# Chemistry, Man and Society

Fourth Edition


Mark M. Jones

Professor of Chemistry,
Vanderbilt University,
Nashville, Tennessee

David O. Johnston

Professor of Chemistry,
David Lipscomb College,
Nashville, Tennessee

John T. Netterville

Superintendent of Schools,
Williamson County Schools,
Franklin, Tennessee

James L. Wood

Consultant,
Resource Consultants, Inc.,
Brentwood, Tennessee





SAUNDERS GOLDEN SUNBURST SERIES

SAUNDERS COLLEGE PUBLISHING

Philadelphia New York Chicago
San Francisco Montreal Toronto
London Sydney Tokyo Mexico City
Rio de Janeiro Madrid

Text Typeface: Century Schoolbook
Compositor: Typothetae Book Composition
Acquisitions Editor: John Vondeling
Project Editor: Patrice L. Smith
Copyeditor: Amy Satran
Managing Editor & Art Director: Richard L. Moore
Design Assistant: Virginia A. Bollard
Text Design: Caliber Design Planning Inc.
Cover Design: Richard L. Moore
Text Artwork: Tom Mallon
Production Manager: Tim Frelick
Assistant Production Manager: Maureen Iannuzzi
Cover Photograph by John Drooyan © 1982

**Library of Congress Cataloging in Publication Data**

Main entry under title:
Chemistry, man, and society.

(Saunders golden sunburst series)
Includes index.

1. Chemistry. I. Jones, Mark Martin, 1928–
QD31.2.C43 1983 540 82-61051

ISBN 0-03-063032-0

CHEMISTRY, MAN AND SOCIETY

ISBN 0-03-063032-0


Library of Congress catalog card number 82-61051.

5 032 987654

**CBS COLLEGE PUBLISHING**
Saunders College Publishing
Holt, Rinehart and Winston
The Dryden Press

# Preface

The fourth edition of *Chemistry, Man, and Society* has been prepared for those students wishing a one- or two-semester course in college chemistry. The fundamental approach of the third edition has been preserved—discoveries are presented along with consequent thought processes that lead to new discoveries and the modern chemical world. No previous knowledge of chemistry is assumed in this presentation. The approach is sufficiently different to challenge and interest the student with a background in high school chemistry, however.

Certainly, there is mystery in chemistry, but to leave the workings of the chemist as a mystery is inevitably to argue that the liberally educated person must be dependent upon the chemist for the chemical decisions that affect society as a whole. This text is based on the belief that the liberal arts student can see and appreciate the chain of events leading from chemical fact to chemical theory and the ingenious manipulation of materials based on the chemical theories. The thoughtful student will then see that the intellectual struggles in chemistry are closely akin to personal intellectual pursuits and will feel that each educated individual can be partner to the ongoing chemical manipulation of the human environment.

The entire presentation in this book is based on stated student needs. Over the years we, through innumerable discussions with our students, have observed an intense interest in the following areas:

1. There is considerable pleasure in understanding a sensible, theoretical explanation of an observed natural phenomenon. Such curiosity appears innate in every person.
2. There is a desire to make highly personal choices as rational as possible. The role of fluoride in tooth structure, for example, is appreciated when understood. A class always perks up when the role of aluminum chloride in a body deodorant is explained, realizing the personal choice that is to be made in the matter.
3. Chemical choices to be made in the immediate environment, in the home, in the automobile, and in the work area are of vital interest. For example, the choice of fuel additives in gasoline proves to be interesting to almost every student.
4. The great environmental questions of a chemical nature also have a compelling interest. The liberal arts student appreciates a sense of awareness in the dilemmas presented in gross pollution problems and in the depletion of natural resources.
5. Energy needs and resources for a growing world population may be the pivotal issue in modern politics. Realizing this, the typical college student is interested in knowing the facts relative to fuels and food.

Several changes refine the fourth edition. These changes fall under the general effort to provide threads of learning that tie the entire presentation into a whole. Structure causes function, chemical properties are periodic, useful and interesting chemical changes, and energy are the threads chosen. Many applications are made in chapter after chapter, and the teacher is provided many leads to add supplementary material of interest. The result is a whole, rather than a collection of pieces. The first six chapters have been rewritten extensively to provide an improved presentation of atomic theory, bonding, the periodic relationship, and chemical reactivity. Periodicity has been given chapter status, and energy has been moved forward for emphasis.

New topics of interest have been added, such as biogenetic engineering, recombinant DNA, acid rain, and the neutron bomb. Additional emphasis has been given to selected consumer products and chemical carcinogens in humans. Extensive rewriting has resulted in an updating of the material throughout the text. Clarification and simplification have been the standing criteria for the revision. Approximately 300 new questions have been added at the ends of the chapters, and each chapter has been introduced with questions that have been proven to catch the interest of the typical reader.

Useful features that have been helpful in previous editions have been maintained and improved. Notes in the margins highlight ideas and add emphasis. Self-tests measure recall and comprehension; matching sets build and maintain vocabulary. The building of vocabulary is also aided by setting forth important new words and terms in boldface type. In review, the illustrative material has been judged to be one of the most important features of the book. The simplicity and clarity of each illustration present an idea in a memorable and understandable way.

We are deeply grateful to all who have contributed to the betterment of each edition of this book. For this edition, we especially acknowledge the indepth review of the manuscript by Professor Calvin VanderWerf (University of Florida) and thank him for sharing his clear and logical insight into the approach and nature of this kind of textbook. We would also like to acknowledge the reviewing contributions of Gregory Choppin (Florida State University) and Thomas Furtsch (Tennessee Technological University).

Only through the expertise of the staff at Saunders College Publishing did these words get into print. Thanks to Amy Satran, Copy Editor, for smoothing the manuscript and adding consistency to the language. Our gratitude goes to Patrice Smith, Senior Project Editor, for bringing the many parts into a whole under challenging circumstances.

To John Vondeling, Associate Publisher, super editor and salesman, we extend our deepest thanks for his encouragement and for his faith in the book through four editions. Thanks, friend.

Much help has come our way, but, of course, the responsibility for the contents of the text rests entirely on us.

As in previous texts, we dedicate this effort to our wives and gratefully acknowledge their support and understanding during the preparation of this manuscript.

Mark M. Jones
David O. Johnston
John T. Netterville
James L. Wood

# Contents Overview

1 The Chemical View of Matter 1
2 Atoms 20
3 Elements in Useful Order—The Periodic Table 49
4 Chemical Bonds 65
5 Some Principles of Chemical Reactivity 98
6 Hydrogen Ion Transfer (Acid-Base) and Electron Transfer (Oxidation-Reduction) Reactions 113
7 Useful Materials from the Earth, Sea, and Air 139
8 Ubiquitous Carbon Atom—An Introduction to Organic Chemistry 164
9 Some Applications of Organic Chemistry 185
10 Man-Made Giant Molecules—The Synthetic Polymers 209
11 Biochemistry—Basic Structures 233
12 Biochemical Processes 261
13 Science and Technology as a New Philosophy 287
14 Energy and Our Society 306
15 Toxic Substances in Our Environment 337
16 Water: Its Use and Misuse 367
17 Air Pollution 390
18 Consumer Chemistry 416
Appendices 476
Answers to Self-Tests and Matching Sets 488
Photo Credits 497
Index 499

# Contents

## 1 The Chemical View of Matter 1

Pure Substances 2
Definitions: Operational and Theoretical 3
Separation of Mixtures into Pure Substances 3
Elements and Compounds 4
Chemical Changes and Physical Changes 6
Why Study Pure Substances and Their Chemical Changes? 8
The Structure of Matter Explains Chemical and Physical Properties 9
Facts, Laws, and Theories 9
The Language of Chemistry 12
Measurement 13

## 2 Atoms 20

The Greek Influence 20
Antoine Laurent Lavoisier: The Law of Conservation of Matter in Chemical Change 21
Joseph Louis Proust: The Law of Constant Composition 22
John Dalton: The Law of Multiple Proportions 23
Dalton's Atomic Theory 23
Dalton's Idea about Atomic Weights: The Follow-up 24
Atoms Are Divisible 27
Electrons 29
Protons 32
Neutrons 32
The Nucleus 34
Atomic Number 35
Isotopes 36
Where Are the Electrons? 38

## 3 Elements in Useful Order—The Periodic Table 49

Elements Described 49
The Periodic Law—The Basis of the Periodic Table 50
Features of the Modern Periodic Table 53
Uses of the Periodic Table 55
The Periodic Table in the Future 61

## 4 Chemical Bonds 65

Ionic Bonds 66
Properties of Ions, Atoms, and Molecules 70
Covalent Bonds 74
Hydrogen Bonding 82
Van der Waals Attractions 85
Metallic Bonding 85
Molecular Structure 86
Hybridization 87
Valence-Shell Electron-Pair Repulsion Theory 90

## 5 Some Principles of Chemical Reactivity 98

Rate of Chemical Reaction 102
Reversibility of Chemical Reactions 104
Quantitative Energy Changes in Chemical Reactions 106
Weight Relationships in Chemical Reactions 107

## 6 Hydrogen Ion Transfer (Acid-Base) and Electron Transfer (Oxidation-Reduction) Reactions 113

Formation of Solutions 113
Ionic Solutions (Electrolytes) and Molecular Solutions (Nonelectrolytes) 114
Acids and Bases 117
Salts 122
Sodium Chloride (Table Salt) 126
Nitrate Salts 127
Sodium Hydrogen Carbonate 127
Electron Transfer (Oxidation-Reduction) 129

## 7 Useful Materials from the Earth, Sea, and Air 139

Metals and Their Preparation 139
Elements from the Air 148
Glass—Silicon's Domain 150
Chemical Fertilizers—Keys to World Food Problems 154
The Numbers on the Bag 160

## 8 The Ubiquitous Carbon Atom—An Introduction to Organic Chemistry 164

Why Are There So Many Organic Compounds? 164
Chains of Carbon Atoms—The Hydrocarbons 165
Structural Isomers 167

Nomenclature—What's in a Name? 170
Optical Isomers 174
Geometric Isomers 177
Functional Groups 178
Aromatic Compounds 179

## 9 Some Applications of Organic Chemistry 185

Some Uses of Hydrocarbons 185
Gasoline 186
Some Important Compounds of Carbon, Hydrogen, and Oxygen 189
Methanol (Methyl Alcohol) 190
Ethanol (Ethyl Alcohol) 191
Propanols (Propyl Alcohols) 192
Ethylene Glycol and Glycerol (Glycerin) 193
Hydrogen Bonding in Alcohols 193
Methanoic Acid (Formic Acid) 196
Ethanoic Acid (Acetic Acid) 196
Fatty Acids 197
Synthetic Detergents (Syndets) 201
Useful Products from Organic Synthesis Reactions 202

## 10 Man-Made Giant Molecules—The Synthetic Polymers 209

What Are Giant Molecules? 209
Addition Polymers 210
Copolymers 214
Condensation Polymers 218
Silicones 223
Rearrangement Polymers 225
Polymer Additives—Toward an End Use 226
The Future of Polymers 226

## 11 Biochemistry—Basic Structures 233

Carbohydrates 233
Starches and Glycogen 236
Cellulose 237
Dietary Fats and Essential Fatty Acids 238
Proteins, Amino Acids, and the Peptide Bond 241
Enzymes 248
Nucleic Acids 253

## 12 Biochemical Processes 261

Biochemical Energy and ATP 261
Photosynthesis 264

Digestion 268
Glucose Metabolism 271
Synthesis of Living Systems 276

13 Science and Technology as a New Philosophy 287

A Philosophical and Historical Background 287
Technology: Its Triumphs and Problems 291
Technology and the Human Environment 294
Technological Development and Its Environmental Consequences 299
Is Technology Escaping Control? 300

14 Energy and Our Society 306

Fundamental Principles of Energy 308
Fossil Fuels 311
Coal Gasification 314
Electricity Production 316
Nuclear Energy 319
Controlled Fusion 329
Solar Energy 330

15 Toxic Substances in Our Environment 337

Dose 337
Corrosive Poisons 338
Metabolic Poisons 340
Neurotoxins 349
Mutagens 353
Teratogens 357
Carcinogens 358
Hallucinogens 361
Alcohols 364

16 Water: Its Use and Misuse 367

Water Reuse 367
Water Purification in Nature 368
The Scope of Water Pollutants 369
Water Purification: Classical and Modern Processes 379
Distillation 382
Freezing 382
Ion Exchange 383
Electrodialysis 384
Reverse Osmosis 385

## 17 Air Pollution 390

Do Air Pollutants Solo or Aggregate? 393
A Major Air Pollutant—Sulfur Dioxide 395
Major Air Pollutants—Nitrogen Oxides 400
A Major Air Pollutant—Carbon Monoxide 404
Major Air Pollutants—Certain Hydrocarbons 405
Ozone—A Secondary Pollutant and a Sunscreen 406
Carbon Dioxide—An Air Pollutant . . . Or Is It? 408
The Automobile—A Special Case of Air Pollution 409
What Does the Future Hold? 413

## 18 Consumer Chemistry 416

Part I—Chemicals in Foods 417
Part II—Medicines 428
Part III—Beauty Aids 442
Part IV—Automotive Products 455
Part V—The Chemistry of Photography 462

## Appendix A The International System of Units (SI) 476

Units of Length 476
Units of Mass 477
Units of Volume 478
Units of Energy 478
Other SI Units 478

## Appendix B Temperature Scales 479

## Appendix C Factor-Label Approach to Conversion Problems 481

## Appendix D Calculations with Chemical Equations 483

## Answers to Self-Tests and Matching Sets 488

## Photo Credits 497

## Index 499

# 1 The Chemical View of Matter

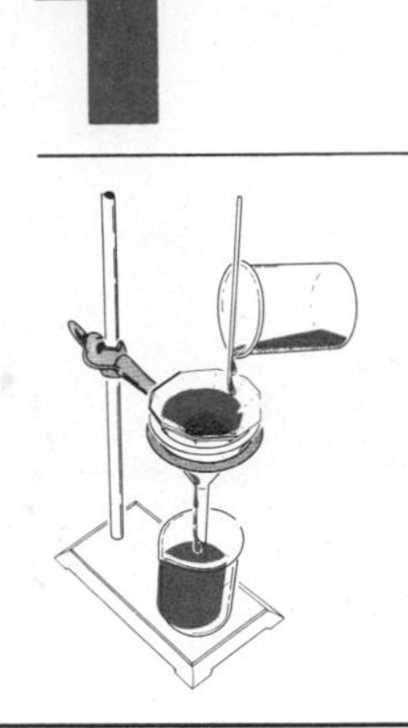

Look around! How many different kinds of things do you see? Broaden your vision and imagine how many different kinds of things exist. Too many to count, right? These things—all of them—that we can see and touch and weigh are made of ***matter.*** Although an uncountable number of things exist, is there any order or simplicity in the makeup of matter? The science of ***chemistry*** addresses this fundamental question.

Most of the things we use in our daily life are very different from the materials which are an obvious part of our natural surroundings. Practically everything we use has been changed from a natural state of little or no utility to one of very different appearance and much greater utility. The processes by which the materials found in nature can be changed, and a detailed description of such changes, are highly intriguing. Wherever we look we see matter undergoing change. This adds another dimension to the science of chemistry: the *changes* that matter undergoes.

An understanding of our environment is necessary if we are to know the consequences of our acts on our environment and, even more critical, if we are to undertake the successful repair of previous damage. Since life itself involves very intricate sequences of changes, each of us has a more personal reason for valuing knowledge of such processes—a desire to gain some insight into the manner in which specific kinds of matter in our food, air, and water can influence our individual health, happiness, and behavior.

**Chemical change alters the kinds of matter present.**

While all changes in matter are of concern, our attention here will be focused on a particular kind of change called ***chemical change.*** In any chemical change the starting material is changed into a different kind of matter. Matter may also undergo another kind of change which does not produce a new kind of matter, but simply results in a new form of the same material. This kind of change is a ***physical change.***

What causes changes to occur in matter? It is ***energy!*** Examples of energy are heat, light, sound, and electricity. Energy is not the same as matter even though they are closely related and associated. Energy has the ability to move matter (engines, eardrums, and motors, for example). Matter is converted into energy in nuclear reactors and nuclear bombs. Energy can infiltrate matter and manifest itself through the actions of the matter. A sample of hot water contains more energy than the same sample when it is cold. Some forms of energy can exist apart from matter; examples are light and radiant heat. It appears that all forms of energy are generated by changes in matter, and that matter, in turn, can absorb energy to produce other physical and chemical changes. Indeed, one definition of energy is that which can produce change in matter. It follows, then, that a study of chemistry involves still another dimension: the *energy* associated with chemical changes.

It is difficult, in a few words, to establish the exact bounds of chemistry. Even so, it will be helpful to think of ***chemistry*** as ***the study of the kinds of matter and the changes of one kind of matter into another with the associated energy changes.***

Since the feature used to recognize a chemical change is the production of a different kind of matter, it is necessary to be able to classify matter into different kinds. In a natural state the kinds of matter are usually mixed together and the separation of such mixtures has to precede their systematic classification. After an examination of the methods of separating such mixtures into their components, we can appreciate some of the problems involved in an accurate definition of the terms "kinds of matter" and "chemical change."

**Common examples of**
**a. chemical change: burning, rusting, souring**
**b. physical change: melting, boiling, reshaping**

## Pure Substances

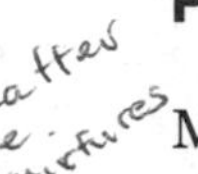

Most natural samples of matter are complex mixtures. Sometimes it is easy to see the various ingredients in a mixture, such as the bits of sand and clay in a sample of soil, or the glittering crystals (mica and quartz) and dark areas (feldspar and magnetite) in a piece of granite. More often, however, it is not apparent to the casual observer that a mixture is made up of distinctly separate ingredients. For example, only by experimentation can we establish that the air we breathe is a mixture.

Bits of iron and sulfur mixed

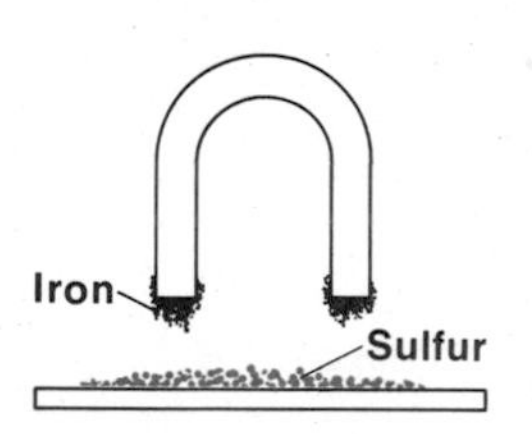


**Purification separates the kinds of matter.**

When a mixture is separated into its components, the components of the mixture are said to be *purified.* However, most efforts at separations are incomplete in a single operation or step, and repetition of the purification process results in a better separation. Ultimately in such a procedure the experimenter arrives at pure substances, samples of matter that cannot be further purified. For example, if sulfur and iron powder are ground together to form a mixture, the iron can be separated from the sulfur by repeated stirrings of the mixture with a magnet. When the mixture is stirred the first time and the magnet removed, much of the iron is removed with it, leaving the sulfur in a higher state of purity. However, after just one stirring the sulfur may still have a dirty appearance due to a small amount of iron that remains. Repeated stirring with the magnet, or perhaps the use of a very strong magnet, will finally leave a bright yellow sample of sulfur that apparently cannot be purified further by this technique. In this purification process a property of the mixture, its color, is a measure of the extent of purification. After the bright yellow color is obtained, it could be assumed that the sulfur has been purified. Drawing a conclusion based on one property of the mixture may be dangerous because other methods of purification might change some other properties of the sample. It is safe to call the sulfur a pure substance only when all possible methods of purification fail to change its properties. This assumes that all pure substances have a set of properties by which they can be recognized, just as a person can be recognized by a set of characteristics. ***A pure substance, then, is a kind of matter with properties that cannot be changed by further purification.***

**About five million pure substances have been identified.**

There are some naturally occurring pure substances. Rain is very nearly pure water, except for small amounts of dust and air. Gold, diamond, and sulfur are also found in very pure form. These substances are special cases. Man, a complex assemblage of mixtures, lives in a world of mixtures—eating them, wearing them, living in houses made of them, and making most of his tools out of them.

**Most materials in nature are mixtures; a few are relatively pure substances.**

Although naturally occurring pure substances are not common, it is possible to produce many pure substances from natural mixtures. Relatively pure substances are now very common as a consequence of the development of modern purification techniques. Common examples are sugar, table salt (sodium chloride), copper, sodium

bicarbonate, nitrogen, dextrose, ammonia, uranium, and carbon dioxide—to mention just a few. In all, about five million pure substances have been identified and cataloged.

## Definitions: Operational and Theoretical

The preceding definition given for a pure substance is an ***operational*** definition, or a definition using specific experiments or operations. That is, if further purification efforts are unsuccessful in changing the properties of a substance, it is said to be a pure substance. It is evident then that operational definitions result from performing operations or tests on matter and summarizing the results in a statement. For example, pure sulfur is a yellow substance that boils at 833°F and has a density of 129 pounds per cubic foot. When all the properties of pure sulfur have been listed, we find that the pure substance has been characterized in a way that distinguishes it from any other pure substance.

**Chemistry begins with observations and experiments.**

A pure substance also can be defined in theoretical terms; that is, in terms of the molecules, atoms, and subatomic particles that compose it. Both types of definitions are important in the study of chemistry and both will be used in this text. The ***theoretical*** definition will follow the development of the theory on which it is based.

### Self-Test 1-A*

1. Four common materials that could not be pure substances are:
   a. Polyester ____ c. Steel ____
   b. glass ____ d. Cement ____
2. Which must come first, the operational definition or the theoretical definition?
   B ____
3. Four common materials that are very nearly pure substances are:
   a. Gold ____ c. Sulfate ____
   b. RAIN ____ d. Diamond ____
4. The properties of two different pure substances could all be identical. True ( ) or False (X)

---

## Separation of Mixtures into Pure Substances

The separation of mixtures is usually more difficult than the magnetic separation of iron and sulfur described previously. Most beginning chemistry students would find it a bewildering task to separate a piece of granite into pure substances. Indeed, a trained chemist would find this a difficult assignment. Since each of the pure substances in the granite has a set of properties unlike those of any other pure substance, it should be possible to use these properties to separate the pure substances, just as the attraction of iron to a magnet is used to separate it from sulfur.

Many different methods have been devised to separate the pure substances in a mixture. In each case, differing properties of the pure substances are exploited to effect

*Use these self-tests as a measure of how well you have learned the material. Take a test only after careful reading of the material preceding the test. Do not return to the text during the self-test, but reread entire sections carefully if you do poorly on the self-test on those sections. The answers to the self-tests are at the end of the text.

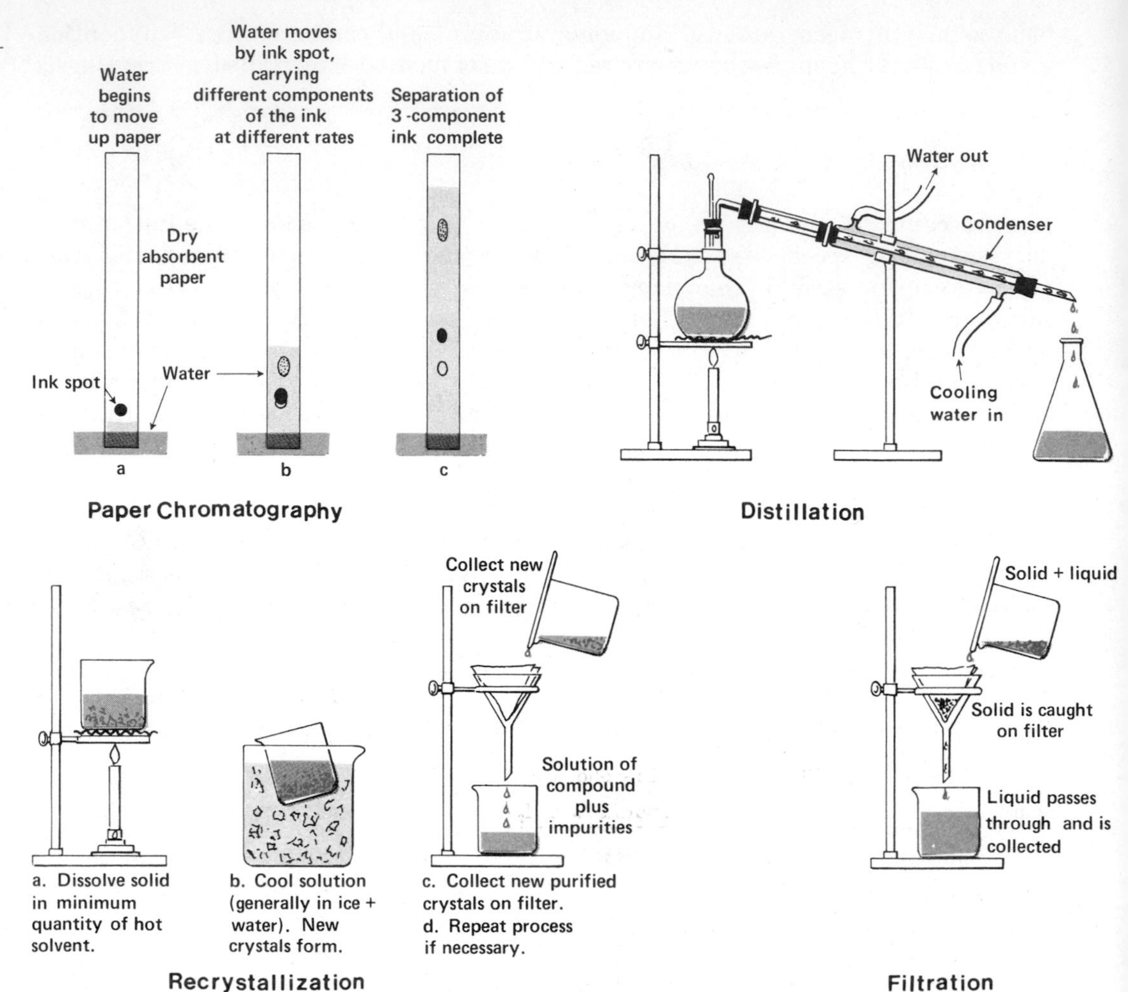
Water begins to move up paper
Water moves by ink spot, carrying different components of the ink at different rates
Separation of 3-component ink complete
Dry absorbent paper
Ink spot
Water
a
b
c
Paper Chromatography
Water out
Condenser
Cooling water in
Distillation
a. Dissolve solid in minimum quantity of hot solvent.
b. Cool solution (generally in ice + water). New crystals form.
Collect new crystals on filter
Solution of compound plus impurities
c. Collect new purified crystals on filter.
d. Repeat process if necessary.
Recrystallization
Solid + liquid
Solid is caught on filter
Liquid passes through and is collected
Filtration

**Figure 1–1** Four methods of purifying mixtures of elements and compounds. **Paper chromatography.** Owing to the absorbent character of paper, water moves against gravity and carries the ink dyes along its path. If the ink dyes move at different rates because of differing attraction to the paper, they will be separated in the developed chromatogram. **Distillation.** Sodium chloride dissolves in water to form a clear solution. When heated above the boiling point (indicated by thermometer), water will vaporize and pass into the condenser. Cool water injected into the glass jacket of the condenser circulates over the inner tube, causing the steam to liquefy and collect in the flask. In this simple example pure water collects in the receiving flask, while the salt remains in the boiling flask. **Recrystallization.** This can be used to separate some solid mixtures. **Filtration.** The separation of a solid from a liquid by filtration.

Scanning electron micrograph of particles of asbestos filtered from a sample of air by a small pore filter.

the separation. Figure 1–1 illustrates four commonly used methods: chromatography, distillation, recrystallization, and filtration.

## Elements and Compounds

Experimentally, pure substances can be classified into two categories: those that can be broken down by chemical change into simpler pure substances and those that cannot. Table sugar (sucrose), a pure substance, will decompose when heated in the oven, leaving carbon, another pure substance, and evolving water. No chemical

**TABLE 1–1 Some Common Elements***

**Metals**

(A metal is a good conductor of electricity, can have a shiny or lustrous surface, and in the solid form usually can be deformed without breaking.)

| Name | Symbol | Properties of Pure Element |
|---|---|---|
| Iron<br>Latin, *ferrum* | Fe | strong, malleable, corrodes |
| Copper<br>Latin, *cuprum* | Cu | soft, reddish-colored, ductile |
| Sodium<br>Latin, *natrium* | Na | soft, light metal, very reactive, low melting point |
| Silver<br>Latin, *argentum* | Ag | shiny, white metal, relatively unreactive, good conductor of electricity |
| Gold<br>Latin, *aurum* | Au | heavy, yellow metal, very unreactive, ductile, good conductor |
| Chromium | Cr | resistant to corrosion, hard, bluish-gray |

**Nonmetals**

(A nonmetal is often a poor conductor of electricity, normally lacks a shiny surface, and is brittle in crystal-solid form.)

| Name | Symbol | Properties of Pure Element |
|---|---|---|
| Hydrogen | H | colorless, odorless, occurs as a very light gas ($H_2$), burns in air |
| Oxygen | O | colorless, odorless gas ($O_2$), reactive, constituent of air |
| Sulfur | S | odorless, yellow solid ($S_8$), low melting point, burns in air |
| Nitrogen | N | colorless, odorless gas ($N_2$), rather unreactive |
| Chlorine | Cl | greenish-yellow gas ($Cl_2$), very sharp choking odor, poisonous |
| Iodine | I | dark purple solid ($I_2$), sublimes easily |

*Chemists usually use the symbol rather than the name of the element. In addition to denoting the element, the chemical symbol has a very specialized meaning which is described later in this chapter. A complete list of the elements with the symbols can be found inside the back cover of this book.

operation has ever been devised that will decompose carbon into simpler pure substances. Obviously sucrose and carbon belong to two different categories of pure substances. Only 89 substances found in nature cannot be reduced chemically to simpler substances; 17 others are available artificially. These 106 substances are called ***elements.*** Pure substances that can be decomposed into two or more different pure substances are referred to as ***compounds.*** Even though there are presently only 106 known elements, there appears to be no practical limit to the number of compounds that can be made.

**These are operational definitions of "element" and "compound."**

Elements are the basic building blocks of the universe and the world in which we live. Table 1–1 lists the properties of some of the common elements; a complete list of the elements is found inside the back cover of this text. A number of elements are found as the elementary substance in nature; examples include gold, silver, oxygen, nitrogen, carbon (graphite and diamond), copper, platinum, sulfur, and the noble gases (helium, neon, argon, krypton, xenon, and radon). Many more elements, however, are found chemically combined with other elements in the form of compounds.

**Figure 1–3 lists the most common elements in the universe and on earth.**

Elements in compounds no longer show all of their original, characteristic properties, such as color, hardness, and melting point. Consider ordinary sugar, which is properly called sucrose, as an example. It is made up of three elements: carbon (which is usually a black powder), hydrogen (the lightest gas known), and oxygen (a gas necessary for respiration). The compound sucrose is completely unlike any of the three elements; it is a white crystalline powder which, unlike solid carbon, is readily soluble in water.

**Compounds have a fixed composition of the elements they contain.**

**Figure1–2** Distillation. Some of the most useful purification techniques copy processes in nature and date back to alchemical times. Distillation allows a more volatile substance to be separated from a less volatile one. In this case, alcohol is partially separated by evaporation from water and other ingredients. One distillation can produce a mixture that is 40 percent alcohol from one that is only 12 percent. Further distillations would produce an even better separation.

A careful distinction should be made between a *compound* of two or more elements and a *mixture* of the same elements. The two gases hydrogen and oxygen can be mixed in all proportions. However, these two elements can and do react chemically to form the compound water. Not only does water exhibit properties peculiar to itself and different from those of hydrogen and oxygen, but it also has a definite percentage composition by weight (88.8% oxygen and 11.2% hydrogen). This is a second distinct difference between compounds and mixtures: ***Compounds have a definite percentage composition by weight of the combining elements.***

## Chemical Changes and Physical Changes

Matter can exist in three states: solid, liquid, or gas, depending upon conditions.

All changes in matter can be distinguished as chemical or physical changes. A chemical change occurs when substances are used up in the production of different ones. For example, the burning of carbon (as charcoal) to form carbon dioxide is a chemical change. A physical change produces no new substances; it only alters the original state of the elements or compounds. An example is the boiling of water, a change of liquid water into gaseous water. Some additional examples of chemical and physical changes are the following:

| Chemical Changes | Physical Changes |
|---|---|
| Rusting of iron | Evaporation of a liquid |
| Burning gasoline in an automobile engine | Melting of iron |
| Preparation of caramel by heating sugar | Drawing metal wire |
| Preparation of iron from its ores | Freezing of water |
| Solution of copper in nitric acid | Crystallizing salt from sea water |
| Souring and ripening of fruit | Grinding or pulverizing a solid |
| | Cutting and shaping wood |
| | Bending, breaking, molding |

The properties of a new substance identify it. There are physical properties and there are chemical properties. A *chemical property* of an element or compound describes its ability to undergo chemical change. It follows, then, that to determine a chemical property we must attempt a chemical change. *Physical properties,* on the other hand, are characteristics of a substance which can be determined without chemically changing the element or compound into some other element or compound. Physical properties of a pure substance include such things as melting temperature, boiling temperature, density (weight of the substance in a given volume), color, physical state, crystalline form, and magnetic properties.

**Nuclear changes, which are basic to radioactivity, atomic bombs, hydrogen bombs, and solar energy, are a third type of change. Nuclear changes involve relatively large amounts of energy.**

It should be emphasized that not all of the properties of a sample need to change in order to establish that a chemical change has occurred. For example, if table salt is

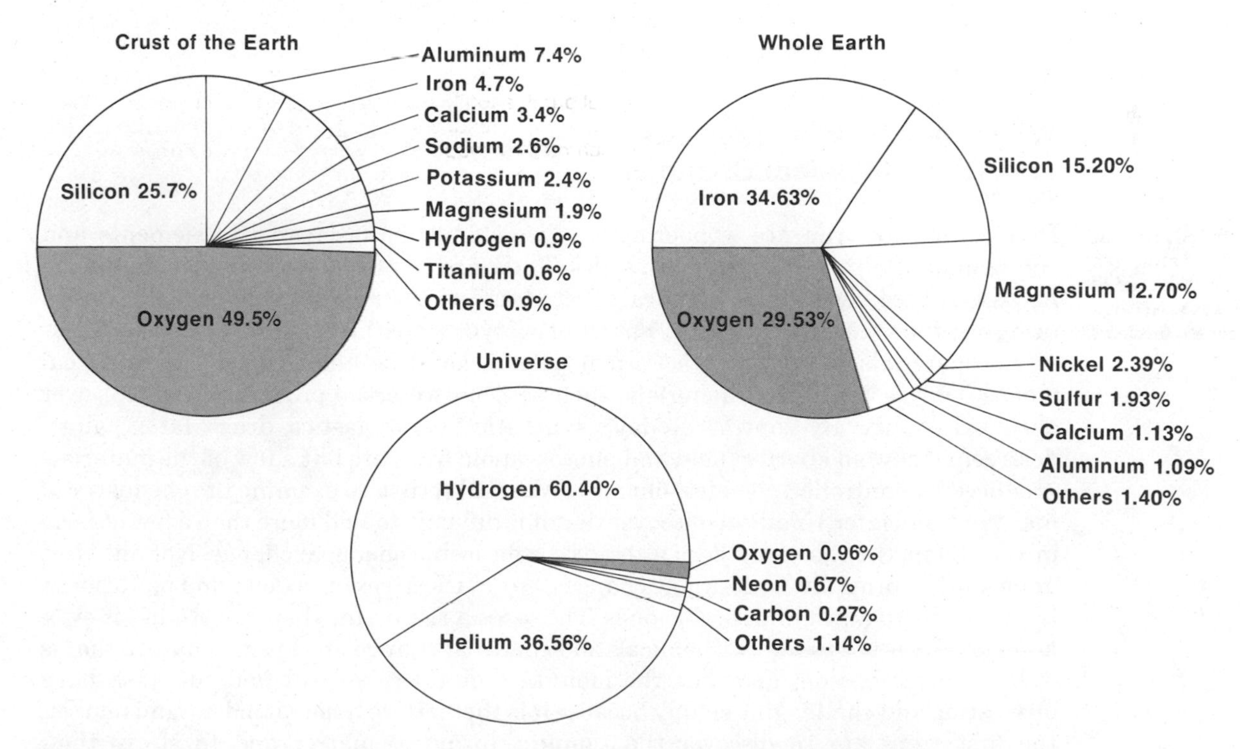


**Figure 1–3** Relative abundance (by mass) of the most common elements in the earth's crust, the whole earth, and the universe. Note that the earth's crust differs significantly from the cosmic array of elements.

The remaining solid is another salt, sodium sulfate. Hydrogen chloride gas also results.

treated with sulfuric acid and heated strongly, the solid product of this reaction is much like table salt in its appearance, but it has a different melting point and different chemical properties. A change in only one significant property usually indicates that a newly formed substance has resulted from a chemical change.

While most changes can easily be classified on the basis of whether a different substance is produced or not, some changes are difficult to put into such neat categories. Examples include those which accompany the dissolving of a solid in a liquid. We often apply a simple test to determine the type of change. If the solid can be recovered unchanged by the evaporation of the liquid, the solution process was a physical change. Thus, table salt or sugar dissolves in water, but either solid can be recovered by careful evaporation, so such solutions are said to be formed by physical changes although the interaction between constituents is considerable. An example of a chemical change resulting from solution is observed when the gas hydrogen chloride dissolves in water. In this case, the properties of the solution are very different from those of the constituents, and evaporation of the solution is not a feasible way to recover the starting substances completely. In general, however, solution processes are of such complexity that their detailed discussion is inappropriate at this point.

The classification of solution changes illustrates some difficulties about exact definitions as used in chemistry. Like the stamp collector arranging stamps and the cook arranging foods in the kitchen, we shall find our work easier if we group similar things and phenomena into categories. This makes it easier to organize knowledge. Of course, we try to define the limits of our categories so there is no question about what fits into each classification, but nature sometimes presents situations that are very hard to classify. There are borderline cases, and all efforts to be specific and limiting will usually meet with an item or idea that defies exact classification. However, a few exceptions do not stop us from originating simple categories for the things we observe.

## Why Study Pure Substances and Their Chemical Changes?

A knowledge of chemical changes allows us to carry them out purposefully to achieve desired results.

Perhaps by now you are wondering why we should be interested in elements and compounds and their chemical properties. There are two basic reasons. The first is the belief that the knowledge of chemical substances and chemical changes will allow us to bring about desired changes in the nature of everyday life. Two hundred years ago most of the materials surrounding a normal person could be changed only by physical means. Only a few useful materials, such as iron, were the product of control over chemical change. By contrast, today's synthetic fibers, plastics, drugs, latex paints, detergents, new and better fuels, and photographic films are but a few of the materials produced by controlled chemical change. (We shall return to examine the chemistry of many of these later.) You will discover that it is difficult to find more than a few objects in your home that have not been altered by a desirable chemical change. Not only is it important to bring about desirable changes, but in the areas of toxicity and pollution it is important to avoid undesirable ones. The second reason for chemical studies is even more important—curiosity. Chemicals and chemical change are a part of nature that is open to investigation, and, like the mountain climber, we will find this task both interesting and challenging simply because it is there. If we hope to understand matter, the first steps are to discover the simplest forms of matter and to study their interactions.

## The Structure of Matter Explains Chemical and Physical Properties

For reasons which are partly theoretical and partly practical, we are deeply interested in the structure of matter—that is, the minute parts of matter and how these parts are fitted together to make larger units. Why does an element or compound have the properties it has? Why does one element or compound undergo a change that another element or compound will not undergo? Inanimate matter is the way it is because of the nature of its parts. A watch is what it is because of the nature of its individual parts. So is a car, a refrigerator, and the salt in your salt shaker. The individual parts (smaller than the whole) are what determine the nature (actions and properties) of the whole. The most basic parts of matter, as we shall see, are very small. If we even hope to understand the nature of matter, it is absolutely necessary that we have some understanding of these minute parts and how they are related to each other.

A very large portion of today's research in chemistry is aimed at sorting out and elucidating the structure of matter. Indeed, the basic theme of this text is the relationship between the structure of matter and its properties. This theme of structure and related properties is of great interest because if we know exactly how and with what strength the minute parts of matter are put together, we can discover exact relationships between structure and properties. Armed with this understanding, we can make changes that result in new substances and predict the properties of these substances. Such knowledge can save many months of trial and error which otherwise may be required to prepare a product with the desired qualities. While this day of predicting chemical changes based on structural characteristics has not completely arrived, such significant advances have been made that the practice of modern chemistry would not be possible without such knowledge.

**Submicroscopic structures help to explain chemistry.**

Samples of matter large enough to be seen and felt and handled, and thus large enough for ordinary laboratory experiments, are called ***macroscopic*** samples, in contrast to ***microscopic*** samples, which are so small that they have to be viewed with the aid of a microscope. The structure of matter that really interests us, however, is at the ***submicroscopic*** level. Our senses have no direct access into this small world of structure, and any conclusions about it will have to be based on circumstantial evidence gathered in the macroscopic and microscopic worlds.

**Scope of chemistry.**

We can now extend our concept of the science of chemistry. It is that science which investigates the properties and changes of pure substances. Chemistry is also deeply concerned with structure, both ***macrostructure*** and ***submicrostructure,*** in an effort to give plausible reasons for properties and change, with emphasis on chemical change.

## Facts, Laws, and Theories

**A scientific fact can be verified independently of any particular observer.**

A ***scientific fact*** is an observation about nature that can usually be reproduced at will. For example, carbon in some forms will readily burn in the presence of air. If you have any doubt about this fact it is easy enough to set up an experiment that will readily demonstrate the fact anew. You would only need some carbon, air, and a source of heat. The repeatability of a scientific fact distinguishes it from a historical fact, which obviously cannot be reproduced. Of course, some scientific facts are also historical facts—such as the movement of heavenly bodies—and are not repeatable at will.

Often a large number of related scientific facts can be summarized into broad, sweeping statements called natural ***laws.*** The law of gravity is a classic example of a natural law. This law, that all bodies in the universe have an attraction for all other bodies which is directly proportional to the product of their masses and inversely related to the square of their separation distance, summarizes in one sweeping statement an enormous number of facts. It implies that any object lifted a short distance from the surface of the earth will fall back if released. Such a natural law can only be established in our minds by inductive reasoning; that is, you conclude that the law applies to all possible cases, since it applies in all of the cases studied or observed. A well-established law allows us to predict future events. When convinced of the generality of a scientific law, we may reason deductively, based on our belief that if the law holds for all situations, it will surely hold for the events in question.

**A scientific law summarizes a large number of related facts.**

**A scientific law predicts what *will* happen. A governmental law describes what people *should do.***

The same procedure is used in the establishment of chemical laws, as can be seen from the following example. Suppose an experimenter carried out hundreds of different chemical changes in closed, leakproof containers, and suppose further that he weighed the containers and their contents before and after each of the chemical changes. Also, suppose that in every case he found that the container and its contents weighed exactly the same before and after the chemical change occurred. Finally, suppose that he repeated the same experiments over and over again, obtaining the same results each time, until he was absolutely sure that he was dealing in reproducible facts. It can be understood then that the experimenter would reasonably conclude: ***"All chemical changes occur without any detectable loss or gain in weight."*** This is indeed a basic chemical law and serves as one of the foundations of modern scientific theory.

After a natural law has been firmly established, its explanation must be sought. Chemists are not satisfied until they have explained chemical laws logically in terms of the submicroscopic structure of matter. This is indeed a difficult process, and until recently its progress has been painfully slow because of our lack of direct access into the submicroscopic structure of matter with our physical senses. All we can do is collect information in the macroscopic world in which we live, and then try, by circumstantial reasoning, to visualize what the submicroscopic world must be like in order to explain our macroscopic world. Such a visualization of the submicroscopic world is called a ***theoretical*** model. If the theoretical model is successful in explaining a number of chemical laws, a major scientific theory is built around it. The atomic theory and the electron theory of chemical bonding are two such major theories, and both will be discussed in relation to chemical laws in later chapters.

**Theories are ideas or models used to explain facts and laws.**

**Chemical theories use the concepts of atoms to explain chemical observations.**

Consider again the chemical law concerning the conservation of weight in chemical changes. What is a possible theoretical model that could explain this law? If we assume that matter is made up of atoms, and that these atoms are grouped in a

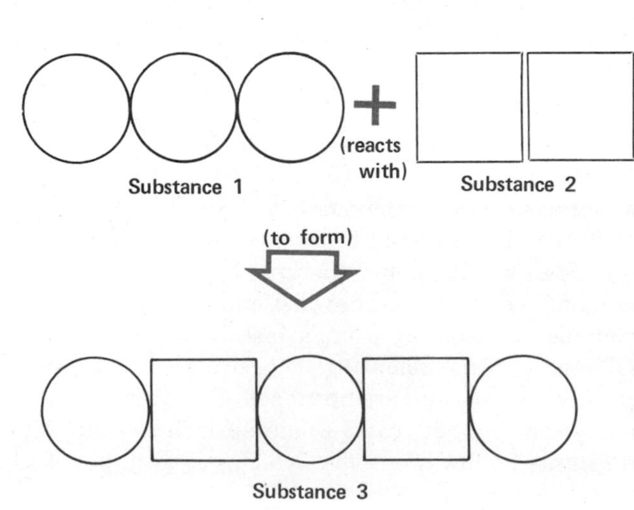


**Figure 1–4** Atomic explanation for weight conservation in chemical changes. If substances are composed of atoms, if chemical changes are atomic rearrangements in which atoms are not destroyed or changed into other atoms, and if atoms have fixed weights, the products in a reaction must weigh the same as the starting substances. Hence the law of conservation of weight is explained in terms of a theoretical model.

particular way in a given pure substance, we can reason that a chemical change is simply the rearrangement of these atoms into new groupings, and, consequently, into new substances. If the same atoms are still there, they should have the same individual characteristic weight, and hence the law of conservation of weight is explained (Fig. 1–4).

For a scientific theory to have much value, it must not only explain the pertinent facts and laws at hand, but it must also be able to explain new facts and laws that are obviously related. If the theory cannot consistently perform in this manner, it must be revised until it is consistent, or, if this is not possible, it must be completely discarded. You must not allow yourself to think that this process of trying to understand nature's secrets is nearing completion. The process is a continuing one.

The word theory is often used in a different sense from the one discussed above. If a student is absent from the chemistry class, his neighbor may say, "I do not know why he is absent, but my theory is that he is sick and unable to come to class." This use of the word theory is similar to the concept of the general scientific theory in that there is an element of speculation in both uses. But the speculative guess of the student about his absent friend is vastly different from the broad theoretical picture that is able to explain a number of laws. The reader should be alert for the considerable amount of confusion that has resulted from the different meanings associated with this word. In

**Figure 1–5** Direct observation stops at the microscopic level. Convinced of structure beyond the microscopic level, the chemist employs circumstantial evidence to construct the world of molecules, atoms, and subatomic parts in the mind's eye.

this book the word ***hypothesis*** is used when speaking of a speculation about a particular event or set of data, reserving the word ***theory*** for the broad imaginative concepts that have gained wide acceptance.

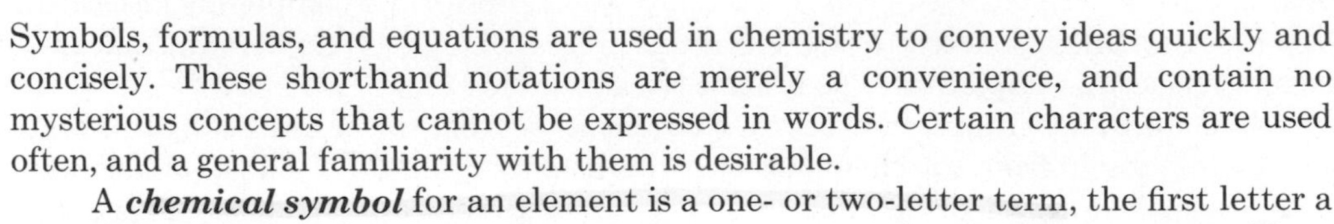

## The Language of Chemistry

"chemistry"

Symbols, formulas, and equations are used in chemistry to convey ideas quickly and concisely. These shorthand notations are merely a convenience, and contain no mysterious concepts that cannot be expressed in words. Certain characters are used often, and a general familiarity with them is desirable.

**Chemical symbols are abbreviations for the different elements.**

A ***chemical symbol*** for an element is a one- or two-letter term, the first letter a capital and the second a lowercase letter. The symbol is a sign for three concepts. First, the symbol stands for the element in general. H, O, N, Cl, Fe, Pt are shorthand notations for the elements hydrogen, oxygen, nitrogen, chlorine, iron, and platinum, respectively, and it is useful and timesaving to substitute these symbols for the words themselves in describing chemical changes. Some symbols originate from Latin words (such as Fe, from *ferrum,* the Latin word for iron); others came from English, French, or German names. Second, the chemical symbol stands for a single atom of the element. The ***atom*** is the smallest particle of the element that can enter into chemical combinations. Third, the elemental symbol stands for a mole of the atoms of the element. The ***mole*** is a term in chemical usage (it is derived from the Latin for "a pile of" or "a quantity of") that means the quantity of a substance that contains 602 sextillion identical particles. Just as a dozen apples would be 12 apples, a mole of atoms would be 602,300,000,000,000,000,000,000 atoms or $6.023 \times 10^{23}$ atoms. How big is this number? A mole of textbooks like this one would cover the *entire surface of the continental states* to a height of 190 miles! To match the population density on Earth, a mole of people would require 150 trillion planets. It takes 134,000 years for a mole of water drops to flow over Niagara Falls at a flow rate of 112,500,000 gallons per minute. A mole is a very large number. Yet, it turns out that a mole of very small atoms is usually a convenient amount for laboratory work. Thus, the symbol Ca can stand for the element calcium, or a single calcium atom, or a mole of calcium atoms. It will be evident from the context which of these meanings is implied.

**A *mole* contains $6.023 \times 10^{23}$ particles.**

**A mole of water molecules in the liquid state occupies only about four teaspoonfuls. Are molecules small. . . or are they small?**

Atoms can unite (bond together) to form molecules. A ***molecule*** is the smallest particle of an element or a compound that can have a stable existence in the close presence of like molecules. One or more of the same kind of atom can make a molecule of an element. For example, two atoms of hydrogen will bond together to form a molecule of ordinary hydrogen, and eight sulfur atoms will form a single molecule. Subscripts in a chemical formula show the number of atoms involved: $H_2$ means a hydrogen molecule is composed of two atoms, and $S_8$ means a sulfur molecule is composed of eight atoms. The noble gases, such as helium, He, have monatomic molecules (monatomic = one atom).

**$H_2SO_4$**
**4 atoms of oxygen**
**1 atom of sulfur**
**2 atoms of hydrogen**

When unlike atoms combine, as in the case of water ($H_2O$) or sulfuric acid ($H_2SO_4$), the formulas tell what atoms and how many of each are present in a molecule of the compound. For example, $H_2SO_4$ molecules are composed of two hydrogen atoms, one sulfur atom, and four oxygen atoms. A **formula** can stand not only for the substance itself, but also for one molecule of the substance or for a mole of such molecules, depending on the context.

When elements or compounds undergo a chemical change, the formulas, arranged in the form of a ***chemical equation,*** can present the information in a very concise fashion. For example, carbon can react with oxygen to form carbon monoxide. Like most solid elements, carbon is written as though it had one atom per molecule; oxygen exists as diatomic (two-atom) molecules, and carbon monoxide molecules contain two atoms, one each of carbon and oxygen. Furthermore, one oxygen molecule will combine with two carbon atoms to form two carbon monoxide molecules. All of this information is contained in the equation

$$2C + O_2 \longrightarrow 2CO$$

The arrow is often read "yields"; the equation then states the following information:

**Chemical equations summarize information on chemical reactions in a concise fashion.**

a. carbon plus oxygen yields carbon monoxide
b. two atoms of carbon plus one diatomic molecule of oxygen yield two molecules of carbon monoxide
c. two moles of carbon atoms plus one mole of diatomic oxygen molecules yield two moles of carbon monoxide

The number written before a formula, the ***coefficient,*** gives the amount of the substance involved, while the ***subscript*** is a part of the definition of the pure substance itself. Changing the coefficient only changes the amount of the element or compound involved, whereas changing the subscript would necessarily involve changing from one substance to another. For example, 2CO means either two molecules of carbon monoxide or two moles of these molecules.

## Measurement

The heart and soul of reliable scientific facts, and, consequently, the basis for laws and theories, is the accurate measurement of weights, volumes, times, lengths, and other quantities. Perhaps more than any other person, Antoine Lavoisier, a French chemist who lived in the late 1700's, put chemistry on a quantitative basis by very accurately measuring the weights of materials involved in chemical reactions. Because of the significance of his work, Lavoisier has been called the "Father of Modern Chemistry." The reproducible data which he obtained led directly to the formulation of several laws concerning chemical change.

Since 1960, a coherent system of units known as the International System of Units, abbreviated SI and bearing the authority of the International Bureau of Weights and Measures, has been in effect and is gaining acceptance among scientists. It is an extension of the metric system which began in 1790, and assigns each physical quantity a unique SI unit. We shall use the more familiar metric system in most of our discussions; many metric units are identical with SI units.

The metric system has one great advantage over older systems: it is simple. Its simplicity lies in its organized structure and its decimal counting, which require relatively few terms to be remembered. For example, units such as those of length are defined in such a way that a large unit is ten (or some power of ten) times larger than a smaller unit. A ***meter,*** a unit of length equal to 39.4 inches, is 100 centimeters, and a

**Accuracy of English and metric systems of measurement is the same, but metric is easier to use.**

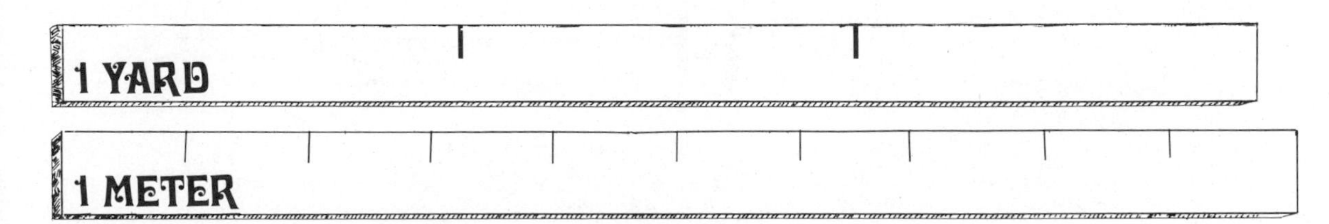


**Figure 1–6** One meter = 39.4 inches.

**Figure 1–7** The volume of a liter is slightly more than that of a quart. One liter = 1.06 quarts.

*Mass* is a measure of the amount of matter in a body, whereas *weight* is a measure of the attraction of the body for the earth. The weight of the body varies with its distance from the earth, whereas the mass remains constant. Unless otherwise stated, the term "weight" refers to the earth-surface weight.

centimeter is 10 millimeters. Making a conversion from one unit of length to another in the metric system merely involves shifting the decimal point. Compare this to the complexity of the conversion of miles to inches, wherein one would probably employ the factors 12 and 5,280. Instead of having a number of different root words for length, as in the English system (league, mile, rod, yard, foot, inch, mil), the metric system has just one, the meter. With suitable prefixes, the meter can be expressive of a unit useful to the watchmaker (millimeter = 0.001 meter) or to the distance runner (kilometer = 1,000 meters).

While the convenience of the metric system can scarcely be overstated, it should be emphasized that the science of chemistry need not rest on this system of measurement. Measurements made in the English system can be and are just as accurate. However, in the metric system only four basic units will be required in this course: a unit of length, the ***meter;*** a unit of volume, the ***liter;*** a unit of mass (weight), the

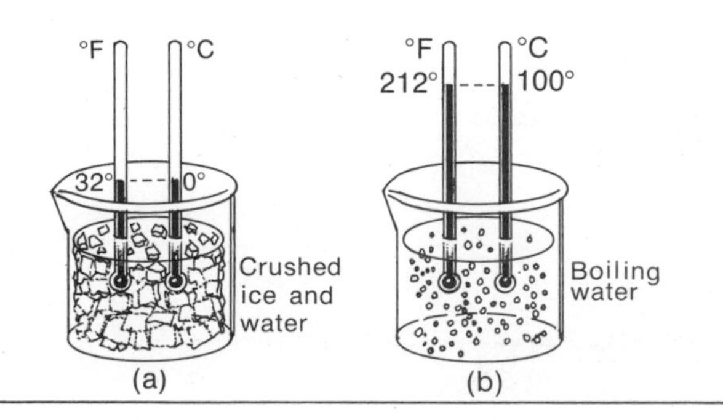


**Figure 1–8** Thermometer scales are defined in terms of the expansion of common materials such as mercury, and in terms of fixed reference points such as the changes of state of water and other common materials. It is only a matter of preference and convenience whether one scale or another is used.

**Figure 1–9** Units of measure are selected for convenience; the millimeter for the small parts of this tiny radio (above), and the kilometer for measuring intercontinental distances. The metric system simply adds the ease of using multiples of 10.

***gram;*** and a unit of temperature, the ***Celsius degree.*** English equivalents of the first three are:

1 meter = 39.4 inches (Fig. 1–6)
1 liter = 1.06 quarts (Fig. 1–7)
1 gram = 0.0352 ounce

Temperature scales are defined in terms of the behavior of samples of matter. For example, the familiar Fahrenheit scale defines the temperature at which water freezes to be 32°F and that at which it boils to be 212°F. Thus there are 180 (212 − 32) Fahrenheit degrees between the freezing and boiling points of water. Most scientists prefer the Celsius scale, which defines the freezing point of water to be 0°C and the boiling point 100°C. There are 100 Celsius degrees in this same temperature range (Fig. 1–8).

Whether you cook English or metric, moist chewy brownies will be delicious either way (Table 1–2).

**TABLE 1–2 Recipe for Perfect Brownies in English and Metric Measurements**

| Ingredient | English | Metric |
|---|---|---|
| Unsweetened chocolate squares | 2.0 oz. | 57 g |
| Butter or margarine | 0.50 cup | 120 ml |
| Sugar | 1.0 cup | 240 ml |
| Eggs | 2 | 2 |
| Vanilla | 1.0 teaspoon | 5 ml |
| Sifted enriched flour | 0.50 cup | 120 ml |
| Chopped walnuts | 0.50 cup | 120 ml |
| Oven | 325°F | 163°C |
| Pan | 8 × 8 × 2 in. | 20 × 20 × 5.0 cm |

1cal = 4.184 J

Energy is measured in joules, calories, British thermal units (BTU's) and similar units. Each is defined as the amount of energy required to produce a specified change in a pure substance. A ***calorie*** is the amount of energy required to raise the temperature of one gram of water from 14.5°C to 15.5°C. To raise one gram of water one degree Celsius requires 4.184 ***joules*** (J). A ***BTU*** is required to raise the temperature of one pound of water one degree Fahrenheit.

For the interested reader and those who need to understand additional aspects of the metric system to support laboratory work, more information is presented in Appendices A, B, and C.

## Self-Test 1-B

1. The two large divisions into which elements can be divided are:

   a. ______ and b. ______

   A half-dozen examples of each are:

   | | |
   |---|---|
   | a. ______ | a. ______ |
   | b. ______ | b. ______ |
   | c. ______ | c. ______ |
   | d. ______ | d. ______ |
   | e. ______ | e. ______ |
   | f. ______ | f. ______ |

2. How many elements are presently listed in the periodic table? 106
3. A compound has properties that are combinations of the elemental properties. True (x) or False ( )
4. Four chemical changes not listed in the chapter are:

   a. ______

   b. ______

   c. ______

   d. ______
5. Four physical changes not listed in the chapter are:

   a. ______

   b. ______

   c. ______

   d. ______
6. A chemical change always produces a new ~~elemen~~ mixture
7. ______ structures explain chemical properties.
8. Put in order of decreasing size: 1. microscopic, 2. molecular, and 3. macroscopic.

   a. 3, b. 1, c. 2
9. Arrange from most abstract to general to specific: laws, facts, and theories.

   a. 3, b. 2, c. 1

10. Consider the equation: $2Na + 2HCl \longrightarrow H_2 + 2NaCl$. Explain what is meant by the symbols:

a. Na ~~2Na = 2 ATOM~~ Sodium

b. 2Na 2 ATOM of " "

c. HCl 1 Hydrogen + 1 clorine

d. $\longrightarrow$ yeilds

e. $H_2$ ______

f. 2NaCl ______

## Matching Set

| | | | |
|---|---|---|---|
| ______ | 1. produces a new type of matter | **a** | filtration |
| ______ | 2. air | **b** | chemical change |
| ______ | 3. unchanged by further purification | **c** | element |
| ______ | 4. used to separate a solid from a liquid | **d** | properties of pure substance |
| ______ | 5. cannot be reduced to simpler substances | **e** | mixture |
| ______ | 6. scientific theory | **f** | used to explain facts and laws |
| ______ | 7. symbol for iron | **g** | $O_3$ |
| ______ | 8. mole of atoms | **h** | CO |
| ______ | 9. molecule containing three oxygen atoms | **i** | Fe |
| ______ | 10. carbon monoxide | **j** | one meter |
| ______ | 11. 100 centimeters | **k** | $6.023 \times 10^{23}$ atoms |

## Questions

**1.** Name as many materials as you can that you have used during the past day that were not chemically changed by artificial means.

**2.** In pottery making, an object is shaped and then baked. Which part of this process is chemical and which part is physical?

**3.** Identify the following as physical or chemical changes. Justify in terms of the operational definitions for these types of changes.

a. formation of snowflakes
b. rusting of a piece of iron
c. ripening of fruit
d. fashioning a table leg from a piece of wood
e. fermenting grapes

**4.** If physics is the study of matter and energy, why can it be said that the study of chemistry is a special case within the general study of physics?

**5.** Name a physical change food undergoes as you eat it.

**6.** Would it be possible for two pure substances to have exactly the same set of properties? Give reasons for your answer.

**7.** Chemical changes can be both useful and destructive to humanity's purposes. Cite a few examples of each kind with which you have had personal experience. Also give observed evidence that each is indeed a chemical change and not a physical change.

**8.** For many years water was thought to be an element. Explain.

**9.** Name two pure substances that are used at the dinner table. Identify each as an element or compound.

**10.** Classify each of the following as a physical property or a chemical property. Justify in terms of operational definitions.

a. density
b. melting temperature
c. substance decomposes into two elements upon heating.
d. electrical conductivity of a solid
e. the substance does not react with sulfur
f. ignition temperature of a piece of paper

**11.** Classify each of the following as an element, compound or mixture. Justify each answer.

a. mercury
b. milk
c. pure water
d. a tree
e. ink
f. iced tea
g. pure ice
h. carbon

**12.** Which of the materials listed in Question 11 can be pure substances?

**13.** Explain how the operational definition of a pure substance allows for the possibility that it is not actually pure.

**14.** Why do theoretical definitions come after operational definitions in a particular concept?

**15.** Is it possible for the properties of iron to change? What about the properties of steel? Explain your answer.

**16.** Suggest a method for purifying water slightly contaminated with a dissolved solid.

**17.** Did most purification techniques arise from theory or practice?

**18.** Distinguish between theory and law as used in chemistry.

(a) Which has a better chance of being true? Why?
(b) Which summarizes? Which explains?

**19.** What is the most abundant element in the universe?

**20.** Name an element that is a solid at room temperature. A gas. A liquid.

**21.** Given the following sentence, write a chemical reaction using chemical symbols that convey the same information. "One nitrogen molecule containing two nitrogen atoms per molecule reacts with three hydrogen molecules containing two hydrogen atoms per molecule to produce two ammonia molecules containing one nitrogen and three hydrogen atoms per molecule."

**22.** Aspirin is a pure substance, a compound of carbon, hydrogen, and oxygen. If two manufacturers produce equally pure aspirin samples, what can be said of the relative worth of the two products?

**23.** Is it possible to have a mixture of two elements and also to have a compound of the same two elements? Explain. Can you think of an example?

**24.** How is the salt content of the sea related to the purity of rain water? What method of purification does nature employ in the purification of rain water?

**25.** Give an example of a chemical fact.

**26.** Give an example of a chemical law.

**27.** Give an example of an assumption made to explain an observed fact.

**28.** How many times do you think a given experiment should give a result before a scientific fact is established? How many failures would you require before rejecting the "fact"?

**29.** As you look up from this page, which are most abundant in your field of view: mixtures, compounds, or elements?

**30.** Suppose a mother and her children discover the family car missing on returning from an afternoon ball game, even though the father rode to work with a neighbor that morning. The mother says to the children, "Don't worry, Dad must have had an unexpected need for the car and got it after lunch." Would you call the statement made by the mother a theory or a hypothesis? Why?

**31.** Is it a fact that a water molecule is made up of two atoms of hydrogen and one atom of oxygen? Explain.

**32.** From the molecular formulas given below, tell what kind of atoms and how many of each kind are present in a molecule. (You may look up the names of any elements whose symbols you do not know in the list printed inside the cover.)

$SO_3$, $HCl$, $NH_3$, $H_2S$

**33.** How many atoms compose a molecule of $N_2O_5$? $H_3PO_4$? $C_{12}H_{22}O_{11}$?

**34.** Use bolts, taps, and bolt-tap combinations to illustrate some properties that distinguish elements from compounds and these, in turn, from mixtures.

**35.** Describe in words the chemical process which is summarized in the following equation:

$$2Na + Cl_2 \longrightarrow 2NaCl$$

**36.** How many *atoms* are present in each of the following:

a. one mole of He
b. one mole of $Cl_2$
c. one mole of $O_3$

**37.** Name ten elements and give the symbol for each one.

**38.** Describe the meaning of each symbol and number in the chemical equation:

$$2NO_2 \longrightarrow N_2O_4$$

**39.** Would you think that tea in tea bags is a pure substance? Use the process of making tea to make an argument for your answer. How would your argument apply to instant tea?

**40.** Which contains more atoms?

a. one mole of water, $H_2O$
b. one mole of hydrogen, $H_2$
c. one mole of oxygen, $O_2$

**41.** Do elements sometimes retain their physical properties in the formation of compounds? Give two examples to support your argument.

**42.** Chemical formulas are to compounds as chemical equations are to ______________.

**43.** What is the most fundamental assumption relative to structure and properties in chemical theory?

**44.** Consider the equation:

$$2C_7H_{16} + 21O_2 \longrightarrow 14CO_2 + 14H_2O$$

| $2C_7H_{16}$ | $21O_2$ | $14CO_2$ | $14H_2O$ |
|---|---|---|---|
| *Heptane (a component of gasoline)* | *Oxygen* | *Carbon dioxide* | *Water* |

Write out in words as much of the information presented in this equation as you can decipher.

**45.** In the normal usage of the terms atom and molecule, which is composed of the other?

**46.** The number 12 is to a dozen, and 144 is to a gross, as $6.023 \times 10^{23}$ is to a(n) ______________.

**47.** If you had a mole of elephants, how many moles of elephant ears would you have? A mole of $O_3$ molecules contains how many moles of oxygen atoms?

**48.** Which English unit is closest to a liter? A meter?

**49.** How tall are you in meters?

**50.** What is your weight in kilograms?

**51.** Which is colder: 0°C or 0°F?

**52.** Convert the distance of your last auto trip into kilometers.

**53.** Library assignment. Look up other purification methods such as sublimation, extractions, and zone refining, and tell in each case how the method is used in the separation of pure substances.

**54.** Which is larger, one BTU or one calorie?

# 2 Atoms

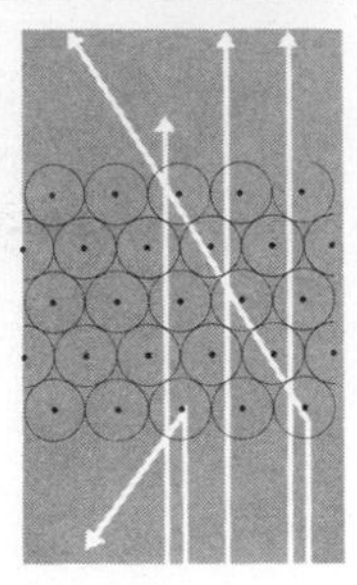

As you focus on a dot over an "i," can you visualize the thousands and thousands of individual, very small atoms in the dot? Trying to fathom the minuteness of the atom is as deeply challenging to the human mind as trying to fathom the wholeness of the universe. One lures the mind to unseen smallness; the other to unseen largeness. If atoms are too small for us to observe, how do we know they exist? And if there were some way we could observe an individual atom, would there be some way we could probe inside it?

Answers to these questions are the focus of this chapter. Assembling the pieces of the puzzle has taken more than two thousand years, but much of the work has been done in the present era. What difference does it make what is inside atoms, or even if they exist? The drive to discover is a strong force in certain individuals and cannot be discounted. Simply, it is thrilling to discover what makes nature tick. A knowledge of the atom leads to a deep and enlightened understanding of bonding, chemical reactivity, light, and other intriguing phenomena. Perhaps the most powerful outcome of knowledge about atoms is the ability to predict accurately the properties of matter.

Democritus

**Since the writings of Leucippus and Democritus have been destroyed, we know about their ideas only from recorded opposition to atoms and from a lengthy poem (55 B.C.) by the Roman poet, Lucretius.**

## The Greek Influence

The ancient Greeks recorded the first theory of atoms. Leucippus and his student, Democritus (460–370 B.C.), argued for the concept of atoms. Democritus used the word ***atom*** to describe the ultimate particles of matter, particles that could not be divided further. He reasoned that in the division of a piece of matter, such as gold, into smaller and smaller pieces, one would ultimately arrive at a tiny particle of gold that could not be further divided and still retain the properties of gold. The atoms that Democritus envisioned representing different substances were all made of the same basic material. His atoms differed only in shape and size.

Democritus used his concept of atoms to explain the properties of substances. For example, the high density and softness of lead could be caused by lead atoms packing very close together like marbles in a box and moving easily one over the other. Iron was known to be a less dense metal that is quite hard. Democritus argued that the properties of iron resulted from atoms shaped like corkscrews, atoms that would entangle in a rigid but relatively lightweight structure. Although his concept of the atom was limited, Democritus did explain in a simple way some well-known phenomena such as the drying of clothes, how moisture appears on the outside of a cold glass of water, how an odor moves through a room, and how crystals grow from a solution. He imagined the scattering or collecting of atoms as needed to explain the events he saw. All atomic theory has been built on the assumption of Leucippus and Democritus: atoms, which we cannot see individually, are the cause of the phenomena that we can see.

Plato (427–347 B.C.) and Aristotle (384–322 B.C.) led the arguments against the atom by asking to be shown atoms. They also argued that the idea of atoms was a challenge to God. If atoms could be used to explain nature, there would be no need for God. For centuries most of those in the mainstream of enlightened thought rejected or ignored the atoms of Democritus.

Ideas about atoms drifted in and out of philosophical discussions for about 2,200 years without playing a major role. Galileo (1564–1642) reasoned that the appearance of a new substance through chemical change involved a rearrangement of parts too small to be seen. Francis Bacon (1561–1626) speculated that heat might be a form of motion by very small particles. Robert Boyle (1627–1691) and Isaac Newton (1642–1727) used atomic concepts to interpret physical phenomena.

**Although the idea was proposed three and a half centuries earlier by Roger Bacon, it was not until 1620 that Francis Bacon wrote his book, *New Organon,* which put experimental science in the most refined and scholarly terms and made it possible for other scholars to accept it.**

It was John Dalton (1766–1844), an English schoolteacher, who forcefully revived the idea of the atom. By Dalton's time experimental results had gained a position of greater respect than authoritative opinions. More clearly than any before him, Dalton was able to explain general observations, experimental results, and laws relative to the composition of matter. Dalton was particularly influenced by the experiments of two Frenchmen, Antoine Lavoisier (1743–1794) and Joseph Louis Proust (1754–1826). We shall look at the major contributions of these two experimentalists before we examine Dalton's theory.

## Antoine Laurent Lavoisier: The Law of Conservation of Matter in Chemical Change

There are many reasons why Antoine Laurent Lavoisier has been acclaimed the "Father of Chemistry." He clarified the confusion over the cause of burning. He wrote an important textbook of chemistry, *Elementary Treatise on Chemistry.* He was the first to use systematic names for the elements and a few of their compounds. While he made still other contributions, his most notable achievement was to show the importance of very accurate weight measurements of chemical changes. His work began the process of establishing chemistry as a quantitative science.

Lavoisier weighed the chemicals in such changes as the decomposition of mercury (II) oxide by heat into mercury and oxygen.

$$2HgO \longrightarrow 2Hg + O_2$$

Mercury (II) Oxide → Mercury + Oxygen

Very accurate measurements showed that the total weight of all the chemicals involved remained constant during the course of the chemical change. Similar measurements on many other chemical reactions led Lavoisier to the summarizing statement now known as the ***Law of Conservation of Matter:*** *Matter is neither lost nor gained during a chemical reaction.* In other words, if one weighed all of the products of a chemical reaction—solids, liquids, and gases—the total would be the same as the weight of the reactants. Substances can be destroyed or created in a chemical reaction, but matter cannot. As a further example of the law of conservation of matter, consider Figure 2–1.

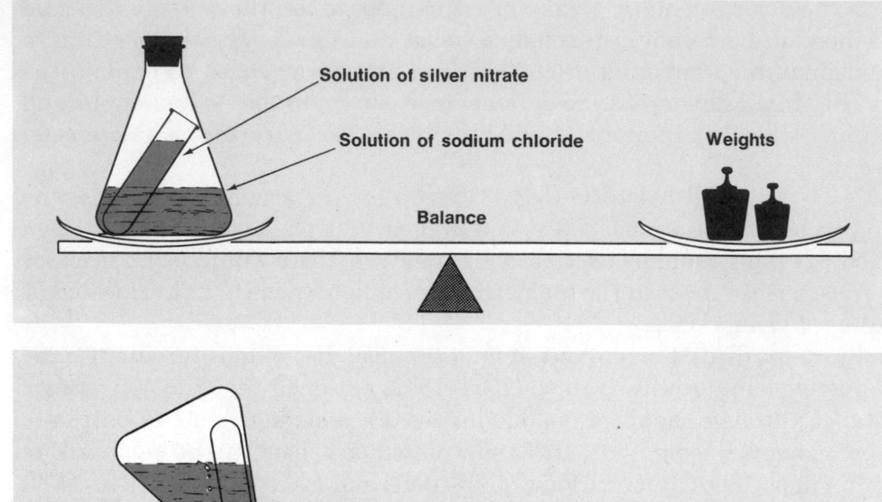

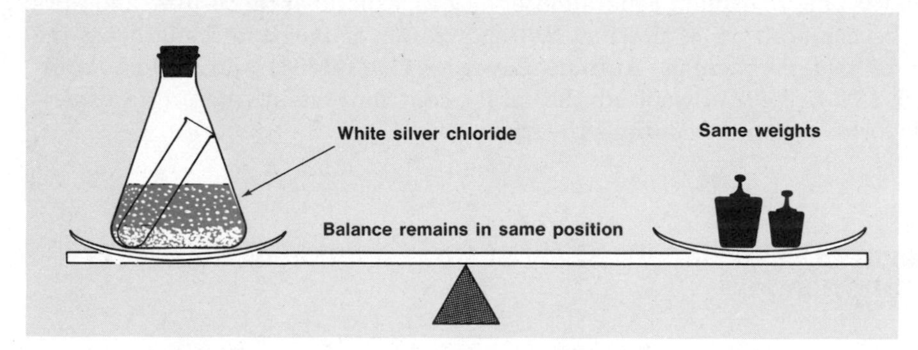


**Figure 2–1** Mixing a solution of sodium chloride with a solution of silver nitrate produces a new substance, solid silver chloride, but the total weight of matter remains the same.

**The Personal Side**

With all of his success, Lavoisier had his problems and disappointments. His highest goal, that of discovering a new element, was never achieved. He lost some of the esteem of his colleagues when he was accused of saying the work of someone else was his own. In 1768, he invested half a million francs in a private firm retained by the French government to collect taxes. He used the earnings (about 100,000 francs a year) to support his research. Although Lavoisier was not actively engaged in tax-collecting, he was brought to trial as a "tax-farmer" during the French Revolution. Lavoisier, along with his father-in-law and other tax-farmers, was guillotined on May 8, 1794, just two months before the end of the revolution.

## Joseph Louis Proust: The Law of Constant Composition

Following the lead of Lavoisier, several chemists investigated the quantitative aspects of compound formation. One such study, made by Proust in 1799, involved copper carbonate. Proust discovered that, regardless of how copper carbonate was prepared in the laboratory or how it was isolated from nature, copper carbonate always contained five parts of copper, four parts of oxygen, and one part of carbon by weight. His careful analyses of this and other compounds led to the belief that a given compound has an unvarying composition. These and similar discoveries are summarized by the ***Law of Constant Composition:*** *In a compound, the constituent elements are always present in a definite proportion by weight.*

**The Personal Side**

Proust's generalization has been verified many times for many compounds since its formulation, but its acceptance was delayed by controversy. Comte Claude Louis Berthollet (1748–1822), an eminent French chemist and medical doctor, believed and strongly argued that the nature of the final product was determined by the amount of reacting materials one had at the beginning of the reaction. The running controversy between Proust and Berthollet reached major proportions, but more careful measurements supported Proust. Proust showed that Berthollet had made inaccurate analyses and had purified his compounds insufficiently—two great errors in chemistry.

Unlike Lavoisier, Proust saved his head during the French Revolution. Proust fled to Spain, where he lived in Madrid and worked as a chemist under the sponsorship of Charles IV, King of Spain. When Napoleon's army ousted Charles IV, Proust's laboratory was looted and his work came to an end. Later, Proust returned to his homeland, where he lived out his life in retirement.

---

Pure water, a compound, is always made up of 11.2% hydrogen and 88.8% oxygen by weight. Pure table sugar, another compound, always contains 42.11% carbon by weight. Contrast these with 14-carat gold, a mixture which should be at least 58% gold, from 14 to 28% copper, and 4 to 28% silver by weight. This mixture can vary in composition and still be properly called 14-carat gold, but a compound that is not 11.2% hydrogen and 88.8% oxygen is not water.

## John Dalton: The Law of Multiple Proportions

John Dalton

John Dalton made a quantitative study of different compounds made from the same elements. Such compounds differed in composition from each other, but each obeyed the law of constant composition. Examples of this concept are the compounds carbon monoxide, a poisonous gas, and carbon dioxide, a product of respiration. Both compounds contain only carbon and oxygen. Carbon monoxide is made up of carbon and oxygen in proportions by weight of 3 to 4. Carbon dioxide is made up of carbon and oxygen in proportions by weight of 3 to 8. Note that for equal amounts of carbon (three parts), the ratio of oxygen in the two compounds is a ratio of 8 to 4, or 2 to 1.

In 1803, after analyzing compounds of carbon and hydrogen such as methane (in which the ratio of carbon to hydrogen is 3 to 1 by weight) and ethylene (in which the ratio of carbon to hydrogen is 6 to 1 by weight), and compounds of nitrogen and oxygen, Dalton first clearly enunciated the ***Law of Multiple Proportions:*** *In the formation of two or more compounds from the same elements, the weights of one element that combine with a fixed weight of a second element are in a ratio of small whole numbers (integers) such as 2 to 1, 3 to 1, 3 to 2, or 4 to 3.*

**Methane is the main component of natural gas. Ethylene is the only component of polyethylene.**

## Dalton's Atomic Theory

Why do the law of conservation of matter, the law of constant composition, and the law of multiple proportions exist? How can they be explained? John Dalton employed the idea of atoms and endowed them with properties that enabled him to explain these chemical laws.

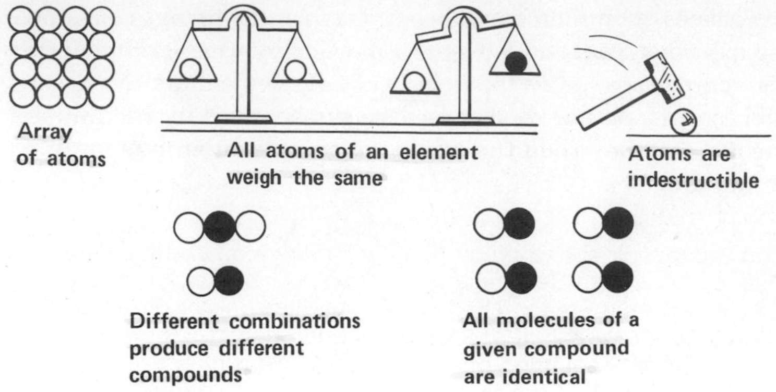

Array of atoms

All atoms of an element weigh the same

Atoms are indestructible

Different combinations produce different compounds

All molecules of a given compound are identical

**Figure 2–2** Features of Dalton's atomic theory.

**The Personal Side**

While Lavoisier is considered the father of chemical measurement, Dalton is considered the father of chemical theory. Dalton, a gentle man and a devout Quaker, gained acclaim because of his work. He made careful measurements, kept detailed records of his research, and expressed them convincingly in his writings. However, he was a very poor speaker and was not well received as a lecturer. When Dalton was 66 years old, some of his admirers sought to present him to King William IV. Dalton resisted because he would not wear the court dress. Since he had a doctor's degree from Oxford University, the scarlet robes of Oxford were deemed suitable, but a Quaker could not wear scarlet. Dalton, being colorblind, saw scarlet as gray, so he was presented in scarlet to the court but in gray to himself. This remarkable man was, in fact, the first to describe color blindness. He began teaching in a Quaker school when only 12 years old, discovered a basic law of physics, the law of partial pressure of gases, and helped found the British Association for the Advancement of Science. He kept over two hundred thousand notes on meteorology. Despite his accomplishments he shunned glory, and maintained he could never find time for marriage.

---

The major points of Dalton's theory, presented in modernized statements, are:

1. Matter is composed of indestructible particles called atoms.
2. All atoms of a given element have the same properties such as size, shape, and weight,* which differ from the properties of atoms of other elements.
3. Elements and compounds are composed of definite arrangements of atoms, and chemical change occurs when the atomic arrays are rearranged.

Dalton's theory was successful in explaining the three laws of chemical composition and reaction. See Figure 2–3.

## Dalton's Idea about Atomic Weights: The Follow-up

John Dalton's idea about unique atomic weights for the atoms of the different elements generated interest in searching for the atomic weight characteristic of each element. Not until after 1860 did a consistent set of atomic weights appear, although several notable attempts were made before that, including the first table prepared by Dalton. In September 1860, many of the most brilliant minds in chemistry met in Karlsruhe, Germany, to discuss the inconsistencies in the atomic weights proposed at

*Contrary to Dalton's belief, all of the atoms of the same element do not have the same weight. The idea of isotopes is introduced later in this chapter.

that time. There were differences of opinion on whether the formula of water was HO or $H_2O$, whether hydrogen gas was $H_2$ or H, and whether oxygen gas was $O_2$ or O. Water is 88.8% oxygen and 11.2% hydrogen by weight—a firmly established experimental fact by that time. If water is HO, as Dalton argued (based on his belief that the simplest formula is likely to be the correct one), then the weight of an oxygen atom should be about eight times that of a hydrogen atom:

$$\frac{\text{weight of an oxygen atom}}{\text{weight of a hydrogen atom}} = \frac{88.8}{11.2} = \frac{7.9}{1}$$

If the formula for water is $H_2O$, as the scientist Amedeo Avogadro (1776–1856) had proposed in 1811, then one oxygen atom is 88.8% of the molecule but two hydrogen atoms are 11.2%. Each hydrogen atom would be $\frac{1}{2}$(11.2%) or 5.6%. With the formula

| Law | Statement of Law | Explanation of Law |
|---|---|---|
| Law of Conservation of Matter | Matter is neither lost nor gained in a chemical change. | A chemical change is the result of a new arrangement of the same atoms present initially; hence, the weight is the same before and after the change. |

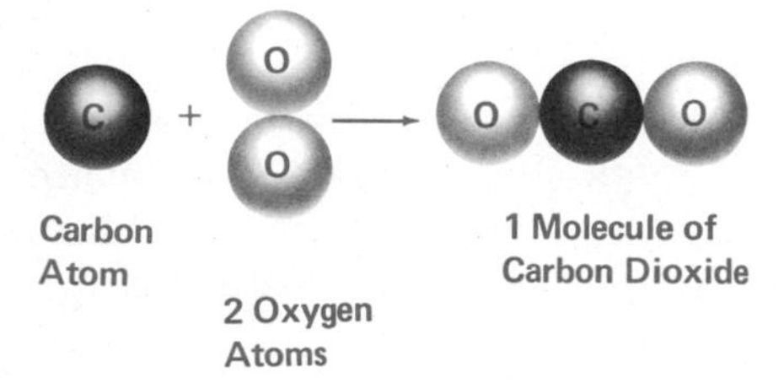


| Law | Statement of Law | Explanation of Law |
|---|---|---|
| Law of Constant Composition | When two or more elements combine to form a given compound, the ratio of the weights of the elements involved is always the same. | The smallest unit of a compound is a molecule. It has a fixed ratio of atoms, hence a fixed ratio of weights. Any larger sample of this compound would merely represent a multiple of the weights in the same ratio. |

Carbon monoxide

| Law | Statement of Law | Explanation of Law |
|---|---|---|
| Law of Multiple Proportions | In the formation of two or more compounds from the same elements, the weights of one element that combine with a fixed weight of a second element are in a ratio of integers such as 2:1, 3:1, 3:2, or 4:3. | If, for example, the first compound has a ratio of one atom of C to one atom of O (above), and a second compound has a ratio of one atom of C to two atoms of O (below), then for a fixed number of atoms of C, the ratio of atoms of O (and the weights of O) is a ratio of integers: 1:2. |

Carbon dioxide

**Figure 2–3** John Dalton's explanation of three laws of chemistry in terms of atoms.

$H_2O$, then, an oxygen atom would be about 16 times heavier than a hydrogen atom:

$$\frac{\text{weight of an oxygen atom}}{\text{weight of a hydrogen atom}} = \frac{88.8}{5.6} = \frac{15.9}{1}$$

Near the end of the meeting at Karlsruhe, Stanislao Cannizzaro (1826–1910) argued for the ideas of Avogadro. In spite of his arguments, which later proved to be correct, the confusion about atomic weights was not resolved during the conference. At the close of the meeting, however, a friend of Cannizzaro named Angelo Pavesi distributed copies of a paper written by Cannizzaro two years earlier. Several years later, the chemical community finally accepted $H_2O$ as the formula of water, and a consistent set of atomic weights was generally agreed upon and used.

The fact that an oxygen atom is about 16 times heavier than a hydrogen atom does not tell us the weight of either atom. These are relative weights in the same way that a grapefruit may weigh twice as much as an orange. This information gives neither the weight of the grapefruit nor that of the orange. However, if a specific number is *assigned* as the weight of any particular atom, this fixes the numbers assigned to the weights of all other atoms. The standard for comparison of atomic weights was for many years the weight of the oxygen atom, which was taken as 16.0000 atomic weight units. This allowed the lightest atom, hydrogen, to have an atomic weight of 1.008, or approximately one.

The modern set of atomic weights (inside the back cover) is an outgrowth of the set of weights begun in the 1860's. The present atomic weight scale, adopted in 1961, is based on assigning the weight of a particular kind of carbon atom, the carbon-12 atom, as exactly 12 atomic weight units. On this scale, an atom of Mg with an atomic weight of about 24 has twice the weight of a carbon-12 atom. An atom of titanium, Ti, with an atomic weight of 48 has four times the weight of a carbon-12 atom.*

Stanislao Cannizzaro (1826–1910), Italian chemist whose work resulted in the clarification of the atomic weight scale.

*What we are talking about here is really atomic mass. Atomic weight, an uncorrected misnomer from the past, persists today in, for example, the "Tables of Atomic Weights" published in most chemistry textbooks. If they were really atomic weights, they would change value wherever the force of gravity changes on earth (less at the equator, more at the poles). Instead, we have only one table for the whole world, which means atomic weights are really atomic masses. In this text, we shall use both terms: atomic weight because it is a practice of chemistry and a term possibly more familiar to students, and atomic mass where it seems necessary to clarify the thought.

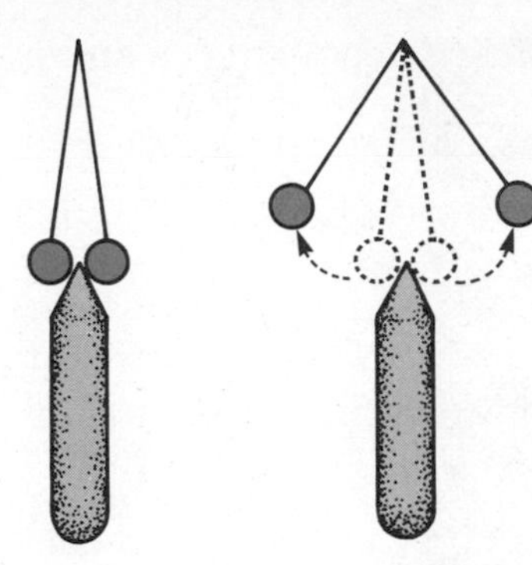

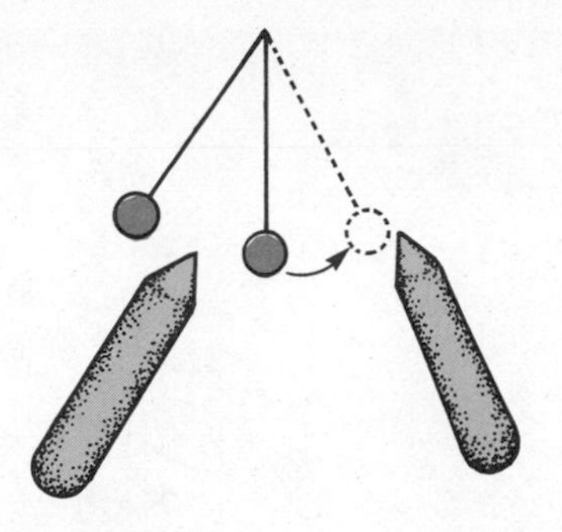

Neutral

Rubber rod rubbed with silk or wool and touched to pith balls.

Rubber rod rubbed with silk and touched to pith balls repels balls that have the same charge as the rod.

Rubber rod rubbed with wool attracts pith balls that have a charge opposite to that on the rod.

**Figure 2–4** Effects of charged matter on other charged matter. Like charges repel. Unlike charges attract.

## Atoms Are Divisible

Dalton's concept of the indivisibility of atoms was severely challenged by the subsequent discoveries of radioactivity and cathode rays, and was even in conflict with some previously known electrical phenomena such as static electrical charges.

Electrical charge was first observed and recorded by the ancient Egyptians, who noted that amber, when rubbed, attracted light objects. A bolt of lightning, a spark between a comb and hair in dry weather, and a shock on touching a doorknob are all results of the discharge of a buildup of electrical charge.

The two types of electrical charge had been discovered by the time of Benjamin Franklin (1706–1790). He named them positive (+) and negative (−) because they appear as opposites, in that they can neutralize each other. The existence and nature of the two kinds of charge, and their effects on each other, can be shown with a simple electroscope (Fig. 2–4). When a hard rubber rod is rubbed vigorously with silk and allowed to touch the lightweight balls, the balls spring apart immediately. The touching allowed the rod and the balls to share the same type of charge (positive). If the rod is then brought near one of the balls, the ball moves away from the rod. This movement indicates that *like charges repel.*

If the same rod is now rubbed vigorously with wool and brought near the charged balls, they move toward the rod. The opposite type of charge is now on the rod. The generalization is: *Unlike charges attract, and like charges repel.*

Radioactivity is a spontaneous process in which some natural materials give off very penetrating radiations. Henri Becquerel (1852–1908) discovered this property in natural uranium and radium ores in 1896. His student, Marie Curie (1867–1934), isolated the radioactive element radium and some of its pure compounds. It turns out that radioactivity is characteristic of the elements, not the compounds, and that about 25 elements are naturally radioactive.

Radioactive elements commonly emit alpha, beta, and gamma rays, as shown in Figure 2–5. Alpha and beta rays are composed of particles, while gamma rays have no detectable mass. Alpha particles have a mass of 4 on the carbon-12 atomic weight scale, positive charge, and low penetrating power (they will not penetrate skin, for example). In the arrangement shown in Figure 2–5, they are attracted toward the negatively charged plate. Beta particles have a mass of 0.0005 on the carbon-12 atomic weight scale, negative charge (they are attracted toward the positive plate), and enough penetrating power to go through kitchen-strength aluminum foil. Gamma rays are a

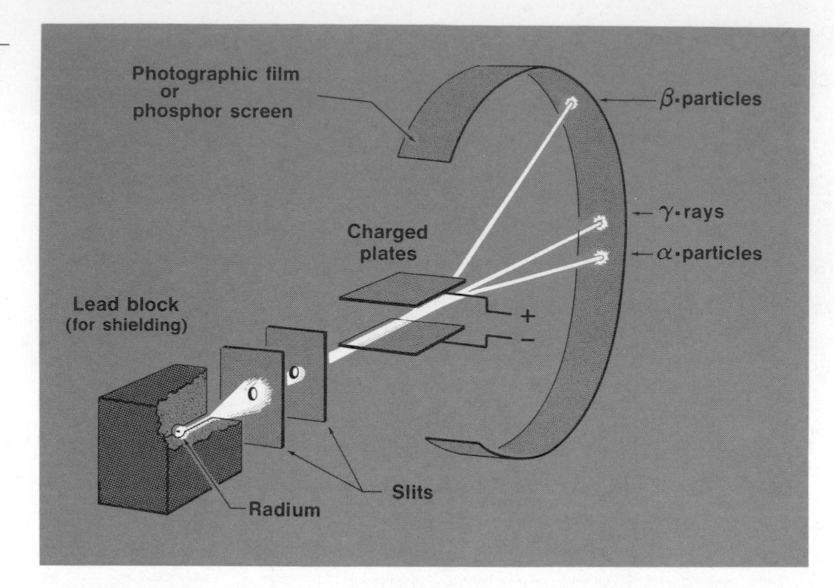


**Figure 2–5**
Separation of alpha, beta, and gamma rays by electrostatic field.

type of electromagnetic energy like light and X rays, but more penetrating. Gamma rays can penetrate a considerable thickness of aluminum, and even thin sheets of lead. They are not deflected at all by charged plates.

**TABLE 2–1 Summary of Properties of Alpha Particles, Beta Particles, and Gamma Rays**

| | Charge | Relative Mass | Symbols |
|---|---|---|---|
| Alpha particle | positive (+2) | 4 | $\alpha$, ${}^{4}_{2}\alpha$, ${}^{4}_{2}He$ |
| Beta particle | negative (−1) | 0.0005 | $\beta$, ${}^{0}_{-1}\beta$ |
| Gamma ray | neutral (0) | 0 | $\gamma$, ${}^{0}_{0}\gamma$ |

Cathode rays, as we shall see below, are similar to beta rays in that they are composed of negatively charged particles; their charge and mass are identical to those of beta particles.

The discoveries of natural radiation, cathode rays, and electrical charge are evidence that atoms can be divided and may even divide spontaneously. The smallest atom is 1836 times more massive than the beta or cathode-ray particle. Therefore, beta (and cathode-ray) particles appear to be subatomic in origin.

## Self-Test 2-A

1. Two Greek philosophers who were influential in advocating the concept of atoms were ______________ and ______________.
2. The Greek approach to the "discovery" of atoms can best be described as:
   a. experimentation
   b. philosophy (play on logic)
   c. direct observation of atoms
   d. consistent explanation of well-known, established laws of nature
   e. deductive reasoning
3. The law of conservation of matter states that matter is neither lost nor ______________ in a ______________ reaction.

4. The law of multiple proportions explains the existence of compounds like ____________ and ____________.
5. a. Assume that you are a chemist of many years ago. Your field of study is compounds composed of nitrogen and oxygen only. There are several that you know about. One contains 16 grams of oxygen for every 14 grams of nitrogen, while another contains 32 grams of oxygen for every 14 grams of nitrogen. Your assistant discovers what he claims is a new compound of nitrogen and oxygen. Upon analysis, the compound is found to contain 8 g of O for every 14 g of N. Has your assistant discovered a new compound or is it one of the others? ____________
   b. What is the ratio by weight of oxygen in these compounds for a given weight of N?

   ____________
   c. What fundamental law of chemistry is illustrated by a comparison of the compounds? ____________
6. According to Dalton's atomic theory, what happens to atoms during a chemical change? Select one:
   a. atoms are made into new and different kinds of atoms
   b. atoms are lost
   c. atoms are gained
   d. atoms are recombined into different arrangements
7. According to Dalton's atomic theory, a compound has a definite percentage by weight of each element because
   a. All atoms of a given element weigh ____________.
   b. All molecules of a given compound contain a definite number and kind of

   ____________.
8. Like charges ____________; unlike charges ____________.
9. The three types of radiation from a radioactive element such as radium are ____________, ____________, and ____________, of which ____________ pass through an electrostatic field without being deflected.

---

## Electrons

The first ideas about electrons came from experiments with cathode-ray tubes. A forerunner of neon signs, fluorescent lights, and TV picture tubes, a typical cathode-ray tube is a partially evacuated glass tube with a piece of metal sealed in each end (Fig. 2–6). The pieces of metal are called electrodes; the one given a negative charge is called the ***cathode,*** and the one given a positive charge is called the ***anode.***

**Cathode rays are streams of the negatively charged particles called *electrons*.**

If a high electrical voltage is applied to the electrodes, an electrical discharge can be created between them. This discharge appears to be a stream of particles emanating from the cathode. This cathode ray will cause gases and fluorescent materials to glow, and will heat metal objects in its path to red heat. Cathode rays travel in straight lines and cast sharp shadows. Unlike light, however, cathode rays are attracted toward a positively charged plate. This led to the conclusion that cathode rays are negatively charged.

Careful microscopic study of a screen that emits light when struck by cathode rays shows that the light is emitted in tiny, random flashes. Thus, not only are cathode rays negatively charged, but they are composed of particles, each one of which

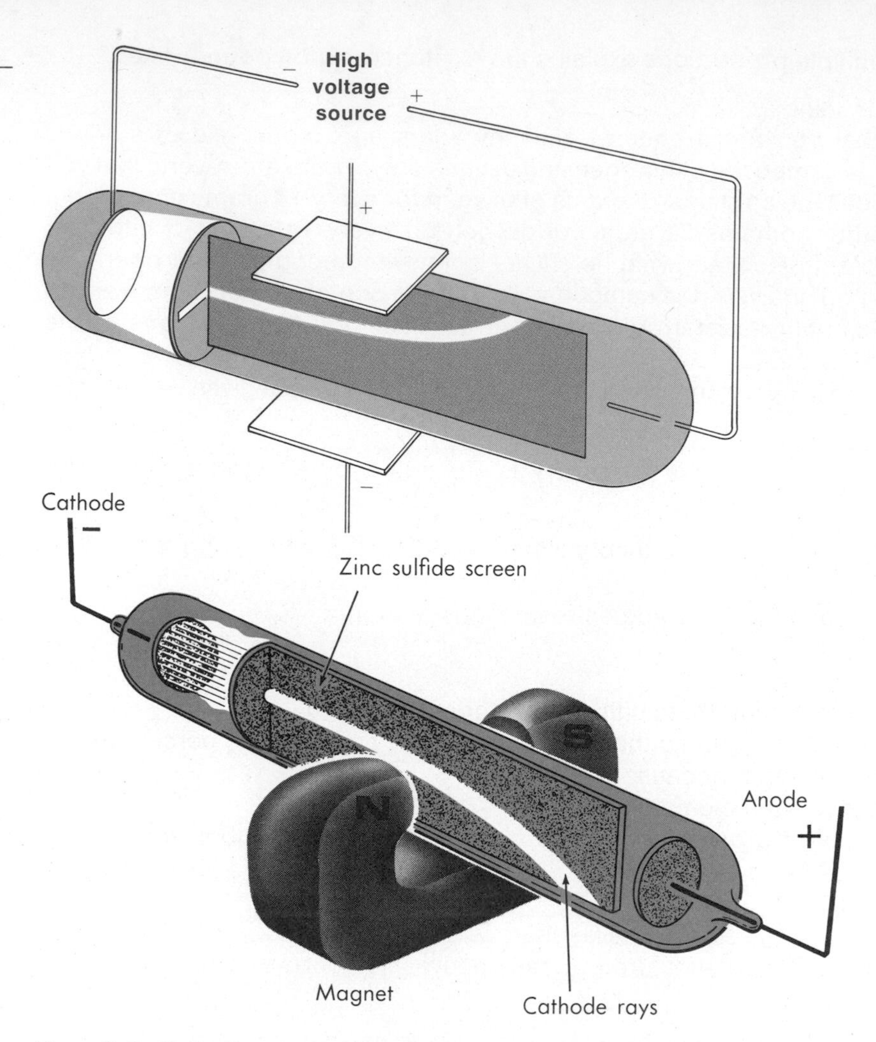


**Figure 2–6** Deflection of a cathode ray by an electric field and by a magnetic field. When an external electric field is applied, the cathode ray is deflected toward the positive pole. When a magnetic field is applied, the cathode ray is deflected from its normal straight path into a curved path.

produces a flash of light upon collision with the material of the screen. The cathode-ray particles became known as ***electrons.***

The charge and mass of an electron were determined by a combination of experiments by Sir Joseph John Thomson in 1897 and by Robert Andrews Millikan in 1911. Both scientists were awarded Nobel prizes, Thomson in 1906 and Millikan in 1923.

**Thomson discovered the charge-to-mass ratio of the electron.**

By using a specially designed cathode-ray tube (Fig. 2–7), Thomson applied electrical and magnetic fields to the rays. Using the basic laws of electricity and magnetism, he determined the charge-to-mass ratio of the electrons. He was able to measure neither the charge nor the mass of the electron, but he established the ratio between the two numbers and made it possible to calculate either one if the other could ever be measured. What Thomson did for the concept of the electron is like showing that a peach weighs 40 times more than its seed. What is the weight of the peach? What is the weight of the seed? Neither is known, but if it can be determined by other means that the peach weighs 120 grams, then the weight of the seed, by ratio, must be 3 grams.

**Electrons are present in all of the elements.**

An important part of Thomson's experimentation was his use of twenty different metals for cathodes and of several gases to conduct the discharge. Every combination

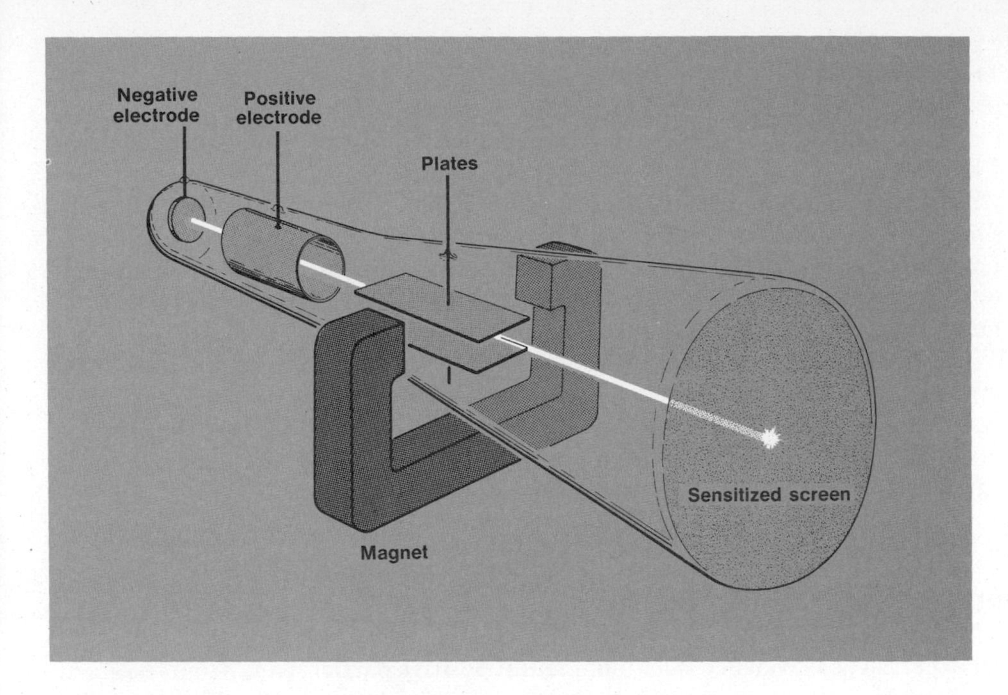


**Figure 2–7** J. J. Thomson experiment. Electric field, applied by plates, and magnetic field, applied by magnet, cancel each other's effects to allow cathode ray (electron beam) to travel in straight line.

of metals and gases yielded the same charge-to-mass ratio for the cathode rays. This led to the belief that electrons are common to all of the metals used in the experiments, and probably to all atoms in general.

**Millikan measured the charge on an electron.**

Millikan measured the fundamental charge of matter—the charge on an electron. A simplified drawing of his apparatus is shown in Figure 2–8. The experiment consisted of measuring the electrical charge carried by tiny drops of oil that are suspended in an electrical field. By means of an atomizer, oil droplets were sprayed into the test chamber. As the droplets settled slowly through the air, high-energy X rays were passed through the chamber to charge the droplets negatively (the X rays caused air molecules to give up electrons to the oil). By using a beam of light and a small telescope, Millikan could study the motion of a single droplet. When the electrical

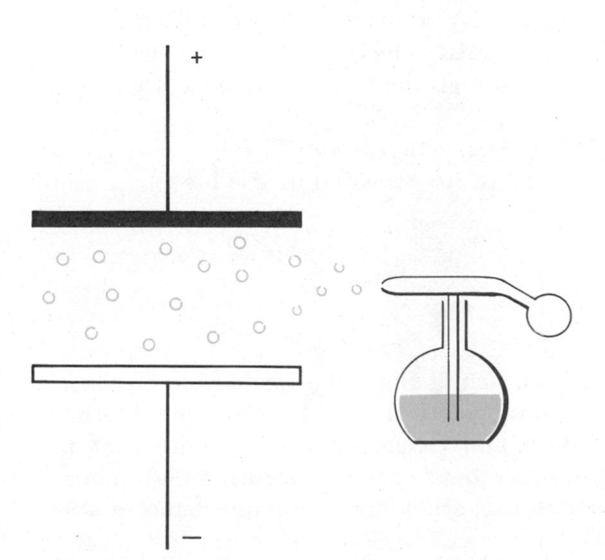


**Figure 2–8** Charged oil drops, suspended as a result of opposing gravitational and electrostatic forces, provided Millikan with the means of calculating the charge on the electron.

charge on the plates was increased enough to balance the effect of gravity, a droplet could be suspended motionless. At this point, the gravitational force would equal the electrical force. Measurements made in the motionless state, when inserted into equations for the forces acting on the droplet, enabled Millikan to calculate the charge carried by the droplet.

Only a whole number of electrons may be present in a sample of matter.

Millikan found different amounts of negative charge on different drops, but the charge measured each time was always a whole-number multiple of a very small basic unit of charge. The *largest* common divisor of all charges measured by this experiment was $1.60 \times 10^{-19}$ coulomb (the coulomb is a charge unit). Millikan assumed this to be the fundamental charge, which is the charge on the electron.

With a good estimate of the charge on an electron and the ratio of charge-to-mass as determined by Thomson, the very small mass of the electron could be calculated. The mass of an electron is $9.11 \times 10^{-28}$ g. On the carbon-12 relative scale, the electron would have a weight of 0.000549 atomic weight units. The negative charge on an electron of $-1.60 \times 10^{-19}$ coulomb is set as the standard charge of $-1$.

## Protons

The first experimental evidence of a fundamental positive particle came from the study of canal rays. A special type of cathode-ray tube produces canal rays (Fig. 2–9). The cathode is perforated, and the tube contains a gas at very low pressure. When high voltage is applied to the tube, cathode rays can be observed between the electrodes as in any cathode-ray tube. On the other side of the perforated cathode, a different kind of a ray is observed. These rays are attracted to a negative plate brought alongside the rays. The rays must therefore be composed of positively charged particles. Each gas used in the tube gives a different charge-to-mass ratio for the positively charged particles. When hydrogen gas was used, the largest charge-to-mass ratio was obtained, indicating that hydrogen provides the positive particles with the smallest mass. This particle was considered to be the fundamental positively charged particle of atomic structure, and was called a ***proton*** (from Greek for "the primary one").

Canal rays are streams of positive ions derived from the gases present in the discharge tube.

Experiments on canal rays were begun in 1886 by E. Goldstein and further work was done later by W. Wien. The production of canal rays is caused by high-energy electrons moving from the negative cathode to the positive anode, hitting the molecules of gases occupying the tube. Electrons are knocked from some atoms by the high-energy electrons, leaving each molecule with a positive charge. The positively charged molecules are then attracted to the negative electrode. Since the electrode is perforated, some of the positive particles go through the holes or channels (hence the name canal rays).

An *ion* is an atom or a group of atoms carrying a charge.

The mass of the proton is $1.67261 \times 10^{-24}$ g, which is 1.00727 relative weight on the carbon-12 scale. The charge of $+1$ on the proton is equal in size but opposite in effect to the charge on the electron.

## Neutrons

The third type of particle in an atom was hard to find. Since the particle has no charge, the usual methods of detecting small individual particles could not be used. Methods using such instruments as cloud chambers and Geiger counters require that the radiation either be charged particles (alpha or beta particles, for example) or high-energy radiation (gamma rays, for example). Neither method could detect a slow-moving, neutral particle.

The current interest in neutrons with regard to the neutron bomb is discussed in Chapter 14.

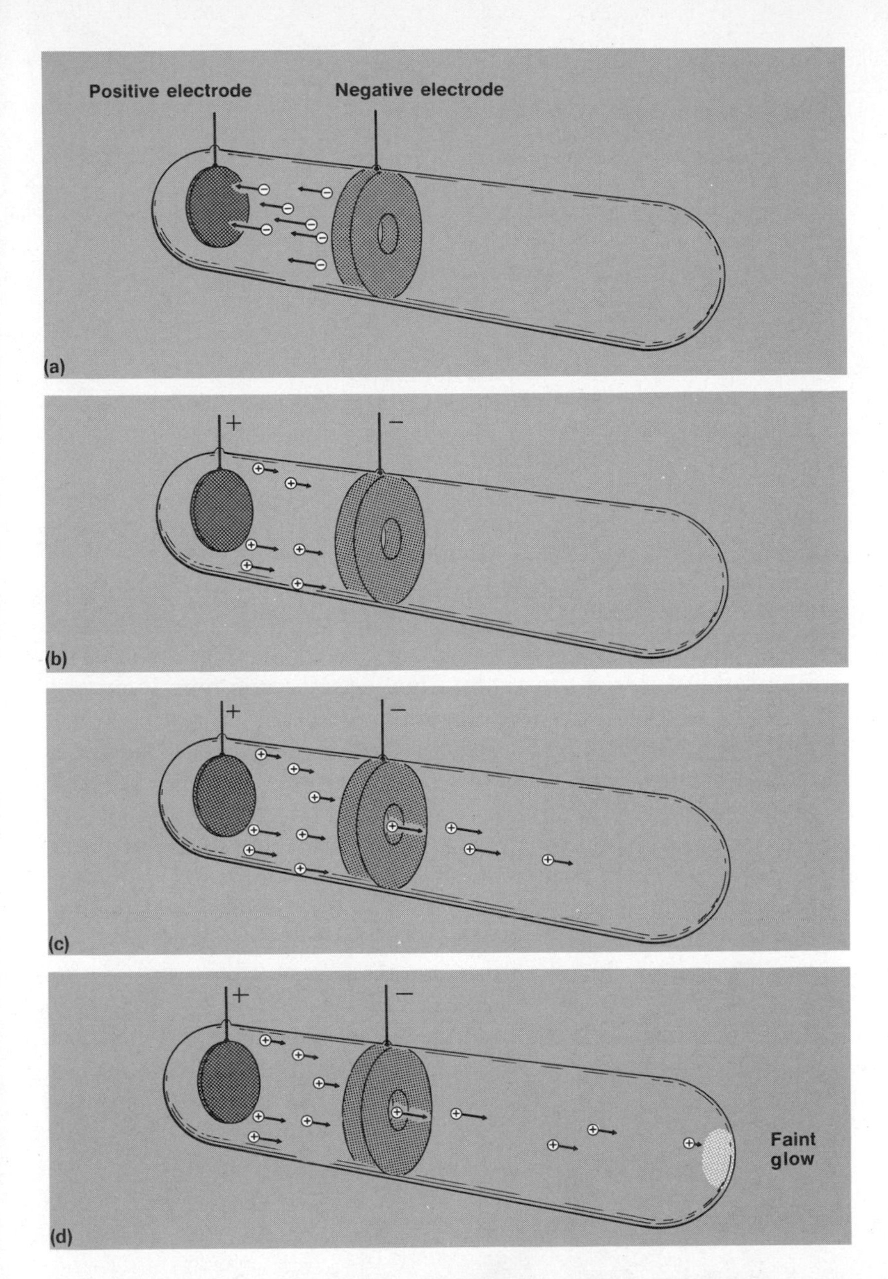


**Figure 2–9** (*a*) Electrons rush from the negative electrode to the positive electrode due to high voltage. (*b*) Electrons collide with gas molecules to produce positive ions that are accelerated toward the negative electrode. (*c*) Some of the positive ions escape capture by the electrode and rush through the opening due to their kinetic energy. (*d*) Some of the positive ions in the positive ray collide with gas molecules to produce a characteristic glow, and strike the end of the glass tube to produce a luminous spot.

It remained for James Chadwick, in 1932, to devise a clever experiment that produced neutrons by a nuclear reaction and then detected them by having the neutrons knock hydrogen ions, a detectable species, out of paraffin.

**Paraffin is the material that seals home-canned strawberry preserves.**

A neutron has no electrical charge and has a mass of $1.67492 \times 10^{-24}$ g, which is a relative weight of 1.00867 on the carbon-12 scale.

**TABLE 2–2 Summary of Properties of Electrons, Protons, and Neutrons**

| | Relative Charge | Relative Mass | Location |
|---|---|---|---|
| Electron | −1 | 0.00055 | outside the nucleus |
| Proton | +1 | 1.00727 | nucleus |
| Neutron | 0 | 1.00867 | nucleus |

## The Nucleus

**Alpha particles are scattered by the nuclei of the gold atoms.**

When Ernest Rutherford and his students directed alpha particles toward a very thin sheet of gold foil in 1909, they were amazed to find a totally unexpected result (Fig. 2–10). As they had expected, the paths of most of the alpha particles were only slightly changed as they passed through the gold foil. The extreme deflection of a few of the alpha particles was a surprise. Some even "bounced" back toward the source. Rutherford expressed his astonishment by stating that he would have been no more surprised if someone had fired a 15-inch artillery shell into tissue paper and then found it in flight back toward the cannon.

What allowed most of the alpha particles to pass through the gold foil in a rather straight path? According to Rutherford's interpretation, the atom is mostly empty space and, therefore, offers little resistance to the alpha particles (Fig. 2–11).

What caused a few alpha particles to be deflected? According to Rutherford's interpretation, concentrated at the center of the atom is a **nucleus** containing most of the mass of the atom and all of the positive charge. When an alpha particle passes near the nucleus, the positive charge of the nucleus repels the positive charge of the alpha particle; the path of the smaller alpha particle is deflected. The closer an alpha particle comes to a target nucleus, the more it is deflected. Those alpha particles that meet a nucleus head on are "bounced" back toward the source by the strong positive-positive repulsion, since the alpha particles do not have enough energy to penetrate the nucleus.

Rutherford's calculations, based on the observed deflections, indicate that the nucleus is a very small part of an atom. An atom occupies about a million million times

Lord Rutherford (Ernest Rutherford, 1871–1937) was Professor of Physics at Manchester when he and his students discovered the scattering of $\alpha$ particles by matter. Such scattering led to the postulation of the nuclear atom. For this work he received the Nobel prize in 1908. In 1919, Rutherford discovered and characterized nuclear transformations.

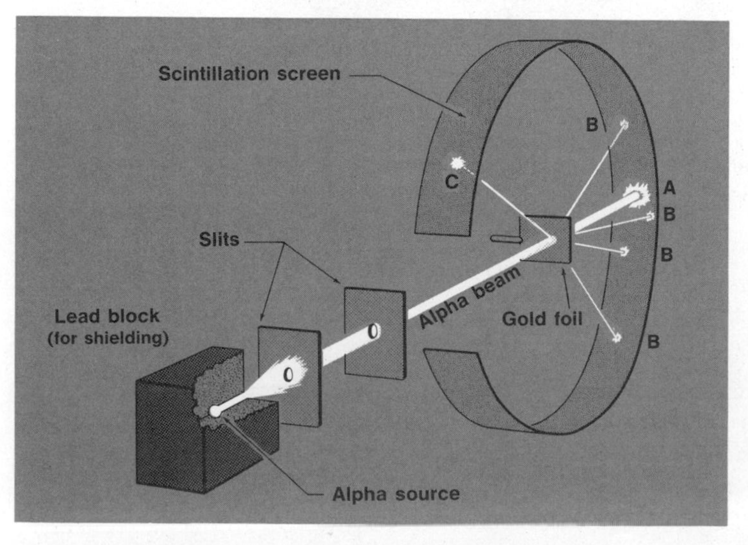


**Figure 2–10** Rutherford's gold foil experiment. A circular scintillation screen is shown for simplicity; actually, a movable screen was employed. Most of the alpha particles pass straight through the foil to strike the screen at point A. Some alpha particles are deflected to points B, and some are even "bounced" backwards to points such as C.

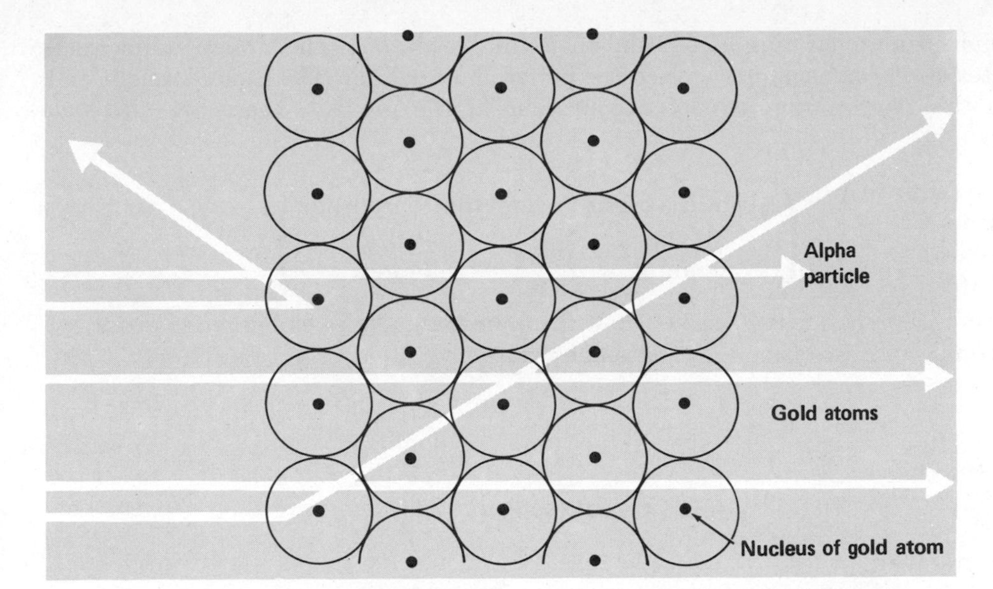


**Figure 2–11** Rutherford's interpretation of how alpha particles interact with atoms in a thin gold foil. Actually, the gold foil was about 1000 atoms thick. For illustration purposes, points are used to represent the gold nuclei and the pathwidths of the alpha particles are drawn larger than scale.

more space than does a nucleus; the radius of an atom is about 10,000 times greater than the radius of its nucleus. Thus, if a nucleus were the size of a baseball, then the edges of the atom would be about one third of a mile away. And most of the space in between would be absolutely empty.

Since the nucleus contains most of the mass and all of the positive charge of an atom, the nucleus must be composed of the most massive atomic particles, the protons and neutrons. The electrons are distributed in the near-emptiness outside the nucleus.

Alpha-particle scattering can be explained if the nucleus occupies a very small volume of the atom.

Truly, Rutherford's model of the atom was one of the most dramatic interpretations of experimental evidence to come out of this period of significant discoveries.

## Atomic Number

The ***atomic number*** of an element indicates the number of protons in the nucleus of the atom, which is the same as the number of electrons outside the nucleus. The two types of particles must be present in equal numbers for the atom to be neutral in charge. Note that the periodic table of the elements, inside the back cover, is an arrangement of the elements consecutively according to atomic number. Beginning with the atomic number 1 for hydrogen, there is a different atomic number for each element.

The lightest atom is the hydrogen atom.

$$\begin{pmatrix}\text{Number of}\\ \text{electrons}\\ \text{per atom}\end{pmatrix} = \begin{pmatrix}\text{Number of}\\ \text{protons}\\ \text{per atom}\end{pmatrix} = \begin{pmatrix}\text{Atomic number}\\ \text{of the}\\ \text{element}\end{pmatrix}$$

The ***atomic mass*** of a particular atom is the sum of the masses of the protons, neutrons, and electrons in that atom. Since an electron has such a small mass, the atomic mass is essentially the sum of the masses of the protons and neutrons in the nucleus. Both protons and neutrons have masses of approximately 1.0 on the atomic weight scale. Hydrogen, with an atomic weight of 1, must be composed of one proton (and no neutrons) in the nucleus, and one electron outside:

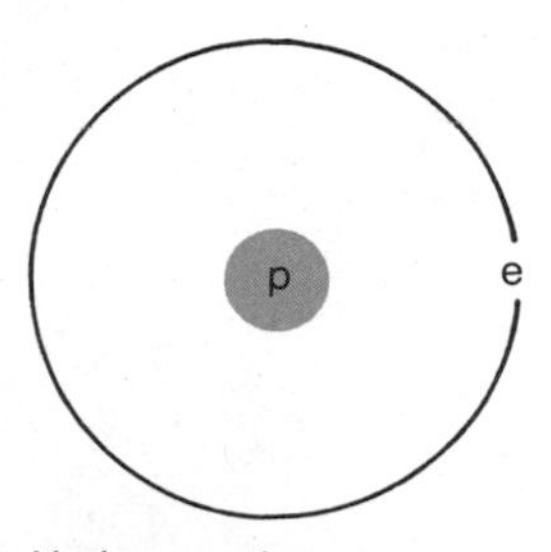


Hydrogen atom

Helium has an atomic number of 2 and an atomic weight of 4. The atomic number of 2 indicates two protons and two electrons per atom of helium. The atomic weight of 4 means that, in addition to the two protons in the nucleus, there are also two neutrons.

$$\begin{pmatrix}\textbf{Approximate}\\ \textbf{number of}\\ \textbf{neutrons}\\ \textbf{per atom}\end{pmatrix} = \begin{pmatrix}\textbf{Atomic weight}\\ \textbf{of the}\\ \textbf{element}\end{pmatrix} - \begin{pmatrix}\textbf{Atomic number}\\ \textbf{of the}\\ \textbf{element}\end{pmatrix}$$

A notation frequently used to show the atomic mass (also called **mass number**) and atomic number of an atom uses subscripts and superscripts to the left of the symbol:

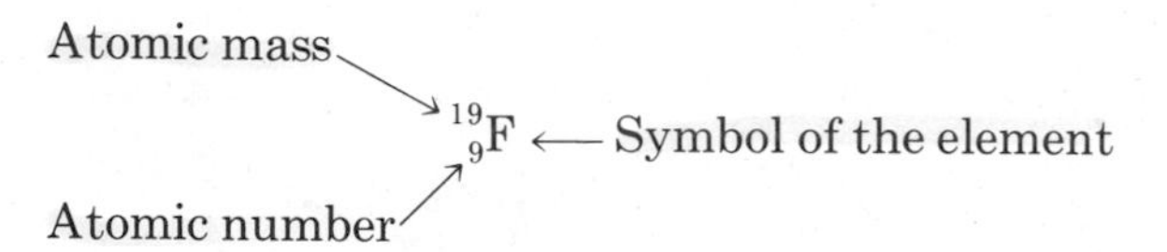


Symbolism refers to individual isotopes.

For an atom of fluorine, $^{19}_{9}F$, the number of protons is 9, the number of electrons is also 9, and the number of neutrons is $19 - 9 = 10$.

## Isotopes

Many of the elements, when analyzed by a special type of canal-ray tube called a *mass spectrometer* (Fig. 2–12), are found to be composed of atoms of different masses (Fig. 2–13). Atoms of the same element having different atomic masses are called ***isotopes*** of that element.

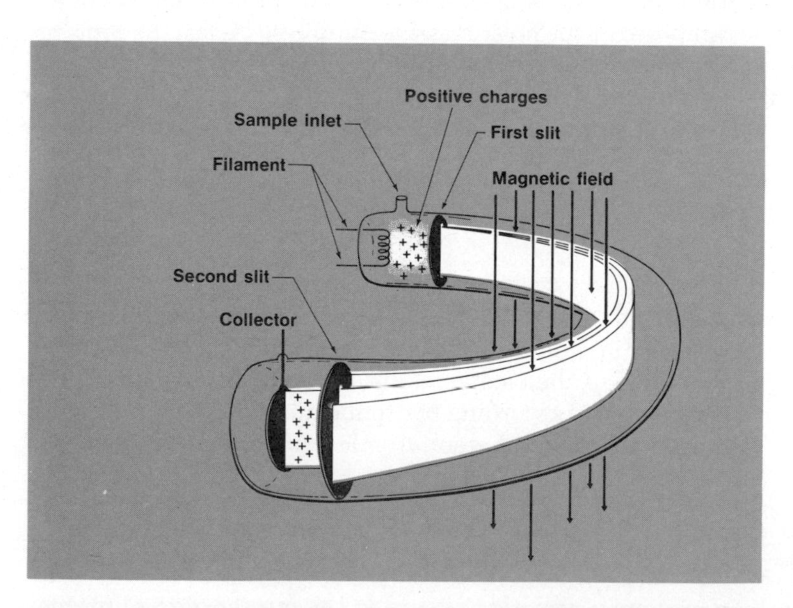


**Figure 2–12** Mass spectrometer. Sample to be studied is injected near filament. Electrodes (not shown) subject sample to electron beam that ionizes a part of the sample. Electrodes are arranged to accelerate positive ions toward first slit. The positive ions that pass the first slit are immediately put into a magnetic field perpendicular to their path and follow a curved path determined by the charge-to-mass ratio of the ion. A collector plate, behind the second slit, detects charged particles passing through the second slit. The relative magnitudes of the electrical signals are a measure of the numbers of the different kinds of positive ions.

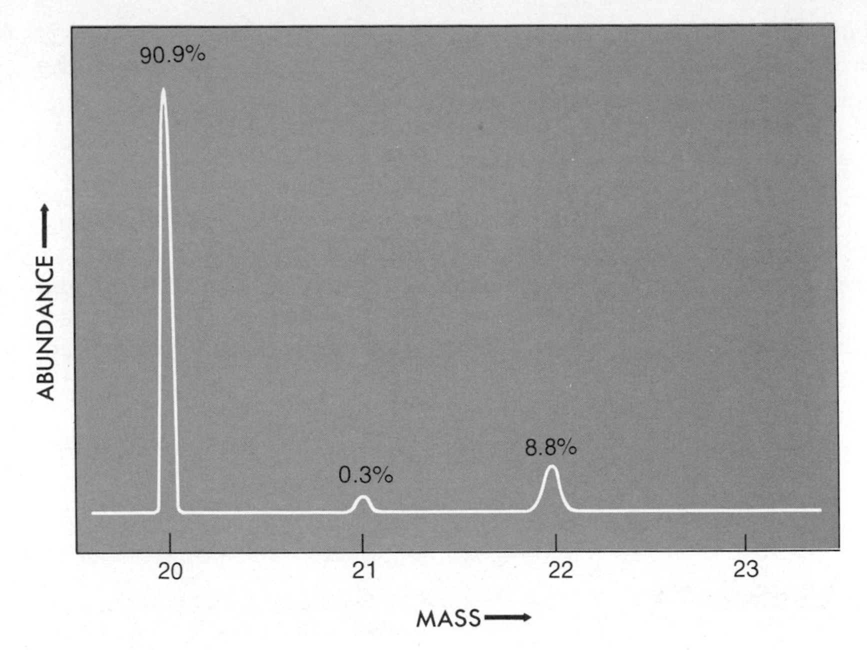


**Figure 2–13** Mass spectrum of neon (+1 ions only). The principal peak corresponds to the most abundant isotope, neon-20.

The element neon is a good example to consider. A natural sample of neon gas is found to be a mixture of three isotopes of neon:

$^{20}_{10}\text{Ne}$ $^{21}_{10}\text{Ne}$ $^{22}_{10}\text{Ne}$

The fundamental difference between isotopes is the different number of neutrons per atom. All atoms of neon have 10 electrons and 10 protons; about 90% of the atoms have 10 neutrons, some have 11 neutrons, and others have 12 neutrons. Because they have different numbers of neutrons, they must have different masses. Note that all the isotopes have the same atomic number, and indeed all behave the same chemically. They are all neon.

**Isotopes are atoms of the same element having different numbers of neutrons.**

There are only 106 known elements, yet more than 1,000 isotopes have been identified. Some elements have many isotopes; tin, for example, has ten natural isotopes. Hydrogen has three isotopes, and they are the only three which are generally referred to by different names: $^{1}_{1}\text{H}$ is called protium, $^{2}_{1}\text{H}$ is called deuterium, and $^{3}_{1}\text{H}$ is called tritium. Tritium is radioactive. The natural assortment of isotopes, each having its own distinctive atomic mass, results in fractional atomic weights for many elements.

**The weighted average of the atomic weights of the isotopes in a natural mixture is the noninteger atomic weight of the element.**

## Self-Test 2-B

1. Isotopes of an element are atoms that have nuclei with the same number of ____________ but different numbers of ____________.

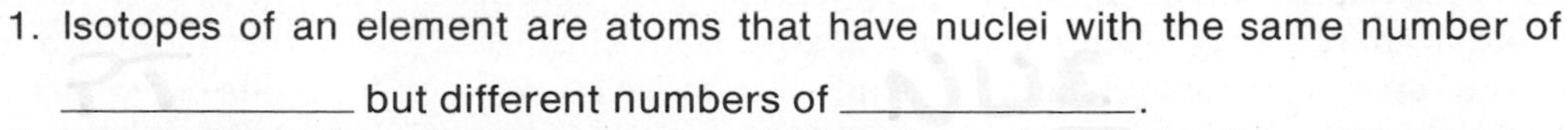

2. The nucleus of an atom occupies a relatively (large/small) fraction of the volume of the atom.
3. The positive charges in an atom are concentrated in its ____________.
4. The negatively charged particles in an atom are ____________; the positively charged particles are ____________; and the neutral particles are ________.

5. In a neutral atom there are equal numbers of ______________ and ______________.
6. The number of protons per atom is called the ______________ number of the element.
7. The mass of the proton is ______________ times that of the electron.
8. An atom of arsenic, ${}^{75}_{33}As$, has ______________ electrons, ______________ protons, and ______________ neutrons.
9. Positive (canal) rays obtained with different gases are (different/identical), while the cathode rays obtained using different cathodes are (different/identical).
10. Cathode rays are composed of a universal constituent of matter named ______________.
11. The two fundamental particles revealed by studies using gas discharge (cathode-ray) tubes are the ______________ and ______________.
12. All atoms of a given element are exactly alike (true/false).

---

## Where Are the Electrons?

Two major theories have been presented concerning the position, movement, and energy of electrons in an atom. The Bohr theory of the hydrogen atom was put forth in 1913 by Niels Bohr. This theory was extended and modified by Erwin Schrödinger, Werner Heisenberg, Louis de Broglie and others in 1926. The newer theory is referred to as the quantum mechanical theory, the wave mechanical theory, or Schrödinger's theory.

### The Bohr Model

Energy of matter in motion is *kinetic* energy.

Energy stored in matter is *potential* energy.

In Bohr's concept, electrons revolve around a nucleus in definite orbits, much as planets revolve around the sun. He equated classical mathematical expressions for the force tending to keep the electron traveling in a straight line and the force tending to pull the electron inward (the positive-to-negative attraction between a proton and an electron). The total energy of the atom is the kinetic energy of the electrons plus the potential energy due to the electron's separation from the nucleus. In a revolutionary sort of way, Bohr suggested that electrons stay in rather stable orbits and can have only certain energies within a given atom. According to Bohr, an electron can travel in one orbit for a long period of time, or in another orbit some distance away for a long period of time, but cannot stay for any measurable time between the two orbits. A rough analogy is provided by considering books in a bookcase. Books may rest on one shelf or on another shelf for very long periods of time but cannot rest between shelves. In moving a book from one shelf to another shelf, the potential energy of the book changes by a definite amount. When an electron moves from one orbit to another, its energy changes by a definite amount, called a ***quantum*** of energy.

Bohr assumed that an atom can exist only in certain energy states.

Packets of light energy are called *photons,* or *quanta.*

Energy is required to separate objects attracted to each other. For example, energy is required to lift a rock from the earth, or to separate two magnets, or to pull a positive charge away from a negative charge. Bohr suggested that a very definite and characteristic amount of energy is required to move an electron from one energy level to another one. The energy added to move an electron farther from its nucleus is stored in the system as potential energy (energy of position). Thus, the electron has more energy when it is in an orbit further from the nucleus, and less energy when it is in an orbit close to the nucleus. When the electron passes from an outer orbit to an inner orbit, energy is emitted from the atom, generally in the form of light energy.

Prism
White Light
Slit
Red
Violet
Screen
Red
Orange
Yellow
Green
Blue
Violet

**Figure 2–14** Spectrum from white light produced by refraction in a glass prism. The different colors blend into one another smoothly.

**Dispersed light produces a *spectrum*.**

Bohr used the idea of electrons moving up and down a "bookcase" of energy levels corresponding to orbits to explain the observable bright-line emission spectrum of hydrogen. A ***spectrum*** is the display produced by an instrument designed to separate or disperse light into its component colors. The spectrum of white light is the rainbow display of separated colors shown in Figure 2–14. An ***emission*** spectrum is produced when the light emitted by atoms energized by a flame or an electric arc is allowed to pass through a narrow vertical slit and then through a prism of glass or quartz. If sunlight or light from a white-hot solid is dispersed, all of the colors of the rainbow are seen. This is a ***continuous emission spectrum.*** However, if the light from an energized gaseous element is dispersed, a set of colored lines is produced, the lines being separated by black spaces. This is a ***bright-line emission spectrum*** (Fig. 2–15). Each line is a pure color, and is really an image of the slit in that particular color. A quantum of light of any given color has a characteristic energy that is different from the energy of a quantum of any other color.

According to Bohr, the light forming the lines in the bright-line emission spectrum of hydrogen come from electrons moving toward the nucleus after having first been energized and pushed to orbits farther from the nucleus (Fig. 2–16). A movement between two particular orbits involves a definite quantum of energy. Each time an electron moves from one orbit to another orbit closer to the nucleus, a quantum of light having a characteristic energy (and color) is emitted. Transitions toward the nucleus between two outer orbits emit quanta having smaller characteristic energies. Transitions from an outer orbit to orbits near the nucleus emit quanta having larger characteristic energies. Each line corresponds to its own particular energy of the same-sized quanta of light.

Niels Bohr (1885–1962) received his doctor's degree the same year that Rutherford announced his discovery of the atomic nucleus, 1911. After studying with Thomson and Rutherford in England, Bohr formulated his model of the atom. Bohr returned to the University of Copenhagen and, as Professor of Theoretical Physics, directed a program that produced a number of brilliant theoretical physicists. He received the Nobel prize for physics in 1922.

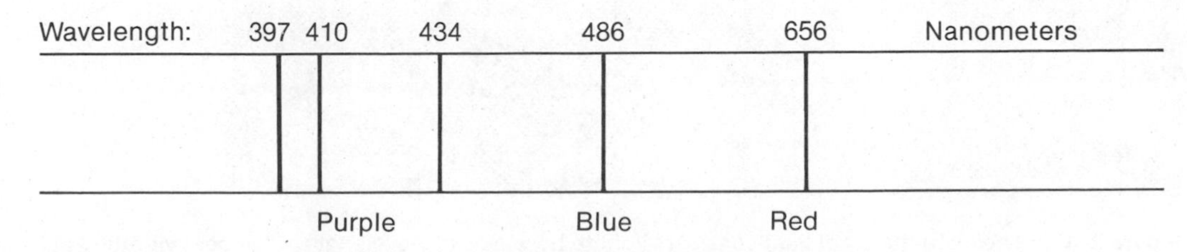


**Figure 2–15** The visible portion of the hydrogen bright-line emission spectrum. Frequency and energy increase to the left; wavelength increases to the right. There are 7 more lines to the left in the ultraviolent range and 13 more lines to the right in the infrared range.

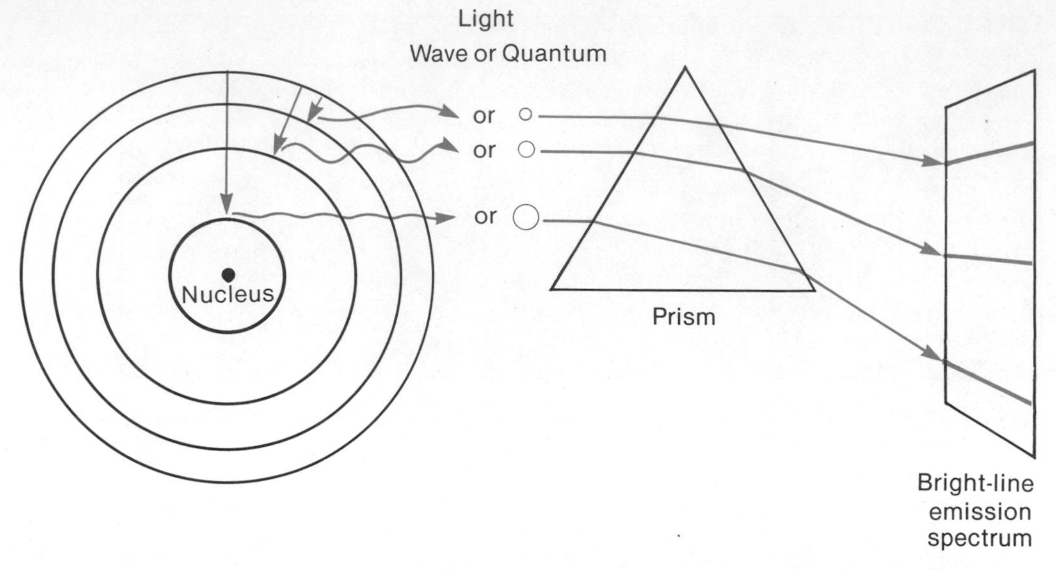


**Figure 2–16** The formation of light according to the atomic theory. Electrons previously energized to higher energy levels make transitions back toward the nucleus (three transitions are shown). The decrease in potential energy during a transition is transformed into light. Transitions between different energy levels produce different-sized quanta of light. When the quanta are dispersed by being passed through a prism, the bright lines can be seen on film or on a spectroscope. Billions of transitions per second make each line bright enough (sufficient same-sized quanta emitted) for each line to be detected.

Not only could Bohr explain the cause of the lines in the bright-line emission spectrum of hydrogen, but he also calculated the expected wavelengths of the lines. He expressed the results of his calculations in the alternate view of the nature of light, its

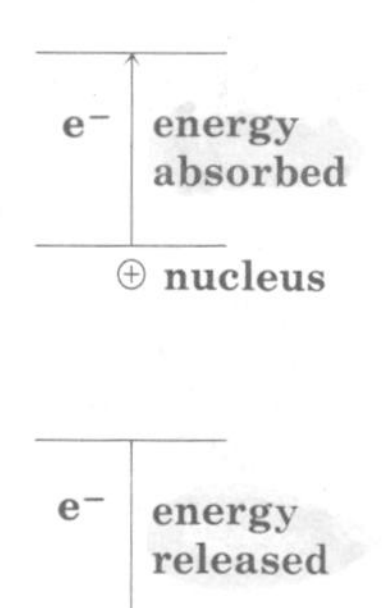


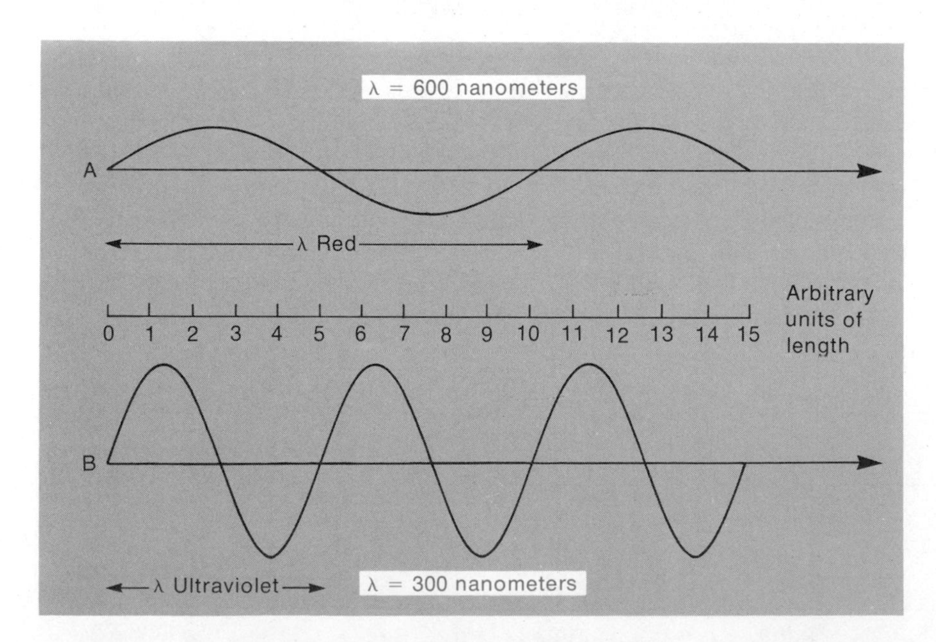


**Figure 2–17** The wave theory of light considers light to be waves of wavelength λ (lambda) vibrating at right angles to their path of motion. Red light, *A*, completes one vibration or wave in the same distance and time that involve two complete waves of ultraviolet light, *B*. Ultraviolet light is of shorter wavelength than violet light. A nanometer is $10^{-9}$ meter.

wave nature. The wave properties of light—wavelength, frequency, and speed—are considered in Figure 2–17 as they apply to any phenomenon possessing wave properties (water waves, sound waves from violin strings, radio waves, and light).

With brilliant imagination, Bohr applied a little algebra and some classical mathematical equations of physics to his tiny solar-system model of the hydrogen atom. The unprecedented requirement was that only a few allowable paths (quantized orbits) are available in which electrons can move stably around the nucleus. A further requirement was that energy differences (quanta) existed between any two orbits. Execution of the mathematics of the model produced the predicted wavelengths of the lines in the hydrogen spectrum, some of which are shown in Table 2–3. Note the close agreement between the measured values and the values predicted by the calculations of the Bohr theory. He calculated the wavelengths of all lines which had been observed with remarkably close agreement. Niels Bohr had tied the unseen (the interior of the atom) with the seen (the observable lines in the hydrogen spectrum)—a fantastic achievement.

The Bohr theory was accepted almost immediately after its presentation, and Bohr was awarded the Nobel prize in physics in 1922 for his contribution to the understanding of the hydrogen atom.

**The Personal Side**

Bohr had a close call in his escape from Denmark when Hitler's forces ravished the country. Having done all he could to get Jewish physicists to safety, Bohr was still in Denmark when Hitler's army suddenly occupied the country in 1940. In 1943, to avoid imprisonment, he escaped to Sweden. There he helped to arrange the escape of nearly every Danish Jew from Hitler's gas chambers. He was later flown to England in a tiny plane, in which he passed into a coma and nearly died from lack of oxygen.

He went on to the United States, where until 1945 he worked with other physicists on the atomic bomb development at Los Alamos. His insistence upon sharing the secret of the atomic bomb with other allies, in order to have international control over nuclear energy, so angered Winston Churchill that he had to be restrained from ordering Bohr's arrest. Bohr worked hard and long on behalf of the development and use of atomic energy for peaceful purposes. For his efforts, he was awarded the first Atoms for Peace prize in 1957. He died in Copenhagen on November 18, 1962.

One way to decorate the interior of atoms other than hydrogen is to use the orbits devised by Bohr for hydrogen and insert the proper number of electrons. While this is at best a rough approximation, we shall find it is adequate in Chapter 4 for explaining such phenomena as some kinds of bonding and the formation of ions.

Orbits are sometimes called shells or just energy levels.

Let us build up some atoms in Tinker Toy fashion. First, we need a few rules in order to play the game. Recall that the atomic number is the number of electrons (and protons) per atom of the element. Consistent with the ionization energies discussed in

**TABLE 2–3 Agreement between Bohr's Theory and the Lines of the Hydrogen Spectrum**

| Changes in Energy Levels | Wavelength Predicted by Bohr's Theory (nm) | Wavelength Determined from Laboratory Measurement (nm) | Spectral Region |
|---|---|---|---|
| 2 ⟶ 1 | 121.6 | 121.7 | ultraviolet |
| 3 ⟶ 1 | 102.6 | 102.6 | ultraviolet |
| 4 ⟶ 1 | 97.28 | 97.32 | ultraviolet |
| 3 ⟶ 2 | 656.6 | 656.7 | visible red |
| 4 ⟶ 2 | 486.5 | 486.5 | visible blue-green |
| 5 ⟶ 2 | 434.3 | 434.4 | visible blue |
| 4 ⟶ 3 | 1876 | 1876 | infrared |

NOTE: These lines are typical; other lines could be cited as well, with equally good agreement between theory and experiment. The unit of wavelength is the nanometer (nm), $10^{-9}$ meter.

Chapter 3, the maximum number of electrons per orbit is $2n^2$, where $n$ is the number of the orbit. Orbits are numbered with integers, beginning with 1 for the orbit closest to the nucleus. As practice, use the formula to check these numbers.

| Orbit | Maximum Number of Electrons |
|---|---|
| 1 | 2 |
| 2 | 8 |
| 3 | 18 |
| 4 | 32 |
| 5 | 50 |

A general, overriding rule to the numbers above is that the outside orbit can have no more than eight electrons. When electrons are placed in orbits as close to the nucleus as possible, the electrons are said to be in their ***ground state.***

You might like to follow along in Figure 2–18 (the Bohr Model column) as the building-up process is described. Hydrogen, with atomic number 1, has one electron. In its ground state, this electron is in the first orbit. The two electrons of helium are in its first orbit since the first orbit can have a maximum of two electrons.

For all atoms of other elements, two electrons are in the first orbit, and the other electrons of the atoms are assorted into higher-numbered energy levels. In atomic-number order, lithium through neon, two electrons are placed in the first orbit (which fill it), and into the second orbit are placed one, two, three, and so on to eight electrons (for Ne). Eight electrons fill the second orbit.

Sodium (Na), with 11 electrons, has the first two orbits filled with 2 and 8 electrons, respectively, and has 1 electron in the third orbit. Each succeeding element in atomic-number order, magnesium through argon, adds one more electron to the third orbit of its atoms.

At argon (Ar), the maximum of eight electrons in the outside orbit comes into play. When 19 electrons are present, as in an atom of potassium (K), the first orbit has 2 electrons, the second orbit has 8 electrons, and the third orbit could have the other 9 electrons (maximum of 18 electrons) if it were not the outside orbit. So to accommodate 19 electrons, there are two choices: 2-8-9 or 2-8-8-1. The first choice violates the requirement of no more than eight in the outside orbit. The second is the proper choice. Calcium (Ca) with 20 electrons per atom has an electronic arrangement of 2-8-8-2.

Beginning with scandium (Sc), atomic number 21, and continuing through zinc (Zn), atomic number 30, ten electrons are added to the third orbit to complete its maximum of 18. Zinc has the electronic arrangement 2-8-18-2.

You might pause in your reading here, and predict the ground-state electronic arrangement of gallium (Ga), atomic number 31, and rubidium (Rb), atomic number 37, using this system.

## The Wave Mechanical Model

Electrons are described by both particle and wave theories.

The Bohr model failed when applied to elements other than hydrogen because it could not account exactly for the line spectra of atoms with more than one electron. It was also weak in explaining why the periods (the horizontal rows) of the periodic table vary considerably in length.

After Bohr's work, a more modern highly sophisticated mathematical theory of the atom was developed by Schrödinger, Heisenberg, Dirac, and others. In this theory, electrons are treated as having both a particle and a wave nature. The locations of the electrons are treated as probabilities, without seeking to locate the exact spot for an electron at a given time. This approach suggested that the Bohr theory sought more precision in describing the atom than nature would allow.

A Frenchman, Louis de Broglie, was the first to suggest (in 1924) that electrons and other small particles should have wave properties. In this respect, he said, electrons should behave like light, a suggestion which scientists of the time found hard to accept. However, in a few years separate experiments by George Thomson (son of J. J. Thomson) in England and Clinton Davisson in the United States justified de Broglie's hypothesis (Fig. 2–19). The electron microscopes found in many research laboratories today are built and operated on our understanding of the wave nature of the electron.

It should not be surprising to find that matter can be treated by both wave and particle theories (the duality of matter), since its convertible counterpart—energy—

Placement of Electrons in Ground State

| Element | Atomic Number | Bohr Model | Wave Mechanical Model |
|---|---|---|---|
| Hydrogen (H) | 1 | 1p 1)e | $1s^1$ |
| Helium (He) | 2 | 2p 2n 2)e | $1s^2$ |
| Lithium (Li) | 3 | 3p 4n 2)e 1)e | $1s^2\,2s^1$ |
| Beryllium (Be) | 4 | 4p 5n 2)e 2)e | $1s^2\,2s^2$ |
| Boron (B) | 5 | 5p 6n 2)e 3)e | $1s^2\,2s^2\,2p^1$ |
| Carbon (C) | 6 | 6p 6n 2)e 4)e | $1s^2\,2s^2\,2p^2$ (or $2p_x^1\,2p_y^1$) |
| Nitrogen (N) | 7 | 7p 7n 2)e 5)e | $1s^2\,2s^2\,2p^3$ (or $2p_x^1\,2p_y^1\,2p_z^1$) |
| Oxygen (O) | 8 | 8p 8n 2)e 6)e | $1s^2\,2s^2\,2p^4$ (or $2p_x^2\,2p_y^1\,2p_z^1$) |
| Fluorine (F) | 9 | 9p 10n 2)e 7)e | $1s^2\,2s^2\,2p^5$ |
| Neon (Ne) | 10 | 10p 10n 2)e 8)e | $1s^2\,2s^2\,2p^6$ |
| Sodium (Na) | 11 | 11p 12n 2)e 8)e 1)e | $1s^2\,2s^2\,2p^6\,3s^1$ |
| Magnesium (Mg) | 12 | 12p 12n 2)e 8)e 2)e | $1s^2\,2s^2\,2p^6\,3s^2$ |
| Aluminum (Al) | 13 | 13p 14n 2)e 8)e 3)e | $1s^2\,2s^2\,2p^6\,3s^2\,3p^1$ |
| Silicon (Si) | 14 | 14p 14n 2)e 8)e 4)e | $1s^2\,2s^2\,2p^6\,3s^2\,3p^2$ (or $3p_x^1\,3p_y^1$) |
| Phosphorus (P) | 15 | 15p 16n 2)e 8)e 5)e | $1s^2\,2s^2\,2p^6\,3s^2\,3p^3$ (or $3p_x^1\,3p_y^1\,3p_z^1$) |
| Sulfur (S) | 16 | 16p 16n 2)e 8)e 6)e | $1s^2\,2s^2\,2p^6\,3s^2\,3p^4$ (or $3p_x^2\,3p_y^1\,3p_z^1$) |
| Chlorine (Cl) | 17 | 17p 18n 2)e 8)e 7)e | $1s^2\,2s^2\,2p^6\,3s^2\,3p^5$ |
| Argon (Ar) | 18 | 18p 22n 2)e 8)e 8)e | $1s^2\,2s^2\,2p^6\,3s^2\,3p^6$ |
| Potassium (K) | 19 | 19p 20n 2)e 8)e 8)e 1)e | $1s^2\,2s^2\,2p^6\,3s^2\,3p^6\,4s^1$ |
| Calcium (Ca) | 20 | 20p 20n 2)e 8)e 8)e 2)e | $1s^2\,2s^2\,2p^6\,3s^2\,3p^6\,4s^2$ |

**Figure 2–18** Electron arrangements of the first 20 elements. The nuclear contents of a typical isotope are shown.

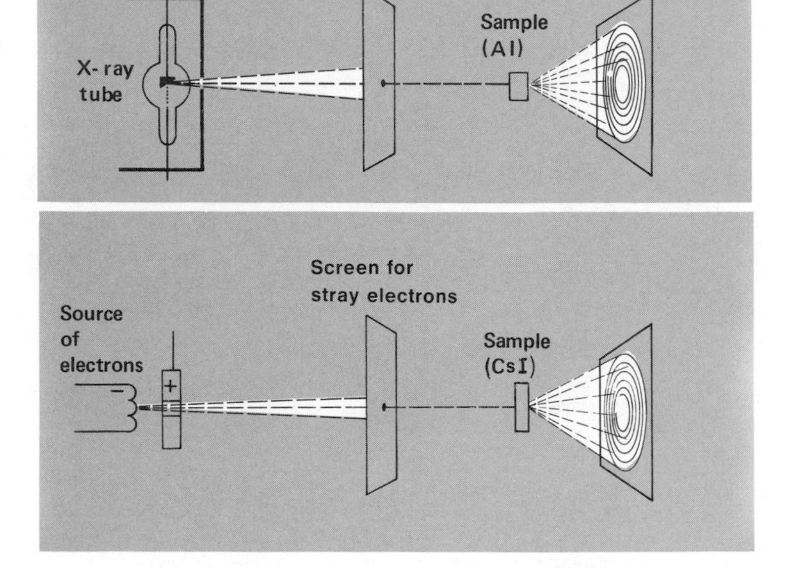


**Figure 2–19** Similar patterns are shown by light, X rays, and electrons as each is diffracted. Diffraction is the bending and spreading of wave motion around edges. The effect is prominent when the wavelength is large compared to the size of the obstacle and small when the wavelength is short compared to the size of the obstacle. Similar effects from light, X rays, and electrons indicate a property common to all: each has a wave nature.

has been successfully treated by both theories for a long time. Keep in mind that we do not really know if matter or light is a wave or a particle. However, because there are limits on what we can visualize in our physical world, in talking about something like subatomic behavior we are forced to use physical models based on known behavior, rather than more sophisticated models that would describe some type of intermediate behavior with which we are unfamiliar in our macroscopic world.

The wave theory of the atom was developed in the 1920's, principally by Erwin Schrödinger. The most fundamental aspects of the theory are the mathematical wave equations used to describe the electrons in atoms. Solutions to the equations are called wave functions or ***orbitals.*** Calculations involving the wave equations are complicated and time-consuming, but we do not need to do the elaborate calculations in order to use the results.

The principal result is a series of orbitals. Orbitals are different in their type, energy, and likely configuration in space (related to the probability of finding the electron there). The types of orbitals are distinguished by the letters ***s, p, d,*** and ***f.*** These letters were derived from terms in spectroscopy (sharp, principal, diffuse, and fundamental, respectively), and emphasize again that atomic theory developed very closely with atomic spectra.

The orbitals of the wave theory are simply subdivisions of the Bohr orbits. Each orbit has an s orbital. Beginning with the second orbit, each orbit has also a set of three

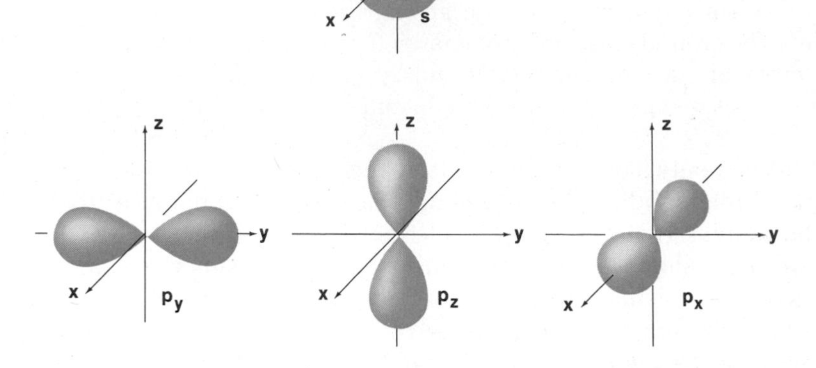


**Figure 2–20** Spatial orientations of s and p orbitals for hydrogen-like atoms. The x, y, and z subscripts denote p orbitals with different orientations in space with reference to the imaginary axes.

p orbitals. The third orbit and all orbits thereafter also have a set of five d orbitals. The fourth orbit and all orbits thereafter also have a set of seven f orbitals. We shall need only s and p orbitals for the explanations given in this book, since the electrons in the outer orbits of the various atoms are in the s and p orbitals of those atoms. These outer s and p electrons are the ones most involved with other atoms.

Only the **probability** of finding an electron in a given volume of space around the nucleus can be calculated from the Schrödinger equation. In order to portray the probabilities of finding an electron, usually the surface of a region in space (similar to the surface of a balloon) is plotted that will enclose the volume in which the electron will be expected to be found 90% of the time.

**How are orbitals visualized?**

As shown in Figure 2–20, the shape of an s orbital is always spherical. The shape of a p orbital is like that of a dumbbell. The d and f orbitals have more complicated shapes. The shapes of the orbitals are about the best that we can do to relate the wave theory of electron probabilities in a pictorial way, for we are trying to visualize where in space a given electron will be. The geometries associated with the orbitals help to explain the structures of molecules that result from the interaction and combinations of atoms.

**Maximum of two electrons per orbital.**

The first Bohr orbit is synonymous with the 1s orbital. (The 1 indicates the orbit; the s indicates the type of orbital.) According to the wave theory, an orbital may be occupied by a *maximum* of two electrons. Hence, the 1s orbital (and also the first orbit) can have a maximum of two electrons. When the electron of a hydrogen atom is as close to its nucleus as it can stably be (in its ***ground state***), the electron is in a 1s orbital, expressed as $1s^1$. The two electrons of helium can be in the same orbital and are in the 1s orbital in their ground state, $1s^2$

**How are electrons assorted into orbitals?**

Since the second Bohr orbit can have a maximum of eight electrons, four orbitals are required to accommodate the eight electrons. The four orbitals are a 2s orbital and a set of three 2p orbitals designated $2p_x$, $2p_y$, and $2p_z$ (Fig. 2–20). When the three electrons of a lithium atom are in their ground state, two electrons are in the 1s orbital and the other electron is in the 2s orbital, $1s^2\ 2s^1$ (Fig. 2–18). The five electrons of a boron atom are distributed with two in the 1s orbital, two in the 2s orbital, and one in a 2p orbital (either $2p_x$, $2p_y$, or $2p_z$, since these orbitals have the same energy unless the atom is in a strong magnetic field), $1s^2\ 2s^2\ 2p^1$.

If there are two, three, or four electrons in a set of p orbitals, the electrons are spread among the orbitals as much as possible. This gives a more stable arrangement of electrons from spin and angular momentum considerations. What this means is that no orbital in a set will have two electrons until each orbital in the set has one electron. Thus, the six electrons of carbon distribute as $1s^2\ 2s^2\ 2p_x^1\ 2p_y^1$. The seven electrons of nitrogen have the configuration $1s^2\ 2s^2\ 2p_x^1\ 2p_y^1\ 2p_z^1$, and the eight electrons of oxygen are arranged in the ground state as $1s^2\ 2s^2\ 2p_x^2\ 2p_y^1\ 2p_z^1$.

In Figure 2–18, wave-mechanical electronic arrangements are shown through atomic number 20. Note the ground state order of energy for the orbitals: 1s 2s 2p 3s 3p 4s. How does the number of electrons in the 2s and 2p orbitals compare with the number of electrons in the second orbit? The 3s and 3p and the third orbit?

What, then, is an atom really like? We have Dalton's concept of an atom as a hard sphere similar to a small billiard ball. We have Bohr's concept of the atom as a small three-dimensional solar system with a nucleus and electrons in paths called orbits. In the modern theory, we have more detail in that orbits now have suborbits called orbitals and we are given approximate spaces where electrons exert their greatest influence in an atom. Why present all three theories? First, an understanding of the simpler Dalton and Bohr theories helps us to understand the more complicated, more detailed modern theory of the atom. Second, all three theories help us to understand the phenomena that we observe. We simply use whatever detail is necessary to explain what we see. For example, the simpler Dalton concept adequately explains many properties of the gaseous, liquid, and solid states. Most bonding between atoms of the light elements can be explained by application of the orbits of Bohr. The shapes of molecules and the arrangement of atoms with respect to each other can best be explained by the orbital representations of the modern theory. In the explanations given in this text, we shall follow the principle that simplest is best.

## Self-Test 2-C

1. Under some conditions light has properties of __________ and under other conditions exhibits the properties of __________.
2. When light is dispersed into the different colors composing the light, a __________ is produced.
3. According to Bohr's theory, light of characteristic wavelength is produced as an electron passes from an orbit (closer to/farther from) the nucleus to an orbit (closer to/farther from) the nucleus.
4. Which of the following led to the modern theory of the atom and was not included in the Bohr theory?
   a. concept of the nucleus
   b. quantum theory
   c. particle nature of the electron
   d. wave nature of the electron
5. According to de Broglie, every moving particle has not only mass and velocity but also a characteristic __________.
6. The maximum number of electrons in the $n = 3$ energy level is __________, and the maximum number in any orbital is __________.
7. Consider the meaning of the representations of the orbitals shown in Figure 2–20.
   a. Are the representations those of the paths of electrons? __________
   b. Are the representations the containers of electrons? __________
   c. Do the representations show where an electron is most likely to be found? __________

## Matching Set

| | | | |
|---|---|---|---|
| ______ | 1. atomic mass | a | attract |
| ______ | 2. unlike electrical charges | b | equal to number of protons in nucleus |
| ______ | 3. $2n^2$ | c | demonstrated wave nature of electron |
| ______ | 4. nucleus | d | cathode ray particle |
| ______ | 5. electron | e | neutrons plus protons |
| ______ | 6. $^{22}Ne$ and $^{20}Ne$ | f | a small, definite amount of energy |
| ______ | 7. atomic number | g | proton and an electron |
| ______ | 8. quantum | h | uncharged elementary particle |
| ______ | 9. gamma ray | i | contains most of the mass in an atom |
| ______ | 10. particles in an H atom | j | maximum number of electrons in an orbit |
| ______ | 11. neutron | k | predicted wave nature of electron |
| ______ | 12. G. P. Thomson C. J. Davisson | l | probable location for electrons in an atom |
| ______ | 13. de Broglie | m | a form of radiant energy |
| ______ | 14. orbital | n | isotopes |

## Questions

**1.** What kinds of evidence did Dalton have for atoms that the early Greeks (Democritus, Leucippus) did not have?

**2.** How does Dalton's atomic theory explain:
a. the law of conservation of matter?
b. the law of constant composition?
c. the law of multiple proportions?

**3.** The laws of chemical change presented in this chapter are often referred to as empirical laws. What does "empirical" mean? How does empirical differ from theoretical?

**4.** Although there may not be a very reliable way to check the conservation of matter in a large explosion of dynamite, what leads us to believe that the law of conservation of matter is obeyed?

**5.** Describe the potential energy and kinetic energy relationships as a rock tumbles off a cliff.

**6.** What experimental evidence indicates that
a. cathode rays have considerable energy?
b. cathode rays have mass?
c. cathode rays have charge?
d. cathode rays are a fundamental part of all matter?
e. two isotopes of neon exist?
f. atoms are destructible?

**7.** Describe in detail Rutherford's gold-foil experiment under the following headings:
a. experimental setup
b. observations
c. interpretations

**8.** Why was Thomson's charge-to-mass ratio determination for electrons very significant although he did not determine either the charge or the mass of the electron?

**9.** What part do electrons play in producing positive rays?

**10.** How do the following discoveries indicate that the Daltonian model of atoms is inadequate?
a. cathode rays
b. positive rays
c. nucleus
d. natural radioactivity
e. isotopes

**11.** Characterize the three types of emissions from naturally radioactive substances as to charge, relative mass, and relative penetrating power.

**12.** Explain what the following terms mean:
a. isotopes of an element
b. atomic number
c. an alpha emitter

**13.** If electrons are a part of all matter, why are we not electrically shocked continually by the abundance of electrons about and in us?

**14.** There are more than 1000 kinds of atoms, each with a different weight. Yet there are only 106 elements. How does one explain this in terms of subatomic particles?

**15.** What is a practical application of cathode-ray tubes?

**16.** A common isotope of lithium (Li) has a mass of 7. The atomic number of lithium is 3. What are the constituent particles in its nucleus?

**17.** An element has 12 protons in its nucleus. How many electrons do the atoms of this element possess?

**18.** An isotope of atomic mass 60 has 33 neutrons in its nucleus. What is its atomic number, and what are the name and chemical symbol of the element?

**19.** An isotope of cerium (Ce) has 88 neutrons in its nucleus. How many protons plus neutrons does this nucleus contain?

**20.** The element iodine (I) occurs naturally as a single isotope of atomic mass 127; its atomic number is 53. How many protons and how many neutrons does it have in its nucleus?

**21.** An element with an atomic number of 8 is found to have three isotopes, with atomic masses of 16, 17, and 18. How many protons and neutrons are present in each nucleus? What is the element?

**22.** Suppose Millikan had determined the following charges on his oil drops:
$1.33 \times 10^{-19}$ coulomb
$2.66 \times 10^{-19}$ coulomb
$3.33 \times 10^{-19}$ coulomb
$4.66 \times 10^{-19}$ coulomb
$7.92 \times 10^{-19}$ coulomb
What do you think his value for the electron's charge would have been?

**23.** Suppose an isotope of aluminum has an atomic mass of 27.0. How many protons, neutrons, and electrons are in an atom of this isotope? What is the charge on the nucleus?

**24.** What is a quantum? What is a photon?

**25.** Discuss, in quantum terms, how a ladder works.

**26.** Distinguish between atomic number and atomic weight.

**27.** Distinguish between a continuous spectrum and a bright-line spectrum under the two headings:
a. general appearance
b. source

**28.** How does the Bohr theory explain the many lines in the spectrum of hydrogen although the hydrogen atom contains only one electron?

**29.** Helium, neon, argon, krypton, xenon, and radon form a group of similar elements in that they form very, very few compounds. From their atomic structures, suggest a reason for this similarity in relative inactivity.

**30.** Compare your view of a valley as you walk from a mountain top to the floor of the valley with the progression of atomic theory.

**31.** The law of conservation of matter is to a chemical change as the law of constant composition is to a chemical ____________.

**32.** (True/False)
a. If compounds conform to the law of multiple proportions, they must necessarily conform to the law of constant composition.
b. If compounds conform to the law of constant composition, they must necessarily conform to the law of multiple proportions.

**33.** What is constant about a compound?
a. the weight of a sample of the compound.
b. the weight of one of the elements in samples of the compound
c. the ratio by weight of the elements in the compound

**34.** If pure water is 88.8% oxygen and 11.2% hydrogen by weight,
a. is it likely to have *only* 88.8 g of oxygen in 110 g of water?
b. is it likely to have *exactly* 22.2 g of oxygen in 25.0 g of water?

**35.** In John Dalton's concept of atoms, were they more like billiard balls, cotton puff balls, tennis balls, or small solar systems?

**36.** How are fluorescent lights and TV picture tubes related to the study of the atom?

**37.** If you found the number of wheels received by an assembly plant to be twice the number of motors, what type of vehicle would you assume to be assembled there? Of what chemical law does this remind you?

**38.** Which is the empirical (observable) fact:
a. water is 88.8% oxygen by weight, or
b. water molecules contain one atom of oxygen each?

**39.** In recent years we have found that pure substances, such as some plastics, do vary in composition and that some elements can be decomposed (nuclear fission). What does this say to you about concepts and progress in science?

**40.** If different materials when heated give off different and characteristic colors of light, what can you assume about the structure and kinds of light?

**41.** Why is it impossible to produce a positive charge without producing a negative charge at the same time?

**42.** Krypton is the name of Superman's home planet and also that of an element. Look up the element, krypton, and list its symbol, atomic number, atomic weight, and electronic arrangement.

**43.** Explain in your own words why alpha particles are deflected in one direction in an electrical field while beta particles are deflected in the opposite direction.

**44.** Read about lasers and seek similarities between the explanation for the generation of laser light and the explanation for the production of bright-line elemental spectra.

**45.** Without looking at Figure 2–18 (except for checking later), write out the placement of electrons in their ground state
a. into orbits according to the Bohr theory for atoms having 6, 10, 13, and 20 electrons.
b. into orbitals according to the wave mechanical theory for atoms having 7, 13, 16, and 20 electrons.

# Elements in Useful Order— The Periodic Table

# 3

About now in this adventure with chemistry, we have the first meaningful opportunity to discuss the ***periodic table*** (or ***chart***), which you have probably noticed on the classroom wall. The careful study of the properties of the elements (defined and partially described in Chapter 1) led to the formulation of the periodic table, and the established periodic table furnishes evidence for atomic structural theory (Chapter 2).

What is there to know about the periodic table? Why is a so-called periodic table on a wall of most science classrooms and labs? Is it just a portrait of chemistry to adorn a wall, or is it useful? Why is the name "periodic" appropriate? Why is the table so arranged, and what are its important features? Does the table give order to the 106 known elements?

We shall discover in this chapter that the periodic table is important because it summarizes, correlates, and predicts a wealth of chemical information. In essence, the periodic table does bring order to 106 individual elements. Elements in an orderly arrangement provide the same benefits as your class notes arranged in a logical order, your room neatly and orderly arranged, or the goods arranged into departments in a store—ease of use and facilitation of understanding. From the standpoint of its logic, the periodic table can be of great help to a student of chemistry. As you read and study this chapter, look for both how and what chemical information is summarized, correlated, simplified, and predicted by the orderly arrangement of the elements in the periodic table.

**Why arrange things in order?**

## Elements Described

If we are to find and see order among the elements, we must have some general acquaintance with them. A few of the chemical and physical properties of twenty of the elements are summarized in Table 3–1. This format has been chosen so you can compare the properties more easily. The properties chosen for comparison are density, hardness, and relative reactivity.

The mass of a substance in a given volume is its ***density.*** For example, one milliliter of water at room temperature contains about one gram of matter; its density is one gram per milliliter (1 g/ml), while lithium has a density of 0.534 g/ml. A piece of lithium will float on water as it reacts with the water. Most gases have very low densities, about 0.001 g/ml, while many metals have densities much greater than that of water.

***Hardness*** is a relative term used to describe solids. Diamond, a form of the element carbon, is the hardest substance known, while talc, like that used in talcum powder, is among the softest of the solids. ***Reactivity*** is often a useful term to describe an element. If an element is described as very reactive, that means it may react vigorously with air upon exposure, with moisture on your fingertips if you touch it, or with other elements. An element called unreactive may not even form *any* compounds!

***Metals*** have high reflectivity (known as metallic luster), the ability to be bent and drawn into wire without shattering, and higher densities than nonmetals. They also conduct heat and electricity well and react with nonmetals. Since metals conduct heat so well, in a cool environment metals feel colder to the touch than do nonmetals. The metal is conducting heat from your skin and you feel cooler.

***Nonmetals*** are insulators, that is, they are extremely poor conductors of heat and electricity. Their crystals are brittle and tend to shatter easily. Therefore nonmetals cannot be drawn into wire or beaten into shapes like metals. Many nonmetals are gases at room temperature. Their densities are usually less than the densities of metals, and they react readily with metals and other nonmetals.

Although you probably know the properties of some of the elements listed in Table 3–1, our primary purpose is not to learn these properties, but rather to use them to search out any trends and similarities among the elements. Do you see any trends or similarities among the elements listed in the table? The elements in Table 3–1 are listed in atomic number order, but is there another, better arrangement for them?

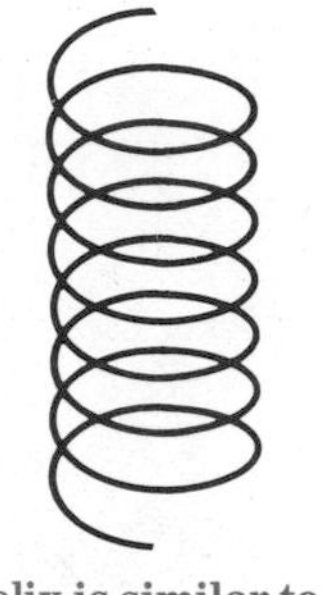

**A helix is similar to a coiled spring.**

## The Periodic Law—The Basis of the Periodic Table

The periodic law did not occur to anyone until 1869, although considerable information was available concerning the then-known elements. Parts of the complete idea had occurred as early as 1817 when Johann Wolfgang Döbereiner saw trends and similarities among several groups of three elements each, which he called *triads*. By 1862, A. Beguyer de Chancourtois saw similarities in elements along vertical lines when the

**TABLE 3–1 Some Properties of 20 Elements**

| Element | Atomic Number | Description | Compound Formation*: with Cl (or Na) | Compound Formation*: with O (or Mg) |
|---|---|---|---|---|
| Hydrogen (H) | 1 | colorless gas; reactive | HCl | $H_2O$ |
| Helium (He) | 2 | colorless gas; unreactive | none | none |
| Lithium (Li) | 3 | soft metal; low density; very reactive | LiCl | $Li_2O$ |
| Beryllium (Be) | 4 | harder metal than Li; low density; less reactive than Li | $BeCl_2$ | BeO |
| Boron (B) | 5 | both metallic and nonmetallic; very hard; not very reactive | $BCl_3$ | $B_2O_3$ |
| Carbon (C) | 6 | brittle nonmetal; unreactive at room temperature | $CCl_4$ | $CO_2$ |
| Nitrogen (N) | 7 | colorless gas; nonmetallic; not very reactive | $NCl_3$ | $N_2O_5$ |
| Oxygen (O) | 8 | colorless gas; nonmetallic; moderately reactive | $Na_2O$, $Cl_2O$ | MgO |
| Fluorine (F) | 9 | greenish-yellow gas; nonmetallic; extremely reactive | NaF, ClF | $MgF_2$, $OF_2$ |
| Neon (Ne) | 10 | colorless gas; unreactive | none | none |
| Sodium (Na) | 11 | soft metal; low density; very reactive | NaCl | $Na_2O$ |
| Magnesium (Mg) | 12 | harder metal than Na; low density; less reactive than Na | $MgCl_2$ | MgO |
| Aluminum (Al) | 13 | metal as hard as Mg; less reactive than Mg | $AlCl_3$ | $Al_2O_3$ |
| Silicon (Si) | 14 | brittle nonmetal; not very reactive | $SiCl_4$ | $SiO_2$ |
| Phosphorus (P) | 15 | nonmetal; low melting point; white solid; reactive | $PCl_3$ | $P_2O_5$ |
| Sulfur (S) | 16 | yellow solid; nonmetallic; low melting point; moderately reactive | $Na_2S$, $SCl_2$ | MgS |
| Chlorine (Cl) | 17 | green gas; nonmetallic; extremely reactive | NaCl | $MgCl_2$, $Cl_2O$ |
| Argon (Ar) | 18 | colorless gas; unreactive | none | none |
| Potassium (K) | 19 | soft metal; low density; very reactive | KCl | $K_2O$ |
| Calcium (Ca) | 20 | harder metal than K; low density; less reactive than K | $CaCl_2$ | CaO |

* The chemical formulas shown are lowest ratios. The molecular formula for $AlCl_3$ is $Al_2Cl_6$, and for $P_2O_5$ is $P_4O_{10}$.

elements were arranged in order of their atomic weights along a helix. A most interesting insight occurred in 1866 when John Newlands arranged elements in the order of their atomic weights and observed that every eighth element had similar properties. Newlands coined the "Law of Octaves" for which he was harshly ridiculed by his peers. All of these early ideas were incomplete and gained no lasting support.

On the evening of February 17, 1869, at the University of St. Petersburg (now Leningrad) in Russia, a 35-year-old professor of general chemistry, Dmitri Ivanovich Mendeleev (1834–1907), was writing a chapter for his soon-to-be-famous textbook on chemistry. He had the properties of each element written on a separate card for each element. While he was shuffling the cards trying to gather his thoughts before writing his manuscript, Mendeleev realized that if the elements were arranged in the order of their atomic weights, there was a trend in properties that repeated itself several times! Thus the periodic law and table were born, although only 63 elements had been discovered by 1869 (the noble gases, He, Ne, Ar, Kr, Xe, and Rn, were not discovered until after 1893), and the clarifying concept of the atomic number was not known until 1913.

Other spellings observed for Mendeleev's name in English: Mendeleef and Mendeleyeff.

Within a month, Mendeleev had prepared a paper and had delivered it before the Russian Chemical Society. His idea and textbook achieved great success, and he rose to a position of prestige and fame as he continued to teach at St. Petersburg. In 1890, he resigned from the university during an episode of student unrest against the government, in which he sided with the students.

By 1871, Mendeleev published a more elaborate periodic table (Fig. 3–1). This version was the forerunner of the modern table currently seen in classrooms and textbooks.

| | Group I $R_2O$ RCl | Group II RO $RCl_2$ | Group III $R_2O_3$ $RCl_3$ | Group IV $RO_2$ $RCl_4$ | Group V $R_2O_5$ $RH_3$ | Group VI $RO_3$ $RH_2$ | Group VII $R_2O_7$ RH | Group VIII $RO_4$ |
|---|---|---|---|---|---|---|---|---|
| 1 | H = 1 | | | | | | | |
| 2 | Li = 7 | Be = 9.4 | B = 11 | C = 12 | N = 14 | O = 16 | F = 19 | |
| 3 | Na = 23 | Mg = 24 | Al = 27.3 | Si = 28 | P = 31 | S = 32 | Cl = 35.5 | |
| 4 | K = 39 | Ca = 40 | — = 44 | Ti = 48 | V = 51 | Cr = 52 | Mn = 55 | Fe = 56, Co = 59<br>Ni = 59, Cu = 63 |
| 5 | (Cu = 63) | Zn = 65 | — = 68 | — = 72 | As = 75 | Sb = 78 | Br = 80 | |
| 6 | Rb = 85 | Sr = 87 | ?Yt = 88 | Zr = 90 | Nb = 94 | Mo = 96 | — = 100 | Ru = 104, Rh = 104<br>Pd = 106, Ag = 108 |
| 7 | (Ag = 108) | Cd = 112 | In = 113 | Sn = 118 | Sb = 122 | Te = 125 | I = 127 | |
| 8 | Cs = 133 | Ba = 137 | ?Di = 138 | ?Ce = 140 | — | — | — | — — — — |
| 9 | (—) | — | — | — | — | — | — | |
| 10 | — | — | ?Er = 178 | ?La = 180 | Ta = 182 | W = 184 | — | Os = 195, Ir = 197<br>Pt = 198, Au = 199 |
| 11 | (Au = 199) | Hg = 200 | Tl = 204 | Pb = 207 | Bi = 208 | — | — | |
| 12 | — | — | — | Th = 231 | — | U = 240 | — | — — — — |

**Figure 3–1** An 1871 version of Mendeleev's periodic table.

Dmitri Mendeleev (1834–1907). Born in Siberia, Mendeleev rose to Professor of Chemistry at St. Petersburg (now Leningrad) and then to director of the Russian Bureau of Weights and Measures. Although a prolific writer, a versatile chemist and inventor, and a popular teacher, the fame of this brilliant scientist rests on his discovery of the periodic law.

Two features of the 1871 version were especially interesting. Empty spaces were left in the table, and there was a problem with the positions of tellurium (Te) and iodine (I).

The empty spaces showed the genius and daring of Mendeleev. He left the empty spaces to retain the rationale of ordered arrangement based upon periodic recurrence of the properties. For example, in atomic weight order are copper (Cu), zinc (Zn), and then arsenic (As). If As had been placed next to Zn, As would have fallen under aluminum (Al). But As forms compounds similar to those formed by phosphorus (P) and antimony (Sb), not Al. Mendeleev reasoned that two as yet undiscovered elements existed, and moved As over two spaces to the position below P. The two missing elements were soon discovered: gallium (Ga) in 1875 and germanium (Ge) in 1886. The other gaps in his 1871 periodic table were later filled by discovered elements.

Mendeleev aided the discovery of the new elements by predicting their properties with remarkable accuracy, and even suggesting the geographical regions in which minerals containing the elements could be found. The properties of a missing element were predicted by consideration of the properties of its neighboring elements in the table. He had learned from Döbereiner, perhaps, that the density of an element is approximately the arithmetical average of the density of the lighter element above the missing element and the density of the heavier element just below. An example of Mendeleev's prediction of the properties of an undiscovered element is shown in Table 3–2. The term *eka* comes from Sanskrit, and means "one"; thus, ekasilicon means one place away from silicon. He also predicted the properties of ekaboron (scandium) and ekaaluminum (gallium).

The empty spaces in the table and Mendeleev's predictions of the properties of missing elements stimulated a flurry of prospecting for elements in the 1870's and 1880's. As a result, gallium (Ga) was discovered in 1875; scandium (Sc), samarium (Sm), holmium (Ho), and thulium (Tm) in 1879; gadolinium (Gd) in 1880; neodymium (Nd) and praseodymium (Pr) in 1885; and germanium (Ge) and dysprosium (Dy) in 1886.

If Mendeleev had followed the atomic weight order precisely, some elements with similar properties would not have been in the same column, or group. In the 1869 table, tellurium (Te) with an atomic weight of 128 was placed one position ahead of iodine, which has a lower atomic weight of 127. On the basis of its chemical properties, Te belonged with Sb, S, and O, and I belonged with F, Cl, and Br.

Mendeleev believed the atomic weight of tellurium was in error, but this was later shown not to be the case. In the 1871 table, the weight of Te had been changed from 128 to 125—an example of the unwise practice of changing data to fit a theory. The record is not clear as to why he changed the value.

Other reversed pairs in the modern periodic table are U before Np, Ar before K, Co before Ni, and Th before Pa. Upon realization of the atomic number concept in 1913, the question was resolved.

**TABLE 3–2 Some of Mendeleev's Predicted Properties of Ekasilicon and the Corresponding Observed Properties of Germanium.**

| | Ekasilicon (Es) | Germanium (Ge) |
|---|---|---|
| Atomic weight | 72 | 72.6 |
| Color of element | gray | gray |
| Density of element (g/ml) | 5.5 | 5.36 |
| Formula of oxide | $EsO_2$ | $GeO_2$ |
| Density of oxide (g/ml) | 4.7 | 4.703 |
| Formula of chloride | $EsCl_4$ | $GeCl_4$ |
| Density of chloride (g/ml) | 1.9 | 1.887 |
| Boiling point of chloride (°C) | under 100 | 86 |

About nine months after Mendeleev delivered his paper before the Russian Chemical Society, Julius Lothar Meyer (1830–1895), a German medical doctor and professor of chemistry at the University of Tübingen, prepared a table very similar to Mendeleev's. Apparently, both men were unaware of each other's work, yet both had left gaps for undiscovered elements. Meyer's table was based primarily on the repeatable trends in physical properties as the property is plotted against the atomic weight. Meyer grouped elements in subfamilies so, for example, zinc (Zn), cadmium (Cd), and mercury (Hg) with similar chemical properties could be separated from their chemical cousins magnesium (Mg), calcium (Ca), strontium (Sr), and barium (Ba). Mendeleev's table was superior because Meyer did not predict the properties of the undiscovered elements, and he did not rectify atomic-weight-position errors.

Building on the work of Mendeleev, Meyer, and others and using the clarifying concept of the atomic number, we are now able to state the modern periodic law: *When elements are arranged in the order of their atomic numbers, their chemical and physical properties show repeatable trends.*

Refer again to Table 3–1 and note how the trend in properties from lithium (Li) to neon (Ne) matches the trend from sodium (Na) to argon (Ar). The pattern in the properties of the elements, then, is *periodic;* hence the name periodic law or table. Other familiar periodic phenomena include the average daily temperature, which is periodic with time in a temperate climate. Low temperatures in January give way to high temperatures in July and low temperatures again in December. The trend repeats each year, not with exactly the same numbers but with the same pattern of change. Drowsiness follows a trend each twenty-four hours. Cash flow follows a cycle related to pay day. Hunger pains are periodic several times each day. A shingled roof has the same pattern over and over and is, therefore, periodic.

So, to build up a periodic table according to the periodic law, line up the elements in a horizontal row in the order of their atomic numbers. Every time you come to an element with similar properties to one already in the row, start a new row. The columns, then, will contain elements with similar properties.

## Features of the Modern Periodic Table

A modern, popular version of the periodic table is shown in Figure 3–2. Note the following features.

The vertical columns are called ***groups.***

The horizontal rows are called ***periods.***

The letters A and B distinguish ***families*** of elements. For example, Group IA is the alkali metal family, and the closely related Group IB is sometimes called the coinage metal family.

The groups of elements are cataloged into four categories. The A groups are the ***representative*** elements. As we shall see, simple atomic theory represents these elements well. The B groups and Group VIII are the ***transition*** elements that link the two areas of representative elements. The ***inner transition*** elements are the lanthanide series, which fits between La and Hf, and the actinide series, which fits between Ac and Ku. The ***noble gases*** are unique and comprise a group to themselves.

### Self-Test 3-A

1. In which group of the periodic table is Mg ______________, Pd ______________, Cl ______________, Ga ______________, Ag ______________?

2. In which period of the periodic table is Li ____________, Mo ____________, Nd ____________, U ____________, Br ____________?
3. Which are transition elements (T), which are representative elements (R), which are noble gases (N), and which are inner transition elements (I)? Be ____________, P ____________, Cr ____________, Kr ____________, Am ____________
4. Who was primarily responsible for formulating the periodic table? ____________
5. According to the periodic law, when the elements are arranged in the order of their ____________ ____________, their properties show periodicity.
6. When a phenomenon shows the same pattern over and over, we say the pattern is ____________.
7. Elements that conduct heat and electricity well are classified as ____________.

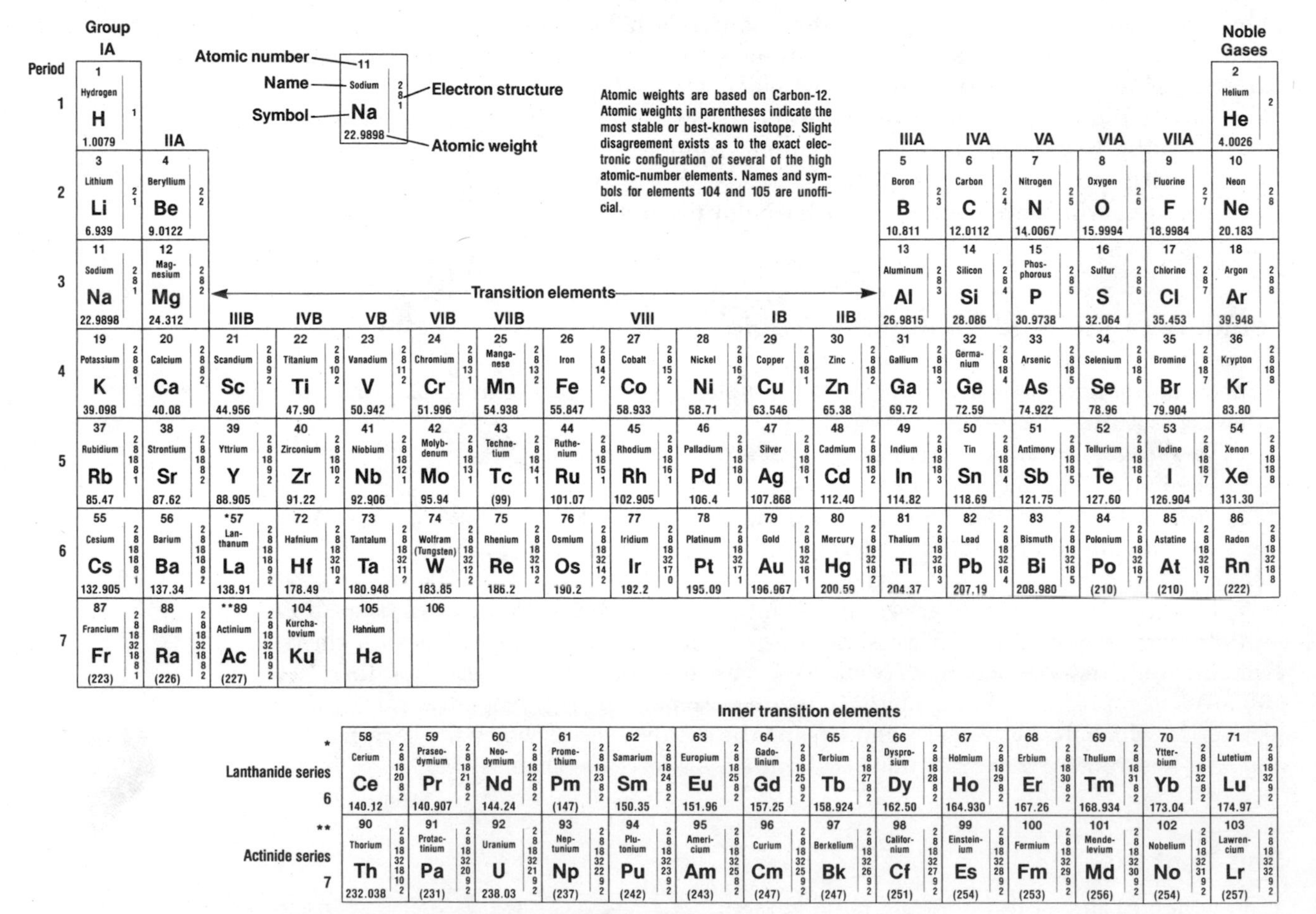


**Figure 3–2** Periodic table of the elements.

## Uses of the Periodic Table

The periodic table is very useful to chemists and students of chemistry in many ways. In addition to being a handy reference for atomic weights, atomic numbers, and whatever other information is printed in each square, the arrangement of the elements—the position of an element in the table—presents a wealth of useful information. Consider the following three ways that the periodic table will be helpful to you as you study chemistry.

### Elements in a Group Have Similar Properties

Refer back to Table 3–1 and note the properties of Li, Na, and K. They are soft metals, have low density, are very reactive, and form chlorides and oxides with formulas of MCl and $M_2O$. Now notice on the modern periodic table (Fig. 3–2) that rubidium (Rb), cesium (Cs), and francium (Fr) are also in the same group, Group IA. What properties would you expect Rb, Cs, and Fr to have? If you predicted soft metals, low density, very reactive, MCl and $M_2O$, you are right. Elements in a group have similar properties, but not the same properties.

Some properties of elements in a group differ by degree in a regular pattern. For example, the melting points beginning with Li and going down the column through Cs are (in °C) 179, 98, 64, 39, and 28, respectively. Lithium reacts slowly with water, sodium reacts faster, potassium still faster, and for the elements at the bottom of the group, just exposure to moist air produces an explosion.

Some properties differ by degree but not in a regular pattern. For example, the densities (in g/ml) of the solids Li through Cs are 0.53, 0.97, 0.86, 1.53, and 1.87, respectively. Elements become more metallic at the bottom of a group and more nonmetallic at the top of a group.

Some properties are the same for every member of a group. Elements generally react with the same other elements and form *some* of the same compounds. This is the most useful and powerful inference that can be made from the periodic table. For example, if the formula for the compound composed of lithium (Li) and chlorine (Cl) is LiCl, then probably there is a compound of Rb and Cl with the formula RbCl; the compound for Rb and Br (in Group VIIA with Cl) would probably have the formula of RbBr. Likewise, if the formula $Na_2O$ is known, then a compound with the formula $K_2S$ predictably exists. This ability to predict formulas from the periodic table has limitations. For example, Na, K, Rb, and Cs all form superoxides (formula $MO_2$), but no superoxide with Li is known. The limitations do not prohibit the use of the periodic table for predicting formulas. In general, elements in the same group of the periodic table form some of the same types of compounds!

**The general properties of metals and nonmetals are described on p. 50.**

Four groups of elements in the periodic table are referred to by the name of the group. These four groups and some of their general properties are described below.

The ***alkali metals*** are Group IA (Li, Na, K, Rb, Cs, and Fr). They are soft metals, have low density, are very reactive, and form MCl and $M_2O$. The name alkali derives from an old word meaning ashes of roasted plants. When alkali metal oxides react with water, alkalis are formed. Alkalis (later in this book to be called bases) are bitter, slick, and neutralize or destroy the effects of acids. A common alkali is sodium hydroxide (NaOH), known commercially as lye. It is formed when sodium oxide ($Na_2O$) reacts with water.

$$Na_2O + H_2O \longrightarrow 2NaOH$$

There are about 15 isotopes of francium, all of which are naturally radioactive.

The ***alkaline earths*** are Group IIA (Be, Mg, Ca, Sr, Ba, and Ra). They are metals harder and less reactive than the alkali metals. They form $MCl_2$ and MO. As the oxides, they are common components of dirt, or earth, and some react with water to form alkalis. For example, CaO (lime) reacts with water to form $Ca(OH)_2$ (slaked lime), a cheap alkali.

The ***halogens*** are Group VIIA (F, Cl, Br, I, and At). Fluorine (F) and chlorine (Cl) are gases at room temperature whereas bromine (Br) is a liquid and iodine (I) is a solid. In the elemental state, each of these elements exists as diatomic molecules ($X_2$). All isotopes of astatine (At) are naturally radioactive and disintegrate quickly. If you could accumulate enough astatine, it would be a solid at room temperature. The name halogen comes from a Greek word and means salt-producing. The most famous salt involving a halogen is sodium chloride (NaCl), table salt. But there are many other halogen salts such as calcium fluoride ($CaF_2$), a natural source of fluorine; potassium iodide (KI), an additive to table salt that prevents goiter; and silver chloride (AgCl), the active photosensitive component of photographic film.

**Aliases:**
**noble gases**
**inert gases**
**rare gases**

The ***noble gases*** are He, Ne, Ar, Kr, Xe, and Rn. All are colorless monatomic gases at room temperature. They are referred to as noble because they generally lack chemical reactivity. The derivations of some of the names of these elements are consistent with their inactivity: argon (from Greek, *argon,* meaning inactive); xenon (from Greek, *xenon,* meaning stranger). Helium (Greek, *helios,* meaning the sun) was discovered by analysis of the sun's light; later it was found on Earth. Neon (Greek, *neos,* meaning new) is a common gaseous filler for "neon" lights. Neon gas glows red when excited in a discharge lamp. Other gases and painted tubes are used to give different colors. Radon (Latin, *nitens,* meaning shining) is naturally radioactive.

**The name, radon, comes from radium (Latin, *radius,* meaning ray). At first, radon was called niton (Latin, *nitens,* meaning shining).**

Until 1962, it was thought that all of the noble gases had absolutely no chemical reactivity. On some older periodic tables, the noble gas column was headed "inert gases." Many reasons were presented to explain why the noble gases were inactive and why they never would react. Beginning in 1962, the situation began to change. A Canadian, Neil Bartlett, prepared the compound $O_2PtF_6$. Realizing that xenon might also form a similar compound, he discovered the first noble gas compound, $XePtF_6$. His discovery was followed quickly by the work of other scientists at Argonne National Laboratory, who made some thirty compounds involving the heavier members of the noble gases combined with fluorine or oxygen. Some of the first prepared compounds were $KrF_2$, $KrF_4$, $XeF_2$, $XeF_6$, $XeO_3$, $XeO_4$, and $RnF_4$. No compounds with He, Ne, or Ar have yet been reported.

**Hydrogen—the element without a home on the periodic table.**

Hydrogen probably should be in a group by itself, although you may see H in both Group IA and Group VIIA in some periodic tables. Hydrogen forms compounds with formulas similar to those of the alkali metals, but with vastly different properties, such as NaCl and HCl; $Na_2O$ and $H_2O$. Hydrogen also forms compounds similar to those of the halogens: NaCl and NaH (sodium hydride); $CaBr_2$ and $CaH_2$ (calcium hydride).

## Similar Trends Occur in Successive Periods

What should be remembered about the periodic nature of certain properties is where the low values and high values occur. For example, from Table 3–1, the number of chlorines combining with Li through Ne is 1-2-3-4-3-2-1-0, respectively. The same trend occurs for Na through Ar (Fig. 3–3). These numbers correspond to an old term called *combining power,* or ***valence*** (Latin, *valens,* meaning strength). If each chlorine has a combining power of one, then Li has a combining power of one (LiCl, one Li to one Cl), Be a combining power of two ($BeCl_2$), boron (B) a combining power of three ($BCl_3$), and so on. Elements with the greatest combining power are near the middle of the periodic table; elements with the smallest combining power are positioned at the far

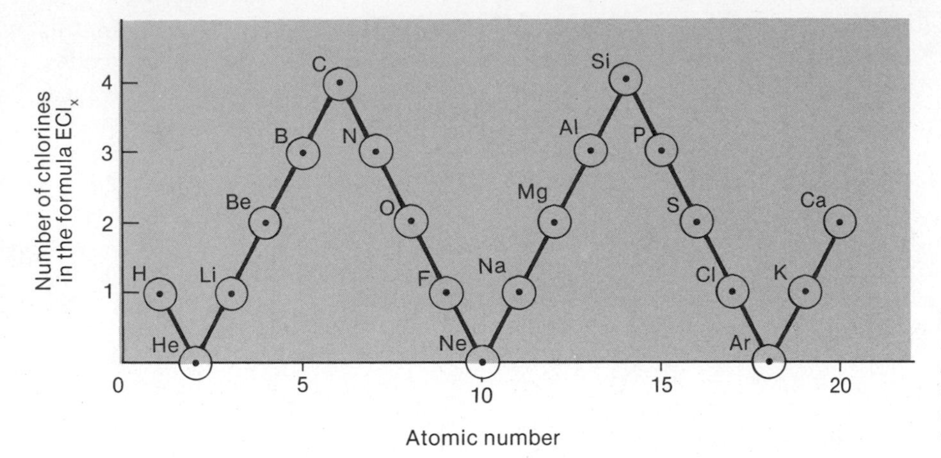


**Figure 3–3** The number of chlorines in the formula $ECl_x$, where $E$ represents the element and $x$ is the subscript on chlorine. The number of chlorines is the combining power or valence of E.

left and the far right of the periodic table. A modern, atomic-theory interpretation of combining power is a fundamental part of the next chapter.

From left to right across each period, metallic character gives way to nonmetallic character (Fig. 3–4). The elements with the most metallic character are at the lower left part of the periodic table near francium (Fr). The elements with the most nonmetallic character are found near the upper right portion of the periodic table near fluorine (F).

The heavy line on the periodic table that begins at boron (B) and staircases down to astatine (At) roughly separates the metals and the nonmetals. Most of the elements (about 80%) lie to the left of this line and are considered metals. The elements positioned along the line are considered ***semimetals*** or ***metalloids.*** Their properties are intermediate between those of metals and nonmetals. For example, silicon (Si), germanium (Ge), and arsenic (As) are ***semiconductors,*** and boron (B) conducts electricity well only at high temperatures.

***Semiconductors*** **conduct electricity less than metals such as silver and copper, but more than insulators such as sulfur; semiconductors are components of transistors.**

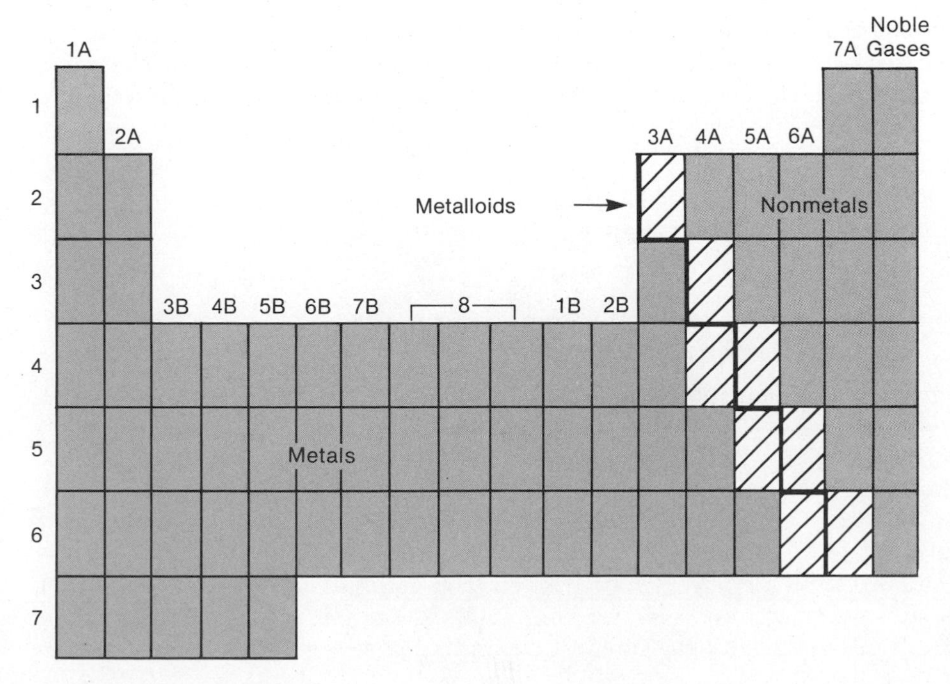


**Figure 3–4** The location of metals, nonmetals, and semimetals (metalloids) in the periodic table.

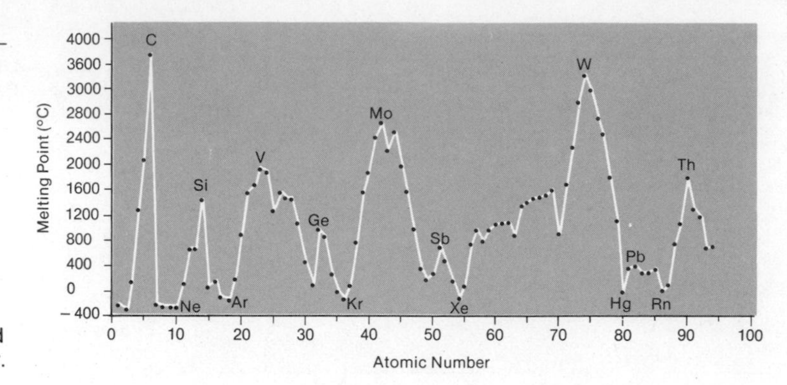


**Figure 3–5** The periodic nature of the melting points of the elements when plotted versus atomic number.

Notice the periodic patterns of melting points of the elements when plotted versus atomic number (Fig. 3–5), and of boiling points when plotted versus atomic number (Fig. 3–6). The trends are not smooth, but a general pattern is obvious. In which groups of the periodic table are the elements with the lowest melting and boiling points? Which groups have the highest? Does this information correlate well with the general information given in Table 3–1?

Atomic volumes show periodicity with atomic number (Figs. 3–7 and 3–8). The volume of a mole of atoms in the solid state can be obtained by dividing the atomic weight (g/mole) by the density of the solid (g/ml). A plot of such atomic volumes versus atomic weight was a main exhibit in support of Meyer's periodic table. Figure 3–8 gives a better general perspective of the trends across and down the periodic table. Why do atoms get larger from top to bottom of a group? Do you suppose it has something to do with more layers making a larger onion? Yes, by analogy, the larger atoms simply have more energy levels (orbits) inhabited by electrons than do the smaller atoms. The theoretical reasons for why there is a decrease in size across a period from left to right are beyond the level of this textbook, but you may see the paradox of adding electrons and getting smaller atoms.

**Larger atoms have more orbits occupied by electrons.**

Periodic relationships are also seen when the first ionization energies of the elements are plotted against atomic number (Fig. 3–9). The ***first ionization energy*** is the energy required to remove the first electron from an atom. The energy required to remove the second electron is the second ionization energy; removal of the third electron requires the third ionization energy, and similar terminology applies for the

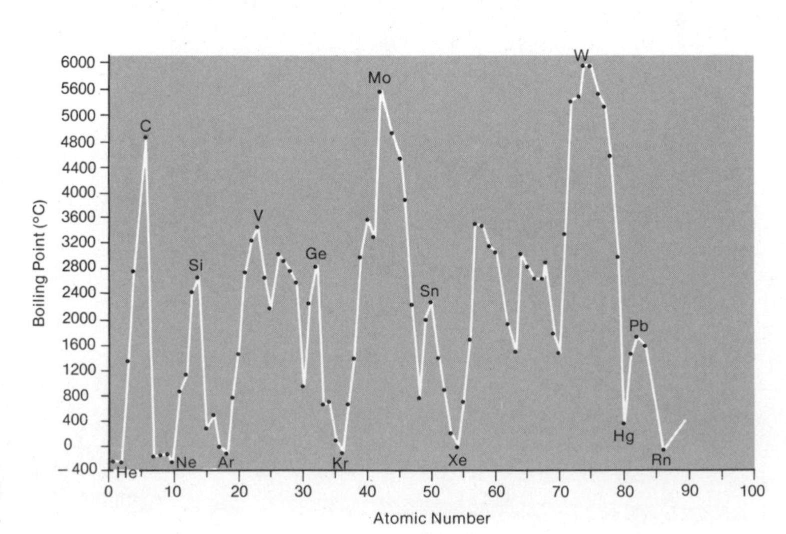


**Figure 3–6** The periodic nature of the boiling points of the elements when plotted versus atomic number.

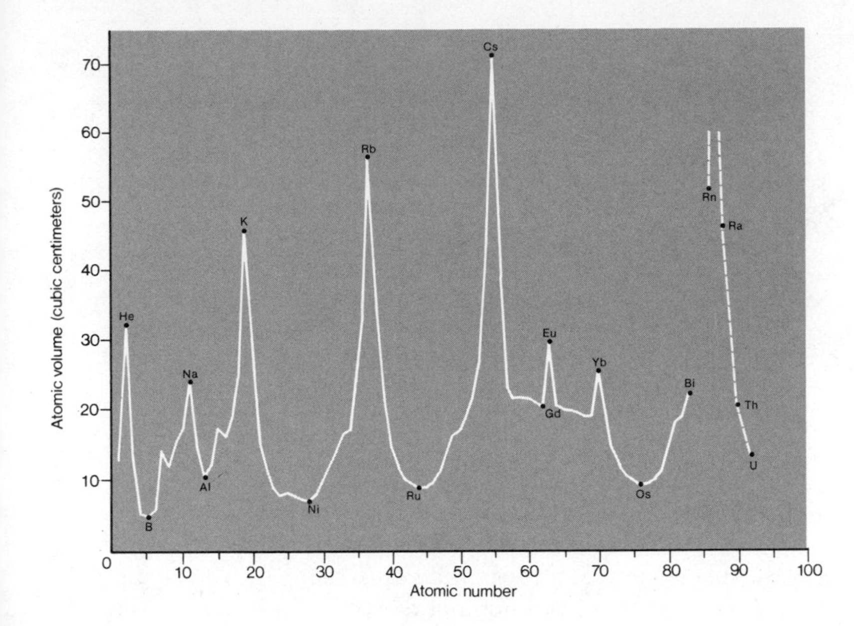


**Figure 3–7** The periodic nature of the atomic volumes of the elements when plotted versus atomic number.

fourth and other electrons. Ionization energies can be determined experimentally for some elements by inserting the gaseous element into a cathode-ray tube and increasing the voltage until a surge of current occurs (the first ionization energy), increasing the

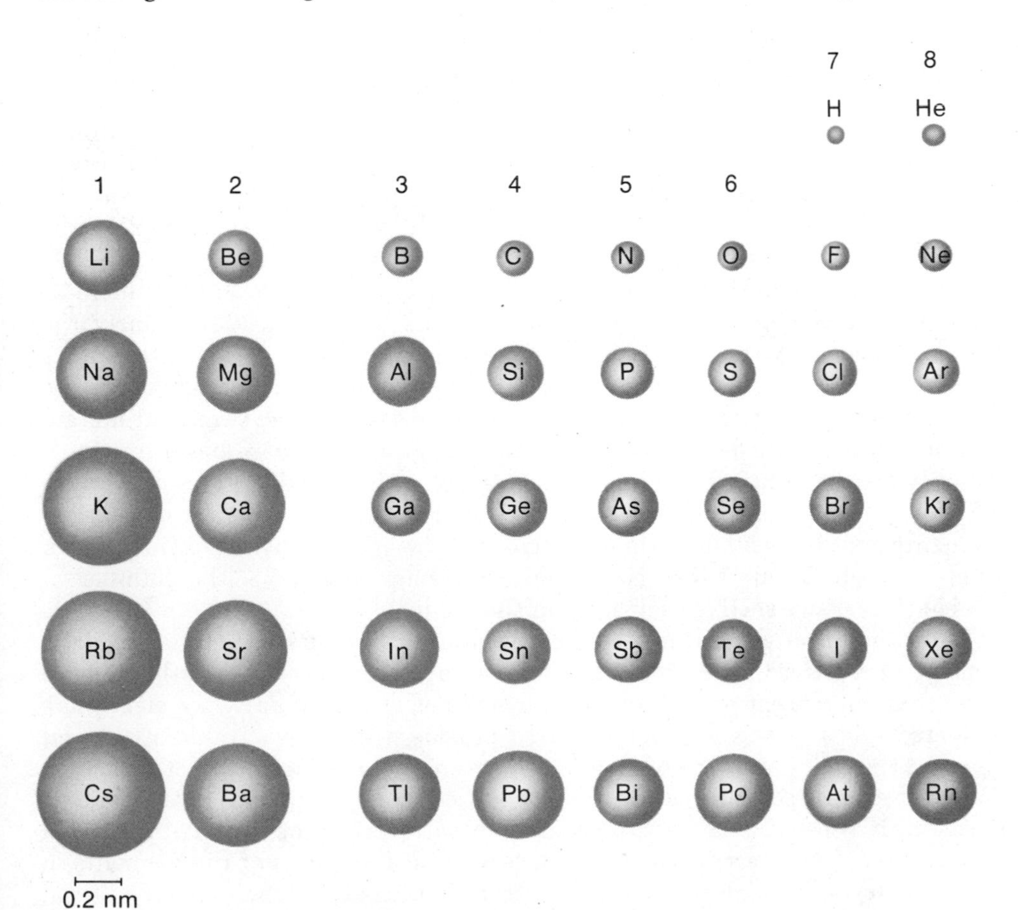


**Figure 3–8** Atomic radii of the A group elements. Atomic radii increase as one goes down a group and in general decrease in going across a row in the periodic table. Hydrogen has the smallest atom, cesium the largest.

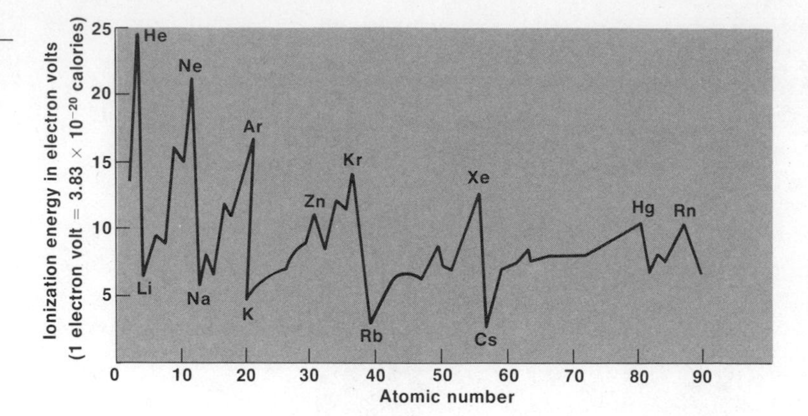


**Figure 3–9** First ionization energies of the elements.

voltage further until a second surge of current occurs (the second ionization energy), and so on. According to Figure 3–9, for which group of elements is it easiest to remove an electron? For which group of elements is it most difficult to remove an electron? Is it easier to remove electrons from metals or from nonmetals? If you answered Group IA, noble gases, and metals, respectively, you are correct. The theoretical reasons for your answers will be given in the next chapter, where ionization energies give a basis for understanding the use of electrons in bonding atoms to atoms.

In subsequent chapters, look for other trends in elemental properties and how they relate to the periodic table. Among these will be the types of bonding, electronegativity, and the number of valence electrons.

## The Periodic Table and Atomic Theory Support Each Other

Atomic theory and the periodic table support the validity of each other. Atomic theory justifies the arrangement of the elements in the periodic table. The periodic table provides observational evidence for the growing understanding of trends in atomic structure. Chronologically, the periodic table was established prior to the development of our modern atomic theory and made its general acceptance possible.

Why do elements in the same group in the periodic table have similar chemical behavior? The answer is because all of the elements in a group (particularly the representative elements, the A groups, and the noble gases) have atoms with similar structural features. In Figure 3–2, the ground-state positions of electrons in orbits are given for each of the 106 atoms. The electronic structures for the first twenty atoms are repeated nearby in Figure 3–10. Note in Group IA that each element has atoms with one and only one electron in the outermost occupied orbit (called the ***valence shell***). Group IIA elements all have two electrons in the valence shells of their atoms. Group IIIA elements and atoms all have three electrons in the valence shell, and the pattern continues through Group VIIA. Note that the group number is the number of electrons in the valence shell of each atom in the group.

**The *valence shell* is the outermost occupied orbit.**

**Eight electrons in the valence shell provide a stable electronic arrangement.**

What is the structural feature of the noble gases that results in their having little or no chemical reactivity? The noble gases each have eight electrons in the valence shell of each atom [except for helium (He), which has a total of only two electrons]. Eight electrons in the valence shell seem to provide a balanced, stable structural arrangement that minimizes the tendency of an atom to react with other atoms.

Why are there repeatable patterns of properties across the periods in the periodic table? Again, it is because there is a repeatable pattern in atomic structure. Each period begins with one electron in the valence shell of the atoms of the elements in Group IA. Each period builds up to eight electrons in the valence shell, and the period ends. This pattern repeats across periods two through six. As more elements are made by nuclear machinery, period seven may end someday. When it does, the last element in period seven will be element number 118, with eight electrons in the valence shells of its atoms. That is the prediction of the periodic table.

Atomic theory and the periodic table complement each other perfectly. One verifies the other, and vice versa.



## The Periodic Table in the Future

The periodic table ties together well what is known about familiar elements, and it predicts accurately properties of unfamiliar elements. It is an indispensable memory aid. It makes intelligent and informal guessing easy, especially when it comes to predicting chemical formulas. All of these benefits are very important to students of chemistry.

Beyond these benefits, perhaps the most elegant contribution of the periodic table toward understanding nature is its stimulation of research. We are already aware of how Mendeleev's gaps stimulated the search for new elements. Going a step further, there is now active research under way to make elements beyond element 106. (There is a paradox here in using huge nuclear machines that cover many acres to try to shoot alpha particles and other very small particles into unseen, very small nuclei.) Part of the stimulus for this reasearch is to see whether the prepared elements have the properties predicted by the periodic table. Without the periodic table, this reaching out would not be occurring. At least 16 elements have already been produced or discovered since 1939, bringing the number of known elements to 106. Prediction on elements 110 through 118 are that they will be very stable but still radioactive. Element 118 is expected to be a noble gas.

The periodic table will continue to stimulate the making of new compounds as it has in the past. An example from the past is sodium perbromate ($NaBrO_4$). Sodium perchlorate ($NaClO_4$) and sodium periodate ($NaIO_4$) had been known for some time. The fact that $NaBrO_4$ could not be prepared was puzzling since Br is in the same group (Group VIIA) of the periodic table as Cl and I, and all three elements should form similar compounds. If not, why not? All attempts failed (at least seven papers appeared

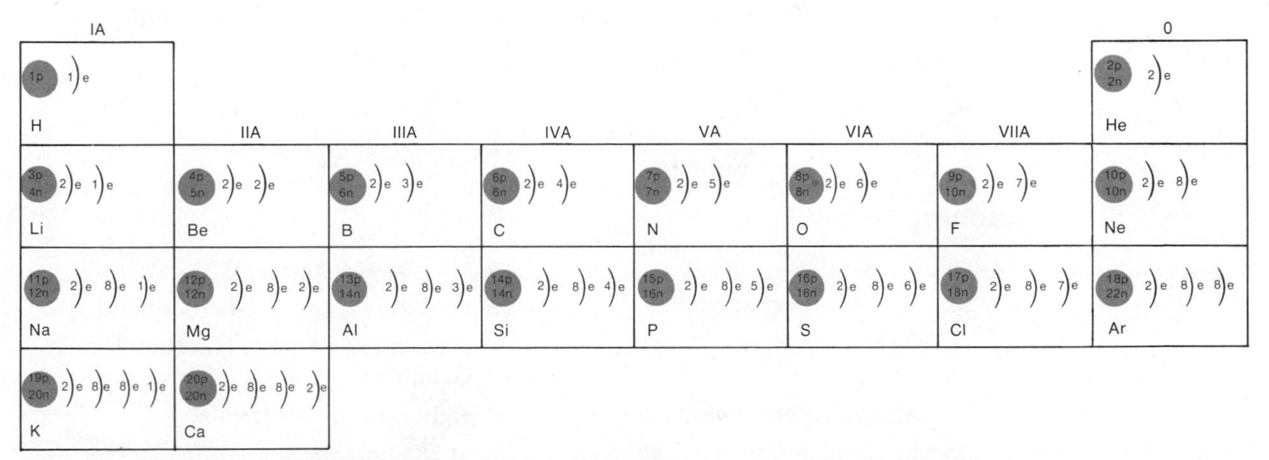


**Figure 3–10** Electron arrangements of the first 20 elements. Above each column, or group, is the group number. The nuclear contents of a typical isotope are shown.

in the chemical literature detailing why $NaBrO_4$ would never be made) until 1968, when another new compound, $XeF_2$, was reacted with sodium bromate ($NaBrO_3$) in water. Faith in the correctness and predictive powers of the periodic table stimulated the research, as it will continue to do.

The periodic table is not a panacea for the chemist, but it is an important correlating unit for tying the properties and relationships of the elements together. For its place in *your* future, we propose this hypothesis: the periodic table will be your most lasting memory of the chemistry you study in this course. As you look at that portrait of chemistry on your classroom wall, what do you see now that you did not see a while ago? You now see where you may not have seen before—a requirement for the appreciation of real art—and of science.

## Self-Test 3-B

1. What is the combining power and formula for the chloride of: Ga? ______ ______, Ba? ______ ______, Se? ______ ______, I? ______ ______
2. Classify each of the following as a metal, nonmetal, or metalloid. Si ______, Ce ______, Cl ______, Cs ______, Ca ______, O ______, H ______, Ge ______
3. How many electrons are in the valence shell of Na ______, Ca ______, F ______, Cl ______, O ______, Al ______, C ______?
4. The amount of energy required to remove an electron from an atom is called the ______ energy.
5. Which element in each pair has the greater ionization energy? He or O ______, Na or F ______, Ca or Br ______, K or S ______
6. Which element in each pair has the larger atoms? Li or K ______, F or Br ______, Na or S ______, B or In ______
7. In which groups of the periodic table would elements with the following electron configurations be found? 2-8-1 ______, 2-8-4 ______, 2-8-8-2 ______

## Matching Set

| | | |
|---|---|---|
| ______ | 1. periodic | a electronic arrangement 2-8-2 |
| ______ | 2. ionization energy | b generally a gas or a brittle solid |
| ______ | 3. larger atoms | c greater for Group VIIA than for Group IA |
| ______ | 4. two valence electrons | d eight valence electrons |
| ______ | 5. a noble gas | e at the bottom of a group |
| ______ | 6. a metal | f praseodymium (Pr) |

_______ 7. a nonmetal
_______ 8. a halogen
_______ 9. an inner transition element
_______ 10. valence shell

**g** electronic arrangement 2-8-1
**h** seven valence electrons
**i** repeated pattern
**j** ruthenium (Ru)
**k** outermost occupied orbit

## Questions

**1.** State the periodic law.

**2.** How did the discovery of the periodic law lead to the discovery of elements?

**3.** How was the atomic weight 72, which appears in Mendeleev's periodic table of 1871 (Fig. 3–1), for a then-unknown element evaluated?

**4.** Omitting argon, write the formulas for a bromide of each of the elements with atomic numbers 11 through 20.

**5.** From their positions in the periodic table, predict which will be more metallic: (a) beryllium (Be) or boron (B); (b) beryllium (Be) or calcium (Ca); (c) calcium (Ca) or potassium (K); (d) arsenic (As) or germanium (Ge); (e) arsenic (As) or bismuth (Bi).

**6.** Use the information on the periodic chart to answer the following:

a. the nuclear charge on cadmium (Cd)
b. the atomic number of arsenic (As)
c. the atomic mass (or mass number) of an isotope of bromine (Br) having 46 neutrons
d. the number of electrons in an atom of barium (Ba)
e. the number of protons in an isotope of zinc (Zn)
f. the number of protons and neutrons in an isotope of strontium (Sr), atomic mass (or mass number) of 88
g. an element forming compounds similar to those of gallium (Ga)

**7.** In a general way, how do average daily temperatures over the past three years at your location relate to properties and electronic structures of the elements when the elements are taken in the order of their atomic numbers?

**8.** Given one formula, based on the positions of the elements in the periodic table, predict the other formula.

a. $BaCl_2$; formula for Sr and Br
b. $Na_2S$; formula for K and Se
c. $Al_2O_3$; formula for Ga and S
d. $NCl_3$; formula for P and Br

**9.** Sodium metal reacts violently with water and forms hydrogen gas in the process. Magnesium metal will react with water only when the water is very hot. Copper metal does not react with water. Suppose you find a bottle containing a lump of metal in a liquid and a label, "Cesium (Cs) Metal." Based on your knowledge of the periodic table, what danger is there, if any, of disposing of the metal by throwing it into a barrel of water?

**10.** How many elements are present in each period?

**11.** Write the symbols of the halogen family in the order of increasing size of their atoms.

**12.** Why is cesium (Cs) a larger atom than lithium (Li)?

**13.** What similarities do you observe in the elements in Group IIA?

**14.** What general electronic arrangement is conducive to chemical inactivity?

**15.** How are the elements in a group related to each other?

**16.** Write the names and symbols of the alkaline earth elements.

**17.** Write the symbols for the family of elements that have three electrons in the valence shells of their atoms.

**18.** What is common about the electron structures of the alkali metals?

**19.** Pick the electron structures below that represent elements in the same chemical family.

(a) $1s^2\ 2s^1$
(b) $1s^2\ 2s^2\ 2p^4$
(c) $1s^2\ 2s^2\ 2p^2$
(d) $1s^2\ 2s^2\ 2p^6\ 3s^2\ 3p^4$
(e) $1s^2\ 2s^2\ 2p^6\ 3s^2 3p^6$
(f) $1s^2\ 2s^2\ 2p^6\ 3s^2\ 3p^6\ 4s^2$
(g) $1s^2\ 2s^2\ 2p^6\ 3s^2\ 3p^6\ 4s^1$
(h) $1s^2\ 2s^2\ 2p^6\ 3s^2\ 3p^6\ 3d^1 4s^2$

**20.** In how many different principal energy levels do electrons occur in period 1, period 3, and period 5?

**21.** Complete the following table:

| *Atomic No.* | *Name of Element* | *Electron Structure* | *Period* | *Metal or Nonmetal* |
|---|---|---|---|---|
| 6 | | | | |
| 12 | | | | |
| 17 | | | | |
| 37 | | | | |
| 42 | | | | |
| 54 | | | | |

**22.** How do the electronic structures of transition elements differ from those of "regular" (representative) elements?

**23.** How many electrons does the last element in each period have in its valence shell?

**24.** Answer this question without referring to the periodic table. Element number 55 is in Group IA, period 6. Describe its valence shell when all electrons are in the ground state; in how many orbits are there electrons?

**25.** If element 36 is a noble gas, what groups would you expect elements 35 and 37 to occur in?

**26.** Oxygen and sulfur are very different elements in that one is a colorless gas and the other a yellow crystalline solid. Why, then, are they both in Group VIA?

**27.** Suppose the popular press reports the discovery of a large deposit of pure sodium metal in northern Canada. What is your reaction as an informed citizen?

**28.** Argon is frequently used to fill tungsten-filament electric light bulbs. The argon does not glow, but it does serve a function. What is it?

**29.** You have probably noticed that balloons filled with helium lose their gas faster than balloons filled with air. Provide a reason.

**30.** Many elements are known to form compounds with hydrogen. Letting E be an element in any group, the following table represents the possible formulas of such compounds.

| *Group* | *IA* | *IIA* | *IIIA* | *IVA* | *VA* | *VIA* | *VIIA* |
|---|---|---|---|---|---|---|---|
| | EH | $EH_2$ | $EH_3$ | $EH_4$ | $EH_3$ | $H_2E$ | HE |

Following the pattern in the table, write the formulas for the hydrogen compounds of (a) Na, (b) Mg, (c) Ga, (d) Ge, (e) As, (f) Cl.

**31.** Why do you suppose that Mendeleev did not predict the existence of the noble gases?

**32.** Write the symbol for an alkali metal, a lanthanide, an alkaline earth, a halogen, an actinide, and a transition metal (first series).

**33.** Below are some selected properties of lithium (Li) and potassium (K). Without looking up the numbers, estimate values for the corresponding properties of sodium (Na).

| | *Lithium* | *Potassium* |
|---|---|---|
| Atomic weight | 6.9 | 39.1 |
| Density (g/cm$^3$) | 0.53 | 0.86 |
| Melting point (°C) | 180 | 63.4 |
| Boiling point (°C) | 1330 | 757 |

**34.** Give the names and symbols for two elements most like selenium (Se), atomic number 34.

**35.** Predict some chemical and physical properties for the element francium (Fr, atomic number 87).

**36.** What is the likelihood of discovering another family of elements such as the noble gases?

**37.** Tin (II) chloride ($SnCl_2$) and tin (IV) chloride ($SnCl_4$) are known compounds of tin. From the positions of tin (Sn) and thallium (Tl) in the periodic table, predict the two expected chlorides of thallium.

**38.** Complete the following table by writing the predicted formula.

| *Element* | *F* | *O* | *Cl* | *S* | *Br* | *Se* |
|---|---|---|---|---|---|---|
| Na | | | | | | |
| K | | | | | | |
| B | | | | | | |
| Al | | | | | | |
| Ga | | | | | | |
| C | | | | | | |
| Si | | | | | | |

**39.** Look up the properties of the other halogens and use them to predict the following properties of astatine: melting point, boiling point (1 atm), density in the gaseous, liquid, and solid states, valence(s) toward oxygen and toward hydrogen, solubility of NaAt in water at 25°C and 100°C.

**40.** If element 118 is ever produced, what will be its position in the periodic table?

**41.** Compare the first ionization energies of oxygen and xenon using Figure 3–9. Explain Neil Bartlett's reasoning when he expected xenon to form a compound with $PtF_6$ similar to $O_2PtF_6$.

# Chemical Bonds 

What holds matter together? In other words, why does glue stick, or what causes pieces of hard candy to stick together? Why is a diamond so hard; why is cotton soft? Or, in reverse, why do things break or fall apart? Why is table salt so brittle? Why does paint peel? Why do some substances melt at a rather low temperature, while others melt at higher temperatures?

These and other questions can be answered logically and be consistent with experimental evidence if we will think of matter as one atom bound to another. Granted, it is a little hard to consider the Empire State Building or the Washington Monument or a living organism as a conglomeration of atoms bonded one to the other. But large pieces of matter, even the Rocky Mountains, conform to the same fundamental principles of nature as a small crystal of sugar or salt.

**Chemical bonds hold atoms, molecules, and ions together.**

Most of the reasons for matter bonding to matter (or atom to atom) can be summarized by two concise notions discussed in Chapter 2.

1. Unlike charges attract.
2. Electrons tend to exist in pairs.

Couple these two ideas (one empirical; one theoretical) with the proximity requirement that only the outer electrons of the atoms (the *valence electrons*) interact directly from atom to atom, and you have the basic concepts that explain how atoms bond to each other. Just how the different atoms use these principles to bond atom to atom is the subject of this chapter. We shall see that the action of an atom in the formation of a bond is dictated by its atomic structure and generalized by its position in the periodic table.

The various actions of the atoms cause the formation of five major types of chemical bonds. The types of bonds, along with some common materials in which they occur, are:

1. Ionic bonding — salts, such as table salt (sodium chloride); and metal oxides, such as lime, iron rust, ruby, and sapphire
2. Covalent bonding — molecular compounds, such as water, methane, and sucrose; and polymers, such as polyethylene
3. Hydrogen bonding — water, ammonia, and large molecules in living organisms
4. Van der Waals attractions — liquid helium and solid $CO_2$
5. Metallic bonding — metals and alloys

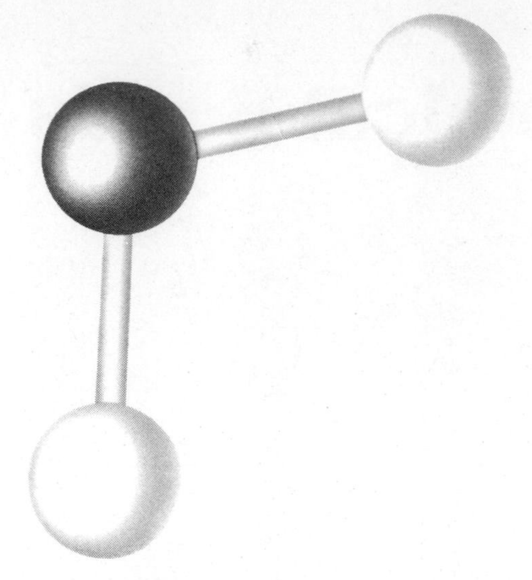

**Figure 4–1** A water molecule is often represented by a ball-and-stick model. This model tells which atoms are bonded together and the angle involved, but gives no information as to why the atoms are bonded in a particular pattern, the relative sizes of the atoms, or the actual distances between them.

As always, it is the properties of the substances that dictate and verify the related theories. It is properties such as chemical reactivity, volatility (ability to pass into the gaseous state), melting point, electrical conductivity, and color that often give some indication of how atoms are bonded to each other. For example, since melting involves a situation in which atoms or molecules become less firmly bound to their neighbors, a high melting point implies that a solid is held together by very stable chemical bonds. As we shall see shortly, compounds composed of a network of tightly bound ions or atoms tend to have relatively high melting points. The volatility of a substance also indicates how strongly molecules are attracted to each other. For example, in the case of carbon dioxide, $CO_2$, we must assume that the bonding between molecules (intermolecular bonding) is slight, since it takes relatively little energy to break up the solid $CO_2$.

$CO_2$ changes readily from a solid to a gas (*sublimes*).

In the ensuing discussion of chemical bonds, major emphasis will be placed on accounting for the properties of a given substance by the bonds that hold that substance together.

Bonding theories must explain the observed behavior of chemicals.

## Ionic Bonds

### Ions and Ion Formation

There is a large category of compounds that forms rather hard and brittle crystalline solids with relatively high melting points. When melted or in solution, they conduct electricity well, but they do not conduct when solid. If in solution or melted, these compounds often react quickly with each other. Compounds with these properties are known as ***ionic compounds.*** Examples of ionic compounds are sodium chloride (NaCl), magnesium fluoride ($MgF_2$), and calcium oxide (CaO).

All of the properties of these compounds can be explained if the compounds are assumed to be composed of charged atoms (called ***ions***) rather than neutral atoms. X ray and mass spectrographic studies of these kinds of compounds strongly confirm that ions exist.

An *ion* is a charged atom or group of atoms.

How then do atoms become ions, and which atoms are most likely to form ions?

Since electrons constitute the outermost parts of the atom, it is reasonable to assume that electrons—and only electrons—are manipulated to form ions. Electrons can be *removed* from an atom, with the result that part of the positive nuclear charge is

There is a periodic table on p. 54 and on the inside of the back cover.

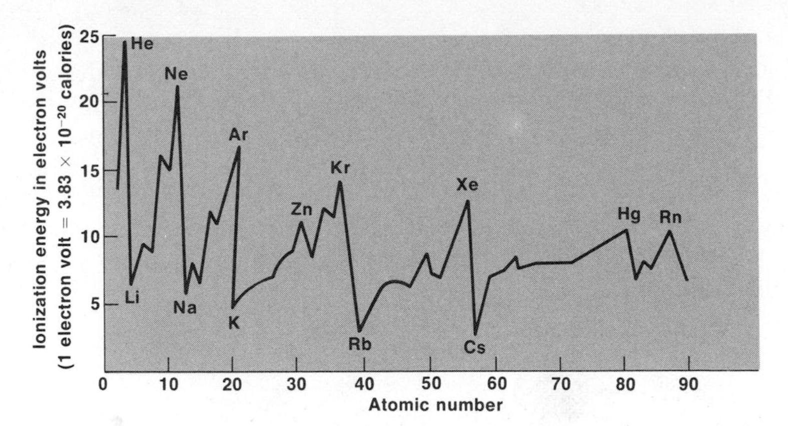


**Figure 4–2** First ionization energies of the elements.

not neutralized and the atom becomes an ion with a positive charge. Electrons can also be *gained* by an atom, with the result that excess negative charge has been added to the atom, making it a negative ion. Metals form positive ions in the presence of nonmetals, and nonmetals form negative ions in the presence of metals.

**Ionization energies are measured experimentally and correspond to the reaction: atom ⟶ positive ion + electron.**

Recall that metals constitute the left side of the periodic table, and that they have one, two, or three valence electrons. Energy is required to remove these electrons from atoms. This is the ionization energy that was shown to be periodic in Chapter 3 (lower ionization energies on the left of the periodic table, higher on the right; see Fig. 4–2 for review). Ionization energy must be added to an atom, and the energy can come either from an outside source (heat or electricity, for example) or from the energy given off when nonmetals receive and pair electrons. Since metals have the lower ionization energies, their electrons are easier to remove than the electrons of nonmetals.

**Metals lose electrons to form positive ions.**

**The electrons easiest to remove are the outermost electrons, the valence electrons. Which is easier to remove, the peel or the seed of an orange?**

The number of electrons removed from metals to form positive ions is the number of valence electrons per atom, as confirmed by the ionization energies listed in Table 4–1. Atoms of Group IA metals (Li, Na, and K as well as Rb, Cs, and Fr) have only one

**Table 4–1 shows the amounts of energy needed to remove one or more electrons from various atoms.**

**TABLE 4–1 Ionization Energies of Selected Gaseous Atoms**

An electron volt (ev) is the energy acquired by an electron when accelerated by a potential difference of 1 volt. For each element, electrons must be removed to the heavy vertical line in order to attain a noble gas electronic configuration.

| Atomic Number | Atom | Ionization Energies (ev) | | | | | | | |
|---|---|---|---|---|---|---|---|---|---|
| | | *1st* | *2nd* | *3rd* | *4th* | *5th* | *6th* | *7th* | *8th* |
| 1 | H | 13.6 | 54.5 | | | | | | |
| 2 | He | 24.6 | 54.4 | | | | | | |
| 3 | Li | 5.4 | 75.6 | 122.4 | | | | | |
| 4 | Be | 9.3 | 18.2 | 153.9 | 217.7 | | | | |
| 5 | B | 8.3 | 25.1 | 37.9 | 259.3 | 340.1 | | | |
| 6 | C | 11.3 | 24.4 | 47.9 | 64.5 | 392.0 | 489.8 | | |
| 7 | N | 14.5 | 29.6 | 47.4 | 77.5 | 97.9 | 551.9 | 666.8 | |
| 8 | O | 13.6 | 35.1 | 54.9 | 77.4 | 113.9 | 138.1 | 739.1 | 871.1 |
| 9 | F | 17.4 | 35.0 | 62.6 | 87.2 | 114.2 | 157.1 | 185.1 | 953.6 |
| 10 | Ne | 21.6 | 41.1 | 64 | 97.2 | 126.4 | 157.9 | | |
| 11 | Na | 5.1 | 47.3 | 71.7 | 98.9 | 138.6 | 172.4 | 208.4 | 264.2 |
| 12 | Mg | 7.6 | 15.0 | 80.1 | 109.3 | 141.2 | 186.9 | 225.3 | 266.0 |
| 13 | Al | 6.0 | 18.8 | 28.4 | 120.0 | 153.8 | 190.4 | 241.9 | 285.1 |
| 14 | Si | 8.1 | 16.3 | 33.5 | 45.1 | 166.7 | 205.1 | 264.4 | 303.9 |
| 15 | P | 10.6 | 19.7 | 30.2 | 51.4 | 65.0 | 220.4 | 263.3 | 309.3 |
| 16 | S | 10.4 | 23.4 | 35.0 | 47.3 | 72.5 | 88.0 | 281.0 | 328.8 |
| 17 | Cl | 13.0 | 23.8 | 39.9 | 53.5 | 67.8 | 96.7 | 114.3 | 348.3 |
| 18 | Ar | 15.8 | 27.6 | 40.9 | 59.8 | 75.0 | 91.3 | 124.0 | 143.5 |
| 19 | K | 4.3 | 31.8 | 46 | 60.9 | 82.6 | 99.7 | 118 | 155 |
| 20 | Ca | 6.1 | 11.9 | 51.2 | 67 | 84.4 | 109 | 128 | 147 |

electron easily removed. To remove a second electron requires at least seven times more energy than to remove one electron. Group IA metal atoms have one valence electron. According to Table 4–1, atoms of Group IIA metals (Be, Mg, and Ca shown plus Sr, Ba, and Ra) have two easier-to-remove electrons, and thus two valence electrons. Atoms of Group IIIA metals (B and Al, for example) have three easier-to remove electrons, and hence three valence electrons.

The removal of electrons from metals and the consequent formation of positive ions can be depicted in varying degrees of detail. For Group IA, using sodium (Na, atomic number 11) as the example:

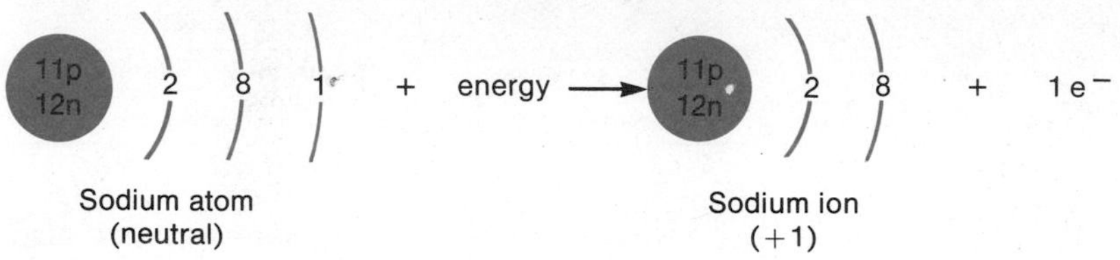

11p 12n 2 8 1 + energy ⟶ 11p 12n 2 8 + 1 $e^-$

Sodium atom (neutral)  Sodium ion (+1)

or

$$\underset{2\text{-}8\text{-}1}{Na} + \text{energy} \longrightarrow \underset{2\text{-}8}{Na^+} + e^-$$

or, simply

$$Na \longrightarrow Na^+ + e^-$$

Likewise for Group IIA metals, using Mg as the example:

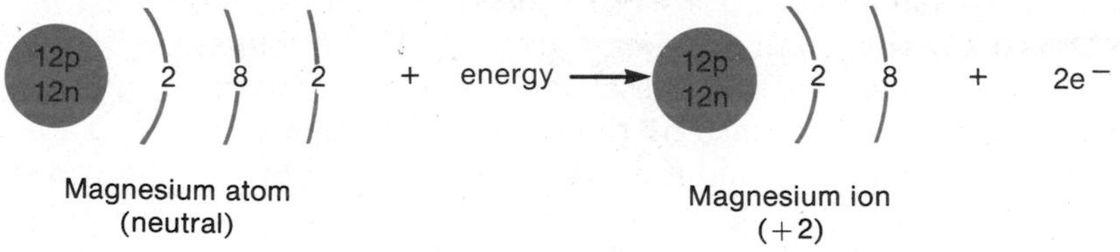

12p 12n 2 8 2 + energy ⟶ 12p 12n 2 8 + 2$e^-$

Magnesium atom (neutral)  Magnesium ion (+2)

or

$$\underset{2\text{-}8\text{-}2}{Mg} + \text{energy} \longrightarrow \underset{2\text{-}8}{Mg^{2+}} + 2e^-$$

or, simply

$$Mg \longrightarrow Mg^{2+} + 2e^-$$

In a similar fashion for Group IIIA metals, using Al as the example:

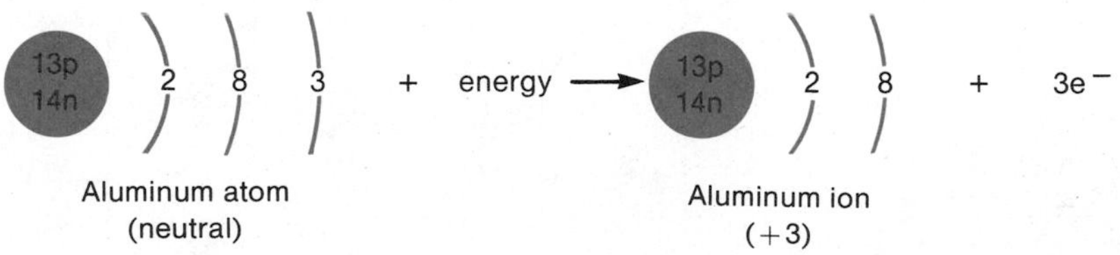

13p 14n 2 8 3 + energy ⟶ 13p 14n 2 8 + 3$e^-$

Aluminum atom (neutral)  Aluminum ion (+3)

or

$$\underset{2\text{-}8\text{-}3}{Al} + \text{energy} \longrightarrow \underset{2\text{-}8}{Al^{3+}} + 3e^-$$

or, simply

$$Al \longrightarrow Al^{3+} + 3e^-$$

**TABLE 4–2 Electronic Configurations of the Noble Gases and Ions with Identical Configurations**

| Species | Configuration |
|---|---|
| He, $Li^+$, $Be^{2+}$, $H^-$ | 2 |
| Ne, $Na^+$, $Mg^{2+}$, $F^-$, $O^{2-}$ | 2-8 |
| Ar, $K^+$, $Ca^{2+}$, $Cl^-$, $S^{2-}$ | 2-8-8 |
| Kr, $Rb^+$, $Sr^{2+}$, $Br^-$, $Se^{2-}$ | 2-8-18-8 |
| Xe, $Cs^+$, $Ba^{2+}$, $I^-$, $Te^{2-}$ | 2-8-18-18-8 |

**Stable ions often have noble gas electronic configurations.**

The more detailed depiction of positive ion formation points out an interesting coincidence for many, if not most, ions. For the ions formed from metals in Group IA, Group IIA, and Group IIIA (B and Al only), each ion has eight electrons in its new outermost orbit. This is exactly the number of valence electrons in the atoms of the noble gases (except He, of course). It appears that when some atoms achieve a noble gas electronic arrangement, they become satisfied, unreactive, and definitely not interested in transferring any more electrons. Some examples are shown in Table 4–2. Stable positive ions that do not conform to the noble gas electronic arrangement are most of the transition metal ions and positive ions formed from elements to the right of the transition metals in Periods 4, 5, and 6.

In summary, positive ions are formed when metal atoms lose one electron (Group IA), two electrons (Group IIA), or three electrons (Group IIIA). The resulting ions often have the same electronic arrangement as a noble gas.

**Why not a +8 stable ion?**

The difficulty of removing successive electrons from the same atom is verified by the ionization energies in Table 4–1. As more and more electrons are removed from a single atom such as a boron (B) atom, the net charge builds up with the loss of each electron. The unneutralized charge helps to hold the remaining electrons more securely. For this reason (no other stabilizing forces considered), a $B^{3+}$ ion is more difficult to make and less likely to occur than a $Li^{+1}$ ion.

**Electrons in an orbital spin in opposite directions.**

Likewise, moving across a period in the periodic table, the lower ionization energies of the metals give way to the higher ionization energies of the nonmetals. It is much more difficult to remove electrons from nonmetals than from metals. The energies available in ordinary chemical changes are simply not great enough to form positive ions from nonmetallic atoms. Instead, two driving forces cause nonmetals to take electrons from metals, and thus make nonmetal atoms into nonmetal negative ions. One driving force is the urge for atoms to mimic the eight valence electrons of the noble gas atoms. Another driving force, and probably a more fundamental one, is the stable arrangement of a pair of electrons. Recall from Chapter 2 that two electrons occupy an orbital. The compatibility of two electrons in an orbital is thought to be caused by the mutual stability afforded by opposite spins on the two electrons. A moving charge (rotation or otherwise) produces a magnetic field about it. When the moving charges spin in opposite directions, compatible magnetic fields are produced that attract the charges toward each other.

$s = +½$

$s = -½$

A nonmetal atom, then, strives to pair all of its valence electrons, and thereby achieves a noble gas electronic arrangement of eight valence electrons. Thus a nonmetal can add to its five, six, or seven valence electrons three, two, or one electron(s) and have the stable eight valence electrons.

**Nonmetal atoms gain electrons to form negative ions.**

Consider an atom of fluorine (F, atomic number 9). Its electronic arrangement is 2-7 (or $1s^2\ 2s^2\ 2p_x^2\ 2p_y^2\ 2p_z^1$). A fluorine atom needs one electron to pair with its one unpaired electron, and this single electron completes the stable eight valence electrons. The gain of electrons by the nonmetals, like the loss of electrons by the metals, can be depicted in various degrees of detail.

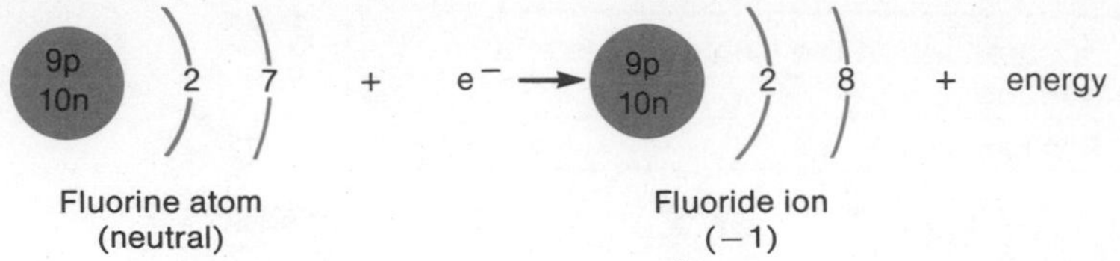

9p 10n ) 2 ) 7 + $e^-$ → 9p 10n ) 2 ) 8 + energy

Fluorine atom (neutral) — Fluoride ion (−1)

$F + e^- \longrightarrow F^- + \text{energy}$
2-7 2-8

$:\ddot{\underset{..}{F}}\cdot + e^- \longrightarrow :\ddot{\underset{..}{F}}:^-$

$F + e^- \longrightarrow F^-$

Note that an electron is on the left side of each equation, and the electron is therefore gained. The third depiction is very informative, and involves ***electron dot formulas.*** Only the valence electrons are represented around the symbol for the element. The electrons are placed at north, south, east, and west positions adjacent to the symbol. The method clearly shows the pairing of electrons as well as the attainment of eight valence electrons when the negative ion is formed.

Next consider how an oxygen atom becomes an oxide ion. The electronic arrangement of oxygen (O, atomic number 8) is 2-6 (or $1s^2\ 2s^2\ 2p_x^2\ 2p_y^1\ 2p_z^1$), which in electron dot representation is $\cdot\ddot{\underset{\cdot}{O}}:$. How an oxygen atom gains two electrons to become an oxide ion ($O^{2-}$) can be depicted in several ways as before.

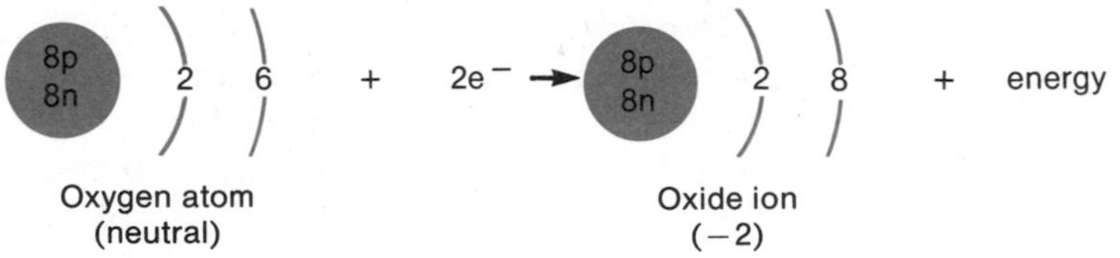

8p 8n ) 2 ) 6 + $2e^-$ → 8p 8n ) 2 ) 8 + energy

Oxygen atom (neutral) — Oxide ion (−2)

$O + 2e^- \longrightarrow O^{2-} + \text{energy}$
2-6 2-8

$\cdot\ddot{\underset{\cdot}{O}}: + 2e^- \longrightarrow :\ddot{\underset{..}{O}}:^{2-}$

$O + 2e^- \longrightarrow O^{2-}$

Other common negative ions are listed in Table 4–2. The types of ions formed by the various groups of elements are shown on the periodic table in Figure 4–3. As the difficulty of removing successive electrons from an atom of a metal increases, it is increasingly more difficult to add successive electrons to a nonmetal atom. The negative charge built up on a negative ion repels an incoming electron. For this reason, highly charged negative ions (−4, −5, −6, etc.) are not stable and do not exist in stable compounds. An atom with four valence electrons has no pronounced tendency to lose or gain electrons or to form ions.

Table 4–2 is on p. 69.

Why not a −8 stable ion?

In summary, nonmetals in the presence of metals gain one, two, or three electrons to form negative ions, which have all valence electrons paired and have the stable eight-electron arrangement of noble gases.

## Properties of Ions, Atoms, and Molecules

Ions and their parent atoms have some different properties. Besides the difference of being charged or neutral, there are other attendant, important differences. For ex-

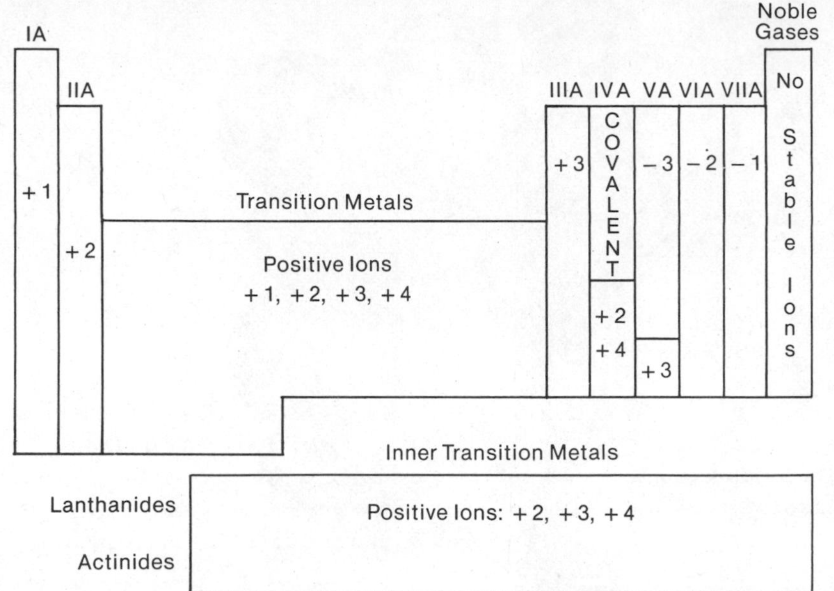


**Figure 4–3** The periodic table and the formation of ions.

ample, sodium atoms react quickly with water to produce hydrogen gas. Sodium ions and water produce no hydrogen gas. Chlorine atoms are poisonous to the human body. Chloride ions are not considered poisonous. In fact, sodium chloride ($Na^+$ and $Cl^-$ ions) is palatable on tomatoes, whereas Na and Cl atoms would blow up the tomato and poison the eater. Hydrogen gas is combustible; oxygen gas supports combustion; hydrogen and oxygen bonded into water molecules put out a fire. The bonding and charging of atoms change their nature. Compounds that contain ionic bonds (formed by the *transfer* of electrons) are classified as ***salts,*** in contrast to molecular compounds in which all the atoms are held together by *sharing* electrons (covalent bonds, p. 74).

The *electrical conductivity* of melted ionic compounds is based on the movement of free ions to oppositely charged poles when an electrical field is imposed. The movement of the ions transports charge, or electric current, from one place to another. In a rigid solid, the immobile ions are not free to move, and the solid does not conduct electricity.

**Electrical conductivity by ionic substances is illustrated in Figure 6–3 on p. 115.**

The *hardness* of ionic compounds is caused by the strong bonding between ions of unlike charge. The strong bonds require much energy to separate the ions and allow the freer movement of the melted state. Much energy means *higher melting points,* which are characteristic of ionic compounds.

Ionic compounds are *brittle* because the structure of the solid is a regular array of ions. Take, for example, the structure of sodium chloride (NaCl) (Fig. 4–4). Each $Na^+$ ion is surrounded by six $Cl^-$ ions, and each $Cl^-$ ion is, in turn, surrounded by six $Na^+$ ions. Each ion is attracted to all of the oppositely charged ions and repelled by all of the identically charged ions in the structure. If a plane of ions is shifted just one ion's distance in any direction, identically charged ions are now next to each other; there is repulsion—no attraction—and the crystalline solid breaks. Sodium chloride cannot be hammered into a thin sheet. It shatters instead.

***Ionic bonds* are the attractions between ions of opposite charge.**

**An ionic structure is a regular geometrical array of ions.**

There is no unique pairing of ions; hence ionic compounds have *no molecules.*

All chemical compounds are inherently *neutral.* For ionic compounds, this means there must be present as many unit positive charges as unit negative charges. This is a theoretical justification for the usual chemical formulas seen for ionic compounds. For example, a calcium ion ($Ca^{2+}$) requires two fluoride ions ($F^-$) to have exactly the same

**Ionic compounds have no molecules.**

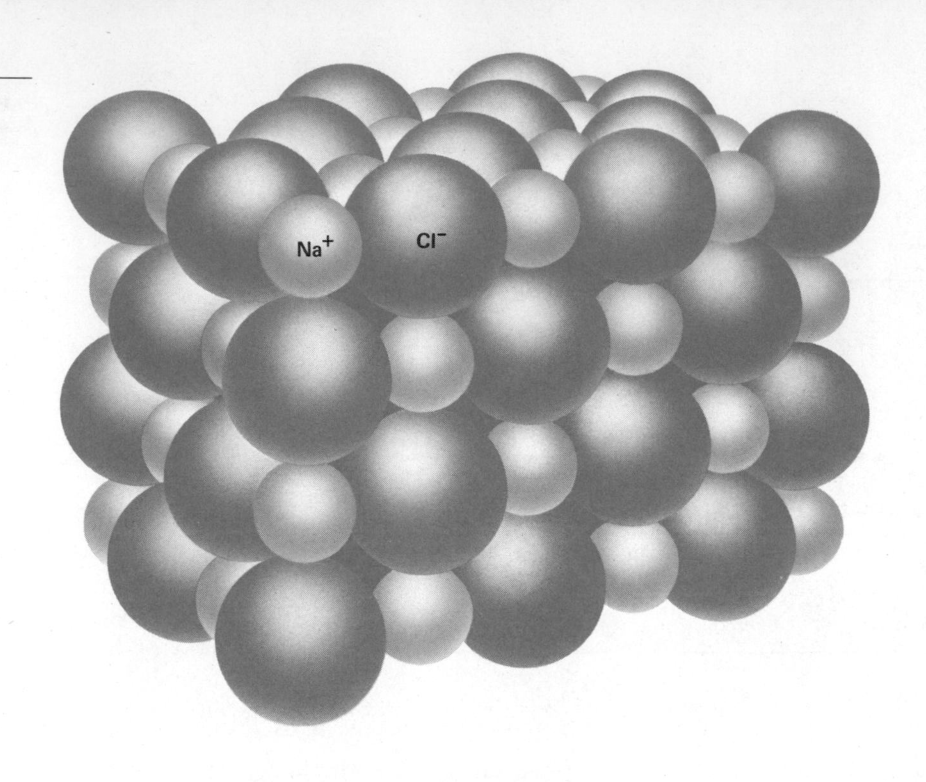


○ $Na^+$ ion

● $Cl^-$ ion

(b)

**Figure 4–4** Structure of sodium chloride crystal. (a) Model showing relative sizes of the ions; (b) ball-and-stick model showing cubic geometry.

Chemical formulas of ionic compounds result from the number of electrons lost or gained by the reacting atoms and the neutrality of the compound, which requires equal numbers of positive and negative charges.

One nanometer, nm, is $10^{-9}$ meter.

amount of each kind of charge. The chemical formula is $CaF_2$. An easy way to arrive at the correct formula is to write the symbols and their attendant charges. Then crisscross the numbers, reduce the subscripts to the lowest ratio, and you have the proper chemical formula. A few examples are given below.

Working setup: $Ca^{2+}_{1}\ F^{-}_{2}$ $\qquad$ $Al^{3+}_{2}\ O^{2-}_{3}$ $\qquad$ $Mg^{2+}_{2}\ S^{2-}_{2}$

Chemical formula: $CaF_2$ $\qquad$ $Al_2O_3$ $\qquad$ $MgS$

*Ion sizes* are different from parent atom sizes. Positive ions are smaller than the atoms from which they were made; negative ions are larger than their atoms (Fig. 4–5). A sodium atom with its single outer electron has a radius of 0.186 nm. One would expect that when this electron is removed (forming the $Na^+$ ion) the resulting ion would be smaller. This decrease in size results because there are now only 10 electrons attracted to a charge of +11 on the nucleus, and these electrons are pulled closer to the nucleus

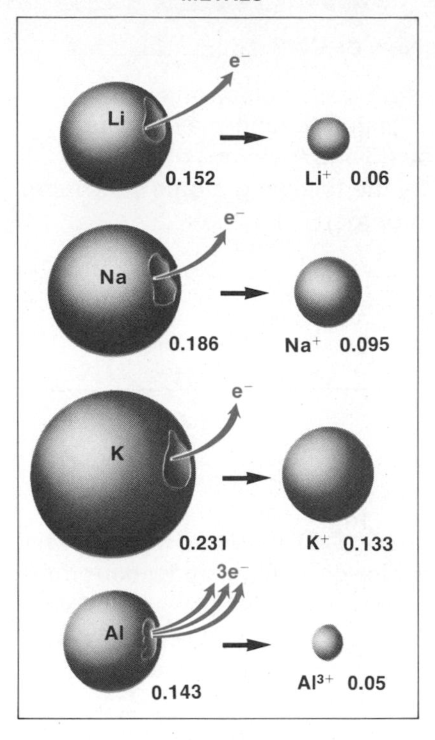


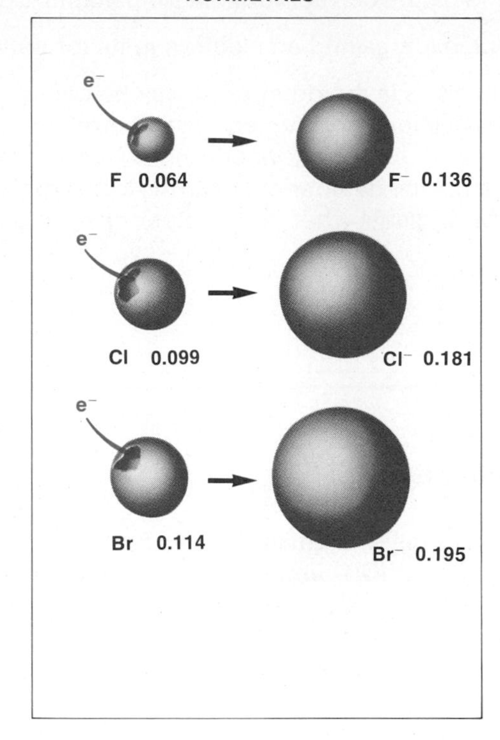


**Figure 4–5** Relative sizes of selected atoms and ions. Numbers given are atomic or ionic radii in nanometers.

by this charge imbalance. The same type of phenomenon is observed for all metal ions.

The metal ions with multiple charges ($Al^{3+}$, $Fe^{2+}$, and so on) are much smaller than the corresponding metal atom, because of still greater surplus positive charge on the nucleus.

Nonmetals gain electrons to form negative ions that are *larger* than the corresponding atoms. This phenomenon results from the addition of electrons to the outer orbit of an atom without increasing the charge on the nucleus. The repulsion of the electrons causes the expansion.

The sizes of ions are important because the strength of the forces that hold ions together in ionic compounds depends on the sizes (and charges) of the ions involved. If two ions have the same charge, the smaller ion would have a more concentrated charge which could get closer to another ion and form a stronger bond.

## Self-Test 4-A

1. Charged atoms are called ____________.
2. The attraction between positive and negative ions produces a(n) ____________ bond.
3. A sodium atom loses ____________ electron(s) in achieving a noble gas configuration.
4. Look at Figure 4–2. Which element requires the most energy for ionization? ____________ The least? ____________

5. What is the correct formula for calcium iodide (Ca and I)? ________
6. Which ion gained an electron in its formation: $Na^+$ or $Cl^-$? ________
7. Electrons in the outer orbit may be called ________ electrons.
8. Positive ions are formed from neutral atoms by (losing/gaining) electrons.
9. Negative ions are formed from neutral atoms by (losing/gaining) electrons.
10. Predict the number of electrons lost or gained by the following atoms in forming ions. Indicate whether the electrons are gained or lost.

Rb ________ ________ S ________ ________

Ca ________ ________ Mg ________ ________

K ________ ________ Br ________ ________

## Covalent Bonds

What holds together substances such as carbon monoxide (CO), methane ($CH_4$), water ($H_2O$), sand ($SiO_2$), ammonia ($NH_3$), carbon tetrachloride ($CCl_4$), and about four million other compounds in which all of the elements are nonmetals? They are all very poor conductors of electricity in the melted state. Remember that all of the nonmetals have higher ionization energies, and none of them are prone to form positive ions to balance possible negative ions.

**The drive to attain the stable eight valence electrons of a noble gas is known as the *octet rule*. The rule is particularly applicable to carbon and a few other nonmetals.**

The driving forces of electron pairing and the stable eight-electron (***octet***) arrangement of the noble gases can be accommodated by *sharing* pairs of electrons between atoms of elements in Groups IVA, VA, VIA, and VIIA. The sharing of electrons between two atoms produces a ***covalent bond.*** The strength of the bond comes from the interaction of an orbital of one atom with an orbital of another atom (Fig. 4–6). The shared electrons are held to each other by the pairing forces, and the electron pairs are held to the two nuclei by the attractions between unlike charges.

### Single Covalent Bonds

**A *covalent bond* results from sharing an electron pair.**

A single covalent bond is formed when two atoms share a single pair of electrons. The simplest examples are the diatomic (two-atom) molecules such as $H_2$ (hydrogen), $F_2$ (fluorine), and $Cl_2$ (chlorine).

A hydrogen atom has one electron. If a hydrogen atom could share its electron with another atom that has an unpaired valence electron, a stable pairing of the two electrons can be achieved and the hydrogen atom can then have the electronic structure of helium, a noble gas. This arrangement can be achieved by two hydrogen atoms sharing their single electrons. The electron dot formula for the $H_2$ molecule is

**To break a bond requires energy; when bonds are formed, energy is released.**

$$2H\cdot \longrightarrow H{:}H + \text{energy}$$

ATOMS MOLECULE

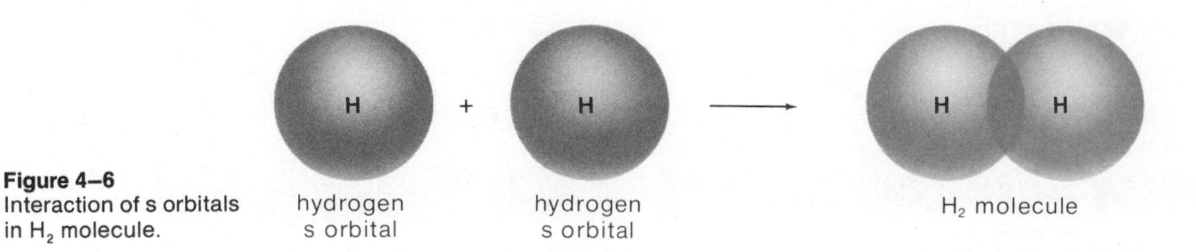


**Figure 4–6**
Interaction of s orbitals in $H_2$ molecule.

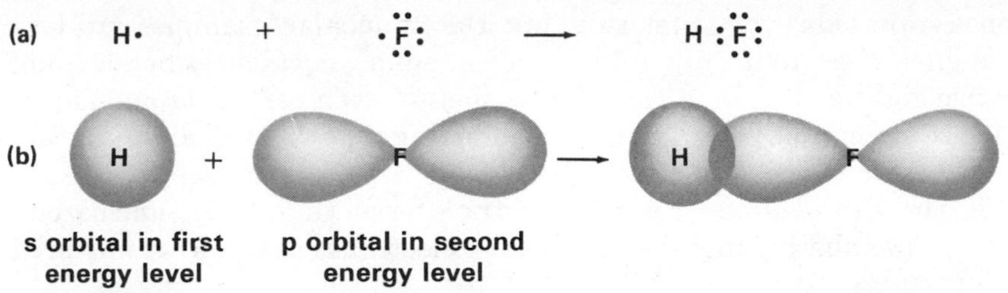

(a) H• + •F: ⟶ H:F:

(b) H + F ⟶ H F

s orbital in first energy level — p orbital in second energy level

**Figure 4–7** Single bond formation in HF. (a) The electron dot representation; (b) the orbital interaction representation.

Since each fluorine atom has one unpaired electron (: F ·) ($1s^2\,2s^2\,2p_x^2\,2p_y^2\,2p_z^1$), two fluorine atoms can share an electron each and form a single covalent bond and an $F_2$ molecule.

2: F · ⟶ : F : F : + energy

ATOMS FLUORINE MOLECULE

Only the pair of electrons represented between the two symbols (the two F's) are bonding electrons. The other six pairs of electrons are called ***nonbonding*** valence electrons.

Before reading any further, you might draw the electron dot structures of $Cl_2$, $Br_2$, and $I_2$.

When hydrogen (H×) and fluorine (· F :) combine to form HF (H×· F :), the s orbital of hydrogen interacts with a half-filled p orbital of fluorine (Fig. 4–7) to form a single covalent bond.

In a water molecule, two HO single covalent bonds are formed. An oxygen atom has six valence electrons, of which two are unpaired · O : ). It needs two more electrons to pair up its electrons and produce the stable eight of a noble gas. Two hydrogen atoms supply the two electrons.

The x's and ·'s distinguish the sources of the identical electrons.

: O · + 2H× ⟶ : O ×· H + energy
H

WATER

An ammonia ($NH_3$) molecule has three NH single covalent bonds. A nitrogen atom has five valence electrons, of which three are unpaired (· N ·). The atom needs three more electrons to pair up its electrons and give it the stable eight. Three hydrogen atoms supply the three electrons to form the $NH_3$ molecule.

The choice of H : O : with H above, over H : O : H will be explained later in this chapter.

· N · + 3H× ⟶ H ×· N ·× H + energy
H

AMMONIA

A molecule of $BF_3$ has three BF single covalent bonds. A boron atom has only three valence electrons (×B×) and each fluorine atom has seven valence electrons, one unpaired (· F :). When a boron atom and three fluorine atoms share electrons to form a molecule of $BF_3$, the boron has all of its electrons paired, but it has only six electrons (three pairs) in its valence shell.

The combining power of old OH (p. 56) is now understood as the number of charges per ion (ionic compounds) or the number of pairs of bonding valence electrons (covalent compounds).

: F :
×B× + 3 · F : ⟶ : F ×· B ·× F : + energy

Since $BF_3$ is a stable molecule, the requirement to have all valence electrons paired must supersede the requirement to have eight electrons in the valence shell. Yes,

The octet rule has exceptions.

pairing sometimes supersedes the octet rule, but the molecular examples are few. Carbon occurs in more than four million known compounds, covalently bonded and requiring the stable eight electrons in the valence shell of each carbon atom. This is reason enough to use the simple rule of eight, remembering that occasionally there are exceptions.

Before reading on, draw the electron dot structures for methane, $CH_4$, and carbon tetrachloride, $CCl_4$, remembering that the four valence electrons of carbon are unpaired ($\cdot\dot{\underset{\cdot}{C}}\cdot$).

Have you ever seen the chemical formula for trisodium phosphate (TSP, $Na_3PO_4$) or for blue vitriol [copper (II) sulfate, $CuSO_4$]? These are common substances sold in hardware stores and elsewhere for cleaning floors ($Na_3PO_4$) and killing algae in ponds ($CuSO_4$). Both of these substances have the properties of ionic compounds. When $Na_3PO_4$ is dissolved in water, sodium ions ($Na^+$) and phosphate ions ($PO_4^{3-}$) are formed. Copper (II) sulfate forms copper ions ($Cu^{2+}$) and sulfate ions ($SO_4^{2-}$) in water. Since the $PO_4^{3-}$ and $SO_4^{2-}$ ions are composed of nonmetals only, the P—O and S—O bonds are covalent bonds. If the P and S atoms are surrounded by the oxygen atoms in electron dot structures, the experimentally correct structure is represented. The arrows point out the electrons transferred from the metal (Na, Cu) atoms to the $PO_4^{3-}$ (addition of three electrons) and $SO_4^{2-}$ (addition of two electrons). The bonds marked *"coordinate covalent"* are formed by one atom supplying both electrons for the shared bond.

**The bonds between $Na^+$ and $PO_4^{3-}$ and between $Cu^{2+}$ and $SO_4^{2-}$ are ionic bonds.**

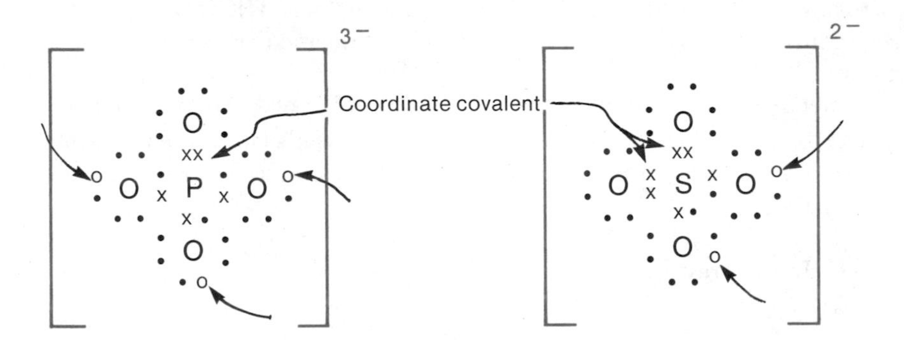


The phosphate ion and the sulfate ion are examples of ***polyatomic*** (many-atom) ***ions,*** which are held intact by covalent bonds. A few common examples are listed in Table 4–3.

Before continuing on, draw the electron dot structure for the perchlorate ion, $ClO_4^-$.

## Multiple Bonding

***One*** **pair of electrons is *one* covalent bond.**

When an atom has fewer than seven electrons in its valence shell, it can form covalent bonds in two ways. The atom may share a single electron with each of several other atoms which can contribute a single electron each. This leads to ***single*** covalent bonds. But the atom can also share two (or three) pairs of electrons with a single other atom. In this case there will be two (or three) bonds between these

**TABLE 4–3 A Few Polyatomic Ions**

| | | | | | |
|---|---|---|---|---|---|
| Ammonium | $NH_4^+$ | Hypochlorite | $ClO^-$ | Chromate | $CrO_4^{2-}$ |
| Acetate | $CH_3CO_2^-$ | Chlorate | $ClO_3^-$ | Silicate | $SiO_3^{2-}$ |
| Nitrate | $NO_3^-$ | Perchlorate | $ClO_4^-$ | Phosphate | $PO_4^{3-}$ |
| Nitrite | $NO_2^-$ | Carbonate | $CO_3^{2-}$ | Arsenate | $AsO_4^{3}$ |
| Hydroxide | $OH^-$ | Sulfate | $SO_4^{2-}$ | | |

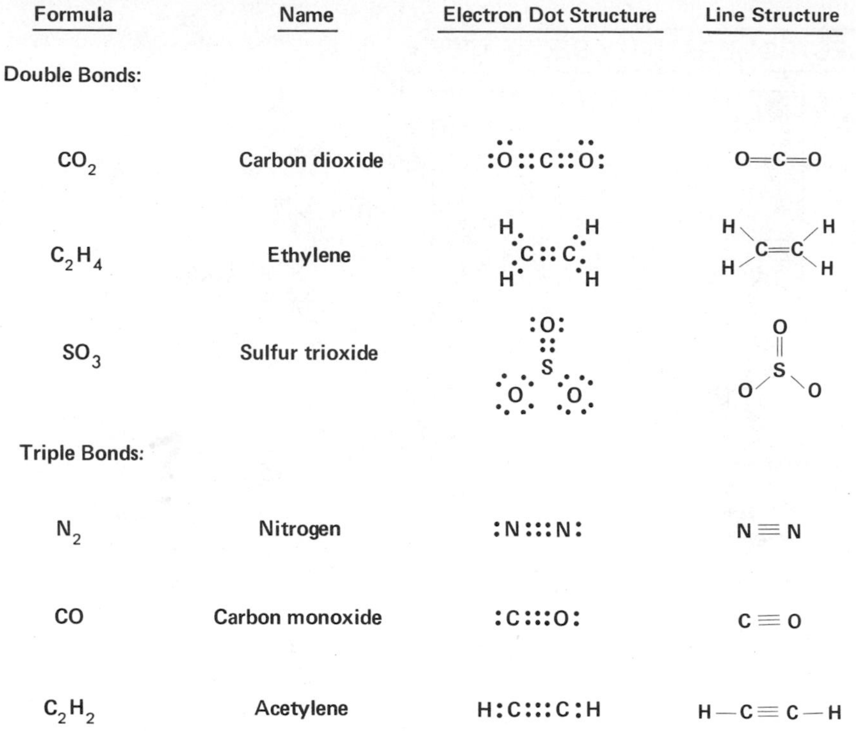

| Formula | Name | Electron Dot Structure | Line Structure |
|---|---|---|---|
| **Double Bonds:** | | | |
| $CO_2$ | Carbon dioxide | :Ö::C::Ö: | O=C=O |
| $C_2H_4$ | Ethylene | H. .H / C::C / H. .H | H\C=C/H, H/ \H |
| $SO_3$ | Sulfur trioxide | :O: :: S / :O: :O: | O ‖ S / O O |
| **Triple Bonds:** | | | |
| $N_2$ | Nitrogen | :N:::N: | N≡N |
| CO | Carbon monoxide | :C:::O: | C≡O |
| $C_2H_2$ | Acetylene | H:C:::C:H | H—C≡C—H |

**Figure 4–8** Electron dot structures of some molecules containing multiple bonds. Line structures are shown for comparison.

two atoms. When two shared pairs of electrons join together the same two atoms, we speak of a ***double bond,*** and when three shared pairs are involved, the bond is called a ***triple bond.*** Examples of these bonds are found in many compounds such as those shown in Figure 4–8.

**A *double bond* consists of two electron pairs shared between two atoms.**

As we can see from these structures, molecules may contain several types of bonds. Thus, ethylene (Fig. 4–8) contains a double bond between the carbon atoms and single bonds between the hydrogen atoms and the carbon atoms. For convenience, an electron pair bond is often indicated by a dash as follows:

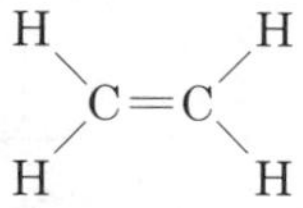

```
H       H
 \     /
  C=C
 /     \
H       H
```

The $H_2$ molecule with a single bond is shown as H—H; ethylene, with a double bond, is shown as $H_2C{=}CH_2$; and diatomic nitrogen, with a triple bond, is shown as N≡N. Note that in each of these cases the stable eight rule is obeyed if the shared electrons can be counted as belonging to both atoms.

**A line between two atoms, as in H—H, represents a bonding pair of electrons.**

Single, double, and triple bonds differ in length and strength. Triple bonds are shorter than double bonds, which in turn are shorter than single bonds. Bond energies normally increase with decreasing bond length due to greater orbital interaction. ***Bond energy*** is the amount of energy required to break a mole of the bonds. Some typical bond lengths and energies are listed in Table 4–4.

**TABLE 4–4 Some Bond Lengths and Bond Energies**

| Bond type | C—C | C=C | C≡C | N—N | N=N | N≡N |
|---|---|---|---|---|---|---|
| Bond length (nm) | 0.154 | 0.134 | 0.120 | 0.140 | 0.124 | 0.109 |
| Bond energy (kcal/mole) | 83 | 146 | 200 | 40 | 100 | 225 |

kcal/mole = thousands of calories necessary to break $6.02 \times 10^{23}$ bonds.

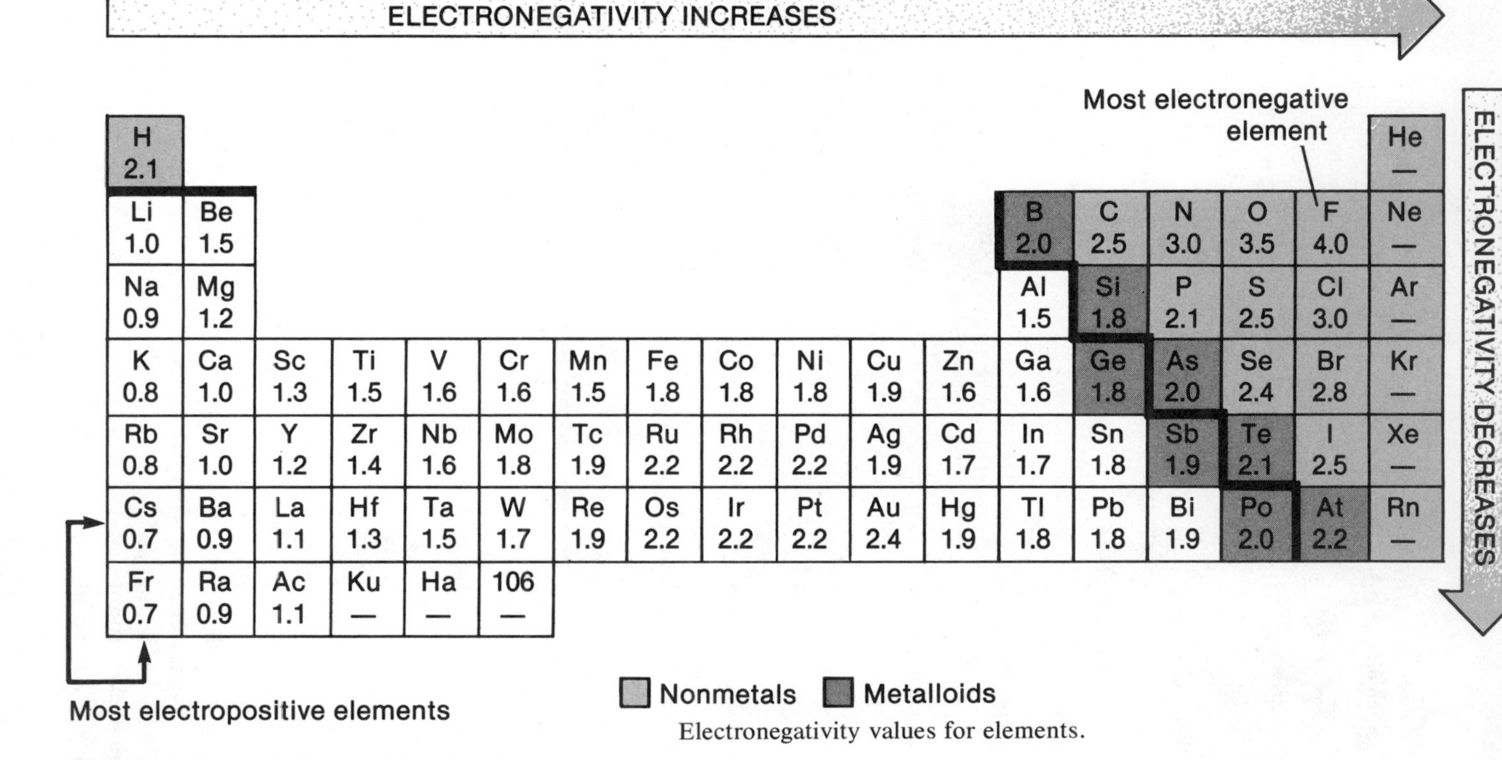


Electronegativity values for elements.

**Figure 4–9** Some electronegativity values in a periodic table arrangement.

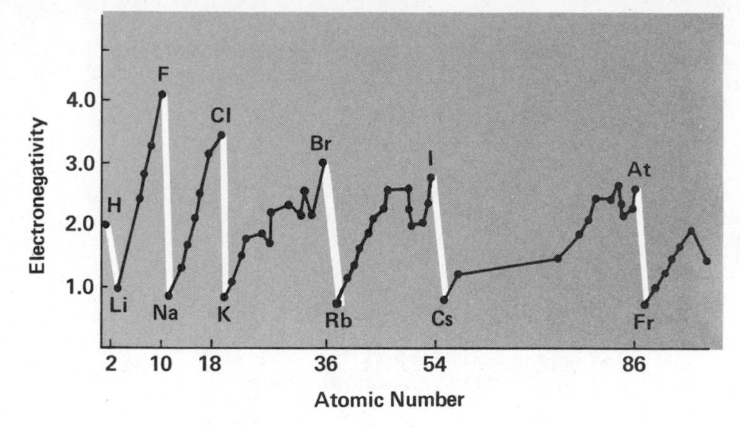


**Figure 4–10** Periodic nature of the electronegativities when plotted versus atomic number.

## Polar Bonds

In a molecule like $H_2$ or $F_2$, where both atoms are alike, there is equal sharing of the electron pair. Where two unlike atoms are bonded, however, the sharing of the electron pair is unequal, and results in a shift of electric charge toward one partner. Recall that the more nonmetallic an element is, the more that element attracts electrons. Nonmetallic character increases across the periodic table toward fluorine (F), the strongest of the nonmetals. The attraction for the electrons in a chemical bond can be expressed on a quantitative basis and is called ***electronegativity.*** A popular set of electronegativities devised by Linus Pauling and based on bond energy differences assigns fluorine the highest value of 4.0 and lesser values to the other elements (Fig. 4–9). The electronegativities generally increase along a diagonal line drawn from francium (Fr) to fluorine.

**In a *polar bond,* there is an unequal sharing of the bonding electrons.**

**Linus Pauling (1901– ) has been awarded two Nobel prizes: in 1954 for his work on molecular structure, and in 1963 for his efforts on nuclear disarmament. Only he and Marie Curie have received two Nobel prizes.**

The periodic nature of electronegativity is clearly demonstrated by Figure 4–10. A comparison of Figures 4–10 and 4–11 reveals that in any period of the periodic table, the smallest atom with the greatest nuclear charge (atomic number) is generally the most electronegative.

**The electronegativity of an atom is a measure of its ability to attract electrons to itself in a compound. The most electronegative atom is fluorine.**

When two atoms are bonded covalently and the electronegativities of the two atoms are the same, there is an equal sharing of the bonding electrons, and the bond is a ***nonpolar*** covalent bond. The bonds in $H_2$, $F_2$, and $CI_4$ (C and I have the same electronegativity, 2.5) are nonpolar.

Two atoms with different electronegativities bonded covalently form a ***polar*** covalent bond. The bonds in HF, NO, $SO_2$, $H_2O$, $CCl_4$, and $BeF_2$ are polar.

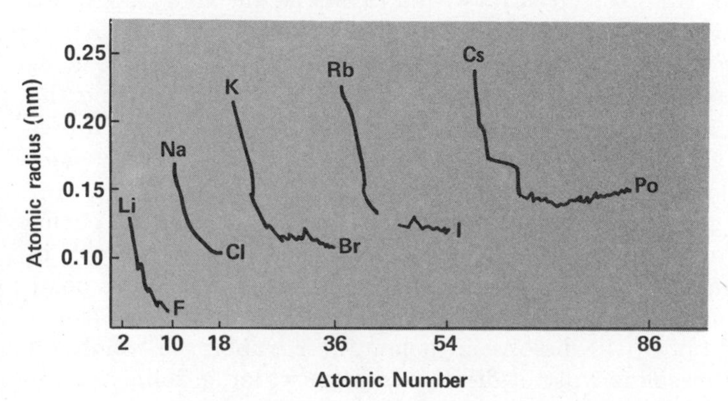


**Figure 4–11** Periodic nature of atomic radii when plotted versus atomic number. A nanometer (nm) is $10^{-9}$ meter.

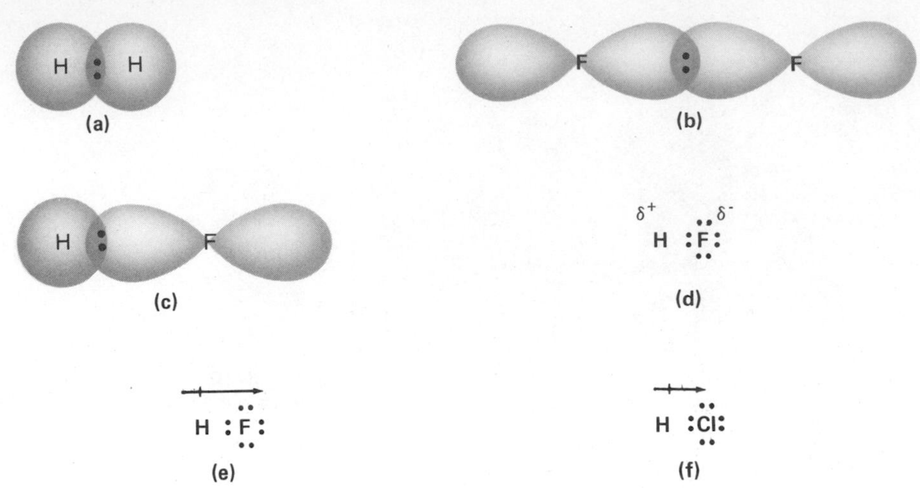


**Figure 4–12** Polar bonds in HF and HCl. (a) Symmetrical distribution of electrons in $H_2$ results in the center of negative charge being identical with the center of positive charge. This is symbolized by the electron dots placed in the overlap area. (b) Overlap of p orbitals in $F_2$ also results in symmetrical distribution of charge. (c) In HF, the electron pair is displaced toward the fluorine nucleus since fluorine is more electronegative than hydrogen. Note the electron dots to the right of the overlap area that conveys the idea of polarity (separation of charge). (d) $\delta+$ (delta positive, meaning fractional positive charge) and $\delta-$ (delta negative, meaning fractional negative charge) are used to indicate poles of charge. In (e) and (f) an arrow is used to indicate electron shift, the arrow having a "plus" tail to indicate partial positive charge on the hydrogen atom. Note that the longer arrow in the HF structure indicates a greater degree of polarity than in HCl. This should not be confused with the greater bond length in HCl.

In a molecule of HF, for example, the bonding pair of electrons is more under the control of the highly electronegative fluorine atom than of the less electronegative hydrogen atom (Fig. 4–12).

When covalent bonds join different atoms, the bonds are generally polar in that one of the atoms has distorted the electron distribution toward itself due to its greater electronegativity. Thus, polar covalent bonds occur in practically every molecule that has different kinds of atoms covalently bonded.

A very common bond that we shall discuss frequently in the remainder of this text is the C—H bond. Since carbon has an electronegativity of 2.5 and hydrogen 2.1, the C—H bond is only slightly polar. The arrangement of C—H bonds around a carbon atom generally makes $—CH_2—$ and $—CH_3$ groups essentially nonpolar.

The polar bonds in beryllium difluoride ($BeF_2$), water ($H_2O$), carbon tetrachloride ($CCl_4$), and chloroform ($CHCl_3$) are indicated in Figure 4–13 by arrows in the direction of the electron shift, pointing toward the more electronegative atom.

**The bases of the structures of molecules are discussed in the last section of this chapter.**

**Water is a polar molecule.**

If a substance has polar bonds, it may have polar molecules or it may have nonpolar molecules. It all depends on the structure of the molecule, discussed in the last section of this chapter. If the electron shifts within the molecule balance out (are symmetrical), the substance has ***nonpolar molecules.*** (See $BeF_2$ and $CCl_4$ in Fig. 4–13.) Or, to say it another way, if the centers of positive and negative charge coincide, the substance is nonpolar. On the other hand, if the electron shifts within the molecule do not balance out (are asymmetrical), the substance has ***polar molecules*** ($H_2O$ and $CHCl_3$ in Fig. 4–13). In other words, if the centers of positive and negative charge do not coincide, the substance is polar.

Whether a substance is polar or nonpolar can have a great effect on the chemical reactivity of the substance and its solubility in various liquids. For example, an old rule of thumb is that ***like dissolves in like:*** polar substances dissolve in polar liquids; nonpolar substances dissolve in nonpolar liquids. Therefore, if rubbing alcohol (2-propanol or isopropyl alcohol) will dissolve in polar water, rubbing alcohol is a polar substance. Likewise, if gasoline will not dissolve in polar water, gasoline is nonpolar and will dissolve in nonpolar carbon tetrachloride.

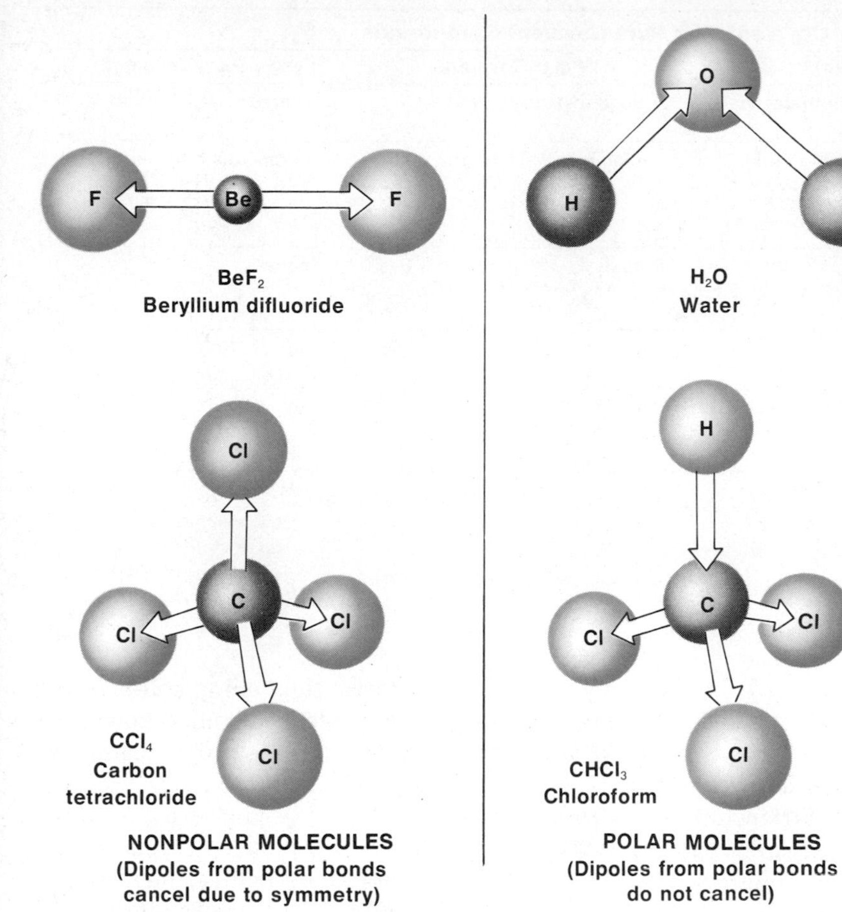


**Figure 4–13** Polar bonds may or may not result in polar molecules. The polar bonds in beryllium difluoride and carbon tetrachloride are arranged about the center atom in such a way as to cancel out the polar effect. In contrast, the polar bonds in water and chloroform molecules do not cancel as a result of the molecular shape but combine to give a polar molecule.

A simple experimental method for detecting polar molecules is represented in Figure 4–14.

Features of ionic, polar covalent, and nonpolar or pure covalent compounds are summarized in Table 4–5.

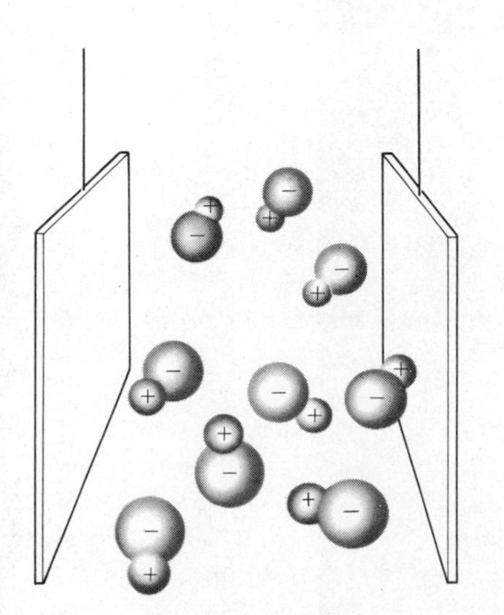


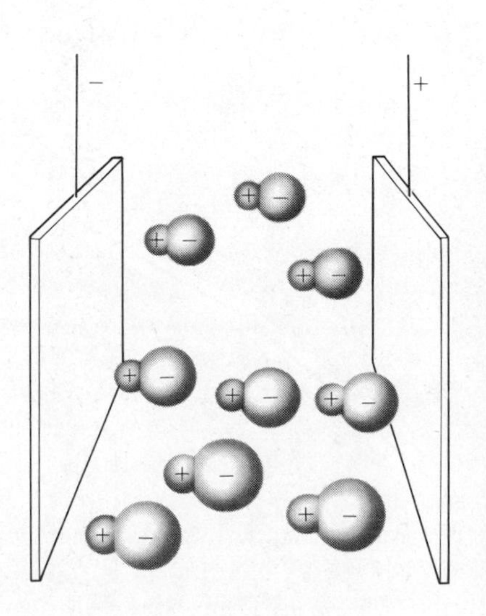


**Figure 4–14** Physical evidence for both the existence of polar molecules and the degree of polarity is provided by a simple electrical capacitor. The capacitor is composed of two electrically conducting plates with nonconducting material (an electrical insulator) between. The storage of charge by the capacitor is least when there is a vacuum between the plates; charge storage is more effective when nonpolar substances are placed between the plates, and is most effective with polar substances between the plates. In the study of individual molecules, the material is placed between the plates in the gas phase.

TABLE 4-5 Characteristics of Ionic, Polar Covalent, and Pure Covalent Compounds

| Type of Bond | Ionic | Polar Covalent | Pure Covalent |
|---|---|---|---|
| **Disposition of the electrons** | transferred from metal to nonmetal | partially transferred | shared |
| **Elements involved** | Groups IA, IIA, transition, inner transition metals with Groups VA, VIA, VIIA nonmetals | nonmetals with nonmetals: IVA, VA, VIA, VIIA | nonmetals with nonmetals: IVA, VA, VIA, VIIA |
| **Electronegativity difference** | great (more than 2) | small | none |
| **Conductance of electricity as a solid** | no | no | no |
| **Conductance of electricity as a liquid (melted solid)** | yes | no | no |
| **Molecules** | no | yes | yes |

## Hydrogen Bonding

For a series of molecular substances with similar structures, the boiling points ordinarily increase as the molecular weights increase. For example, the boiling points of fluorine ($F_2$), chlorine ($Cl_2$), bromine ($Br_2$), and iodine ($I_2$) increase rather regularly with increasing molecular weight (Figure 4–15).

Strongly electronegative atoms compete for a single hydrogen atom to give hydrogen bonds.

The general relationship between boiling points and molecular weights also holds for hydrogen chloride (HCl), hydrogen bromide (HBr), and hydrogen iodide (HI). However, the boiling point of hydrogen fluoride (HF), the lightest member of this series of compounds, is abnormally high (Fig. 4-16). Irregularities similar to this are also found in other compounds in which hydrogen is bonded to fluorine,

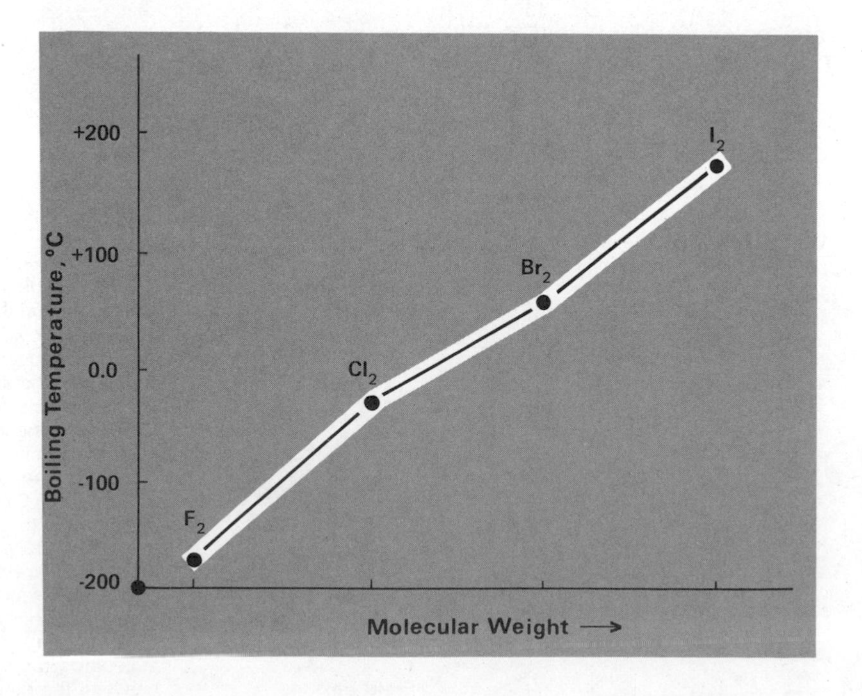


**Figure 4–15** Boiling points of $F_2$, $Cl_2$, $Br_2$, and $I_2$, as a function of molecular weight.

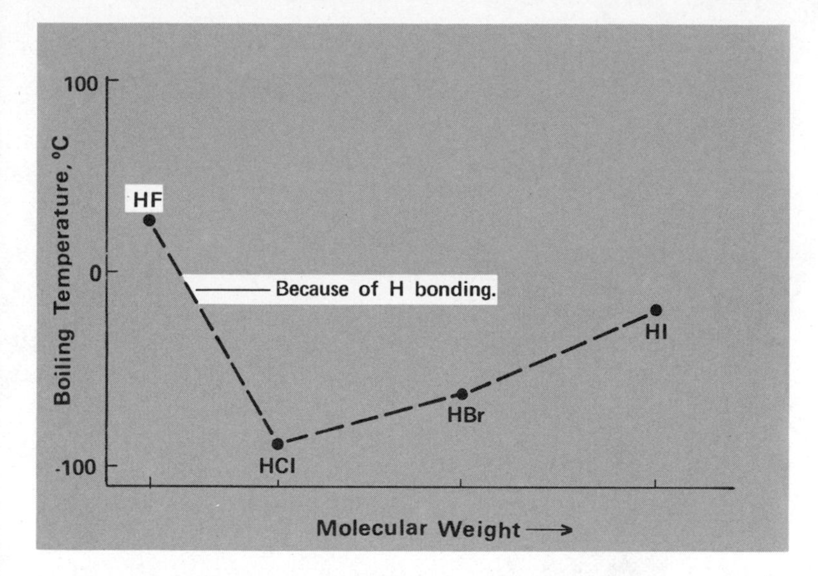


**Figure 4–16** Boiling points of HF, HCl, HBr, and HI, plotted against molecular weights.

oxygen, and nitrogen. The increased attraction between molecules containing H—F, H—O, or H—N bonds is termed ***hydrogen bonding.***

The explanation for hydrogen bonding is to be found in the extremely large electronegativity differences of the H—F, H—O, and H—N bonds, the polarity being due to the extreme electronegativity of fluorine, oxygen, and nitrogen. Consider the bonding in liquid HF. Since unlike-charged ends of these molecules should attract each other, we expect HF molecules to be associated with one another. This association is illustrated in Figure 4–17. The increased association between HF molecules compared with that found in HCl offers a ready explanation for the unusually high boiling point of HF.

**For a substance to boil, its molecules must gain enough energy to break loose from each other.**

Water provides another good example of hydrogen bonding. Water molecules are not linear but rather are angular, with two nonbonding (unshared) electron pairs located toward one end of the molecule and the partially positive hydrogen atoms located toward the opposite end. The two polar bonds in this geometry result in a distinctly polar molecule.

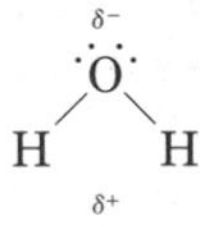

$\delta^-$
O
H H
$\delta^+$

**The $\delta^-$ and $\delta^+$ represent partial charges.**

In liquid and solid water, where the molecules are close enough to interact, the hydrogen atom on one of the water molecules is attracted to the nonbonding elec-

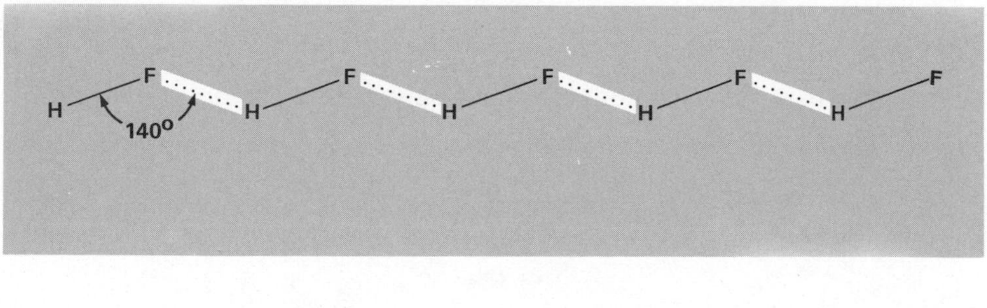

H F H F H F H F H F
140°

**Figure 4–17** Hydrogen bonding in HF.

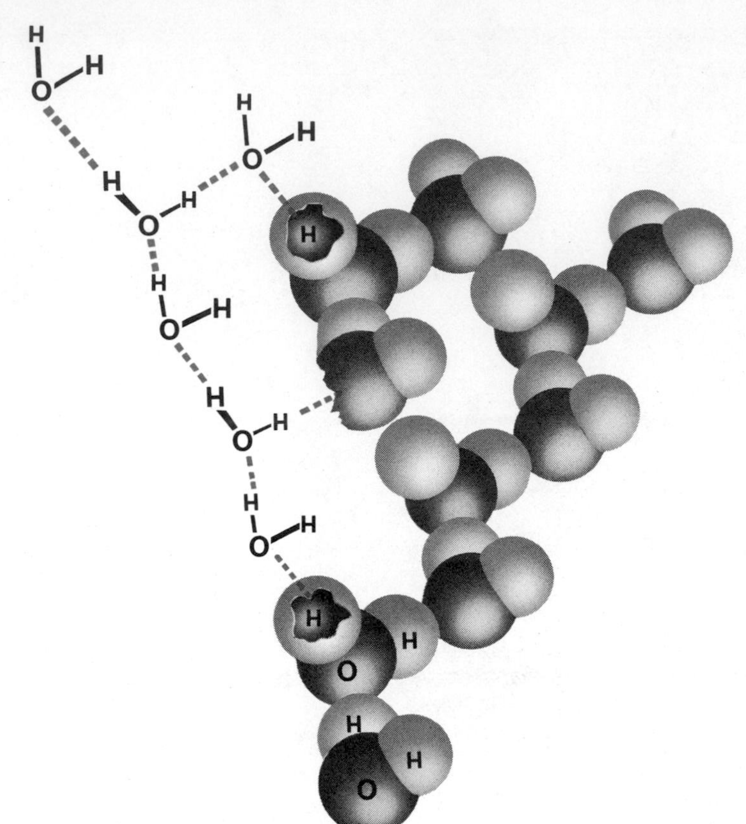


**Figure 4–18** Hydrogen bonding in the structure of ice. The hydrogen bonds are indicated by the dashed lines. In liquid water the hydrogen bonding is not as extensive as it is in ice.

trons on the oxygen atom of an adjacent water molecule (Fig. 4–18). This is possible because of the small size of the hydrogen atom. The result of this association is called hydrogen bonding because the slightly positive hydrogen atom acts as a sort of a bridge to hold two molecules together in much the same way as electron pairs hold atoms together in molecules. Hydrogen bonds are much weaker than ordinary covalent bonds.

**Why does ice float?**

The extent of hydrogen bonding in liquid water varies with temperature. In ice, hydrogen bonding is extensive, resulting in a very open structure for ice. Consequently, at ordinary pressures ice is *less* dense than water. As ice is melted and the molecules gain more energy and move about, this bonding begins to break down. Not all the hydrogen bonds are broken, however, and large aggregates of water molecules exist in liquid water even near 100°C. As water is heated, thermal agitation disrupts the hydrogen bonding until, in water vapor, there is only a small fraction of the number of hydrogen bonds that are found in liquid or solid water.

**Water has an abnormally high boiling point because of hydrogen bonding between the molecules.**

Look for hydrogen bonding in proteins, DNA, ice cream, cotton, hand moistening creams, hair, nylon, humectants, and alcohols later in the text.

## Self-Test 4-B

1. a. An example of a molecule containing covalent bonding where the electrons are equally shared between the atoms is ______________; b. one where they are unequally shared is ______________.
2. The number of electrons shared in a triple covalent bond is ______________.

3. There are ____________ covalent bonds in an ammonia ($NH_3$) molecule.
4. Which atom cannot form a double bond, fluorine or oxygen? ____________
5. a. How many valence (bonding) electrons are thought to be involved in molecules containing covalently bound atoms of period 2 and 3 elements? ____________

   b. This is known as the ____________ rule. c. Is it true most of the time or all of the time? ____________
6. Which is the most electronegative of all of the elements? ____________
7. Hydrogen bonding very probably occurs when hydrogen is bound to atoms of ____________, ____________, or ____________.
8. Which molecule ($H_2O$, $H_2$, $O_2$, $CCl_4$) is a polar molecule? ____________

## Van der Waals Attractions

The bonding between chemical species that cannot be explained by ionic, covalent, or hydrogen bonding is often due to a weak attraction known as ***van der Waals attraction.*** In helium, for example, the two electrons are nonbonding, and yet there is a slight attraction between two helium atoms. This slight attraction allows helium to become a solid at −272°C under high pressure.

**Van der Waals attractions explain the existence of solid helium.**

Van der Waals interactions can arise from a variety of causes. For instance, as two atoms or molecules approach each other, intermolecular interactions will cause a temporary shifting of their electron clouds. An uneven electron distribution in an atom makes the atom itself temporarily polar, producing a dipole (meaning "two poles"). The temporary poles on two adjacent atoms can interact with each other, resulting in a momentary attractive force (Fig. 4–19).

This type of dipole-dipole interaction is called van der Waals attraction, after the Dutch physicist who suggested its possibility. The existence of the solid state of many nonpolar molecular substances (such as oxygen, nitrogen, and helium) can be explained by invoking van der Waals attractions.

## Metallic Bonding

**Loosely held electrons are found in metals. They can move freely and are not confined to the area between any particular pair of atoms.**

Metals have some properties totally unlike those of other substances. For example, most metals are good electrical conductors; they are shiny solids and have relatively high melting points (with a few notable exceptions such as mercury). Any theory of

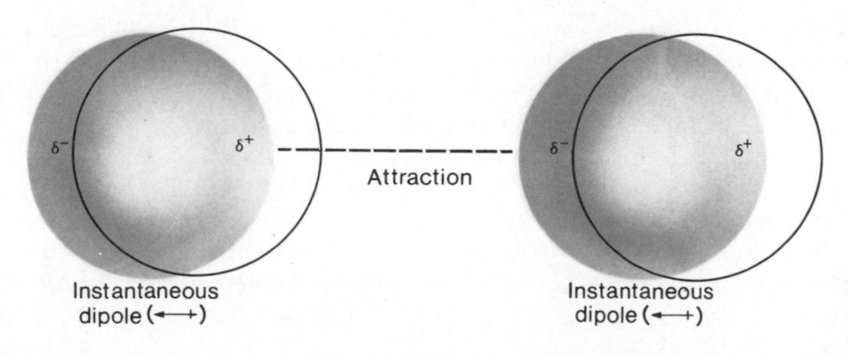


**Figure 4–19** An illustration of van der Waals attraction. One instantaneous dipole interacts with another in a neighboring atom.

the bonding of metal atoms must be consistent with these properties. Structural investigations of metals have led to the conclusion that metals are composed of regular arrays or lattices of metal ions in which the bonding electrons are loosely held. The loosely held electrons can be made to move rather easily through the lattice upon application of an electric field. In this way the metal acts as a conductor of electricity. As a consequence of this movement of bonding electrons, we cannot really write a satisfactory description of the bonding in a metal using an electron dot structure.

## Molecular Structure

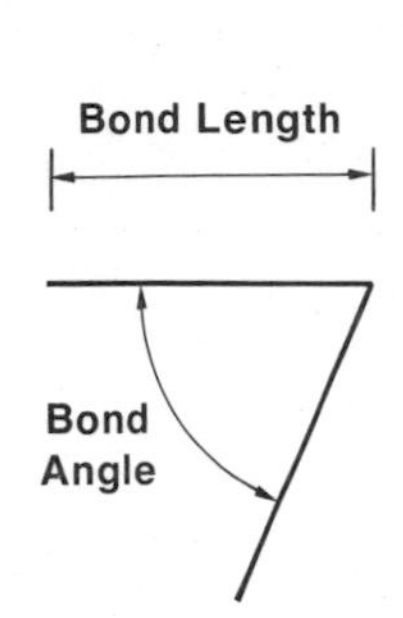


The modern theory of the atom is effective in explaining bonding and periodicity of the elements. It is also capable of explaining related chemical phenomena, such as molecular structure. The structure of a molecule is determined by the arrangement of the atoms in the molecule with respect to each other, that is, by the angles between and the lengths of its covalent bonds. The angle formed by two intersecting lines drawn from the two nuclei of the attached atoms through the nucleus of the central atom is called the ***bond angle.*** Molecular structures are established by many different experimental techniques. Here we shall see what molecular structures are predicted by atomic theory and what modifications are needed for the theoretical and the experimental findings to agree.

### Molecular Structures Predicted by Atomic Orbitals

The starting point for a discussion of molecular structures is the set of atomic orbitals, since these determine the preferred regions in space for electrons which will be involved in the bonding. Expected molecular structures when s and p atomic orbitals interact are shown in Figure 4–20.

Since an s orbital is spherically symmetrical, there is no preferred direction for the bonding with the other atom. There is only one s orbital per valence shell and only one bond can be formed from this kind of orbital. Hydrogen, with its one electron in an s orbital, can form only one covalent bond, and there is no specified direction from the hydrogen atom.

There are three p orbitals per valence shell. The electrons in a set of p orbitals concentrate charge along the x, y, and z axes, which are at 90° angles to each other. Any two atoms bonded to a third atom by interacting p orbitals would be expected to form bonds enclosing an angle of 90°.

Is this predicted bond angle actually found in molecules? Some molecules, such as those of $H_2S$ and $H_2Se$, come close. However, instead of the expected bond angle of 90°, when s orbitals of two hydrogens interact with two p orbitals of S or Se, the angles are 92° for $H_2S$ and 91° for $H_2Se$. It is rare for the bond angles of molecules to agree this closely with the angles predicted by combinations of simple atomic orbitals. In $H_2O$, for example, the bond angle is 104.5°.

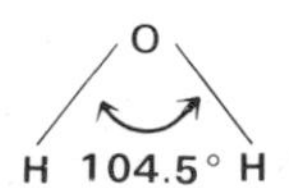


Before discarding the theory, we should look a little further. With a slight adjustment, the modern atomic theory can be brought into best agreement with experimentally derived molecular structures of relatively simple molecules. The theory is also available for use with complex molecules that contain large numbers of atoms.

***Hybridization*** **combines atomic orbitals to give new orbitals with different properties.**

The method of adjusting the pure atomic orbitals is called ***hybridization.*** This process brings a modified theoretical model into agreement with experimental data. Although hybridization is a mathematical process, we can gain an insight into its methods and its usefulness through a qualitative examination of its features. An alternative approach to this same problem is the ***valence-shell electron-pair repulsion theory.***

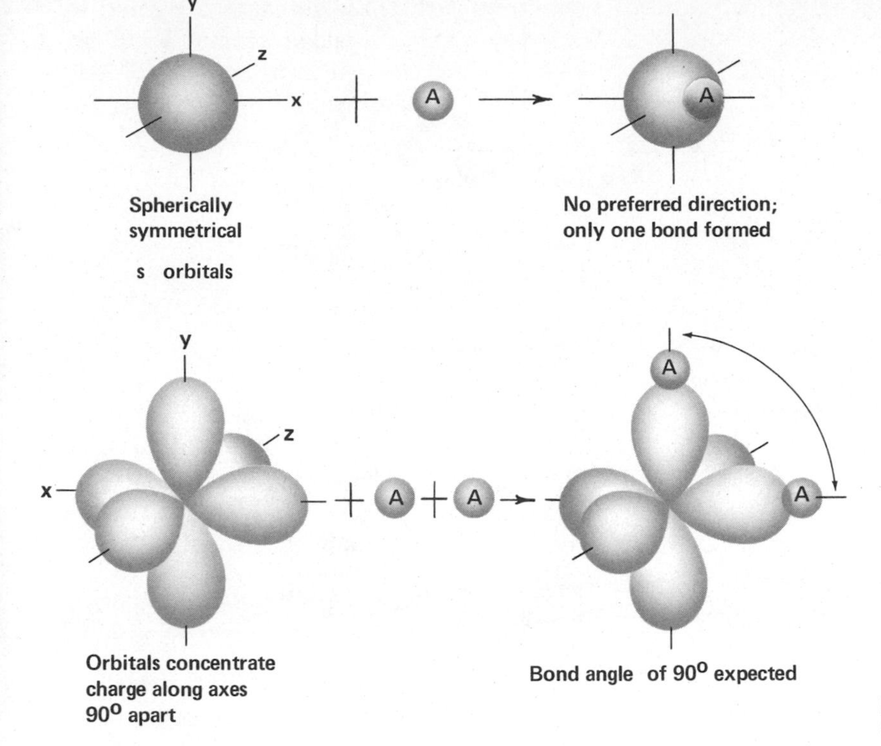


**Figure 4–20** Predicted molecular structures when s and p orbitals are used to form bonds. **A** represents any atom to be joined to the central atom.

Both of these theories will be examined, along with some of their applications and some of the experimental observations used as a guide in their development.

## Hybridization

The tetrahedral structure of species such as $CH_4$ and $NH_4^+$ requires an adjustment of atomic theory. The isolated carbon atom in its ground state has the electron configuration 2-4, with the four electrons in the second orbit arranged with two in the 2s orbital, one in the $2p_x$ orbital, one in the $2p_y$ orbital, and none in the $2p_z$ orbital, which we will write in a shorthand notation as $2s^2$, $2p_x^1$, $2p_y^1$, $2p_z^0$. The letters x, y, and z are used to denote the orientation of the three p orbitals along lines that are perpendicular to each other. If the unpaired electrons in the 2p orbitals are used for bonding, this leads us to expect that carbon would form two bonds at an angle of 90°. Actually carbon commonly forms four single bonds, and each bond angle is 109.5°. This angle directs each bond from carbon toward a corner of a regular tetrahedron.

**The *ground state* of an atom is its lowest possible energy state. Higher-energy electron configurations result when energy is added, such as in an experiment in spectroscopy.**

To explain these four equivalent bonds, we assume that prior to or during the formation of the bonds, four orbitals of the carbon atom are hybridized (or mixed) to form a new set of four equivalent orbitals. The carbon atom promotes one of the electrons from the 2s orbital to the empty $2p_z$ orbital to give the electronic arrangement $2s^1$, $2p_x^1$, $2p_y^1$, $2p_z^1$ for the outer electrons. These four orbitals are rearranged mathematically according to specific rules to obtain four new equivalent hybrid orbitals. They are designated ***sp³*** orbitals, since they are made from one s and three p orbitals. The new orbitals are directed toward the corners of a tetrahedron (Fig. 4–21). The four valence electrons are distributed equally among the hybrid orbitals (one electron to each orbital). The energy required for the hybridization is less than that provided by the formation of additional, stronger bonds with the hybrid orbitals.

**A tetrahedron is a four-sided structure.**

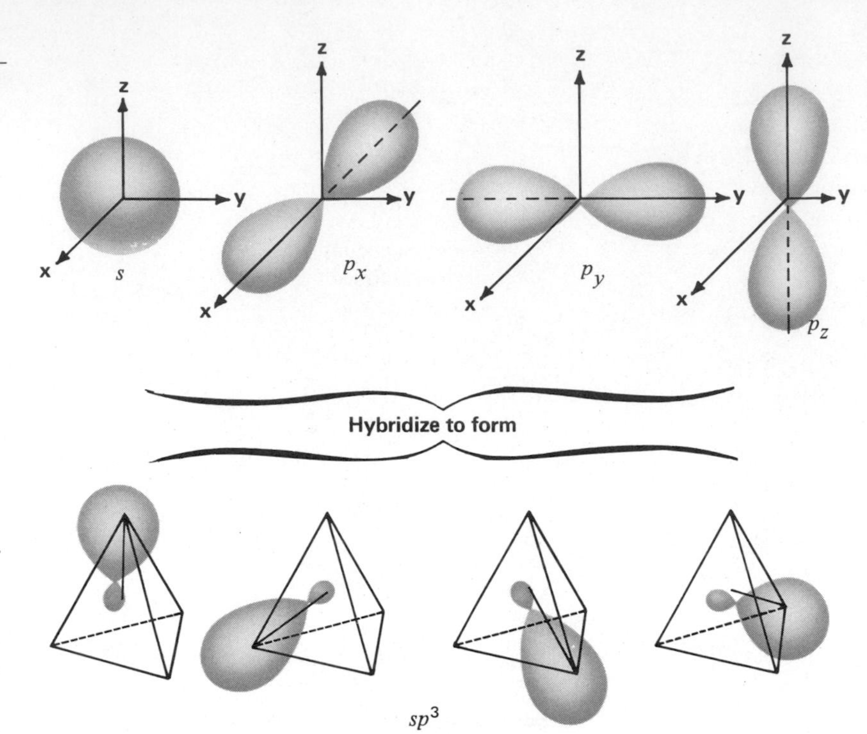


**Figure 4–21** Hybridization of one s and three p orbitals into four $sp^3$ orbitals results in new spatial representations of the electron density patterns.

Thus, we have a more favorable energy change than would result from bonding with p orbitals alone.

Other sets of bonding orbitals with geometries very closely in accord with established molecular geometries can be derived in a similar fashion. Some of these are listed in Figure 4–22. A hybridization analogy in orbitals, plants, and animals is illustrated in Figure 4–23.

As examples of the use of hybrid orbitals, consider tetrahedral molecules such as $CH_4$, $SiCl_4$, and $CCl_4$. In all of these, the atomic orbitals of the central atom can be

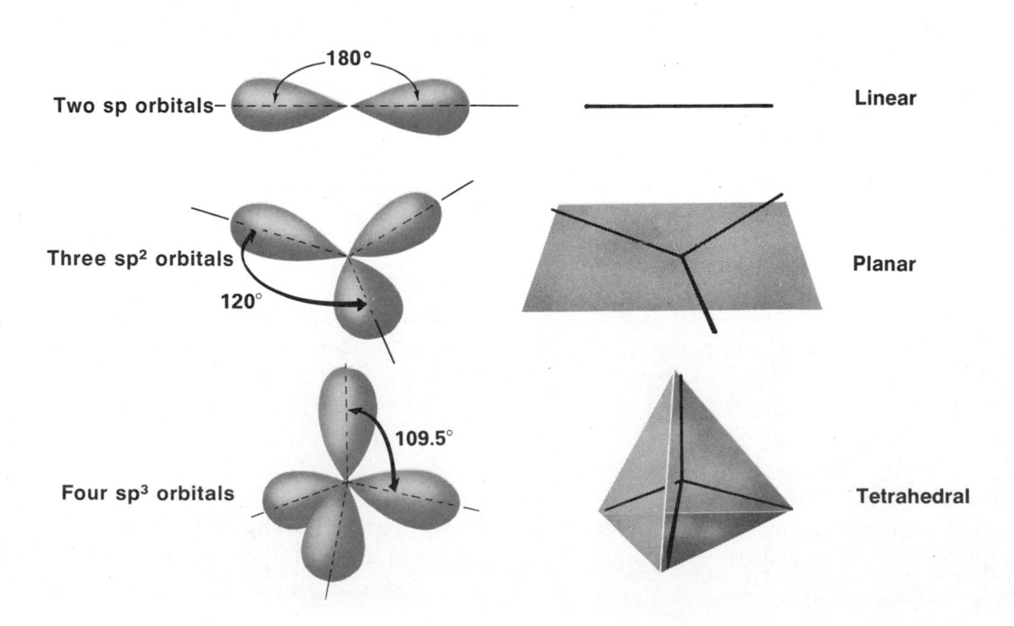


**Figure 4–22** Spatial orientations of some hybrid orbitals.

described as $sp^3$ hybrids. For molecules such as $H_2O$ and $NH_3$, $sp^3$ hybrid orbitals can also be postulated if it is assumed that the ***nonbonding pairs*** of electrons (those on the O or N which are not involved in bonds to H) are also placed in hybridized orbitals. The experimentally determined bond angles in these molecules are not quite the same as normal tetrahedral bond angles (as shown in Fig. 4–24); however, the agreement is relatively close.

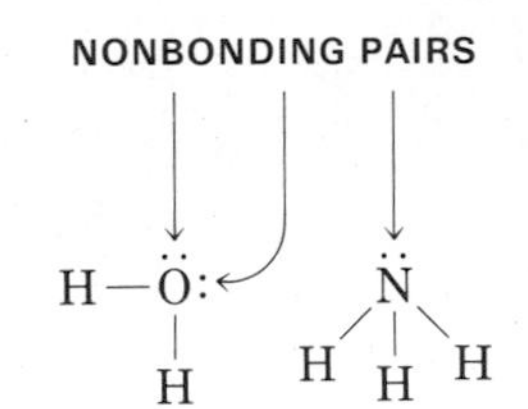


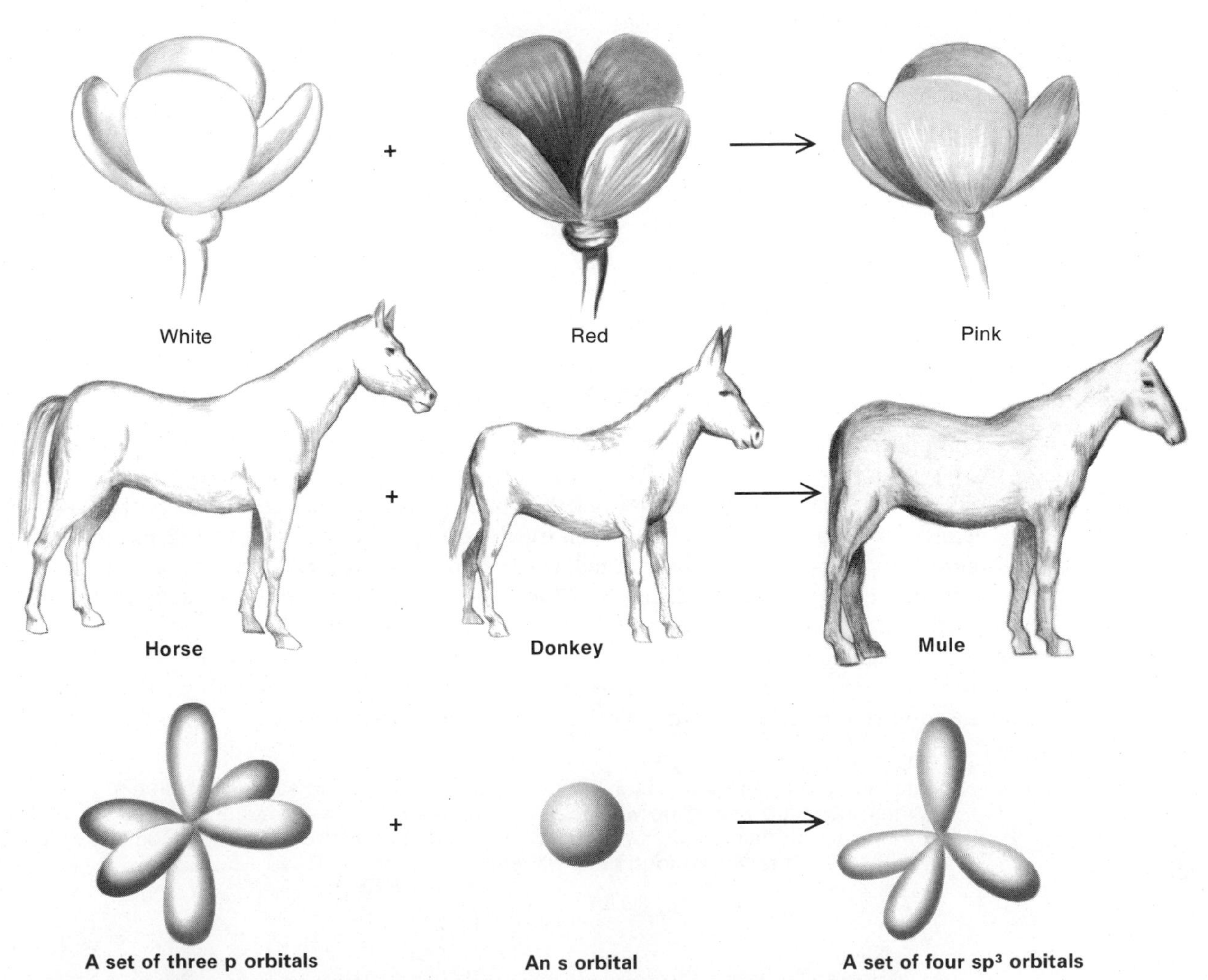


**Figure 4–23** Hybridization is a familiar concept in nature. A mixture of two similar but different entities produces an entity with intermediate characteristics. Biological hybrids are analogous to hybrid orbitals in the sense of being the result of mixtures, yet hybrid orbitals are obtained much more exactly by mathematical calculations.

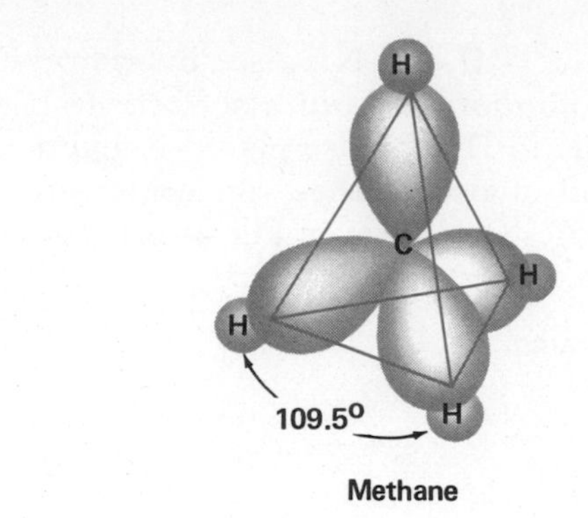


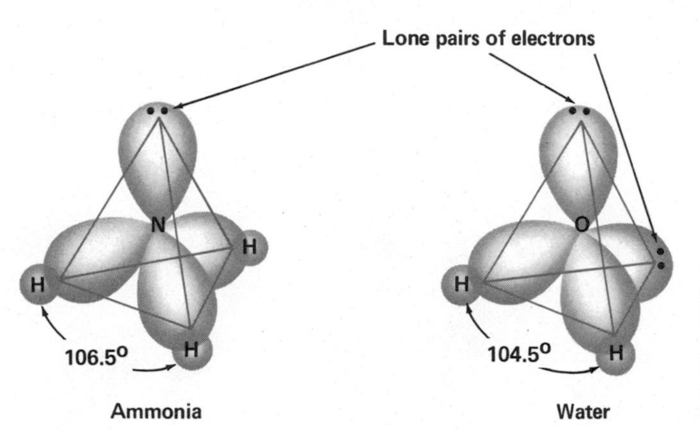


**Figure 4–24** Molecules containing $sp^3$ orbitals. Methane has only bonding pairs of valence electrons; water and ammonia have bonding and nonbonding pairs of valence electrons.

## Valence-Shell Electron-Pair Repulsion Theory

There is another, even simpler, notion that can be used to predict molecular structures of covalent molecules. The basic idea is that electron pairs repel each other. In fact, this theory assumes that electron pairs in the valence shell of a central atom behave like a group of electrically charged balloons that are connected to a central point by strings. If similarly charged, the balloons would tend to be as far apart from each other as possible. In a similar manner, electron-pair repulsion leads to the arrangements shown in Figure 4–25.

These ideas allow us to predict the shapes of molecules if the underlying electronic structure is known. It is particularly simple when every pair of electrons in the valence shell is a bonding pair. Examples can be seen in $BCl_3$, $SiCl_4$, and $SF_6$.

**$BCl_3$**

Atomic boron has three outer orbit electrons in its ground state electronic structure. When these are paired up with the unpaired electrons of three chlorine atoms, :Cl·, three electron-pair bonds are formed. This leads to the structure shown below, which actually is the structure found experimentally.

Cl
120°
B
Cl Cl

The use of one s and two p orbitals to form $sp^2$ hybrid orbitals also leads to the same structure.

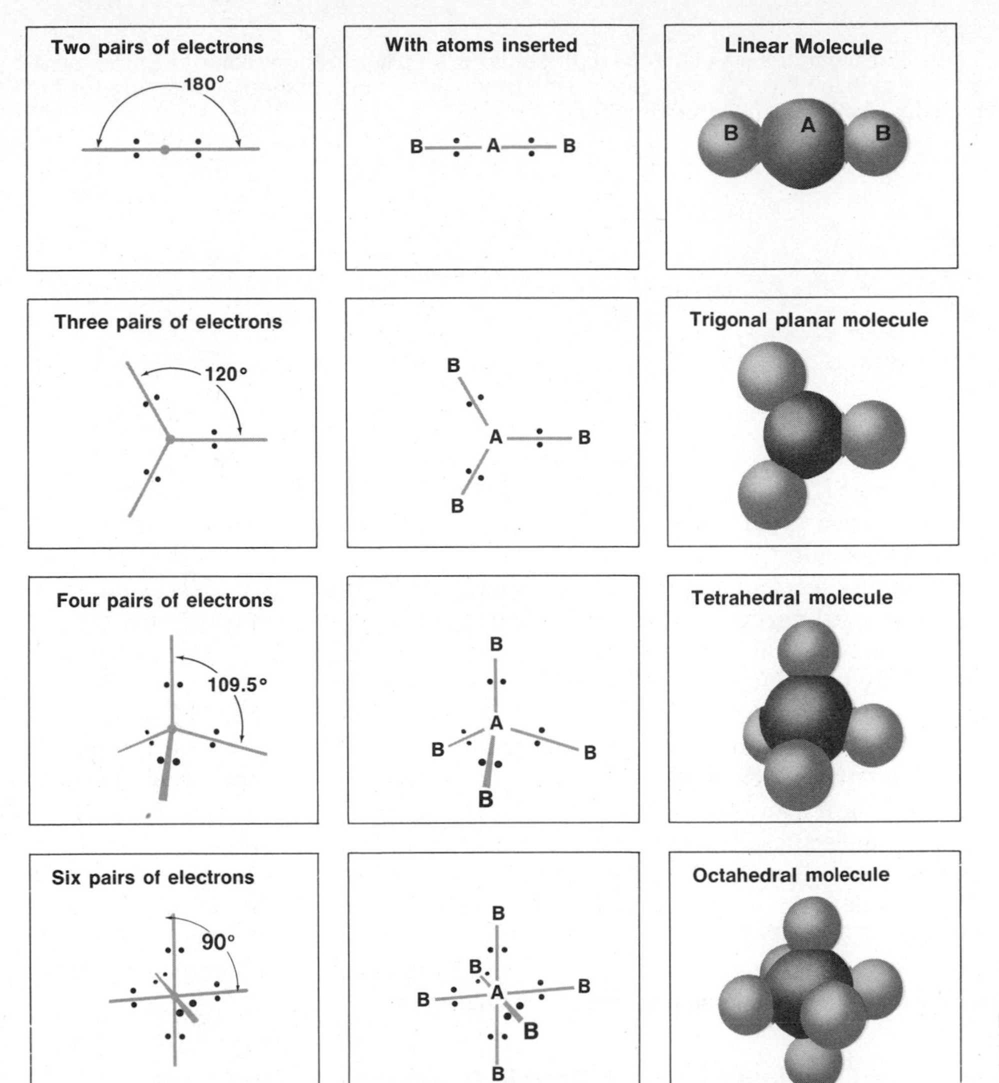


**Figure 4–25** Arrangements of electron pairs according to the electron-pair repulsion theory.

## $SiCl_4$

Atomic silicon has four electrons in its outer orbit. When these are paired with the unpaired electrons of four chlorine atoms, :C̈l·, four electron-pair bonds are formed. These four pairs of electrons in the valence shell will repel each other to form the predicted tetrahedral structure.

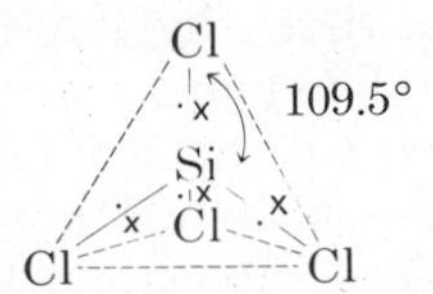


The combination of one s and three p orbitals to form $sp^3$ hybrids leads to the same structure. The actual structure is, indeed, tetrahedral. Other tetrahedral structures are sulfate ($SO_4^{2-}$), phosphate ($PO_4^{3-}$), and perchlorate ($ClO_4^-$) polyatomic ions.

**$SF_6$**

The sulfur atom has six electrons in the outer orbit prior to formation of $SF_6$. Six electron pairs in the valence shell lead to the prediction of an octahedral structure for the molecule, which is the structure found.

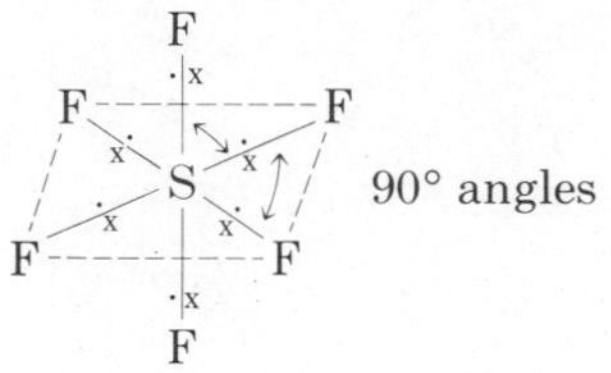

F F F S F F F

90° angles

At this point you might ask how the presence of nonbonding pairs of electrons (in the outermost shell of the central atom) affects the disposition of the other electron pairs. In brief, a pair of electrons occupies somewhat more volume when it is not involved in bonding.

**Water consists of bent, polar molecules.**

Nonbonding electron pairs help to explain the structures of $NH_3$ and $H_2O$ molecules discussed earlier and shown in Figure 4–24. It can be postulated, for example, that the $NH_3$ and $H_2O$ molecules have bond angles (106.5° in $NH_3$ and 104.5° in $H_2O$) slightly less than the tetrahedral angle of 109.5° because the lone, nonbonding pairs of electrons exert more repulsion than do the bonding pairs. This spreads the nonbonding pairs further apart and squeezes the bonding pairs closer together.

From the typical examples given here, it is obvious that both the electron-pair repulsion theory and the hybridization theory are consistent with experimental structures for molecules. The methods differ, however, in their approach. Hybridization utilizes the mathematical combination of atomic orbitals to form hybrid orbitals properly oriented to give the established bond angles. On the other hand, electron-pair repulsion theory assumes that electron pairs simply move as far away from each other as is possible while maintaining the proper bond length. The nonbonding pairs of electrons exert more repulsion than do bonding pairs.

## Structures of Molecules with Double Bonds and Triple Bonds

The electron dot structure for ethylene, $C_2H_4$, is

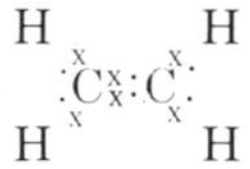

H H
C::C
H H

Four electrons form a double bond between the two carbon atoms. Each carbon atom furnishes two unpaired electrons for the double bond. This means two orbitals from each carbon atom are involved in the double bond.

Experimentally, it is observed that ethylene is a *planar* molecule (all six atoms are in the same plane) and that all bond angles are close to 120°. This kind of structure around a carbon atom can exist if the 2s and *two* of the 2p orbitals of carbon undergo hybridization to form three $sp^2$ hybrid orbitals. The hybridization can come about in the following way. Each carbon atom can be excited to the electronic arrangement $2s^1$, $2p_x^1$ $2p_y^1$ $2p_z^1$. When the 2s and two of the p orbitals are mixed, three equivalent orbitals are produced. The three $sp^2$ orbitals are all in the same plane and form angles of 120° with each other (refer back to Fig. 4–22).

From two s orbitals

From two p orbitals (end-to-end interaction)

From an s orbital and a p orbital

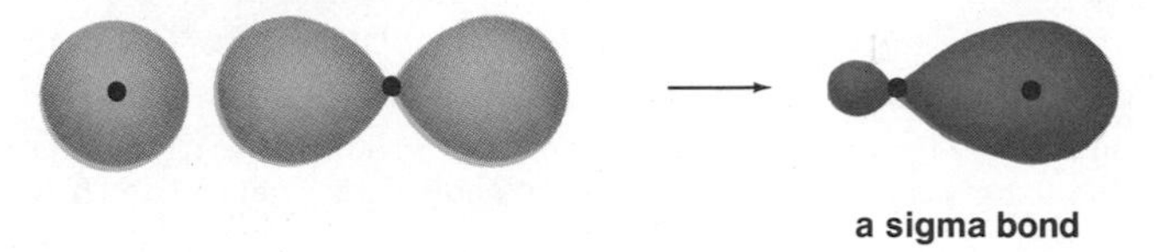


From two p orbitals (lateral interaction)

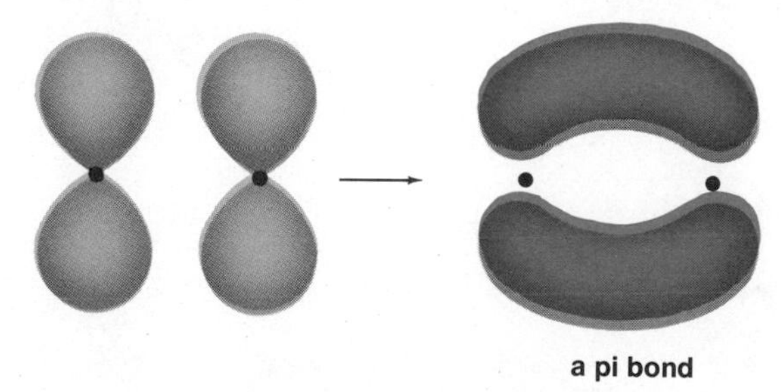


**Figure 4–26** Ways of interacting s and p orbitals to form sigma ($\sigma$) and ($\pi$) bonds. Sigma bonds can also be formed by interacting the end of an $sp^2$ hybrid orbital with the end of a p orbital (similar to the interaction of two p orbitals) or with an s orbital (similar to the interaction of an s orbital and a p orbital).

The interactions of s, p, and $sp^2$ orbitals to form sigma and pi bonds are illustrated in Figure 4–26. As shown in Figure 4–27, a **sigma ($\sigma$) bond** is formed between the two carbon atoms when the end of an $sp^2$ orbital of one carbon atom interacts with the end of an $sp^2$ orbital of the other carbon atom. The second bond of the double bond between the two carbon atoms is formed by a lateral (or side-by-side) interaction of the two p orbitals not involved in the $sp^2$ hybridizations. The lateral interaction is called a **pi ($\pi$) bond.** It prevents rotation about the bond axis. The two $CH_2$ groups of ethylene cannot rotate freely with respect to each other because the rotation would reduce or eliminate the orbital interaction of the pi bond. As we shall see later, this rotational limitation about double bonds is quite important in the chemistry of organic compounds.

**Sigma ($\sigma$) bond: end-to-end interaction. Pi ($\pi$) bond: side-by-side interaction.**

**A double bond is made up of one sigma bond and one pi bond.**

A pi bond, with less effective overlap of orbitals, is easier to break than a sigma bond. However, a double bond, consisting of a sigma bond *and* a pi bond, is more difficult to break than a single bond.

**A pi bond is easier to break than a sigma bond.**

Acetylene, $C_2H_2$, has a triple bond composed of one sigma and two pi bonds (Figure 4–28).

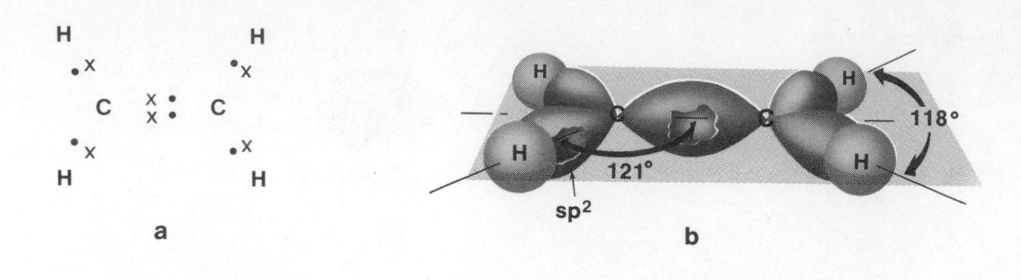


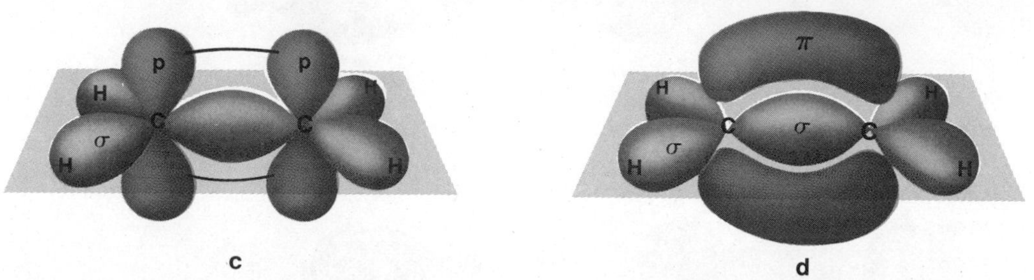


**Figure 4–27** Structure and bonding in ethylene, $C_2H_4$. (a) Electron dot structure; (b) sigma ($\sigma$) bond formation; (c) assembly of p orbitals to form a pi ($\pi$) bond; (d) representation of the sigma ($\sigma$) and pi ($\pi$) bonds.

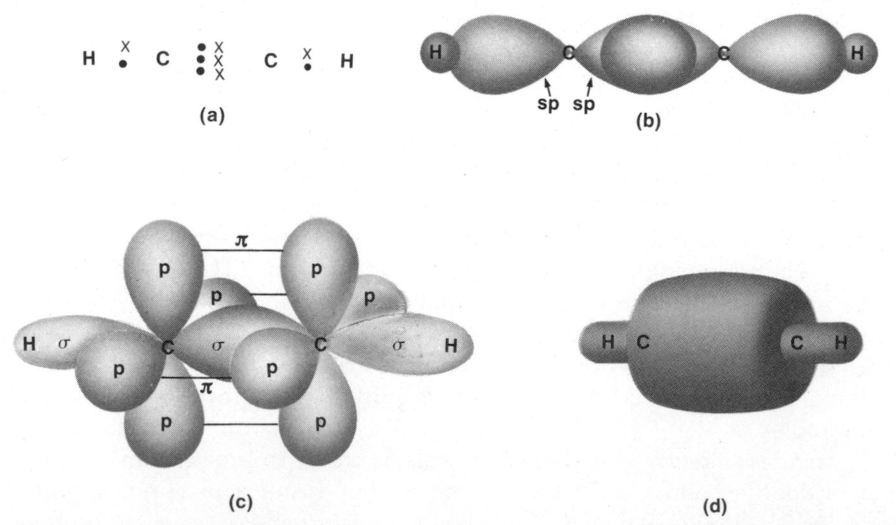


**Figure 4–28** Structure and bonding of acetylene, $C_2H_2$. (a) Electron dot structure; (b) sigma ($\sigma$) bond formation; (c) assembly of p orbitals to form two pi ($\pi$) bonds; (d) representation of the electron density pattern, which is cylindrically symmetric in the molecule.

## Self-Test 4-C

1. A double bond is composed of one ____________ bond and one ____________ bond.
2. A triple bond is composed of one ____________ bond and two ____________ bonds.
3. An $sp^3$ hybrid orbital is one of a set of orbitals that can be formed by mixing one ____________ orbital and three ____________ orbitals.
4. (Bonding/nonbonding) pairs of electrons repel more than (bonding/nonbonding) pairs of electrons.
5. Around which kind of bond (single, double, triple) can atoms rotate on an axis drawn through the nuclei of the two atoms bonded? ____________
6. When three atomic orbitals are combined in hybridization, how many new orbitals are produced? ____________
7. Give the molecular shapes expected for the hypothetical molecules $AX_2$ ____________, $BZ_3$ ____________, $DQ_6$ ____________. (There are no nonbonding valence electrons in these structures, and all of the bonds are single bonds.)

## Matching Set I

| | | | |
|---|---|---|---|
| ______ | 1. double covalent bond | **a** | explanation for covalent bonding |
| ______ | 2. ionic bonds | **b** | shared electrons |
| ______ | 3. ionization energy | **c** | requires O, F, or N |
| ______ | 4. metallic bonding | **d** | positive ions attracted to negative ions |
| ______ | 5. van der Waals attractions | **e** | ionic compounds |
| ______ | 6. noble gas | **f** | electrons free to move |
| ______ | 7. covalent bonds | **g** | element with eight valence-shell electrons |
| ______ | 8. NaCl | **h** | covalent compound |
| ______ | 9. metal ion | **i** | attraction between nonpolar neutral particles |
| ______ | 10. hydrogen bonds | **j** | measures gaseous atom's hold on electron |
| ______ | 11. $NH_3$ | **k** | smaller than parent atom |
| ______ | 12. orbital interaction | **l** | four electrons shared |
| ______ | 13. single covalent bond | **m** | two electrons shared |

## Matching Set II

| | | | |
|---|---|---|---|
| ______ | 1. pi bond | **a** | an electrically neutral arrangement of covalently bonded atoms |
| ______ | 2. sigma bond | **b** | arrangement of parts |
| ______ | 3. $sp^3$ orbital | **c** | four-sided polygon |
| ______ | 4. hybridization | **d** | lateral interaction of p orbitals |

| | | |
|---|---|---|
| ______ | 5. molecule | e one pi bond and one sigma bond |
| ______ | 6. double bond | f mixing of orbitals |
| ______ | 7. symmetry | g end-to-end interaction of orbitals |
| ______ | 8. tetrahedral | h hybrid orbital |

## Matching Set III

| | | |
|---|---|---|
| ______ | 1. $BeCl_2$ (no nonbonding valence electrons) | a linear |
| ______ | 2. $H_2O$ | b trigonal planar |
| ______ | 3. $SiCl_4$ | c bent |
| ______ | 4. $SF_6$ | d tetrahedral |
| ______ | 5. $BCl_3$ | e octahedral |

## Questions

**1.** Diamond (a form of carbon) has a melting point of 3500°C, whereas carbon monoxide (CO) has a melting point of −207°C. What does this suggest about the kinds of bonding found in these two substances?

**2.** Write the electronic configuration for the element potassium (atomic number 19). What will be the electronic configuration when a $K^+$ ion is formed?

**3.** Is $Ca^{3+}$ a possible ion under normal chemical conditions? Why?

**4.** Write the symbols for the six elements with the highest ionization potentials, selecting one from each period of the periodic table.

**5.** Match the electronic configurations that would be expected to lead to similar chemical behavior. The numbers denote the numbers of electrons in the orbits.

| | |
|---|---|
| a. 2-2 | d. 2-8-2 |
| b. 2-5 | e. 2-1 |
| c. 2-8-8-1 | f. 2-8-5 |

**6.** Fluorine (atomic number 9) has an electronic configuration of 2-7. How many electrons will be involved in the formation of a single covalent bond?

**7.** What kind of bond (ionic, pure covalent, polar covalent) is likely to be formed by the following pairs of atoms?

a. a Group IA element with a Group VIIA element
b. a Group VIA element with a Group VIIA element
c. two chlorine atoms
d. an element with low electronegativity and an element with high electronegativity
e. two elements with about the same electronegativity
f. two elements with the same electronegativity

**8.** How many electrons would there be in an iodide ion, $I^-$?

**9.** Write the electron dot structures for the fluoride ion, $F^-$, the chloride ion, $Cl^-$, and the bromide ion, $Br^-$.

**10.** Draw the electron dot structure for water. Based on bonding theory, why is water's formula not $H_3O$?

**11.** Define the term *bond energy.*

**12.** Draw electron dot structures for the following molecules:

| | | |
|---|---|---|
| a. $NF_3$ | g. $N_2H_4$ | m. $SiH_4$ |
| b. $CCl_4$ | h. $CH_3OH$ | n. IBr |
| c. $C_2Cl_2$ | i. $Br_2$ | o. $ClO_3^-$ |
| d. $OF_2$ | j. HCl | p. $SO_3^{2-}$ |
| e. $H_2S$ | k. $BCl_3$ | q. $NH_4^+$ |
| f. CO | l. $PH_3$ | r. $OH^-$ |
| | | s. $AsO_4^{3-}$ |

**13.** The members of the nitrogen family, N, P, As, and Sb, form compounds with hydrogen: $NH_3$, $PH_3$, $AsH_3$, and $SbH_3$. The boiling points of these compounds are

| | |
|---|---|
| $SbH_3$ | −17°C |
| $AsH_3$ | −55°C |
| $PH_3$ | −87.4°C |
| $NH_3$ | −33.4°C |

Comment on why $NH_3$ doesn't follow the downward trend of boiling points.

**14.** Match the substances listed below with the type of bonding responsible for holding units in the solid together.

| | |
|---|---|
| solid krypton (Kr) | ionic |
| ice | covalent |
| diamond | metallic |
| $CaF_2$ | hydrogen bonding |
| iron | van der Waals |

**15.** Predict the general kind of chemical behavior (that is, loss, gain, or sharing of electrons) you would expect from atoms with the following electron arrangements:
a. 2-8-1
b. 2-7
c. 2-4

**16.** Show how two fluorine atoms can form a bond by the interaction of their half-filled p orbitals.

**17.** Select the *polar* molecules from the following list and explain why they are polar:
$N_2$, HCl, CO, NO

**18.** How many bonds join the two atoms in each of the following?
$CN^-$, $Cl_2$, $S_2$

**19.** Boron trichloride has the electron dot formula
:Cl:
:Cl:B:Cl:. What does this tell you about the octet rule even for Period 2 elements?

**20.** Use your chemical intuition and suggest a reaction that might occur between boron trichloride (Question 19 above) and ammonia, H:N:H.
H

**21.** Sketch separately the spatial distribution of electronic charge associated with an s and a p atomic orbital.

**22.** What is the meaning of the notation $sp^3$?

**23.** What is the structure around a central atom having the following hybridization?
a. sp b. $sp^2$ c. $sp^3$

**24.** What is the meaning of hybridization?

**25.** Account for the fact that carbon in methane, $CH_4$, forms four equivalent bonds.

**26.** What is a sigma bond; a pi bond?

**27.** Give the basic points of the valence-shell electron-pair repulsion theory.

**28.** Explain the 106.5° H—N—H bond angles of $NH_3$, the ammonia molecule.

**29.** How many atomic positions must be specified in order to define a bond angle?

**30.** $BeCl_2$ is a compound known to contain polar Be—Cl bonds, yet the $BeCl_2$ molecule is not polar. Explain.

**31.** What happens to the third p orbital in $sp^2$ hybridization?

**32.** If ionic bonds are represented by white, how would pure covalent bonds and polar covalent bonds be represented by black and grey?

**33.** How is an ionic bond formed? How is a covalent bond formed?

**34.** A compound will not conduct electricity when melted, and it melts at 46°C, a low melting point. What type of bond holds atom to atom in this compound?

**35.** What ions would probably be formed by Br, Al, Ba, Na, Ca, Ga, I, S, O, Mg, K, At, Fr, all Group IA metals, all Group VIIA nonmetals?

**36.** How are ionic solids held together?

**37.** A compound will conduct electricity when melted, but it is rather hard to melt. What type of bonds are in this compound?

**38.** A substance is composed of carbon only, or of two nonmetals. The substance has a high melting point and is very hard. What kind of bonds hold the atoms of the substance together?

**39.** When you melt a molecular solid, the molecules separate somewhat from each other and are free to move over each other in a random fashion. Why should it be much more difficult to melt ionic solids?

**40.** In which case would hydrogen bonding be most extensive: a. liquid water, b. water vapor, c. ice?

**41.** Which type of bond would not be employed between the freely moving unit parts of a liquid?

**42.** Predict the formulas of compounds consisting of the following elements.

| Nonmetal \ Metal | *Ba* | *Al* | *K* |
|---|---|---|---|
| *Cl* | | | |
| *O* | | | |
| *S* | | | |
| *N* | | | |
| *I* | | | |

**43.** Do you suppose any type of glue holds things together without chemical bonds?

**44.** Which is harder to break, an ordinary covalent bond or a hydrogen bond?

**45.** What is the direction of energy transfer in a
a. bond-making process?
b. bond-breaking process?

**46.** Liquid water consists of water molecules held together by covalent O—H bonds, and the water molecules are held together loosely by hydrogen bonds. When water boils, which type of bond breaks first?

**47.** Since energy is required to break bonds, do you expect water molecules to break apart at some elevated temperature above its normal boiling point?

# 5 Some Principles of Chemical Reactivity

Now that we are acquainted with atoms and have them bonded, what else can atoms do? One very important role of atoms is their activity during chemical change: which atoms react with which atoms, how many atoms of one kind are required to react with one atom of another kind, what drives atoms to react. Knowledge of this kind allows one to predict the outcome when two chemicals are put together under prescribed conditions and to prepare for safety measures when handling the chemicals. Custom-made chemical products come about occasionally by trial and error, but more often by knowing the fundamental principles that control and guide chemical reactivity. In this chapter we shall examine a few of the basic principles of chemical reactivity.

**Reactants disappear, and products are formed.**

In millions of chemical reactions, the principles of the periodic table and bonding discussed in earlier chapters undergird, predict, and explain which chemicals react with each other and what products are likely to be formed. For example, when potassium (K) is placed with chlorine (Cl), a predictable reaction occurs between the ***reactants*** (the metal K and the nonmetal Cl) to form the predictable ***product,*** potassium chloride (KCl).

$$2K + Cl_2 \longrightarrow 2KCl$$

REACTANTS PRODUCT

The ratio of one K to one Cl in the formula of the product, KCl, is predictable because K is in Group IA of the periodic table and has a normal charge of +1. Chlorine is in Group VIIA of the periodic table and has a normal charge of −1. One ion of each neutralizes the charge of the other ion, and the compound is neutral as observed.

We expect this reaction to occur as shown because metals generally react with nonmetals and form compounds typical of their groups in the periodic table.

Furthermore, we would expect a nonmetal to react with a nonmetal because they would form a covalent compound. For example, phosphorus (P) reacts with chlorine (Cl). Both are nonmetals. The reaction produces two products, either

$$P_4 + 6Cl_2 \longrightarrow 4PCl_3$$

or

$$P_4 + 10Cl_2 \longrightarrow 4PCl_5$$

Which reaction occurs depends primarily upon how much chlorine is present for each mole of phosphorus. If less Cl is present, the product is $PCl_3$; if more Cl is present, the product is $PCl_5$; ordinarily, a mixture of $PCl_3$ and $PCl_5$ is formed. This reaction is typical of those involving two nonmetals in that more than one product is both possible and usual. The law of multiple proportions has many applications among nonmetal–nonmetal compounds.

**Figure 5–1** Chemical changes produce new substances, generally with properties very different from the starting substances. Bright, shiny nuts and bolts are changed to crumbly, dull rust by iron reacting with oxygen.

Some nonmetal–nonmetal reaction products can be eliminated by the application of bonding theories. For example, by drawing valence-shell electron dot structures, you can see that $PCl_3$ and $PCl_5$ would give all paired valence-shell electrons whereas $PCl_2$ would leave an unpaired electron.

:Cl:P:Cl:
:Cl:
$PCl_3$

:Cl: :Cl:
:Cl: P
:Cl: :Cl:
$PCl_5$

:Cl:P· ← unpaired electron
:Cl:
$PCl_2$
**(does not exist)**

Although atomic structure and the combining capacities of the elements help us to predict the formulas of the products of a chemical reaction, the final proof, of course, is found by conducting and studying the reaction in the laboratory.

The formulas of the products of a given reaction are sometimes predictable and sometimes unpredictable, but some general results are so predictable that we can classify many of the reactions to be found in this book into four types.

In a ***combination*** or ***synthesis*** chemical reaction, two substances are united into one different substance. The general form of the chemical equation is

$$A + B \longrightarrow AB$$

Examples:

$$2Mg + O_2 \longrightarrow 2MgO$$

$$2Na + Cl_2 \longrightarrow 2NaCl$$

$$2H_2 + O_2 \longrightarrow 2H_2O$$

$$SO_3 + H_2O \longrightarrow H_2SO_4$$

If the substance is formed from a metal ion and a nonmetal ion, the ratio of elements in the formula of the compound formed is consistent with balancing the charges on the ions.

In a ***decomposition*** chemical reaction, a compound is broken down into two or more different substances. The starting material must be a compound, and the products may be elements or compounds. The reaction may be considered the reverse of a combination reaction. The general form of the equation is

$$AB \longrightarrow A + B$$

Examples:

$$2H_2O \longrightarrow 2H_2 + O_2$$

$$2HgO \longrightarrow 2Hg + O_2*$$

$$2KClO_3 \longrightarrow 2KCl + 3O_2\dagger$$

In a ***single replacement*** chemical reaction, an element substitutes for another element in a compound. The products of the reaction are the replaced element and a compound containing the replacing element. Metals substitute for metals, that is, positive ion formers substitute for positive ions; nonmetals substitute for nonmetals, that is, negative ion formers substitute for negative ions. The general form of the equation for single replacement reactions is

$$A + BC \longrightarrow AC + B$$

Examples:

**Single replacement reactions are types of oxidative-reduction reactions to be discussed in the next chapter.**

$$Zn + 2HCl \longrightarrow ZnCl_2 + H_2$$

$$2Na + 2HOH \longrightarrow 2NaOH + H_2$$

$$Zn + CuSO_4 \longrightarrow ZnSO_4 + Cu$$

$$Cl_2 + 2KI \longrightarrow 2KCl + I_2$$

**Single replacement reactions are reminiscent of the game "King on the Mountain."**

Whether an element will replace another element depends on the relative tendencies of the two elements to form ions. By placing elements in solutions of various compounds, an ***activity series*** of the elements has been determined experimentally. In the partial activity series for the metals (or positive ion formers) listed below, the metal to the left replaces any metal on the right.

| | K | Ca | Na | Mg | Al | Zn | Cr | Fe | Ni | Pb | H | Cu | Ag | Hg | Au |
|---|---|---|---|---|---|---|---|---|---|---|---|---|---|---|---|
| usual ion formed: | +1 | +2 | +1 | +2 | +3 | +2 | +3 | +3 | +2 | +2 | +1 | +2 | +1 | +2 | +3 |

Thus, potassium (K) replaces any element in the list by forming potassium ions and the free element. Another example is zinc (Zn) metal replacing copper (Cu) ions in a solution of copper sulfate.

$$Zn + CuSO_4 \longrightarrow Cu + ZnSO_4$$

However, copper metal will not replace zinc ions in a solution of zinc sulfate.

**The charges on the ions are predicted by the periodic table and the list of polyatomic ions on p. 76.**

$$Cu + ZnSO_4 \not\longrightarrow \textit{No reaction}$$

In the new compound formed, both ions have their usual charges, and the formula of the compound has the proper ratio of ions to make the compound neutral. For example, aluminum (Al) will replace nickel (Ni) as follows.

*In 1774, this reaction was the basis for the discovery of $O_2$ by Joseph Priestley and a partial basis for the discovery of the law of conservation of matter by Antoine Lavoisier.

†This is the usual preparation of oxygen in an academic laboratory.

$Al + NiSO_4 \longrightarrow \underbrace{Al^{3+} SO_4^{2-}}_{} + Ni \longleftarrow$ Correct replacement

$Al_2(SO_4)_3 \longleftarrow$ Correct new formula

$2Al + 3NiSO_4 \longrightarrow Al_2(SO_4)_3 + 3Ni \longleftarrow$ Balanced equation

Likewise, in the nonmetal series (or negative ion formers),

| | F | Cl | Br | O | I | S |
|---|---|---|---|---|---|---|
| usual ion formed: | −1 | −1 | −1 | −2 | −1 | −2 |

fluorine (F) replaces chlorine (Cl), and chlorine replaces bromine (Br).

$F_2 + 2KCl \longrightarrow Cl_2 + 2KF$

$Cl_2 + 2KBr \longrightarrow Br_2 + 2KCl$

Bromine will not replace chlorine.

$Br_2 + 2KCl \not\longrightarrow$ *No reaction*

Metathetical is from Greek, *metathesis*, meaning transposition or interchange.

In ***double replacement*** or ***metathetical*** reactions, the positive ions in two compounds exchange negative ions. Two different compounds are formed. The general form of the chemical equation is

$AB + CD \longrightarrow AD + CB$

Examples:

$HCl + NaOH \longrightarrow NaCl + HOH$*

$AgNO_3 + NaCl \longrightarrow AgCl$ (a precipitate) $+ NaNO_3$

$2HCl + ZnS \longrightarrow ZnCl_2 + H_2S$ (a gas)

In all of the four types of chemical reactions just described, the products must conform to experimental findings. This usually means that the ionic compounds contain ions with normal charges predicted by their position in the periodic table, and the ions combine in correct ratios to form neutral compounds.

Such generalities as the four types of chemical reactions given, the periodic table, atomic structure, and the charges on ions all converge to aid in the prediction of chemical reactions. But there is more—much more, in fact. What, for example, drives one element to replace another? One type of driving force for chemical reactions is the drive to produce products that exist in a lower, more stable energy state than the reactants. Heat is liberated during such reactions. (We will return to this topic later in the chapter and in Chapter 14.) For reactions in solution, a driving force can be the formation of a substance that is much less soluble than any of the reactants (a precipitate), the formation of a gas that escapes the reaction solution, and (or) the formation of a slightly ionized substance such as water.

The formation of water in the first example of a double replacement reaction, AgCl precipitate in the second, and $H_2S$ gas in the last help to drive those reactions to products.

Even if the products can be predicted accurately, what does this say about how fast the process goes? In the next section, the factors that affect the rate of a chemical reaction are discussed.

*An example of a neutralization reaction, to be discussed in the next chapter.

**Figure 5–2** The biochemical processes of decomposition occur more rapidly at higher temperatures. Half the peach shown in this photograph was refrigerated, while the other half was kept warm. The refrigerated half on the right shows little discoloration, whereas the other one shows the typical signs of decay.

## Rate of Chemical Reaction

The whole notion of how fast or how slow chemical reactions proceed can be put on a quantitative basis by the concept of ***reaction rate.*** The rate of a reaction is always defined in terms of the changes in the amounts of chemical substances present per unit of time. Thus, if we consider the burning of sulfur to produce sulfur dioxide,

$$S + O_2 \longrightarrow SO_2$$

we can discuss the rate of the reaction in terms of the amount of $SO_2$ formed per minute or of the amount of S or $O_2$ consumed per minute.

**Raising the temperature speeds up chemical reactions.**

**For many reactions, a temperature rise of 10°C doubles the rate.**

### Effect of Temperature on Reaction Rate

It is possible to alter the rate of a chemical reaction by changing the temperature. If the temperature is raised, the rates of chemical reactions are increased; if the temperature is reduced, the rates are decreased. We make use of this principle in cooking foods (a roast will cook at a faster rate at a higher temperature) and in preserving foods (foods spoil less quickly if refrigerated). Figure 5–2 illustrates the effect of temperature on the reactions that take place in a slice of fruit exposed to air.

**Figure 5–3** Effect of concentration on reaction rate. Steel wool held in the flame of a gas burner is oxidized rapidly. It is in contact with air, which is 20 percent oxygen. When the red hot metal is placed in pure oxygen in the flask, it oxidizes much more rapidly.

## Effect of Concentration on Reaction Rate

It is also possible to alter the rate of a reaction by changing the concentrations of the reactants. For example, in the reaction of sulfur and oxygen given earlier, if air replaces oxygen, the reaction will proceed at a slower rate, since air is a mixture of about one part oxygen and four parts nitrogen. The rusting of iron can be retarded by painting or coating the surface of the metal to cut down on the concentration of the oxygen and moisture at the surface. A demonstration such as that shown in Figure 5–3 contrasts the reaction of iron with oxygen in relatively small and in relatively large concentrations.

**Increasing the concentration of reactants speeds up a reaction.**

A theoretical (molecular) explanation for both the concentration effect and the temperature effect on the rate of chemical reactions is illustrated in Figure 5–4. On a molecular basis, the reaction between sulfur and oxygen in air occurs more slowly than in pure oxygen because there are fewer oxygen molecules per volume of air to react (collide) with sulfur molecules.

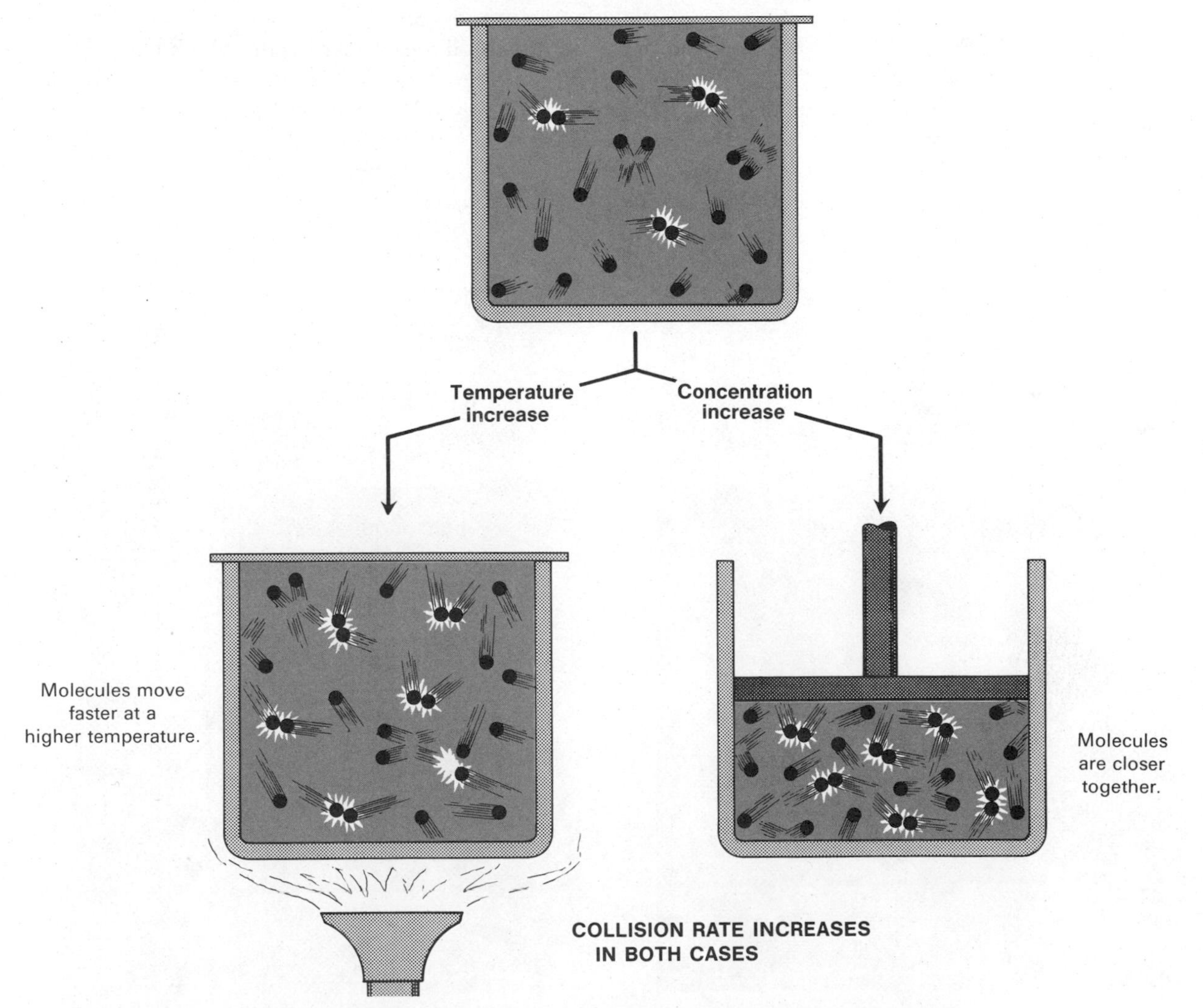


**Figure 5–4** Effects of temperature and concentration on rates of chemical reactions. At the higher temperature, more collisions occur between molecules, and a greater percentage of the collisions produce a chemical reaction. At the higher concentration (no temperature change), more collisions occur, but the percentage of effective collisions remains the same.

## Reversibility of Chemical Reactions

**Chemical reactions are capable of going forward or backward.**

Most chemical processes are capable of being reversed under suitable conditions. When a chemical reaction is reversed, some of the products are converted back into reactants. For example, heating calcium hydroxide will drive off some of the water. This process is the reverse of adding water to quicklime (CaO).

$$Ca(OH)_2 + heat \longrightarrow CaO + H_2O$$

Other methods can be used to reverse chemical reactions. For example, if we put calcium hydroxide in a vacuum, there will soon be water vapor in the space around the solid. It is easier to reverse chemical reactions when they are associated with small heat changes. For the compound hydrogen iodide, HI, it is possible to break the HI molecules apart into hydrogen and iodine

$$2HI + heat \longrightarrow H_2 + I_2$$

at slightly elevated temperatures. On the other hand, water is decomposed into hydrogen and oxygen only by the use of considerable amounts of electrical energy. The reaction that is the reverse of the electrolytic decomposition of water is also shown in Figure 5–5. Hydrogen when burned in air produces water vapor, which can be condensed rapidly.

The fact that many reactions can be approached from either direction leads to the conclusion: *Chemical reactions are generally reversible.*

There are many reversible reactions that are important to human life. One of these is involved in the transport of atmospheric oxygen from the lungs to the various

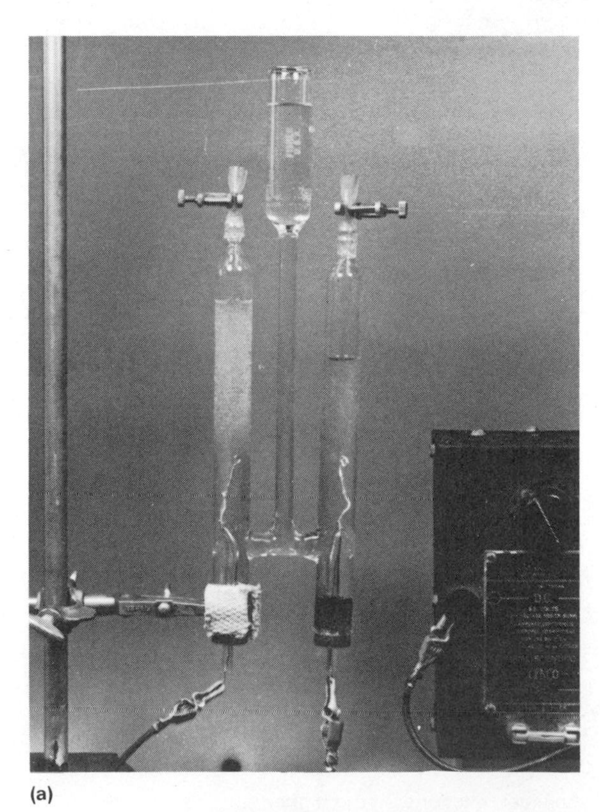

(a)

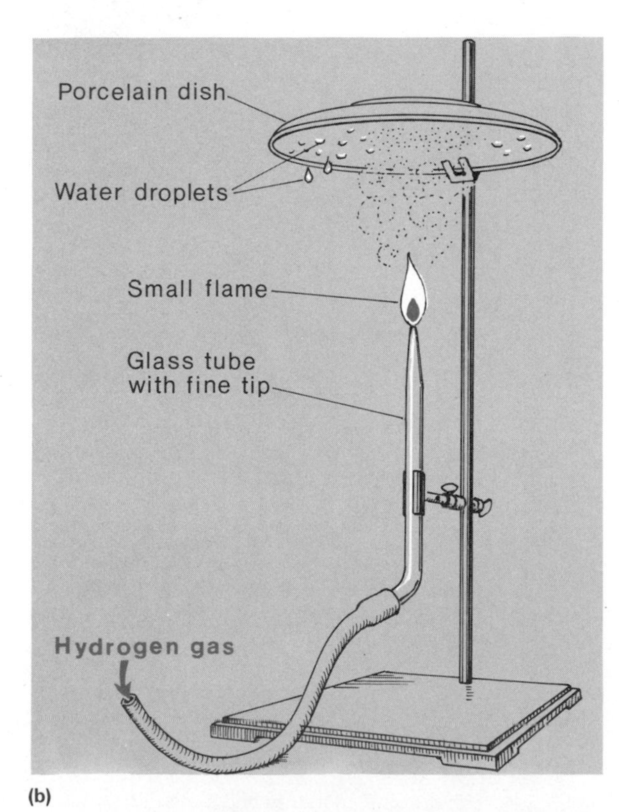


(b)

**Figure 5–5** (a) Electrical energy is required to decompose water into hydrogen (*right tube*) and oxygen (*left tube*). Note that there are two volumes of hydrogen produced for each volume of oxygen.
(b) Hydrogen and oxygen burning to produce water in the gaseous state. The water is condensed on the cooler porcelain dish.

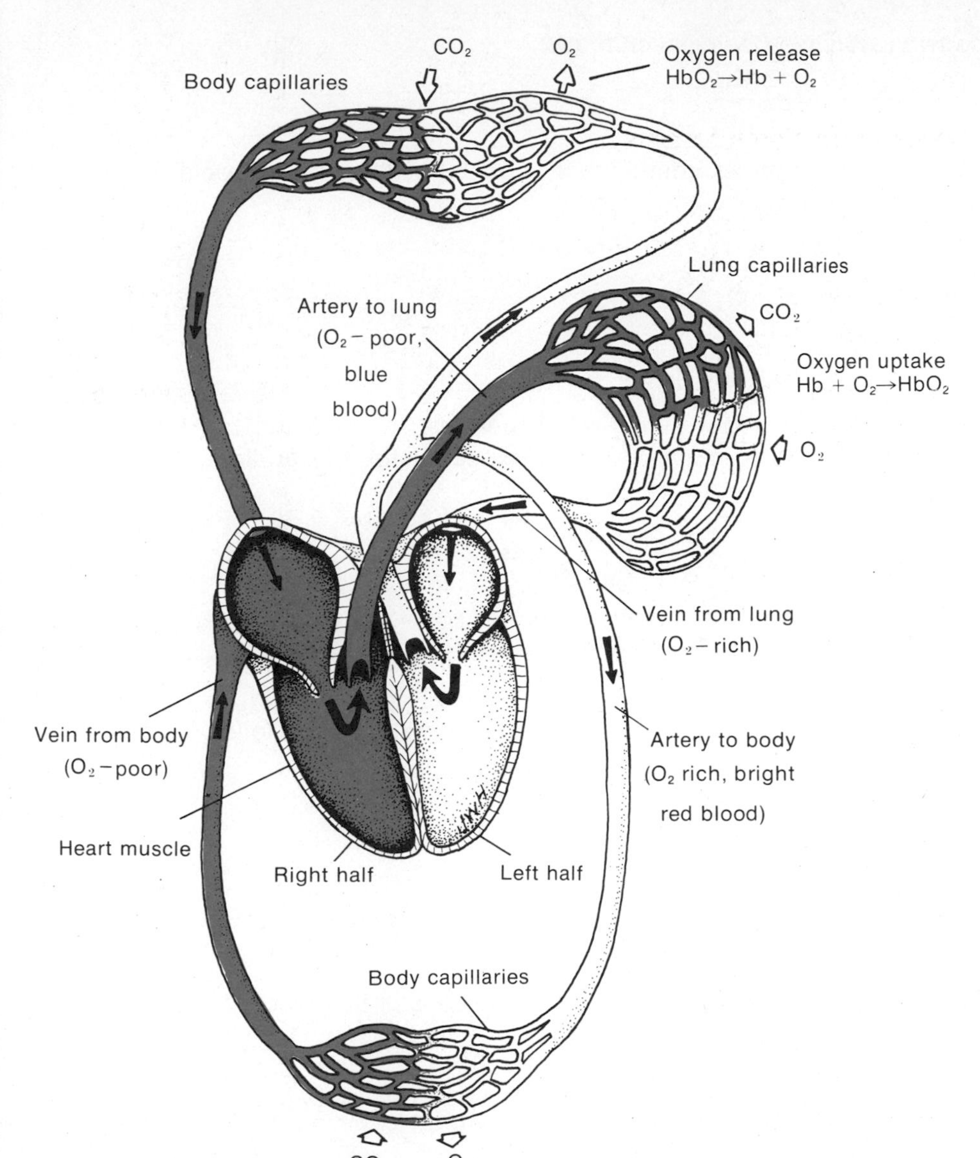


**Figure 5–6** Simplified diagram of human circulation. The heart (shown in front view) is divided into two parallel halves. The right half pumps oxygen-poor blood to the lungs; the left half pumps oxygen-rich blood to the body. Hb = hemoglobin; $HbO_2$ = oxyhemoglobin.

parts of the body. This task is carried out by hemoglobin, a complex compound found in the blood. This substance takes up oxygen while in the lungs to form oxyhemoglobin.

Hemoglobin + $O_2 \rightleftharpoons$ Oxyhemoglobin

The oxyhemoglobin is then carried by the bloodstream to the various parts of the body, where it releases the oxygen for use in metabolic processes, as shown in Figure 5–6.

**The double arrows, $\rightleftharpoons$, indicate a reversible reaction.**

## Self-Test 5-A

1. Which factor affecting reaction rate—temperature or concentration—is most closely related to freezing foods to prevent spoilage? temp
   Why? rapidity if heat↑
2. If water can be produced by burning hydrogen in oxygen, what are the products of the decomposition of water? Hydrogen and oxygen

3. Name two reversible chemical changes.

Water

oxygen in breathing

4. Identify each reaction as combination, decomposition, single replacement, or double replacement.

a. $Al(OH)_3 + 3HNO_3 \rightleftharpoons Al(NO_3)_3 + 3HOH$ double rep.

b. $4Fe + 3O_2 \rightleftharpoons 2Fe_2O_3$ comb.

c. $2Fe + 6HCl \longrightarrow 2FeCl_3 + 3H_2$ singl rep

d. $2Ag_2O \longrightarrow 4Ag + O_2$ decom

5. Selenium is directly below sulfur in the periodic table. Given the sulfur compounds below, write formulas that are expected for the analogous selenium compounds.

a. Oxides
Sulfur: $SO_2$
$SO_3$

Selenium: ________

________

b. Fluorides
Sulfur: $SF_4$
$SF_6$

Selenium: ________

________

c. Oxyacids
Sulfur: $H_2SO_3$
$H_2SO_4$

Selenium: ________

________

6. With the aid of the periodic table, complete and balance the following chemical equations.

a. $Na + F_2 \longrightarrow$
b. $Mg + Cl_2 \longrightarrow$
c. $Ra + O_2 \longrightarrow$
d. $CaCl_2 + AgNO_3 \longrightarrow$
e. $K + H_2CO_3 \longrightarrow$
f. $Li + HClO_4 \longrightarrow$
g. $Ba + S \longrightarrow$
h. $Mg(OH)_2 + HCl \longrightarrow$

7. Which should burn faster: (1) a pound of flour in a sack, or (2) the same flour in dust form in the air of a flour mill? ________

8. What chemical reaction can you name that is constantly reversed in your bloodstream? ________

9. Name two chemical reactions that are not easily reversed. ________ and ________

---

## Quantitative Energy Changes in Chemical Reactions

| REACTANTS | | PRODUCTS | HEAT EFFECT* |
|---|---|---|---|
| $CaO$ Calcium oxide (quicklime) | $+ H_2O$ Water | $\longrightarrow Ca(OH)_2$ Calcium hydroxide (slaked lime) | + 15.6 kcal per mole of $Ca(OH)_2$ |
| $2Na$ Sodium | $+ Cl_2$ (gas) Chlorine | $\longrightarrow 2NaCl$ Sodium chloride (table salt) | + 196.4 kcal (98.2 kcal per mole of NaCl) |
| $H_2$ (gas) Hydrogen | $+ I_2$ (gas) Iodine | $\longrightarrow 2HI$ (gas) Hydrogen iodide | − 12.4 kcal (−6.20 kcal per mole of HI) |

+ Heat means heat is liberated.

− Heat means heat is required.

*Heat energy can be measured in calories (cal). A kilocalorie (kcal) is 1,000 calories. A calorie is the amount of heat required to raise the temperature of 1 g of water 1° Celsius.

In the first reaction, calcium oxide (quicklime) reacts with water to give calcium hydroxide (slaked lime) with the evolution of heat. In the second reaction, metallic sodium reacts with the greenish-yellow gas, chlorine, to give sodium chloride (table salt). If a piece of hot sodium is put into a flask containing chlorine, the sodium burns quickly, liberating a great deal of heat and light, to produce white crystals of sodium chloride. In the last reaction, gaseous hydrogen reacts with gaseous iodine to produce gaseous hydrogen iodide, with the absorption of heat. These facts, along with similar ones, lead to a generalization about chemical reactions: *A given amount of a particular chemical change corresponds to a proportional amount of energy change.* For example, the preparation of one mole of $Ca(OH)_2$ from CaO and $H_2O$ releases 15.6 kcal; for two moles of $Ca(OH)_2$, $2 \times 15.6$ or 31.2 kcal of heat is released.

**Energy changes in chemical reactions are proportional to the amount of reactant (or product).**

Sometimes energy changes in reactions are difficult to observe because of the very slow rate of reaction. An example is the rusting of iron. This is a very important reaction since it has been estimated that the loss from corrosion of iron and steel in the United States is slightly over $60 per person per year. The reaction involved is complicated, but we can represent it by the simplified equation

$$\underset{\text{IRON}}{4Fe} + \underset{\text{MOIST AIR}}{3O_2 + 6H_2O} \longrightarrow \underset{\text{IRON HYDROXIDE (RUST)}}{4Fe(OH)_3} + \underset{\text{HEAT}}{788\text{ kcal}}\ [197\text{ kcal per mole of }Fe(OH)_3]$$

Ordinarily, the rusting of iron occurs so slowly that the liberation of heat is perceptible only with the aid of special instruments. The total amount of heat evolved in rusting is considerable, but it typically takes place over a long period of time.

## Weight Relationships in Chemical Reactions

An important question concerning weight relationships in chemical reactions could be: What weight of aluminum can be produced from one ton of aluminum oxide? Once the atomic weight scale was established, such important calculations could be accomplished readily.

**All calculations involving chemical equations are based on the law of conservation of matter.**

Since the law of conservation of matter states that matter is neither lost nor gained in a chemical reaction, the weight of the reactants in a chemical reaction must be the same as the weight of the products. For example, according to the law of conservation of matter, when hydrogen reacts with chlorine to form hydrogen chloride, the weight of the reactants, hydrogen and chlorine, used in the reaction must equal the weight of the product, hydrogen chloride. The reaction can be written more concisely by using the symbolism of a chemical equation.

**Accounting for atoms requires balancing an equation. To balance a chemical equation:**
1. **Place numbers (coefficients) only *before* formulas. ($2H_2$ means 2 molecules of $H_2$ *and* $2 \times 2 = 4$ atoms of H.)**
2. **Have same number of each kind of atom on each side of the arrow ($\longrightarrow$).**
3. **Do not change subscripts. Each symbol represents one atom. $H_2$ means a molecule composed of two atoms.**

$$H_2 + Cl_2 \longrightarrow 2HCl$$

Note that a coefficient, 2, has been placed in front of the hydrogen chloride so that 2 atoms of hydrogen and 2 atoms of chlorine are represented in both the reactants and the products; that is, none are gained or lost. What is the meaning of the symbolism of the equation? There are two alternative but equally meaningful ways to interpret the equation:

> 1 molecule of hydrogen reacts with 1 molecule of chlorine to form 2 molecules of hydrogen chloride;

or,

> 1 mole of hydrogen molecules reacts with 1 mole of chlorine molecules to form 2 moles of hydrogen chloride molecules.

Once the equation is balanced, the relative number of moles for each substance involved is given by the respective coefficients. Furthermore, since 1 mole weighs 1 gram molecular weight, a set of weights for all substances involved in the reaction can easily be obtained by adding up the atomic weights of the atoms represented by the formula.

For example, the molecular weight of $C_8H_{18}$, a component of gasoline, is 114. This is calculated as follows:

| | Atomic weight | × | No. of atoms | = | Total weight |
|---|---|---|---|---|---|
| C | 12.0 | × | 8 | = | 96.0 |
| H | 1.0 | × | 18 | = | 18.0 |
| | | | Molecular weight | = | 114.0 |

A mole of any item is $6.02 \times 10^{23}$ of them.

The molecular weight is simply the sum of the atomic weights. The gram molecular weight of $C_8H_{18}$ is 114.0 *grams,* and this is the weight of $6.02 \times 10^{23}$ molecules or one mole of $C_8H_{18}$. It follows that two moles of $C_8H_{18}$ molecules weigh 2 moles × 114 grams per mole, or 228 grams.

These facts allow several types of mole calculations to be made for chemical reactions. The following examples will illustrate some of these.

### Example 1

How many moles of nitrogen ($N_2$) are required to react with 6 moles of hydrogen ($H_2$) in the formation of ammonia ($NH_3$)?

a. Write and balance the equation:

Ammonia is used as a crop fertilizer.

$$N_2 + 3H_2 \longrightarrow 2NH_3$$

b. Since 1 mole of $N_2$ reacts with 3 moles of $H_2$, how many moles of $N_2$ will react with 6 moles of $H_2$? The answer is 2 moles of $N_2$. If the number of moles of $H_2$ is doubled, then the number of moles of $N_2$ must be doubled to keep the same ratio of nitrogen and hydrogen that react with each other.

### Example 2

How many moles of nitrogen dioxide ($NO_2$) will be produced by 4 moles of oxygen reacting with sufficient nitric oxide (NO)?

a. Write and balance the equation:

Nitrogen dioxide is a major air pollutant.

$$2NO + O_2 \longrightarrow 2NO_2$$

b. From the balanced equation we see that the number of moles of $NO_2$ produced is twice that of the oxygen reacting. Therefore 4 moles of oxygen would produce 8 moles of $NO_2$.

### Example 3

How many grams of carbon dioxide ($CO_2$) can be produced by burning 2,650 grams of gasoline ($C_8H_{18}$)?

a. Write and balance the equation:

$$2C_8H_{18} + 25O_2 \longrightarrow 16CO_2 + 18H_2O$$

b. The balanced equation states that 2 moles of gasoline produce 16 moles of $CO_2$. Since the molecular weight of $C_8H_{18}$ is 114, two moles would weigh 228 grams. The molecular weight of $CO_2$ is 44, i.e., $(1 \times 12) + (2 \times 16)$, so 16 moles would weigh $16 \times 44 = 704$ grams. Thus, 228 grams of gasoline would produce 704 grams of $CO_2$; that is, the weight of $CO_2$ produced is about three times the weight of gasoline burned. This means that our 2,650 grams of gasoline should produce about 8,000 grams of $CO_2$. To be more exact:

$$\text{grams of } CO_2 = 2650 \cancel{\text{g } C_8H_{18}} \times \frac{704 \text{ g } CO_2}{228 \cancel{\text{g } C_8H_{18}}}$$
$$= 8182 \text{ g } CO_2$$

**Example 4**

How many pounds of aluminum can be obtained from 1.00 ton of pure aluminum oxide, $Al_2O_3$? (This was the opening question in this section.)

a. Write and balance the chemical equation.

$2Al_2O_3 \longrightarrow 4Al + 3O_2$

b. The balanced equation states that 2 moles of $Al_2O_3$ produce 4 moles of Al. Add up the formula weight of $Al_2O_3$ [(27.0 g/mole × 2 moles Al/mole $Al_2O_3$) + (16.0 g/mole × 3 moles O/mole $Al_2O_3$) = 102 g/mole]. Thus, 2 moles would weigh 2 moles × 102 g/mole = 204 g. Four moles of Al weigh 4 moles × 27.0 g/mole = 108 g. Since 108 is about half of 204, a ton of $Al_2O_3$ should produce about half a ton of Al, which is about 1000 pounds. To be more exact:

$$\text{pounds of Al} = 1.00 \cancel{\text{ton } Al_2O_3} \times \frac{108 \cancel{\text{g Al}}}{204 \cancel{\text{g } Al_2O_3}} \times \frac{\text{pound Al}}{454 \cancel{\text{g Al}}} \times \frac{454 \cancel{\text{g } Al_2O_3}}{\cancel{\text{pound } Al_2O_3}} \times \frac{2000 \cancel{\text{pounds}}}{\cancel{\text{ton}}}$$
$$= 1060 \text{ pounds of Al}$$

By observing this problem carefully, perhaps you can see a short cut. If the units given for $Al_2O_3$ and the units sought for Al are the same units (ton, for example), the numbers obtained from the chemical equation need no additional factors. (We did not need the two 454 g/pound factors in the numerical solution above; they cancel.) Thus, the numbers obtained from the chemical equation can have any weight units (tons, pounds, grams, kilograms, etc.) *as long as both weights have the same units.* In other words, the solution to the problem resolves simply into the following setup.

**Other problems of this type are given in Appendix D.**

$$\text{pounds of Al} = 1.00 \cancel{\text{ton } Al_2O_3} \times \frac{108 \cancel{\text{tons Al}}}{204 \cancel{\text{tons } Al_2O_3}} \times \frac{2000 \text{ pounds}}{\cancel{\text{ton}}}$$
$$= 1060 \text{ pounds of Al}$$

---

## Self-Test 5-B

1. The reaction of hydrogen with iodine (releases/absorbs) ____________ energy.
2. If 68 kcal of energy is released in the formation of 18 g (one mole) of water, how much energy would be released in the formation of 36 g of water? 136
3. Balance the following equations:

   a. 2 Mg + 2 $O_2 \longrightarrow$ 2 MgO

   b. 1 Si + 2 $Cl_2 \longrightarrow$ 1 $SiCl_4$

   c. 4 Al + 3 $O_2 \longrightarrow$ 2 $Al_2O_3$

4. An important source of hydrogen, used as a rocket fuel, is the decomposition of water by electrical energy. The reaction is:

   $H_2O \xrightarrow{\text{electrical energy}} H_2 + O_2$

   a. Balance the equation.

   $2H_2O \longrightarrow 2H_2 + O_2$

b. What weight of water is necessary to produce 2.0 g of hydrogen? 18

c. How many grams of oxygen would be produced as a by-product? 16

d. How much water would be necessary to produce 2.0 tons of hydrogen? ______

______ 2 tons H 2 grH / 1 Mol H2O 1 Mol H2O / 18 g H2O

## Matching Set I

______ 1. rate of chemical reaction — a atom count same on both sides of the arrow

______ 2. corrosion — b F, Cl, Br, I, At

______ 3. lower temperature — c react similarly with oxygen

______ 4. chemical family — d atomic number ordering

______ 5. Mg and Ca — e slower chemical reaction

______ 6. periodic table — f formation of rust

______ 7. balanced equation — g would form an oxide similar to $K_2O$

______ 8. Li — h amount of matter reacted in a given time

## Matching Set II

______ 1. $6.02 \times 10^{23}$ — a 18 grams

______ 2. gram molecular weight of $P_4O_{10}$ — b 20

______ 3. weight of one mole of water molecules — c one mole

______ 4. conserved during chemical reaction — d carbon-12 isotope

______ 5. number of atoms in the formula $(NH_4)_3PO_4$ — e 284 grams

______ 6. reference point for atomic weights — f mass or weight

## Questions

**1.** Based on information presented in this chapter on likenesses of elements in groups, predict the formulas of the products of the following reactions.

$Rb + Cl_2 \longrightarrow$ $K + Cl_2 \longrightarrow$
$Ba + O_2 \longrightarrow$ $Mg + Se \longrightarrow$
$Na + Br_2 \longrightarrow$ $Mg + S \longrightarrow$
$Sr + S \longrightarrow$ $Be + S \longrightarrow$

**2.** In the periodic table the elements are arranged in the order of their ______.

**3.** If you were arranging the elements A, B, C, D in a table and came upon an element, G, which had properties similar to those of A, where would you place it?

**4.** Why is it necessary to balance a chemical equation before it can be used in making a calculation?

**5.** Identify four chemical reactions that we use in our daily lives in which energy plays an important role.

**6.** Give an example of a chemical reaction whose rate is fast and one whose rate is slow.

**7.** List three characteristics of all chemical reactions and illustrate each characteristic with an example.

**8.** Using the periodic table, select all chemical elements that can be expected to have chemical properties similar to

a. Ca (calcium), atomic number 20
b. Fe (iron), atomic number 26
c. Sn (tin), atomic number 50
d. S (sulfur), atomic number 16

**9.** In 1968 an Apollo spacecraft cabin fire killed three astronauts. The fact that pure oxygen was used as the cabin atmosphere contributed to the severity of the fire. How?

**10.** The element helium is very unreactive chemically. What type of behavior would you expect for argon? Study their relationship in the periodic table.

**11.** Consider what the relative rates of iron rusting would be in a dry climate as opposed to a damp climate. What principle(s) of reaction rate is (are) involved?

**12.** Fires have been started by water seeping into bags in which quicklime (CaO) was stored. Why would this produce a fire?

**13.** Write and balance the chemical equation for the reaction between water and hot carbon to form gaseous hydrogen ($H_2$) and gaseous carbon monoxide (CO).

**14.** When iron rusts, heat energy is produced. Why does a rusty piece of iron not feel warm?

**15.** How could greater surface area (hot coal dust versus hot lump coal in the presence of air) affect the rate of a chemical reaction?

**16.** Balance the following equations.

a. $SO_2 + O_2 \longrightarrow SO_3$
b. $NO_2 + H_2O \longrightarrow HNO_3 + NO$
c. $K + Br_2 \longrightarrow KBr$
d. $Mg_3N_2 + H_2O \longrightarrow NH_3 + Mg(OH)_2$
e. $P_4 + Cl_2 \longrightarrow PCl_5$
f. $PbO_2 \longrightarrow PbO + O_2$
g. $Al + O_2 \longrightarrow Al_2O_3$
h. $Fe + O_2 \longrightarrow Fe_3O_4$
i. $HgO \longrightarrow Hg + O_2$
j. $Fe + H_2O \longrightarrow Fe_3O_4 + H_2$
k. $Al + H_2SO_4 \longrightarrow Al_2(SO_4)_3 + H_2$
l. $CH_4 + O_2 \longrightarrow CO_2 + H_2O$
m. $C_8H_{18} + O_2 \longrightarrow CO_2 + H_2O$
n. $O_3 \longrightarrow O_2$
o. $C_{12}H_{22}O_{11} + O_2 \longrightarrow CO_2 + H_2O$

**17.** If the term *endothermic* is used to describe a chemical reaction that absorbs energy as it proceeds, what adjective would be used to describe a reaction that produces energy?

**18.** In a balanced chemical equation, which is conserved, (a) molecules or (b) atoms?

**19.** The kinetic-molecular theory states that matter is made up of molecules and that molecules move faster at higher temperatures. Why should molecules move faster at elevated temperatures?

**20.** The electrolysis (electrical decomposition) of water is the reverse of what chemical reaction?

**21.** Firefighters use the methods of controlling the rate of a chemical reaction to combat a fire. Beside each of the firefighting methods listed below, give the rate-controlling factor that is being applied.

a. use of water ______________________________

b. limiting the fuel supply ______________________________

c. use of a fire blanket ______________________________

d. carbon dioxide extinguisher ______________________________

**22.** Which burns faster, a large log or the same log cut into small sticks of wood? What principle of chemical reactivity applies?

**23.** What principle of chemical reactivity applies to the storage of food in a freezer?

**24.** Nitrogen, phosphorus, arsenic, and antimony all form compounds called hydrides. The hydride of nitrogen is very common and is called ammonia. Its formula is $NH_3$. Predict the hydride formulas for phosphorus (P), arsenic (As), and antimony (Sb).

**25.** Add the atomic weights to determine the formula weights:

| | |
|---|---|
| a. $H_2O_2$ | e. $C_6H_{12}O_6$ |
| b. $H_3BO_3$ | f. $Ag_2O$ |
| c. $C_2H_4(OH)_2$ | g. $H_2SO_4$ |
| d. $Fe_2O_3$ | h. $Ca_3(PO_4)_2$ |

**26.** What is the weight (in grams) of one mole of each of the following?

| | |
|---|---|
| a. Xe | e. $NH_3$ |
| b. $C_2H_5OH$ | f. $Na_2S_2O_3(H_2O)_5$ |
| c. $H_2NCH_2COOH$ | g. $MgSO_4(H_2O)_7$ |
| d. Au | h. $CF_2Cl_2$ |

**27.** Complete and balance the following chemical reactions. Identify each type as combination (synthesis), decomposition, single replacement or double replacement.

a. $H_2 + I_2 \longrightarrow$
b. $BaO_2 \longrightarrow BaO +$ ______________
c. $Zn + AuCl_3 \longrightarrow$
d. $Mg + AgNO_3 \longrightarrow$
e. $Zn(OH)_2 + HBr \longrightarrow$
f. $Ga(OH)_3 + HNO_3 \longrightarrow$
g. $Al + Br_2 \longrightarrow$
h. $CaCl_2 + H_2SO_4 \longrightarrow$

**28.** What are some factors that drive reactions toward products?

**29.** What is the molecular weight of $H_2O$? of $H_2SO_4$?

**30.** How many moles of KCl can be made using one mole of potassium and one mole of chlorine?

$2K + Cl_2 \longrightarrow 2KCl$

**31.** How many moles of sulfur trioxide are needed to form 100 moles of sulfuric acid? The equation is

$SO_3 + H_2O \longrightarrow H_2SO_4$

**32.** Hydrogen chloride (HCl) is produced by the action of sulfuric acid ($H_2SO_4$) on sodium chloride (NaCl)

$NaCl + H_2SO_4 \longrightarrow HCl + Na_2SO_4$

How many moles of HCl can be made from 50.0 moles of $H_2SO_4$? Be sure to balance the equation first.

**33.** Chlorine can be made by the electrical decomposition of melted sodium chloride.

$NaCl \xrightarrow{\text{energy}} Na + Cl_2$

How many moles of products can be made from one mole of sodium chloride? Be sure to include both products.

**34.** How many grams of hydrogen are liberated when 75.0 g of sodium metal react with excess water?

$2Na + 2H_2O \longrightarrow 2NaOH + H_2$

**35.** Copper metal can be produced by heating copper sulfide with carbon and air.

$CuS + O_2 + C \longrightarrow Cu + SO_2 + CO_2$

a. Balance the equation.
b. How many moles of oxygen are required for each mole of CuS?
c. How many grams of copper can be produced from 100 g of CuS?

**36.** How many grams of sulfur dioxide ($SO_2$) can be formed by burning 70.0 g of sulfur?

$S + O_2 \longrightarrow SO_2$

**37.** Silver sulfide ($Ag_2S$) is the common tarnish on silver objects. What weight of silver sulfide can be made from 1.00 milligram of hydrogen sulfide ($H_2S$) obtained from a rotten egg?

$4Ag + 2H_2S + O_2 \longrightarrow 2Ag_2S + 2H_2O$

**38.** If a member of Group VA forms a chloride, $XCl_5$, what would you expect the formula of the fluoride to be?

# 6 Hydrogen Ion Transfer (Acid-Base) and Electron Transfer (Oxidation-Reduction) Reactions

Two major classes of chemical reactions that have far-reaching applications in our lives are those of hydrogen ion transfer and electron transfer between molecular or ionic species. ***Acids*** such as vinegar or lemon juice, and ***bases*** such as lime or baking soda, are commonly encountered. Reactions of acids with bases can be defined in terms of the transfer of hydrogen ions from an acid to a base. Chemical oxidation and reduction are equally common. The rusting of iron, the burning of wood, and the bleaching of hair or fabrics are examples of oxidation, while the transformation of iron ore in a blast furnace to iron metal is a classic example of chemical reduction. Such ***redox*** (short for ***oxidation-reduction***) reactions can be described in terms of the transfer of electrons from one chemical to another. If we can understand how hydrogen ions and electrons are transferred, we shall be better able to control many processes of importance.

**Redox is pronounced "ree-dox."**

Acid-base and redox reactions occur in the solid, liquid, or gaseous phase. However, the most commonly encountered acid-base and redox reactions occur in liquid solutions. A logical place to start, then, is with some of the general properties of solutions.

## Formation of Solutions

How many liquid solutions are familiar to you? How about sugar or salt dissolved in water, or oil paints dissolved in turpentine, or grease dissolved in gasoline? In each of these solutions, the substance present in the greater amount, the liquid, is defined as the ***solvent,*** and the substances dissolved in the liquid, the ones present in smaller amounts, are the ***solute(s).*** For example, in a glass of tea, water is the solvent and sugar, lemon juice, and the tea itself are solutes.

**Solution: homogeneous mixture of atoms, ions, or molecules.**

In theoretical terms a solution of sugar in water can be thought of as a collection of sugar molecules dispersed among the water molecules (Fig. 6–1). In this presenta-

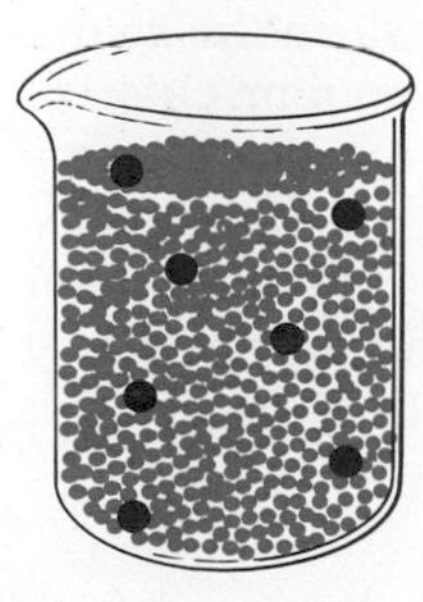


**Figure 6–1** A schematic illustration at the molecular level of sugar solution in water. Large circles represent the sugar molecules and the small circles water. The size of the container and the size of the particles are not to scale.

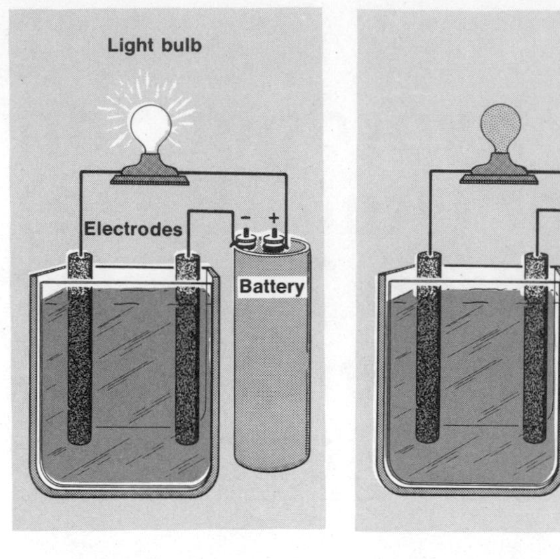


**(a) Solution of table salt**
(an electrolytic solution)

**(b) Solution of table sugar**
(a nonelectrolytic solution)

**(c) Pure water**
(a nonelectrolyte)

**Figure 6–2** A simple test for an electrolytic solution. In order for the light bulb to burn (*a*), electricity must flow from one pole of the battery and return to the battery via the other pole. To complete the circuit, the solution must conduct electricity. A solution of table salt, sodium chloride, results in a glowing light bulb. Hence, sodium chloride is an electrolyte. In (*b*), the light bulb does not glow. Hence, table sugar is a nonelectrolyte. In (*c*), it is evident that the solvent, water, does not qualify as an electrolyte since it does not conduct electricity in this test.

tion of acid-base and redox reactions, most of the chemistry studied will be in water or *aqueous solutions,* where water is the solvent. Generally, one of the species exchanging hydrogen ions or electrons is the solute. Complicating the study somewhat is the fact that solvent molecules themselves can enter into both acid-base and redox reactions. Water molecules can and do exchange hydrogen ions and electrons with solute particles under suitable conditions.

**Aqueous solutions are water solutions.**

## Ionic Solutions (Electrolytes) and Molecular Solutions (Nonelectrolytes)

**Solute particles may be ions or molecules.**

Solutes in aqueous solutions can be classified by their ability or inability to render the solution electrically conductive. When aqueous solutions are examined to see whether they conduct electricity, we find that solutions fall into one of two categories: ***electrolytic*** solutions, which conduct electricity, and ***nonelectrolytic*** solutions, which do not. A simple apparatus such as that shown in Figure 6–2 can be used to determine into which classification a given solution falls.

The conductance of electrolytic solutions is readily explained, since the solute particles in such solutions are ions rather than molecules. Recall that sodium chloride crystals are composed of sodium ions, which are positively charged, and chloride ions, which are negatively charged. When sodium chloride dissolves in water, ***ionic dissociation*** occurs (Fig. 6–7). The resulting solution (Fig. 6–3a) contains positive sodium ions and negative chloride ions dispersed in water. Of course, the solution as a whole is neutral since the total number of positive charges is equal to the total number of negative charges.

**Ionic dissociation is the separation of ions of a solute when the substance is dissolved.**

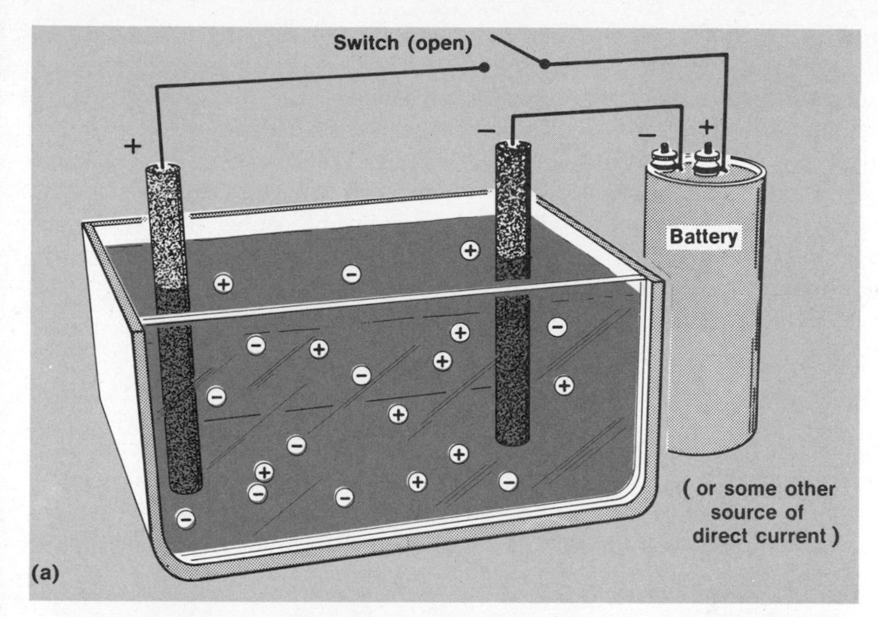


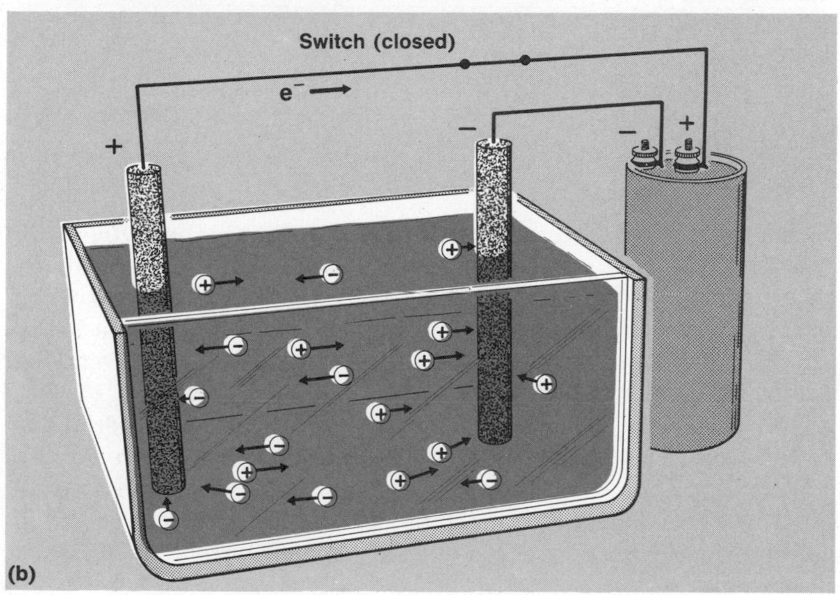


**Figure 6–3** Conductance of electricity by ionic solution. (*a*) The hydrated ions are randomly distributed throughout the salt solution; the net charge is zero. (*b*) Negative electrode attracts positive ions; positive electrode attracts negative ions. If electrons are transferred from negative electrode to positive ions and from negative ions to positive electrode, the circuit is complete, and electricity will flow through the circuit.

$$\underset{\text{(SOLID)}}{Na^+Cl^-} \xrightarrow{\text{water}} \underset{\substack{\text{(AQUEOUS)}\\\text{SODIUM}\\\text{ION}}}{Na^+_{(aq)}} + \underset{\substack{\text{(AQUEOUS)}\\\text{CHLORIDE}\\\text{ION}}}{Cl^-_{(aq)}}$$

The random motions of the sodium and chloride ions are not completely independent. The charges on the particles prevent all of the sodium ions from going spontaneously to one side of the container while all of the chloride ions are going to the other side. However, a net motion of ions occurs when charged electrodes are placed in an electrolytic solution (Fig. 6–3b). If the negative ions give up electrons to one electrode while the positive ions receive electrons from the other electrode, a flow of electrons or electricity is maintained.

**Ions migrate toward oppositely charged electrodes in an electric field.**

Nonelectrolytic solutions composed of solute molecules dispersed throughout solvent molecules are insensitive to negatively and positively charged electrodes unless the voltage is so great that it breaks the molecules into ions.

Sometimes ionic solutions arise when a molecular substance dissolves in water. For example, hydrogen chloride, HCl, is a gas composed of diatomic molecules, each having one hydrogen and one chlorine atom. When hydrogen chloride dissolves in water an ***ionization*** reaction occurs, producing ions from molecules. The resulting solution is composed of hydrogen ions and chloride ions dispersed among the water molecules; consequently it is a conducting solution, and hydrogen chloride in water is properly termed an electrolyte.

**When a molecular solute dissolves in water to produce ions, the process is called ionization.**

$$\underset{\text{MOLECULE}}{HCl} \xrightarrow{\text{water}} \underbrace{H^+_{(aq)} + Cl^-_{(aq)}}_{\text{IONS}}$$

**The symbol $H^+_{(aq)}$ means a hydrogen ion dissolved in water, an aqueous system.**

The hydrogen ion in aqueous systems is not free and unattached. Recall that water molecules are polar. A free proton (isolated hydrogen ion) could not exist in such a medium; it becomes attached to the negative end of one of the water dipoles. In fact, the attraction of water dipoles for the polar HCl molecule probably brings about its ionization in the first place.

$$H^+ + H:\ddot{\underset{\cdot\cdot}{O}}: \atop \quad\quad H \longrightarrow \left[\begin{array}{c} H:\ddot{\underset{\cdot\cdot}{O}}:H \\ H \end{array}\right]^+$$

HYDRONIUM ION

**$H_3O^+$ is the hydronium ion.**

Thus, the hydrogen ion in water is ***hydrated*** and is often referred to as the ***hydronium*** ion, $H_3O^+$ or $H^+(H_2O)$. When one considers hydrogen bonding between water molecules, it is very likely that other water molecules are attached to the molecule to which the proton is attached. The best representation we can give for the hydrogen ion in water then is $H^+(H_2O)_n$, where $n$ is a constantly changing number, perhaps averaging about 4 or 5 in dilute solutions at room temperature.

When sugar, sodium chloride, alcohol, or any other readily soluble material dissolves in water, we can have either a concentrated or a dilute solution. Such a qualitative description of concentration is much less satisfactory and useful than a quantitative description, which tells us just how much of a given substance is dissolved in a specified volume. In chemical work we often express ***concentrations*** in ***moles*** of solute ***per liter*** of solution.

Solutions of known concentration are prepared using volumetric flasks. These are glass vessels with the stems precisely marked to indicate specific volumes, such as 1.000 liter. The procedure involves the steps shown in Figure 6–4.

To show how concentrations are determined, let us consider a case where a 25 g sample of NaCl is carefully weighed, then transferred to a one-liter volumetric flask and dissolved in water. The next step is to add water to the flask until the solution has a total volume of one liter. In order to determine the concentration of such a solution, we need to know the number of moles of NaCl present. The formula weight of NaCl is 23.0 + 35.5 or 58.5. Thus, if we have 58.5 g of NaCl, we have one mole. We have 25 g, so the concentration is

$$\frac{25\ \cancel{\text{grams}}}{58.5\ \cancel{\text{grams}}/\text{mole}} = 0.43 \text{ mole of NaCl per liter of solution}$$

**Molar concentration: number of moles of a substance per liter of solution.**

We usually indicate this as 0.43 M NaCl, where M stands for moles of solute per liter of solution, and is read as "molar."

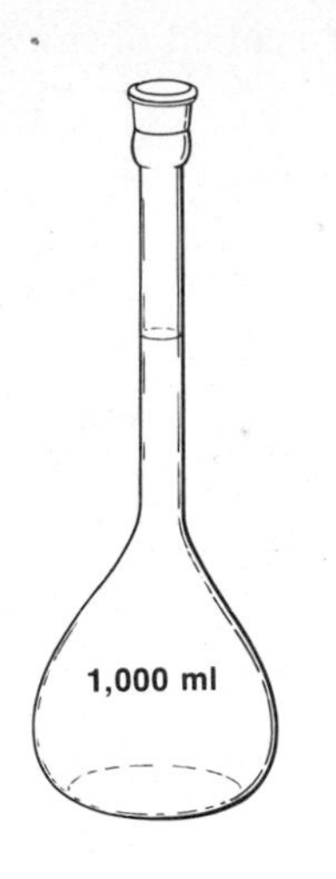


1. Take a volumetric flask

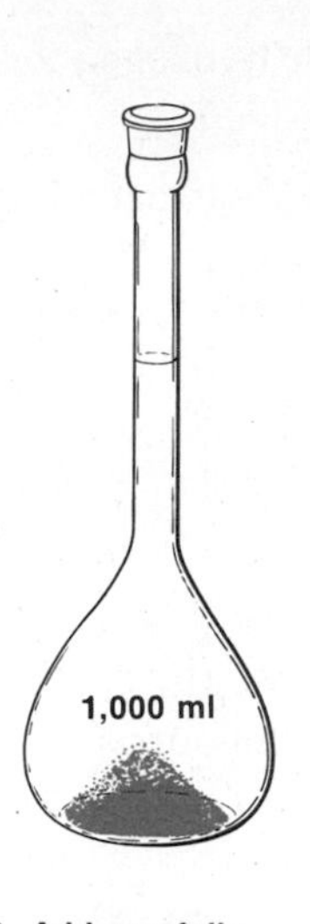


2. Add carefully the weighed amount of solid

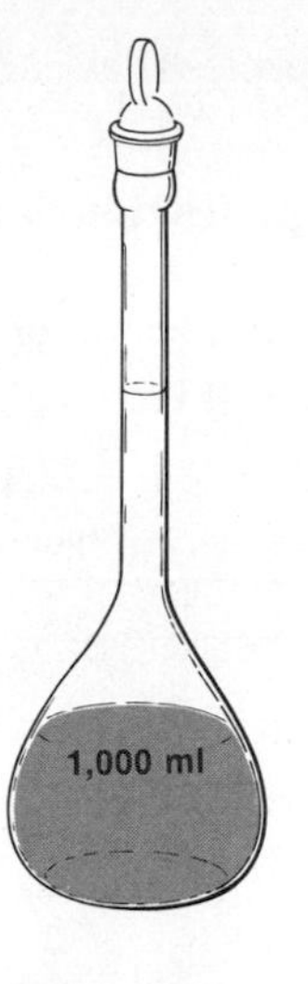


3. Add some water, shake, and dissolve solid

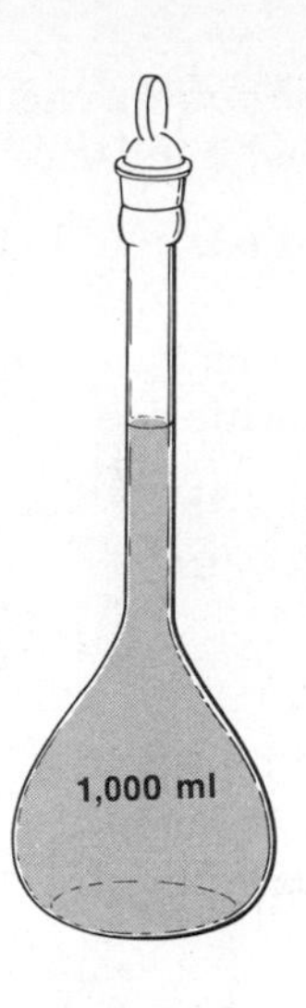


4. Fill flask to one liter mark and shake until homogeneous solution is obtained

**Figure 6–4** Laboratory procedure for the preparation of a solution of known concentration.

Suppose we have 86 g of sucrose (table sugar: $C_{12}H_{22}O_{11}$, molecular weight 342) dissolved in a volume of 500 ml. What is the concentration of sugar in this solution? Since we have $^{86}/_{342}$ mole of sugar, or 0.25 mole, dissolved in $^{500}/_{1000}$ of a liter, the concentration of the sugar solution is

$$\frac{\dfrac{86\ \cancel{g}}{342\ \cancel{g}/\text{mole}}}{\dfrac{500\ \cancel{\text{ml}}}{1000\ \cancel{\text{ml}}/\text{liter}}} = 0.50\ \text{M sucrose}$$

**Molarity = M = $\dfrac{\text{moles of solute}}{\text{liter of solution}}$**

With this knowledge about solutions and their concentrations in mind, let us look at acid-base reactions occurring in solution.

## Acids and Bases

The terms "acid" and "base" have been used by chemists for several hundred years. These names were originally given to substances that showed certain properties. For example, acids have long been characterized as substances that are sour-tasting, corrosive, and able to react with alkaline substances called bases. Acids turn blue litmus red. Bases, on the other hand, have a bitter taste, make the fingers or skin feel slippery on contact, and react with acids. Bases turn red litmus blue. As more and more information was collected on the properties of acids and bases, these simple definitions were refined to the point where current definitions of acids and bases are theoretical in nature.

**Acids taste sour; bases taste bitter.**

**Litmus is an acid-base–sensitive dye obtained from lichens.**

### Hydrogen Ion Transfer and Neutralization

Perhaps the most used definitions of acids and bases are those first given by J. N. Brønsted and T. M. Lowry in 1923.

***Brønsted acid:*** a chemical species that can *donate* hydrogen ions (also called protons or $H^+$ ions) is an acid.

***Brønsted base:*** a chemical species that can *accept* hydrogen ions is a base.

To illustrate these definitions we again consider the reaction between gaseous hydrogen chloride (HCl) and water:

$$\underset{\text{ACID}}{HCl\,(gas)} + \underset{\text{BASE}}{H_2O} \longrightarrow \underbrace{\underset{\text{Hydronium Ion}}{H_3O^+} + \underset{\text{Chloride Ion}}{Cl^-}}_{\text{HYDROCHLORIC ACID}}$$

Examination of the above reaction shows that the HCl molecule has donated a proton ($H^+$ ion) to the water molecule. This behavior is understandable in terms of the electronic structures of the reacting molecules.

$$\underset{\text{BASE}}{H{:}\overset{\delta-}{\ddot{\underset{\cdot\cdot}{O}}}{:}\,\overset{\delta+}{\underset{}{H}}} + \underset{\text{ACID}}{\overset{\delta+}{H}{:}\overset{\delta-}{\ddot{\underset{\cdot\cdot}{Cl}}}{:}} \longrightarrow \left[H{:}\ddot{\underset{\cdot\cdot}{O}}{:}H \atop H\right]^+ + {:}\ddot{\underset{\cdot\cdot}{Cl}}{:}^-$$

The partially positive ($\delta+$) hydrogen atom of the polar HCl molecule is attracted to the partially negative ($\delta-$) end of the polar water molecule, and the hydrogen ion is transferred. This reaction is essentially complete; that is, almost all the HCl is converted to $H_3O^+$ and $Cl^-$. A concentrated (about 12 M) solution of hydrogen chloride in water is mostly a solution of hydronium ($H_3O^+$) ions and chloride ($Cl^-$) ions.

If the ionic solid sodium oxide, $Na_2O$, is dissolved in water, a vigorous reaction produces a solution containing sodium ions ($Na^+$) and hydroxide ions ($OH^-$). In this process, the oxide ion ($O^{2-}$) reacts with polar water to form the hydroxide ion. In this, as in other such aqueous reactions, it is understood that the ions are hydrated (i.e., water molecules are bonded to the ions on a transitory basis).

**Reaction is:**
$Na_2O + H_2O \longrightarrow 2Na^+ + 2OH^-$

$$\underset{\text{BASE}}{{:}\ddot{\underset{\cdot\cdot}{O}}{:}^{2-}} + \underset{\text{ACID}}{H{:}\ddot{\underset{\cdot\cdot}{O}}{:} \atop H} \longrightarrow \underbrace{{:}\ddot{\underset{\cdot\cdot}{O}}{:}H^- + {:}\ddot{\underset{\cdot\cdot}{O}}{:}H^-}_{\text{HYDROXIDE IONS}}$$

There are many other bases that take a hydrogen ion from a water molecule in this way. For example:

$$\underset{\text{SULFIDE ION BASE}}{S^{2-}} + \underset{\text{ACID}}{H_2O} \longrightarrow \underset{\text{HYDROGEN SULFIDE ION}}{HS^-} + \underset{\text{HYDROXIDE ION}}{OH^-}$$

$$\underset{\text{CYANIDE ION BASE}}{CN^-} + \underset{\text{ACID}}{H_2O} \longrightarrow \underset{\text{HYDROGEN CYANIDE}}{HCN} + \underset{\text{HYDROXIDE ION}}{OH^-}$$

According to the Brønsted-Lowry definition, water acts as an acid in these reactions and donates a hydrogen ion to the other molecule or ion, which acts as a base. A species such as water that can either donate or accept hydrogen ions is called ***amphiprotic.*** The existence of amphiprotic species imples that acid-base reactions possess a reciprocal nature; an acid and a base react to form another acid (to which a hydrogen ion has just been added) and another base (from which a hydrogen ion has just been removed). Because water is the most commonly used solvent, it is also the most usual reference compound for acid-base reactions.

Also, one water molecule can transfer a hydrogen ion to another water molecule.

$$2H_2O \rightleftharpoons H_3O^+ + OH^-$$

This reaction takes place to only a very small extent, as is indicated by arrows of unequal length. Since $H_3O^+$ and $OH^-$ are produced in equal amounts when only water is present, pure water is neither acidic nor basic, but is described as ***neutral.***

A chemical species in water solution is commonly spoken of as an acid if it donates hydrogen ions to water and increases the concentration of $H_3O^+$ or $H^+_{(aq)}$. Similarly, a base in water solution is commonly described as a compound whose addition to water increases the concentration of $OH^-$. Since water is not the only possible solvent, these concepts are too narrow for general scientific use; they have been extended by the definitions given above, which focus on the essential feature of such acid-base behavior—that is, the donation or acceptance of a hydrogen ion ($H^+$) in a reaction.

**Acids form $H^+$ ions in water; bases form $OH^-$ ions in water.**

When acids react with bases the properties of both species disappear. The process involved is called ***neutralization.*** To get a more precise picture of acid-base neutralization reactions, we will consider what happens when a solution of hydrochloric acid is mixed with a solution of sodium hydroxide. The hydrochloric acid contains $H_3O^+$ and $Cl^-$ ions; the sodium hydroxide solution contains $Na^+$ and $OH^-$ ions. When these two solutions are mixed, a reaction occurs between $H_3O^+$ and $OH^-$.

**When an acid neutralizes a base, acid and base properties are suppressed.**

$$Na^+ + Cl^- + \underset{\text{ACID}}{H_3O^+} + \underset{\text{BASE}}{OH^-} \longrightarrow H_2O + H_2O + Na^+ + Cl^-$$

If we have an equal number of $H_3O^+$ and $OH^-$ ions, they will react to produce a neutral solution, with the hydronium ions ($H_3O^+$) donating their hydrogen ions to the hydroxide ions ($OH^-$), forming molecules of water. Such reactions are called nuetralization reactions because the acids and bases neutralize each other's properties. If we have more $H_3O^+$ ions than $OH^-$ ions, the extra $H_3O^+$ will make the resulting solution ***acidic.*** If we have more $OH^-$ ions than $H_3O^+$ ions, only a fraction of the $OH^-$ ions will be neutralized, and the extra $OH^-$ ions will make the resulting solution ***basic.***

**Conjugate means kindred or kin to.**

When an acid ionizes, it produces a hydronium ion plus a species called the ***conjugate base*** of that acid. For example:

$$\underset{\text{NITRIC ACID}}{HNO_3} + \underset{\text{WATER}}{H_2O} \longrightarrow \underset{\text{HYDRONIUM ION}}{H_3O^+} + \underset{\text{NITRATE ION, THE CONJUGATE BASE OF THE ACID } HNO_3}{NO_3^-}$$

**Conjugate acids and bases differ by one hydrogen ion. The conjugate acid has one more hydrogen ion than its conjugate base.**

In the same manner we speak of nitric acid, $HNO_3$, as being the ***conjugate acid*** of the base $NO_3^-$.

## Self-Test 6-A

1. When ammonia dissolves in water, the resulting solution conducts electricity. Ammonia in water is therefore a(n) ______________.
2. A chemical species that can accept a hydrogen ion is a(n) ______________.

   A chemical species that can donate a hydrogen ion is a(n) ______________.
3. A compound HA is found to undergo a reaction forming a product $H_2A^+$. Therefore HA is a(n) (acid/base). If compound HA reacted to form $A^-$, then HA would be a(n) (acid/base).

4. If compound HA mentioned above undergoes both reactions described, then HA is termed ______________.
5. A solution that contains equal concentrations of $OH^-$ and $H^+$ ions is termed ______________
6. The word "aqueous" means ______________.
7. In a neutralization reaction, a(n) ______________ reacts with a(n) ______________.

### The Strengths of Acids and Bases

Because the water molecule is itself a weak base, the strongest acid (that is, the best hydrogen ion donor) that can exist in water is the hydronium ion ($H_3O^+$). If strong acids such as sulfuric acid ($H_2SO_4$) or nitric acid ($HNO_3$) are added to water, they donate their hydrogen ions to water molecules, and the resulting solutions contain $H_3O^+$ and $HSO_4^-$, and $H_3O^+$ and $NO_3^-$, respectively. All of these reactions are reversible, at least in principle, but when strong acids are dissolved in a large excess of water to form dilute solutions, virtually all of the strong acid is converted to products.

Strong electrolytes dissociate completely in water.

*Reactants* *Products*

$$HNO_3 + H_2O \longrightarrow H_3O^+ + NO_3^-$$
$$H_2SO_4 + H_2O \longrightarrow H_3O^+ + HSO_4^-$$

| ACID | BASE | CONJUGATE ACID | CONJUGATE BASE |
|---|---|---|---|

Since there are a large number of ions present, the solution conducts electricity well; $HNO_3$ and $H_2SO_4$ are termed ***strong electrolytes.***

Under equilibrium conditions the concentrations of the species in solution remain unchanged even though reactions in the forward and reverse directions continue.

Not all acids lose hydrogen ions as readily to water as do nitric acid and sulfuric acid. Some anions are capable of competing with water for the hydrogen ion being exchanged. The result of this competition is the establishment of an ***equilibrium*** (or balance) between neutral acid molecules and hydronium ions in water solution.

Acetic acid, found in vinegar, is a weak acid; that is, its conjugate base, acetate ion, competes well for a hydrogen ion. The molecular structure of acetic acid ($HC_2H_3O_2$) is as follows:

```
     H
     |      O
     |     //
H — C — C
     |     \
     |      O — H
     H
```

ACIDIC HYDROGEN ATOM

The hydrogen atom bonded to the oxygen in the molecule is the only one that is donated to a base in water solution; for that reason it is designated an acidic hydrogen. When acetic acid is dissolved in water, some ions are produced. However, most of the acetic acid molecules do not donate hydrogen ions to water molecules. The result is a mixture containing a few $H_3O^+$ and $C_2H_3O_2^-$ ions and many $HC_2H_3O_2$ molecules. The reaction is:

$$HC_2H_3O_2 + H_2O \rightleftharpoons H_3O^+ + C_2H_3O_2^-$$

| ACETIC ACID | WATER | HYDRONIUM ION | ACETATE ION |
|---|---|---|---|

Ammonia is the number two commercial chemical in quantity produced, sodium hydroxide is number seven, and nitric acid is number ten. (See inside front cover.)

The relatively few ions in an acetic acid solution do not conduct electricity very effectively; consequently, acetic acid is a ***weak electrolyte.*** (A dilute solution of acetic acid would barely conduct in the apparatus shown in Fig. 6–2.)

Another way of looking at this reaction is to realize that in the mixture there are two bases: the water molecule $H_2O$, and the acetate ion $C_2H_3O_2^-$. Since the reverse reaction dominates, the acetate ion must be a stronger base than the water molecule.

The same kind of considerations can be made for other bases. Ammonia dissolved in water is a weak base. The resulting reaction produces relatively few ions; ammonia is mostly in the molecular form:

$$\underset{\text{AMMONIA}}{NH_3} + \underset{\text{WATER}}{H_2O} \rightleftarrows \underset{\text{AMMONIUM ION}}{NH_4^+} + \underset{\text{HYDROXIDE ION}}{OH^-}$$

Consequently, the relatively few ions present do not conduct electricity well, and ammonia may thus be called a weak electrolyte.

Table 6–1 gives some common acids and bases ranked according to their relative strengths.

## The pH Scale

Because water solutions of dilute acids and bases are used so extensively, it is convenient to have a simple way of designating the acidity or basicity of these solutions. The pH scale was devised for this purpose; it furnishes a number which describes the acidity of a solution. The ***pH*** is defined as the negative logarithm of the concentration of the hydrogen ion, $pH = -\log[H_3O^+]$. The brackets mean moles per liter of hydronium ions. The pH scale, for practical purposes, runs from 0 to 14. A pH of 7 indicates a neutral solution (such as pure water), a pH below 7 indicates an acidic solution, and a pH above 7 indicates a basic solution.

The pH number is related to the concentration of the hydrogen ions in an aqueous solution expressed as the negative power of 10 (logarithm). A hydrogen ion concentration of 0.00001 or $1 \times 10^{-5}$ M corresponds to a pH of 5. When the pH is 7, as in pure water, the hydrogen ion concentration would have to be $1 \times 10^{-7}$ M. Note that for each unit decrease in the pH number, there is a ten-fold increase in the concentration of the hydrogen ion (that is, the acidity).

**At pH = 7, the number of $H_3O^+$ ions equals the number of $OH^-$ ions.**

To understand why the pH of pure water is 7, we will need to reexamine the very slight acid-base reaction between water molecules:

$$H_2O + H_2O \rightleftarrows H_3O^+ + OH^-$$

This ionization reaction produces ions, which should cause water to conduct electricity. But we have said that water is a nonelectrolyte. Actually, with sensitive electrical equipment, pure water can be shown to conduct electricity *very slightly* because of the small amount of ionization. Laboratory measurements reveal that 0.0000001 or $1 \times 10^{-7}$ mole per liter of water is present as $H_3O^+$ and $OH^-$ ions at 25°C. Pure water contains 55.6 moles of water per liter. Consequently, we can see that the actual amount of the water that ionizes is very small compared to the total amount of water present. Pure water, then, which is defined as being neutral, contains $1 \times 10^{-7}$ mole of hydrogen ions (hydronium ions) per liter, and the pH is 7. In pure water the concentrations of

**One liter of water weighs 1000 grams and is 55.6 molar:**

$$\frac{\frac{1000\ \text{g}}{\text{liter}}}{\frac{18\ \text{g}}{\text{mole}}} = \frac{55.6\ \text{moles}}{\text{liter}}$$

**TABLE 6–1 Relative Strengths of Some Acids and Bases**

| Increasing Acid Strength ↑ | Acid | | | | | Conjugate Base | | Increasing Base Strength ↓ |
|---|---|---|---|---|---|---|---|---|
| | perchloric acid | $HClO_4$ | + | $H_2O \rightleftharpoons H_3O^+$ | + | $ClO_4^-$ | perchlorate ion | |
| | hydrochloric acid | $HCl$ | + | $H_2O \rightleftharpoons H_3O^+$ | + | $Cl^-$ | chloride ion | |
| | nitric acid | $HNO_3$ | + | $H_2O \rightleftharpoons H_3O^+$ | + | $NO_3^-$ | nitrate ion | |
| | hydronium ion | $H_3O^+$ | + | $H_2O \rightleftharpoons H_3O^+$ | + | $H_2O$ | water | |
| | hydrofluoric acid | $HF$ | + | $H_2O \rightleftharpoons H_3O^+$ | + | $F^-$ | fluoride ion | |
| | acetic acid | $HC_2H_3O_2$ | + | $H_2O \rightleftharpoons H_3O^+$ | + | $C_2H_3O_2^-$ | acetate ion | |
| | water | $H_2O$ | + | $H_2O \rightleftharpoons H_3O^+$ | + | $OH^-$ | hydroxide ion | |
| | hydroxide ion | $OH^-$ | + | $H_2O \rightleftharpoons H_3O^+$ | + | $O^{2-}$ | oxide ion | |

hydronium ions ($H_3O^+$) and hydroxide ions ($OH^-$) must be equal, since each time a hydronium ion ($H_3O^+$) is produced in the ionization reaction a hydroxide ion ($OH^-$) is also produced. Therefore, in pure water the concentration of the hydroxide ion ($OH^-$) is also $1 \times 10^{-7}$ mole per liter.

Figure 6–6 graphically displays the relationship between the pH number and the concentration of the hydrogen ion. It also gives the approximate pH values for common solutions.

A close examination of Figure 6–6 reveals that basic solutions such as aqueous ammonia have hydrogen ion concentrations less than that of pure water. Note that the pH of a typical sample of aqueous ammonia is 11. This is equivalent to a hydrogen ion concentration of $1 \times 10^{-11}$, or 0.00000000001 mole per liter.

## Salts

### Preparation of Salts

Salts can be formed by acid-base neutralization.

The chemical compounds known as ***salts*** play a vital role in nature, in plant and animal growth and life, and in the manufacture of various chemicals for human use.

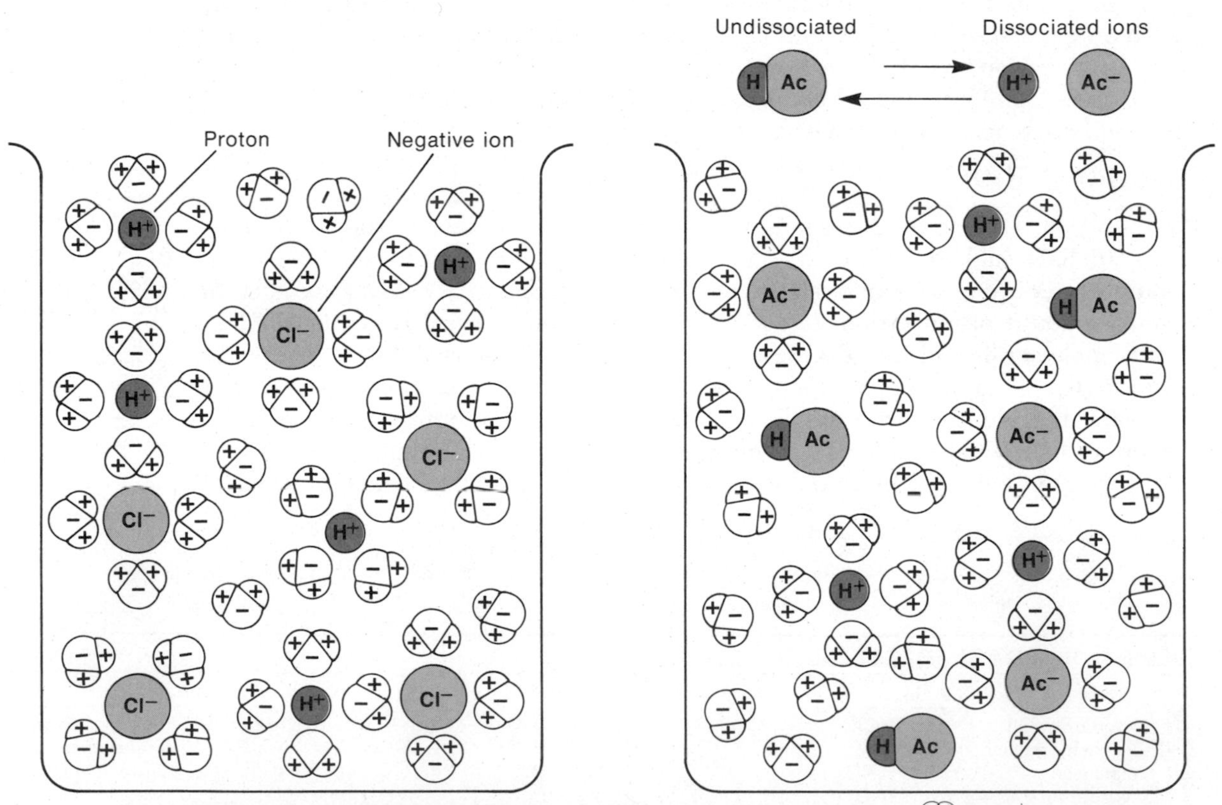


**Figure 6–5** An illustration of what it may be like in an aqueous (water = ) solution of a strong acid (HCl) and an aqueous solution of a weak acid (acetic acid, HAc). The strong acid is all ions; no molecules. The weak acid is mostly molecules with only a few ions.

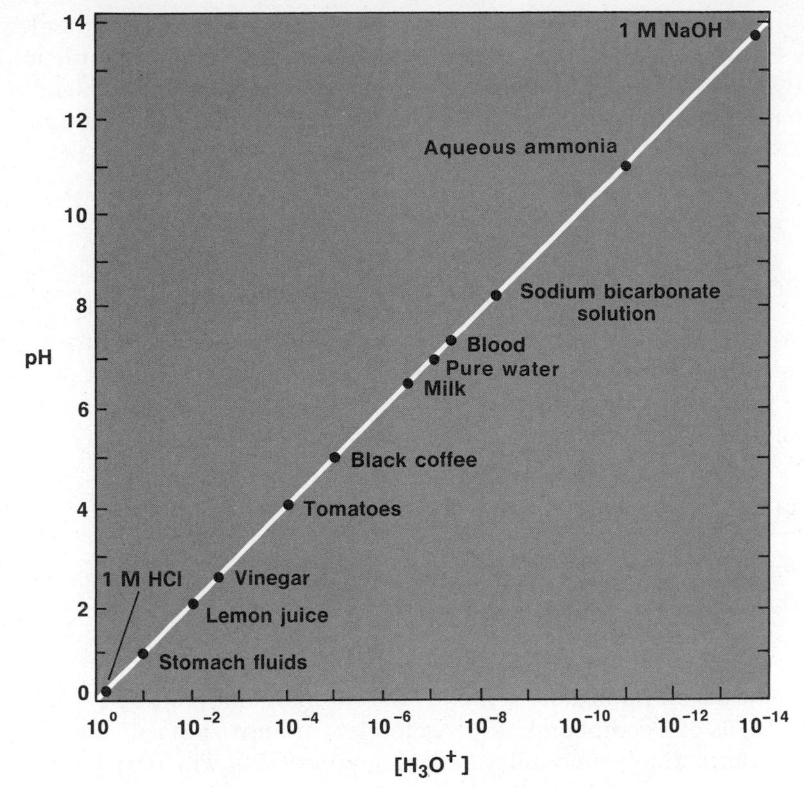


**Figure 6–6** A plot of pH versus hydrogen ion concentration. $[H_3O^+]$. Note that the pH *increases* as the $[H_3O^+]$ decreases. The pH values of some common solutions are given for reference. (A solution in which $[H_3O^+] = 1$ M has a pH of 0 since $1 = 10^0$.)

They can be formed as the products of acid-base neutralizations, as in the following example:

$$\underset{\substack{\text{Potassium Hydroxide}\\ \text{in Water}\\ \text{BASE}}}{(K^+ + OH^-)} + \underset{\substack{\text{Hydrochloric Acid}\\ \text{in Water}\\ \text{ACID}}}{(H_3O^+ + Cl^-)} \longrightarrow \underbrace{K^+ + Cl^-} + 2H_2O$$

Crystallize the Salt by Removal of Solvent, Water

$$\underset{\text{SALT}}{KCl\ (\text{Solid})}$$

Most salts contain ions held together by ***ionic bonding*** (see Chapter 4). Solid potassium chloride, for example, is composed of an equal number of $K^+$ ions and $Cl^-$ ions arranged in definite positions with respect to one another in an ionic structure or lattice (see Fig. 4–4). Since the salt crystal must be electrically neutral, it can have neither an excess nor a deficiency of positive or negative charge.

Let us imagine that we have at our disposal the ions listed below, and let us see what salts could result.

| Ions | | Possible Salts | Salt Name |
|---|---|---|---|
| $Na^+$ | sodium | $NaCl$ | sodium chloride |
| $Ca^{2+}$ | calcium | $NaNO_3$ | sodium nitrate |
| $Cl^-$ | chloride | $Na_2SO_4$ | sodium sulfate |
| $NO_3^-$ | nitrate | $NaC_2H_3O_2$ | sodium acetate |
| $SO_4^{2-}$ | sulfate | $CaCl_2$ | calcium chloride |
| $C_2H_3O_2^-$ | acetate | $Ca(NO_3)_2$ | calcium nitrate |
| | | $Ca(C_2H_3O_2)_2$ | calcium acetate |
| | | $CaSO_4$ | calcium sulfate |

In the examples just given, notice that in order to attain an electrically neutral lattice, it is necessary to balance the charges of the ions. A sodium ($Na^+$) ion requires just one chloride ion ($Cl^-$), and the NaCl lattice contains an equal number of $Na^+$ and $Cl^-$ ions. A sulfate ion ($SO_4^{2-}$) with two negative charges must have its negative charge balanced by two positive charges. This may be done by using two $Na^+$ ions

$$2Na^+ + SO_4^{2-} \longrightarrow Na_2SO_4$$

or one $Ca^{2+}$ ion

In the formula of a salt, the positive and negative charges are equal.

$$Ca^{2+} + SO_4^{2-} \longrightarrow CaSO_4$$

It is possible to form many solid salts by mixing water solutions of different soluble salts with each other. For example, both lead acetate and sodium chloride are soluble in water. If we prepare solutions of these salts and then mix the solutions, we find the insoluble salt, lead (II) chloride, precipitates from the mixture and may be removed by filtration.

$$\underbrace{2Na^+ + 2Cl^-}_{\text{SODIUM CHLORIDE IN SOLUTION}} + \underbrace{Pb^{2+} + 2C_2H_3O_2^-}_{\text{LEAD (II) ACETATE IN SOLUTION}} \longrightarrow \underbrace{PbCl_2}_{\text{SOLID LEAD (II) CHLORIDE}} + \underbrace{2Na^+ + 2C_2H_3O_2^-}_{\text{SODIUM ACETATE IN SOLUTION}}$$

Sodium acetate may be recovered by evaporating the water.

This reaction illustrates an important principle based upon differences in solubility. If the component ions of a compound of low solubility are mixed in solution in great enough concentrations, the compound containing those ions will precipitate from solution.

## Salts in Solution

Although some salts are very soluble in water, others are quite insoluble. Salts are found with a wide range of water solubilities.

An important property of many salts is their solubility in suitable solvents. The amount of a salt that will dissolve in a given quantity of solvent tells us the salt's solubility in that solvent. The preparation of lead (II) chloride shown above was made possible by the differences in solubilities of different salts in the same solvent. As a result of these differences in solubilities, we can make roads out of calcium carbonate (limestone), which is insoluble in water, but not calcium chloride, which is water-soluble.

The two most abundant ions in ocean water are $Cl^-$ and $Na^+$ ions, sufficient to recover about 27 grams of NaCl per kilogram of sea water. To put it another way, there are about 128 million tons of NaCl per cubic mile of sea water, and there are 328,000,000 cubic miles available!

Consider what happens at the ionic level when a sodium chloride crystal is placed in contact with water. We know that sodium chloride is soluble in water. This means that most, if not all, of the attractive forces between the ions in the crystal lattice are somehow overcome in the solution process.

The surface of the salt crystal appears calm when the crystal is placed in water, but on the ionic level there is a great deal of agitation. Water molecules have sufficient polarity to interact strongly with the ions and bond with them. Once this takes place the ion is less strongly bound in the lattice, and so can be removed from the crystal. The crystal lattice now has a gap in it where the ion was removed. Ions that are bonded by solvent molecules are termed solvated ions, and the process of ion-solvent interaction is called ***solvation.*** But just how are these ions solvated? What causes this solvation?

Because of its polar nature, the water molecule is ideally suited to interact with ions.

The answers lie in the structure of the water molecule and its ability to interact with ions. As we saw in Chapter 4, water is a *polar* molecule. The negative end of the molecule is attracted to positive ions, and the positive end is attracted to negative ions. As a result, several water molecules will interact with each ion. Figure 6–7 shows several water molecules solvating a $Na^+$ ion. The positive end of the water molecule will tend to interact with a negative ion; this is shown for the $Cl^-$ ion in Figure 6–7.

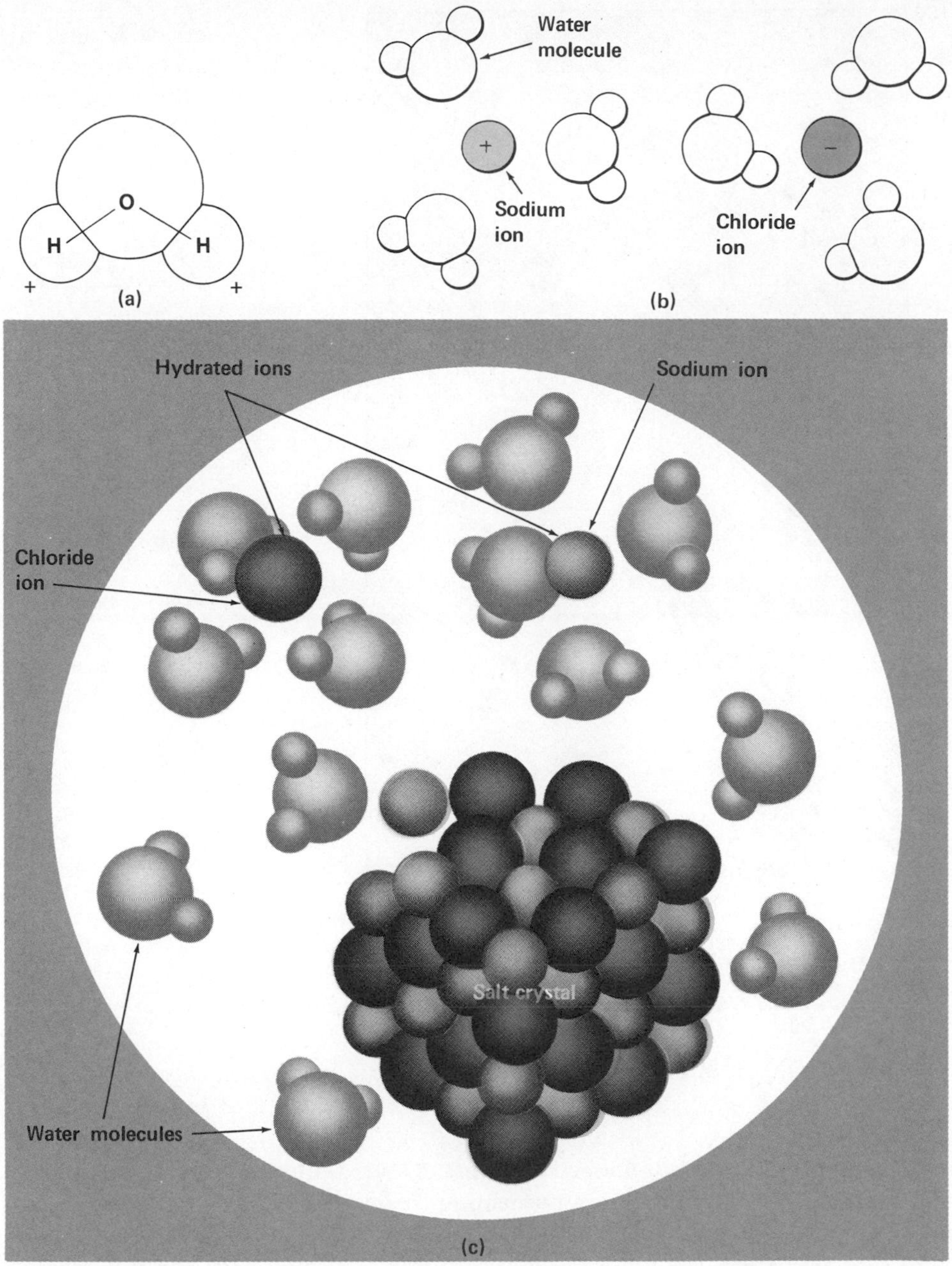

**Figure 6–7** Dissolution of sodium chloride in water. (*a*) Geometry of the polar water molecule. (*b*) Solvation of sodium and chloride ions due to interaction (bonding) between these ions and water molecules. (*c*) Dissolution occurs as collisions between water molecules and crystal ions result in the removal of the crystal ion. In the process the ion becomes completely solvated.

Every salt has what may be termed a "solubility limit" for a given solvent: at a given temperature, a certain number of grams of salt, and *no more,** will dissolve in a certain quantity of solvent. A solution that contains all the dissolved salt that it can hold is termed a ***saturated solution.*** One might ask just why this type of solubility limit is found in nature. The reason for this can be understood if we remember that the oppositely charged ions of the salt solution actually attract one another. If one

**Generally, for a given solvent, a salt will dissolve to a greater extent in the hot solvent than in the cold solvent.**

*Supersaturated solutions can be formed under special conditions, but are not stable in the presence of the solid solute. The presence of the solid solute causes the excess dissolved solute to crystallize and a saturated solution is formed.

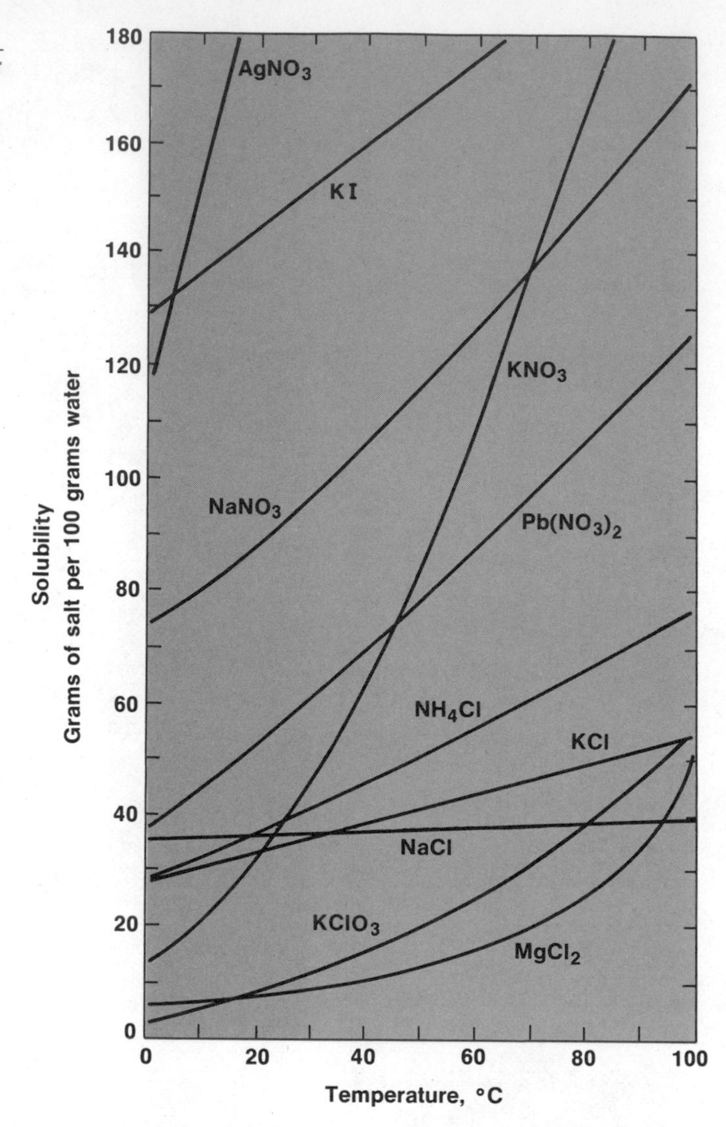


**Figure 6–8** The effect of temperature on the solubility of some common salts in water.

crowds the solution with solvated ions to too great an extent and "ties up too many solvent molecules," the ions will begin re-forming the crystal lattice. This is ***crystallization,*** or solvation in reverse. When undissolved salt is in contact with a saturated solution of that salt, there is a dynamic equilibrium established with the salt crystal being broken down at one point while being formed at another. The effect of temperature on the solubility of some common salts is shown in Figure 6–8.

## Uses of Salts

Salts are used for so many different applications that it would be impossible to list them all. We can, however, look at the uses of several typical salts and see just how versatile they are.

**Many body fluids contain potassium ions ($K^+$) in greater concentration than $Na^+$ ions.**

### Sodium Chloride (Table Salt)

This is the best-known salt and perhaps the most important to us. Physiologically, mammals need this salt for an ionic balance in the bloodstream and for the production of hydrochloric acid in the stomach to aid in digestion of food. Very early in our history, sodium chloride was used in barter, and wars have actually been fought for control of salt deposits. Sodium chloride is commonly used for medicinal purposes, such as in water solution, to wash wounds and to treat burns.

## Nitrate Salts

Salts containing the nitrate ion ($NO_3^-$), such as sodium nitrate ($NaNO_3$), potassium nitrate ($KNO_3$), and ammonium nitrate ($NH_4NO_3$), find special uses in fertilizers and in the manufacture of explosives. Sodium nitrate and potassium nitrate occur in natural deposits, while ammonium nitrate can be prepared by a reaction involving ammonia ($NH_3$) and nitric acid.

**Ammonium nitrate is the number eleven commercial chemical. See inside the front cover.**

$$\underset{\text{BASE}}{NH_3} + \underset{\text{ACID}}{HNO_3} \rightleftarrows \underset{\substack{\text{SALT} \\ \text{(AMMONIUM NITRATE)}}}{NH_4NO_3}$$

Explosions are possible when nitrate salts and combustible materials are mixed. Gunpowder contains $KNO_3$. Ammonium nitrate will explode by itself under certain conditions, and a number of disasters have occurred when large amounts of this salt have exploded in warehouses and in the holds of ships in harbors. Under more controlled conditions, $NH_4NO_3$ decomposes to yield dinitrogen oxide ($N_2O$) and water. Dinitrogen oxide is called laughing gas and is used as an anesthetic in certain forms of surgery.

**Gunpowder contains potassium nitrate ($KNO_3$), sulfur, and charcoal.**

$$NH_4NO_3 \xrightarrow{\text{heat}} \underset{\substack{\text{DINITROGEN} \\ \text{OXIDE} \\ \text{(LAUGHING GAS)}}}{N_2O} + 2H_2O$$

This gas also has a limited solubility in milk and cream and finds a use as a propellent in canned whipped cream. When the valve is opened, the cream and gas escape together, and upon reaching the atmosphere the gas expands, thus "whipping" the cream.

## Sodium Hydrogen Carbonate

Sodium hydrogen carbonate, also called sodium bicarbonate ($NaHCO_3$), has the common name baking soda. The negative ion of the salt, the hydrogen carbonate ion, is a base and is capable of reacting with water in an acid-base reaction.

$$\underset{\text{BASE}}{HCO_3^-} + \underset{\text{ACID}}{H_2O} \rightleftharpoons \underset{\text{ACID}}{H_2CO_3} + \underset{\text{BASE}}{OH^-}$$

When this salt is placed in a mixture containing water, such as biscuit batter, this reaction takes place to a slight extent. The product of the reaction, carbonic acid ($H_2CO_3$), is quite unstable and decomposes under the influence of heat, yielding the products carbon dioxide ($CO_2$) and water. The decomposition of the $H_2CO_3$ causes the reaction of bicarbonate with water to take place to a greater extent. In effect, the equilibrium is shifted to the products side, and the overall reaction is essentially a complete conversion of the bicarbonate ion to carbon dioxide and water. The reaction is

$$HCO_3^- + H_2O \longrightarrow H_2CO_3 + OH^-$$

$$H_2CO_3 \xrightarrow{\text{heat}} H_2O + \underset{\text{(GAS)}}{CO_2}$$

The $CO_2$ gas produced in the reaction causes the biscuit batter to rise. To hasten these reactions, pioneer women often added sour milk or buttermilk, which contain lactic acid, a weak acid. Baking powder contains baking soda along with a weak acid that shifts the equilibrium to the right. Self-rising flour contains a small amount of baking soda and a weak acid to produce a controlled amount of $CO_2$ gas as the mixture is heated in the oven.

The $HCO_3^-$ ion can also act as a weak acid in the presence of a strong base:

$$\underset{\text{ACID}}{HCO_3^-} + \underset{\text{BASE}}{OH^-} \rightleftharpoons \underset{\text{ACID}}{HOH} + \underset{\text{BASE}}{CO_3^{2-}}$$

**The bicarbonate ion is amphiprotic.**

This explains its use as a neutralizer of spilled acids and bases in the laboratory.

## Barium Sulfate

This salt may be prepared by taking advantage of its insolubility in water. If a solution of barium chloride ($BaCl_2$) is mixed with sodium sulfate solution ($Na_2SO_4$), insoluble $BaSO_4$ precipitates:

$$(Ba^{2+} + 2Cl^-) + (2Na^+ + SO_4^{2-}) \longrightarrow BaSO_4\ (\text{solid}) + (2Na^+ + 2Cl^-)$$

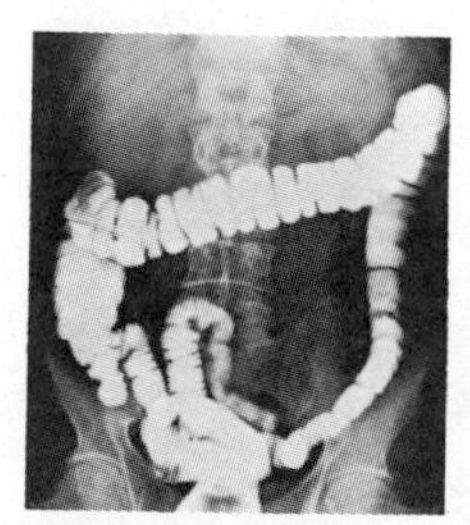

**Figure 6–9** $BaSO_4$ is used in intestinal X-ray studies to minimize the danger of barium poisoning, since the very insoluble salt passes through the body unaltered.

In the diagnosis of stomach ulcers and intestinal disorders X rays are often used. Since these organs are normally transparent to X rays, a suitable material must be placed in the organ to render it partially opaque. Barium is a good element for this purpose, since it is essentially opaque toward the type of X rays used in medicine (Fig. 6–9). Barium ions, however, are quite poisonous and cannot be taken internally. The solution to this problem lies in the extreme insolubility of $BaSO_4$. A suspension of solid $BaSO_4$ is swallowed to introduce barium into the digestive system; since the barium is in an insoluble form, it cannot be absorbed by the body.

## Self-Test 6-B

1. Complete the matching set.

   **pH**
   10 a. acidic
   7 b. basic
   3 c. neutral
2. If the solubility of KI is 140 g per 100 g of water at 20°C, how much KI will dissolve in 1000 g of water? ____________
3. Write a chemical reaction showing water to be an acid. Use ammonia ($NH_3$) as the base.

   $H_2O + NH_3 \longrightarrow$ ____________ + ____________
4. Write a chemical reaction showing water to be a base. Use HCl as the acid.

   $H_2O + HCl \longrightarrow$ ____________ + ____________
5. Which is more acidic, a pH of 6 or a pH of 2? ____________
6. Is a pH of 8 more basic than that of water? ____________
7. The salt barium sulfate poses no threat of barium toxicity because it is ____________ in the stomach and intestinal fluids.
8. When sodium bicarbonate reacts with an acid, the gas ____________ is evolved.
9. Name two uses of nitrate salts. ____________ and ____________

## Electron Transfer (Oxidation-Reduction)



Have you ever watched iron rust or a battery operate? Not much seems to happen. In fact, it's really not very exciting—until you translate your thought to the submicroscopic level where the ever-busy electrons scurry about from one species to another in a mind-boggling frenzy of activity. On the macroscopic level of observation, the evidence for an exchange of electrons is a change in charges on the various species. For example, the iron atoms in shiny iron metal have no residual charge. As a first step in changing iron metal into rust, neutral iron atoms become iron ions with a charge of plus two ($Fe^{2+}$). Each iron atom loses two electrons. The process is called ***oxidation.***

**The loss of electrons is called *oxidation*.**

Oxidation (loss of electrons): $Fe \longrightarrow Fe^{2+} + 2e^-$

Where do these electrons go? They are not stockpiled on the iron, or else they would build up a detectable charge on the rusty iron. Rather, some other species takes the electrons and completes the process. Oxygen is a good candidate. Each oxygen atom will accept two electrons to complete the pairing of its valence electrons, and in the process becomes an oxygen ion with a minus two charge ($O^{2-}$). The process is called ***reduction.***

**The gain of electrons is called *reduction*.**

Reduction (gain of electrons): $O_2 + 4e^- \longrightarrow 2O^{2-}$

In nature, when one species takes electrons, another species has to give the electrons. Or conversely, if one species gives electrons, another species is always there to take the electrons. In an oxidation-reduction reaction, called ***redox*** for short, electrons are conserved. Recall that the formation of an ionic bond is a give-and-take of electrons and is a general example of a redox reaction.

Other examples include:

$2Fe + 3Br_2 \longrightarrow 2FeBr_3 (Fe^{3+} \text{ and } Br^-)$

$2Ca + O_2 \longrightarrow 2CaO\ (Ca^{2+} \text{ and } O^{2-})$

$2Na + Cl_2 \longrightarrow 2NaCl\ (Na^+ \text{ and } Cl^-)$

If the atom has either lost electrons or lost some control over the electrons, we say it has been ***oxidized.*** If it has gained electrons or gained a greater degree of control over the electrons, we say it has been ***reduced.*** In the reactions above, the sodium, hydrogen, and iron atoms have all lost electrons, so these elements have been oxidized. The species that have caused this change ($Cl_2$, $O_2$, and $Br_2$) are called ***oxidizing agents.*** Since these molecules have picked up electrons, they have been reduced, and the species that caused this change (Na, Ca, and Fe) are called ***reducing agents.*** In oxidation-reduction reactions, oxidizing agents become reduced and reducing agents become oxidized.

The term oxidation has its historical background in the fact that the first oxidizing agent whose chemical behavior was thoroughly studied was oxygen. The phenomenon was then named after the element. As the understanding of chemical reactions deepened, it became apparent that the reaction of a metal with oxygen was very similar to its reaction with fluorine, chlorine, or bromine. Thus, in each of the following reactions, the iron atom loses electrons to the other reactant:

**Oxidizing agents gain electrons at the expense of reducing agents.**

$4Fe + 3O_2 \longrightarrow 2Fe_2O_3\ (Fe^{3+} \text{ and } O^{2-})$

$2Fe + 3F_2 \longrightarrow 2FeF_3\ (Fe^{3+} \text{ and } F^-)$

$2Fe + 3Cl_2 \longrightarrow 2FeCl_3\ (Fe^{3+} \text{ and } Cl^-)$

**TABLE 6–2 Some Methods Used to Separate Metals from Their Ores by Reduction**

| Metal | Occurrence | A Reduction Process | Uses of Metal |
|---|---|---|---|
| Cu | $Cu_2S$, chalcocite | air blown through melted ore $Cu_2S + O_2 \longrightarrow 2Cu + SO_2\uparrow$ $(Cu^+ + e^- \longrightarrow Cu)$ | electrical wiring, boilers, pipes, brass (Cu 85%, Zn), bronze (Cu 90%, Sn, Zn), other alloys |
| Na | NaCl, rock salt | electrolysis of fused chloride $2NaCl \longrightarrow 2Na + Cl_2$ $(Na^+ + e^- \longrightarrow Na)$ | coolant in nuclear reactors, orange street lights, making ethyl gasoline |
| Mg | $Mg^{2+}$, sea water | electrolysis of fused chloride $MgCl_2 \longrightarrow Mg + Cl_2$ $(Mg^{2+} + 2e^- \longrightarrow Mg)$ | light alloys such as duralumin (0.5% Mg, Al), Dowmetal H (90.7% Mg, Al, Zn, Mn), flares, some flash bulbs |
| Ca | $CaCO_3$, chalk, limestone, marble | thermal reduction of chloride obtained from carbonate $3CaCl_2 + 2Al \xrightarrow{heat} 3Ca + 2AlCl_3$ $(Ca^{2+} + 2e^- \longrightarrow Ca)$ | bearing metal alloys (0.7% Ca, Pb, Na, Li), storage battery electrodes |
| Al | $Al_2O_3 \cdot H_2O$, bauxite | electrolysis in fused cryolite, $Na_3AlF_6$, at 800–900°C $2Al_2O_3 \longrightarrow 4Al + 3O_2$ $(Al^{3+} + 3e^- \longrightarrow Al)$ | packaging, airplane and automobile parts, alloys, roofing, siding |
| Ag | $Ag_2S$, argentite | reduction by a more active metal $2Ag(CN)_2^- + Zn \longrightarrow Zn(CN)_4^{2-} + 2Ag\downarrow$ $(Ag^+ + e^- \longrightarrow Ag)$ | jewelry, tableware, plating, electrical wiring, photography |
| Zn | ZnS, zinc blende | roasting in air with carbon $ZnS + 2O_2 + C \longrightarrow Zn + SO_2\uparrow + CO_2\uparrow$ $(Zn^{2+} + 2e^- \longrightarrow Zn)$ | galvanizing iron and steel, die-cast auto parts |
| Hg | HgS, cinnabar | roasting in air $HgS + O_2 \xrightarrow{heat} Hg + SO_2\uparrow$ $(Hg^{2+} + 2e^- \longrightarrow Hg)$ | some thermometers, electrical switches, blue-green street lights, amalgams in dentistry, fluorescent lighting |
| Fe | $Fe_2O_3$, hematite | reduction by carbon monoxide $Fe_2O_3 + 3CO \longrightarrow 3CO_2 + 2Fe$ $(Fe^{3+} + 3e^- \longrightarrow Fe)$ | as cast and forged iron, alloyed with C (0.1–1.5%) to make steels (stainless steel has 8% or more Cr) |

After this similarity was noted, the concept of oxidation was generalized to cover all situations in which an atom, ion, or molecule lost electrons. In the same manner the concept of reduction is now used to cover all situations in which an atom, ion, or molecule gains electrons.

**Here, chlorine is the oxidizing agent and gets reduced; sodium is the reducing agent and gets oxidized.**

The formation of a salt by the direct combination of the elements is an example of an oxidation-reduction reaction. When sodium reacts with chlorine to form sodium chloride, the following reactions take place.

$$Na \longrightarrow Na^+ + e^- \qquad \text{(OXIDATION OF SODIUM)}$$

$$Cl_2 + 2e^- \longrightarrow 2Cl^- \qquad \text{(REDUCTION OF CHLORINE)}$$

Since two electrons are gained per chlorine molecule, two sodium atoms must be oxidized for every $Cl_2$ molecule reduced; the overall chemical reaction is

$$2Na + Cl_2 \longrightarrow 2Na^+ + 2Cl^-$$

## Reduction of Metals

**The reduction of metals for making tools is as old as recorded history.**

One of the most practical applications of the oxidation-reduction principle is the separation of metals from their ores. The majority of metals are found in nature in compounds; that is, the metals are in an oxidized state. In order to obtain the metal,

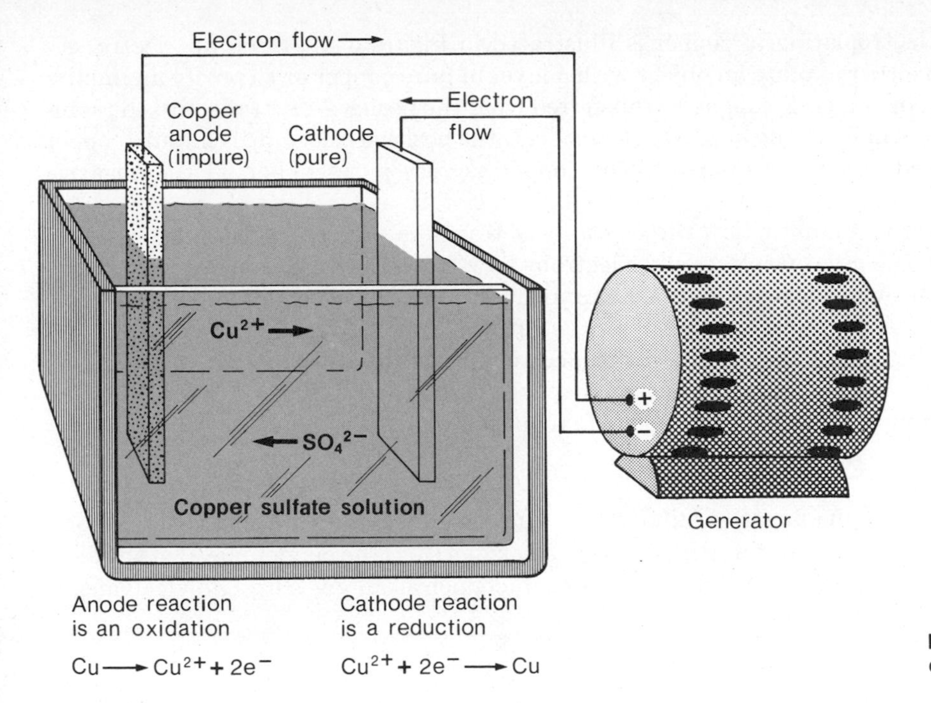


**Figure 6–10** Electroplating from a copper sulfate solution.

it must be reduced from the positive oxidation state to the neutral elemental state. This requires a gain of electrons. In addition to electricity, there are a variety of chemicals that can supply the electrons for the reduction process. Some of the ways that metals can be reduced from their ores are given in Table 6–2.

## Electrolysis

Several metals either are separated from their ores or are purified afterward by electrolysis, as noted in Table 6–2. ***Electrolysis*** is a type of chemical reaction caused by the application of electrical energy.

The suffix *-lysis* means splitting or decomposition; electrolysis is decomposition by electricity.

The principal parts of an electrolysis apparatus are shown in Figure 6–10. Electrical contact between the external circuit and the solution is obtained by means of ***electrodes,*** which are often made of graphite or metal. The electrode at which electrons enter an electrolysis cell is termed the ***cathode,*** and this is the electrode at which reduction takes place. The electrode at which the electrons leave the cell is the ***anode.*** At the anode oxidation takes place. The battery or generator produces a current of electrons that flow toward one electrode (the cathode) and away from the other electrode (the anode). A salt solution can easily be observed to conduct electricity if a light bulb is part of the circuit (Fig. 6–2). When the switch is closed, the positive ions in solution migrate toward the cathode. Soon, evidence of a chemical reaction can be seen at the electrodes. Depending on the substances present in the solution, gases may be evolved, metals deposited, or ionic species changed at the electrodes. The ions that migrate to the electrodes are not necessarily the species undergoing reaction at the electrodes, because sometimes the solvent undergoes reaction more easily. Whatever happens, the chemical reactions that are taking place at the cathode and anode are due to electrons going into and coming out of the solution. The chemical reaction at the cathode is one that furnishes electrons to solution species (reduction). At the anode, electrons are taken from species in solution, so the chemical reaction at the anode is one that gives up electrons (oxidation).

The electroplating of copper is illustrated in Figure 6–10. Such an electrolysis can be used either to plate an object with a layer of pure copper or to purify an impure sample of copper metal; copper is transferred from the positive electrode into the solution and eventually to the negative electrode. If the positive electrode is impure copper to be purified, electrolysis deposits the copper as very pure copper on the negative electrode.

Now let us examine how the electrolysis transfers the copper from the positive electrode to the negative electrode. Electrons flow out of the negative terminal of the generator through the wire, and are pumped into the negative electrode. Somehow this negative charge must be used up at the surface of the electrode.

Consider what happens when the electrons build up on the negative electrode. Since positive copper ions are nearby, they will be attracted to the surface and take electrons from it. Thus, the $Cu^{2+}$ ions are reduced:

**In any electrochemical cell, oxidation occurs at the anode; reduction occurs at the cathode.**

$Cu^{2+} + 2e^- \longrightarrow Cu$ (CATHODE REACTION)

In a similar way the negative sulfate ions migrate to the positive electrode (anode). However, it turns out that it is easier to get electrons from the copper metal of the electrode than it is from the sulfate ions. As each copper atom gives up two electrons, it passes into solution:

$Cu \longrightarrow Cu^{2+} + 2e^-$ (ANODE REACTION)

In effect, then, the copper of the positive electrode (anode) passes into solution; the copper ions in solution migrate to the negative electrode (cathode) and plate out as copper metal. Large amounts of copper are purified in this way each year. Silver and gold purification can be carried out similarly.

**Copper can be plated onto an object by making that object the negative electrode in a cell containing dissolved copper salts.**

If we desire to plate an object with copper, we have only to render the surface conducting and make it the negative electrode in a solution of copper sulfate. It will become coated with copper, with the copper coating growing thicker as the electrolysis is continued. If the object is a metal, it will conduct electricity by itself. If it is a nonmetal, its surface can be lightly dusted with graphite powder to render it conducting.

A potentially very important electrolysis reaction is the electrolysis of water. When electricity is passed into graphite electrodes immersed in a dilute salt solution, water is reduced to hydrogen and hydroxide ions at the cathode:

$2H_2O + 2e^- \longrightarrow H_2\text{ (gas)} + 2OH^-$ (CATHODE REACTION)

At the anode, water is oxidized to oxygen and hydrogen ions:

$2H_2O \longrightarrow O_2 + 4H^+ + 4e^-$ (ANODE REACTION)

The $OH^-$ and $H^+$ ions combine to re-form water. The overall, or net, cell reaction is:

$2H_2O \xrightarrow{\text{electricity}} 2H_2\text{ (gas)} + O_2\text{ (gas)}$

The hydrogen produced by the reduction of water can be stored and used as a fuel—for example, to power rockets into space. Someday, if electricity becomes low enough in cost (see Chapter 14), water may be electrolyzed to produce hydrogen that can then be piped to the point of use, just as natural gas is today.

## Relative Strengths of Oxidizing and Reducing Agents

When a piece of metallic zinc is placed in a solution containing hydrated copper ions ($Cu^{2+}$), an oxidation-reduction reaction occurs:

$Zn + Cu^{2+} \longrightarrow Zn^{2+} + Cu$

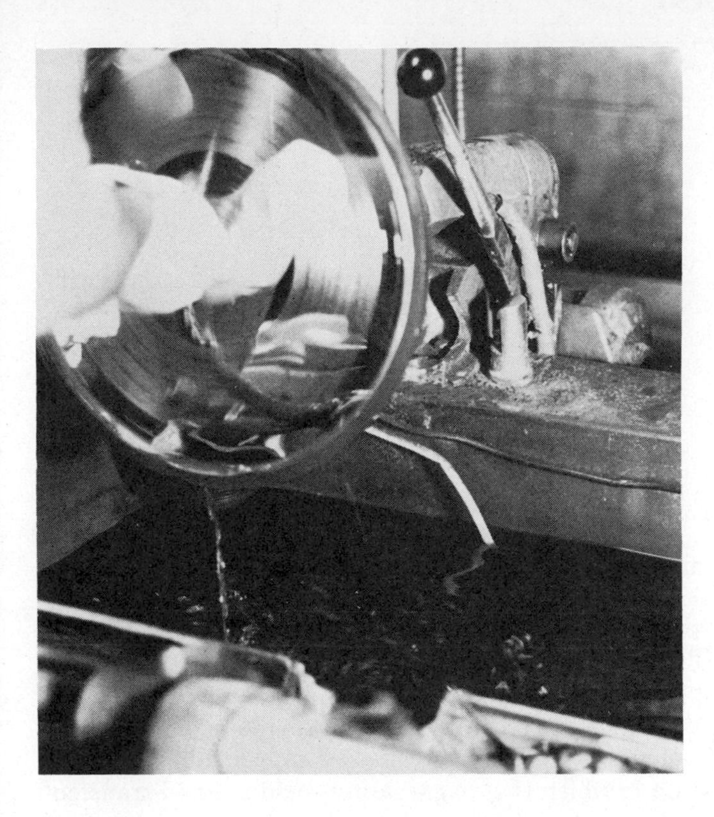

**Figure 6–11** Stamping masters for making high quality phonograph records are made by electroplating nickel onto a plastic record finely coated with silver.

Evidence for this reaction is the deposit of copper on the zinc and the gradual decrease in the intensity of the blue color of the solution, indicating that the $Cu^{2+}$ ions are being removed.

The oxidation of zinc by copper ions can be thought of as a competition between zinc ions ($Zn^{2+}$) and copper ions ($Cu^{2+}$) for the two electrons. Since the reaction proceeds almost to completion, the $Cu^{2+}$ ions obviously win out in the competition. Other metals can compete similarly for electrons.

The ***activity*** of a metal is a measure of its tendency to lose electrons. Zinc is a more active metal than copper on the basis of the experiment just described. This means that given an equal opportunity, the first reaction will take place to a greater extent:

**A copper atom has a greater attraction for electrons than does a zinc atom.**

$$Zn \longrightarrow Zn^{2+} + 2e^-$$

$$Cu \longrightarrow Cu^{2+} + 2e^-$$

Experiments of this type with various pairs of metals and other reducing agents yield an ***activity series*** of the elements, which ranks each oxidizing and reducing agent according to its *strength* or *tendency* for the electron transfer to take place. An iron nail will be partly dissolved in a solution of a copper salt containing $Cu^{2+}$ ions, with copper being deposited on the nail that remains. From this, it is determined that iron, like zinc, is more active than copper. The reaction that takes place is

**The active metals lose electrons more easily; hence, these free metals are not found in nature.**

$$Fe + Cu^{2+} \longrightarrow Fe^{2+} + Cu$$

Now, which is more active, zinc or iron? This question can be answered by placing an iron nail in a solution containing $Zn^{2+}$ ions and, in a separate container, a strip of zinc in a solution containing $Fe^{2+}$ ions. It is found that the zinc strip is eaten away in the solution containing $Fe^{2+}$ ions. The reaction is

$$Zn + Fe^{2+} \longrightarrow Fe + Zn^{2+}$$

**TABLE 6–3 Relative Strengths of Some Oxidizing and Reducing Agents: The Activity Series**

| Oxidizing Agents (Increasing Strength of Oxidizing Agent ↓) | | Reducing Agents (Increasing Strength of Reducing Agent ↑) |
|---|---|---|
| $Na^+ + e^-$ | $\rightleftharpoons$ | Na |
| $Ca^{2+} + 2e^-$ | $\rightleftharpoons$ | Ca |
| $Mg^{2+} + 2e^-$ | $\rightleftharpoons$ | Mg |
| $Zn^{2+} + 2e^-$ | $\rightleftharpoons$ | Zn |
| $Fe^{2+} + 2e^-$ | $\rightleftharpoons$ | Fe |
| $2H^+ + 2e^-$ | $\rightleftharpoons$ | $H_2$ |
| $Cu^{2+} + 2e^-$ | $\rightleftharpoons$ | Cu |
| $Fe^{3+} + e^-$ | $\rightleftharpoons$ | $Fe^{2+}$ |
| $Ag^+ + e^-$ | $\rightleftharpoons$ | Ag |
| $O_2 + 4e^- + 4H^+$ | $\rightleftharpoons$ | $2H_2O$ |
| $F_2 + 2e^-$ | $\rightleftharpoons$ | $2F^-$ |

Nothing happens to the iron nail in the solution of $Zn^{2+}$ ions. We deduce that Zn loses electrons more readily than Fe.

Such an activity series can be extended to include other metals and even nonmetals as well. The concentrations of the ions in solution and other factors often must be considered for accurate work, but for our purposes these will be ignored. Table 6–3 is an activity series of some oxidizing and reducing agents.

Activity: Zn > Fe > Cu

An application of the activity series can be seen in the use of ***cathodic protection*** to reduce corrosion (Fig. 6–12). The metal to be protected, say an iron pipeline, is connected to a more active metal, such as magnesium, by a heavy copper cable. In moist earth, an electrolytic cell is set up in which the magnesium is oxidized and transfers electrons to the iron. This keeps the iron from being oxidized by the air or any other reagent that might otherwise remove electrons from it.

The value of the activity series is that it allows us to predict the feasibility of reactions involving the species in it. Thus, a reducing agent in the table (right-hand column) is able to reduce the oxidized form of any species below it. For example, magnesium can reduce $Cu^{2+}$ to Cu:

$$Mg + Cu^{2+} \longrightarrow Cu + Mg^{2+}$$

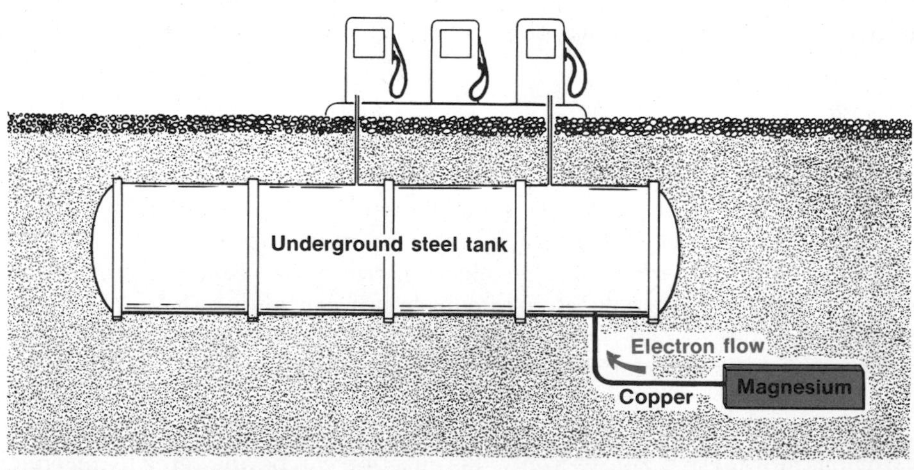


**Figure 6–12** Cathodic protection. If magnesium is connected to the steel tank to be protected, the magnesium is more easily oxidized than the iron or copper connecting wire. The magnesium serves as a sacrificial anode. Hence, the cathode is protected with no points of oxidation occurring on its surface. The anode is the electrode where oxidation occurs; reduction occurs at the cathode. When the magnesium is used up, it is replaced by another block. The replacement is much easier and cheaper than replacing the tank.

Magnesium can also reduce $Ag^+$ to Ag:

$Mg + 2Ag^+ \longrightarrow Mg^{2+} + 2Ag$

Zinc can also reduce silver ions:

$Zn + 2Ag^+ \longrightarrow 2Ag + Zn^{2+}$

Zinc cannot reduce calcium ions, since calcium is above zinc in the series:

$Zn + Ca^{2+} \not\longrightarrow$ *No reaction*

The series also arranges oxidizing agents in the order of their effectiveness. Fluorine, $F_2$, can oxidize water, silver, iron (II) ion, or any species above it in the series; copper (II) ion, $Cu^{2+}$, can oxidize $H_2$, Fe, Zn, and the metals above Cu in the table, because it has a greater tendency to take on electrons than the ions that are formed. Thus:

The activity series predicts possible single replacement reactions (See p. 101).

$Cu^{2+} + Fe \longrightarrow Fe^{2+} + Cu$

$Cu^{2+} + Mg \longrightarrow Mg^{2+} + Cu$

$Cu^{2+} + Ag \longrightarrow$ *No reaction*

## Self-Test 6-C

1. In a redox reaction, the oxidizing agent (gains/loses) electrons and gets (reduced/oxidized); the reducing agent (gains/loses) electrons and gets (reduced/oxidized).
2. Consider the reaction:
   $Ca + 2Ag^+ \longrightarrow Ca^{2+} + 2Ag$
   Calcium is the (oxidizing/reducing) agent. Silver ions ($Ag^+$) are the (oxidizing/reducing) agent.
3. When silver-plated dinnerware is made, silver is deposited by the reaction $[Ag(CN)_2]^- + e^- \longrightarrow Ag + 2CN^-$. The silver is (oxidized/reduced). This reaction takes place at the (cathode/anode) of an electrolysis cell.
4. Consider the activity series in Table 6–3, and write yes or no beside the following reactions, depending on whether or not they will proceed as written.

   $2Ag + Cu^{2+} \longrightarrow Cu + 2Ag^+$ ____________

   $F_2 + Zn \longrightarrow Zn^{2+} + 2F^-$ ____________

   $Fe^{2+} + Ag^+ \longrightarrow Fe^{3+} + Ag$ ____________

   $Mg^{2+} + 2Fe^{2+} \longrightarrow 2Fe^{3+} + Mg$ ____________
5. When electrons are written on the left-hand side of an equation, as in $Cu^{2+} + 2e^- \longrightarrow Cu$, the reaction is a(n) (oxidation/reduction).
6. When electrons are written on the right-hand side of an equation, as in $Zn \longrightarrow Zn^{2+} + 2e^-$, the reaction is a(n) (oxidation/reduction).

### Batteries

One of the most useful applications of oxidation-reduction reactions is in the production of electrical energy. A device that produces an electron flow (current) is called an ***electrochemical cell.*** Although a series of such cells is a ***battery,*** the term battery is commonly used even for single cells such as those we will describe.

Consider the reaction between zinc and copper ions that was previously discussed. If zinc is placed in a solution containing $Cu^{2+}$ ions, the electron transfer takes place

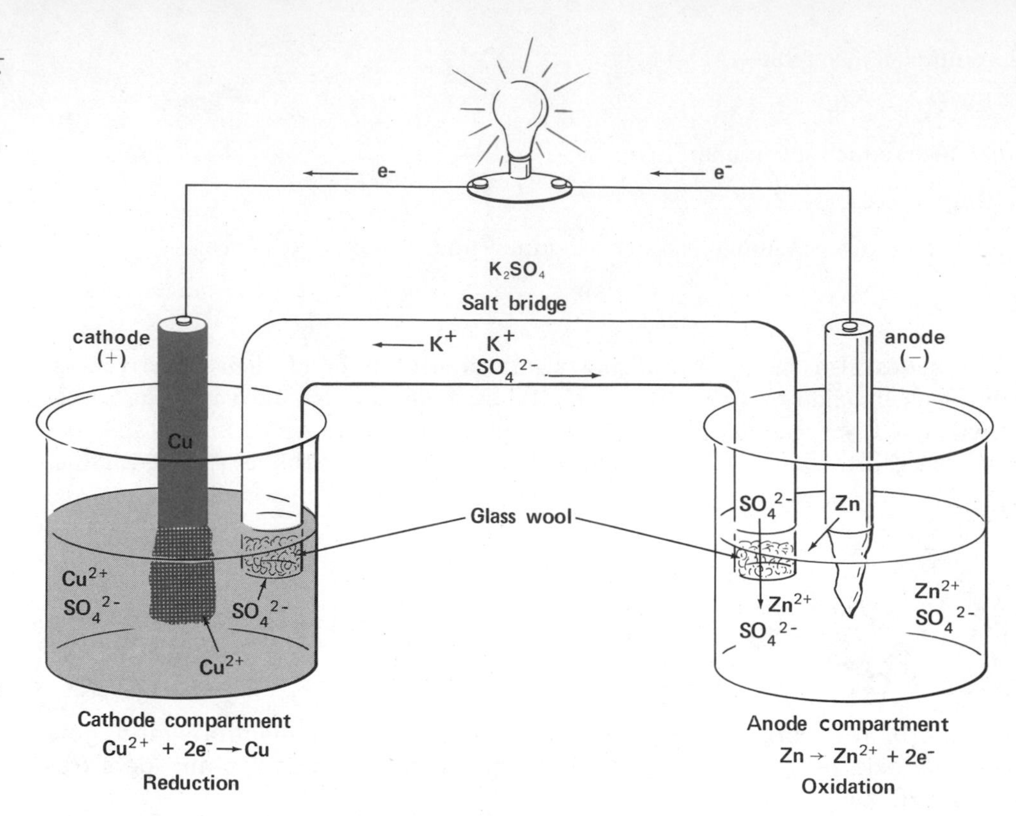


**Figure 6–13** A simple battery involving the oxidation of zinc metal and the reduction of $Cu^{2+}$ ions.

between the zinc metal and the copper ions, and the energy liberated simply causes a slight heating of the solution and the zinc strip. If the zinc could be separated from the copper solution, and the two connected in such a way to allow current flow, the reaction can proceed, but now the electrons are transferred through the connecting wires. Figure 6–13 shows a battery that can be constructed to make use of the energy evolved in the reaction of Zn with $Cu^{2+}$.

The anode reaction is the oxidation of zinc to $Zn^{2+}$ ions.

$$Zn \longrightarrow Zn^{2+} + 2e^-$$

**Historic battery used in the early telegraph.**

The electrons flow from the Zn electrode through the connecting wire, light the lamp in the circuit, and then flow into the copper cathode where reduction of $Cu^{2+}$ ions takes place:

$$Cu^{2+} + 2e^- \longrightarrow Cu$$

The copper is deposited on the copper cathode.

**In commercial batteries, the salt bridge is often replaced by a porous membrane.**

This flow of electrons (negative charge) from the anode to the cathode compartment in the battery must be neutralized electrically. In order to accomplish this, a "salt bridge" is provided that connects the two compartments. The salt bridge contains a solution of a salt such as $K_2SO_4$. The purpose of the salt bridge is to keep the two solutions neutral. Around the cathode the deposition of positive copper ions ($Cu^{2+}$) would tend to cause the solution to become negative owing to the presence of excess negative sulfate ions ($SO_4^{2-}$). Two actions can keep the solution around the cathode neutral: either positive potassium ions ($K^+$) pass into the solution, or negative sulfate ions pass out of the solution and into the salt bridge. Actually, both processes occur. Similarly, around the anode, the solution would tend to become positive because positive zinc ions ($Zn^{2+}$) are put into the solution. Two actions can keep the solution

TABLE 6–4 Characteristics of Some Batteries

| System | Anode (Oxidation) | Cathode (Reduction) | Electrolyte | Typical Operating Voltage Per Cell |
|---|---|---|---|---|
| Dry cell | Zn | $MnO_2$ | $NH_4Cl$-$ZnCl_2$ | 0.9–1.4 |
| Edison storage | Fe | Ni oxides | KOH | 1.2–1.4 |
| Nickel-cadmium–NiCad | Cd | Ni oxides | KOH | 1.1–1.3 |
| Silver cell | Cd | $Ag_2O$ | KOH | 1.0–1.1 |
| Lead storage | Pb | $PbO_2$ | $H_2SO_4$ | 1.95–2.05 |
| Mercury cell | Zn(Hg) | HgO | KOH-ZnO | 1.30 |
| Alkaline cell | Zn(Hg) | $MnO_2$ | KOH | 0.9–1.2 |

around the anode neutral: either negative sulfate ions pass into the solution or positive zinc ions pass out of the solution and into the salt bridge. In all of this exchange, it is necessary for the solution to maintain the same number of positive charges as negative charges. The reaction of zinc atoms and copper ions continues until one or the other is completely consumed.

Many different oxidation-reduction combinations are used in commercial batteries to produce a flow of electrons. A few of the more popular ones are listed in Table 6–4.

## Matching Set

| | | |
|---|---|---|
| ______ | 1. chemicals in automobile lead storage battery | **a** 7 |
| ______ | 2. most active metal in Table 6–3 | **b** gain of electrons |
| ______ | 3. ore of mercury | **c** hydrogen ion donor |
| ______ | 4. pH of pure water | **d** Na |
| ______ | 5. solution | **e** hydrogen ion acceptor |
| ______ | 6. strong acid | **f** homogenous mixture of atoms, molecules, and/or ions |
| ______ | 7. oxidation | **g** $HC_2H_3O_2$ |
| ______ | 8. reduction | **h** Pb, $PbO_2$, $H_2SO_4$ |
| ______ | 9. acid definition | **i** causes solution to conduct |
| ______ | 10. base definition | **j** loss of electrons |
| ______ | 11. weak acid | **k** cinnabar |
| ______ | 12. electrolyte | **l** $H_2SO_4$ |

## Questions

**1.** Define acid-base reactions in terms of hydrogen ions, and oxidation-reduction reactions in terms of electrons.

**2.** Indicate the solute and solvent in (a) coffee, (b) a 5% alcohol in water solution, (c) a 5% water in alcohol solution, and (d) a solution of 50% alcohol and 50% water.

**3.** What is the one test that all aqueous electrolytes must pass?

**4.** Describe a test to determine whether a solution is a weak acid or a strong acid.

**5.** Give an example of ionic dissociation. Give an example of ionization. What is the difference between the two?

**6.** What mobile units must be present in electrolytic solutions?

**7.** Describe a test to determine whether a solution is acidic or basic.

**8.** Why can boric acid ($H_3BO_3$) be used in eyewashes while hydrochloric acid (HCl) is not safe to use?

**9.** Distinguish between the hydrogen ion and the hydronium ion.

**10.** Write a neutralization reaction between lye (NaOH) and muriatic acid (HCl).

**11.** Write the equation for a chemical reaction in which water acts as a Brønsted acid; as a Brønsted base.

**12.** A solution of hydrochloric acid is electrolyzed. The products are hydrogen at the cathode and chlorine at the anode. Write the reactions occurring at each electrode and tell which ions move toward the cathode and which toward the anode.

**13.** What is the difference between ionization and dissociation in the production of mobile ions?

**14.** a. What is the pH of a neutral solution? b. Which pH is more acidic, a pH of 5 or a pH of 2? c. Which pH is more basic, a pH of 5 or a pH of 10?

**15.** Classify each of the following as acids or bases, using the Brønsted-Lowry definitions: $H_2SO_4$, $CO_3^{2-}$, $Cl^-$, $HCO_3^-$, $O^{2-}$, $H_2O$.

**16.** Would liquid Ajax be more likely to have a pH greater or less than 7?

**17.** Two solutions contain 1% acid. Solution A has a pH of 4.6, and solution B has a pH of 1.1. Which solution contains the stronger acid?

**18.** Predict the formulas of salts formed with the following pairs of ions:

$Na^+$ and $SO_4^{2-}$
$Ca^{2+}$ and $I^-$
$Mg^{2+}$ and $NO_3^-$
$Ca^{2+}$ and $PO_4^{3-}$
$K^+$ and $Br^-$

**19.** Moist baking soda is often put on acid burns. Why? Write an equation for the reaction assuming the acid to be hydrochloric (HCl).

**20.** Hydrochloric acid is the acid present in the human stomach. Is this a strong or a weak acid?

**21.** What is the pH of a bicarbonate of soda (sodium bicarbonate) solution?

**22.** Using the table that gives relative oxidizing and reducing strengths, predict whether or not the following reactions would be expected to proceed to the right.

$$Ag + Na^+ \rightleftharpoons Ag^+ + Na$$

$$H_2 + Cu^{2+} \rightleftharpoons Cu + 2H^+$$

$$F_2 + 2Fe^{2+} \rightleftharpoons 2Fe^{3+} + 2F^-$$

$$Mg + 2Ag^+ \rightleftharpoons 2Ag + Mg^{2+}$$

**23.** If a drug consumes 37 times its weight in excess stomach acid, the reaction is one called __________ and the drug must be a(n) __________.

**24.** A special case of solvation is the hydration of ions by polar water molecules. What does this suggest about other polar solvents such as ethyl alcohol?

**25.** Describe what happens when an ionic solid dissolves in water.

**26.** Describe vividly the scenario of HCl molecules being added to water. Compare and contrast this scenario with what happens when acetic acid (HAc) molecules are added to water.

**27.** What ions are present in water solutions of the following salts: $Na_2SO_4$, $CaBr_2$, $Mg(NO_3)_2$?

**28.** Give five practical applications of oxidation-reduction.

**29.** Library assignment: Under what conditions can silver be plated onto steel to make fine silverplate?

**30.** Use Fe and $Fe^{3+}$ to write an oxidation reaction and then, separately, a reduction reaction.

**31.** From the information given in Table 6–4, write the oxidation reaction and the reduction reaction for the silver cell. One of the products is $Cd(OH)_2$ and another is silver metal (Ag).

**32.** What is the purpose of the salt bridge in Figure 6–13? Is a salt bridge always necessary in a battery?

**33.** In oxidation-reduction cells, oxidation always occurs at the (anode/cathode), and reduction occurs at the (anode/cathode).

**34.** Is it possible for the air to be "electrified?" What molecules would have to be ionized in this case?

**35.** A mole of ethyl alcohol weighs 46 g. How many grams of ethyl alcohol would be required to make one liter of 1.5 M solution?

**36.** Look at an automobile battery that has a case you can see through. If it is a 12 volt battery, there will be six cells. Why is "battery" an apt use of this word in view of its classic use in a military sense?

**37.** Which has more grams of solute dissolved per liter of solution, a 0.50 M sucrose (molecular weight 342) solution, or a 0.50 M sodium chloride (molecular weight 58.5) solution?

**38.** In a battery circuit, is the direction of electron flow plus to minus, or minus to plus?

**39.** What is the molar (M) concentration of a solution containing 12.0 g of NaCl dissolved in 500 ml of solution?

**40.** If a hydrochloric acid solution is 0.1 M, how many grams of HCl are dissolved in one liter of this solution?

# Useful Materials from the Earth, Sea, and Air

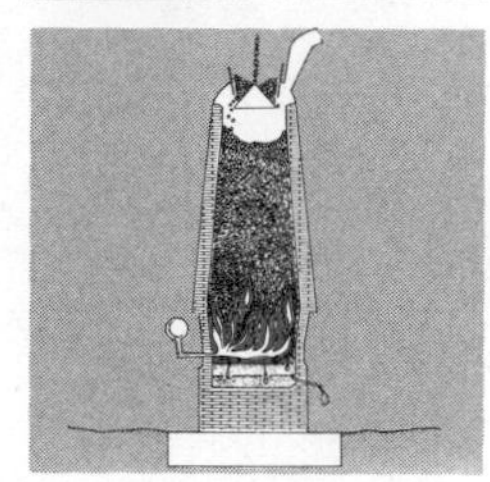

Many of the things we use in everyday life result from chemical reactions that are carried out on an industrial scale. They include iron and steel products; fabricated items of aluminum, magnesium, and copper; glass, medicines, fertilizers, and numerous chemicals used to alter our surroundings. The purpose of this chapter is to discuss some of the chemistry involved in preparing these products from natural resources found in the earth, sea, and air.

***Industrial chemistry*** has taken on some new demands. Not only must it produce a useful arrangement of atoms, but it must also provide for the return of unused atoms to the economy or to nature in a desirable form—that is, the eventual recycling of the atoms. In the process of producing chemicals commercially, energy must not be wasted and pollution must be minimized.

**Recycling decreases our demands on many natural resources.**

## Metals and Their Preparation

Metals occur mostly as compounds in the crust of the earth, though some of the less active metals such as copper, silver, and gold can be found also as free elements. Fortunately, the distribution of elements in the crust is not uniform. Some elements that are not particularly abundant are familiar to us because they tend to occur in very concentrated, localized deposits, called ***ores,*** from which they can be extracted economically. Examples of these are lead, copper, and tin, none of which is among the more abundant elements in the crust of the earth (Fig. 7–1). Other elements that actually form a much larger percentage of the crust are almost unknown to us because concentrated deposits of their ores are less commonly found or the metal is difficult to extract from its ore. An example is titanium, the tenth most abundant element in the crust of the earth. Although the ore of titanium, rutile (mostly $TiO_2$), is common, the use of the metal is rare because it is difficult to reclaim from the ore.

**A continual search is under way for new ore deposits.**

Some common minerals are listed in Table 7–1.

The preparation of metals from their ores, ***metallurgy,*** involves chemical reduction (Chapter 6). Indeed, the concept of oxidation and reduction developed from metallurgical operations. Iron in the ore iron oxide ($Fe_2O_3$) is in the form of $Fe^{3+}$. If we *reduce* $Fe^{3+}$ ions to Fe atoms, we must find a source of electrons. (Recall that oxidation is the loss of electrons and that reduction is the gain of electrons.) Sometimes the desired metal is in solution (e.g., magnesium in the sea), where it exists in the oxidized form ($Mg^{2+}$ ions). To obtain the free metal magnesium we must add electrons to these ions (reduction) to produce neutral atoms.

**Reduction of magnesium:**
$Mg^{2+} + 2e^- \longrightarrow Mg$

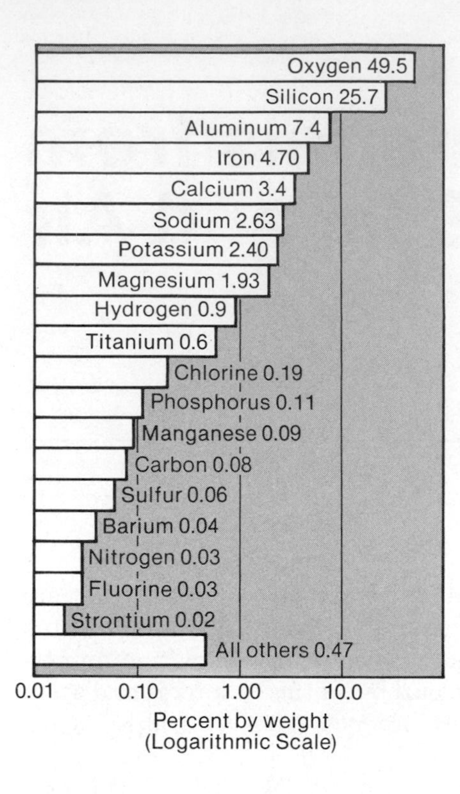


**Figure 7–1** Abundance of the elements in the crust of the earth (percent by weight).

Iron ores are iron compounds. To get iron from the ores, the iron in the compounds must be reduced.

## Iron and Steel

The sources of most of the world's iron are large deposits of the iron oxides in Minnesota, Sweden, France, Venezuela, Russia, Australia, and England. In nature these oxides are frequently mixed with impurities, so the production of iron usually incorporates steps to remove such impurities. Iron ores are then reduced to the metal by using carbon, in the form of coke, as the reducing agent.

The reduction of iron ore is carried out in a blast furnace (Fig. 7–2). The solid material fed into the top of the blast furnace consists of a mixture of an oxide of iron ($Fe_2O_3$), coke (C), and limestone ($CaCO_3$). A blast of heated air is forced into the

**TABLE 7–1 Some Common Metals and Their Minerals**

| Metal | Chemical Formula of Compound of the Element | Name of Mineral |
|---|---|---|
| Aluminum | $Al_2O_3 \cdot xH_2O$ | bauxite |
| Calcium | $CaCO_3$ | limestone |
| Chromium | $FeO \cdot Cr_2O_3$ | chromite |
| Copper | $Cu_2S$ | chalcocite |
| Iron | $Fe_2O_3$ | hematite |
| | $Fe_3O_4$ | magnetite |
| Lead | PbS | galena |
| Manganese | $MnO_2$ | pyrolusite |
| Tin | $SnO_2$ | cassiterite |
| Zinc | ZnS | sphalerite |
| | $ZnCO_3$ | smithsonite |

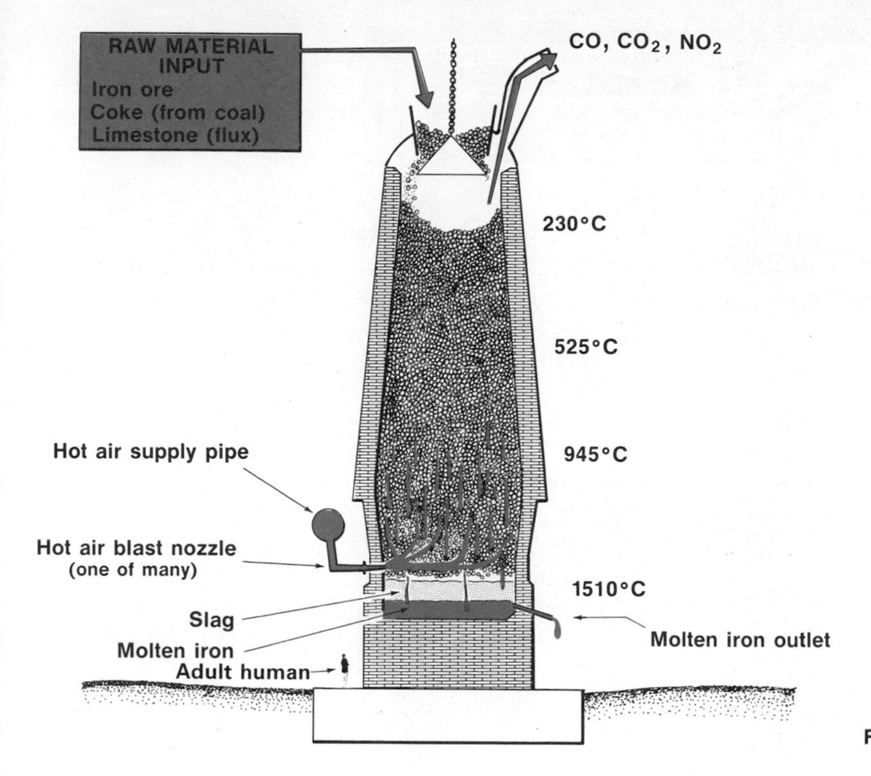


**Figure 7–2** Diagram of a blast furnace.

furnace near the bottom. The heat speeds up the reaction, which is important in making the process economical. The reactions that occur within the blast furnace are

$$2C + O_2 \longrightarrow 2CO + \text{heat}$$

CARBON + OXYGEN → CARBON MONOXIDE

$$Fe_2O_3 + 3CO \longrightarrow 2Fe + 3CO_2 + \text{heat}$$

IRON OXIDE + CARBON MONOXIDE → IRON + CARBON DIOXIDE

Limestone (calcium carbonate) is added to remove the silica ($SiO_2$) impurity.

$$CaCO_3 \xrightarrow{\text{heat}} CaO + CO_2$$

CALCIUM CARBONATE → CALCIUM OXIDE + CARBON DIOXIDE

$$CaO + SiO_2 \longrightarrow CaSiO_3$$

CALCIUM OXIDE + SILICON DIOXIDE → CALCIUM SILICATE

The calcium silicate, or ***slag,*** exists as a liquid in the furnace. Consequently, as the blast furnace operates, two molten layers collect in the bottom. The lower, denser layer is mostly liquid iron that contains a fair amount of dissolved carbon and often smaller amounts of other impurities. The upper, lighter layer is primarily molten calcium silicate with some impurities. From time to time the furnace is tapped at the bottom and the molten iron is drawn off. Another outlet somewhat higher in the blast furnace base can be opened to remove the liquid slag.

As it comes from the blast furnace, the iron contains too much carbon for most uses. If some of the carbon is removed, the mixture becomes structurally stronger

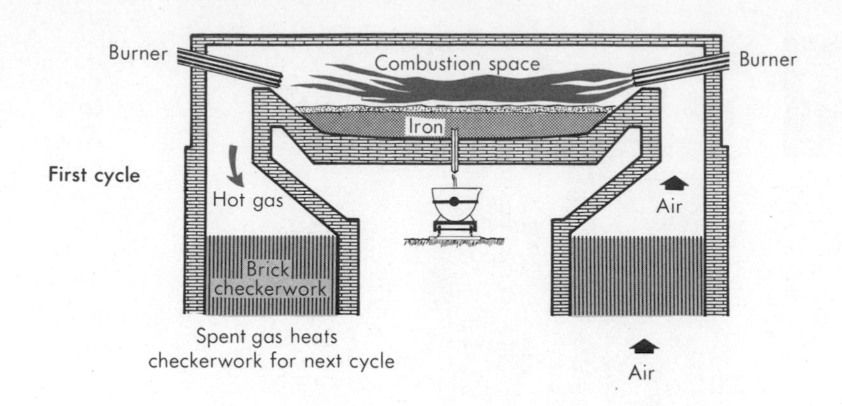


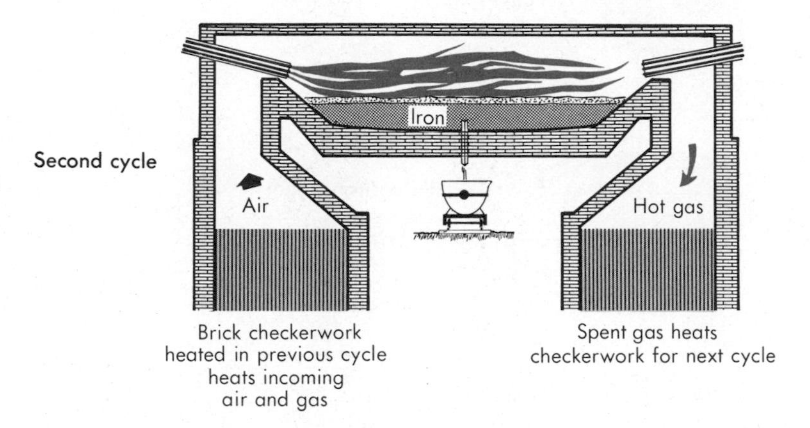


**Figure 7–3** The open hearth furnace for the conversion of iron to steel. Air and fuel are blown in first in one direction and then the other in order that heat, which is costly, might be stored in the brick checkerwork.

**Alloy: a metal consisting of two or more elements.**

and is known as ***steel.*** Steel is an ***alloy*** of iron with a relatively small amount of carbon (less than 1.5%); it may also contain other metals. In order to convert iron into steel, the excess carbon is burned out with oxygen.

There are several techniques for burning the excess carbon (see Figs. 7–3 and 7–4). A recent development that has been very widely adopted is the basic oxygen process (Fig. 7–4). In this process pure oxygen is blown into molten iron through a refractory tube (oxygen gun), which is pushed below the surface of the iron. At elevated temperatures, the dissolved carbon reacts very rapidly with the oxygen to give gaseous carbon monoxide and carbon dioxide, which then escape.

After the carbon content has been reduced to a suitable level, the molten steel is formed into desired shapes. During the processing the steel is subjected to carefully regulated heat treatment to ensure that it has a uniform crystallinity, which in turn determines its pliability, toughness, and other useful mechanical properties.

All of the processes in steelmaking, from the blast furnace to the final heat treatment, use tremendous quantities of energy, mostly in the form of heat. To produce a ton of steel, approximately one ton of coal or its energy equivalent is consumed.

**Several billion dollars worth of valuable materials are destroyed each year by corrosion. A number of techniques are used to reduce this destruction.**

## Corrosion, Rusting, and Stainless Steel

Many metals undergo corrosion when exposed to moist air over a long period of time. The metal is transformed into its oxide or hydroxide. Corrosion of iron is

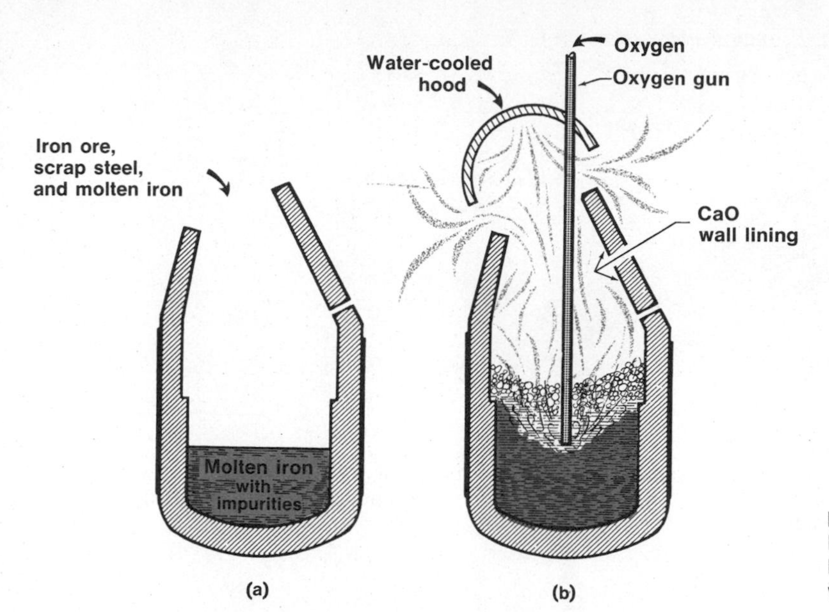


**Figure 7–4** The basic oxygen process furnace. Much of the steel manufactured today is refined by blowing oxygen through a furnace charged with ore, scrap, and molten iron.

called rusting and leads to the transformation of iron into rust ($Fe_2O_3 \cdot xH_2O$). The initial reaction is the oxidation of iron to iron (II) hydroxide:

$$\underset{\text{MOIST AIR}}{2Fe + O_2 + 2H_2O} \longrightarrow \underset{\text{IRON (II) HYDROXIDE}}{2Fe(OH)_2}$$

Iron (II) hydroxide is itself subject to further oxidation in moist air to give iron (III) hydroxide:

$$\underset{\text{MOIST AIR}}{4Fe(OH)_2 + O_2 + 2H_2O} \longrightarrow \underset{\text{IRON (III) HYDROXIDE (RUST)}}{4Fe(OH)_3} \quad \underset{\text{IRON (III) OXIDE TRIHYDRATE (RUST)}}{(\text{or } 2Fe_2O_3 \cdot 3H_2O)}$$

The iron (III) hydroxide loses water readily to form iron (III) oxides with variable amounts of water. The rusting process changes iron back into a compound similar in composition to its ore, and thus undoes all the effort expended in obtaining the metal.

Rusting also occurs when we make an iron object into an electrochemical cell or battery, often without realizing it. This happens when the iron is in contact with copper or nickel or any of the less active metals listed in Table 6–3, or when part of the iron surface becomes wet in contact with air (Fig. 7–5). The oxygen in the air will dissolve in the water and remove electrons from the iron via the reactions

$$2Fe \longrightarrow 2Fe^{2+} + 4e^-$$

$$O_2 + 2H_2O + 4e^- \longrightarrow 4OH^-$$

There are many ways to prevent or reduce corrosion of a metal object. Three of these are: (1) protective coatings, (2) cathodic protection, and (3) alloying (stainless steel is a corrosion-resistant alloy of iron).

**Cathodic protection is discussed on p. 134.**

***Protective coatings*** prevent access of water and atmospheric oxygen to the iron surface. Such coatings may be paint, enamel, grease, or another more resistant metal, such as chromium. This method is successful as long as cracks or holes do not develop in the coating. Galvanized iron has a coating of the more active metal zinc,

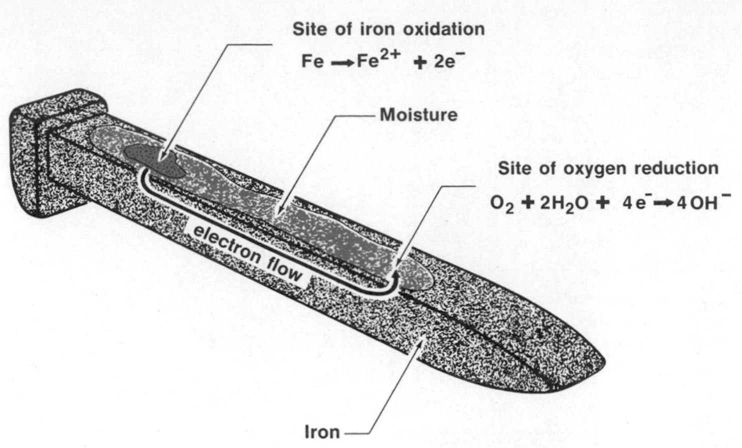


**Figure 7–5** The site of iron oxidation may be different from the point of oxygen reduction owing to the ability of the electrons to flow through the iron. The point of oxygen reduction can be located with an acid-base indicator because of the $OH^-$ ions produced.

A stainless steel that is widely used is the so-called 18-8; 18% chromium, 8% nickel, a trace of carbon, and the rest iron.

which gives good cathodic protection (Table 6–3) and also forms an oxide that is hard and impervious to the air.

***Stainless steels*** are alloys of iron and other metals such as nickel, chromium, or cobalt. They are made by melting iron and the alloying metals together in an electric furnace. The resulting alloys are resistant to corrosion, because in the presence of oxygen they form a very thin, tough, impervious, and adherent layer of metal oxide on their surfaces. The layer is so thin that it is essentially transparent (the luster of the metal is retained). This protects the underlying metal from further contact with the oxygen of the air and renders the objects "stainless," or very resistant to corrosion under normal circumstances.

Stainless steels are used in household items; larger amounts are used in the construction of industrial plants that handle great quantities of hydrochloric acid and other highly corrosive materials.

When aluminum was first made, it was very expensive and rare. A bar of aluminum was displayed next to the Crown Jewels at the Paris Exposition in 1855.

## Aluminum

Aluminum, in the form of $Al^{3+}$ ions, constitutes 7.4% of the crust of the earth. However, because of the difficulty of reducing $Al^{3+}$ to Al, only recently have we learned to isolate and use this abundant element. Aluminum metal is soft and has a low density. Many of its alloys, however, are quite strong. Hence, it is an excellent choice when a lightweight, strong metal is required. In structural aluminum, the high chemical reactivity of the element is offset by the fact that a transparent, hard film of aluminum oxide, $Al_2O_3$, forms over the surface, protecting it from further oxidation:

$$4Al + 3O_2 \longrightarrow 2Al_2O_3$$

The principal ore of aluminum contains the mineral bauxite, a hydrated aluminum oxide, $Al_2O_3 \cdot xH_2O$. Because impurities such as iron oxides in the ore have undesirable effects on the properties of aluminum, these must be removed, generally by the purification of the ore. This is accomplished with the Bayer process, which is based upon the fact that aluminum oxide and aluminum hydroxide react with strong bases. In the Bayer process the mixture of oxides is treated with a sodium hydroxide

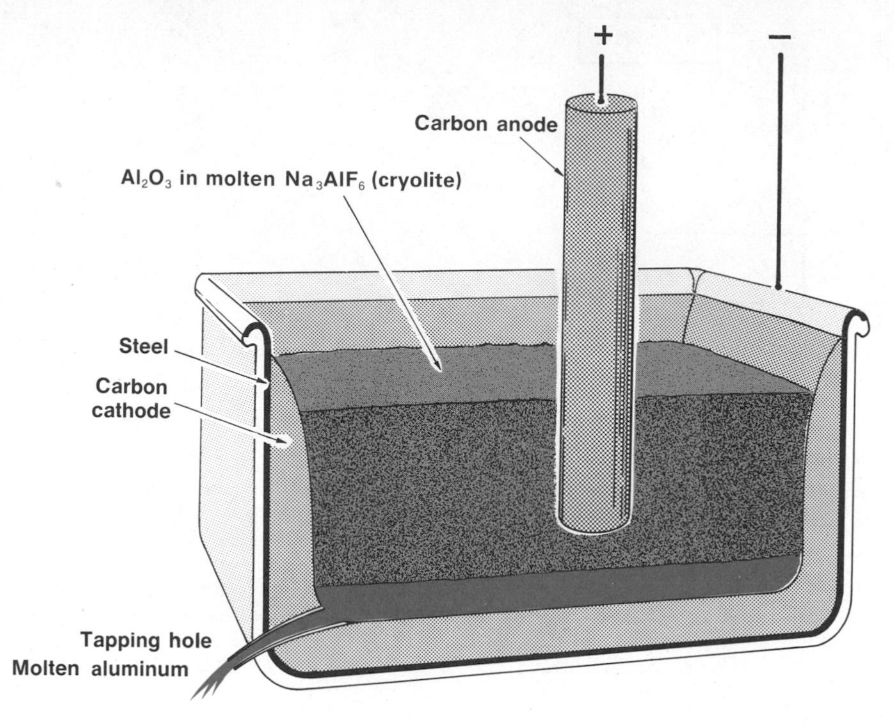


**Figure 7–6** Schematic drawing of a furnace for producing aluminum by electrolysis of a melt of $Al_2O_3$ in $Na_3AlF_6$. The molten aluminum collects at the bottom of the carbon cathode container.

solution, which dissolves aluminum oxide and leaves iron oxide, which is insoluble in the solution.

$$\underset{\text{(SOLID)}}{Al_2O_3 \cdot xH_2O} + \underset{\text{(SOLID)}}{Fe_2O_3} \xrightarrow[\text{solution}]{\text{NaOH}} \underset{\text{(SOLUTION)}}{Al(OH)_4^- + Na^+} + \underset{\text{(SOLID)}}{Fe_2O_3}$$

The mixture is filtered; $Al(OH)_3$ is then carefully precipitated out of the clear solution by the addition of carbon dioxide (an acid), which makes the solution less basic:

$$CO_2 + Al(OH)_4^- \longrightarrow Al(OH)_3\downarrow + HCO_3^-$$

The aluminum hydroxide is heated to transform it into pure anhydrous aluminum oxide:

$$2Al(OH)_3 \xrightarrow{\text{heat}} Al_2O_3 + 3H_2O$$

Aluminum metal is obtained from the purified oxide by electrolysis in molten cryolite (Fig. 7–6 and Table 6–2). Cryolite, $Na_3AlF_6$, has a melting point of 1000°C; the molten compound dissolves considerable amounts of aluminum oxide, which in turn lowers the melting point of the cryolite solution. This mixture of cryolite and aluminum oxide is electrolyzed in a cell with carbon anodes and a carbon cell lining that serves as the cathode on which aluminum is deposited. As the operation of the cell proceeds, the molten aluminum sinks to the bottom of the cell. From time to time the cell is tapped and the molten aluminum is run off into molds.

**The top of the Washington Monument is a casting of aluminum made in 1884.**

Aluminum is used both as a structural metal and as an electrical conductor in high-voltage transmission lines. It competes with copper as an electrical conductor because of the lower cost of aluminum. Larger diameter aluminum wires must be used to offset the lower electrical conductivity of aluminum compared with copper.

**About ten times more energy is needed to produce a ton of aluminum than a ton of steel.**

Can you list all the ways energy in its various forms is used to prepare aluminum from its ore?

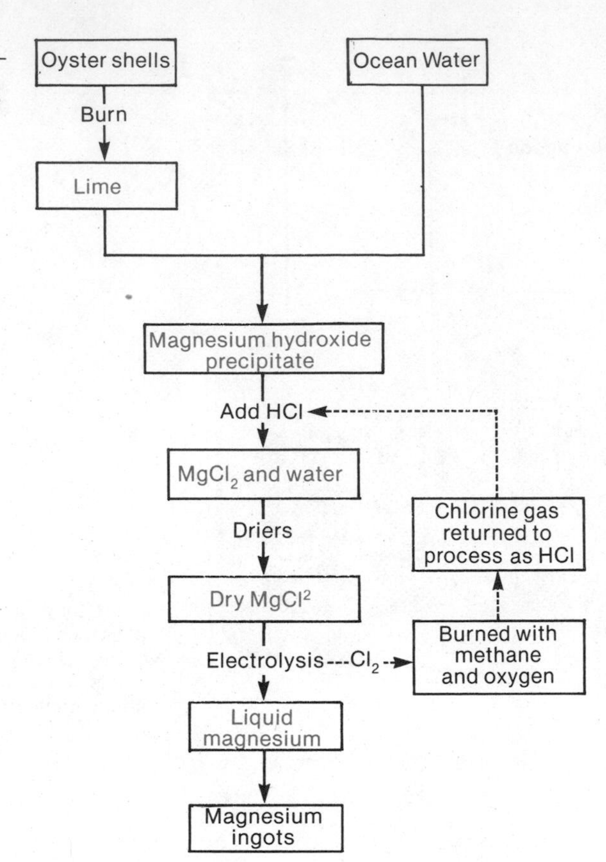


**Figure 7–7** Flow diagram showing how magnesium metal is produced from sea water.

## Magnesium

Magnesium, with a density of 1.74 grams per milliliter, is the lightest structural metal in common use. For this reason it is most often used in alloys designed for light weight and great strength. It is a relatively active metal chemically because it loses electrons easily. Magnesium "ores" include sea water, which has a magnesium concentration of 0.13%, and dolomite, a mineral with the composition $CaCO_3 \cdot MgCO_3$. Because there are six million tons of magnesium present as $Mg^{2+}$ salts in every cubic mile of sea water, the sea can furnish an almost limitless amount of this element.

**There are about 328 million cubic miles of sea water.**

The recovery of magnesium from sea water (Fig. 7–7) begins with the precipitation of magnesium hydroxide by the addition of lime to sea water:

$$CaO + H_2O \longrightarrow Ca^{2+} + 2OH^-$$

$$Mg^{2+} + 2OH^- \longrightarrow Mg(OH)_2$$

The magnesium hydroxide is removed by filtration and then neutralized with hydrochloric acid to form the chloride:

$$Mg(OH)_2 + 2H^+ + 2Cl^- \rightleftarrows Mg^{2+} + 2Cl^- + 2H_2O$$

The water is evaporated; this is followed by the electrolysis of molten magnesium

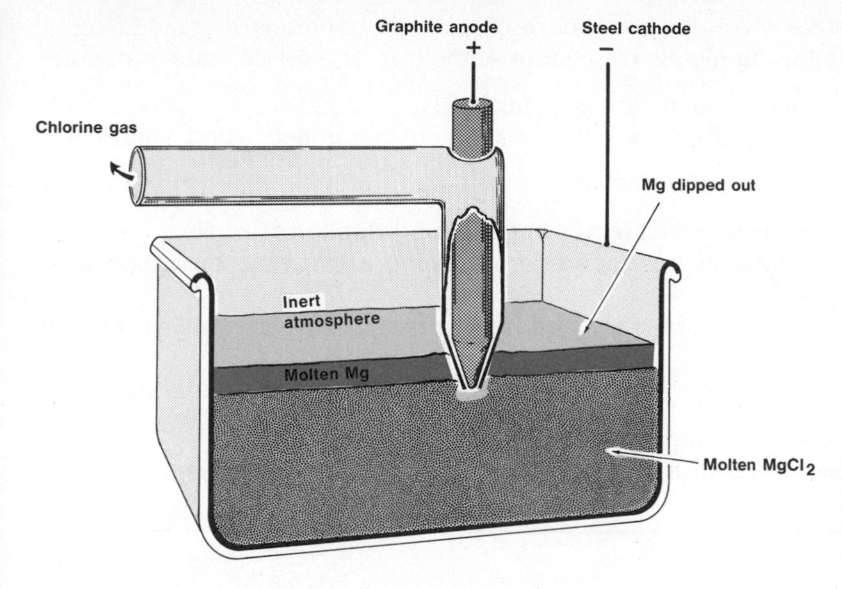


Figure 7–8 A cell for electrolyzing molten $MgCl_2$. The magnesium metal is formed on the steel cathode and rises to the top where it is dipped off periodically. Chlorine gas is formed on the graphite anode and is piped off.

chloride in a huge steel pot that serves as the cathode (Fig. 7–8). Graphite bars serve as the anodes.

$$\underset{\text{(MELTED)}}{Mg^{2+} + 2Cl^-} \xrightarrow{\text{electrolysis}} Mg + Cl_2\uparrow$$

AT THE CATHODE: $Mg^{2+} + 2e^- \longrightarrow Mg$

AT THE ANODE: $2Cl^- \longrightarrow Cl_2\uparrow + 2e^-$

As the melted magnesium forms, it floats to the surface and is removed periodically. The chlorine is recovered and reacted with air and natural gas (methane, $CH_4$) to form hydrochloric acid, which in turn is used in dissolving the magnesium hydroxide:

**To make magnesium one can use sea water, lime from oyster shells, methane from natural gas, and electricity.**

$$4Cl_2 + \underset{\text{METHANE}}{2CH_4} + O_2 \longrightarrow 2CO + 8HCl$$

The lime used to precipitate the magnesium as the hydroxide is obtained by heating limestone or oyster shells:

$$CaCO_3 \xrightarrow{\text{heat}} \underset{\text{LIME}}{CaO} + CO_2$$

The total world production of magnesium is only about 250,000 tons per year; although it is potentially available on a larger scale.

## Self-Test 7-A

1. The most abundant element in the earth's crust is ______________.
2. The most abundant metal in the earth's crust is ______________.
3. In the United States, the largest iron ore deposits are found in the state of ______________.

4. A natural material which is almost pure calcium carbonate is ______.
5. Which of the following metals may occur in the free or metallic state in mineral deposits? Iron, copper, aluminum, magnesium. ______.
6. Which of the metals listed are either produced or purified using electricity? Iron, copper, aluminum, magnesium. ______.
7. Another name for calcium silicate as it applies to production of iron is ______.
8. In order for most metals to be prepared from their ores, they must be (oxidized/reduced).
9. In an electrical refining process for metals, the purest metal will always be found at the (anode/cathode).
10. Which metal is sufficiently concentrated in the oceans to be extracted commercially? Magnesium, aluminum, copper, iron. ______.
11. Most metals are found in the earth as (neutral atoms/positive ions/negative ions).

## Elements from the Air

The atmosphere of the earth is a fantastically large source of the elements nitrogen and oxygen, and certain of the noble gases including argon, neon, and xenon (see Table 7–2).

In order to obtain pure oxygen and nitrogen from the air, water vapor and carbon dioxide must be removed first. This is usually done by precooling the air by refrigeration, or by using silica gel to absorb water and lime to absorb carbon dioxide. Afterward, the air is compressed to a pressure exceeding 100 times normal atmospheric pressure, cooled to room temperature, and allowed to expand into a chamber. This expansion produces a cooling effect (the Joule-Thompson effect) due to the breaking of weak attractive van der Waals bonds between the gaseous molecules. Recall that breaking bonds requires energy so the expanding gas absorbs energy from the surroundings, thus cooling the surroundings and the gas itself. If this expansion is controlled properly the expanding air actually cools to the point of liquefaction (Fig. 7–9). The temperature of the *liquid air* is usually well below the boiling points of nitrogen (−195.8°C), oxygen (−183°C) and argon (−189°C). This liquid air is then allowed to partially vaporize again (Fig. 7–10), and since nitrogen is more volatile than oxygen or argon (N has a lower boiling point), the liquid becomes more concentrated in oxygen and argon. This process, known as the Linde

**TABLE 7–2 The Atmospheric Composition of the Earth at Sea Level**

| | Percentage by Volume |
|---|---|
| Nitrogen | 78.084 |
| Oxygen | 20.948 |
| Argon | 0.934 |
| Carbon dioxide | 0.033* |
| Neon | 0.00182 |
| Helium | 0.00052 |
| Methane | 0.0002 |

*Estimated for 1977.

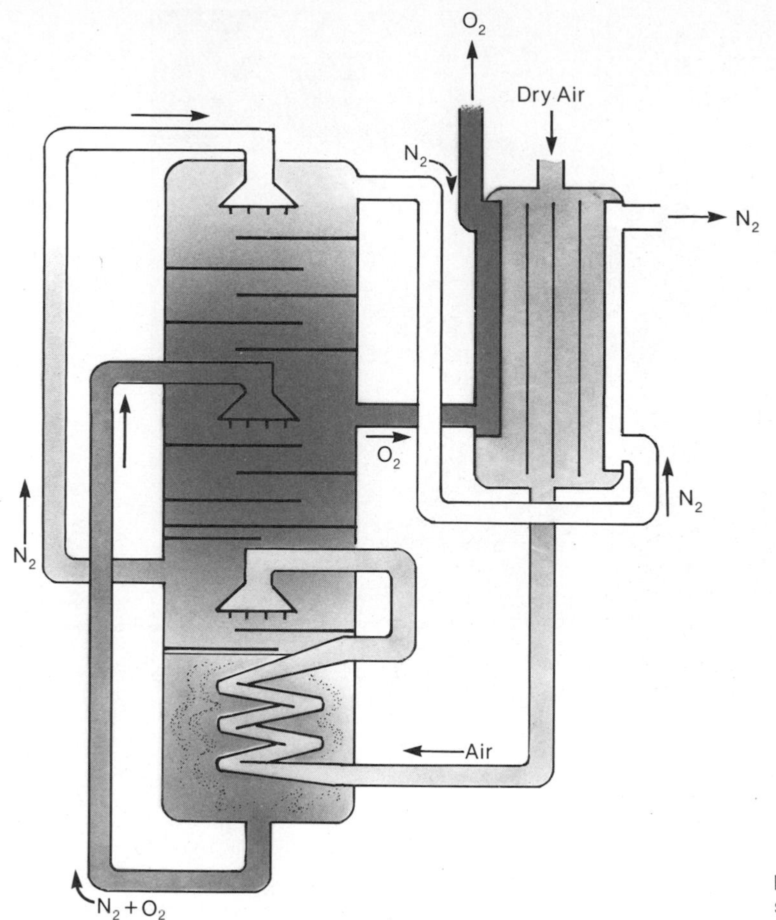


**Figure 7–9** Diagram of a fractionating column for separating oxygen and nitrogen in an air supply.

process, produces high-purity nitrogen (99.5+%) and oxygen having a purity of 99.5%. Further processing produces pure argon, neon (bp −246°C) and even helium (bp−268.9°C), but most helium used in the United States is produced from natural gas wells.

Most of the oxygen produced by the fractionation of liquid air is used in steelmaking, although some finds use in controlled oxidation reactions of other types. Liquid oxygen (LOX) can be shipped and stored at its boiling temperature of −183°C at atmospheric pressure. Substances this cold are called ***cryogens*** (from Greek, *kryos,* meaning icy cold). They represent special hazards since contact produces instantaneous frostbite, and structural materials such as plastics, rubber gaskets, and some metals become brittle and fracture easily at these temperatures. Liquid oxygen can accelerate oxidation reactions to the point of explosion due to the high oxygen concentration. For this reason liquid oxygen must be prevented from contacting substances that will ignite and burn in air.

Special cryogenic containers holding liquid oxygen are actually huge vacuum-walled bottles much like those used to carry hot soup or hot coffee. These containers can be seen outside hospitals or industrial complexes, on highways and railroads, and even aboard ocean-going vessels (Fig. 7–10).

Liquid nitrogen is also a cryogen. It has uses in medicine (cryosurgery), for example in cooling a localized area of skin prior to removal of a wart or other unwanted

**Figure 7–10** Photo of a cargo tanker capable of carrying several thousand gallons of liquified, cryogenic oxygen, at a temperature of −183°C. Although it is extremely cold, its high concentration in the liquid state makes liquid oxygen exceptionally reactive with anything that can burn.

or pathogenic tissue. Since nitrogen is so chemically unreactive it is used as an inert atmosphere for certain applications such as welding, and liquid nitrogen is a convenient source of high volumes of the gas. Due to its low temperature and inertness, liquid nitrogen has found wide use in frozen food preparation and preservation during transit. Containers of nitrogen atmospheres, such as railroad boxcars or truck vans, represent health hazards since they contain little (if any) oxygen to support life, and workers have died when they entered such areas without breathing apparatus.

## Glass—Silicon's Domain

Silicon dioxide, $SiO_2$ (also called silica), occurs naturally in large amounts as sand or more rarely in much larger crystals (quartz). It has a melting point of 1710°C. If the melted material is cooled rapidly, a noncrystalline solid is obtained. Crystalline quartz consists of an extended structure in which each silicon atom is bonded tetrahedrally to four oxygen atoms (Fig. 7–11a), and each oxygen atom is bonded to two silicon atoms. The bonding thus extends throughout the crystal (Fig. 7–11b). When silica is melted, some of the bonds are broken and the units move with respect to each other. When the liquid is cooled, the re-formation of the original solid requires a reorganization that is hard to achieve because of the difficulty the groups experience in moving. The very viscous liquid structure is thus partially preserved on cooling to give the characteristic feature of a ***glass,*** which is an apparently solid material (pseudo-solid) with some of the randomness in structure characteristic of a liquid. This random structure accounts for one of the typical properties of a glass: it breaks irregularly rather than splitting along a plane like a crystal.

**Glass will flow slowly like a liquid. Many colonial homes contain glass window panes thicker at the bottom than at the top.**

By the addition of metal oxides to silica, the melting temperature of the mixture can be reduced from 1710°C to about 700°C. The oxides most often added are

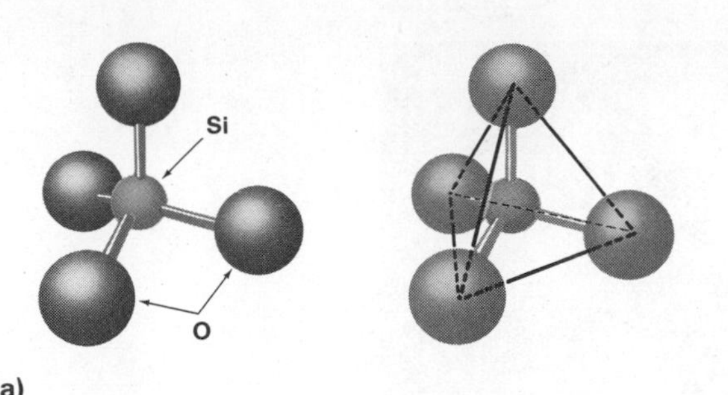


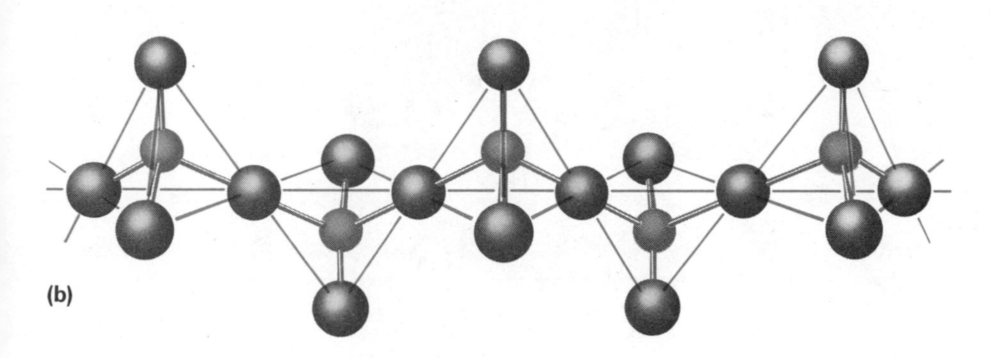


**Figure 7–11** (*a*) Tetrahedral structure of silicon and oxygen in silicates. (*b*) Chain of tetrahedra showing that an oxygen is common at each point of contact between tetrahedra.

sodium oxide (added as $Na_2CO_3$, soda ash) and calcium oxide (added as $CaCO_3$). The metal ions form ionic bonds, which are nondirectional, with oxygen atoms that previously had been bonded rigidly to specific Si atoms. As a result, the so-called soda-lime glass has a lower melting temperature and viscosity than pure $SiO_2$, and can be produced and fabricated more easily.

Sodium carbonate ($Na_2CO_3$) is the number twelve commercial chemical. Calcium oxide is number four. See inside the front cover.

Soda-lime glass will be clear and colorless only if the purity of the ingredients has been controlled carefully. If, for example, too much iron oxide is present, the glass will be green. Other metal oxides produce other colors (see Table 7–3). To some extent, one color can counteract another.

Viscosity is the resistance to flow.

Most glass consists of a mixture of the oxides of Si, Na, and Ca, which are melted together. Colored glass is produced by addition of other metal oxides.

The substances are melted together in a gas- or oil-fired furnace. As they react, bubbles of $CO_2$ gas are evolved.

$$CaCO_3 + SiO_2 \longrightarrow CaSiO_3 + CO_2\uparrow$$

$$Na_2CO_3 + SiO_2 \longrightarrow Na_2SiO_3 + CO_2\uparrow$$

The mixture is heated to about 1500°C to remove the bubbles of $CO_2$. At this temperature the viscosity is low, and the bubbles of entrapped gas easily escape. The mix-

**TABLE 7–3 Substances Used in Colored Glasses**

| Substance | Color |
|---|---|
| Copper (I) oxide | red, green or blue |
| Tin (IV) oxide | opaque |
| Calcium fluoride | milky white |
| Manganese (IV) oxide | violet |
| Cobalt (II) oxide | blue |
| Finely divided gold | red, purple, or blue |
| Uranium compounds | yellow, green |
| Iron (II) compounds | green |
| Iron (III) compounds | yellow |

**Figure 7–12** Craftsman working with molten glass.

ture is cooled somewhat and then is blown into bottles by machines, or is drawn into sheets or molded into other forms (Fig. 7–12).

It is possible to incorporate a wide variety of materials into glass for special purposes. Some examples are given in Table 7–4.

## Silicon

The element silicon is the backbone of our planet. Since the crust of the earth is made up of 49.5% oxygen and 25.7% silicon, there is an ample amount of this ele-

**TABLE 7–4 Special Glasses**

| Special Addition or Composition | Desired Property |
|---|---|
| Large amounts of $PbO$ with $SiO_2$ and $Na_2CO_3$ | brilliance, clarity, suitable for optical structures: crystal or flint glass |
| $SiO_2$, $B_2O_3$, and small amounts of $Al_2O_3$ | small coefficient of thermal expansion: borosilicate glass, "Pyrex," "Kimax," etc. |
| One part $SiO_2$ and four parts $PbO$ | ability to stop (absorb) large amounts of X rays and gamma rays: lead glass |
| Large concentrations of $CdO$ | ability to absorb neutrons |
| Large concentrations of $As_2O_3$ | transparency to infrared radiation |
| Suspended Se particles | red color |

ment. Silicon is a shiny, silvery, brittle element that looks like a metal but doesn't always act like one. Instead of reacting with acids or bases, silicon is nonreactive. When reacted with a halogen such as chlorine or fluorine, instead of forming salts like metals do, silicon forms covalent, gaseous or liquid halides.

$$Si + 2F_2 \longrightarrow SiF_4 \text{ (bp } -86°C)$$

$$Si + 2Cl_2 \longrightarrow SiCl_4 \text{ (bp } 57.6°C)$$

Silicon of about 98% purity can be obtained by heating silica and a form of carbon called coke at 3000°C in an electric arc furnace.

$$SiO_2 + 2C \longrightarrow Si + 2CO$$

Silicon of this purity is alloyed with aluminum and magnesium to increase their hardness and durability, and is used in making silicone polymers (Chapter 10).

High-purity silicon can be prepared by reducing $SiCl_4$ with magnesium.

$$SiCl_4 + 2Mg \longrightarrow Si + 2MgCl_2$$

The magnesium chloride, being water-soluble, is then washed from the silicon. The final purification of the silicon takes place by a melting process called ***zone-refining*** (Fig. 7–13), which produces silicon containing less than 1 part per *billion* of impurities such as boron, aluminum, or arsenic.

One outstanding property of silicon in a high state of purity is its electrical conductivity. Unlike a metal, which easily conducts electricity, and unlike a nonmetal, which fails to conduct electricity, silicon is a ***semiconductor.*** That is, it fails to conduct until a certain electrical voltage is applied, but beyond that it conducts very well. By placing other atoms in a crystal of pure silicon, a process known as ***doping,*** experimenters have found that its conductivity properties can be changed. Doping a silicon crystal with a Group V element such as arsenic produces a crystal with extra electrons. (Arsenic has five valence electrons whereas silicon has four.) This is known as an n-doped semiconductor. Doping silicon with a Group III element such as gallium produces a p-doped semiconductor, since gallium has only three valence electrons and looks positive in the silicon lattice.

In 1947 an electrical device called the ***transistor*** was invented. The simplest device used layers of n-p-n– or p-n-p–doped silicon. Germanium, a Group IV element just below silicon in the periodic table, was also used. Later, scientists used electrical

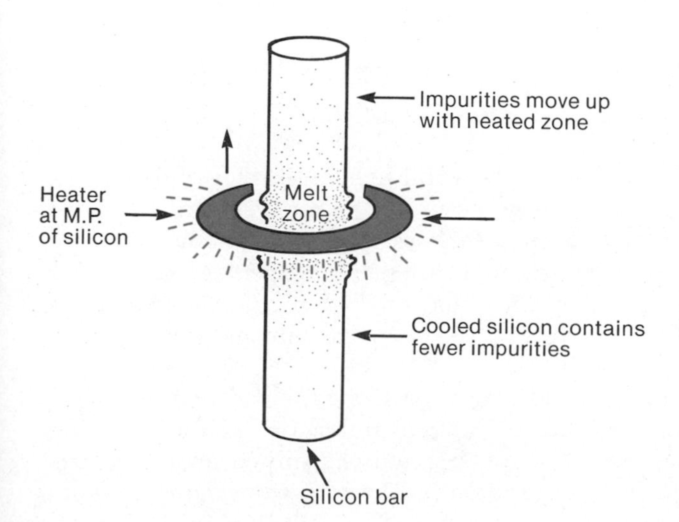


**Figure 7–13** Zone refining. The hot zone moves upward on the silicon bar. As the silicon melts, impurities become mobile and move with the hot molten zone. Repeated passes of the heater produce a crystalline silicon bar with fewer than one part impurities per billion parts of silicon (1 ppb).

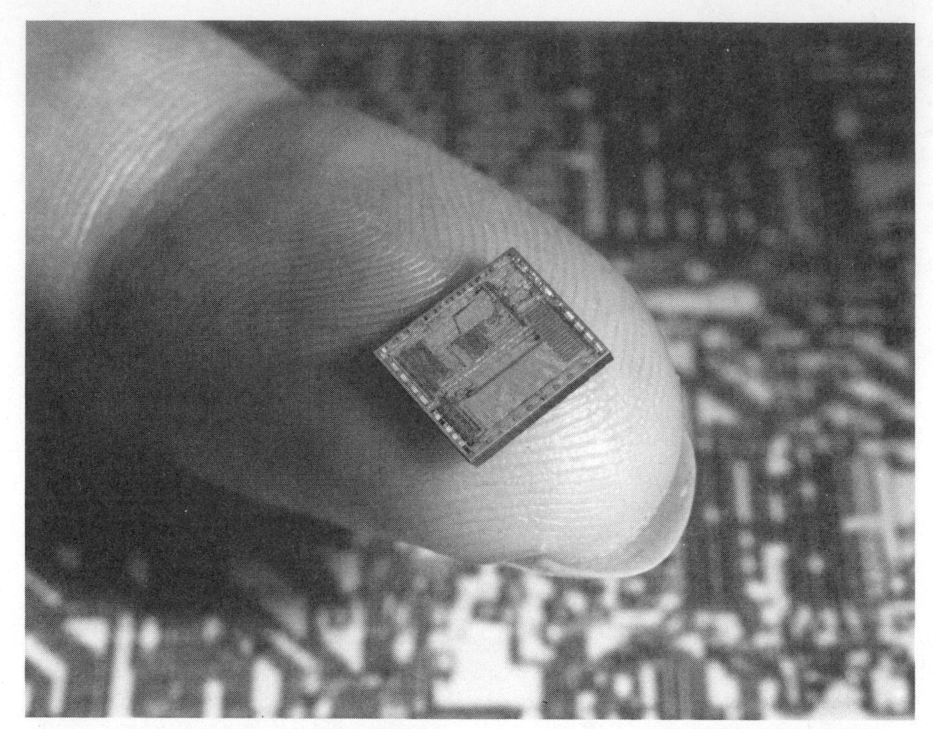

**Figure 7–14** A tiny microcomputer (Intel 8748) fabricated from a single piece of highly purified silicon. Such computers are capable of many millions of computations per second. Their speed and small size have revolutionized computers and their applications.

fields to control conductivity in silicon transistors. These ***field-effect transistors*** (FET's) have been put to good use by engineers designing low-noise amplifiers, receivers, and other forms of electronic equipment.

The most revolutionary application of silicon's semiconductor properties has been the design of ***integrated electrical circuits*** (IC's), computer memories, and even whole computers called ***microprocessors*** on tiny chips of silicon scarcely larger than a millimeter or so in diameter. These devices have begun to permeate our whole society. You will find them in calculators, cameras, watches, toys, coin changers, cardiac pacemaker devices, and many other products. Truly, silicon is both the world we walk on and at the same time our constant companion.

## Chemical Fertilizers—Keys to World Food Problems

Regardless of the method used to predict future world populations, everyone agrees that food production for that population is a major problem that is sure to get worse. Indeed, adequate food production is a problem for most people on the earth today. Food production is complicated, involving the application of management, economics, mechanical and civil engineering, genetics, and chemistry in an area we usually call ***agriculture.*** In this section we shall deal with the chemical nutrients of plants as they relate to food production.

Crops were fertilized with manure, dead fish, or straw for centuries before it was understood that growing plants require 16 different nutrients (Table 7–5). Nine of these, the ***macronutrients,*** are needed in larger amounts, while seven, the ***micronutrients,*** are needed in smaller or trace amounts. Three of the macronutrients,

**TABLE 7–5 Nutrients Needed by Plants**

| Element | Source Compound |
|---|---|
| | *Macronutrients* |
| C | $CO_2$, carbon dioxide |
| H | $H_2O$, water |
| O | $H_2O$, water |
| N | $NH_3$, ammonia; $NH_4NO_3$, ammonium nitrate; $H_2NCONH_2$, urea |
| P | $Ca(H_2PO_4)_2$, calcium dihydrogen phosphate |
| K | KCl, potassium chloride |
| Ca | $Ca(OH)_2$, slaked lime |
| Mg | $MgCO_3$, magnesium carbonate |
| S | elemental sulfur |
| | *Micronutrients* |
| B | $Na_2B_4O_7 \cdot 10H_2O$, borax |
| Cu | $CuSO_4 \cdot 5H_2O$, copper (II) sulfate pentahydrate |
| Fe | $FeSO_4$, iron (II) sulfate |
| Mn | $MnSO_4$, manganese (II) sulfate |
| Zn | $ZnSO_4$, zinc sulfate |
| Mo | $(NH_4)_2MoO_4$, ammonium molybdate |
| Cl | KCl, potassium chloride |

carbon, hydrogen, and oxygen, come from the atmosphere and are available to all plants, while the remainder of the nutrients must come from the soil. Growing plants deplete the available nutrients rather drastically (Table 7–6). Within the space of one or two growing cycles the soil nutrients must be replaced.

**Soil nutrients must be replaced every one or two growing cycles.**

With today's requirements on yields, the use of chemical fertilizers is the only* way to maintain the high productivity of the world's farmland.

There are about 3.5 billion acres currently under cultivation worldwide. It has been calculated that the application of $35 worth of fertilizer per acre would increase crop production by 50%, equivalent to 1.7 billion more acres under cultivation. Of course this would be expensive, requiring about $40 per capita worldwide,

**There is a little less than one acre under cultivation for every individual on the earth.**

**TABLE 7–6 Approximate Amounts of Nutrients Required to Produce 150 Bushels of Corn**

| Nutrient | Approximate Pound per Acre | Source |
|---|---|---|
| Oxygen | 10,200 | air |
| Carbon | 7,800 | air |
| Water | 3,225–4,175 tons | 29–36 in. of rain |
| Nitrogen | 310 | 1,200 lbs. of high-grade fertilizer |
| Phosphorus | 120 (as phosphate) | 1,200 lbs. of high-grade fertilizer |
| Potassium | 245 (as $K_2O$) | 1,200 lbs. of high-grade fertilizer |
| Calcium | 58 | 150 lbs. of agricultural limestone |
| Magnesium | 50 | 275 lbs. of magnesium sulfate (epsom salt) |
| Sulfur | 33 | 33 lbs. of powdered sulfur |
| Iron | 3 | 15 lbs. of iron sulfate |
| Manganese | 0.45 | 1.3 lbs. of manganese sulfate |
| Boron | 0.05 | 1 lb. of borax |
| Zinc | trace | small amount of zinc sulfate |
| Copper | trace | small amount of copper sulfate |
| Molybdenum | trace | trace of ammonium molybdate |

*Organic materials such as feedlot animal wastes, slaughterhouse wastes, and plant wastes (stalks, leaves, etc.) could satisfy much of the soil nutrient demand, but social and sanitary restrictions have retarded their extensive application in this country.

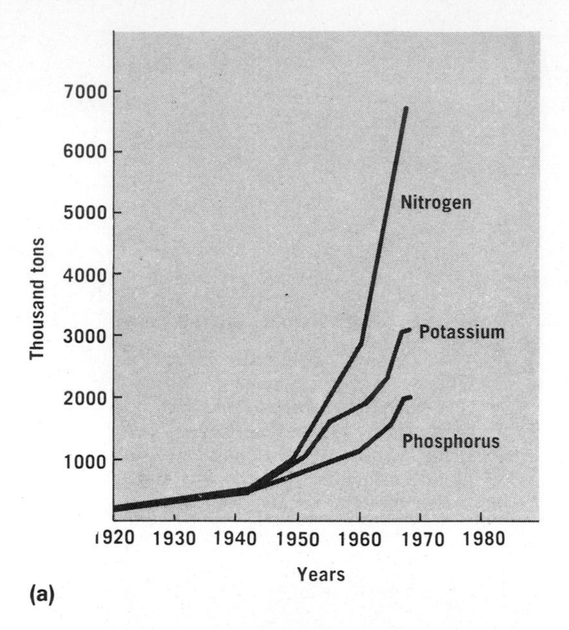


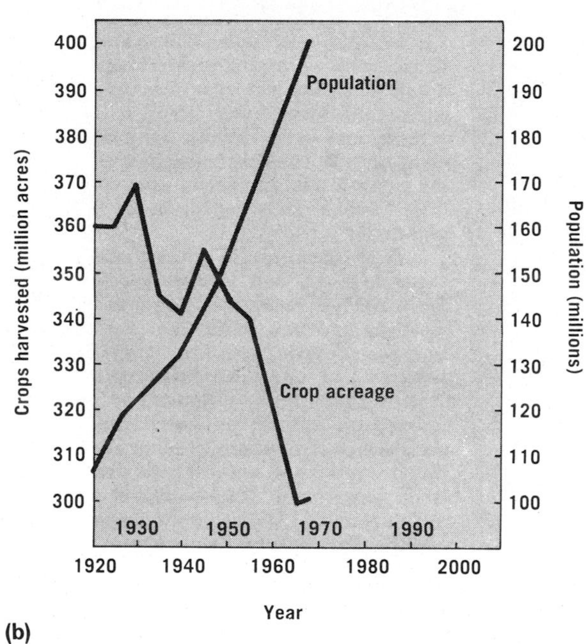


**Figure 7–15** Changes in fertilizer usage, crop acreage, and population in the United States in the last half century. Amounts of phosphorus, potassium, and especially nitrogen have increased markedly (*a*). As the population has increased (*b*), land use for agriculture has diminished sharply.

**The large amount of petroleum needed to make fertilizer causes fertilizer prices to rise with the price of petroleum.**

and might be impossible due to energy limitations since fertilizer manufacture uses a significant portion of our fossil fuels. Poorer countries such as India have almost been priced out of the world fertilizer markets by the recent steep increase in the prices of petroleum products.

## Nitrogen Chemistry

Since the atmosphere is 78% nitrogen by volume, it might appear that this nutrient would be readily available to plants. However, because of the strong nitrogen-nitrogen triple bond in the $N_2$ molecule, gaseous nitrogen is rather unreactive and as such is not directly available to plants as a nutrient. To be useful, the nitrogen

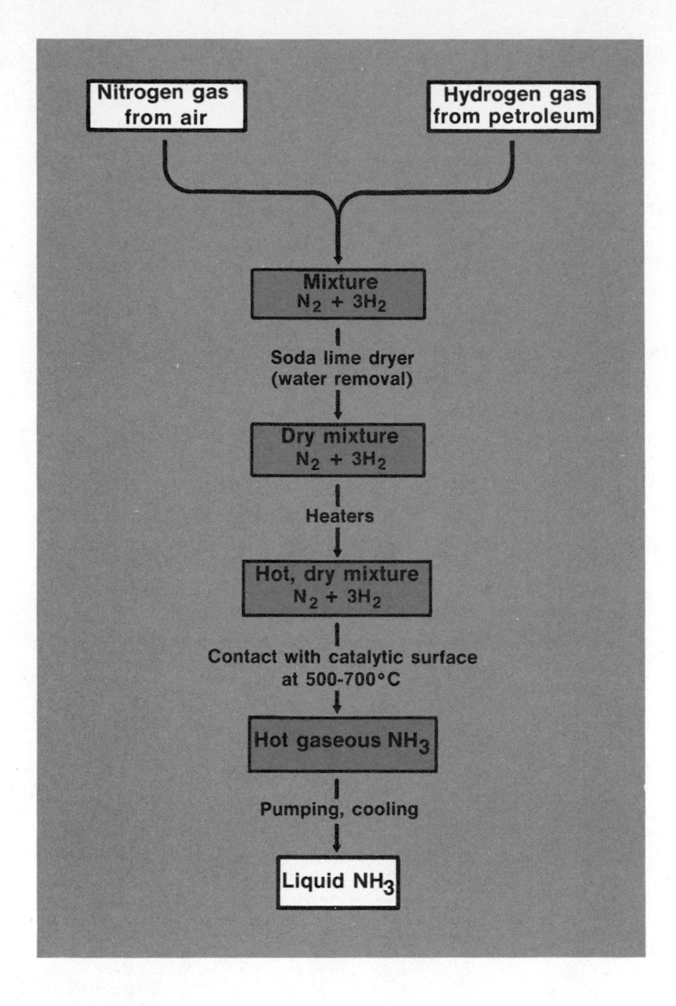


**Figure 7–16** The Haber process for synthesizing ammonia. Ammonia is the number two commercial chemical, and nitrogen is number three. See inside the front cover.

must be ***fixed***—that is, oxidized or reduced to some species containing nitrogen such as nitrate ion, $NO_3^-$ (oxidized form), or ammonia, $NH_3$ (reduced form), either of which is chemically more reactive than the nitrogen molecule. Only a few organisms, such as some bacteria and blue-green algae, are able to fix atmospheric nitrogen. Bacteria of the genus *Rhizobium* can fix nitrogen when they grow in nodules in the roots of legumes such as peas, beans, and alfalfa. Generally more nitrogen is fixed than the plants themselves need; thus, the soil nitrogen is increased. In some parts of the world, crop rotation between legumes and nonlegumes works quite nicely for soil nitrogen requirements. Two of the most important cereal crops, wheat and corn, are not favored by nitrogen-fixing bacteria, and chemists and biologists are working at present on a newly discovered bacterium that grows in the root structures of certain South American grasses, in hopes of establishing a new ***synergistic*** relationship between nitrogen-fixing bacteria and plants.

Fixed nitrogen refers to nitrogen present in chemical compounds, that is, in combinations other than $N_2$.

Synergism is a cooperative action producing an effect greater than the sum of the individual effects.

Commercial fertilizers contain nitrogen fixed by the Haber process, the direct reaction of nitrogen with hydrogen to produce ammonia (Fig. 7–16):

$$N_2 + 3H_2 \rightleftharpoons \underset{\text{AMMONIA}}{2NH_3}$$

Pure nitrogen is obtained by distilling oxygen and other gases from liquid air. Hydrogen is more difficult to obtain. At present, petroleum products such as propane,

$CH_3CH_2CH_3$, are made to react with steam in the presence of catalysts to produce hydrogen:

$$\underset{}{CH_3CH_2CH_3} + \underset{\text{STEAM}}{6H_2O} \xrightarrow{\text{catalysts}} 3CO_2 + 10H_2$$

This is one of the principal reasons why ammonia fertilizer costs are so closely tied to petroleum prices. Hydrogen can also be prepared by the electrolysis of water

$$2H_2O \xrightarrow[\text{KOH}]{\text{electricity}} 2H_2 + O_2$$

and by several other methods, but all of these require great quantities of energy. So, as energy costs continue to rise, food costs will necessarily rise due to the added costs of fertilizers.

**Ammonium nitrate is a solid fertilizer. It is often applied to crops as a solution.**

Nitrogen-containing fertilizers are all based upon ammonia. In the United States, much of the ammonia is injected directly into the ground (Fig. 7–17). Solid fertilizers can also be prepared from ammonia. ***Ammonium nitrate,*** $NH_4NO_3$, contains 35% nitrogen and is the salt of the reaction between ammonia and nitric acid:

**Ammonium nitrate is the number eleven commercial chemical.** **See inside the front cover.**

$$\underset{\text{AMMONIA}}{NH_3} + \underset{\text{NITRIC ACID}}{HNO_3} \longrightarrow \underset{\text{AMMONIUM NITRATE}}{NH_4NO_3}$$

**Figure 7–17** Liquid ammonia is being injected directly into the ground to provide nitrogen for the growing plants.

Ammonium nitrate has an advantage over ammonia in ease of handling, although it is sensitive to violent shock and can explode. Nitric acid used to prepare ammonium nitrate is itself prepared from ammonia by oxidation to nitrogen dioxide followed by reaction with water:

$$4NH_3 + 5O_2 \xrightarrow[\text{Pt catalyst}]{700°C} 4NO + 6H_2O$$

$$2NO + O_2 \longrightarrow 2NO_2$$

$$\underset{\text{NITROGEN DIOXIDE}}{3NO_2} + H_2O \longrightarrow \underset{\text{NITRIC ACID}}{2HNO_3} + \underset{\text{NITRIC OXIDE (RECYCLED)}}{NO}$$

***Urea*** ($NH_2CONH_2$) is probably one of the world's most important chemicals because of its wide use as a fertilizer and as a feed supplement for cattle. Ammonia and carbon dioxide react under high pressure near 200°C to produce first ammonium carbamate, which then decomposes into urea and water:

**Nitric acid is the number ten commercial chemical. See inside the front cover.**

$$2NH_3 + CO_2 \longrightarrow \underset{\text{AMMONIUM CARBAMATE}}{H_2N{-}C({=}O){-}ONH_4} \longrightarrow \underset{\text{UREA}}{H_2N{-}C({=}O){-}NH_2} + H_2O$$

Urea synthesis also produces the compound biuret as a by-product. Biuret must be removed from fertilizer-grade urea because it retards seed germination.

$$\underset{\text{BIURET}}{H_2N{-}C({=}O){-}NH{-}C({=}O){-}NH_2}$$

A slurry of water, urea, and ammonium nitrate is often applied to crops under the name of "liquid nitrogen." Such a solution can contain up to 30% nitrogen and is easy to store and apply.

## Phosphate Rock and Potash

Phosphate rock and potash are two minerals that can be mined, pulverized, and dusted directly onto deficient soil. Often they are specially treated to produce desirable mixing properties. Phosphorus, for example, is found scattered throughout the world in deposits of ***phosphate rock,*** which when treated with sulfuric acid becomes more soluble and hence produces a product of greater phosphorus availability called "superphosphate."

**Phosphoric acid, $H_3PO_4$, is the number nine commercial chemical. It is prepared from phosphate rock. See inside the front cover.**

$$\underset{\text{PHOSPHATE ROCK}}{Ca_3(PO_4)_2} + 2H_2SO_4 + H_2O \longrightarrow \underbrace{Ca(H_2PO_4)_2 + 2CaSO_4}_{\text{"SUPERPHOSPHATE"}}$$

Phosphate rock itself is not useful to a growing plant because of its very low solubility. Recently, it has become evident that phosphate rock demand will eventually exceed supply unless large new deposits are discovered.

**The world has limited deposits of phosphate rock, which is essential to the manufacture of fertilizers.**

Potassium in the form of ***potash,*** $K_2CO_3$, exists in enormous quantities throughout the world. A soluble form of potassium is its chloride (KCl), called muriate of potash or simply potash. Because this compound often occurs with sodium chloride,

**Figure 7–18** Phosphate from agricultural uses is a principal cause of water pollution.

which is toxic to plants, the potash ores must be treated by some process such as recrystallization to separate the two compounds.

## The Numbers on the Bag

People who purchase fertilizers—homeowners with yards to maintain, gardeners, and farmers—usually pay close attention to a set of three numbers that specify the three macronutrients present. These are nitrogen, phosphorus, and potassium. Most fertilizers also contain varying amounts of the micronutrients but these are not so prominently displayed. What do these numbers mean and what is their significance?

A common fertilizer used in gardening might be labeled 6–12–12. Since the mid-1880's, these numbers have meant respectively percentage composition by weight of nitrogen; percentage of available phosphoric oxide, $P_2O_5$; and percentage of water-soluble potassium oxide, $K_2O$, also known as potash. The calculation of these numbers from known compositions of fertilizer mixtures is an exercise in quantitative changes in matter.

For example, pure urea contains 46.7% nitrogen by weight (46.7% N) and no phosphoric oxide or potash, so the designation is 46–0–0.

To obtain the 46.7% N figure, we use the formula weight for urea and the atomic weight of nitrogen. More accurately, we use the total weight of nitrogen in the formula unit:

$$\frac{\text{total weight N in urea}}{\text{formula weight } H_4N_2CO} = \frac{2N}{H_4N_2CO} = \frac{2(14)}{60} = .467 \text{ or } 46.7\% \text{ N}$$

Diammonium phosphate (DAP), another commonly used fertilizer, offers another simple example. The formula for DAP is $(NH_4)_2HPO_4$. The percentage N in DAP can be calculated as

$$\frac{\text{total weight N in DAP}}{\text{formula weight DAP}} = \frac{2N}{(NH_4)_2HPO_4} = \frac{2(14)}{131.97} = .212 \text{ or } 21.2\% \text{ N}$$

The phosphoric content as $P_2O_5$ must be computed indirectly as follows. First the percentage of phosphorus is calculated.

$$\frac{\text{total weight P in DAP}}{\text{formula weight DAP}} = \frac{P}{(NH_4)_2HPO_4} = \frac{30.97}{131.97} = .235 \text{ or } 23.5\% \text{ P}$$

Then the phosphoric percentage is converted to $P_2O_5$ by multiplying by a factor that relates a given quantity of phosphorus to $P_2O_5$.

$$23.5\% \text{ P} \times \frac{\text{formula weight } P_2O_5}{2 \text{ atomic weight P}} = 23.5\% \text{ P} \times \frac{141.94}{61.94} = 53.8\% \ P_2O_5$$

Although this seems strange and cumbersome, it is not unlike multiplying boxes of shoes by two in order to obtain the total number of shoelaces. A number of years ago efforts were made to simplify the phosphorus content in fertilizers to percentage of phosphorus, but these efforts were resisted in the United States. European countries use the simpler method.

Potassium chloride (KCl) is commonly used in fertilizers to furnish potassium. Calculation of the potassium as $K_2O$ requires similar arithmetic. Using pure KCl,

$$\frac{\text{total weight K in KCl}}{\text{formula weight KCl}} = \frac{39.1}{74.6} = .524 \text{ or } 52.4\% \text{ K}$$

Converting this to $K_2O$ requires multiplication by the conversion factor $K_2O/2K$:

$$52.4\% \text{ K in KCl} \times \frac{\text{formula weight } K_2O}{2 \text{ atomic weight K}} = 52.4\% \text{ K} \times \frac{94.2}{78.2} = 63.1\% \ K_2O$$

The most modern technology in multicomponent fertilizers involves mixing urea and DAP with KCl to produce products with compositions such as 17–17–17 or 23–11–11. The production of these components is highly energy-intensive, and therefore increases in cost as energy costs go up.

## The Trace Nutrient Iron

Although they are needed in smaller amounts, the need for the micronutrients listed in Table 7–4 is as urgent as the need for macronutrients. For example, iron is an essential component of the catalyst involved in the formation of chlorophyll, the green plant pigment. When the soil is iron deficient, or when too much "lime," $Ca(OH)_2$*, is present in the soil, iron availability will decrease. This condition is usually present when plant leaves lighten in color or even turn yellow. Often a gardener or lawn worker will apply phosphate and lime to adjust soil acidity, only to see the green plants turn yellow. What is happening is that both phosphate and the hydroxide from the lime tie up the iron and make it unavailable to the plants.

**Lime, as CaO or $Ca(OH)_2$, is the number four commercial chemical. See inside the front cover.**

**The addition of lime, a basic substance, raises the pH of the soil.**

$$Fe^{3+} + \underset{\text{PHOSPHATE}}{2PO_4^{3-}} \longrightarrow \underset{\text{TIGHTLY BOUND COMPLEX}}{Fe(PO_4)_2^{3-}}$$

$$Fe^{3+} + 3OH^- \longrightarrow \underset{\text{INSOLUBLE HYDROXIDE}}{Fe(OH)_3}$$

These problems can be overcome, since iron can be ***chelated*** with certain organic molecules that hold the iron tightly but will release the iron at the root structure. The word chelate comes from the Greek, *chela,* meaning claw.

*Actually, lime is calcium oxide, CaO, while hydrated lime, called slaked lime, is $Ca(OH)_2$. Slaked lime is sold as "lime."

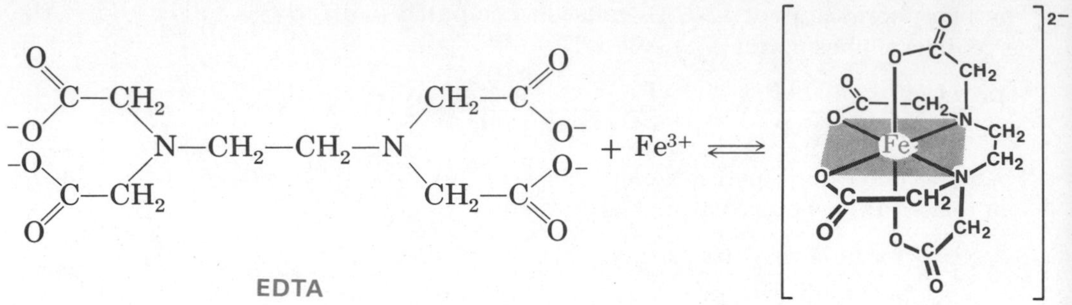

EDTA $+ Fe^{3+} \rightleftharpoons$ $[\ldots]^{2-}$

**Figure 7–19** EDTA structure and its iron (III) chelate.

Certain large negative ions, such as ethylenediaminetetraacetate (EDTA), will react with an iron ion and chelate it, holding the iron tightly. Since the resulting chelate is an ion, the iron can move through the soil to the root structures.

Chelate structures are important in many aspects of chemistry. Chlorophyll is a chelate of magnesium and hemoglobin is an iron chelate. In Chapter 15 chelating agents such as EDTA will be discussed in relation to metal poisoning.

## Self-Test 7-B

1. The principal element in glass is ______________.
2. Flint or crystal glass contains a large amount of the metal ______________.
3. The fertilizer produced by the Haber process is ______________.
4. Three macronutrients that must be replaced in the soil and constitute the main ingredients in chemical fertilizers are ______________, ______________, and ______________.
5. The principal source of hydrogen used to make ammonia is (water/petroleum hydrocarbons).
6. Superphosphate is manufactured from phosphate rock and (nitric/sulfuric) acid.
7. Potassium is usually added to the soil in the form of ($K_2O$/$KCl$/$KNO_3$).
8. The green color of glass is usually due to the presence of ______________.
9. Which is more water-soluble, $MgCl_2$ or silicon? ______________

## Matching Set

| | | | |
|---|---|---|---|
| ______ | 1. copper | **a** | a nutrient necessary for the manufacture of chlorophyll |
| ______ | 2. aluminum | **b** | calcium fluoride added to glass |
| ______ | 3. milk glass | **c** | a nutrient in short supply |
| ______ | 4. magnesium | **d** | used in making transistors |
| ______ | 5. oyster shells | **e** | source of carbon for plants |
| ______ | 6. atmosphere | **f** | made from nitrogen and hydrogen |
| ______ | 7. ammonia | **g** | a limitless supply in sea water |

_______ 8. iron

_______ 9. phosphate

_______ 10. silicon

**h** supply calcium hydroxide for magnesium production

**i** purified electrolytically

**j** the most abundant metal in the earth's crust

## Questions

**1.** Name three metals that you would expect to find free in nature. Name three that you would not.

**2.** What is the primary reducing agent in the production of iron from its ore?

**3.** Why is CaO necessary for the production of iron in a blast furnace?

**4.** What is the chemical difference between iron and steel?

**5.** Both iron and magnesium will oxidize in the air. Why is the oxidation of iron a much greater problem than the oxidation of magnesium?

**6.** How is it possible that both oxidation and reduction can occur at different points on the same piece of iron?

**7.** Give examples of three ways in which metals can be protected from corrosion.

**8.** Describe the solution used in a commercial cell for the electrolytic reduction of aluminum.

**9.** Why is it so important to purify industrial quantities of copper electrolytically to a level above 99.9% pure?

**10.** What chemical is obtained from oyster shells in the production of magnesium from sea water? What is the role of this chemical in the process?

**11.** Explain how the structures of glass and a liquid are similar.

**12.** What oxide is the main ingredient in glass?

**13.** Give reactions involved in the preparation of:

$Ca(OH)_2$ from $CaCO_3$
$NH_3$ from $N_2$ and $H_2$
$HNO_3$ from $NH_3$
$SiF_4$ from Si

**14.** A typical soda-lime glass has a composition reported as 70% $SiO_2$, 15% $Na_2O$, and 10% CaO. What ratio of weights of sand ($SiO_2$), sodium carbonate ($Na_2CO_3$), and calcium carbonate ($CaCO_3$) must be melted together to make this glass? The carbonates are decomposed by heat to evolve carbon dioxide gas.

**15.** What is the maximum weight (in pounds) of magnesium that can be obtained from 1000 pounds of sea water? (See p. 146.)

**16.** Explain why fertilizer costs are directly tied to petroleum costs.

**17.** Of the three major plant macronutrients, nitrogen, potassium, and phosphorus, which is in the most danger of running out? Explain.

**18.** Calculate the percentage of $P_2O_5$ in monoammonium phosphate, $NH_4H_2PO_4$.

**19.** Calculate the percentage of N in potassium nitrate, $KNO_3$.

**20.** What acid is used to make superphosphate fertilizer from phosphate rock?

**21.** Calculate the percentage of nitrogen in ammonium nitrate, $NH_4NO_3$.

**22.** What number would you find on a bag of 50% ammonium nitrate and 50% potassium chloride?

# 8 The Ubiquitous Carbon Atom—An Introduction to Organic Chemistry

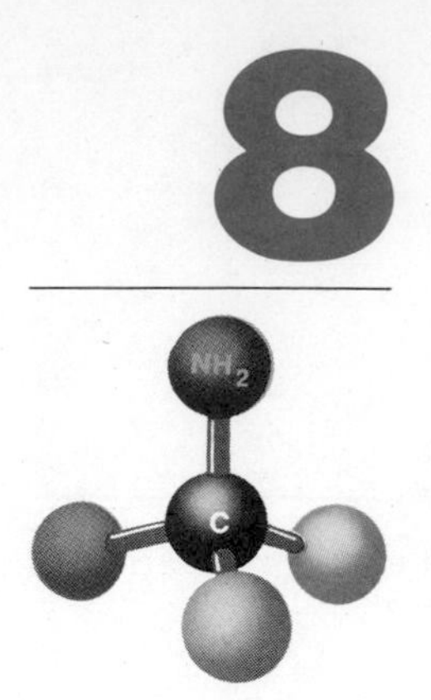


**Carbon and its compounds are vital to life on this planet.**

The importance of carbon compounds to life on earth cannot be overestimated. Consider what the world would be like if all the carbon and carbon compounds were removed suddenly. The result would be somewhat like the barren surface of the moon! Many of the little everyday things often taken for granted would be quite impossible without this versatile element. In an ordinary pencil, for example, the "lead" in the pencil (made from graphite, an elementary form of carbon), the wood, the rubber in the eraser, and the paint on the surface are all carbon or carbon compounds. The paper in this book, the cloth in its cover, and the glue holding it together are also made of carbon compounds. All of the clothes one wears, including the leather in shoes, would not exist without carbon. If carbon compounds were removed from the human body, there would be nothing left except water and a small residue of minerals, and the same is true of all forms of living matter. Fossil fuels, foods, and most drugs are essentially made of carbon compounds. In addition, many carbon compounds such as plastics and detergents, which are not directly connected with the life processes, play a vital role in our lives.

**Organic chemistry is the study of the compounds of carbon.**

Several million different carbon compounds have been studied and described in the chemical literature, and thousands of new ones are reported every year. Although there are 88 other naturally occurring elements, the number of known carbon compounds is many times greater than that of the known compounds which contain no carbon. The very large and important branch of chemistry devoted to the study of carbon compounds is ***organic chemistry.*** The name "organic" is actually a relic of the past, when chemical compounds produced from once-living matter were called "organic" and all other compounds were called "inorganic."

## Why Are There So Many Organic Compounds?

**The large number of carbon compounds is due to**
1. **stability of chains of carbon atoms**
2. **occurrence of isomers**
3. **reactivity of functional groups**

The enormous number of organic compounds has intrigued chemists for over a hundred years. The atomic theory, as developed earlier for all atoms, describes a structure for the carbon atom which explains this multiplicity of carbon compounds. The peculiar structure of this atom allows it *to form covalent bonds with other carbon atoms in a seemingly endless array of possible combinations.* A simple organic molecule may contain a single carbon-carbon bond, whereas a complex molecule may contain literally thousands of such bonds. A few other elements are capable of forming stable bonds between like atoms. These include such elements as nitrogen, $N_2$, oxygen, $O_2$, and sulfur, $S_8$, to name a few. But only S, Sn, Si, and P can form long-chain molecules.

An additional factor in the large number of carbon compounds lies in the stability of carbon chains. The carbon chains are not normally subject to attack by water or, at ordinary temperatures, by oxygen. Chains formed by atoms of other elements undergo reaction with either water or oxygen, or both, much more easily than do carbon chains.

A further reason for the large number of organic compounds is the ability of a given number of atoms to combine in more than one molecular pattern and, hence, produce more than one compound. Such compounds, each of which has molecules containing the same number and kinds of atoms, but arranged differently relative to each other, are called ***isomers.*** For example, the molecular structure represented by A—B—C is different from the molecular structure A—C—B, as is C—A—B; these three species are isomers. If we consider the number of possible ways the digits one through nine can be ordered to make nine-digit numbers, we can begin to imagine how a single group of atoms could possibly form hundreds of different molecules. Carbon, with its ability to bond to other carbon atoms, is especially well suited to form isomers.

**_Isomers_ are two or more different compounds with the same number of each kind of atom per molecule.**

A final factor explaining the large number of organic compounds is the ability of the carbon atom to form strong covalent bonds with atoms of numerous other elements, such as nitrogen, oxygen, sulfur, chlorine, fluorine, bromine, iodine, silicon, boron, and even many metals. As a result there are large classes of organic compounds. A ***functional group,*** a particular combination of atoms, appears in each member of a class. For example, all ***organic acids*** have a carboxyl group attached to another carbon atom.

**A dash in a formula represents a single bond; two electrons are shared. The double dash represents a double bond; four electrons are shared.**

$$\left(-C\begin{matrix} \nearrow O \\ \searrow OH \end{matrix}\right)$$ CARBOXYL GROUP, A FUNCTIONAL GROUP

## Chains of Carbon Atoms—The Hydrocarbons

Carbon is intermediate among the elements in its ability to attract electrons to it in a covalent bond. Thus, a carbon atom is unable to remove electrons completely from metals, and even fluorine (the best attractor of electrons among the elements) is unable to remove an electron completely from a carbon atom. As a result, carbon atoms tend not to form ionic bonds but rather to share electrons in the formation of covalent bonds. In addition to the tendency for carbon to form covalent bonds with many other atoms, there is also a remarkable inclination for carbon atoms to form relatively strong covalent bonds with each other.

**Carbon has an intermediate electronegativity.**

Only two elements, hydrogen and carbon, are required to explain the existence of literally thousands of compounds known as ***hydrocarbons.*** The simplest hydrocarbon is methane, $CH_4$. Methane is tetrahedral and has four C—H bonds, as shown in Figure 8–1.

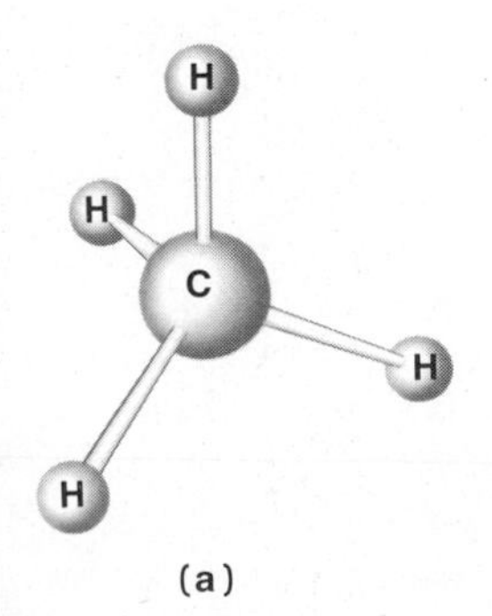


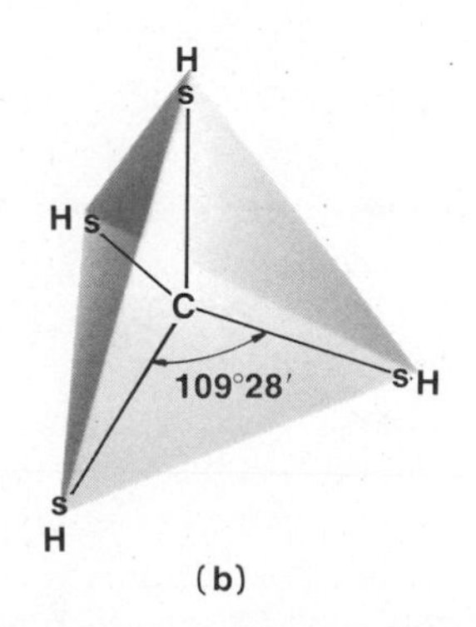

**Figure 8–1** Methane. (*a*) Ball-and-stick model showing tetrahedral structure. (*b*) Geometry of regular tetrahedron. (*c*) Model of methane, $CH_4$, showing relative size of atoms in relationship to interatomic distances.

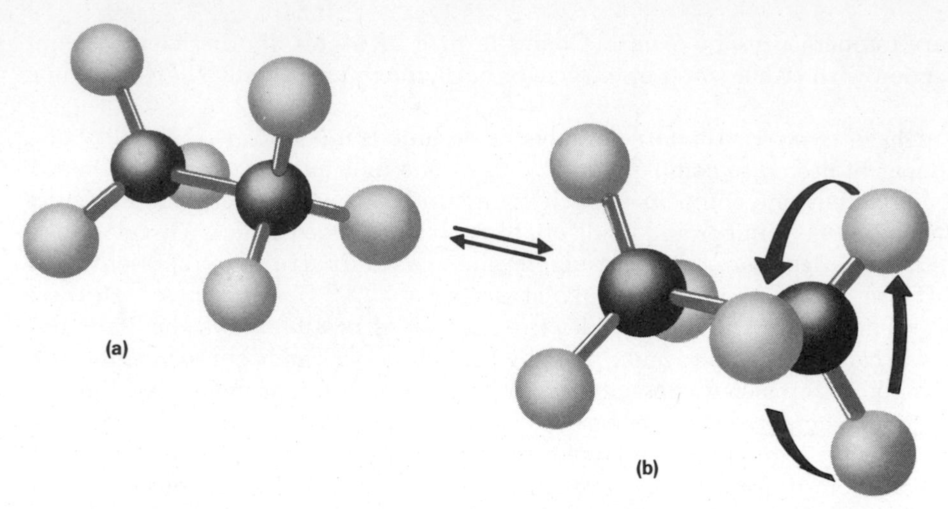


**Figure 8–2** Two possible rotational forms of the ethane molecule. The hydrogen atoms in the methyl ($CH_3$) groups may be in an eclipsed position (*a*), staggered (*b*), or in any intermediate position. In ethane the two methyl groups can rotate easily about the carbon-carbon bond.

A hydrocarbon with two carbon atoms is ethane, $C_2H_6$. Ethane has six C—H bonds and one C—C bond. The bonding in ethane is illustrated by the following formulas:

$$\begin{array}{c}\mathrm{H}\,\mathrm{H}\\ \mathrm{H}\!:\!\ddot{\mathrm{C}}\!:\!\ddot{\mathrm{C}}\!:\!\mathrm{H}\\ \ddot{\mathrm{H}}\,\ddot{\mathrm{H}}\end{array} \quad \text{or} \quad \begin{array}{ccccccc} & & \mathrm{H} & & \mathrm{H} & & \\ & & | & & | & & \\ \mathrm{H} & — & \mathrm{C} & — & \mathrm{C} & — & \mathrm{H} \\ & & | & & | & & \\ & & \mathrm{H} & & \mathrm{H} & & \end{array}$$

Even though the molecule is represented as flat, in reality the bonds to each carbon atom are in a tetrahedral arrangement, as shown in Figure 8–2. Also shown in Figure 8–2 is the rotation that takes place around the carbon-carbon single bond. Rotation of this type is a common feature in many molecules and becomes important as the molecular size increases.

By applying what we have learned, it is a simple matter to extend the concept of carbon-carbon bonding to a three-carbon molecule such as that of propane ($C_3H_8$).

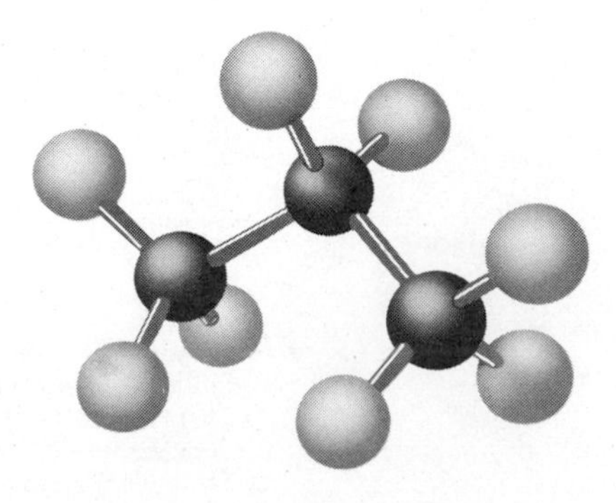

Propane

**Figure 8–3** Ball-and-stick model of propane.

```
 H H H               H   H   H
 ·· ·· ··            |   |   |
H:C:C:C:H   or   H — C — C — C — H
 ·· ·· ··            |   |   |
 H H H               H   H   H
```

In Figure 8–3, note that the three carbon atoms in propane do not lie in a straight line because of the tetrahedral bonding about each carbon atom. Also, because of the rotation about the two C—C single bonds, the molecule is "flexible."

It is apparent that these bonding concepts can be extended to a four-carbon molecule and to a limitless number of larger hydrocarbon molecules. Actually, many such compounds are known; some, such as natural rubber, are known to contain over a thousand carbon atoms in a chain.

## Structural Isomers

When we try to write the structure for butane, $C_4H_{10}$, we soon discover that two structures are possible.

Structural isomers have the same molecular formulas but a different pattern of bonds.

```
                                        H
                                        |
     H   H   H   H               H  H — C — H  H
     |   |   |   |               |      |      |
 H — C — C — C — C — H       H — C ———— C ———— C — H
     |   |   |   |               |      |      |
     H   H   H   H               H      H      H
```

| | n-BUTANE | METHYLPROPANE (ISOBUTANE) |
|---|---|---|
| MELTING POINT | −138.3°C | −160°C |
| BOILING POINT | − 0.5°C | − 12°C |
| DENSITY (at 20°C) | 0.579 g/ml | 0.557 g/ml |

The two formulas represent two distinctly different compounds. Both are well known, each with its own particular set of properties. We must conclude, then, that molecular formulas such as $C_4H_{10}$ are sometimes ambiguous, and that structural formulas are necessary. Since these are different compounds, they have different names, whose derivation will be explained later in this chapter. However, all hydrocarbons that have four carbon atoms can be generally referred to as ***butanes.***

If no carbon atom is attached to more than two other carbon atoms, the carbon chain is said to be a ***straight-chain*** structure. Actually, as shown in Figure 8–4, the carbon chain is bent (109.5°) at each carbon atom, but it is called a straight chain because the carbon atoms are bonded together in succession one after the other. You might note that many molecular shapes are possible for n-butane because of the possible rotational motions about the single bonds. These arrangements (called ***conformations***) do not constitute different molecules. Because of the ease of bond rotation, a sample of a single pure hydrocarbon contains all of its conformations, that is, molecules rotated into all of their possible shapes.

If one carbon atom is bonded to either three or four other carbon atoms in a molecule, the molecule is said to have a ***branched chain.*** Isobutane is an example of a branched-chain hydrocarbon (Fig. 8–4). Isobutane and n-butane are called ***structural isomers*** because both molecules contain exactly the same number and kinds of atoms, $C_4H_{10}$, but the molecules have different atom-to-atom bonding sequences. Structural isomerism can be compared to the results you might expect from a child building many different structures with the same collection of building blocks, and using all of the blocks in each structure.

In all branched-chain structures, at least one carbon atom is bonded to three or four other carbon atoms.

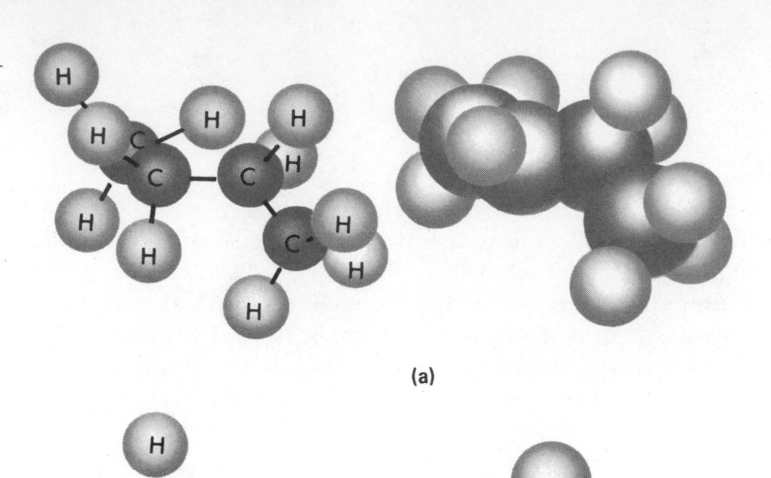


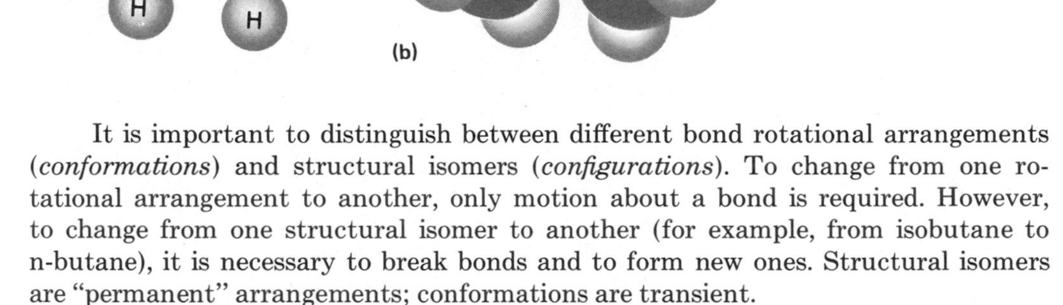


**Figure 8–4** The isomeric butanes, $C_4H_{10}$. (*a*) Normal butane, usually written n-butane. (*b*) Methylpropane (isobutane).

**Rotational forms resulting from the twisting around C—C single bonds are called conformations; they are not isomers.**

It is important to distinguish between different bond rotational arrangements (*conformations*) and structural isomers (*configurations*). To change from one rotational arrangement to another, only motion about a bond is required. However, to change from one structural isomer to another (for example, from isobutane to n-butane), it is necessary to break bonds and to form new ones. Structural isomers are "permanent" arrangements; conformations are transient.

The two butanes (and all hydrocarbon molecules) are essentially nonpolar, since the C—C bonds are nonpolar, and the slightly polar C—H bonds are symmetrically arranged to cancel each other out. The forces holding these molecules together in the liquid, therefore, are van der Waals attractions, which depend upon the surface area of a molecule and the closeness of approach of the molecules to each other. In general, a branched-chain isomer has a lower boiling point than a straight-chain isomer, since the branched-chain isomer does not permit intermolecular distances as short as those of a straight chain and has less surface area. Both of these mean less intermolecular attraction. Melting points of isomers generally do not follow the same pattern, since they also depend upon the ease with which the molecules fit into a crystalline array.

Consider the isomeric pentanes, $C_5H_{12}$. There are three of these:

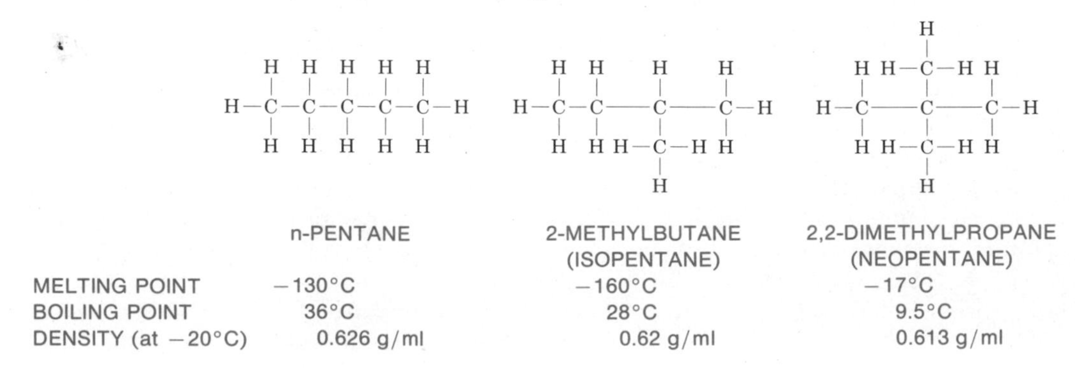

| | n-PENTANE | 2-METHYLBUTANE (ISOPENTANE) | 2,2-DIMETHYLPROPANE (NEOPENTANE) |
|---|---|---|---|
| MELTING POINT | −130°C | −160°C | −17°C |
| BOILING POINT | 36°C | 28°C | 9.5°C |
| DENSITY (at −20°C) | 0.626 g/ml | 0.62 g/ml | 0.613 g/ml |

**TABLE 8–1 Structural Isomers of Some Hydrocarbons**

| Formula | Isomers Predicted | Found |
|---|---|---|
| $C_6H_{14}$ | 5 | 5 |
| $C_7H_{16}$ | 9 | 9 |
| $C_8H_{18}$ | 18 | 18 |
| $C_{15}H_{32}$ | 4,347 | — |
| $C_{20}H_{42}$ | 366,319 | — |
| $C_{30}H_{62}$ | 4,111,846,763 | — |

All three isomers are predicted by bonding theory, and the theory predicts no other possible isomers for $C_5H_{12}$, since there are no other ways to unite the 17 atoms and have all valence electrons paired. This is the octet rule for carbon. These three isomers of pentane are well known, and no others have ever been found.

**According to the octet rule, each atom (except H) shares or controls eight valence shell electrons. Hydrogen shares only two electrons.**

Table 8–1 gives the number of isomers predicted for some larger molecular formulas, starting with $C_6H_{14}$. Every predicted isomer, *and no more,* has been isolated and identified for the $C_6$, $C_7$, and $C_8$ groups. However, not all of the $C_{15}$'s and $C_{20}$'s have been produced, but there is sufficient belief in the theory to presume that if enough time and effort were spent, all of the isomers could eventually be produced. Structural isomerism certainly helps to explain the vast number of carbon compounds.

Structural isomers can also exist in molecules with double and triple carbon-carbon bonds. For example, two of the six isomers of $C_4H_8$, 1-butene and *trans*-2-butene, have the following structures:

**Can you draw another isomer of $C_4H_8$ (there are four others)?**

```
  H  H  H  H                          H
  |  |  |  |                          |
H—C=C—C—C—H               H          C—H
        |  |                \       /|
        H  H                 C=C    H
                            /   \
                        H—C—H    H
                          |
                          H
```

| | 1-BUTENE | *TRANS*-2-BUTENE |
|---|---|---|
| MELTING POINT | −185.4°C | −106.0°C |
| BOILING POINT | −6.3°C | 1.0°C |
| DENSITY (at −20°C) | 0.641 g/ml | 0.649 g/ml |

The number placed before the name butene indicates the position number of the double bond. *Trans*-2-butene is an example of a geometric isomer that results from double bonds within a molecule. Geometric isomers will be considered later. Note that the properties of 1-butene and 2-butene definitely indicate two *different* compounds. 1-butyne and 2-butyne illustrate structural isomerism due to the positioning of triple bonds.

```
        H  H          H        H
        |  |          |        |
H—C≡C—C—C—H      H—C—C≡C—C—H
        |  |          |        |
        H  H          H        H
```

| | 1-BUTYNE | 2-BUTYNE |
|---|---|---|
| MELTING POINT | −125.8°C | −32.2°C |
| BOILING POINT | 8.1°C | 27°C |
| DENSITY (at −20°C) | 0.65 g/ml | 0.69 g/ml |

There are large numbers of single-bonded ring structures which contain three or more carbon atoms in the rings.

Hydrocarbons containing a double bond have isomers with single-bonded ring-type structures. For example, in addition to the isomers given before for $C_4H_8$, an isomer exists that has a cyclic structure with single bonds.

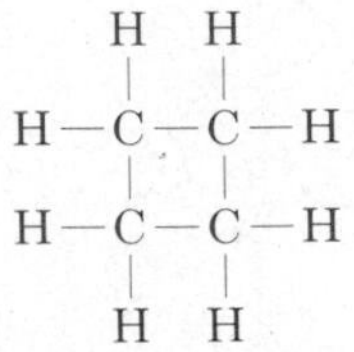

```
    H   H
    |   |
H — C — C — H
    |   |
H — C — C — H
    |   |
    H   H
```

CYCLOBUTANE

| | |
|---|---|
| MELTING POINT | −50°C |
| BOILING POINT | 12°C |
| DENSITY (AT 0°C) | 0.703 g/ml |

*Alkanes:* hydrocarbons with single bonds only; name ending -ane.

*Alkenes:* hydrocarbons with one or more double bonds; name ending -ene.

*Alkynes:* hydrocarbons with one or more triple bonds; name ending -yne.

Hydrocarbons with double and triple bonds (called ***alkenes*** and ***alkynes,*** respectively) undergo a special type of combination reaction called ***addition*** reactions. Straight-chain hydrocarbons with only single bonds (called ***alkanes***) do not undergo addition reactions. A step-by-step example of an addition reaction is discussed in the next chapter, and addition reactions find important application in addition polymerization (Chapter 10). Because alkenes and alkynes can add on other molecules (without eliminating an atom or two), they are a type of ***unsaturated compounds,*** whereas alkanes, which can add an atom only by eliminating one or more of the atoms they already contain, are a type of ***saturated compounds.***

Saturated and unsaturated hydrocarbons do undergo some similar reactions. One type shared by all hydrocarbons is ***combustion reactions,*** in which the hydrocarbon reacts with oxygen, that is, burns in air. Energy is produced, and if the amount of air is sufficient the products are carbon dioxide ($CO_2$) and water ($H_2O$).

*Unsaturated* compounds contain double or triple bonds; *saturated* compounds do not contain such bonds.

$$2CH_3{-}CH_3 + 7O_2 \longrightarrow 4CO_2 + 6H_2O + \text{heat}$$

$$CH_2{=}CH_2 + 3O_2 \longrightarrow 2CO_2 + 2H_2O + \text{heat}$$

$$2HC{\equiv}CH + 5O_2 \longrightarrow 4CO_2 + 2H_2O + \text{heat}$$

## Nomenclature—What's in a Name?

With so many organic compounds, a system of common names quickly fails owing to the shortage of unique names. As organic chemistry grew in complexity, a system of nomenclature developed that made use of numbers as well as names. Much attention has been given to the problems of naming organic compounds, and several international conventions have been held to work out a satisfactory system that can be used throughout the world. The International Union of Pure and Applied Chemistry has given its approval to a very elaborate nomenclature system (*IUPAC system*), and this system is now in general use.

A few of these IUPAC names will be needed for our discussion, and an appreciation for the basic simplicity of the approach in naming organic compounds is desirable. Inscrutable names, such as some of those encountered on medicine bottles, are replaced by systematic names that are descriptive of the molecules involved. The names of a few of the hydrocarbons are presented in Table 8–2.

For branched-chain hydrocarbons it becomes necessary to name submolecular groups. The $-CH_3$ group is called the methyl group; this name is derived from methane by dropping the -ane and adding -yl. Any of the nine other hydrocarbons listed

**The -yl ending indicates an attached group such as $-CH_3$, methyl.**

**TABLE 8–2 The First Ten Straight-Chain Saturated Hydrocarbons**

| Name | Formula | Structural Formula |
|---|---|---|
| Methane | $CH_4$ | $\begin{array}{ccccc} & & H & & \\ & & \vert & & \\ H & - & C & - & H \\ & & \vert & & \\ & & H & & \end{array}$ |
| Ethane | $C_2H_6$ | $\begin{array}{ccccccc} & & H & & H & & \\ & & \vert & & \vert & & \\ H & - & C & - & C & - & H \\ & & \vert & & \vert & & \\ & & H & & H & & \end{array}$ |
| Propane | $C_3H_8$ | $\begin{array}{ccccccccc} & & H & & H & & H & & \\ & & \vert & & \vert & & \vert & & \\ H & - & C & - & C & - & C & - & H \\ & & \vert & & \vert & & \vert & & \\ & & H & & H & & H & & \end{array}$ |
| n-Butane | $C_4H_{10}$ | $\begin{array}{ccccccccccc} & & H & & H & & H & & H & & \\ & & \vert & & \vert & & \vert & & \vert & & \\ H & - & C & - & C & - & C & - & C & - & H \\ & & \vert & & \vert & & \vert & & \vert & & \\ & & H & & H & & H & & H & & \end{array}$ |
| n-Pentane | $C_5H_{12}$ | $\begin{array}{ccccccccccccc} & & H & & H & & H & & H & & H & & \\ & & \vert & & \vert & & \vert & & \vert & & \vert & & \\ H & - & C & - & C & - & C & - & C & - & C & - & H \\ & & \vert & & \vert & & \vert & & \vert & & \vert & & \\ & & H & & H & & H & & H & & H & & \end{array}$ |
| n-Hexane | $C_6H_{14}$ | $\begin{array}{ccccccccccccccc} & & H & & H & & H & & H & & H & & H & & \\ & & \vert & & \vert & & \vert & & \vert & & \vert & & \vert & & \\ H & - & C & - & C & - & C & - & C & - & C & - & C & - & H \\ & & \vert & & \vert & & \vert & & \vert & & \vert & & \vert & & \\ & & H & & H & & H & & H & & H & & H & & \end{array}$ |
| n-Heptane | $C_7H_{16}$ | $\begin{array}{ccccccccccccccccc} & & H & & H & & H & & H & & H & & H & & H & & \\ & & \vert & & \vert & & \vert & & \vert & & \vert & & \vert & & \vert & & \\ H & - & C & - & C & - & C & - & C & - & C & - & C & - & C & - & H \\ & & \vert & & \vert & & \vert & & \vert & & \vert & & \vert & & \vert & & \\ & & H & & H & & H & & H & & H & & H & & H & & \end{array}$ |
| n-Octane | $C_8H_{18}$ | $\begin{array}{ccccccccccccccccccc} & & H & & H & & H & & H & & H & & H & & H & & H & & \\ & & \vert & & \vert & & \vert & & \vert & & \vert & & \vert & & \vert & & \vert & & \\ H & - & C & - & C & - & C & - & C & - & C & - & C & - & C & - & C & - & H \\ & & \vert & & \vert & & \vert & & \vert & & \vert & & \vert & & \vert & & \vert & & \\ & & H & & H & & H & & H & & H & & H & & H & & H & & \end{array}$ |
| n-Nonane | $C_9H_{20}$ | $\begin{array}{ccccccccccccccccccccc} & & H & & H & & H & & H & & H & & H & & H & & H & & H & & \\ & & \vert & & \vert & & \vert & & \vert & & \vert & & \vert & & \vert & & \vert & & \vert & & \\ H & - & C & - & C & - & C & - & C & - & C & - & C & - & C & - & C & - & C & - & H \\ & & \vert & & \vert & & \vert & & \vert & & \vert & & \vert & & \vert & & \vert & & \vert & & \\ & & H & & H & & H & & H & & H & & H & & H & & H & & H & & \end{array}$ |
| n-Decane | $C_{10}H_{22}$ | $\begin{array}{ccccccccccccccccccccccc} & & H & & H & & H & & H & & H & & H & & H & & H & & H & & H & & \\ & & \vert & & \vert & & \vert & & \vert & & \vert & & \vert & & \vert & & \vert & & \vert & & \vert & & \\ H & - & C & - & C & - & C & - & C & - & C & - & C & - & C & - & C & - & C & - & C & - & H \\ & & \vert & & \vert & & \vert & & \vert & & \vert & & \vert & & \vert & & \vert & & \vert & & \vert & & \\ & & H & & H & & H & & H & & H & & H & & H & & H & & H & & H & & \end{array}$ |

in Table 8–2 can give rise to a similar group. For example, the propyl group would be $-C_3H_7$. As an illustration of the use of the group names, consider this formula:

**An *aliphatic compound* is any hydrocarbon that has an open chain of single-bonded carbon atoms.**

$$
\begin{array}{ccccccccccccc}
 & & H & & H & & H & & H & & H & & \\
 & & | & & | & & | & & | & & | & & \\
H & - & C & - & C & - & C & - & C & - & C & - & H \\
 & & | & & | & & | & & | & & | & & \\
 & & H & & H & & H & & H-C-H & & H & & \\
 & & & & & & & & | & & & & \\
 & & & & & & & & H & & & &
\end{array}
$$

**position of group on chain**
↓
**2-*methylpentane***
↑ group attached to chain ↖ longest chain of C atoms

The *longest* carbon chain in the molecule is five carbon atoms long; hence, the root name is pentane. Furthermore, it is a methylpentane (written as one word) because a methyl group is attached to the pentane structure. In addition, a number is needed because the methyl group could be bonded to either the second or third carbon atom.

$$
\begin{array}{ccccccccccccc}
 & & H & & H & & H & & H & & H & & \\
 & & | & & | & & | & & | & & | & & \\
H & - & C & - & C & - & C & - & C & - & C & - & H \\
 & & | & & | & & | & & | & & | & & \\
 & & H & & H & & H-C-H & & H & & H & & \\
 & & & & & & | & & & & & & \\
 & & & & & & H & & & & & &
\end{array}
$$

3-METHYLPENTANE

$$
\begin{array}{ccccccccccccc}
 & & H & & H & & H & & H & & H & & \\
 & & | & & | & & | & & | & & | & & \\
H & - & C & - & C & - & C & - & C & - & C & - & H \\
 & & | & & | & & | & & | & & | & & \\
 & & H & & H & & H & & H-C-H & & H & & \\
 & & & & & & & & | & & & & \\
 & & & & & & & & H & & & &
\end{array}
$$

2-METHYLPENTANE

**Positional numbers of groups should have the smallest possible sum.**

Note that 2-methylpentane is the same as 4-methylpentane since the latter would be the same molecule turned around; the accepted rule requires numbering from the end of the carbon chain that will result in the smallest numbers. Therefore, 2-methylpentane is the correct name.

Any number of substituted groups can be handled in this same fashion. Consider the name and the formula for 3,3,4,6-tetramethyl-5,5-diethyloctane.

$$
\begin{array}{ccccccccccc}
 & & CH_3 & & H & & C_2H_5 & & H & & \\
 & & | & & | & & | & & | & & \\
CH_3CH_2 & - & C & - & C & - & C & - & C & - & CH_2CH_3 \\
 & & | & & | & & | & & | & & \\
 & & CH_3 & & CH_3 & & C_2H_5 & & CH_3 & &
\end{array}
$$

If a double bond appears in a hydrocarbon, then the root name, which indicates the number of carbon atoms, must be modified to reflect the double bond structure and its position. Changing -ane to -ene indicates the presence of the double bond, and a number is used to indicate its position. For example:

**No number is necessary for ethene or propene because there is only one possible position for the double bond.**

$CH_2{=}CH_2$ ETHENE (COMMON NAME: ETHYLENE)

$CH_2{=}CHCH_3$ PROPENE

$CH_2{=}CHCH_2CH_3$ 1-BUTENE

$CH_3CH{=}CHCH_3$ 2-BUTENE

$$
\begin{array}{ccc}
 & & CH_3 \\
 & & | \\
CH_3 & - & C{=}CHCH_2CH_3
\end{array}
$$

2-METHYL-2-PENTENE

If a triple bond is present, the -ane is changed to -yne. Examples:

$H-C{\equiv}C-H$ ETHYNE (COMMON NAME: ACETYLENE)

$CH_3CH_2C{\equiv}CH$ 1-BUTYNE

$CH_3-C{\equiv}C-CH_3$ 2-BUTYNE

When other groups are present in an organic compound, the names are developed on the same basis as those of the hydrocarbons, as can be seen from the following examples:

| | |
|---|---|
| $CH_3CH_2CH_2Br$ | 1-BROMOPROPANE |
| $CH_3CHBrCH_3$ | 2-BROMOPROPANE |
| $BrC{\equiv}CH$ | BROMOETHYNE |
| $CH_3{-}CH(CH_3){-}CH_2Br$ (structure below) | 1-BROMO-2-METHYLPROPANE |
| $CH_3CHBrCH_2Br$ | 1,2-DIBROMOPROPANE |

```
      CH3
       |
CH3 —  C — H
       |
      CH2Br
```

In formulas such as $CH_3CH_2CH_2Br$, elements directly following a carbon are bonded to that carbon. $CH_3CH_2CH_2Br$ is the same as

```
    H   H   H
    |   |   |
H — C — C — C — Br
    |   |   |
    H   H   H
```

## Self-Test 8-A

1. The branch of chemistry that deals with compounds of carbon is known as ____________ chemistry.
2. The structure of the $CH_4$ molecule is described as ____________.
3. How many covalent bonds are in a molecule of butane? ____________

   Does it make a difference which butane is considered? ____________
4. A straight-chain hydrocarbon, such as pentane, actually has all of its carbon atoms in a straight line. (True/False)
5. How many different isomers of $C_5H_{12}$ are shown below?

```
(a)   H   H   H   H
      |   |   |   |
  H — C — C — C — C — H
      |   |   |   |
      H   H   H   |
                  |
              H — C — H
                  |
                  H

(b)       H
          |
      H — C — H
          |
          |   H   H
          |   |   |
      H — C — C — C — H
          |   |   |
          H   H   |
                  |
              H — C — H
                  |
                  H

(c)       H   H   H   H
          |   |   |   |
      H — C = C — C — C — H
                  |   |
                  H   |
                      |
                  H — C — H
                      |
                      H

(d)   H        H        H
      |        |        |
  H — C ————— C ——————— C — H
      |        |        |
      |        H        |
  H — C — H        H — C — H
      |                 |
      H                 H

(e)           H               H
              |               |
      H   H — C — H   H — C — H
      |       |               |
  H — C ————— C ————————————— C — H
      |       |               |
      H       H               H

(f)   H   H   H   H
      |   |   |   |
  H — C — C — C — C — H
      |   |   |   |
      H   H   |   H
              |
          H — C — H
              |
              H
```

6. When the name of a compound ends in -ene (for example, butene), what structural feature is indicated? ____________

7. Name the compound shown on the right:

______________

```
                          H
                          |
                        H-C-H
                          |
        H   H   H         |    H   H
        |   |   |         |    |   |
      H-C - C - C ------- C -- C - C-H
        |   |   |         |    |   |
        H   |   H         H    H   H
            |
          H-C-H
            |
            H
```

8. The formula for the ethyl group is ______________.

---

## Optical Isomers

Stereoisomers have the same atoms and the same bonds but the atoms are arranged in space differently. Optical isomers are one type of stereoisomers.

In the preceding section, it was pointed out that in order to change one *structural* isomer into another, at least two carbon atoms have to change the atoms to which they are bonded. However, it is possible for some sets of atoms to form two isomeric molecules, both of which have the same atoms bonded to each other. One type of this kind of isomerism is ***optical isomerism.***

Optical isomerism is possible when a molecular structure is asymmetric (without symmetry). One common example of an asymmetric molecule is one containing a tetrahedral carbon atom bonded to four *different* atoms or groups of atoms. Such a carbon atom is called an ***asymmetric*** carbon atom; an example is the carbon atom in the molecule CBrClIH.

Figure 8–5 shows the two ways to arrange four different atoms in the tetrahedral positions about the central carbon atom. These result in two nonsuperimposable, mirror-image molecules that are optical isomers.

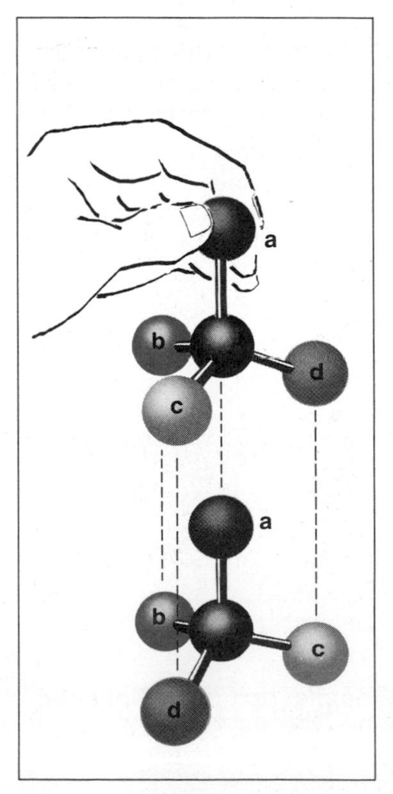


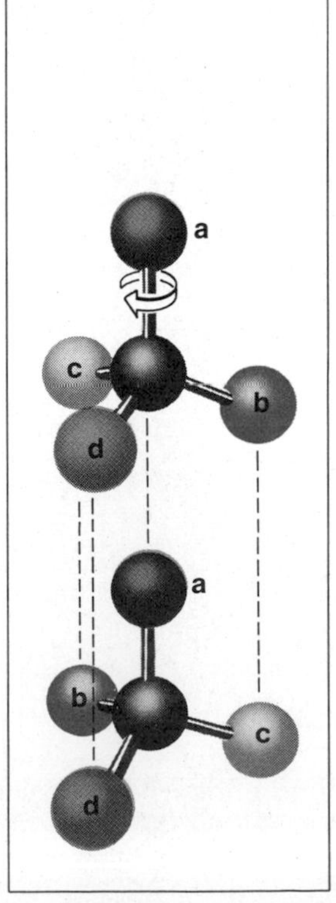


**Figure 8–5** Optical isomers. Four different atoms, or groups of atoms, are bonded to tetrahedral center atoms so that the upper isomeric form cannot be turned in any way and exactly match the lower structure. The upper structure and the lower structure are nonsuperimposable mirror images. See also Figure 8–6.

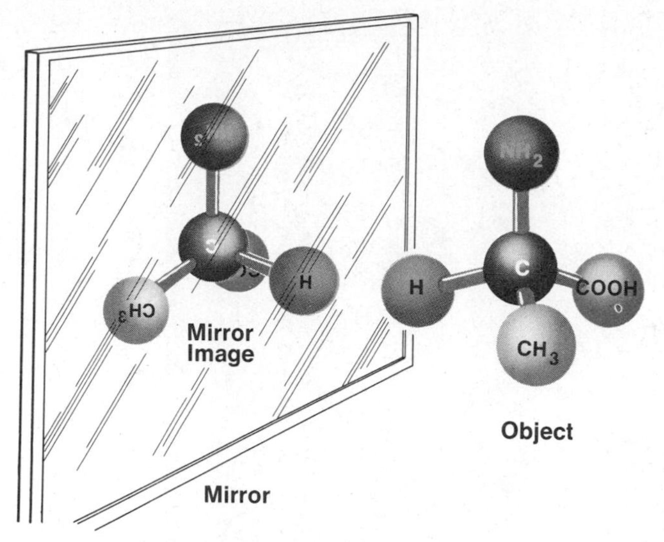


**Figure 8–6** Optical isomers of the amino acid, alanine.

$$\begin{array}{c} COOH \\ | \\ H_2N-C-H \\ | \\ CH_3 \end{array}$$

The D-form is the nonsuperimposable mirror image of the L-form (see Fig. 8–5).

There are many examples of nonsuperimposable mirror images in the macroscopic world. Consider right- and left-hand gloves, for instance. They are mirror images of one another and are nonsuperimposable. In Figure 8–6, this mirror image relationship is shown for isomeric forms of alanine, the molecules of which contain a tetrahedral carbon atom surrounded by an amino group ($-NH_2$), a methyl group ($-CH_3$), an acid group ($-COOH$), and a hydrogen atom. Note in Figure 8–6 that the carbon atoms in the methyl and acid groups are not asymmetric since these atoms are not bonded to four different groups.

The properties of some optical isomers are almost identical. Different compounds whose molecules are *mirror images* of one another have the same melting point, the same boiling point, the same density, and many other identical physical and chemical properties. However, they always differ in one physical property: they rotate the plane of ***polarized*** light in opposite directions. According to the wave theory of light, a light wave traveling through space vibrates at right angles to its path (Fig. 8–7). A group of such rays traveling together vibrate in random directions, all of which are at right angles to the path of travel. If such a group of waves is passed through a polarizing crystal, such as Iceland spar (a form of $CaCO_3$), or through a sheet of Polaroid material, the light is split into two rays and the waves emerging along the incoming axis will vibrate in only one plane perpendicular to the light path. Such light is said to be ***plane polarized.*** When plane-polarized light is passed through a solution of D-lactic acid, the light is still polarized, but the plane of vibration is rotated somewhat in one direction. If the other lactic acid isomer is substituted (L-lactic acid), just the opposite rotation of the light is obtained.

**D- and L- simply indicate that two structures are possible around an asymmetric C atom. The D- and L- notations do not indicate which way the substance will rotate the plane-polarized light.**

D: R′ — C (H, R, OH)

L: R′ — C (HO, R, H)

PUT MORE OXIDIZED GROUP SUCH AS ALDEHYDE (—CHO) AT TOP

Optical isomers can also differ in biological properties. An example is the hormone adrenalin (or epinephrine). Adrenalin is one of a pair of optical isomers. C* designates the asymmetric carbon atom. Only the isomer that rotates plane-polarized light to the left is effective in starting a heart that has stopped beating momentarily, or in giving a person unusual strength during times of great emotional stress. The other isomer is inactive.

$$\begin{array}{c} OH \\ | \\ C \\ H-C \quad C-OH \\ H-C \quad C-H \\ C \\ H \quad\quad H \\ H-C^*-C-N \\ OH \quad H \quad CH_3 \end{array}$$

*Adrenalin (epinephrine)*

It is also interesting to note that during the contraction of muscles the body produces only the L-form of lactic acid and not the D-form. The concentration of this lactic acid in the blood is associated with the feeling of tiredness, and a period of rest is necessary to reduce the concentration of this chemical by oxidation.

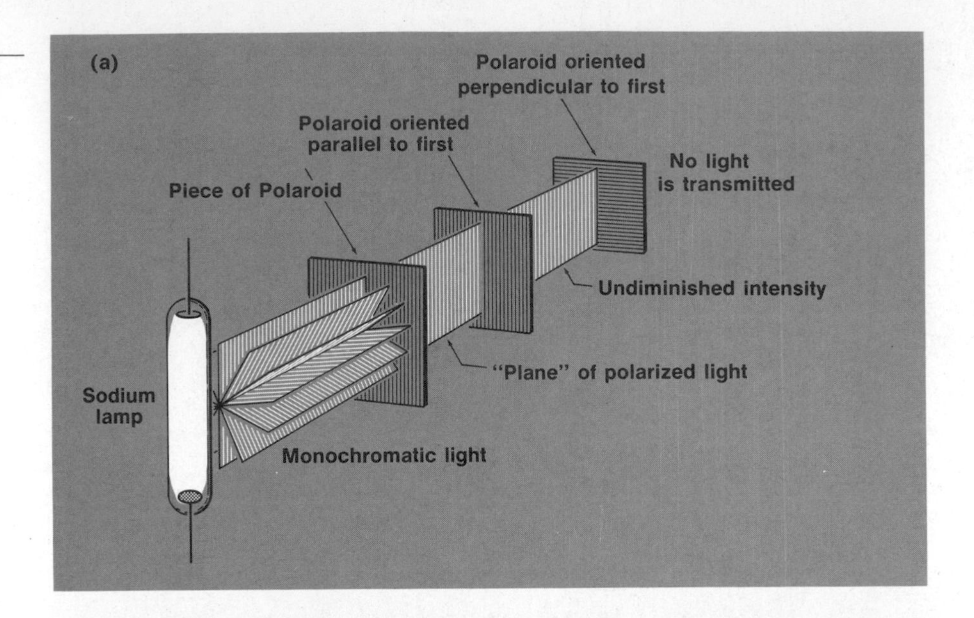


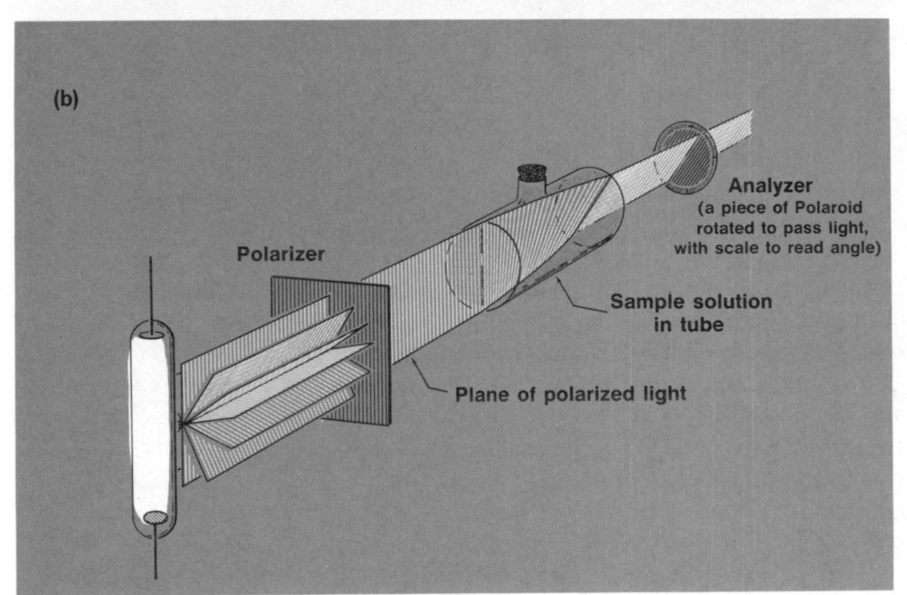


**Figure 8–7** Rotation of plane-polarized light by an optical isomer. (*a*) A sodium lamp provides a monochromatic yellow light. The original beam is nonpolarized; it vibrates in all directions at right angles to its path. After passing through a Polaroid filter, the light is vibrating in only one direction. This polarized light will pass through another Polaroid filter if the filter is lined up properly but will not pass through the third Polaroid filter if it is at right angles to the other two. The direction of the Polaroid filters determines the direction of polarization. (*b*) The plane of polarized light is rotated by a solution of an optically active isomer. The analyzer can be a second Polaroid filter that can be rotated to find the angle for maximum transmission of light. If the solution rotates the plane of polarized light, the analyzer will not be at the same angle as the polarizer for maximum transmission.

Large organic molecules may have many asymmetric carbon atoms within the same molecule. At each such carbon atom there exists the possibility of *two* arrangements of the molecule. The total number of possible molecules, then, increases exponentially with the number of asymmetric centers. With two asymmetric carbon atoms there are $2^2$ or four possible structures; for three, there are $2^3$ or eight possible structures. It should be emphasized that each of the eight isomers can be made from the *same* set of atoms with the *same* set of chemical bonds. Glucose, a simple blood sugar also known as dextrose, contains four asymmetric carbon atoms per molecule. Thus, there are $2^4$ (16) isomers in the family of stereoisomers to which glucose belongs. Obviously, then, the concept of optical isomerism helps to explain the vast number of carbon compounds.

```
       O
       ‖
       C—H
       |
   H—C*—OH
       |
  HO—C*—H
       |
   H—C*—OH
       |
   H—C*—OH
       |
  HO—C—H
       |
       H
```

GLUCOSE (C* = asymmetric carbon atom)

## Geometric Isomers

Geometric isomers are another type of stereoisomers.

Where carbon-carbon double bonds exist in a molecule, geometric isomerism is possible. A double bond between carbon atoms does not allow free rotation, and this lack of rotation provides a structural basis for geometric isomerism.

Consider the compound ethene, $C_2H_4$. Its six atoms lie in the same plane, with bond angles of approximately 120°:

```
H       H
 \     /
  C=C
 /     \
H       H
```

To have geometric isomers, each carbon connected by the double bond must have two unlike groups attached.

If two chlorine atoms replace two hydrogen atoms, one on each carbon atom of ethene ($H_2C{=}CH_2$), the result is CHCl=CHCl. Experimental evidence confirms the existence of two compounds with this general arrangement. If the two chlorine atoms are close together, this is characteristic of one isomer (the ***cis*** isomer), and if they are far apart, another isomer (the ***trans*** isomer) is indicated. Both compounds are called 1,2-dichloroethene (the 1 and 2 indicate that the two chlorine atoms are attached to different carbon atoms). They are distinguished from each other by the prefixes *cis* and *trans*. Note that the two isomeric compounds have significant differences in their properties.

```
Cl       Cl            H        Cl
  \     /                \     /
   C=C                    C=C
  /     \                /     \
H        H             Cl       H
```

| | *CIS*-1,2-DICHLOROETHENE | *TRANS*-1,2-DICHLOROETHENE |
|---|---|---|
| MELTING POINT | −80.5°C | −50°C |
| BOILING POINT | 60.1°C | 48.4°C |
| DENSITY (AT 15°C) | 1.291 g/ml | 1.265 g/ml |

The third possible isomer 1,1-dichloroethene (a *structural* isomer of the *cis* and *trans* isomers), does *not* have *cis* and *trans* structures.

As a general rule, *trans* isomers have higher melting points than *cis* isomers because of the greater ease with which the *trans* molecules can fit into the lattice and form strong intermolecular bonds.

When there is a carbon-carbon double bond in an organic molecule, the possibility exists for *cis* and *trans* isomers. Sometimes a number of such bonds can be found in the same molecule, giving rise to numerous isomeric compounds.

## Functional Groups

Alcohols contain the —OH functional group.

Carbon forms covalent bonds with a number of elements besides hydrogen. As a result, certain groups of atoms called ***functional groups*** appear over and over again in different organic compounds. Consider, for example, the —OH group, called the hydroxyl group. A large number of molecules contain an —OH group and have properties characteristic of a class of organic compounds called ***alcohols.***

The —OH group has the same combining power as an atom of hydrogen. This means that in any of the hydrocarbons considered thus far, any of the hydrogen atoms could be replaced by an —OH group to form an alcohol. A single hydrocarbon molecule can give rise to a number of alcohols if there are different isomeric positions for the —OH group. Three different alcohols result when a hydrogen atom is replaced by an —OH group in n-pentane, depending on which hydrogen atom is replaced (Table 8–3). When one or more functional groups appear in a molecule, the IUPAC name reveals the functional group name and position. For example, the name of an alcohol will use the root of the name of the hydrocarbon to which it corresponds to indicate the number of carbon atoms, and the suffix -ol to denote an alcohol. As before, a number is used to indicate the position of the alcohol group.

Some major functional groups that we shall consider in following chapters are shown in Table 8–4. The symbol —R stands for any hydrocarbon group such as methyl ($-CH_3$) or ethyl ($-C_2H_5$).

**TABLE 8–3 Alcohols Derived From Pentane ($C_5H_{12}$)**

```
  H   H   H   H   H                                  H   H   H   H   H
  |   |   |   |   |                                  |   |   |   |   |
H-C - C - C - C - C-H <-                  gives    H-C - C - C - C - C-OH
  |   |   |   |   |                                  |   |   |   |   |
  H   H   H   H   H                                  H   H   H   H   H
          substitution of an —OH for an end hydrogen      1-PENTANOL

  H   H   H   H   H                                  H   H   H   H   H
  |   |   |   |   |                                  |   |   |   |   |
H-C - C - C - C - C-H                     gives    H-C - C - C - C - C-H
  |   |   |   |   |                                  |   |   |   |   |
  H   H   H   H   H                                  H   H   H   O   H
              ↑                                                  H
          substitution of an —OH for a 2-carbon hydrogen      2-PENTANOL

  H   H   H   H   H                                  H   H   H   H   H
  |   |   |   |   |                                  |   |   |   |   |
H-C - C - C - C - C-H                     gives    H-C - C - C - C - C-H
  |   |   |   |   |                                  |   |   |   |   |
  H   H   H   H   H                                  H   H   O   H   H
          ↑                                                  H
          substitution of an —OH for a 3-carbon hydrogen      3-PENTANOL
```

**TABLE 8–4 Classes of Organic Compounds Based on Functional Groups***

| General Formulas of Class Members | Class Name | Typical Compound | Compound Name | Common Use of Sample Compound |
|---|---|---|---|---|
| $R{-}OH$ | alcohol | $H{-}\underset{H}{\overset{H}{C}}{-}OH$ | methanol (wood alcohol) | solvent |
| $R{-}\overset{\overset{O}{\|}}{C}{-}H$ | aldehyde | $H{-}\overset{\overset{O}{\|}}{C}{-}H$ | methanal (formaldehyde) | preservative |
| $R{-}\overset{\overset{O}{\|}}{C}{-}OH$ | carboxylic acid | $H{-}\underset{H}{\overset{H}{C}}{-}\overset{\overset{O}{\|}}{C}{-}OH$ | ethanoic acid (acetic acid) | vinegar |
| $R{-}\overset{\overset{O}{\|}}{C}{-}R'$ | ketone | $H{-}\underset{H}{\overset{H}{C}}{-}\overset{\overset{O}{\|}}{C}{-}\underset{H}{\overset{H}{C}}{-}H$ | propanone (acetone) | solvent |
| $R{-}O{-}R'$ | ether | $C_2H_5{-}O{-}C_2H_5$ | diethyl ether (ethyl ether) | anesthetic |
| $R{-}O{-}\overset{\overset{O}{\|}}{C}{-}R'$ | ester | $CH_3{-}CH_2{-}O{-}\overset{\overset{O}{\|}}{C}{-}CH_3$ | ethyl ethanoate (ethyl acetate) | solvent in fingernail polish |
| $R{-}N{<}^{H}_{H}$ | amine | $H{-}\underset{H}{\overset{H}{C}}{-}N{<}^{H}_{H}$ | methylamine | tanning (foul odor) |
| $R{-}\overset{\overset{O}{\|}}{C}{-}\overset{H}{N}{-}R'$ | amide | $CH_3{-}\overset{\overset{O}{\|}}{C}{-}N{<}^{H}_{H}$ | acetamide | plasticizer |

*R stands for an H or a hydrocarbon group such as $CH_3{-}$, $C_2H_5{-}$, etc.
R′ could be a different group from R.

## Aromatic Compounds

All of the hydrocarbons we have discussed up to this point have localized electronic structures; that is, the bonding electrons are essentially fixed between two atomic centers as in C—C or C=C bonds. For a large group of organic compounds known as ***aromatic*** compounds, this type of complete electron localization is not found. Rather, these compounds have some delocalized electrons, which are spread over several atoms or even the entire molecule, a feature that leads to some interesting chemical properties.

**Delocalized electrons can occupy orbitals on several nuclei.**

The simplest aromatic compound is *benzene* ($C_6H_6$). The molecular structure of benzene is a ring of carbon atoms in a plane, with one hydrogen atom bonded to each carbon atom. The bonds between these carbon atoms are shorter than single bonds, but longer than double bonds. The measured bond angles are 120°.

**Benzene is the number sixteen commercial chemical.**

One way to account for the bonding and structure of the benzene molecule is to use two of the three $sp^2$ hybrid orbitals of each carbon to form ***sigma*** bonds between the carbon atoms and to use the six remaining $sp^2$ orbitals to form six sigma bonds with six hydrogens.* The six p orbitals (one on each carbon) not involved in the formation

*At this point, you may wish to review sigma and pi bonds in Chapter 4.

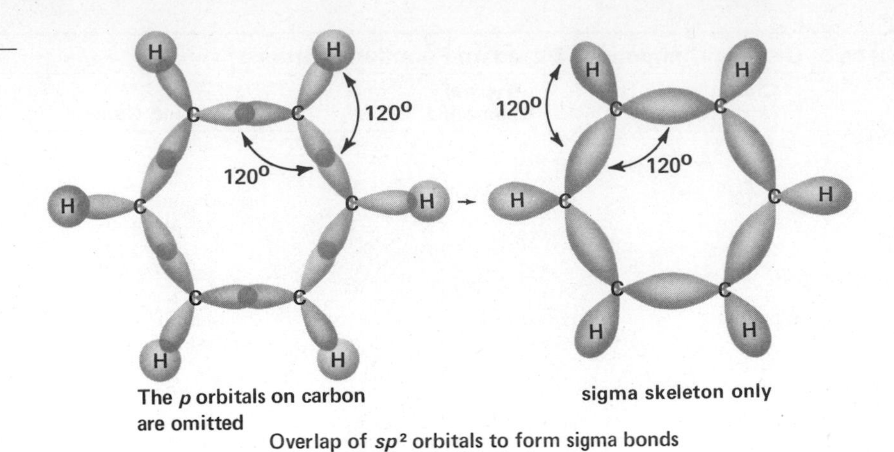


Overlap of $sp^2$ orbitals to form sigma bonds

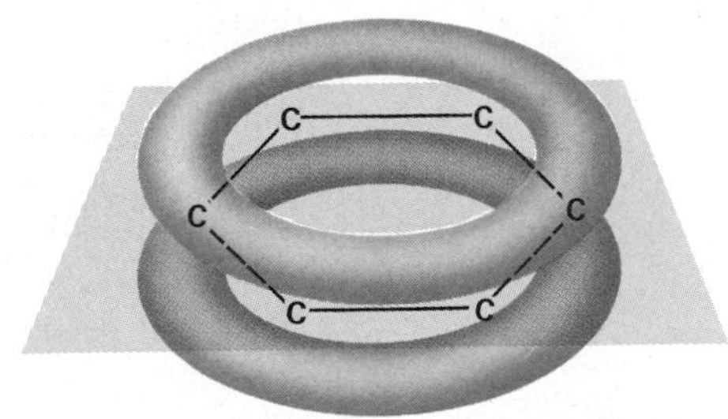


Lateral overlap of six *p* orbitals to form pi bonds

**Figure 8–8** Bonding in an aromatic compound, benzene, $C_6H_6$.

of $sp^2$ orbitals can overlap laterally to form ***pi*** bonds. These pi bonds are not exactly the same as those discussed earlier for double and triple bonds because there is a ring system of p orbitals in an aromatic molecule. Each p orbital overlaps with the p orbitals on both neighboring carbon atoms. This causes the pi bonds to be less well fixed (or less localized) between the three pairs of p orbitals. Figure 8–8 illustrates this type of pi bonding. The pi bonds are averaged evenly around the ring; we say that they are ***delocalized.***

**Aromatic compounds have electrons delocalized over the orbitals of carbon atoms arranged in a ring.**

Various symbols have been used to represent benzene. The first two of the following structures suggest two different types of carbon-carbon bonding within the same ring, with three single bonds and three double bonds; however, there is only one type of bond. Hence, the structure on the right is preferred. The hydrogen atoms are not shown.

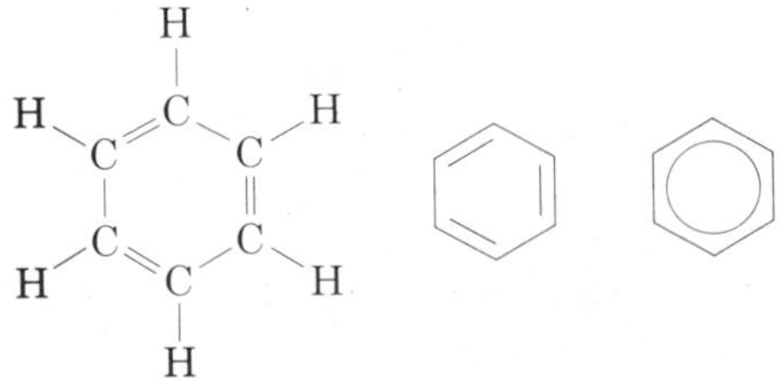

REPRESENTATIONS OF THE BENZENE RING

A large number of aromatic compounds related to benzene contain various atoms, hydrocarbon subgroups, and functional groups replacing ring hydrogen atoms. These molecules exhibit a wide variety of properties and differ greatly in their chemical reactivity. Some interesting examples of isomerism are also possible. Consider the three different compounds with the formula $C_8H_{10}$ found in some coal tars. There are several names given to these isomers:

Xylenes rank number twenty-three among commercial chemicals.

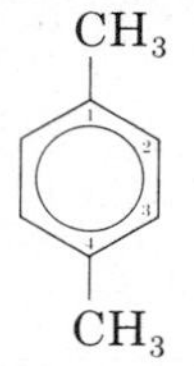

1,4-DIMETHYLBENZENE (*PARA*-XYLENE) mp 13.3°C

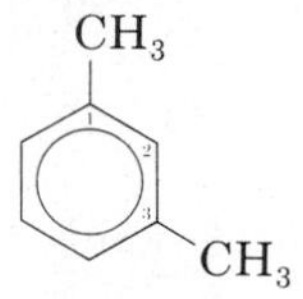

1,3-DIMETHYLBENZENE (*META*-XYLENE) mp −47.9°C

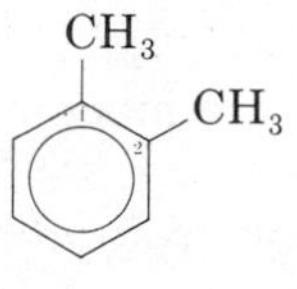

1,2-DIMETHYLBENZENE (*ORTHO*-XYLENE) mp −25°C

Each of these isomers has two methyl groups substituted for hydrogen atoms on the ring. The prefixes ***para, meta-,*** and ***ortho-*** are used if there are only two groups on the benzene ring.

If more than two groups occur, a number system is most useful. Consider the following compounds:

**Nomenclature in aromatic compounds:**

***ortho***—**groups on adjacent carbon atoms (1,2 positions)**
***meta***—**groups have one carbon atom between them (1,3 positions)**
***para***—**groups have two carbon atoms between them (1,4 positions)**

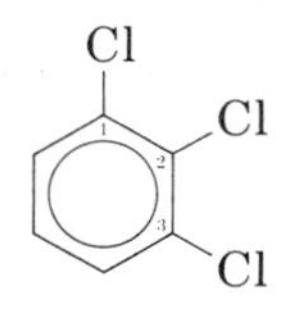

1,2,3-TRICHLOROBENZENE

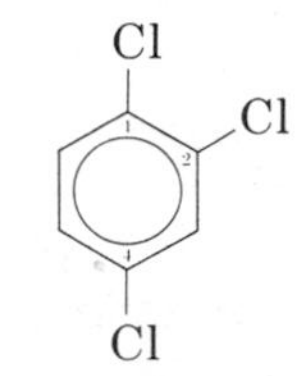

1,2,4-TRICHLOROBENZENE

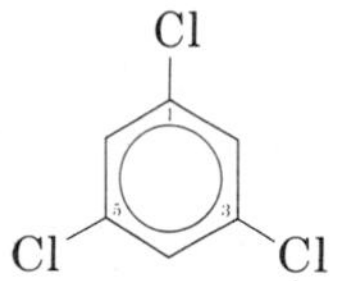

1,3,5-TRICHLOROBENZENE

There are no other ways of drawing these three isomers, and only three trichlorobenzenes have been isolated in the laboratory.

Some rings have nitrogen, oxygen, or sulfur atoms in place of a few carbon atoms in the ring. Only aromatic ring structures incorporating nitrogen will be encountered later in the text. Examples are pyridine and pyrimidine.

PYRIDINE

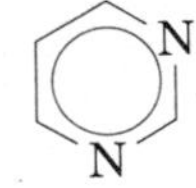

PYRIMIDINE

Not all ring structures are aromatic. The bonds in the alkane cyclohexane are all localized single bonds.

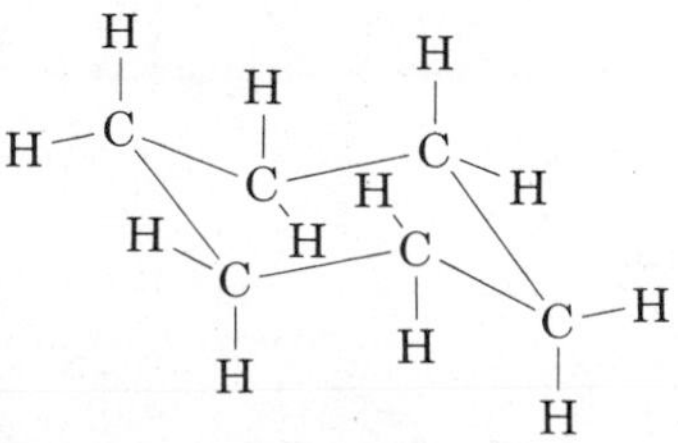

CYCLOHEXANE

There are some specific reasons for the large number of carbon compounds.

No wonder carbon is ubiquitous; there are over four million recorded compounds containing carbon. These millions of carbon compounds exist for the following reasons:

1. The ability of carbon to form covalent bonds to other carbon atoms almost without limit.
2. The ability of a given number of carbon atoms to combine in more than one molecular pattern—isomers.
3. The ability of carbon to form stable covalent bonds to a large number of other atoms—functional groups.
4. The stability of carbon chains in the presence of substances such as oxygen and water, which usually destroy chains of other atoms.

Structural differences explain how two compounds can have the same chemical composition by weight, and yet have different physical and chemical properties. Some order can be brought out of chaos with an understanding of isomerism and a systematic approach to the possible molecular structures.

## Self-Test 8-B

1. In order to have optical isomers in carbon compounds, a carbon atom must have ____________ different groups attached.
2. In what physical property do optical isomers that are mirror images differ? ____________
3. How many optical isomers would be possible if five asymmetric carbon atoms are contained in a structure? ____________
4. The benzene ring has both localized electrons (sigma bonds) and ____________ electrons (pi bonds).
5. How many atoms does the symbol, [benzene ring], represent? ____________
6. Name the following compound. ____________

   [benzene ring with $CH_3$, $CH_3$, $CH_3$ substituents]

7. All ring structures contain only carbon atoms in the ring and have delocalized electrons (True/False)

## Matching Set

| | | |
|---|---|---|
| ______ | 1. organic chemistry | a $-\overset{\overset{\displaystyle O}{\|\|}}{C}-O-H$ |
| ______ | 2. isomers | b occurs when pi bonds overlap over three or more atomic centers |
| ______ | 3. functional group | c compound such as $C_3H_8$ |
| ______ | 4. hydrocarbon | d lateral overlap of p orbitals |
| ______ | 5. methyl group | e chemistry of carbon compounds |

_______ 6. asymmetric carbon atom
_______ 7. delocalized electrons
_______ 8. sigma bond
_______ 9. pi bond

**f** $-CH_3$
**g** same number and kinds of atoms differently arranged
**h** end-to-end overlap of p orbitals
**i** has four different groups attached

## Questions

**1.** *Saturated hydrocarbons* are so named because they have the maximum amount of hydrogen present for a given amount of carbon. The saturated hydrocarbons have the general formula $C_nH_{2n+2}$ where $n$ is a whole number. What are the names and formulas of the first four members of this series of compounds?

**2.** What is the simplest aromatic compound?

**3.** Using the periodic table and electron dot formulas, illustrate the bonding in the compound cyclopropane, $C_3H_6$.

**4.** Draw the structural formula for each of the five isomeric hexanes, $C_6H_{14}$.

**5.** Write the structural formulas for:
a. 2-methylbutane
b. ethylpentane
c. 4,4-dimethyl-5-ethyloctane
d. methylbutane
e. 2-methyl-2-hexene
f. 3-methyl-3-hexanol

**6.** Give the names for:

```
(a)      H     H     H
         |     |     |
     H—C———C———C—H
         |     |     |
         H   H—C—H   H
               |
               H

(b)      H  H      H       H  H
         |  |      |       |  |
     H—C—C———C———C—C—H
         |  |      |       |  |
         H  H    H—C—H     H  H
                   |
                   H

(c)            H          H
               |          |
         H   H—C—H    H—C—H   H
         |     |          |     |
     H—C———C=======C———C—H
         |                      |
         H                      H

(d)      H          H
         |          |
     H—C—C≡C—C—H
         |          |
         H          H

(e)                H
                   |
         H  H    H—C—H   H
         |  |      |      |
     H—C—C———C———C—OH
         |  |      |      |
         H  H      H      H
```

**7.** How can optical isomers be distinguished from each other experimentally?

**8.** a. Which arrangement has a mirror image that is nonsuperimposable?
b. Explain the term "asymmetric carbon atom."

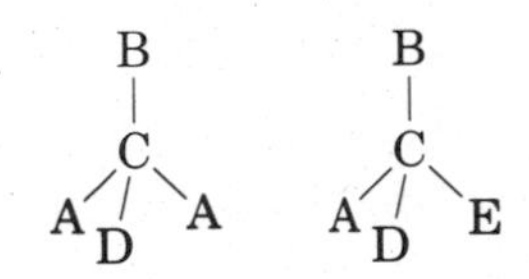


**9.** If for a pair of optical isomers, a solution of the D-isomer rotates plane-polarized light clockwise and the L-isomer rotates the light counterclockwise, predict how a mixture containing equal amounts of the two isomers would affect the light.

**10.** How many optical isomers can there be for a molecular structure containing eight asymmetric carbon atoms?

**11.** What unique bond is present in an alkyne hydrocarbon?

**12.** Which do you think would be better to use for medicinal purposes, pure adrenalin obtained from natural products (found in nature) or the pure compound as synthesized in the laboratory? Why?

**13.** Distinguish between the classical and modern use of the word "organic" in chemistry.

**14.** Look up the formula for ethyl alcohol (use the index), and explain how the formula for this compound indicates that it is a member of the alcohol family.

**15.** What structural feature characterizes aromatic compounds?

**16.** Indicate the functional groups present in the following molecules:

a. $CH_3CH_2CH_2COOH$

b. $CH_3CH_2NH_2$

c.
```
CH3CHCH2CH2COOH
   |
   NH2
```

d.
```
CH3CHCH2COOH
   |
   OH
```

e.
```
CH3CCH2CH2COOH
   ||
   O
```

f.
```
CH3CHCH2OH
   |
   NH2
```

**17.** Name the following compounds:

a. $CH_3CH_2CH_2COOH$,

b. $CH_3CH_2CH_2OH$

c. $CH_3CH_2CH_2C{\equiv}CH$,

**18.** Write structural formulas for butanoic acid, aminomethane, 2-butanol, and 3-aminopentane.

**19.** Give an example of:

a. an alkane
b. an amine
c. a carboxylic acid
d. an ether
e. an ester
f. an alkene
g. an alkyne
h. an alcohol
i. a ketone

**20.** Write structural formulas for two compounds that can have each of the molecular formulas listed:

a. $C_5H_{12}O$
b. $C_3H_6O$
c. $C_5H_{10}O_2$

**21.** Draw the *cis* and *trans* isomers for:

a. 1,2-dibromoethene
b. 1-bromo-2-chloroethene

**22.** How many trichloroethylene structures are possible?

**23.** Draw the structure of the compound 1,1,1-trichloroethane.

**24.** Define ubiquitous and explain how this word is descriptive of the carbon atom.

**25.** Among the possibilities of structural, geometrical, and (or) optical isomers, what type(s) of isomers can isoprene form? (Isoprene is the fundamental structural unit of rubber; refer to the index.)

**26.** a. Carbon and hydrogen have almost the same electronegativities. On the basis of this information and the tetrahedral structure around each carbon atom in a hydrocarbon such as octane ($C_8H_{18}$), would octane be polar or nonpolar?

b. Would octane be likely to dissolve in polar water?

**27.** When you see this symbol, [benzene ring symbol], in a chemistry book, what does it represent?

**28.** What two elements compose hydrocarbons?

**29.** Use the index to locate structures of glucose, epinephrine, dimethyldichlorosilane, histidine, and norepinephrine. Which of these substances have asymmetric carbon atoms and, therefore, can be one of a pair of optical isomers?

**30.** Are more organic or inorganic compounds known?

**31.** Why is carbon the "central" element in organic compounds?

**32.** How many carbon atoms are in a molecule of heptane?

**33.** What does the word "asymmetric" mean?

**34.** Describe the bonding in benzene.

**35.** What are the elements in sugar?

**36.** What functional groups are found in glucose?

**37.** How many trichloroheptane structures are possible?

**38.** What is the structural formula for 1-pentene?

**39.** What type of bond is always in an alkene?

**40.** Draw a structural formula for a molecule containing:

a. alcohol and amine functional groups
b. ether and ester functional groups
c. a double bond and a triple bond
d. a triple bond and an amine group

**41.** Draw the three dichlorobenzene structures and name them.

**42.** Draw the three trichlorobenzene structures and name them.

**43.** Biphenyl, [biphenyl structure], is a molecule that can be chlorinated to make chlorinated biphenyl. The family of chlorinated biphenyls is called polychlorinated biphenyls or PCB's. Draw the structures of several polychlorinated biphenyls. There are 209 possible structures.

**44.** Do you think any hydrocarbons can exist as optical isomers? Give reasons for your answer.

# Some Applications of Organic Chemistry

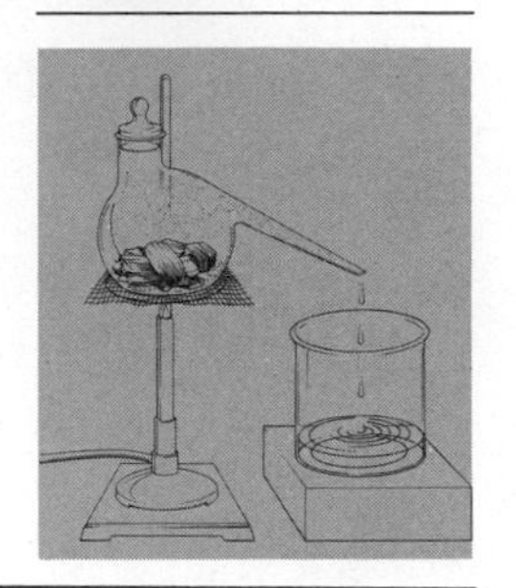

The preparation of new and different organic compounds through chemical reactions is called ***organic synthesis.*** Millions of organic compounds have been synthesized in the laboratories of the world during the past 150 years. Prior to 1828, it was widely believed that chemical compounds synthesized by living matter could not be made without living matter—a "vital force" was necessary for the synthesis. In 1828, a young German chemist, Friedrich Wöhler, destroyed the vital force myth and opened the door to modern organic syntheses. Wöhler heated a solution of silver cyanate and ammonium chloride, neither of which had been derived from any living substance. From these he prepared urea, a major animal waste product found in urine.

$$\underset{\text{SILVER CYANATE}}{AgOCN} + \underset{\text{AMMONIUM CHLORIDE}}{NH_4Cl} \longrightarrow \underset{\text{SILVER CHLORIDE (PRECIPITATE)}}{AgCl} + \underset{\text{AMMONIUM CYANATE}}{NH_4OCN}$$

$$\underset{\text{AMMONIUM CYANATE}}{NH_4OCN} \xrightarrow{\text{heat}} \underset{\text{UREA}}{H_2N\overset{\overset{\displaystyle O}{\|}}{C}NH_2}$$

The notion of a mysterious vital force declined as other chemists began to synthesize more and more organic chemicals without the aid of a living system. Soon it was shown that chemists could do more than imitate the products of living tissue; they could form unique materials of their own design.

Advances in understanding the structure of organic compounds gave organic synthesis a tremendous boost. Knowing the structure of compounds, the organic chemist could predict by analogy with simpler molecules what reactions might take place when organic reagents were used. Very elegant and reliable schemes of synthesis could now be constructed.

In this chapter emphasis will be given to some important organic compounds that not only are useful in themselves but are necessary for the synthesis of many other organic compounds. The dependence of synthesis upon a knowledge of structure will be pointed out from time to time.

Friedrich Wöhler (1800–1882) was Professor of Chemistry at the University of Berlin and later at Göttingen. His preparation of the organic compound urea from the inorganic compound ammonium cyanate did much to overturn the theory that organic compounds must be prepared in living organisms. One of the first to study the properties of aluminum, he was the first to isolate the element beryllium and is known for many other outstanding contributions to chemistry.

## Some Uses of Hydrocarbons

Complex mixtures of hydrocarbons, compounds containing only carbon and hydrogen, occur in enormous quantities in nature as petroleum, natural gas, and coal. These materials were formed from organisms that lived millions of years ago. After their death, they became covered with layers of sediment and ultimately were subjected to high temperatures and pressures in the depths of the earth's crust. In the absence of free oxygen, these conditions converted once-living tissue into petroleum and coal. After petroleum and natural gas are brought to the surface of the earth, they can be separated into various fractions with different boiling points by the use of fractional distillation (Fig. 9–1). Once distilled, the fractions are refined further and made to undergo various types of chemical change in order to make desirable substances.

Coal is one natural source of aromatic compounds.

By heating coal or wood in the absence of air (destructive distillation), coal tar can be separated from the coke or charcoal (carbon). A ton of a typical soft coal produces about 140 pounds of coal tar, which is about one-half pitch, the rest being composed of such chemicals as naphthalene, benzene, phenol, cresols, toluene, and xylenes.

"Bottled gas" is actually stored in the liquid state under pressure. In order for bottled gas to be burned in stoves or furnaces, it must be vaporized.

The naturally occurring compounds not only are important in themselves, but they also serve as starting materials for making numerous organic compounds that do not occur in nature. From the simplest hydrocarbons come such diverse consumer products as plastic dishes, acrylic and polyester fibers for textiles, vinyl and latex paints, neoprene rubber, and such industrial products as Teflon and cattle feed.

Natural gas consists primarily of low molecular weight hydrocarbons. It is predominantly methane ($CH_4$), but ethane ($C_2H_6$), propane ($C_3H_8$), and butane ($C_4H_{10}$) are also present. This mixture is conveyed in long pipelines from the areas in which it occurs to cities where it is used as a fuel.

Energy is obtained by burning the constituents of natural gas in air:

$$CH_4 + 2O_2 \longrightarrow CO_2 + 2H_2O + 213 \text{ kcal}$$

$$C_2H_6 + 3\tfrac{1}{2}O_2 \longrightarrow 2CO_2 + 3H_2O + 372.8 \text{ kcal}$$

The energy released in these reactions can be used to heat homes, run electric power plants, or power special internal combustion engines.

Propane and butane, principal components of bottled gas, can be liquefied by the use of moderate pressure and cooling. Because propane is more volatile than butane, bottled gas is prepared so that it contains more propane in colder climates and more butane in warmer ones.

The enormous increase in our use of organic compounds in the last century forces us to recognize that the sources of these materials are limited and subject to eventual exhaustion. Thus, petroleum sources, from which we draw so much of our energy and which serve as starting materials for the synthesis of industrial organic compounds, cannot be expected to last beyond 50 to 100 years at the most. The transformation of many of the natural sources of organic compounds into disposable convenience items which soon end up in a garbage dump represents an extremely short-sighted use of irreplaceable natural resources. Most of us realize that we must begin ultimately to recycle a much larger percentage of the material we use in our daily lives, but few seem ready to begin effective efforts in this direction now. It seems probable that the next decade will force us to change many of our casual attitudes toward the utilization of our natural sources of organic compounds.

## Gasoline

From the time petroleum was first discovered in the United States in 1859 until 1900 when the automobile became popular, most oil was refined to yield kerosene, a mixture of hydrocarbons in the $C_8$ to $C_{13}$ range. This liquid was used principally as a fuel for lamps.

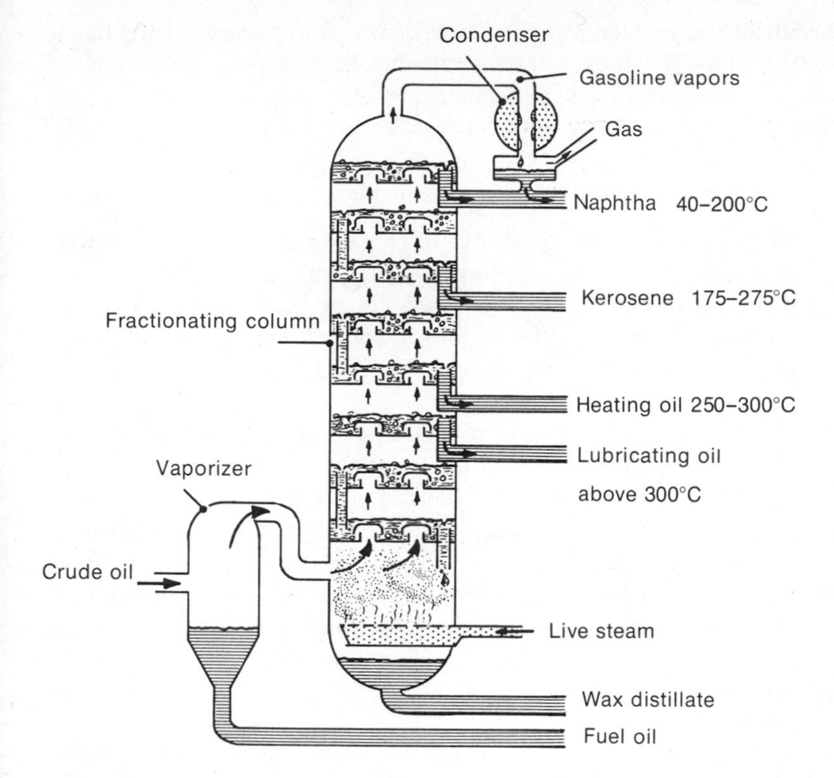


**Figure 9–1** A diagram of a fractionating column for distilling petroleum. Notice that the higher boiling substances condense at the lower levels and the lower boiling substances do not condense until the higher, cooler levels.

The internal combustion engines used in early automobiles were designed to burn a more volatile mixture of hydrocarbons, the $C_6$ to $C_{10}$ fraction, which became known as ***gasoline.*** These lower molecular weight hydrocarbons mix readily with air in simple carburetors and burn fairly completely.

**Our supplies of coal, and especially petroleum, are limited.**

With the increasing popularity of the automobile, petroleum refiners had to shift the output of a barrel of crude oil from a reasonably large fraction of kerosene to almost no kerosene and a much greater fraction of gasoline (Table 9–1). This dramatic increase in the amount of gasoline from a barrel of crude oil was accomplished by the discovery of chemical processes that convert nongasoline molecules into ones that burn well in an automobile engine.

**Mixtures of hydrocarbons in the $C_6$ to $C_{10}$ range are known as *gasolines*.**

**A barrel of crude oil is 42 gallons.**

The first conversion discovered was ***thermal cracking,*** a process in which long-chain hydrocarbon molecules were heated under pressure in the absence of air. This

**TABLE 9–1 Division of a Barrel of Crude Oil**

| | 1920 % | 1967 % |
|---|---|---|
| Gasoline | 26.1 | 44.8 |
| Kerosene | 12.7 | 2.8 |
| Jet fuel | – | 7.6 |
| Heavy distillates | 48.6 | 22.2 |
| Other (asphalt, etc.) | 12.6 | 22.6 |
| total | 100.0 | 100.0 |

heating strains the bonds of the molecule so that it may break or "crack," thus forming two smaller molecules, some of which will be in the gasoline range.

$$\underset{\text{AN ALKANE}}{C_{16}H_{34}} \xrightarrow[\text{heat}]{\text{pressure}} \underset{\text{AN ALKANE IN THE GASOLINE RANGE}}{C_8H_{18}} + \underset{\text{AN ALKENE}}{C_8H_{16}}$$

**Cracking breaks larger molecules into smaller ones.**

Later, cracking was carried out in the presence of certain catalysts that allowed the processes to proceed at lower pressures and to produce even higher yields of gasoline. Today, these catalysts include specially processed clays. Each refiner of petroleum has his own special methods, which offer different advantages in cost and type of crude oil handled. The hydrocarbon molecules produced are much the same regardless of the methods used.

Cracking of petroleum fractions brought about not only an increase in the quantity of gasoline available from a barrel of crude oil, but also an increase in *quality*. That is, gasoline from a cracking process can be used at higher efficiency (in a high-compression engine) than can "straight run" gasoline, because the molecular structures of the hydrocarbons in the cracked gasoline allow them to oxidize more smoothly at high pressure. When the burning of the gasoline-air mixture is too rapid or irregular, ignition occurs in the combustion chamber too early, resulting in a small detonation which is heard as a "knock" in the engine. This knocking can spell trouble to the owner of an automobile because it will eventually lead to the breakdown of the internal parts of the automobile's engine.

**Knocking in gasoline engines is a sign of improper combustion.**

An arbitrary scale for rating the relative knocking properties of gasolines has been developed. Normal heptane, typical of straight run gasoline, knocks considerably and is assigned an octane rating of 0,

$$\underset{\text{n-HEPTANE}}{CH_3CH_2CH_2CH_2CH_2CH_2CH_3} \quad \text{(octane rating = 0)}$$

whereas 2,2,4-trimethylpentane (isooctane) is far superior in this respect and is assigned an octane rating of 100.

$$CH_3-\overset{\overset{\large CH_3}{|}}{\underset{\underset{\large CH_3}{|}}{C}}-CH_2-\overset{\overset{\large CH_3}{|}}{\underset{\underset{\large H}{|}}{C}}-CH_3 \quad \text{(octane rating = 100)}$$

2,2,4-TRIMETHYLPENTANE

**The octane scale measures the ability of a mixture to burn without knocking in a gasoline engine.**

The octane rating of a gasoline is determined by first using the gasoline in a standard engine and recording its knocking properties. This is compared to the behavior of mixtures of n-heptane and isooctane, and the percentage of isooctane in the mixture with identical knocking properties is called the octane rating of the gasoline. Thus, if a gasoline has the same knocking characteristics as a mixture of 9% n-heptane and 91% isooctane, it is assigned an octane rating of 91. This corresponds to a regular grade of gasoline. Since the octane rating scale was established, fuels have been developed that are superior to isooctane, so the scale has been extended well above 100.

**A catalyst increases the speed of a reaction without being consumed in the reaction.**

A primary source of the present high-octane gasolines is a chemical process known as ***catalytic reforming.*** Under the influence of certain catalysts, such as finely divided platinum, straight-chain hydrocarbons with low octane numbers can be re-formed into their branched-chain isomers, which have higher octane numbers.

$$CH_3CH_2CH_2CH_2CH_3 \xrightarrow[\text{heat}]{\text{platinum}} CH_3CH_2CH(CH_3)CH_3$$

n-PENTANE 2-METHYLBUTANE

The knock caused by combustion of gasoline can be avoided by either of two procedures. The first is to add "antiknock" agents to a motley mixture of hydrocarbons. The best-known antiknock additive is tetraethyllead, $(C_2H_5)_4Pb$. Tetraethyllead is extremely toxic and, as new car owners know, is being limited in use. The Environmental Protection Agency has approved a newer antiknock mixture, which is mostly 2-methyl-2-propanol (also called *tertiary* butyl alcohol). The second procedure is to use only hydrocarbons with structures known to burn satisfactorily in a high-compression engine. Branched-chain structures are particularly desirable in this respect, and gasolines with appreciable amounts of aromatic hydrocarbons (20–40%) are also less prone to knock. Some petroleum is just naturally better supplied with aromatic compounds. For example, California crude has a higher percentage of aromatic compounds than Pennsylvania crude. Regardless of the source, petroleum fractions such as naphtha, kerosene, and gas oil are mixtures of open-chain, branched-chain, and single-bonded-ring hydrocarbons. When these stocks, in the vapor phase, are passed over a copper catalyst at about 650°C, a high percentage of the original material is converted into a mixture of aromatic hydrocarbons from which benzene, toluene, xylenes, and similar compounds may be separated by fractional distillation. For example, n-hexane is converted into benzene

$$CH_3CH_2CH_2CH_2CH_2CH_3 \longrightarrow C_6H_6$$

n-HEXANE BENZENE

and n-heptane is changed into toluene.

$$CH_3CH_2CH_2CH_2CH_2CH_2CH_3 \longrightarrow C_6H_5CH_3$$

n-HEPTANE TOLUENE

## Some Important Compounds of Carbon, Hydrogen, and Oxygen

Carbon, hydrogen, and oxygen can be combined to form an enormous number of compounds. As we have seen, considerable order is introduced into the study of these compounds when they are divided into classes on the basis of the functional groups they contain. The ***alcohols,*** the ***organic acids*** and their derivatives (compounds that can be made from them), and the ***esters*** and ***soaps*** are important compounds of carbon, hydrogen, and oxygen that find wide application in our everyday lives.

### Alcohols

When a hydroxyl (—OH) group is attached to a nonaromatic carbon skeleton of a hydrocarbon (an R group), the resulting R—OH compound has properties common to a class of compounds called ***alcohols.*** Formulas and names for some important substances of this type are given in Table 9–2.

**Alcohols are compounds with the structure R—O—H, where R is a hydrocarbon group.**

**TABLE 9–2 Some Important Alcohols**

| Formula | IUPAC Name | Common Name |
|---|---|---|
| H<br>\|<br>H—C—OH<br>\|<br>H | methanol | methyl alcohol<br>(wood alcohol) |
| H H<br>\| \|<br>H—C—C—OH<br>\| \|<br>H H | ethanol | ethyl alcohol<br>(grain alcohol) |
| H H H<br>\| \| \|<br>H—C—C—C—OH<br>\| \| \|<br>H H H | 1-propanol | n-propyl alcohol |
| H H H<br>\| \| \|<br>H—C—C—C—H<br>\| \| \|<br>H O H<br>H | 2-propanol | isopropyl alcohol<br>(rubbing alcohol) |
| H<br>\|<br>H—C—OH<br>\|<br>H—C—OH<br>\|<br>H | 1,2-ethanediol | ethylene glycol<br>(permanent antifreeze) |
| H<br>\|<br>H—C—OH<br>\|<br>H—C—OH<br>\|<br>H—C—OH<br>\|<br>H | 1,2,3-propanetriol | glycerol<br>(glycerin) |

## Methanol (Methyl Alcohol)

**Methanol is number eighteen in the list of commercial chemicals.**

Methanol was originally called wood alcohol since it was obtained by the destructive distillation of wood. It is the simplest of all alcohols and has the formula $CH_3OH$. In the older method for the production of wood alcohol, hardwoods such as beech, hickory, maple, or birch are heated in the absence of air in a retort (Fig. 9–2). Methanol that is 92–95% pure can be obtained by fractional distillation of the resulting liquid.

In 1923, the price of wood alcohol in the United States was 88¢ per gallon. In that year German chemists discovered how to produce this useful compound synthetically. Methanol is formed when carbon monoxide, CO, and hydrogen are heated at a pressure of 200 to 300 atmospheres over a catalyst of mixed oxides (90% ZnO–10% $Cr_2O_3$).

$$CO + 2H_2 \xrightarrow[\text{300°C, pressure}]{ZnO-Cr_2O_3} CH_3OH$$

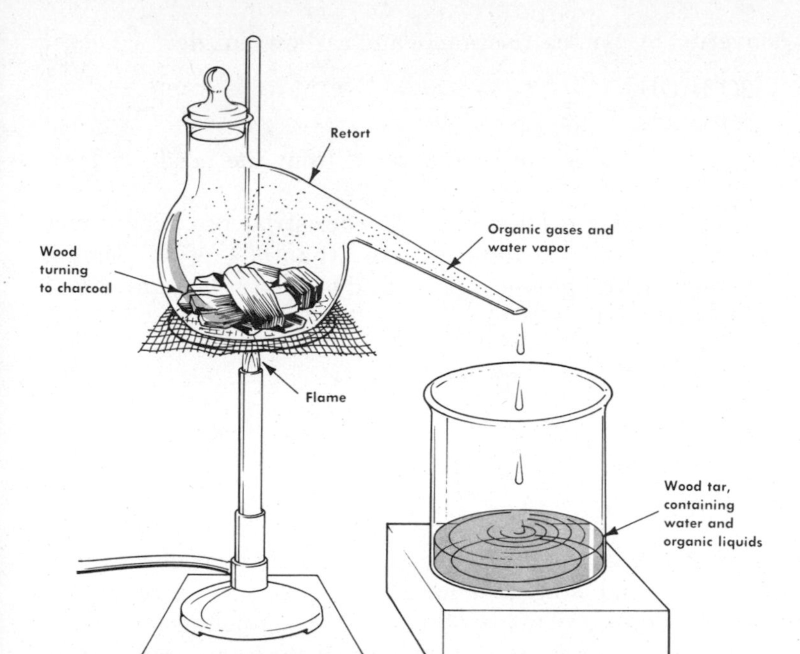


**Figure 9–2** Destructive distillation of wood.

As a result of this synthetic process, German industrialists were able to sell pure methanol at 20¢ per gallon. Even a high tariff was not able to save the wood distillers in their outdated operations. The synthetic product soon dominated the market in the United States.

Industrial chemists are continually searching for less costly ways to prepare important chemicals such as methanol.

The production of synthetic methanol in the United States is over 4.0 billion pounds per year. About half of this is used in the production of formaldehyde (used in plastics, embalming fluid, germicides, and fungicides), 30% in the production of other chemicals, and smaller amounts for jet fuels, antifreeze mixtures, solvents, and as a denaturant (a poison added to ethanol to make it unfit for beverages). Methanol is a *deadly poison;* it causes blindness in less than lethal doses. Many deaths and injuries have resulted when this alcohol was mistakenly substituted for ethanol in beverages.

## Ethanol (Ethyl Alcohol)

Ethanol (ethyl alcohol) is called grain alcohol because it can be fractionally distilled from the fermented mash made from corn, rice, barley, and other grains. Fermentation is a breakdown of starch in the grain by means of enzymes. Enzymes are catalysts that are complex organic molecules produced by living cells. If the enzyme diastase is mixed with ground grain and water, and the mixture is allowed to stand at 40°C for a period of time, the starch in the grain will be changed into maltose.

$$\underset{\text{STARCH}}{2(C_6H_{10}O_5)_n} + nH_2O \xrightarrow{\text{diastase}} \underset{\text{MALTOSE (A SUGAR)}}{nC_{12}H_{22}O_{11}}$$

The subscript $n$ in the formula for starch indicates that starch is made up of many $C_6H_{10}O_5$ units.

Brewers call the resulting mixture of maltose and water the ***wort.*** The wort is diluted and mixed with yeast, and held at a temperature of 30°C for 40–60 hours. The living yeast cells secrete the enzymes maltase and zymase. The maltase causes the sugar, maltose, to hydrolyze into a simple sugar, glucose:

$$\underset{\text{MALTOSE}}{C_{12}H_{22}O_{11}} + H_2O \xrightarrow{\text{maltase}} \underset{\text{GLUCOSE}}{2C_6H_{12}O_6}$$

The glucose, in turn, is converted by zymase to ethanol and carbon dioxide:

$$\underset{\text{GLUCOSE}}{C_6H_{12}O_6} \xrightarrow{\text{zymase}} 2CO_2 + \underset{\text{ETHANOL}}{2C_2H_5OH}$$

A solution of 95% ethanol and 5% water can be recovered from the mash by fractional distillation.

**Over 2 billion pounds of ethanol are produced synthetically in the United States each year.**

Synthetic ethanol is produced on a large scale for industrial use. The direct chemical addition of water to ethylene accounts for more than 80% of all ethanol production. Under high pressure in the presence of a catalyst and a large excess of water vapor, ethylene produces ethanol:

**Ethylene is number six in the list of commercial chemicals.**

$$\underset{\text{ETHYLENE}}{H_2C{=}CH_2} + HOH \xrightarrow[300\,°C]{70\text{ atm.}} \underset{\text{ETHANOL}}{H{-}\overset{H}{\underset{H}{C}}{-}\overset{H}{\underset{H}{C}}{-}OH}$$

Pure ethanol is termed 200 proof (that is, twice the percentage of alcohol). Apart from the alcoholic beverage industry, ethanol is used widely in solvents and in the preparation of chloroform, ether, and many other organic compounds.

Some of the most commonly encountered alcoholic beverages and their characteristics are presented in Table 9–3.

## Propanols (Propyl Alcohols)

When one considers the possible structures for propyl alcohol, it is apparent that two isomers are possible.

$$\underset{\substack{\text{1-PROPANOL}\\ \text{n-PROPYL ALCOHOL}}}{H{-}\overset{H}{\underset{H}{C}}{-}\overset{H}{\underset{H}{C}}{-}\overset{H}{\underset{H}{C}}{-}OH} \qquad \underset{\substack{\text{2-PROPANOL}\\ \text{ISOPROPYL ALCOHOL}}}{H{-}\overset{H}{\underset{H}{C}}{-}\overset{H}{\underset{OH}{C}}{-}\overset{H}{\underset{H}{C}}{-}H}$$

Of the two propanols, 1-propanol is the more expensive; it is prepared by the oxidation of simple hydrocarbons. It finds uses as a solvent and as a raw material in the manufacture of other organic compounds.

**Propylene is number fourteen in the list of commercial chemicals.**

The hydration of propylene yields 2-propanol (isopropyl alcohol), which is rubbing alcohol. It has greater germicidal activity than the other simple alcohols and is used as an antiseptic.

**TABLE 9–3 Common Alcoholic Beverages**

| Name | Source of Fermented Carbohydrate | Amount of Ethyl Alcohol |
|---|---|---|
| Beer | barley, wheat | 3.2–9% |
| Wine | grapes or other fruit | 12% maximum, unless fortified* |
| Brandy | distilled wine | 40–45% |
| Whiskey | barley, rye, corn, etc. | 45–55% |
| Rum | molasses | ~45% |

*The growth of yeast is inhibited at alcohol concentrations over 12% and fermentation comes to a stop. Beverages with a higher concentration are prepared either by distillation or by fortification with alcohol that has been obtained by the distillation of another fermentation product.

## Ethylene Glycol and Glycerol (Glycerin)

More than one alcohol group (—OH) can be present in a single molecule. Glycerol and ethylene glycol, the base of permanent antifreeze, are examples of such compounds.

Permanent antifreeze is ethylene glycol.

```
      H                  H
      |                  |
  H—C—OH             H—C—OH
      |                  |
  H—C—OH             H—C—OH
      |                  |
      H              H—C—OH
                         |
                         H
```

1,2-ETHANEDIOL
ETHYLENE GLYCOL

1,2,3-PROPANETRIOL
GLYCEROL (GLYCERIN)

Glycerol has many uses in foods and tobacco as a digestible and nontoxic humectant (gathers and holds moisture), and in the manufacture of drugs and cosmetics. It is also used in the production of nitroglycerin and numerous other chemicals. Perhaps the most important compounds of glycerol are its natural esters (fats and oils), which we shall discuss later in this chapter.

## Hydrogen Bonding in Alcohols

The physical properties of water, methanol, ethanol, the propanols, ethylene glycol, and glycerol offer another interesting example of the effects of ***hydrogen bonding*** between molecules in liquids. In Table 9–4 the boiling points for these compounds are listed.

Since boiling involves overcoming the attractions between liquid molecules as they pass into the gas phase, a higher boiling point indicates stronger intermolecular forces holding the molecules together. Another factor is also present: as the molecules become larger, higher boiling points result, since more energy is required to change the longer-chain molecules from the liquid to the gaseous phase, owing in part to the larger van der Waals attraction. A graph showing the boiling points of the normal alcohols (straight carbon chains with the —OH group on an end carbon) as a function of chain length is given in Figure 9–3.

Hydrogen bonding is responsible for the fact that an alcohol has a higher boiling point than its parent hydrocarbon.

Methanol, like water, has an —OH group, and some hydrogen bonding is to be expected, as shown in Figure 9–4. Hydrogen bonding explains why methanol (molecular weight 32) is a liquid, whereas propane ($C_3H_8$, molecular weight 44), an even heavier molecule, is a gas at room temperature. Methanol has only one hydro-

**TABLE 9–4 Boiling Points for Some —OH Compounds**

| | | |
|---|---|---|
| Water | HOH | 100°C |
| Methanol | $CH_3OH$ | 65.0° |
| Ethanol | $CH_3CH_2OH$ | 78.5° |
| 1-Propanol | $CH_3CH_2CH_2OH$ | 97.4° |
| 2-Propanol | $CH_3CHOHCH_3$ | 82.4° |
| Ethylene glycol | $CH_2OHCH_2OH$ | 198° |
| Glycerol | $CH_2OHCHOHCH_2OH$ | 290° |

Note: The parent hydrocarbon of methanol is methane (bp −164°C); of ethanol and ethylene glycol, ethane (bp −88.6°C); and of the propanols and glycerol, propane (bp −42.1°C).

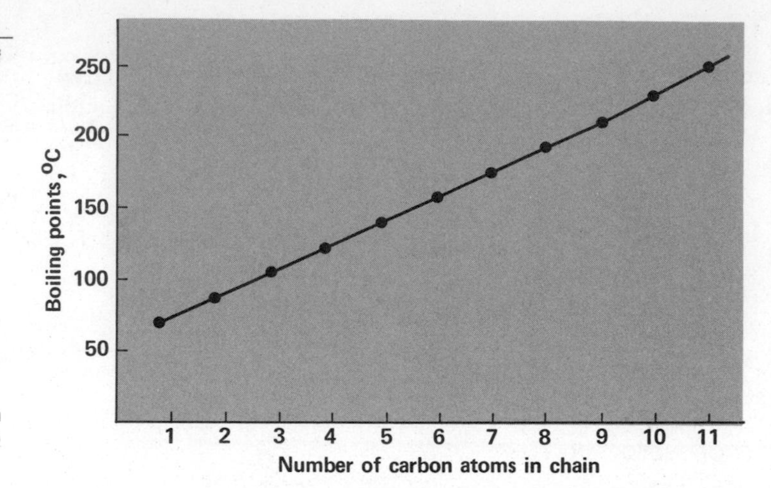


**Figure 9–3** Boiling points of stright-chain alcohols (—OH group on an end carbon).

gen through which it can hydrogen bond, while water can hydrogen bond from either of its two hydrogen atoms. Thus, water, with more extensive intermolecular bonding, has the higher boiling point even though it is made up of lighter molecules.

Methanol and ethanol oxidize easily to acids, and so corrode the radiator.

Both methanol and ethanol can be used as antifreeze, but they tend to distill out of the coolant at the temperatures of a hot gasoline engine. Protection against freezing is then lost over a period of time. Ethylene glycol is equally effective (molecule for molecule) in lowering the freezing point of water, and its high boiling point (198°C) makes it a permanent antifreeze. This property makes ethylene glycol more desirable, even though it takes almost twice as much ethylene glycol by weight as methanol to provide the same amount of protection for a car's cooling system. Ethylene glycol with suitable additives to protect the radiator system is sold under a number of brand names. The higher boiling point of ethylene glycol is readily explained in terms of the two —OH groups per molecule and the enhanced possibility for hydrogen bonding. Glycerol, with three —OH groups per molecule, has an even higher boiling point, as well as a very high viscosity (resistance to flow).

Since the —OH group in an organic molecule causes at least that area of the molecule to be polar, such molecules will be attracted to other polar molecules such as water molecules. As a result of these attractions, the lower molecular weight alcohols are quite soluble in water. In the higher molecular weight alcohols, the nonpolar hydrocarbon chain decreases significantly their solubility in water (see Table 9–5).

## Synthesis with Alcohols

Alcohols can serve as the starting substances for the synthesis of many other types of organic compounds. Oxidation of alcohols may yield aldehydes, ketones, or acids

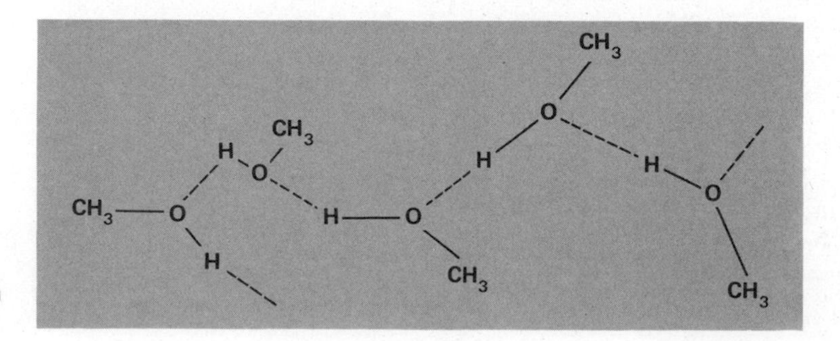


**Figure 9–4** Hydrogen bonding in methanol.

**TABLE 9–5 Water Solubilities of Some Alcohols**

| IUPAC Name | Formula | Solubility (g/100 g $H_2O$) | Comment |
|---|---|---|---|
| Methanol | $CH_3OH$ | $\infty$* | soluble in all proportions |
| Ethanol | $CH_3CH_2OH$ | $\infty$ | same |
| 1-Propanol | $CH_3CH_2CH_2OH$ | $\infty$ | same |
| 2-Propanol | $CH_3CHOHCH_3$ | $\infty$ | effect of —OH still strong |
| 1-Butanol | $CH_3CH_2CH_2CH_2OH$ | 7.9 | hydrocarbon chain effect now apparent |
| 2-Methyl-1-propanol | $(CH_3)_2CHCH_2OH$ | 10.0 | shorter overall chain length |
| 2-Butanol | $CH_3CH_2CHOHCH_3$ | 12.5 | effect of —OH group stronger in this position |
| 2-Methyl-2-propanol | $(CH_3)_3COH$ | $\infty$ | nonpolar effect is diminished |
| 1-Pentanol | $CH_3CH_2CH_2CH_2CH_2OH$ | 2.3 | effect of the long hydrocarbon chain |

*This symbol means "infinitely" soluble, or that there is no limit to solubility.

depending on the starting compound and the amount of oxygen added. For example, the oxidation of ethanol can be used to make acetaldehyde and acetic acid:

$$CH_3CH_2OH \xrightarrow{\text{oxidation}} CH_3CH{=}O \xrightarrow{\text{oxidation}} CH_3C(=O){-}OH$$

ETHANOL ACETALDEHYDE ACETIC ACID

In organic compounds, oxidation is often accompanied by the loss of hydrogen or the addition of oxygen to a molecule.

The oxidation of 2-propanol provides the ketone, acetone:

$$CH_3CH(OH)CH_3 \xrightarrow{\text{oxidation}} CH_3C(=O)CH_3$$

2-PROPANOL ACETONE

Workable oxidants are hot copper oxide (CuO), potassium dichromate ($K_2Cr_2O_7$) with sulfuric acid, or potassium permanganate ($KMnO_4$).

Substitution reactions produce alkyl halides from simple alcohols:

$$CH_3CH(OH)CH_3 + HBr \longrightarrow CH_3CH(Br)CH_3 + H_2O$$

2-PROPANOL 2-BROMOPROPANE

## Organic Acids

Earlier an acid was defined as a species that has a tendency to donate hydrogen ions, or protons. We shall now consider the carboxylic acids, which contain the carboxyl group, $-C(=O)OH$. The electronegative character of the C=O group tends to drain electron density away from the region between the oxygen and hydrogen atoms. The partial positive charge assumed by the hydrogen then makes it possible for polar water molecules to remove the hydrogen ions from some of the carboxyl groups. The strength of an organic acid depends on the group that is attached to the carboxyl

Organic acids are compounds of the type $R-C(=O)-OH$. They are generally weak acids.

group. If the attached group has a tendency to pull electrons away from the carboxyl group, the acid is a stronger acid. For example, trichloroacetic acid is a much stronger acid than acetic acid:

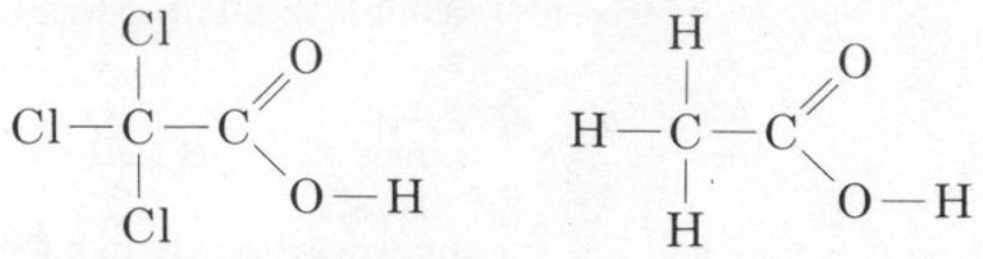

$$\mathrm{Cl{-}C(Cl)_2{-}C({=}O){-}O{-}H} \qquad \mathrm{H{-}C(H)_2{-}C({=}O){-}O{-}H}$$

TRICHLOROACETIC ACID  ACETIC ACID

The highly electronegative chlorine atoms withdraw electron density from the region of the carboxyl group and make the loss of the hydrogen ion even easier than in acetic acid. However, trichloroacetic acid is still weaker than a strong mineral acid such as sulfuric acid.

The ionization of carboxylic acids in water is as follows:

$$\mathrm{R{-}\overset{O}{\overset{\|}{C}}{-}OH + H_2O \rightleftharpoons R{-}\overset{O}{\overset{\|}{C}}{-}O^- + H_3O^+}$$

They are ***neutralized*** by bases to form salts:

$$\mathrm{R{-}\overset{O}{\overset{\|}{C}}{-}OH + Na^+ + OH^- \longrightarrow \underset{\text{A SALT}}{R{-}\overset{O}{\overset{\|}{C}}{-}O^-Na^+} + H_2O}$$

## Methanoic Acid (Formic Acid)

The simplest organic acid is methanoic acid, also called formic acid, in which the carboxyl group is attached directly to a hydrogen atom.

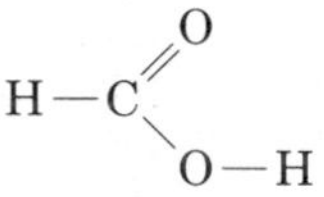

$$\mathrm{H{-}C({=}O){-}O{-}H}$$

METHANOIC ACID
FORMIC ACID

**Weak bases such as ammonia, $NH_3$, which would neutralize formic acid, are used in the treatment of insect bites.**

This acid is found in ants and other insects and is part of the irritant that produces itching and swelling after a bite.

Formic acid may be prepared from its sodium salt, which is readily prepared by heating carbon monoxide (CO) with sodium hydroxide (NaOH):

$$\mathrm{CO + NaOH \xrightarrow[\text{6–10 min.}]{200^\circ C} \underset{\text{SODIUM FORMATE}}{HCOO^-Na^+}}$$

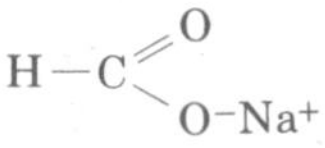

$$\mathrm{H{-}C({=}O){-}O^-Na^+}$$

Sodium Formate

If the resulting salt is mixed with a mineral acid, formic acid can be distilled from the mixture:

$$\mathrm{\underset{\text{SODIUM FORMATE}}{HCOO^-Na^+} + \underset{\text{HYDROCHLORIC ACID}}{H_3O^+ + Cl^-} \longrightarrow \underset{\text{FORMIC ACID}}{HCOOH} + \underset{\text{SODIUM CHLORIDE}}{Na^+Cl^-} + H_2O}$$

## Ethanoic Acid (Acetic Acid)

Ethanoic (acetic) acid is the most widely used of the organic acids. It is found in vinegar, an aqueous solution containing 4–5% acetic acid. Flavor and colors are imparted to vinegars by the constituents of the alcoholic solutions from which they are made. Ethanol in the presence of certain bacteria and air is oxidized to acetic acid:

$$\underset{\text{ETHANOL}}{CH_3CH_2OH} + \underset{\text{OXYGEN}}{O_2} \xrightarrow{\text{bacteria}} \underset{\substack{\text{ETHANOIC ACID} \\ \text{(ACETIC ACID)}}}{CH_3COOH} + \underset{\text{WATER}}{H_2O}$$

The bacteria, called mother of vinegar, form a slimy growth in a vinegar solution. The growth of bacteria can sometimes be observed in a bottle of commercially prepared vinegar after it has been opened to the air.

Acetic acid is an important starting substance for making textile fibers, vinyl plastics, and other chemicals, and is a convenient choice when a cheap organic acid is needed.

## Fatty Acids

A fatty acid contains a carboxyl group attached to a long hydrocarbon chain. The chains often contain only single carbon-carbon bonds, but may contain carbon-carbon double bonds as well. Examples are stearic acid, palmitic acid, and oleic acid.

$$CH_3CH_2CH_2CH_2CH_2CH_2CH_2CH_2CH_2CH_2CH_2CH_2CH_2CH_2CH_2CH_2CH_2C{\overset{\displaystyle O}{\diagup\!\!\diagup}\atop\underset{\displaystyle OH}{\diagdown}}$$

STEARIC ACID, $CH_3-(CH_2)_{16}-COOH$

$$CH_3CH_2CH_2CH_2CH_2CH_2CH_2CH_2CH_2CH_2CH_2CH_2CH_2CH_2CH_2C{\overset{\displaystyle O}{\diagup\!\!\diagup}\atop\underset{\displaystyle OH}{\diagdown}}$$

PALMITIC ACID, $CH_3-(CH_2)_{14}-COOH$

$$CH_3CH_2CH_2CH_2CH_2CH_2CH_2CH_2CH{=}CHCH_2CH_2CH_2CH_2CH_2CH_2CH_2C{\overset{\displaystyle O}{\diagup\!\!\diagup}\atop\underset{\displaystyle OH}{\diagdown}}$$

OLEIC ACID, $CH_3(CH_2)_7CH{=}CH(CH_2)_7COOH$

Fatty acids may be obtained from animal and vegetable fats or oils. The carbon chain in fatty acids is generally 8 to 18 carbon atoms in length.

Stearic acid is obtained by the hydrolysis of animal fat, palmitic acid results from the hydrolysis of palm oil, and oleic acid is obtained from olive oil. These reactions are given in the next two sections. Stearic and palmitic acids are especially important in the manufacture of soaps.

## Self-Test 9-A

1. Which of the following hydrocarbons would be expected to have the highest octane rating?

$$\underset{\text{(a)}}{CH_3CH_2CH_2CH_2CH_2CH_2CH_3} \qquad \underset{\text{(b)}}{CH_3CH_2-\overset{\displaystyle CH_3}{\overset{|}{C}H}-CH_2CH_2CH_3}$$

$$\underset{\text{(c)}}{CH_3-\underset{\displaystyle CH_3}{\underset{|}{\overset{\displaystyle CH_3}{\overset{|}{C}}}}-\underset{\displaystyle H}{\underset{|}{\overset{\displaystyle CH_3}{\overset{|}{C}}}}-CH_3}$$

2. Name the following compounds:

a. $CH_3OH$ ____________

b. $CH_3CH_2OH$ ____________

c. $HCOOH$ ____________

d. $CH_3CH_2\underset{|}{C}HCH_3$ (with OH on the third carbon) ____________

$\quad\quad\quad\quad OH$

e. $CH_3COOH$ ____________

f. $CH_2-CH_2$ (each carbon bearing OH) ____________

$\quad OH \quad OH$

g. $CH_3CHO$ ____________

h. $CH_3CHCH_3$ (with Br on the middle carbon) ____________

$\quad\quad Br$

3. Two methods by which ethanol is made on a large scale are:

a. ____________

b. ____________

4. An example of a carboxylic acid that contains a long hydrocarbon chain is ____________.

5. Identify the functional groups present in each of the following molecules:

a. $R-OH$ ____________

b. $R-\overset{O}{\overset{\|}{C}}-OH$ ____________

c. $R-\overset{O}{\overset{\|}{C}}-H$ ____________

d. $R-\overset{O}{\overset{\|}{C}}-R'$ ____________

## Esters

In the presence of strong mineral acids, organic acids react with alcohols to form compounds called ***esters.*** For example, when ethyl alcohol is mixed with acetic

**TABLE 9–6 Some Alcohols, Acids, and Their Esters**

| Alcohol | Acid | Ester | Odor of the Ester |
|---|---|---|---|
| $CH_3CHCH_2CH_2OH$<br>$\vert$<br>$CH_3$<br>ISOPENTYL ALCOHOL | $CH_3COOH$<br>ACETIC ACID | $CH_3CHCH_2CH_2-O-CCH_3$<br>$\vert \quad\quad\quad\quad \|$<br>$CH_3 \quad\quad\quad\quad O$<br>ISOPENTYL ACETATE | banana |
| $CH_3CHCH_2CH_2OH$<br>$\vert$<br>$CH_3$<br>ISOPENTYL ALCOHOL | $CH_3CH_2CH_2CH_2COOH$<br>n-VALERIC ACID | $CH_3CHCH_2CH_2-O-C-CH_2CH_2CH_2CH_3$<br>$\vert \quad\quad\quad\quad \|$<br>$CH_3 \quad\quad\quad\quad O$<br>ISOPENTYL n-VALERATE | apple |
| $CH_3CH_2CH_2CH_2OH$<br>n-BUTYL ALCOHOL | $CH_3CH_2CH_2COOH$<br>n-BUTYRIC ACID | $CH_3CH_2CH_2CH_2-O-C-CH_2CH_2CH_3$<br>$\|$<br>$O$<br>BUTYL n-BUTYRATE | pineapple |
| $CH_3CHCH_2OH$<br>$\vert$<br>$CH_3$<br>ISOBUTYL ALCOHOL | $CH_3CH_2COOH$<br>PROPIONIC ACID | $CH_3CHCH_2-C-CH_2CH_3$<br>$\vert \quad\quad \|$<br>$CH_3 \quad\quad O$<br>ISOBUTYL PROPIONATE | rum |

acid in the presence of sulfuric acid, ethyl acetate is formed. This reaction is a dehydration in which sulfuric acid acts as a catalyst and dehydrator.

$$CH_3CH_2O\boxed{H + HO}\overset{O}{\overset{\|}{C}}CH_3 \xrightarrow{H_2SO_4} CH_3CH_2O\overset{O}{\overset{\|}{C}}CH_3 + H_2O$$

ETHYL ACETATE

Ethyl acetate is a common solvent for lacquers and plastics, and is often used as fingernail polish remover.

Some of the odors of common fruits are due to the presence of mixtures of volatile esters (Table 9–6). In contrast, higher molecular weight esters often have a distinctly unpleasant odor.

**Organic esters are compounds of the type**

$$R-O-\overset{O}{\overset{\|}{C}}-R'$$

**formed by the reaction of organic acids and alcohols.**

## Fats, Oils, and Soaps

Fats and oils are esters of glycerol (glycerin) and a fatty acid. R, R′, and R″ stand for the hydrocarbon chains of the acids in the following equation.

**Fats and oils are esters of fatty acids and glycerol. Fats are solids, and oils are liquids.**

$$\begin{matrix} CH_2-OH \\ | \\ CH-OH \\ | \\ CH_2-OH \end{matrix} + \begin{matrix} HO-\overset{O}{\overset{\|}{C}}-R \\ HO-\overset{O}{\overset{\|}{C}}-R' \\ HO-\overset{O}{\overset{\|}{C}}-R'' \end{matrix} \rightleftharpoons \begin{matrix} CH_2-O-\overset{O}{\overset{\|}{C}}-R \\ | \\ CH-O-\overset{O}{\overset{\|}{C}}-R' \\ | \\ CH_2-O-\overset{O}{\overset{\|}{C}}-R'' \end{matrix} + 3H_2O$$

GLYCEROL (ONE MOLECULE) — FATTY ACID (THREE MOLECULES THAT MAY OR MAY NOT BE THE SAME) — FAT OR OIL (ONE MOLECULE) — WATER (THREE MOLECULES)

The term ***fat*** is usually reserved for solid glycerol esters (butter, lard, tallow) and ***oil*** for liquid esters (castor, olive, linseed, tung, and so forth). The term ***lipid*** includes fats, oils, and fat-soluble compounds.

**Lipids are soluble in fats and oils.**

Saturation (all single bonds with maximum hydrogen content) in the carbon chain of the fatty acids is usually found in solid or semi-solid fats, whereas unsaturated fatty acids (containing one or more double bonds) are usually found in oils. Hydrogen can be catalytically added to the double bonds of an oil to convert it into a semi-solid fat. For example, liquid soybean and other vegetable oils are hydrogenated to produce cooking fats and margarine.

Consumers in Europe and North America have historically valued butter as a source of fat. As the population increased, the advantages of a substitute for butter became apparent, and efforts to prepare such a product began about a hundred years ago. One problem that arose was the fact that common fats are almost all *animal* products with very pronounced tastes of their own. Analogous compounds from vegetable oils, which are bland or have mixed flavors, were generally *unsaturated* and consequently *oils.* A solid fat could be made from the much cheaper vegetable oils if an inexpensive way could be discovered to add hydrogen across the double bonds. After extensive experiments, many catalysts were found, of which finely divided nickel is among the most effective. The nature of the process can be illustrated by the reaction

**Catalytic hydrogenation can convert a liquid oil into a solid fat.**

$$\begin{array}{l} H_2C-O-\underset{\underset{O}{\|}}{C}-(CH_2)_7CH{=}CH(CH_2)_7CH_3 \\ HC-O-\underset{\underset{O}{\|}}{C}-(CH_2)_7CH{=}CH(CH_2)_7CH_3 \\ H_2C-O-\underset{\underset{O}{\|}}{C}-(CH_2)_7CH{=}CH(CH_2)_7CH_3 \end{array} \xrightarrow[\sim 200^\circ C]{H_2,\ Ni} \begin{array}{l} H_2C-O-\underset{\underset{O}{\|}}{C}-(CH_2)_7CH_2CH_2(CH_2)_7CH_3 \\ HC-O-\underset{\underset{O}{\|}}{C}-(CH_2)_7CH_2CH_2(CH_2)_7CH_3 \\ H_2C-O-\underset{\underset{O}{\|}}{C}-(CH_2)_7CH_2CH_2(CH_2)_7CH_3 \end{array}$$

TRIOLEIN, A LIQUID OIL — TRISTEARIN, A SOLID FAT

Oils commonly subjected to this process include those from cottonseed, peanut, corn germ, soybean, coconut, and safflower seeds. In recent years, as it became apparent that saturated fats may encourage diseases of the heart and arteries, soft margarines and cooking oils (which still contain some of the unhydrogenated fatty acid) have been placed on the market.

**Soaps can be made by treating fats or oils with sodium hydroxide.**

Naturally occurring fats and oils can be hydrolyzed in strongly basic solutions to form glycerol and salts of the fatty acids. Such hydrolysis reactions are called ***saponification*** reactions; the sodium or potassium salts of the fatty acids formed are ***soaps.*** Pioneers prepared their soap by boiling animal fat with an alkaline solution obtained from the ashes of hard wood. The resulting soap could be "salted out" by adding sodium chloride, making use of the fact that soap is less soluble in a salt solution than in water.

$$\begin{array}{l} CH_3(CH_2)_{16}COO-CH_2 \\ CH_3(CH_2)_{16}COO-CH \\ CH_3(CH_2)_{16}COO-CH_2 \end{array} + 3NaOH \longrightarrow 3CH_3(CH_2)_{16}COO^-Na^+ + \begin{array}{l} HO-CH_2 \\ HO-CH \\ HO-CH_2 \end{array}$$

TRISTEARIN (GLYCERYL TRISTEARATE) — SODIUM STEARATE (A SOAP) — GLYCEROL

The cleansing action of soap can be explained in terms of its molecular structure. Material that is water-soluble can be readily removed from the skin or a surface by simply washing with an excess of water. To remove a sticky sugar syrup from one's hands, the sugar is dissolved in water and rinsed away. Many times the material to be removed is oily and water will merely run over the surface of the oil. Since the skin has natural oils, even substances such as ordinary dirt that are not oily themselves can cover the skin in a greasy layer. The cohesive forces (forces between like molecules tending to hold them together) within the water layer are too large to allow the oil and water to intermingle (Fig. 9–5). When present in an oil-water system, soap molecules such as sodium stearate

$$CH_3CH_2CH_2CH_2CH_2CH_2CH_2CH_2CH_2CH_2CH_2CH_2CH_2CH_2CH_2CH_2CH_2C\begin{array}{l}{/\!\!/}O \\ \backslash O^-Na^+\end{array}$$

will move to the interface between the two liquids. The carbon chain, which is a nonpolar organic structure, will readily mix with the nonpolar grease molecules whereas the highly polar $-COO^-Na^+$ group enters the water layer because it becomes hydrated (Fig. 9–5b). The soap molecules will then tend to lie across the oil-water interface. The grease is then broken up into small droplets by agitation, each droplet surrounded by hydrated soap molecules (Fig. 9–5c). The surrounded oil droplets can-

**Soap, water, and oil together form an *emulsion,* with the soap acting as the emulsifying agent.**

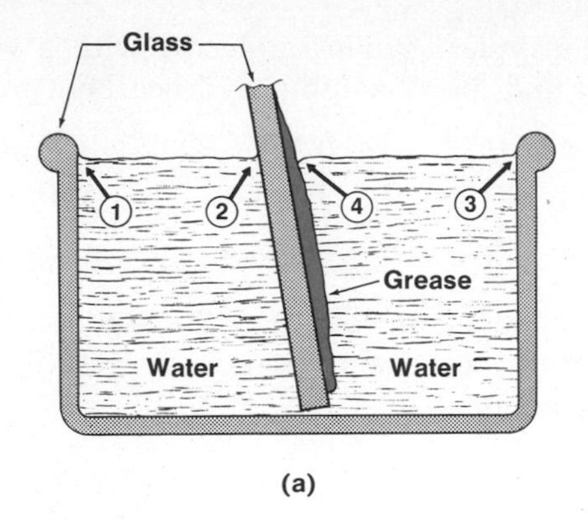


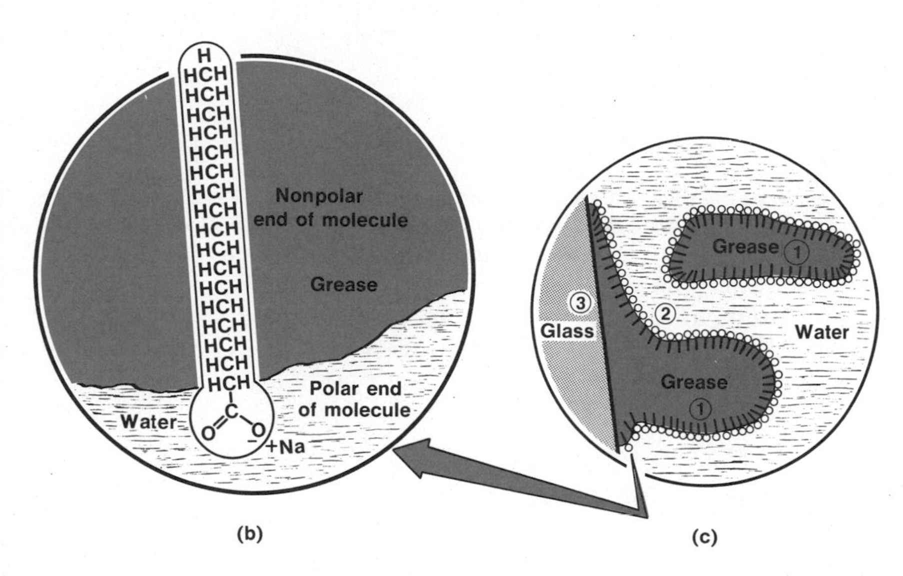


**Figure 9–5** The cleansing action of soap. (*a*) A piece of glass coated with grease inserted in water gives evidence for the strong adhesion between water and glass at 1, 2, and 3. The water curves up against the pull of gravity to wet the glass. The relatively weak adhesion between oil and water is indicated at 4 by the curvature of the water away from the grease against the force tending to level the water. (*b*) A soap molecule, having oil-soluble and water-soluble ends, will become oriented at an oil-water interface such that the hydrocarbon chain is in the oil (with molecules that are electrically similar, nonpolar) and the $COO^-Na^+$ group is in the water (highly charged polar groups interacting electrically). (*c*) In an idealized molecular view, a grease particle, 1, is surrounded by soap molecules, which in turn are strongly attracted to the water. At 2 another droplet is about to break away. At 3 the grease and clean glass interact before the water moves between them.

not come together again since the exterior of each is covered with $-COO^-Na^+$ groups that interact strongly with the surrounding water. If enough soap and water are available, the oil will be swept away, forming a clean and water-wet surface.

## Synthetic Detergents (Syndets)

**Ordinary soaps give precipitates with hard water ions such as $Ca^{2+}$, $Mg^{2+}$, and $Fe^{3+}$.**

If $Ca^{2+}$ or $Mg^{2+}$ ions are present in water, an ordinary soap will precipitate as an insoluble salt of these ions and the water is said to be ***hard.*** The ring on the bathtub and the scum in the washer are visible signs of this precipitate. Only after all these interfering ions have been precipitated can the added soap cause the water to form suds. Water softeners remove the hard water ions, replacing them with ions such as $Na^+$ which do not interfere with the soap's action, but this is often an expensive process. Synthetic detergents (***syndets***) are similar to soaps in that they are composed of molecules having water-soluble and oil-soluble ends, but syndets have an advantage over soaps in hard water in that they do not form precipitates with $Ca^{2+}$

or $Mg^{2+}$. The synthesis of a typical syndet molecule with a water-soluble sodium sulfate ($-SO_3^- Na^+$) group and an oil-soluble hydrocarbon group is shown below.

$$CH_3(CH_2)_{11}OH \xrightarrow{H_2SO_4} CH_3(CH_2)_{11}OSO_3H \xrightarrow{NaOH} CH_3(CH_2)_{11}OSO_3^- Na^+$$

LAURYL ALCOHOL — LAURYL HYDROGEN SULFATE — SODIUM LAURYL SULFATE (AN INGREDIENT OF SOME SHAMPOOS)

The most widely used syndets are sodium salts of substituted benzenesulfonic acids. Normally, a long hydrocarbon chain, designated $-R$, is attached to the benzene ring by a substitution reaction.

$SO_3^-Na^+$

R

SODIUM SALT OF A BENZENESULFONIC ACID

Syndets are effective cleansing agents, but they pose problems in water treatment and purification. By careful choice of starting materials, chemists have been able to produce syndets which, once "used," can be broken down into simpler molecules by the action of bacteria. R must be a straight-chain hydrocarbon for proper breakdown. These biodegradable syndets have been used almost exclusively in the United States since about 1965.

## Useful Products from Organic Synthesis Reactions

Many chemists are engaged in the synthesis of organic compounds. In educational and industrial laboratories throughout the world they prepare new and different compounds on a small scale. If the new compound has commercial value, the preparation is subsequently adapted for full-scale plant operations (Fig. 9–6).

The thousands of chemical changes required to synthesize the many known organic compounds have a few characteristics in common:

1. Several chemical changes are usually required to synthesize a single organic compound. Each chemical change is called a step, and each step produces an intermediate compound that is used in the next step. The final step produces the desired end product.
2. Normally only one functional group undergoes change in each step. The rest of the molecule remains intact and unchanged. This is known as the ***principle of minimum structural change.***
3. From one principal starting substance can come many diverse products. The kind of product obtained depends upon the reactants and the conditions imposed.
4. The more steps in the synthesis, the lower the percentage yield of the final product. The starting substance and each intermediate are only partially converted to the next intermediate because of equilibrium considerations, side reactions, or both, which convert some of the starting substance into undesirable products. If an intermediate product must be removed and purified before proceeding to the next step, some additional material is lost. The principal purification methods are recrystallization and extraction for solids and distillation for liquids.

**Organic syntheses are usually carried out so that only one part of the molecule changes in each reaction step.**

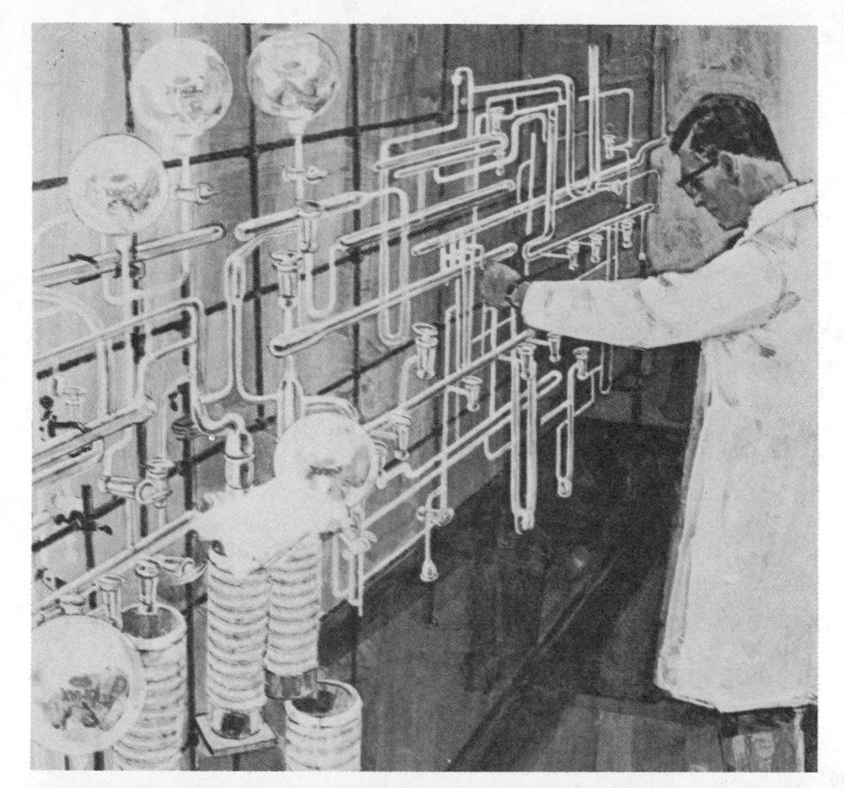

**Figure 9–6** Organic syntheses begin on a small scale in the laboratory (*top*) but graduate to a very large scale if they become commercially important. The top view shows a laboratory bench. The bottom view shows an automatic, modern chemical plant. In both cases the chemical reactions are the same; the scale is different.

## Addition Reactions

Many important organic compounds are formed by ***addition reactions.*** These reactions get their name from the net effect of the reaction: addition of the reactants to each other. For example, chlorine reacts with ethylene to produce 1,2-dichloroethane.

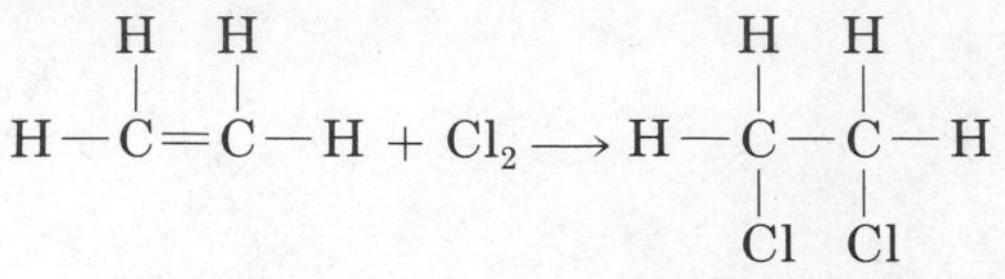

$$\mathrm{H{-}\underset{\phantom{|}}{\overset{H}{\overset{|}{C}}}{=}\overset{H}{\overset{|}{C}}{-}H + Cl_2 \longrightarrow H{-}\underset{\underset{Cl}{|}}{\overset{H}{\overset{|}{C}}}{-}\underset{\underset{Cl}{|}}{\overset{H}{\overset{|}{C}}}{-}H}$$

ETHYLENE 1,2-DICHLOROETHANE

**Carbon-carbon double bonds react more readily than carbon-carbon single bonds.**

This is an addition reaction since two chlorine atoms have been added to the ethylene molecule. The mechanism for this reaction makes use of a special property of the C=C bond. Since there are four electrons localized between the two carbon atoms, they can interact with a chlorine molecule and cause its electron distribution to be distorted. This electron distortion is called ***polarization*** and can result in the breaking of the Cl—Cl bond and the formation of a bond between a carbon and a chlorine atom.

**$\delta^+$ denotes a partial positive charge.**

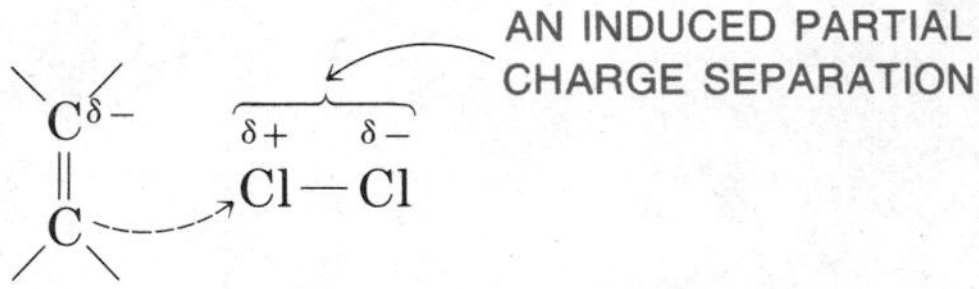

AN INDUCED PARTIAL CHARGE SEPARATION

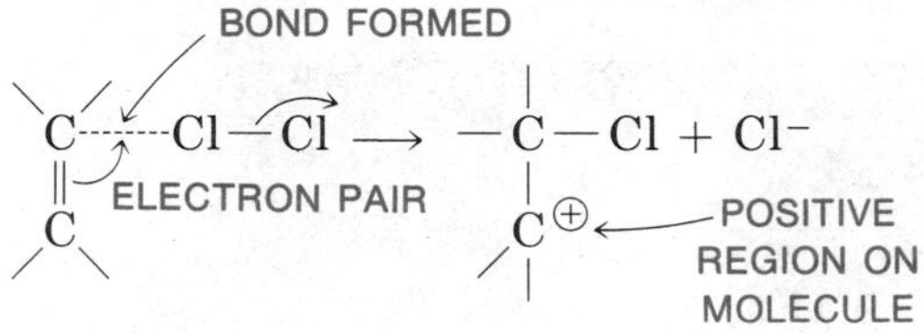

$$\mathrm{C^{\delta-}{=}C \cdots \overset{\delta+}{Cl}{-}\overset{\delta-}{Cl}}$$

BOND FORMED

$$\mathrm{C{=}C \cdots Cl{-}Cl \longrightarrow {-}\overset{|}{C}{-}Cl + Cl^- \quad (C^{\oplus})}$$

ELECTRON PAIR

POSITIVE REGION ON MOLECULE

The resulting positive region on the molecule then attracts the negative chloride ion:

$$\mathrm{{-}\overset{|}{C}{-}Cl \;\; {-}\underset{|}{C^{\oplus}} + Cl^- \longrightarrow {-}\overset{|}{C}{-}Cl \;\; {-}\underset{|}{C}{-}Cl}$$

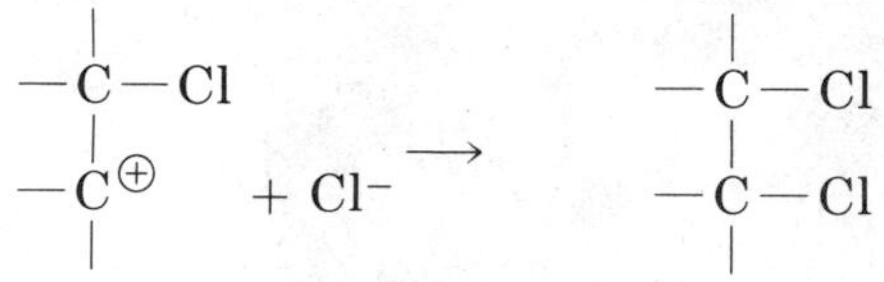

DICHLORO PRODUCT

Other addition reactions take place by different mechanisms.

Addition reactions are used to prepare some polymers, an important class of compounds discussed in the next chapter.

## Substitution Reactions of Aromatic Compounds

When the aromatic compound benzene reacts with chlorine in the presence of a suitable catalyst such as iron (III) chloride, a ***substitution reaction*** takes place in which one of the benzene hydrogen atoms is replaced by a chlorine atom. The mechanism for this reaction is similar to that of addition, except that a positively charged chlorine ion reacts with the aromatic benzene ring, which is electron-rich. A positive hydrogen ion leaves the aromatic ring and becomes associated with the negative chloride ion as HCl.

$$Cl_2 + FeCl_3 \longrightarrow FeCl_4^- + Cl^+$$

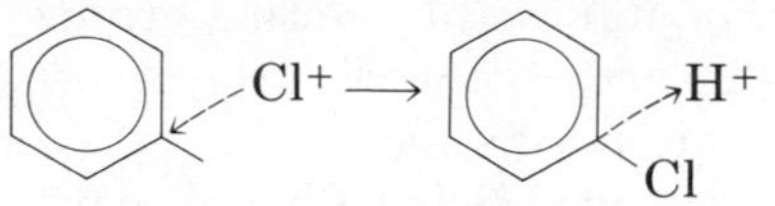

$$C_6H_6 + Cl^+ \longrightarrow [C_6H_6Cl]^+ \longrightarrow C_6H_5Cl + H^+$$

$$H^+ + FeCl_4^- \longrightarrow FeCl_3 + HCl$$

The overall reaction can be written as

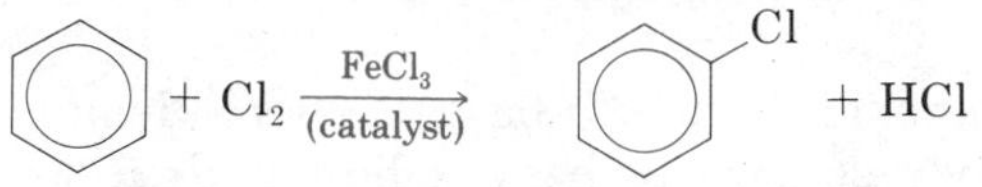

$$C_6H_6 + Cl_2 \xrightarrow[\text{(catalyst)}]{FeCl_3} C_6H_5Cl + HCl$$

BENZENE CHLOROBENZENE

**Substitution reactions are like replacement reactions, discussed in Chapter 5.**

Chlorobenzene is manufactured on an enormous scale and is widely used in industrial processes to prepare other organic compounds.

## Aspirin and Oil of Wintergreen

The series of steps required to synthesize aspirin and oil of wintergreen from chlorobenzene is outlined in Figure 9–7. Each structure represents the beginning or end of a step in the synthesis. Only the principal organic substance is shown for each step. Other products such as NaCl (a coproduct with phenol) and water (a coproduct with sodium phenoxide) are sometimes important in the synthesis because they have to be removed to avoid interference with subsequent steps. However, in giving

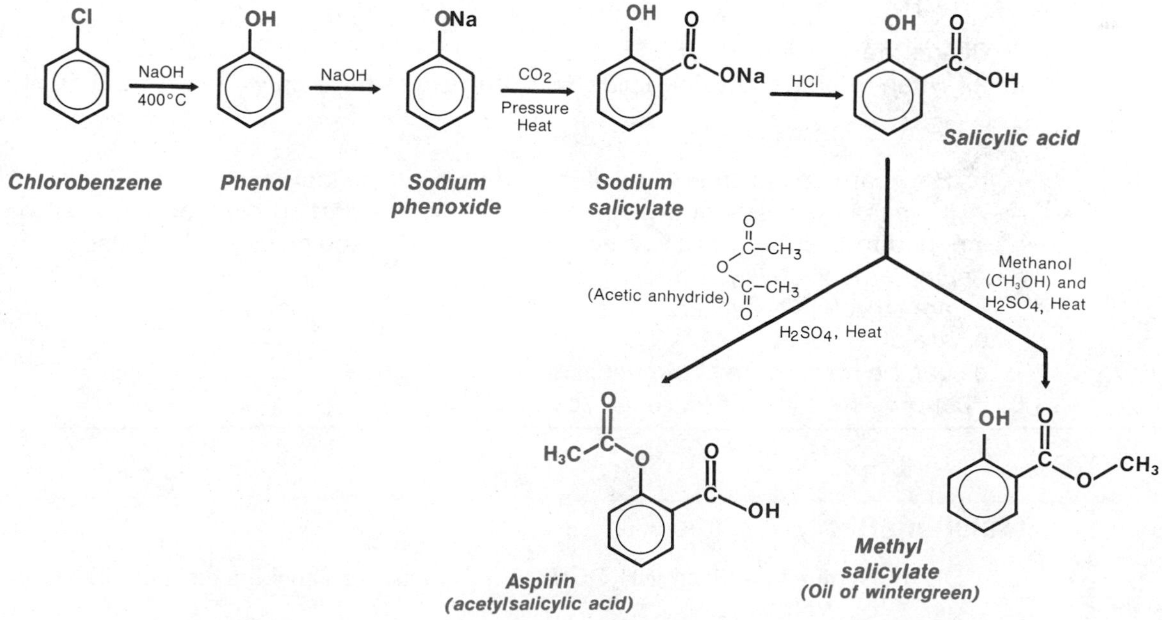

Chlorobenzene $\xrightarrow[400°C]{NaOH}$ Phenol $\xrightarrow{NaOH}$ Sodium phenoxide $\xrightarrow[\text{Pressure, Heat}]{CO_2}$ Sodium salicylate $\xrightarrow{HCl}$ Salicylic acid

Salicylic acid + (Acetic anhydride) $\xrightarrow{H_2SO_4,\ \text{Heat}}$ Aspirin (acetylsalicylic acid)

Salicylic acid + Methanol ($CH_3OH$) and $H_2SO_4$, Heat $\longrightarrow$ Methyl salicylate (Oil of wintergreen)

**Figure 9–7** Preparation of aspirin and oil of wintergreen. A discussion of the syntheses is given in the text.

Any given organic compound can usually serve as the starting substance for the synthesis of many other organic compounds.

a broad outline of the synthetic process, the coproducts are generally omitted; only those products made in a previous step and required for subsequent steps are included. In Figure 9–7 the step-by-step structural changes can be followed by noting groups in color. Conditions and additional reactants for each step are written with the arrow. These conventions are generally used to summarize organic syntheses.

Some intermediates are useful compounds in their own right. Phenol, commonly called carbolic acid, is used to prepare plastics such as Bakelite, drugs, dyes, and other compounds. Phenol also has medical application as a topical anesthetic for some types of lesions and in the treatment of mange and colic in animals. Methyl salicylate, or oil of wintergreen, is used as a flavoring agent and as a component of rubbing alcohol for sore muscles.

The conversion of phenol into sodium phenoxide is an acid-base reaction. Phenol, with an acidic hydrogen in the hydroxyl group, reacts with a base, sodium hydroxide, to give a salt, sodium phenoxide, and water. The reaction of salicylic acid to form oil of wintergreen is an ***esterification.*** The organic acid reacts with an alcohol in the presence of strong mineral acid to produce an ester, methyl salicylate, and water.

Obviously, it is beyond the scope of this text to give an extensive overview of organic synthesis. You can be assured, however, that if the "miracle cancer drug" is found or whenever any other new and useful compounds are discovered, their production will involve step-by-step molecular modifications to produce the desired product. An exciting aspect of organic synthesis is the prediction of desired properties of new molecular arrangements and then the testing of the theoretical properties. With so many possibilities yet to be discovered, one can only guess at the potential power of organic synthesis.

## Self-Test 9-B

1. Complete the following equation:

   $CH_3CH_2CH_2OH + CH_3CH_2COOH \xrightarrow{H_2SO_4}$

2. Define the term soap. ______________
3. a. When referring to edible lipids, what is the difference between a fat and an oil?

   ______________

   b. How can the melting points of most edible oils be increased? ______________
4. In a typical synthesis of organic compounds, the starting compounds must be stripped to their atoms before new compounds can be made. (True/False)
5. Aspirin and oil of wintergreen
   a. are structurally similar
   b. are both acids
   c. can be made from chlorobenzene
6. Phenol is the same as (carbonic/carbolic) acid.

## Matching Set

| | | |
|---|---|---|
| ______ 1. Synthesized from $NH_4OCN$ by Wöhler | | **a** measures knocking behavior in engine |
| ______ 2. RCOOH | | **b** breaks hydrocarbons into smaller molecules |
| ______ 3. R—OH | | **c** antiknocking agent |

_______ 4. $R-O-\underset{\underset{O}{\|}}{C}-R'$ | d aldehyde

_______ 5. octane rating | e carboxylic acid

_______ 6. RCHO | f soap

_______ 7. cracking | g urea

_______ 8. $Pb(C_2H_5)_4$ | h alcohol

_______ 9. sodium stearate | i ester

## Questions

**1.** Hydrocarbons are generally separated by what purification technique?

**2.** Would you expect $CH_3CH_2CH_2CH_2CH_2CH_2CH_2CH_2CH_3$ to be an important useful constituent of bottled gas? Explain your answer. Would it be useful in gasoline? Explain.

**3.** Indicate what products would be formed in the reaction of the following:
a. methanol and acetic acid
b. 1-propanol and stearic acid
c. ethylene glycol and acetic acid

**4.** Explain how hydrogen bonding could play a significant role in fixing the boiling point of acetic acid.

**5.** Write structural formulas for the four alcohols with the composition $C_4H_9OH$.

**6.** Wood alcohol is a deadly poison that can be made from what deadly gas?

**7.** Draw a structural formula for each of the following:
a. an alcohol
b. an organic acid
c. an ester
d. glycerol

**8.** What is the formula for tetraethyllead?

**9.** What is meant by each of the following terms?
a. proof rating of an alcohol
b. octane rating of a gasoline
c. denatured alcohol

**10.** Beginning with petroleum, outline the steps and write the chemical equations for the production of ethyl acetate.

**11.** What is a major structural difference between a motor oil and a vegetable oil?

**12.** How does a soap cleanse?

**13.** What ions make water hard?

**14.** Discuss the fundamental characteristics of the following:
a. synthetic detergents
b. soaps

**15.** Would you use the sodium salt of a fatty acid to wash your hands? Explain.

**16.** Which would you expect to boil at a higher temperature, $CH_3CH_2CH_2CH_3$ or $CH_3CH_2OH$? Why?

**17.** Would you say that hydrogen bonding is stronger or weaker in alcohol when contrasted with that in water?

**18.** Why will the addition of strong alkali aid in unstopping a greasy sink drain?

**19.** Draw structural formulas for:
a. aspirin
b. oil of wintergreen
c. phenol

**20.** What ester has the smell of bananas?

**21.** What chemical reactions can be used to distinguish between:
a. $C_2H_5OH$ and $CH_3COOH$
b. $CH_3COOC_2H_5$ and $CH_3COOH$

**22.** What would be the product if *para*-dichlorobenzene were heated with NaOH at 400°C?

**23.** Tell how you could prepare the following compounds:
a. chlorobenzene
b. urea
c. methanol

**24.** The product of oxidation of a primary alcohol is a(n) ______________, which can be oxidized to a(n) ______________.

**25.** What hydrocarbons are gases in their natural state?

**26.** Pure ethyl alcohol is what proof?

**27.** What is the use of tetraethyllead in gasoline?

**28.** Describe two ways to prevent knocking in gasoline.

**29.** Explain the common names for methanol and ethanol.

**30.** Why is a fatty acid so named?

**31.** Is pure synthetic ethanol different from pure grain alcohol? Explain.

**32.** Which propanol is used as rubbing alcohol?

**33.** How many hydrogen bonds are possible per molecule of methanol? of ethylene glycol? of glycerol?

**34.** Would you expect glycerin to be water-soluble? Why?

**35.** Which structural group hydrogen bonds most readily: alcohol, carboxylic acid, or ester?

**36.** What is the primary chemical in permanent antifreeze?

**37.** What structural features and properties make ethylene glycol a desirable antifreeze agent?

**38.** What is the acid in vinegar?

**39.** The herbicide 2,4-D has the chemical name of 2,4-dichlorophenoxyacetic acid. Its structure is

$$\mathrm{Cl{-}C_6H_3(Cl){-}O{-}\underset{H}{\overset{H}{C}}{-}C(=O){-}O{-}H}$$

Draw two other isomers of this molecule.

**40.** After reading this chapter and the last one, what new thoughts do you have when you view a lump of coal or a drop of petroleum?

# Man-Made Giant Molecules—The Synthetic Polymers

# 10

A plastic is a substance that will flow under heat and pressure, and hence is capable of being molded into various shapes. All plastics are polymers, but not all polymers are plastic.

It is impossible for most Americans to get through a day without using a dozen or more materials based on synthetic ***polymers.*** Many of these materials are *plastics* of one sort or another. Examples of these include plastic dishes and cups, combs, automobile steering wheels and seat covers, telephones, pens, plastic bags for food and wastes, plastic pipes and fittings, plastic water-dispersed paints, false eyelashes and wigs, a wide range of synthetic fibers for clothing, synthetic glues, and flooring materials. In fact, these materials are so widely used they are usually taken for granted. All these materials are composed of ***giant molecules.***

This "flood of plastic objects" did not arise accidentally; it slowly became necessary over a period of 30 or 40 years because (1) natural resources dwindled and so many materials became scarce, (2) with rising labor costs, many items could be made less expensively by molding than by whittling, shaving, sawing, and gluing, and (3) the new materials were so superior in properties that they did the job better.

Some of our most useful polymer chemistry has resulted from copying giant molecules found in nature. Rayon is remanufactured cellulose. Synthetic rubber is copied from natural latex rubber. As useful as they may be, however, polymer chemistry is not restricted to nature's models. Nylon, Dacron, and polycarbonates are a few examples of synthetic molecules that do not have exact duplicates in nature. We have gone to school on nature and extended our knowledge to new situations.

The purpose of this chapter is to investigate the structural chemistry of polymers to see just why they have such useful properties. Are these properties the result of stronger bonds, or groups of molecules acting together, or is there some other explanation? As we shall see, giant molecules were observed first in nature, and then copied and improved upon by the chemist. In the next chapter we will study some of nature's polymers and their functions.

## What Are Giant Molecules?

A *macromolecule* is a molecule with a very high molecular weight.

Many chemists were reluctant to accept the concept of giant molecules, but in the 1920's a persistent German chemist, Hermann Staudinger (1881–1965; Nobel prize, 1953), championed the idea and introduced a new term, ***macromolecule,*** for these giant molecules. Staudinger devised experiments that yielded accurate molecular weights, and, in addition, he synthesized "model compounds" to test his theory. One of his first model compounds was prepared from styrene, a chemical made from ethylene and benzene.

Styrene is the number twenty-two commercial chemical.

$$H_2C{=}CH{-}C_6H_5$$

STYRENE

Under the proper conditions, styrene molecules use the "extra" electrons of the double bond to undergo a ***polymerization*** reaction to yield polystyrene, a material composed of giant molecules. The word ***polymer*** means "many parts" (Greek, *poly* = many, *meros* = parts). The molecules of styrene are the ***monomers*** (Greek, *mono* = one); they provide the recurring units in the giant molecule analogous to identical railroad cars coupled together to make a long train.

A polymer has molecules composed of a large number of similar units.

The macromolecule polystyrene is represented as a long chain of monomer units bonded to each other. Each unit is bonded to the next by a strong covalent bond. The polymer chain is not an endless one; some polystyrenes made by Staudinger were found to have molecular weights of about 600,000, corresponding to a chain of about 5,700 styrene units. The polymer chain can be indicated as

$$R{-}CH_2{-}\underset{C_6H_5}{CH}{-}\left(CH_2{-}\underset{C_6H_5}{CH}{-}\right)_n CH_2{-}\underset{C_6H_5}{CH}{-}R$$

where R represents some terminal group, often an impurity, and $n$ is a large number.

Polystyrene is a clear, hard, colorless solid at room temperature. Since it can be molded easily at 250°C, the term ***plastic*** has become associated with it and similar materials. Polystyrene has so many useful properties that its commercial production, which began in Germany in 1929, today exceeds 5 billion pounds per year. It is used to make combs, bowls, toys, electrical parts, and many other items such as tough plastics used for radio and TV cabinets and synthetic rubber for tires.

Synthetic polymers are commonly called *plastics* when in a solid form.

There are two broad categories of plastics. One, when heated repeatedly, will soften and flow; when it is cooled, it hardens. Materials that undergo such reversible changes when heated and cooled are called ***thermoplastics;*** polystyrene is one example. The other type is plastic when first heated, but when heated further it forms a set of interlocking bonds. When reheated, it cannot be softened and reformed without extensive degradation. These materials are called ***thermosetting plastics*** and include such familiar names as Bakelite and rigid-foamed polyurethane, a polymer which is finding many new uses as a construction material.

Thermoplastic polymers can be repeatedly softened by merely heating.

Thermosetting polymers form cross-linking bonds when heated and then become rigid.

In order to gain a better understanding of polymers it is necessary to look at representative examples of the different types of polymerization processes.

## Addition Polymers

In the previous section it was noted that some polymers, such as polystyrene, are made by adding monomer to monomer to form a polymer chain of great length. Perhaps the easiest addition reactions to understand chemically are those involving monomers containing double bonds. The simplest monomer of this group is ethylene, $C_2H_4$. When ethylene (ethene) is heated under pressure in the presence of oxygen, polymers with

Ethylene (ethene) is the number six commercial chemical.

molecular weights of about 30,000 are formed. In order to enter into reaction, the double bond of an ethylene molecule must be broken. This forms ***reactive sites*** composed of unpaired electrons at either end of the molecule.

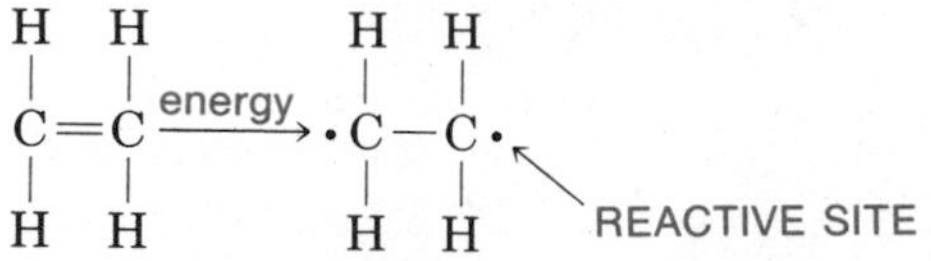

$$\begin{matrix} H & H \\ | & | \\ C & = C \\ | & | \\ H & H \end{matrix} \xrightarrow{\text{energy}} \begin{matrix} H & H \\ | & | \\ \cdot C & - C \cdot \\ | & | \\ H & H \end{matrix}$$

REACTIVE SITE

The partial breaking of the double bond can be accomplished by physical means such as heat, ultraviolet light, X rays, and high-energy electrons. This *initiation* of the polymerization reaction can also be accomplished with chemicals such as organic peroxides. These initiators, which are very unstable, break apart into pieces with unpaired electrons. These fragments (called ***free radicals***) are ravenous in trying to find a "buddy" for their unpaired electrons. They react readily with molecules containing carbon-carbon double bonds.

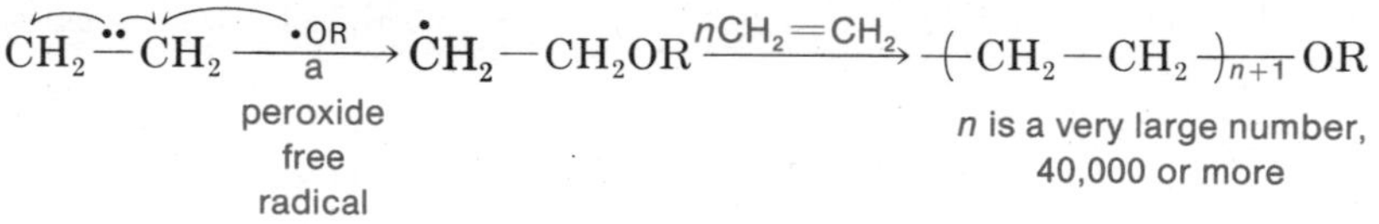

$$CH_2 \overset{\cdot\cdot}{-} CH_2 \xrightarrow[\text{a}]{\cdot OR} \dot{C}H_2 - CH_2OR \xrightarrow{nCH_2=CH_2} \left( CH_2 - CH_2 \right)_{n+1} OR$$

a peroxide free radical

$n$ is a very large number, 40,000 or more

The extension of the polyethylene chain shown above comes from the unpaired electron bonding to an electron in an unreacted ethylene molecule. This leaves another unpaired electron to bond with yet another ethylene molecule. For example,

$$ROCH_2 - CH_2\cdot + CH_2 \overset{\cdot\cdot}{-} CH_2 \longrightarrow ROCH_2 - CH_2 - CH_2 - CH_2\cdot$$

Polyethylenes formed under various pressures and catalytic conditions have different molecular structures and hence different physical properties. For example, chromium oxide as a catalyst yields almost exclusively the linear polyethylene shown in the margin. Actually, a methyl group is attached to about every eighth or tenth carbon in the chain. If ethylene is heated to 230°C at a pressure of 200 atm, irregular branches result. Under these conditions, free radicals undoubtedly attack the chain at random positions, thus causing the irregular branching.

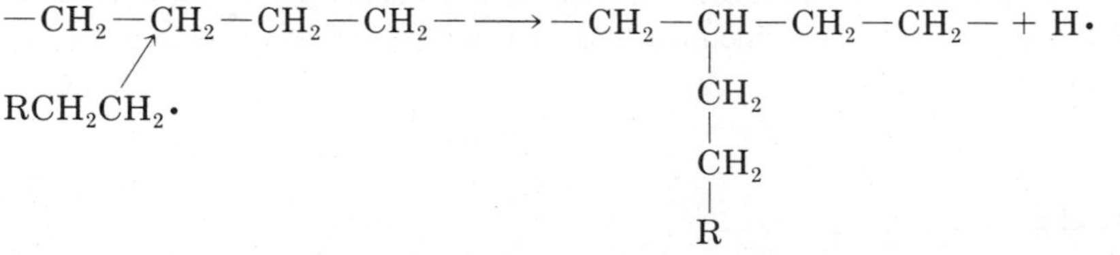

$$-CH_2 - CH_2 - CH_2 - CH_2 - \longrightarrow -CH_2 - \underset{\underset{\underset{\underset{R}{|}}{CH_2}}{\underset{|}{CH_2}}}{\underset{|}{CH}} - CH_2 - CH_2 - + H\cdot$$

$RCH_2CH_2\cdot$

BRANCHED POLYMER CHAINS

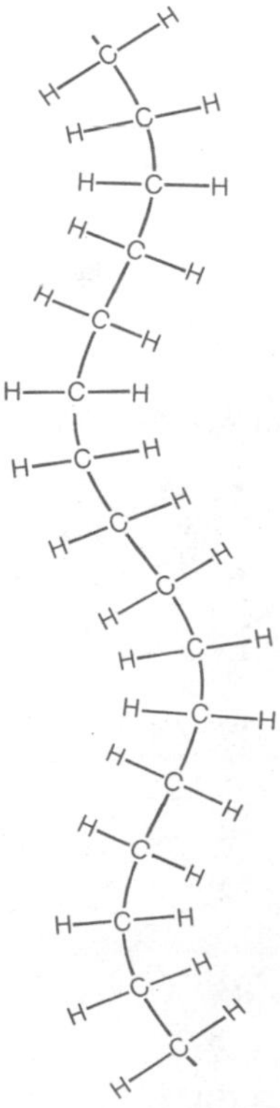

The molecules in linear polyethylene can line up with one another very easily, yielding a tough, high-density crystalline compound that is useful in making toys, bottles, and structural parts. The polyethylene with irregular branches is less dense, more flexible, and not nearly as tough as the linear polymer, since the molecules are generally farther apart and their arrangement is not as precisely ordered. This material is used for squeeze bottles and other similar applications.

There is a large group of derivatives of ethylene that undergo addition polymerization. Table 10–1 summarizes some information on these materials.

**A portion of a linear polyethylene molecule. Ethylene is unsaturated; polyethylene is saturated.**

**TABLE 10–1 Ethylene Derivatives That Undergo Addition Polymerization**

| Formula | Monomer Name | Polymer Name | Uses |
|---|---|---|---|
| $CH_2{=}CH_2$ | ethylene | polyethylene | coats milk cartons, wire insulation, bread wrappers, toys, films |
| $C_6H_5{-}HC{=}CH_2$ | styrene | polystyrene | synthetic rubber, combs, toys, bowls, packaging, appliance parts |
| $CH_2{=}CHCl$ | vinyl chloride (number twenty-one commercial chemical) | polyvinylchloride (PVC) | as a vinyl acetate copolymer in phonograph records, credit cards, rain wear, pipes, adhesives, films |
| $CH_2{=}CH{-}O{-}C({=}O){-}CH_3$ | vinyl acetate | polyvinylacetate | latex paint |
| $CH_2{=}CH{-}CN$ | acrylonitrile | polyacrylonitrile (PAN) | rug fibers, high-impact plastics |
| $CH_2{=}CH{-}CH{=}CH_2$ | divinyl (1,3-butadiene) | buna rubbers | tires and hoses |
| $CH_2{=}C(CH_3){-}C({=}O){-}O{-}CH_3$ | methyl methacrylate | polymethyl methacrylate (Plexiglas, Lucite) | transparent objects, lightweight "pipes" |
| $CF_2{=}CF_2$ | tetrafluoroethylene | polytetrafluoroethylene (TFE) (Teflon) | insulation, bearings, nonstick fry pan surfaces |

**Rubber is vulcanized by heating it with sulfur, which forms links between the polymer chains.**

*VULCANIZATION:*

$$>C{=}C< \; + \; S_8 \; + \; >C{=}C< \;\longrightarrow\; {-}C{-}C{-} \text{ linked through } {-}S{-}(S)_6{-}S{-} \text{ to } {-}C{-}C{-}$$

*Sulfur, or $S_8$ molecule*

## Synthetic "Natural" Rubber—A Tailor-Made Addition Polymer

A very interesting application of stereochemical control over polymerization is the manufacture of synthetic rubber. When several structures are possible and only one is desired, stereochemical control must be exercised.

Natural rubber, a product of the *Hevea brasilieusis* tree, is a hydrocarbon with the composition $C_5H_8$, and when it is decomposed in the absence of oxygen it yields the monomer isoprene:

$$CH_2{=}C(CH_3){-}CH{=}CH_2$$

ISOPRENE

Natural rubber occurs as latex (an emulsion of rubber particles in water) that oozes from rubber trees when they are cut. Precipitation of the rubber particles yields a gummy mass that is not only elastic and water-repellent but also very sticky, especially when warm. In 1839, after 10 years' work on this material, Charles Goodyear (1800–1860) discovered that heating gum rubber with sulfur produced a material that was no longer sticky, but still elastic, water-repellent, and resilient.

***Vulcanized rubber,*** as Goodyear called his product, contains short chains of sulfur atoms that bond together the polymer chains of the natural rubber and reduce its unsaturation (Fig. 10–1). The sulfur chains help to align the polymer chains,

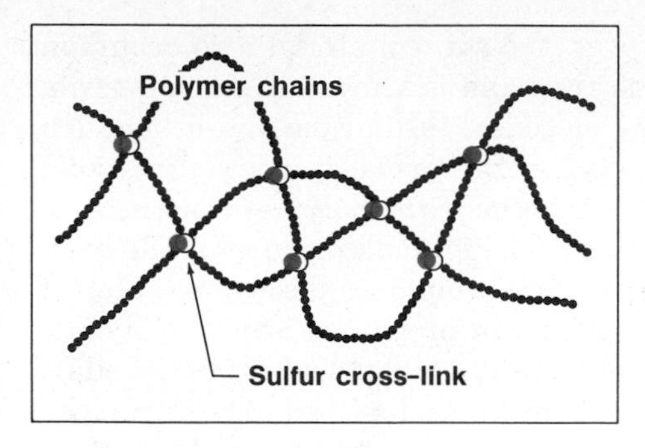


a. Before stretching

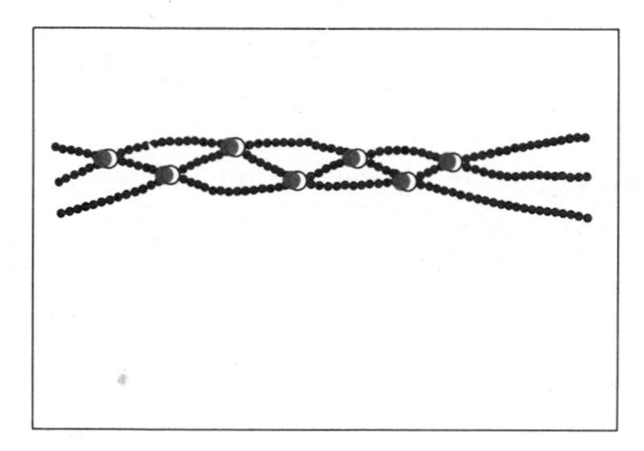

b. Stretched

**Figure 10–1** Stretched vulcanized rubber will spring back to its original structure, an electomeric property.

so the material does not undergo a permanent change when stretched, but springs back to its original shape and size when the stress is removed. Substances that behave this way are called ***elastomers.***

**Stereoregulating: controlling the arrangement of monomer units in the polymer.**

In later years chemists searched for ways to make a synthetic rubber so we would not be completely dependent on imported natural rubber during emergencies, such as during the first years of World War II. In the mid-1920's, German chemists polymerized butadiene (obtained from petroleum and structurally similar to isoprene, but without the methyl group side chain). The product was buna rubber, so named because it was made from butadiene (Bu—) and catalyzed by sodium (—Na).

$CH_2{=}CH{-}CH{=}CH_2$
*Butadiene*

The behavior of natural rubber (polyisoprene), it was learned later, is due to the specific arrangement within the polymer chain. We can write the formula for polyisoprene with the $CH_2$ groups on opposite sides of the double bond (the *trans* arrangement)

$$\begin{array}{llllll}
-CH_2 & H \quad CH_3 & & CH_2-CH_2 & & H \\
\quad\diagdown\; C{=}C \;\diagup & & \quad\diagdown\; C{=}C \;\diagup & & \quad\diagdown\; C{=}C \;\diagup & \\
CH_3 & CH_2-CH_2 & & H \quad CH_3 & & CH_2-
\end{array}$$

POLY-*TRANS*-ISOPRENE (THE $-CH_2-CH_2-$ GROUPS ARE *TRANS*)

or with the $CH_2$ groups on the same side of the double bond (the *cis* arrangement, from Latin meaning "on this side").

$$\begin{array}{llllll}
-CH_2 & CH_2-CH_2 & & CH_2-CH_2 & & CH_2- \\
\quad\diagdown\; C{=}C \;\diagup & & \quad\diagdown\; C{=}C \;\diagup & & \quad\diagdown\; C{=}C \;\diagup & \\
CH_3 & H \quad CH_3 & & H \quad CH_3 & & H
\end{array}$$

POLY-*CIS*-ISOPRENE (THE $-CH_2-CH_2-$ GROUPS ARE *CIS*)

Natural rubber is poly-*cis*-isoprene.

Natural rubber is poly-*cis*-isoprene. However, the *trans* material also occurs in nature, in the leaves and bark of the sapotacea tree, and is known as *gutta-percha*. It is used as a thermoplastic for golf ball covers, electrical insulation, and other such applications. Without an appropriate catalyst, polymerization of isoprene yields a solid that is like neither rubber nor gutta-percha. Neither the *trans* polymer nor the randomly arranged material is as good as natural rubber (*cis*) for making automobile tires.

In 1955, chemists at the Goodyear and Firestone companies discovered, almost simultaneously, how to use stereoregulation catalysts to prepare synthetic poly-*cis*-isoprene. This material is, therefore, structurally identical to natural rubber. Today, synthetic poly-*cis*-isoprene can be manufactured cheaply and is used almost equally well (there is still an increased cost) when natural rubber is in short supply. More than 2.4 million tons of synthetic rubber are produced in the United States yearly. Table 10–2 gives a typical rubber formulation as it might be used in a tire.

## Copolymers

After examining Table 10–1, you might well wonder what would happen if a mixture of two monomers were polymerized. If we polymerize pure monomer A, we get a ***homopolymer;*** poly A:

—AAAAAAAAAA—

Likewise, if pure monomer B is polymerized, we get a homopolymer, poly B:

—BBBBBBBBBB—

**TABLE 10–2 A Rubber Formulation**

| Ingredient | Name | Percent | Formula | Function |
|---|---|---|---|---|
| Rubber | poly-*cis*-isoprene | 62.0 | $-CH_2-H_2C(H_3C)C{=}C(H)CH_2-CH_2-$ | elastomer |
| Activators | zinc oxide<br>stearic acid | 2.7<br>0.6 | $ZnO$<br>$C_{18}H_{37}COOH$ | activates vulcanizing agents; stearic acid acts as a lubricant in processing |
| Vulcanizing agent | sulfur | 1.5 | $S_8$ | crosslinks polymer chains |
| Filler | carbon black | 30.5 | C | provides strength and abrasion resistance |
| Accelerator | dibenzthiozole disulfide | 1.1 | (benzothiazolyl)–C–S–S–C–(benzothiazolyl) | catalyzes vulcanization |
| Antioxidant | alkylated diphenylamine | 1.1 | $C_8H_{17}-C_6H_4-N(H)-C_6H_4-C_8H_{17}$ | inhibits attack by oxygen or ozone in the air |
| Processing oil | hydrocarbon oil | 0.5 | $C_nH_{2n+2}$ | plasticizer |

In contrast, if the monomers A and B are mixed and then polymerized, we get ***copolymers*** such as the following:

**A copolymer is made by polymerizing two or more different monomers together.**

—AABABAAABB—

—AABABABABB—

—BABABBAABA—

In such polymers the order of the units is often completely random, in which case the properties of the copolymer will be determined by the ratio of the amount of A to the amount of B and the reaction conditions during polymerization.

It is possible to produce copolymers that have long chains of similar monomers in their structures. These are called ***block copolymers,*** and can be represented as

—AAAAAABBBBBBBBBAAAAAA—

To overcome brittleness in polypropylene, both random and block copolymers with ethylene are made. The block copolymers are more resistant to impact. They are molded into articles such as toys by a process called injection molding, in which the molten plastic flows under pressure into the mold. Random copolymers are more transparent, whereas the homopolymer is used for filaments in such applications as carpets and rope.

A copolymer can have useful properties that are different from and often superior to those of the polymers of its pure constituents. As an example, consider synthetic rubbers again. During World War II it was apparent to our military planners that we would be hard-pressed if our rubber supplies from Asia were cut off by Japan. A crash program was begun to develop synthetic rubber that would be as good as natural rubber. The Germans had earlier polymerized styrene, but this is a hard thermoplastic with little elasticity. They had also polymerized butadiene to make the first synthetic rubber (buna rubber), although it was not very serviceable. American chemists found, however, that a 1 to 6 copolymer of styrene and butadiene possessed properties closer to those of natural rubber.

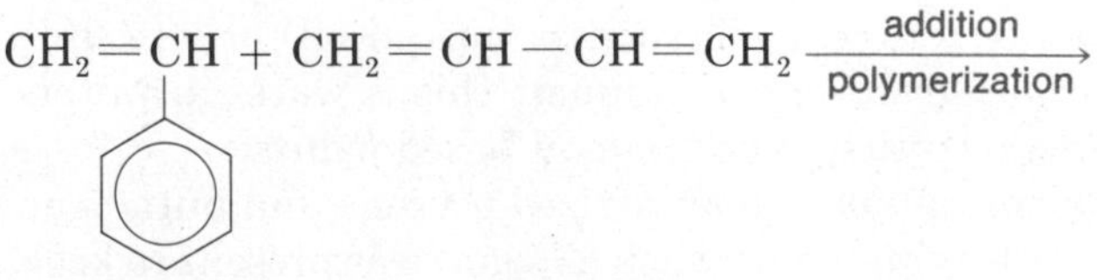

$$\underset{\text{STYRENE}}{CH_2{=}CH(C_6H_5)} + \underset{\text{BUTADIENE}}{CH_2{=}CH{-}CH{=}CH_2} \xrightarrow[\text{polymerization}]{\text{addition}}$$

$$-CH_2CH{=}CHCH_2CH_2\underset{C_6H_5}{CH}CH_2CH{=}CHCH_2CH_2CH{=}CHCH_2-$$

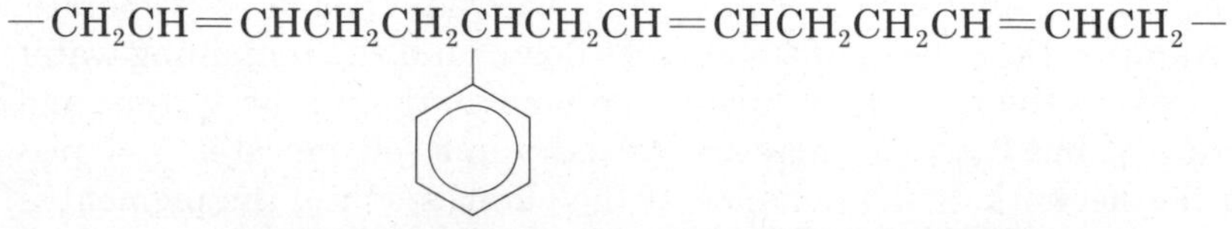

SBR COPOLYMER (STYRENE-BUTADIENE RUBBER)

**The largest group of synthetic polymers is used to make rubber—about 10 billion pounds per year. About 40% is used for tires.**

The double bonds remaining in the polymer chain allow them to undergo vulcanization like natural rubber polymer chains.

Copolymers are also used extensively in the plastics industry. For example, high-impact material for radio cabinets, tools, handles, and anything that might be dropped or struck is made of a copolymer of acrylate, butadiene, and styrene. Fibrous fillers and reinforcing agents add strength. A pure form of SBR has even replaced the latex in chewing gum.

**Figure 10–2** Photographers photographing a painter painting—in this picture the John Quincy Adams birthplace in Quincy, Massachusetts. The Sears Great American Home Series of advertisements illustrates the importance of surface coatings as preservatives.

## Addition Polymers and Paints

All paints involve polymers in one form or another. Popular latex and acrylic paints contain addition polymers that serve as ***binders.*** A paint binder forms a molecular network to hold the ***pigment*** (coloring agent) in place and to hold the paint to the painted surface. In oil-based paints, a drying oil (such as linseed oil) or a resin is the binder. All paints also have a volatile solvent or thinner; this is water in water-based paints and turpentine (or mineral spirits, or both) in oil-based paints.

**In 1970, nearly 830 million gallons of coatings were sold by U.S. companies. The $3 billion sales represented 5.5% of sales of all chemicals and allied products during 1970.**

Early latex paints were emulsions of partly polymerized styrene and butadiene in water (Fig. 10–3). Some type of emulsifying agent (such as soap) was present to keep the small drops of nonpolar styrene and butadiene dispersed in the polar water.

**The first commercial water-based latex paint was Glidden's Spred Satin, introduced in 1948.**

Immediately after the application of a latex paint, the water begins to evaporate. When some of the water is gone, the emulsion breaks down, and the remaining water evaporates quickly, leaving the paint film. Further polymerization of the styrene and butadiene follows slowly, but the paint appears to be dry in a few minutes. The pigment is trapped in the network of the polymer. If the paint is white, the pigment is probably titanium dioxide, $TiO_2$, which has replaced the poisonous compound "white lead," $Pb(OH)_2 \cdot 2PbCO_3$, that was used in older paints.

**Water-based latex paints reduce fire hazards and air pollution associated with the handling and application of oil-based paints.**

The styrene-butadiene resin is the least expensive binder material used, but it has a relatively long curing period, relatively poor adhesion, and a tendency to yellow with age. Polyvinylacetate is only a little more expensive and is an improvement over the styrene-butadiene resin. It quickly captured 50% of the latex market for interior paints. Another type with rapidly growing popularity, though about one third more expensive, includes the acrylic resins and the "acrylic latex" paints. These are more washable and much more resistant to light damage. They are especially useful as exterior paints.

**Monomer of polyvinylacetate:**

$$\mathrm{H_2C{=}CH{-}O{-}\overset{\displaystyle O}{\overset{\|}{C}}{-}CH_3}$$

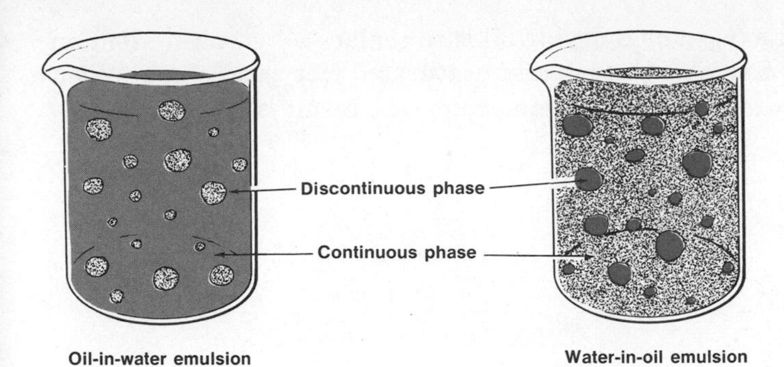


**Figure 10–3** Two kinds of emulsions. An emulsion is composed of two immiscible liquids, one dispersed as tiny droplets in the other. An emulsifying agent is required to stabilize an emulsion.

The fluoropolymers, similar to Teflon, are especially promising as surface coatings because of their great stability. Fluorine atoms are substituted for hydrogen atoms in the organic structure. Metals covered with polyvinylidene fluoride carry up to a 20-year guarantee against failure from exposure.

**Acrylic polymers have a sheen that allows latex paint to compete in the exterior gloss market traditionally monopolized by oil-based coatings. Acrylics adhere well and control corrosion. Acrylics are polymers of acrylonitrile,**

$$\begin{matrix} H & & CN \\ & C=C & \\ H & & H \end{matrix}$$

In the past few years, paint manufacturers have begun to blend linseed oil emulsions with latex emulsions in order to take advantage of the penetrating ability of the triglyceride molecules in the linseed oil. Some "latex" paints now contain as much as 75% linseed oil emulsion, but still have the desirable characteristic of latex. Table 10–3 gives a summary of various additives to emulsion paints and the rationale for their use.

The drying of modern oil-based paints involves much more than the evaporation of the mineral spirits or turpentine solvent. The chemical reaction between a drying oil and oxygen from the air completes the drying process. Common drying oils are soybean, castor, coconut, and linseed oils; the most widely used is linseed, which comes from the seed of the flax plant. All of these oils are glyceryl esters of fatty acids, as discussed in Chapter 8. Hydrolysis of a typical linseed oil yields the following assortment of fatty acids:

**See Tables 10–1 and 10–4.**

**Mineral spirits are petroleum fractions of moderate volatility.**

| | | |
|---|---|---|
| 4–7% | palmitic acid (16 C atoms) | (saturated) |
| 2–5% | stearic acid (18 C atoms) | (saturated) |
| 9–38% | oleic acid (18 C atoms) | (unsaturated) |
| 3–43% | linoleic acid (18 C atoms) | (unsaturated) |
| 25–58% | linolenic acid (18 C atoms) | (unsaturated) |

**TABLE 10–3 Additives Used in Emulsion Paints**

| | |
|---|---|
| Dispersing agents for pigments | Example: tetrasodium pyrophosphate ($Na_4P_2O_7$). The principle of like-charged particles repelling. |
| Protective colloids and thickeners | A thicker paint is slower to settle and drips and runs less. A protective colloid tends to stabilize the organic-water interface in the emulsion. Examples: sodium polyacrylates, carboxymethylcellulose, clays, gums. (Same mechanism as soap dispersing oil in water, Chapter 9.) |
| Defoamers | Foaming presents a serious problem if not corrected. Chemicals used: tri-n-butylphosphate, n-octyl alcohol and other higher alcohols, silicone oil. |
| Coalescing agents | As the water evaporates and the paint dries, an agent is needed to stick the pigment particles together. As the resin film forms, the agent evaporates. Coalescing agents must volatilize very slowly. Examples: hexylene glycol and ethylene glycol. |
| Freeze-thaw additives | Freezing will destroy the emulsion. Antifreezes such as ethylene glycol are used. |
| pH Controllers | The effectiveness of the ionic or molecular form of the emulsifier depends on the acid or alkaline conditions (pH). The wrong pH will break down the emulsion. Most paints tend to be too acidic. Ammonia, $NH_3$, is added to neutralize the acid. |

Typical molecule in linseed oil.

- Carbon atom
- Oxygen atom

The chemical action of oxygen on a drying oil is to replace a hydrogen atom on a carbon atom next to a C=C double bond in an unsaturated fatty acid chain. When oxygen reacts with two fatty acids on two oil molecules, the result is crosslinking between the two molecules.

part of one molecule

$$\begin{matrix} -CH_2-CH_2-CH{=}CH-CH_2- \\ \\ -CH_2-CH_2-CH{=}CH-CH_2- \end{matrix} + O_2 \longrightarrow \begin{matrix} -CH_2-CH-CH{=}CH-CH_2- \\ | \\ O \text{ ether linkage} \\ | \\ -CH_2-CH-CH{=}CH-CH_2- \end{matrix} + H_2O$$

part of another molecule

The polymeric network produced by the crosslinking hardens the paint, traps the pigment, and secures the paint in the crevices of the painted surface.

Metal ions such as $Zn^{2+}$, $Co^{2+}$, $Fe^{3+}$, $Mn^{2+}$, and $Ca^{2+}$ are added to oil-based paints to catalyze the drying process. These ions decompose peroxides (compounds containing the —O—O group) formed during the crosslinking process and precipitate free acids as salts of the ions.

## Condensation Polymers

### Polyesters

A chemical reaction in which two molecules react by splitting out or eliminating a small molecule is called a ***condensation reaction.*** For example, acetic acid and ethyl alcohol will react, splitting out a water molecule, to form ethyl acetate, an ***ester.***

$$\underset{\text{ACETIC ACID}}{CH_3\overset{O}{\overset{\|}{C}}-OH} + \underset{\text{ETHANOL}}{HOCH_2CH_3} \xrightarrow[\text{catalyst}]{H^+} \underset{\text{ETHYL ACETATE (AN ESTER)}}{CH_3\overset{O}{\overset{\|}{C}}-OCH_2CH_3} + H_2O$$

In a condensation polymerization, molecules are linked when they react to split out a small molecule such as water. The backbone of the polymer contains functional groups.

This important type of chemical reaction does not depend upon the presence of a double bond in the reacting molecules. Rather, it requires the presence of two kinds of functional groups on two different molecules. If each reacting molecule has *two* functional groups, both of which can react, it is then possible for condensation reactions to lead to a long-chain polymer. If we take a molecule with two carboxyl groups, such as terephthalic acid, and another molecule with two alcohol groups, such as ethylene glycol, each molecule can react at each end. The reaction of one acid group of terephthalic acid with one alcohol group of ethylene glycol initially produces an ester molecule with an acid group left over on one end and an alcohol group left over on the other:

The esterification of a dialcohol and a diacid involves two positions on each molecule.

$$\underset{\text{TEREPHTHALIC ACID}}{HO-\overset{O}{\overset{\|}{C}}-C_6H_4-\overset{O}{\overset{\|}{C}}-OH} + \underset{\text{ETHYLENE GLYCOL}}{HO-CH_2-CH_2-OH} \longrightarrow$$

$$\underset{\text{(AN ESTER)}}{HO-\overset{O}{\overset{\|}{C}}-C_6H_4-\overset{O}{\overset{\|}{C}}-O-CH_2-CH_2-OH} + H_2O$$

Subsequently, the remaining acid group can react with another alcohol group, and the alcohol group can react with another acid molecule. The process continues until an extremely large polymer molecule, known as a ***polyester,*** is produced with a molecular weight in the range of 10,000–20,000.

**Terephthalic acid is the number twenty-four commercial chemical.**

$$HO-\overset{O}{\overset{\|}{C}}-C_6H_4-\overset{O}{\overset{\|}{C}}-OCH_2CH_2O\left(-\overset{O}{\overset{\|}{C}}-C_6H_4-\overset{O}{\overset{\|}{C}}-OCH_2CH_2O\right)_n-$$

$$\cdot\ -\overset{O}{\overset{\|}{C}}-C_6H_4-\overset{O}{\overset{\|}{C}}-OCH_2CH_2OH$$

POLY(ETHYLENE GLYCOL TEREPHTHALATE)

Poly(ethylene glycol terephthalate) is used in making polyester textile fibers marketed under such names as "Dacron" and "Terylene," and films such as "Mylar." The film material has unusual strength and can be rolled into sheets one-thirtieth the thickness of a human hair. In film form this polyester is often used as a base for magnetic recording tape and for packaging frozen food. Dacron (and Teflon) tubes substitute for human blood vessels in heart bypass operations. The inert, nontoxic, nonallergenic, noninflammatory, non-blood-clotting natures of these polymers make them excellent substitutes.

**A typical polyester is produced from a dialcohol and a diacid.**

## Condensation Polymers and Baked-On Paints

If you have ever had a car repainted, perhaps you have seen the baking oven in which the paint is dried (Fig. 10–4). Automobile finishes, and those on major appliances (such as refrigerators, washing machines, and stoves), require very tough, adherent paints in order to withstand abuse. The tough coating is produced by extensively crosslinked condensation polymers.

A popular type of baked-on paint is the ***alkyd*** variety. The term comes from a combination of the words ***alc***ohol and ac***id.*** Alkyds, then, are polyesters with extensive crosslinking. One of the simpler alkyds is formed from the diacid, phthalic acid, and the trialcohol, glycerol.

**When General Motors lacquered the 1923 Oakland with a nitrocellulose lacquer, the protective coatings industry first began its expansion into the use of a wide variety of materials instead of a few naturally occurring oils and minerals.**

**Figure 10–4** The high temperatures in a drying oven cause numerous cross-linking reactions to take place, which increase the surface strength of an alkyd paint.

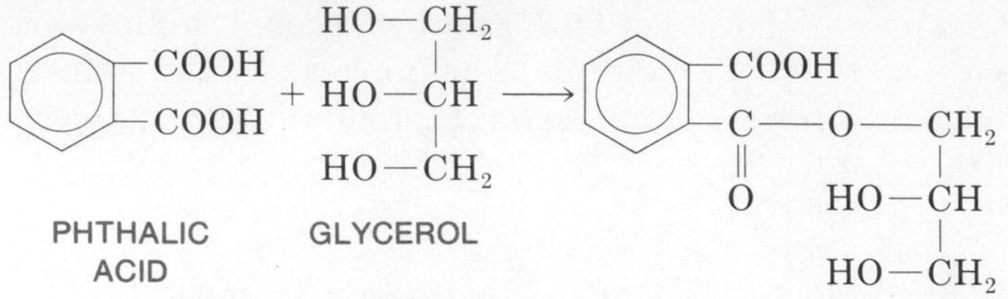

$$C_6H_4(COOH)_2 + HO{-}CH_2{-}CH(OH){-}CH_2{-}OH \longrightarrow C_6H_4(COOH){-}C(=O){-}O{-}CH_2{-}CH(OH){-}CH_2{-}OH$$

PHTHALIC ACID GLYCEROL

The —OH and —COOH groups continue to react with more and more reactant molecules until extensive crosslinking occurs. Heating to about 130°C for about one hour causes maximum crosslinking. A portion of the resin's structure is shown below.

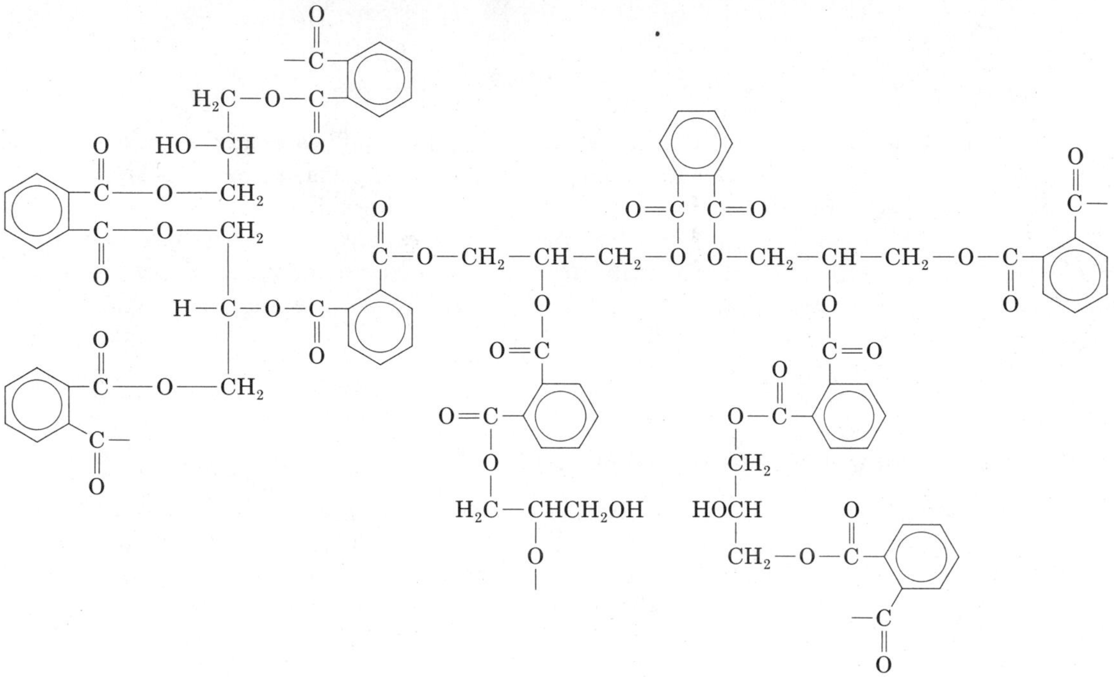

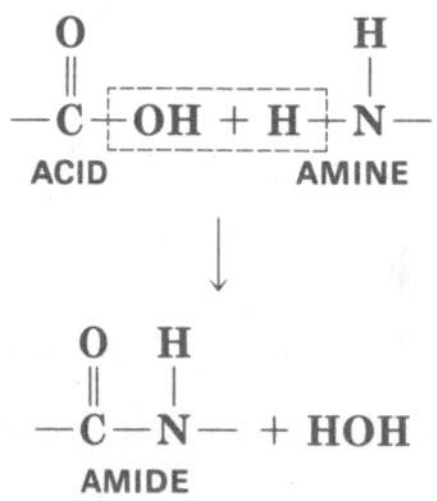

$H_2C{-}O{-}C$ HO—CH $CH_2$ $CH_2$ H—C—O—C $CH_2$ C—O—$CH_2$—CH—$CH_2$—O O—$CH_2$—CH—$CH_2$—O—C O=C C=O O=C O=C C=O O—C $CH_2$ $H_2C{-}CHCH_2OH$ HOCH $CH_2{-}O{-}C$

## Polyamides (Nylons)

$$-\overset{O}{\overset{\|}{C}}\text{-}OH + H\text{-}\overset{H}{\overset{|}{N}}- \;(\text{ACID},\ \text{AMINE}) \longrightarrow -\overset{O}{\overset{\|}{C}}-\overset{H}{\overset{|}{N}}- + HOH\ (\text{AMIDE})$$

Another useful condensation reaction is that occurring between an acid and an amine to split out a water molecule and form an ***amide.*** Reactions of this type yield a group of polymers that perhaps have had a greater impact upon society than any other type. These are the ***polyamides,*** or nylons.

In 1928, the Du Pont Company embarked upon a program of basic research headed by Dr. Wallace Carothers (1896–1937), who came to Du Pont from the Harvard University faculty. His research interests were high molecular weight compounds, such as rubber, proteins, and resins, and the reaction mechanisms that produced these compounds. In February 1935, his research produced a product known as nylon 66, prepared from adipic acid (a diacid) and hexamethylenediamine (a diamine):

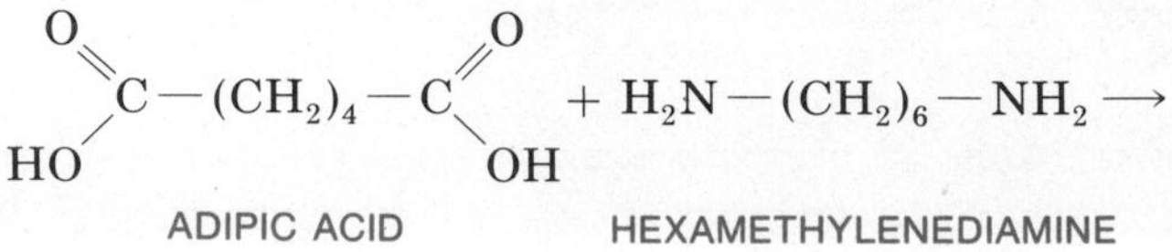

$$HO{-}C(=O){-}(CH_2)_4{-}C(=O){-}OH + H_2N{-}(CH_2)_6{-}NH_2 \longrightarrow$$

ADIPIC ACID HEXAMETHYLENEDIAMINE

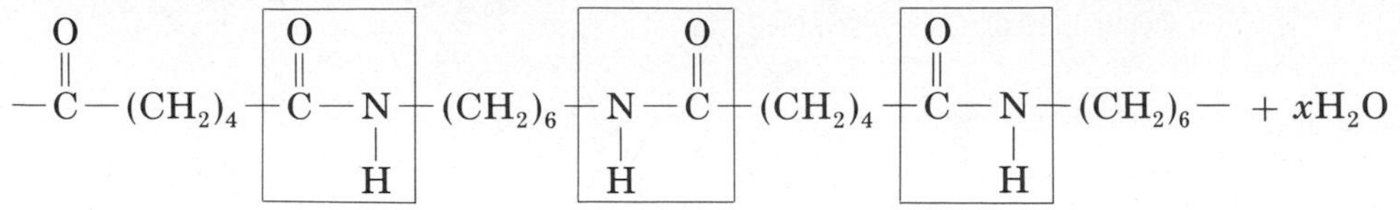

$$-\overset{\overset{\displaystyle O}{\|}}{C}-(CH_2)_4-\boxed{\overset{\overset{\displaystyle O}{\|}}{C}-\underset{\underset{\displaystyle H}{|}}{N}}-(CH_2)_6-\boxed{\underset{\underset{\displaystyle H}{|}}{N}-\overset{\overset{\displaystyle O}{\|}}{C}}-(CH_2)_4-\boxed{\overset{\overset{\displaystyle O}{\|}}{C}-\underset{\underset{\displaystyle H}{|}}{N}}-(CH_2)_6- \; + xH_2O$$

NYLON 66

*(The amide groups are outlined for emphasis.)*

**Common nylon can be made by the reaction of adipic acid and hexamethylenediamine.**

This material could easily be extruded into fibers that were stronger than natural fibers and chemically more inert. The discovery of nylon jolted the American textile industry at almost precisely the right time. Natural fibers were not meeting the needs of twentieth-century Americans. Silk was not durable and was very expensive, wool was scratchy, linen crushed easily, and cotton did not lend itself to high fashion. All four had to be pressed after cleaning. As women's hemlines rose in the mid-1930's silk stockings were in great demand, but they were very expensive and short-lived. Nylon changed all that almost overnight. It could be knitted into the sheer hosiery women wanted, and it was much more durable than silk. The first public sale of nylon hose took place in Wilmington, Delaware (the hometown of Du Pont's main office), on October 24, 1939. The stockings were so popular they had to be rationed. World War II caused all commercial use of nylon to be abandoned until 1945, as the industry turned to making parachutes and other war materials. Not until 1952 was the nylon industry able to meet the demands of the hosiery industry and to release nylon for other uses as a fiber and as a thermoplastic.

Many kinds of nylon have been prepared and tried on the consumer market, but two, nylon 66 and nylon 6, have been most successful. Nylon 6 is prepared from caprolactam, which comes from aminocaproic acid. Notice how aminocaproic acid contains an amine group on one end of the molecule and an acid group on the other end.

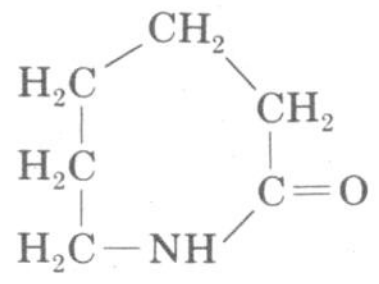

$$\begin{array}{ccc} & CH_2 & \\ H_2C & & CH_2 \\ | & & | \\ H_2C & & C{=}O \\ | & & \\ H_2C-NH & & \end{array}$$

*Caprolactam*

$$H_2N-(CH_2)_5-C\begin{matrix}\nearrow O\\ \searrow OH\end{matrix} \xrightarrow[\text{polymerization}]{-H_2O} -\overset{\overset{\displaystyle H}{|}}{N}-(CH_2)_5-\overset{\overset{\displaystyle O}{\|}}{C}-\overset{\overset{\displaystyle H}{|}}{N}-(CH_2)_5-\overset{\overset{\displaystyle O}{\|}}{C}-$$

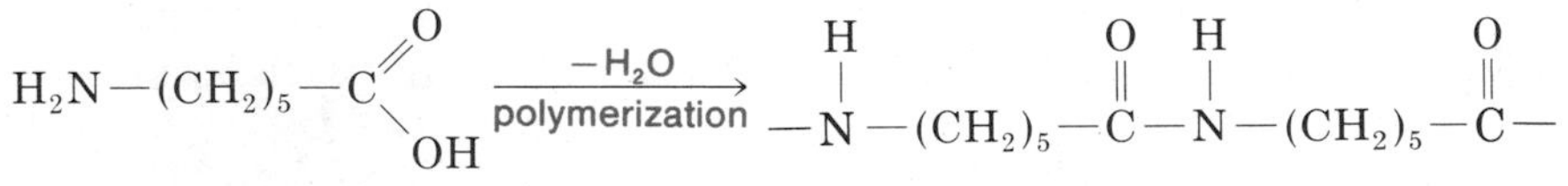

AMINOCAPROIC ACID — PORTION OF NYLON 6

**Hydrogen bonding is important in determining the properties of nylon fibers.**

Figure 10–5 illustrates another facet of the structure of nylon—***hydrogen bonding.*** This type of bonding explains why the nylons make such good fibers. In order to have good tensile strength, the chains of atoms in a polymer should be able to attract one another, but not so strongly that the plastic cannot be initially extended to form the fibers. Ordinary covalent chemical bonds linking the chains together would be too strong. Hydrogen bonds, with a strength about one tenth that of an ordinary covalent bond, link the chains in the desired manner. We shall see later that this type of bonding is also of great importance in protein structures.

## Polycarbonates

The tough, clear polycarbonates constitute another important group of condensation plastics. One type of polycarbonate, commonly called Lexan or Merlon, was first made in Germany in 1953. It is as "clear as glass" and nearly as tough as steel. A one-inch sheet can stop a .38-caliber bullet fired from 12 feet away. Such unusual properties have resulted in Lexan's use in "bullet-proof" windows and as visors in astronauts' space helmets. More than 115 million pounds of polycarbonates were produced in the United States in 1980.

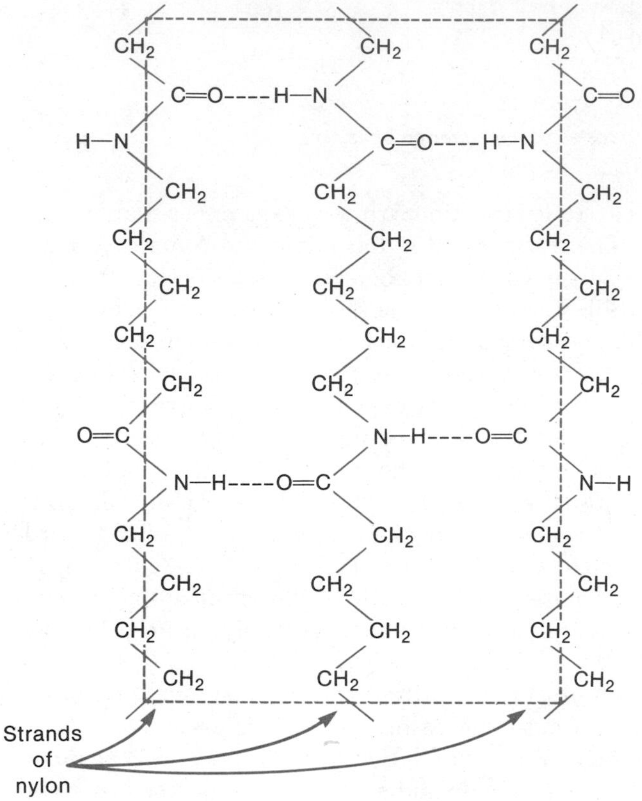

Figure 10–5
Structure and hydrogen bonding in nylon 6.

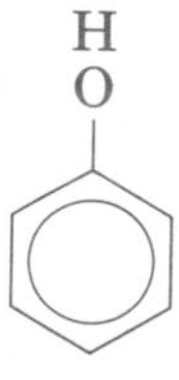

Strands of nylon

Phenol

The polycarbonates are formed by condensing phosgene with a substance containing two phenol structures. A molecule of HCl is condensed out in the formation of a C—O bond to complete the ester (—C(=O)—O—) linkage. Other chlorines then react with other alcohol groups to give a polymer chain containing —O—C(=O)—O— functional groups in the backbone of the chain. A representative portion of Lexan is made as follows.

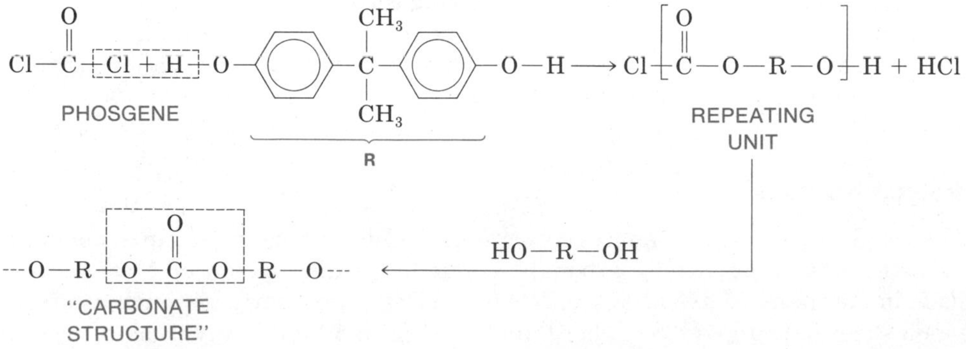

Cl—C(=O)—Cl + H—O—C₆H₄—C(CH₃)₂—C₆H₄—O—H ⟶ Cl—[C(=O)—O—R—O]—H + HCl

PHOSGENE  R  REPEATING UNIT

HO—R—OH

---O—R—O—C(=O)—O—R—O---

"CARBONATE STRUCTURE"

The name polycarbonate comes from the linkage's similarity to an inorganic carbonate ion, $CO_3^{2-}$.

## Self-Test 10-A

1. The individual molecules from which polymers are made are called ______________ ______________.
2. Draw the formulas of the monomers used to prepare the polymers listed below. For example, $CH_2{=}CH_2$ is used to prepare polyethylene.
   a. polypropylene ______________
   b. polystyrene ______________
   c. Teflon ______________
3. Many molecules of a carboxylic diacid reacting with many molecules of a di-alcohol produce a ______________.
4. Natural rubber is a polymer of ______________.
5. When styrene and butadiene are polymerized together, the product is a type of ______________.
6. Nylon is an example of a ______________ polymer.
7. Polyamides are formed when ______________ is split out from the reaction of many organic acid groups and many amine groups.
8. When an acid such as terephthalic acid $HO{-}C({=}O){-}C_6H_4{-}C({=}O){-}OH$ reacts with ethylene glycol, $HOCH_2CH_2OH$, the structure of the resulting polymer is: ______________
9. Polyesters are formed by (addition/condensation) reactions.
10. In paints, the components are dispersed in a liquid called the ______________ and the color is supplied by the ______________.
11. The drying oil used most often in paints is ______________ oil.
12. Water-based or "latex-type" paints are ______________ because two liquids that are ordinarily incompatible are stabilized.
13. Baked-on paints are usually alkyds. This means that they are usually made from polyfunctional ______________ and ______________.

## Silicones

The element silicon, in the same chemical family as carbon, also forms many compounds with numerous Si—Si and Si—H bonds, analogous to C—C and C—H bonds. However, the Si—Si bonds and the Si—H bonds are reactive toward both oxygen and water; hence, there are no useful silicon counterparts to most hydrocarbons. However, silicon does form stable bonds with carbon, and especially oxygen, and this fact gives rise to an interesting group of condensation polymers containing silicon, oxygen, carbon, and hydrogen (bonded to carbon).

**Silane, $SiH_4$, is structurally like methane, $CH_4$, in that both are tetrahedral.**

In 1945, E. G. Rochow, at the General Electric Research Laboratory, discovered that a silicon-copper alloy will react with organic chlorides to produce a whole class of reactive compounds, the ***organosilanes.***

$$\underset{\text{METHYL CHLORIDE}}{2CH_3Cl} + \underset{\text{SILICON-COPPER ALLOY}}{Si(Cu)} \longrightarrow \underset{\text{DIMETHYLDICHLOROSILANE}}{(CH_3)_2SiCl_2} + Cu$$

The chlorosilanes readily react with water and replace the chlorine atoms with hydroxyl (—OH) groups. The resulting molecule is similar to a dialcohol.

$$(CH_3)_2SiCl_2 + 2H_2O \longrightarrow (CH_3)_2Si(OH)_2 + 2HCl$$

**Silicones are polymers held together by a series of covalent Si—O bonds.**

Two dihydroxysilane molecules undergo a condensation reaction in which a water molecule is split out. The resulting Si—O—Si linkage is very strong; the same linkage holds together all the natural silicate rocks and minerals. Continuation of this condensation process results in polymer molecules with molecular weights in the millions:

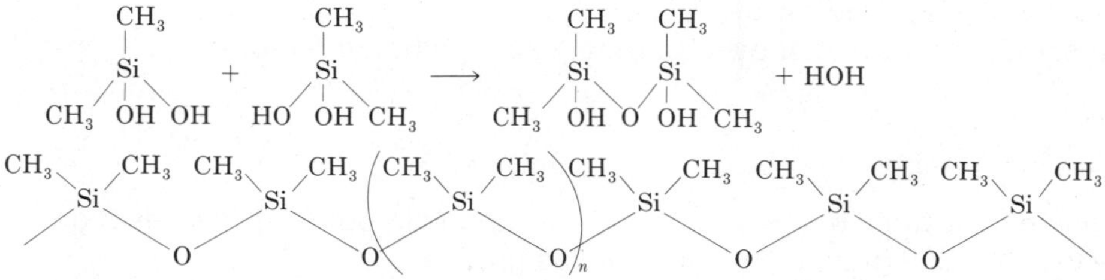

$$(CH_3)_2Si(OH)_2 + HO(CH_3)_2SiOH \longrightarrow HO(CH_3)_2Si{-}O{-}Si(CH_3)_2OH + HOH$$

$$-Si(CH_3)_2-O-Si(CH_3)_2-O\left(-Si(CH_3)_2-O\right)_n-Si(CH_3)_2-O-Si(CH_3)_2-O-Si(CH_3)_2-$$

By using different starting silanes, polymers with different properties result. For example, two methyl groups on each silicon atom results in ***silicone oils,*** which are more stable at high temperatures than hydrocarbon oils and also have less tendency to thicken at low temperatures.

Silicone rubbers are very high molecular weight chains crosslinked by Si—O—Si bonds. Room-temperature-vulcanizing (RTV) silicone rubbers are commercially available; they contain groups that readily crosslink in the presence of atmospheric moisture. The —OH groups are first produced, and then they condense in a crosslinking "cure" similar to the vulcanization of organic rubbers.

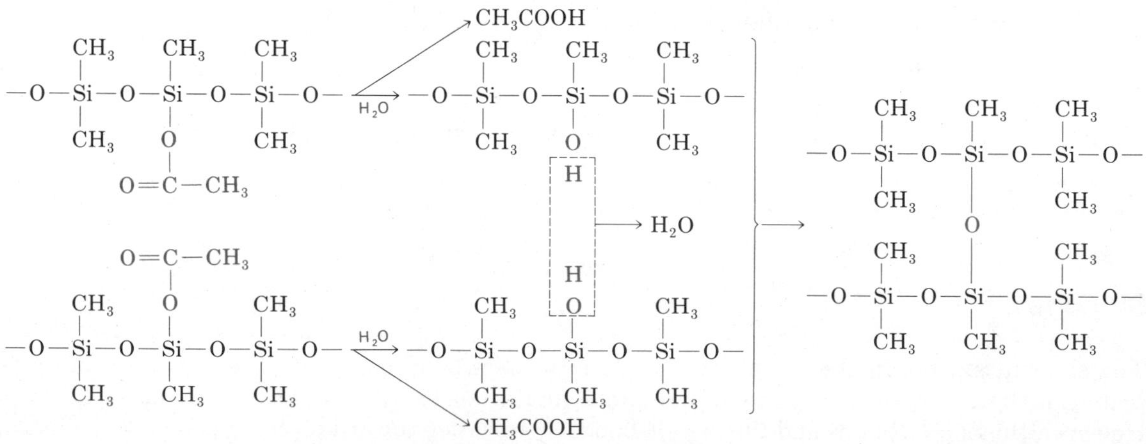

$$-O-Si(CH_3)_2-O-Si(CH_3)(OCOCH_3)-O-Si(CH_3)_2-O- \xrightarrow{H_2O} CH_3COOH + -O-Si(CH_3)_2-O-Si(CH_3)(OH)-O-Si(CH_3)_2-O-$$

$$-O-Si(CH_3)_2-O-Si(CH_3)(OCOCH_3)-O-Si(CH_3)_2-O- \xrightarrow{H_2O} CH_3COOH + -O-Si(CH_3)_2-O-Si(CH_3)(OH)-O-Si(CH_3)_2-O-$$

$$\longrightarrow H_2O + \begin{matrix} -O-Si(CH_3)_2-O-Si(CH_3)-O-Si(CH_3)_2-O- \\ | \\ O \\ | \\ -O-Si(CH_3)_2-O-Si(CH_3)-O-Si(CH_3)_2-O- \end{matrix}$$

**Silicone oils and rubbers find many medical uses.**

Over 3,000,000 pounds of silicone rubber are produced each year in the United States. The uses include window gaskets, o-rings, insulation, sealants for buildings, space ships, jet planes, and even some wearing apparel. The first footprints on the moon

**Figure 10–6** Examples of the use of silicone in the space program. Soles of lunar boots worn by the Apollo astronauts were made of high-strength silicone rubber. A silicone compound was also used for the air-tight seal of the lunar module hatch from which Astronaut Edwin E. Aldrin, Jr., has just emerged in this photo of the first manned landing on the moon on July 20, 1969.

were made with silicone rubber boots, which readily withstood the extreme surface temperatures (Fig. 10–6).

"Silly Putty," a silicone widely distributed as a toy, is intermediate between silicone oils and silicone rubber. It is an interesting material with elastic properties on sudden deformation, but its elasticity is quickly overcome by its ability to flow like a liquid when allowed to stand.

## Rearrangement Polymers

Some molecules polymerize by ***rearrangement*** reactions to yield very useful products. Molecules containing the isocyanate group ($-NCO$), for example, will react with almost any other molecule containing an active hydrogen atom (such as in an $-OH$ or $-NH_2$ group) in a rearrangement process. An example is the reaction of hexamethylene diisocyanate and butanediol. The urethane linkage $\left(\begin{array}{ccccccc} - & \text{N} & - & \text{C} & - & \text{O} & - \\ & | & & \| & & & \\ & \text{H} & & \text{O} & & & \end{array}\right)$ is produced by a shift (or rearrangement) of a hydrogen atom, moving it from the alcohol (butanediol) to a nitrogen atom on the isocyanate group; the linkage ($-N-C-O-$) is similar to, but not the same as, the amide bond ($-N-C-C-$) in nylons.

$$\mathrm{OCN(CH_2)_6NCO + HO(CH_2)_4OH \xrightarrow{} OCN(CH_2)_6{-}\overset{\displaystyle H}{\overset{|}{N}}{-}\overset{\displaystyle O}{\overset{\|}{C}}{-}O{-}(CH_2)_4OH}$$

rearrangement

HEXAMETHYLENE DIISOCYANATE — 1,4-BUTANEDIOL — PRODUCT MOLECULE (A URETHANE)

The continued reaction of the other groups gives rise to a polymer chain—a polyurethane.

$$\mathrm{{-}\overset{\displaystyle O}{\overset{\|}{C}}{-}\overset{\displaystyle H}{\overset{|}{N}}{-}(CH_2)_6{-}\overset{\displaystyle H}{\overset{|}{N}}{-}\overset{\displaystyle O}{\overset{\|}{C}}{-}O{-}(CH_2)_4{-}O{-}\overset{\displaystyle O}{\overset{\|}{C}}{-}\overset{\displaystyle H}{\overset{|}{N}}{-}(CH_2)_6{-}\overset{\displaystyle H}{\overset{|}{N}}{-}\overset{\displaystyle O}{\overset{\|}{C}}{-}O{-}(CH_2)_4{-}O{-}}$$

A PORTION OF POLYURETHANE

**Polyurethanes are structurally similar to many polyamides.**

A polyurethane is structurally similar to a polyamide (nylon). In Europe polyurethanes have applications similar to those of nylon in this country. Polyurethanes have viscosities and melting points that make them useful for foam applications. Foamed polyurethanes are known as "foam rubber" and "foamed hard plastics," depending on the degree of crosslinking.

Table 10–4 lists the trade names and compositions of some of the addition, condensation, and rearrangement polymers on the market today.

## Polymer Additives—Toward an End Use

Few plastics produced today find end uses without some kind of modification. Polyurethanes are a good example. In order for polyurethane to be useful as insulation in refrigerators, refrigerated trucks and railroad cars, and as construction insulation, a ***foaming agent*** is used. For polyesters to have the strength to compete with metals and certain natural materials such as wood, they are usually modified with a ***reinforcing additive.*** Inorganic fibers are usually used for this purpose (Table 10–5).

A plastic such as polystyrene, polyethylene, or polypropylene is often too stiff to have an immediate application. For example, polyethylene makes good bread wrap, but only if it is made flexible by an additive called a ***plasticizer.*** Some plasticizers, such as dioctyl phthalate (DOP), have not been approved for food use but still can be found in samples of living tissue taken from laboratory animals and humans. DOP and other plasticizers also occur in water samples taken from rivers and lakes. The health implications of these findings are not known at this time.

Since most plastics are used where they are exposed to sunlight, almost all contain some form of an additive to absorb the harmful portion of sunlight, ultraviolet (UV) light. These compounds, called ***UV stabilizers,*** can absorb photons in the 290–400 nanometer range of the spectrum, which have sufficient energy to break the chemical bonds found in polymers. If enough bonds are broken, the polymer becomes brittle and will break under stress (Fig. 10–7). Of course, with time, even the stabilizers decay, and the polymer article begins to show visible signs of sunlight degradation.

Table 10–5 summarizes a number of common polymer additives.

## The Future of Polymers

As we have seen in this chapter, the development and use of synthetic polymers is quite recent. Polyethylene, for example, was not discovered until 1933, yet by 1981, its production in the United States amounted to almost twelve billion pounds. Chemists are constantly synthesizing new polymers and finding applications for them. The space age has brought with it the need for new polymers, especially in electronics and as special coatings that can withstand high temperatures without breaking down. Among

**TABLE 10–4 Composition of Some Trade-Name Polymers***

| Trade Name | Composition |
|---|---|
| Acrilan | 85% acrylonitrile plus vinyl acetate or vinyl pyridine |
| Acrylic | at least 85% acrylonitrile |
| Arnel | cellulose triacetate |
| Bakelite | phenol-formaldehyde condensation product |
| Caprolan | nylon 6 |
| Creslan | copolymer of acrylonitrile and acrylamide |
| Dacron | ethylene glycol-terephthalic acid condensation product |
| Delrin | polyacetal |
| Dynel | 60% vinyl chloride plus 40% acrylonitrile |
| Epoxy | phenol-acetone-epichlorohydrin condensation product |
| Formica | phenol-formaldehyde condensation product |
| Fortrel | polyester similar to Dacron |
| Herculon | polypropylene |
| Kodel | polyester; terephthalic acid plus 1,4-cyclohexane dimethanol |
| Lucite | methyl methacrylate |
| Melmac | melamine-formaldehyde condensation product |
| Mylar | polyethylene terephthalate |
| Neoprene | 2-chlorobutadiene |
| Nylon 501 | nylon 66 |
| Nytril | at least 85% vinylidene dinitrile |
| Orlon | originally pure acrylonitrile; now up to 14% of another monomer |
| Plexiglas | methyl methacrylate |
| Polythene | polyethylene |
| Saran | vinylidene chloride-vinyl chloride addition product |
| Spandex | polyurethane; ethylene glycol-diisocyanate condensation product |
| Teflon | polytetrafluoroethylene |
| Terylene | polyester similar to Dacron |
| Vectra | polypropylene |
| Velon | vinylidene chloride-vinyl chloride addition product; see Saran |
| Zantrel | rayon fiber |

$CH_2{=}C(OC_2H_5)_2$
*acetal*

$CH_2{=}CH(CONH_2)$
*acrylamide*

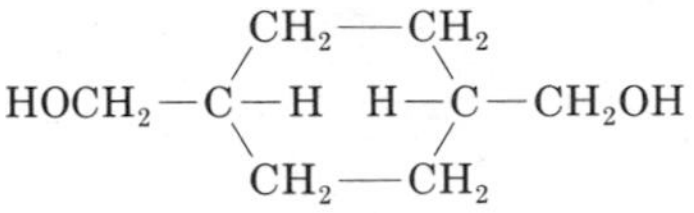

$HOCH_2{-}\overset{}{C}H(CH_2CH_2)_2CH{-}CH_2OH$
*1,4-cyclohexane dimethanol*

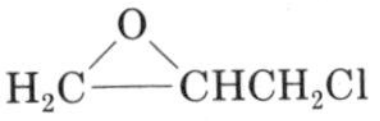

$H_2C\overset{O}{\text{—}}CHCH_2Cl$
*epichlorohydrin*

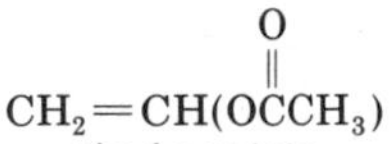

$CH_2{=}CH(O\overset{O}{\overset{\|}{C}}CH_3)$
*vinyl acetate*

$CH_2{=}C(CN)_2$
*vinylidine dinitrile*

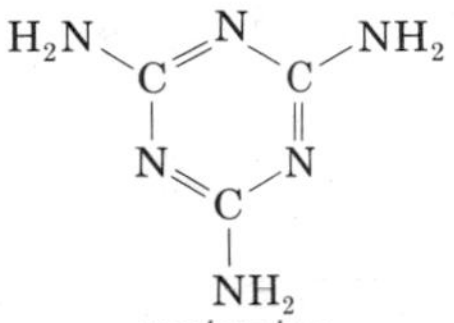

$C_3N_3(NH_2)_3$
*melamine*

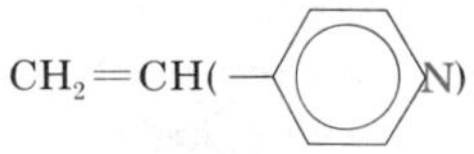

$CH_2{=}CH(C_5H_4N)$
vinyl pyridine

*Structures are given for those monomers not found elsewhere in the text. Consult the index for formulas not given.

the newcomers are the polyimides, prepared from the polycondensation of a diacid anhydride and a diamine. Some of these polymers have very high service temperatures (Fig. 10–8).

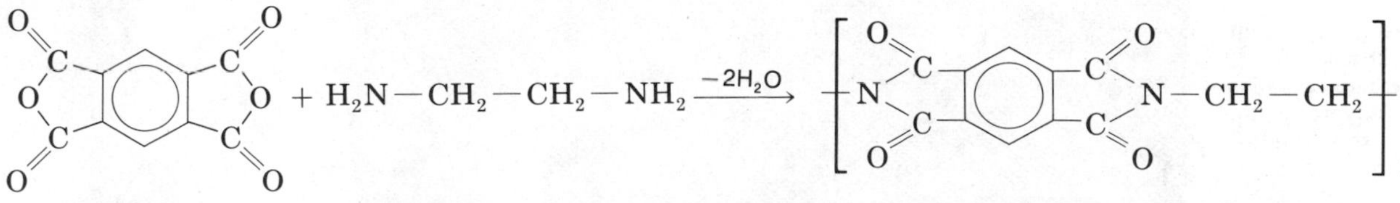

$$C_6H_2(C_2O_3)_2 + H_2N{-}CH_2{-}CH_2{-}NH_2 \xrightarrow{-2H_2O} \left[ {-}N(CO)_2C_6H_2(CO)_2N{-}CH_2{-}CH_2{-} \right]_n$$

PYROMELLITIC ANHYDRIDE — 1,2-DIAMINOETHANE — A POLYIMIDE

Plastic materials are being improved constantly. Some have been made with the strength and rigidity of steel while having only 15–20% of the density of steel. The

**Anhydride: an HOH has been removed from the structure.**

**TABLE 10–5 Polymer Additives**

| Additive | Structure | Use, Comments |
|---|---|---|
| Foaming agent<br>Pentane | $CH_3CH_2CH_2CH_2CH_3$ | Used to foam polyurethane. Dissolved in liquid polymer under high pressure, then heated and placed in mold. |
| Plasticizers<br>Dioctyl phthalate (DOP) | $C_6H_4$ ring bearing $-C(=O)-O-CH_2-CH(C_2H_5)-C_4H_9$<br>and $-C(=O)-O-CH_2-CH(C_2H_5)-C_4H_9$ | Plasticizer in PVC to lend flexibility. Gets into the environment. |
| Dioctyl adipate (DOA) | $CH_2-C(=O)-O-CH_2-CH(C_2H_5)-C_4H_9$<br>$\vert$<br>$CH_2$<br>$\vert$<br>$CH_2$<br>$\vert$<br>$CH_2-C(=O)-O-CH_2-CH(C_2H_5)-C_4H_9$ | Used in plastic films to make them flexible. Has Food and Drug Administration approval for food contact. |
| UV stabilizers<br>Phenylsalicylate | $C_6H_4(OH)-C(=O)-O-C_6H_5$ | Absorbs UV light very efficiently. |
| Carbon black | similar to graphite below, but small particles, less structure | Absorbs UV light and radiates energy as heat. Fine for all-black articles. |
| Reinforcing agents<br>Glass | $SiO_2$ units (see Fig. 7–10) | Used in polyesters and other plastics to improve strength. Found in car bodies, boats, fishing poles, tennis racquets, bicycle frames, radio antennas, etc. |
| Boron | clusters of $B_{12}$ units | |
| Graphite | hexagonal rings of carbon atoms, joined on all sides in a layered arrangement | |

structural strength of such plastics offers the possibility of self-supporting domes for buildings and automobiles that contain more plastic than metal. New low-temperature polymerizations without the use of a solvent are being developed. An application is to make "spray-on" clothes. Simply spray the monomers onto a mannequin—a little

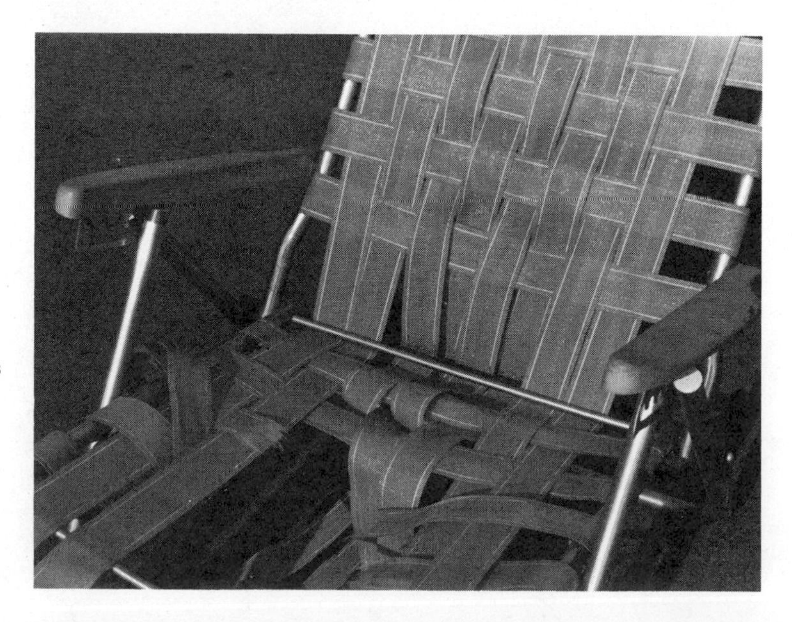

**Figure 10–7** Sunlight (mostly ultraviolet) damages most plastics such as polypropylene webbing. The results are shorter product life and higher costs to the consumer. Plastics containing ultraviolet absorbing chemicals have a longer outdoor life.

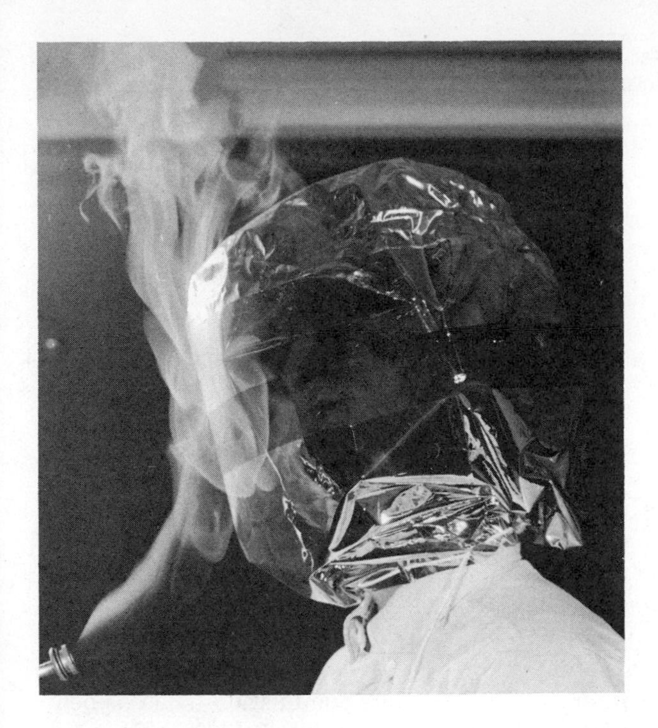

**Figure 10–8** An example of how a polyimide film can withstand, for a short period, the flame of a blowtorch.

more here, a little less there. Then cut along desired lines, add buttons and zippers, and wear. Some plastics are being developed to replace wood fiber in paper for the printed page. These papers offer smooth surfaces without the graininess of paper, an improvement especially in the quality of microfilming.

Because polymers are used so extensively throughout the world today, the problem of waste disposal is inevitable. Plants are in operation in which solid wastes undergo first a magnetic separation to remove iron and steel objects, then a ballistic separation (Fig. 10–9) based on density, since glass and aluminum objects are denser than plastics. Another method is to have consumers separate waste into garbage, plastics, metals, paper, wood, and glass. Plans are being developed to treat the plastics thus separated in two ways. If suitable separation methods could be developed, thermoplastics could be reprocessed into new items (e.g., if all the nylon could be separated from polystyrene).

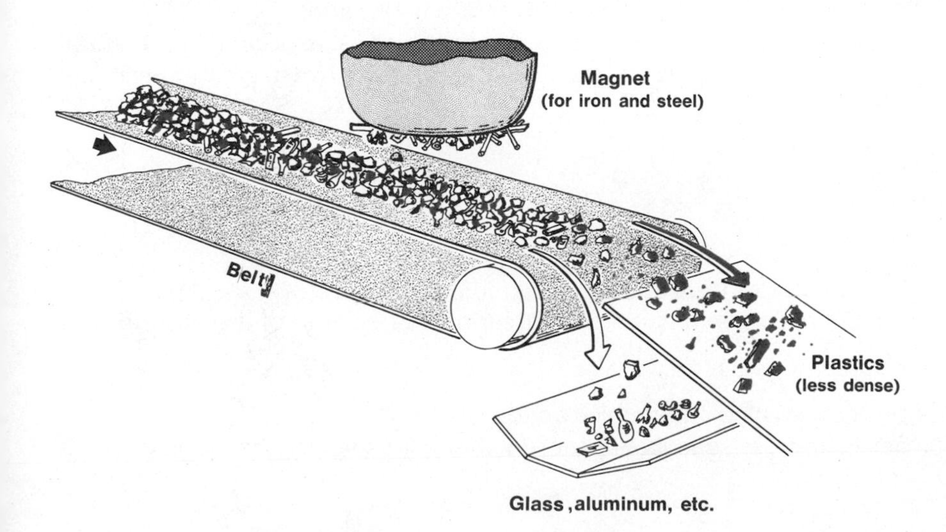


**Figure 10–9** A method of separation of plastics from other wastes prior to recycling the plastics for reuse.

**TABLE 10–6 Energy Requirements to Produce Some Plastics and Metals (Including the Fuel Equivalent of the Monomers)**

| | Million BTU/Ton* |
|---|---|
| Aluminum | 244 |
| Copper | 112 |
| Low-density polyethylene | 106 |
| High-density polyethylene | 96 |
| Polystyrene | 64 |
| Polyvinyl chloride | 49 |
| Steel | 24 |

*Burning one ton of coal produces about 25 million BTU of heat energy. BTU energy units are discussed in Chapter 14.

Thermosetting plastics could not be treated this way, however, because breaking the crosslinking would cause complete molecular degradation. If separation and reuse were not feasible, asphalt could be made from the mixed plastics. In their original composition, combustion units built near cities could actually use plastics as fuels, since they are mostly carbon and hydrogen. There is danger, though, in that some plastics contain elements that could create massive pollution if released into the atmosphere. An example is polyvinylchloride, which on burning yields hydrogen chloride, a very corrosive gas. However, some of the products of incomplete combustion (such as benzene, styrene, and acetylene) could be recycled as raw materials for other chemical syntheses.

**In the United States, 95% of our petroleum is used as fuel; only 5% is used to make products such as medicines, textiles, and plastics.**

The long-range future of plastics looks dismal unless we curtail our ravenous burning of petroleum. Most of the raw materials for plastics come from petroleum, and less from coal. Petroleum and coal are the principal sources of energy in this country. Not only are plastics and energy linked through the raw materials—fossil fuels relationship, but considerable energy is required to purify starting materials and to change them into the desired plastic in the preferred shape. It is the age-old principle that we cannot continue to eat our cake (burn petroleum and coal) and have it, too (use petroleum and coal to make plastics, fibers, and medicines). Table 10–6 illustrates that when we consider polymers in the broadest sense in terms of their energy costs of production, we must include their energy value as if they were used as fuel instead of for some object. Considered this way, plastics, fibers, and other items made of polymers derived from petroleum may have only a short history in man's existence. Wood, paper, and mineral products such as metals and cement appear either renewable or present in the earth's crust in greater abundance than petroleum.

We hope that these and similar problems will be solved as we begin to understand more fully how to use what we have on this planet, and how to live in greater harmony with nature.

## Self-Test 10-B

1. When $(CH_3)_2SiCl_2$ reacts with water, a representative portion of the structure of the polymer obtained is ____________.
2. Stabilizers protect plastics against the action of ____________.
3. A plastic that is too stiff can be rendered more flexible by the addition of a ____________.

4. A silicone polymer contains Si— ______________ bonds.
5. The burning of plastics containing chlorine, such as polyvinyl chloride, produces what toxic gas? ______________
6. (Ultraviolet/visible) light is more destructive to plastics.
7. Which requires more energy per ton to produce, polyethylene or steel? ________ ______________

## Matching Set

| | | |
|---|---|---|
| ______ | 1. nylon | **a** plastic that forms interlocking bonds when heated |
| ______ | 2. block copolymer | **b** rubber |
| ______ | 3. monomer | **c** crosslinking via reaction with sulfur |
| ______ | 4. thermoplastic | **d** —AAAAABBBBBAAAAA— |
| ______ | 5. thermosetting plastic | **e** causes RTV silicone to crosslink |
| ______ | 6. homopolymer | **f** forms polymers of desired structure |
| ______ | 7. polymer with a memory | **g** —AAAAAAAAAAAAAAA— |
| ______ | 8. vulcanize | **h** natural rubber |
| ______ | 9. stereochemical control | **i** a synthetic rubber |
| ______ | 10. styrene-butadiene copolymer | **j** building unit for a polymer |
| ______ | 11. poly-*cis*-isoprene | **k** plastic softened by heat |
| ______ | 12. polyester | **l** a polyamide |
| ______ | 13. moisture | **m** formed from a dialcohol and a diacid |
| ______ | 14. dioctyl phthalate | **n** possibly harmful plasticizer |

## Questions

**1.** In what way is a railroad train like polystyrene?

**2.** Where do you suppose the first synthetic chemist who prepared a polymer got the idea for these giant molecules?

**3.** What property does a polymer have when it is extensively crosslinked?

**4.** Describe on the molecular level the end result of the vulcanization process.

**5.** What is the origin of the word "polymer"?

**6.** Is polystyrene a thermoplastic or thermosetting plastic?

**7.** What property of the molecular structure of rubber allows it to be stretched?

**8.** Explain how polymers could be prepared from each of the following compounds. (Other substances may be used.)

(a) $CH_3-\overset{\displaystyle H}{\overset{|}{C}}=\overset{\displaystyle H}{\overset{|}{C}}-CH_3$

(b) $HO-\overset{\displaystyle O}{\overset{\|}{C}}-CH_2-CH_2-\overset{\displaystyle O}{\overset{\|}{C}}-OH$

(c) $\underset{\displaystyle OH}{\underset{|}{CH_2}}-\underset{\displaystyle OH}{\underset{|}{CH}}-\underset{\displaystyle OH}{\underset{|}{CH_2}}$

(d)  $H_2N-CH_2-C_6H_4-CH_2-NH_2$ (benzene ring drawn between the two $CH_2$ groups)

**9.** What are the monomers used to prepare the following polymers?

(a) $-CH_2CH_2CH_2CH_2CH_2CH_2CH_2CH_2CH_2-$

(b)
$$\begin{array}{l} \;\;CH_3\quad\;\; CH_3\quad\;\; CH_3 \\ \;\;\;| \qquad\quad\; | \qquad\quad\; | \\ -CHCH_2CHCH_2CHCH_2- \end{array}$$

(c) 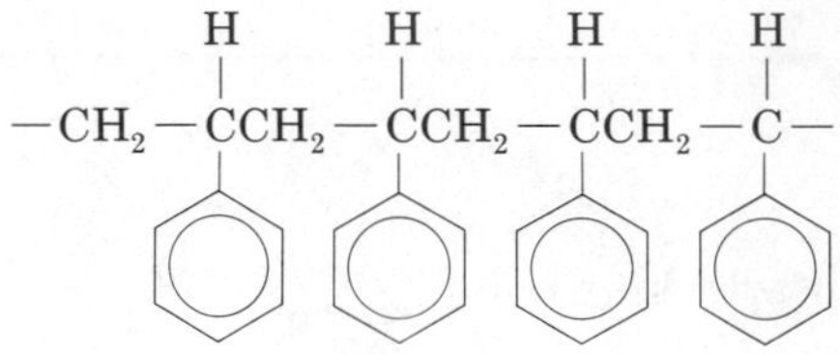
$$\begin{array}{l} \qquad\quad\; H \qquad\quad\;\; H \qquad\quad\;\; H \qquad\quad\;\; H \\ \qquad\quad\; | \qquad\quad\;\;\; | \qquad\quad\;\;\; | \qquad\quad\;\;\; | \\ -CH_2-CCH_2-CCH_2-CCH_2-C- \\ \qquad\quad\; | \qquad\quad\;\;\; | \qquad\quad\;\;\; | \qquad\quad\;\;\; | \\ \qquad\;\; C_6H_5 \qquad C_6H_5 \qquad C_6H_5 \qquad C_6H_5 \end{array}$$

**10.** Write equations showing the formation of polymers by the reaction of the following pairs of molecules:

(a) 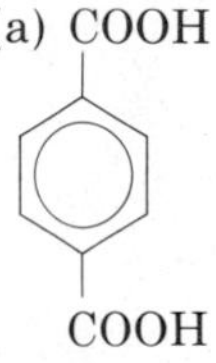 $HOOC-C_6H_4-COOH$ (COOH groups at opposite ring positions) and $HOCH_2CH_2OH$

(b) $HOOCCH_2CH_2COOH$ and $H_2NCH_2CH_2NH_2$

(c) 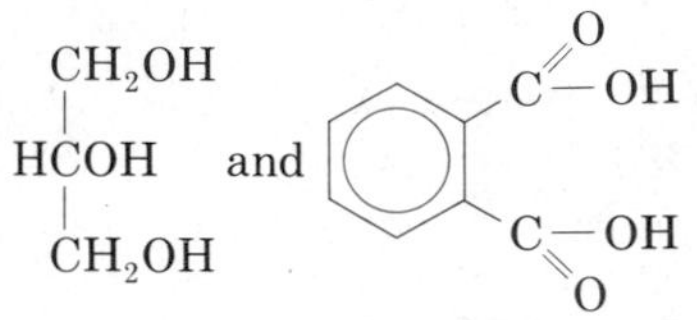
$$\begin{array}{l} CH_2OH \\ | \\ HCOH \\ | \\ CH_2OH \end{array}$$
and $C_6H_4(COOH)_2$ (two $-C(=O)-OH$ groups on adjacent ring positions)

**11.** Is a small molecule eliminated when each monomer unit is added to the chain in addition polymers?

**12.** Give an example of a copolymer.

**13.** You are given two specimens of plastic, A and B, to identify. One is known to be nylon and the other polymethyl methacrylate, but you do not know which is which. Analysis of A shows it to contain C, H, and O, while B contains C, H, O, and N. What are A and B?

**14.** What structural features must a molecule have in order to undergo addition polymerization?

**15.** What is meant by the term "macromolecule"?

**16.** Orlon has a polymeric chain structure of

$$\begin{array}{l} -CH_2-CH-CH_2-CH-CH_2-CH- \\ \qquad\quad\;\; | \qquad\qquad\quad\; | \qquad\qquad\quad\; | \\ \qquad\quad CN \qquad\qquad\; CN \qquad\qquad\; CN \end{array}$$

What is the monomer from which this structure can be made?

**17.** Which white pigment is banned in interior paints? Explain.

**18.** What feature do all condensation polymerization reactions have in common?

**19.** Give an example of the possibilities that exist if a trifunctional acid reacts with a difunctional alcohol.

**20.** What type of chemical change takes place during the drying of oil paints?

**21.** What are the starting materials for nylon 66?

**22.** Suggest a major difference in the bonding of thermosetting and thermoplastic polymers. Which is more likely to have an interlacing (crosslinking) of covalent bonds throughout the structure? Which is most likely to have weak bonds between large molecules?

**23.** What is a major difference between silicone oils and silicone rubbers?

**24.** Explain how a plasticizer can make a polymer more flexible.

**25.** Name one commercial plasticizer found in food wraps.

**26.** Could an oil-based paint "dry" in a vacuum? Explain.

**27.** Would a latex paint "dry" in a vacuum? Explain.

**28.** Describe the properties and structure of Silly Putty.

**29.** Which is more likely to produce a thermosetting polymer, the monomers of Question 10a or 10c?

**30.** Draw representative portions of Acrilan, Delrin, Saran, and Plexiglas. Refer to Table 10–4.

**31.** In what way is the structure of ice like that of a crosslinked polymer? How is it different?

**32.** What single property must a molecule possess in order to be a monomer?

**33.** Which do you think is the source of most polymers used today, green plants or petroleum? Do you think this will ever change? Explain.

**34.** a. Should we stop the burning of petroleum? What are the problems involved?
b. Should we start to develop research on how to change wood and straw into plastics? What are a few of the problems involved?

**35.** Would isoprene make a good motor fuel? Explain.

**36.** What properties of plastics make them superior to metals? What properties of plastics make them inferior to metals?

# Biochemistry—Basic Structures

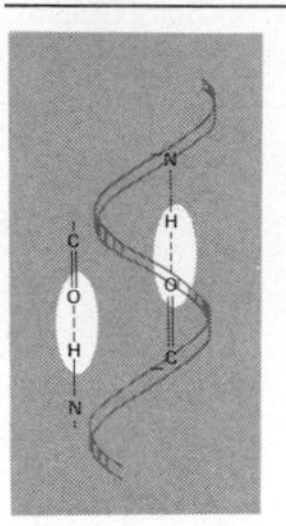

Do you think of yourself as an array of the chemical elements? Probably not, because it is hard to explain your characteristics in terms of the properties of the elements—you are so complex! However, the physical you is chemical in nature and many relationships are now known to exist between your gross properties and the chemicals you contain and ingest. Chapter 11 looks at basic structures in biological systems, Chapter 12 at some of the reactions that occur in biological systems, and Chapter 15 at some of the chemicals that will disrupt the chemistry of a healthy organism. This is the world of ***biochemistry.*** Our knowledge in this exciting field of study is currently expanding at an explosive rate.

The goal of biochemistry is to develop a chemically based understanding of living cells of all types. This includes the determination of the kinds of atoms present, the investigation of how they are joined together to form the larger structural units present in cells, and the study of the chemical reactions by which living cells obtain the energy required for the life processes of growth, movement, and reproduction.

In this chapter emphasis will be placed on fundamental biochemical structures and their preparation. These structures are (1) ***carbohydrates,*** (2) ***fats,*** (3) ***proteins,*** (4) ***enyzmes,*** and (5) ***nucleic acids.***

## Carbohydrates

Carbohydrates contain the elements carbon, hydrogen, and oxygen, with hydrogen atoms and oxygen atoms generally in the ratio of 2 to 1.

*mono*—one
*oligo*—few
*poly*—many

Carbohydrates are composed of the three elements carbon, hydrogen, and oxygen. Three structural groups are prevalent in carbohydrates: alcohol (—OH), aldehyde $\left(\begin{array}{c}\mathrm{O}\\ \|\\ -\mathrm{CH}\end{array}\right)$, and ketone $\left(\begin{array}{c}\mathrm{O}\\ \|\\ -\mathrm{C}-\end{array}\right)$. The carbohydrates can be classified into three main groups: ***monosaccharides*** (Latin, *saccharum,* sugar), ***oligosaccharides,*** and ***polysaccharides.*** Monosaccharides are simple sugars that cannot be broken down into smaller units by mild acid hydrolysis.

When a saccharide undergoes hydrolysis, the larger molecular unit is broken into smaller ones with little change in the overall structure. At the point where the break occurs, water furnishes H· and ·OH groups to bond to the broken ends, forming stable structures in the aqueous medium (Fig. 11–3). The hydrolysis of biochemicals is usually more rapid in acidic or basic media than in a neutral solution. Hydrolysis of a molecule of an oligosaccharide yields two to six molecules of a simple sugar; complete hydrolysis of a polysaccharide produces many monosaccharide units.

Carbohydrates are synthesized by plants from $CO_2$ and $H_2O$, using energy from the sun.

Carbohydrates are synthesized by plants from water and atmospheric carbon dioxide. The process is called ***photosynthesis*** since it is a synthetic reaction that occurs when energized by photons (that is, quanta) of light energy. It requires an

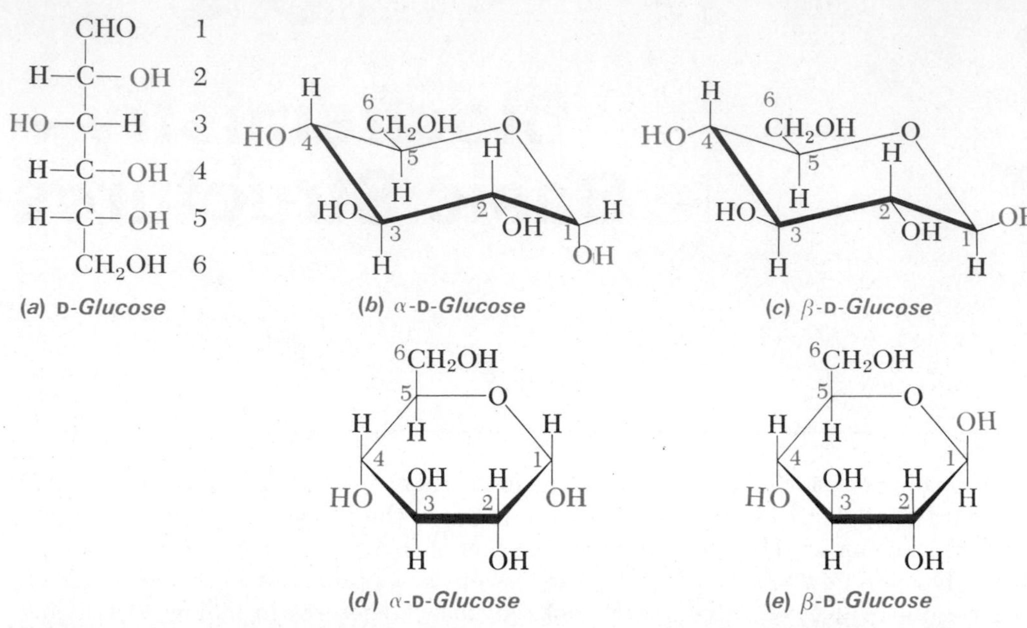


**Figure 11–1** The structures of D-glucose; *d* and *e* are two-dimensional representations of *b* and *c*, respectively. Note the difference in the positions of the —OH groups (*color*) in the $\alpha$ and $\beta$ forms of glucose: the —OH groups on the 1 and 4 carbons are *trans* when the structure is beta ($\beta$), and the —OH groups are *cis* when the structure is alpha ($\alpha$). In both alpha and beta glucose, the —OH group on the number 4 carbon atom must be in the same position.

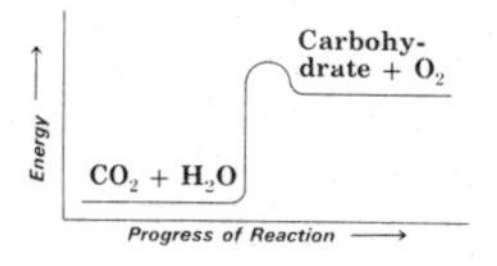


input of energy; consequently, carbohydrates are energy-rich compounds. These compounds serve as important sources of energy for the metabolic processes of plants and animals. Glucose, $C_6H_{12}O_6$, along with some of the other simple sugars are quick energy sources for the cell. Large amounts of energy are stored in polysaccharides such as starch. The stored energy is usable by living cells only if polysaccharides are broken down into monosaccharides.

Some complex carbohydrates are also used by cells for structural purposes. Cellulose, for example, partially accounts for the structural properties of wood.

## Monosaccharides

Approximately 70 monosaccharides are known; 20 of these simple sugars occur naturally. Unlike many organic compounds, these sugars are very soluble in water owing to the numerous —OH groups present, which can form hydrogen bonds with water.

**D-glucose, the most important monosaccharide, is found in fruits, blood, and living cells.**

The most common simple sugar is ***D-glucose.*** This monosaccharide requires three structures for its adequate representation (Fig. 11–1). Structure (*a*), in which the carbon atoms are numbered for later reference, depicts the "straight-chain" structure with the aldehyde group (—CHO) in position 1. The properties of a water solution of D-glucose cannot be explained by this structure alone. At any given time, most of the molecules exist in the ring form, structures (*b*) and (*c*), which results from a molecular rearrangement in which carbon 1 bonds to carbon 5 through an oxygen atom. Both ring structures are possible since the OH group on carbon 1 may form in such a way to point either along the plane of the molecule or out of the plane. It should be emphasized that a solution of D-glucose contains a mixture of three forms in a

CH₂OH — C=O — HO—C—H — H—C—OH — H—C—OH — CH₂OH

(a) *Ketone structure*

HOH₂C¹, O, H, 6, H, OH, 2, 5, H, HO, HO, 3, 4, H, OH, H

(b) *β-Ring structure (Pyranose structure: 6-membered ring with an oxygen atom in the ring)*

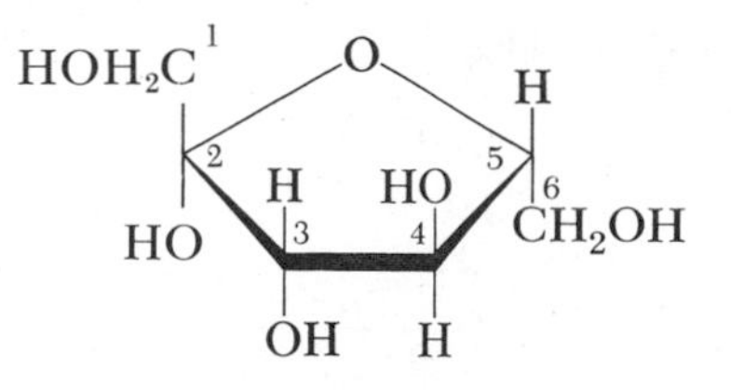


(c) *β-Ring structure (Furanose structure: 5-membered ring with an oxygen atom in the ring)*

**Figure 11–2** The structures of D-fructose. The α-ring structure (not shown) differs from the β-ring structure in that the $CH_2OH$ and OH groups are in reversed positions on carbon 2.

dynamic state of change from one form to another. There is more of the ring forms and much less of the straight-chain form.

Because it is sweet, D-glucose is used in the manufacture of candy and in commercial baking. This simple sugar, also called *dextrose, grape sugar,* and *blood sugar,* is prevalent in fruits, vegetables, blood, and tissue fluids. A solution of D-glucose is fed intravenously when a readily available source of energy is needed to sustain life. As will be discussed later, many polysaccharides including starch are composed of glucose units, and serve as a source of this important chemical upon hydrolysis of the complex structures.

**Glucose has a relative sweetness of 74.3, compared with sucrose having an assigned value of 100.0. The value for fructose is 173.3.**

Another important monosaccharide is ***D-fructose.*** Its structure, which has a ketone group, is given in Figure 11–2.

## Oligosaccharides

The most important oligosaccharides are the disaccharides (two simple sugar units per molecule). Examples include these widely used disaccharide sugars:

***sucrose*** (from sugar cane or sugar beets), which consists of a glucose unit and a fructose unit,
***maltose*** (from starch), which consists of two glucose units, and
***lactose*** (from milk), which consists of a glucose unit and a galactose (an optical isomer of glucose) unit.

The formula for these disaccharides, $C_{12}H_{22}O_{11}$, is not simply the sum of two monosaccharides, $C_6H_{12}O_6 + C_6H_{12}O_6$. A water molecule must be eliminated as two monosaccharides are united to form the disaccharide. The structures for sucrose, maltose, and lactose, along with their hydrolysis reactions, are given in Figure 11–3.

**Disaccharides are molecules containing two simple sugars bound together, such as in sucrose, which contains a glucose and a fructose unit in each molecule. A water molecule is eliminated when the bond forms between the two simple sugars.**

The disaccharides are important as foods. Sucrose is produced in a high state of purity on an enormous scale: the annual production amounts to over 80 million tons per year. Originally produced in India and Persia, sucrose is now used universally as a sweetener. About 40% of the world sucrose production comes from sugar beets and 60% from sugar cane. Sugar provides a high caloric value (1794 kcal per pound); it is also used as a preservative in jams, jellies, and candied fruit.

## Polysaccharides

There is an almost limitless number of possible structures in which monosaccharide units (monosaccharide molecules minus one water molecule at each bond between

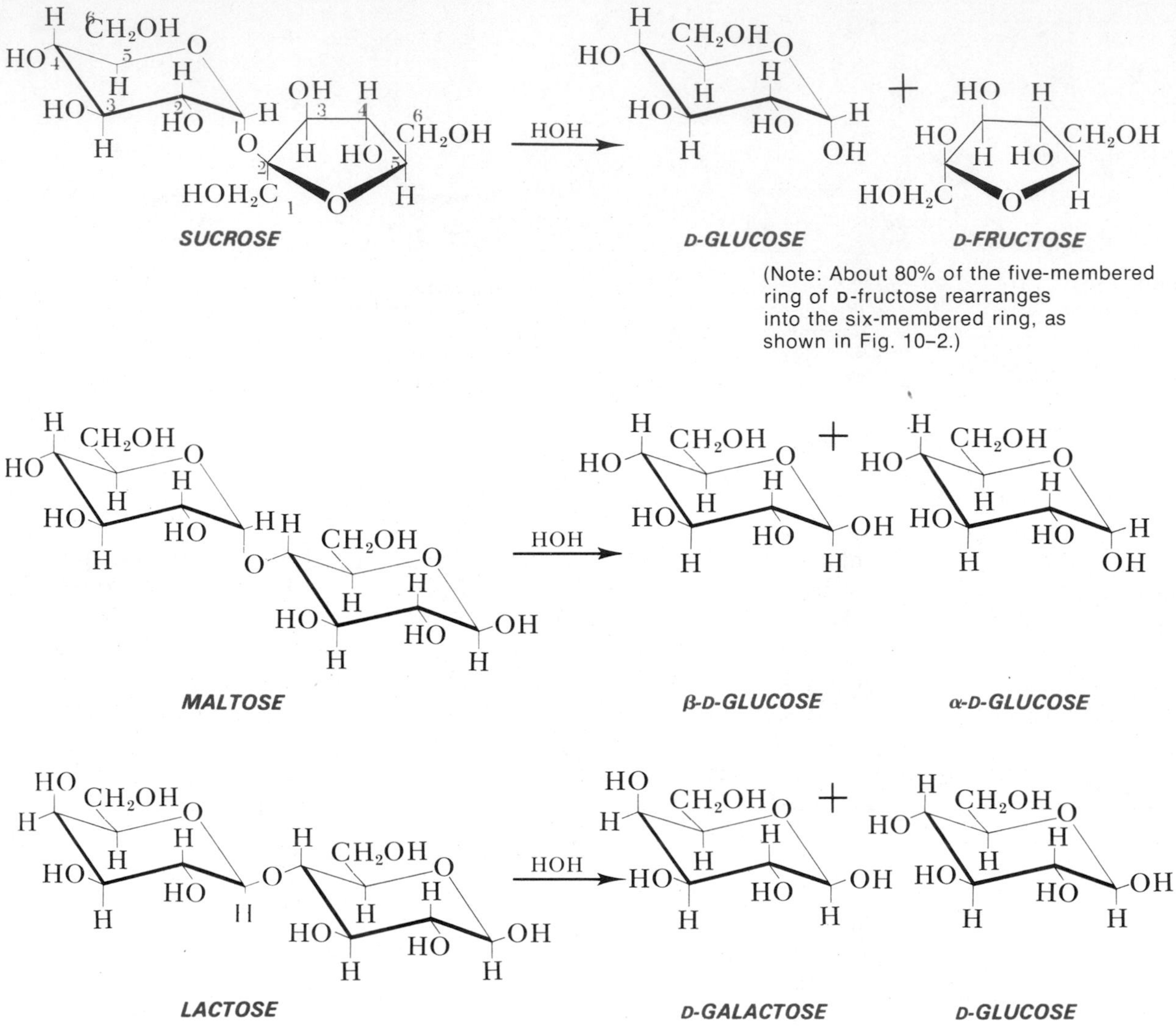

SUCROSE → HOH → D-GLUCOSE + D-FRUCTOSE

(Note: About 80% of the five-membered ring of D-fructose rearranges into the six-membered ring, as shown in Fig. 10–2.)

MALTOSE → HOH → β-D-GLUCOSE + α-D-GLUCOSE

LACTOSE → HOH → D-GALACTOSE + D-GLUCOSE

**Figure 11–3** Hydrolysis of disaccharides (sucrose, maltose, and lactose).

units) can be combined. Molecular weights are known to go above 1,000,000. Apparently nature has been very selective in that only a few of the many possible monosaccharide units are found in polysaccharides.

## Starches and Glycogen

**Starch molecules consist of many glucose units bonded together.**

Starch is found in plants in protein-covered granules. These granules are disrupted by heat, and part of the starch content is soluble in hot water. Soluble starch is ***amylose*** and constitutes 22–26% of most natural starches; the remainder is ***amylopectin.*** Amylose gives the familiar blue-black starch test with iodine solutions; amylopectin turns red on contact with iodine.

Structurally amylose is a straight-chain polymer of α-D-glucose units, each one bonded to the next, just as the two units are bonded in maltose (Fig. 11–3). Molecular

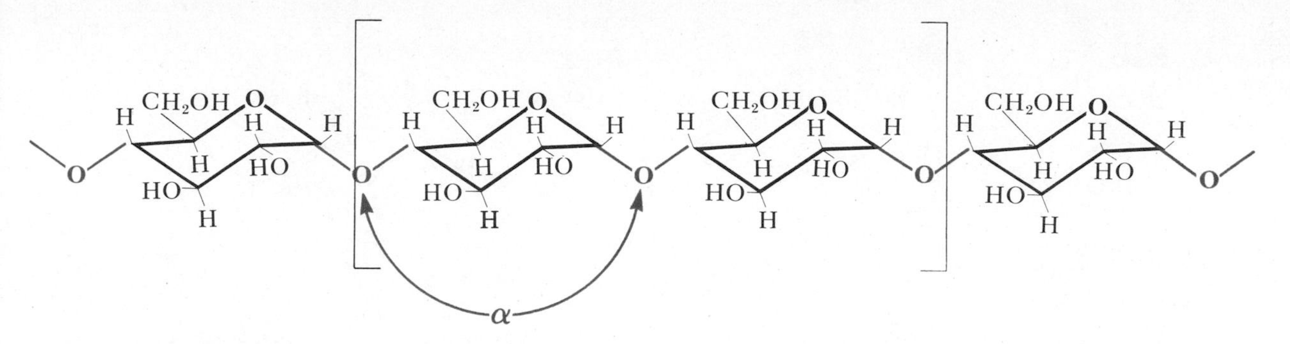


**Figure 11–4** Amylose structure. From 60 to 300 α-D-glucose units are bonded together by **alpha** linkages to form amylose molecules.

weight studies on amylose indicate that the average chain contains about 200 units. The structure of amylose is illustrated in Figure 11–4.

Amylopectin is made up of branched chains of α-D-glucose units (Fig. 11–5). Its molecular weight generally corresponds to about 1000 glucose units. Partial hydrolysis of amylopectin yields mixtures called ***dextrins.*** Complete hydrolysis, of course, yields D-glucose. Dextrins are used as food additives, mucilages, paste, and in finishes for paper and fabrics.

***Glycogen*** serves as an energy reservoir in animals as does starch in plants. Glycogen has a structure similar to that of amylopectin (branched chains of glucose units), except that glycogen is even more highly branched.

## Cellulose

Cellulose is the most abundant polysaccharide in nature. Like amylose, it is composed of D-glucose units. The difference between the structure of cellulose and that

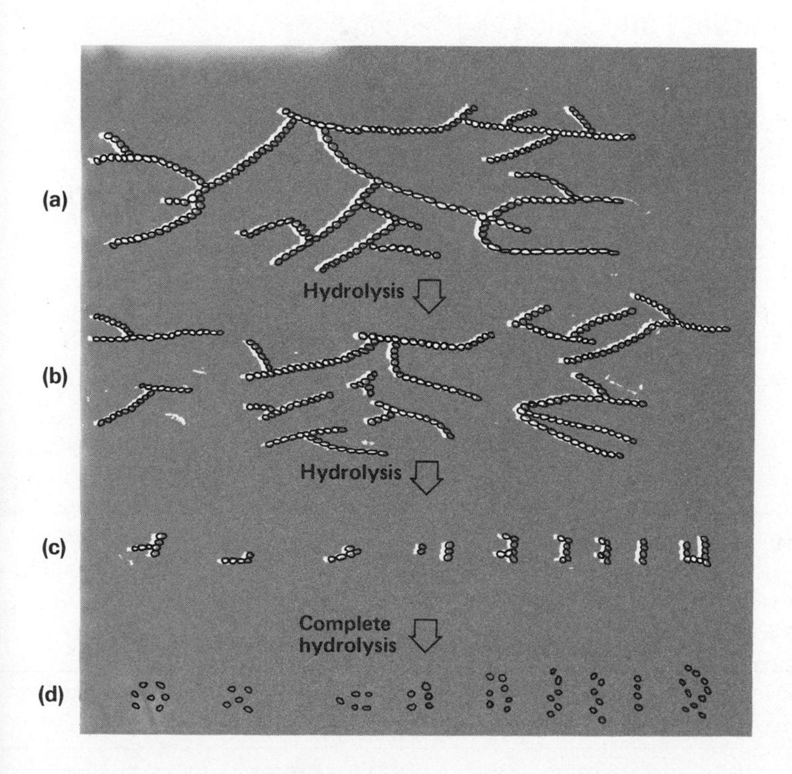


**Figure 11–5** (*a*) Partial schematic amylopectin structure. (*b*) Dextrins from incomplete hydrolysis of *a*. (*c*) Oligosaccharides from hydrolysis of dextrins. (*d*) Final hydrolysis product: D-glucose. Each circle represents a glucose unit.

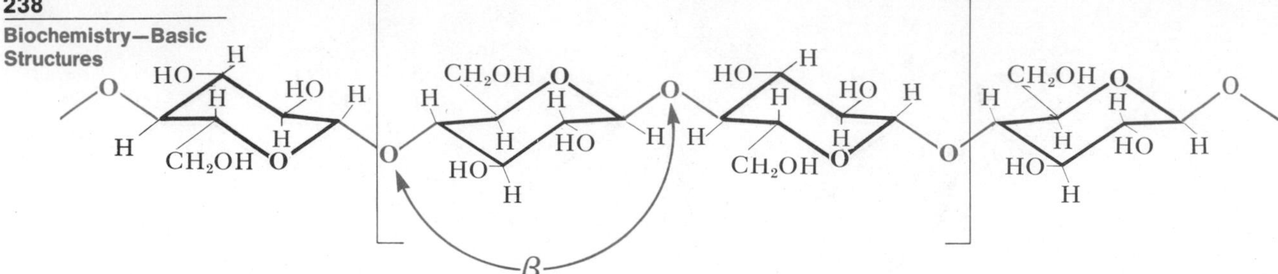


**Figure 11–6** Cellulose structure. About 2800 $\beta$-D-glucose units are bonded together by **beta** linkages to form a cellulose molecule.

of amylose lies in the bonding between the D-glucose units. In cellulose, all of the glucose units are in the $\beta$-ring form in contrast to the $\alpha$-ring form in amylose. (Review the ring forms in Fig. 11–1 and compare the structures in Figs. 11–4 and 11–6.)

**Humans do not have an enzyme to split cellulose into its glucose units.**

The different structures of starch and cellulose account for their difference in digestibility. Human beings and carnivorous animals do not have the necessary enzymes (biochemical catalysts) to break down the cellulose structure as do numerous microorganisms. Cellulose is readily hydrolyzed to D-glucose in the laboratory by heating a suspension of the polysaccharide in the presence of a strong acid. Unfortunately, at present this is not an economically feasible solution to the world's growing need for an adequate food supply.

Paper, rayon, cellophane, and cotton are principally cellulose. A representative portion of the structure of cotton is shown in Figure 11–7. Note the hydrogen bonding between cellulose chains.

## Dietary Fats and Essential Fatty Acids

**Refer to Chapter 9 for introductory material on fats and fatty acids.**

**An ordinary scientific calorie is the amount of heat required to raise 1 g of water 1°C. A food calorie is 1000 scientific calories (or 1 kilocalorie).**

Most diets in the United States gain 40–50% of their calories from fats or oils. This is rather high when compared with diets in most other parts of the world. Natural fats and oils are generally mixtures of various esters of glycerol with more than one kind of fatty acid. In our diets, most of the fatty acids are ***saturated*** fatty acids (Table 11–1). Such fatty acids can be (1) used as a source of energy if the body burns them to $CO_2$ and $H_2O$, (2) stored for possible future use in fat cells, or (3) used as starting materials for the synthesis of other compounds needed by the body. Fats are the most concentrated source of food energy in our diets, as they furnish about 9000 calories/g when burned for energy. The human body can make some fats from carbohydrates and carries out such processes to store the excess energy furnished in the diet.

A high intake of dietary fat has been implicated as one of the factors that can give rise to ***atherosclerosis,*** a complex process in which the walls of the arteries suffer damage and ultimately develop scar tissue and fatty deposits. Atherosclerosis is generally considered to be a precursor to certain types of heart disease and strokes. Atherosclerosis may also be related to the amount of cholesterol in the diet, but the relationship of both dietary fat and cholesterol intake to atherosclerosis does not appear at this time to be a simple one.

It has been known for about 60 years that the human body has a small requirement for certain types of fatty acids (called ***essential fatty acids***), and in recent years the basis for the need for these essential fatty acids has been determined.

**Figure 11–7** The properties of cotton, about 98% cellulose, can be explained in terms of this submicroscopic structure. A small group of cellulose molecules, each with 2000 to 9000 units of D-glucose, are held together in an approximately parallel fashion by hydrogen bonding (- - - -). When several of these ***chain bundles*** cling together in a relatively vast network of hydrogen bonds, a ***microfibril*** results; the microfibril is the smallest microscopic unit that can be seen. The macroscopic ***fibril*** is a collection of numerous microfibrils. The absorbent nature of cotton results from the numerous capillaries that exist between the cellulose chains wherein the smaller water molecules are held by hydrogen bonds.

The essential fatty acids are ***linoleic, linolenic,*** and ***arachidonic*** acids.

$$\underset{18}{CH_3}\underset{17}{CH_2}\underset{16}{CH_2}\underset{15}{CH_2}\underset{14}{CH_2}\underset{13}{CH}=\underset{12}{CH}\underset{11}{CH_2}\underset{10}{CH}=\underset{9}{CH}\underset{8}{CH_2}\underset{7}{CH_2}\underset{6}{CH_2}\underset{5}{CH_2}\underset{4}{CH_2}\underset{3}{CH_2}\underset{2}{CH_2}\underset{1}{C}(=O)OH$$

LINOLEIC ACID ($C_{18}\Delta_{9,12}$)

$$\underset{18}{CH_3}\underset{17}{CH_2}\underset{16}{CH}=\underset{15}{CH}\underset{14}{CH_2}\underset{13}{CH}=\underset{12}{CH}\underset{11}{CH_2}\underset{10}{CH}=\underset{9}{CH}\underset{8}{CH_2}\underset{7}{CH_2}\underset{6}{CH_2}\underset{5}{CH_2}\underset{4}{CH_2}\underset{3}{CH_2}\underset{2}{CH_2}\underset{1}{C}(=O)OH$$

LINOLENIC ACID ($C_{18}\Delta_{9,12,15}$)

$\Delta$ indicates the positions of the double bonds.

$$\underset{20}{CH_3}\underset{19}{CH_2}\underset{18}{CH_2}\underset{17}{CH_2}\underset{16}{CH_2}\underset{15}{CH}=\underset{14}{CH}\underset{13}{CH_2}\underset{12}{CH}=\underset{11}{CH}\underset{10}{CH_2}\underset{9}{CH}=\underset{8}{CH}\underset{7}{CH_2}\underset{6}{CH}=\underset{5}{CH}\underset{4}{CH_2}\underset{3}{CH_2}\underset{2}{CH_2}\underset{1}{C}(=O)OH$$

ARACHIDONIC ACID ($C_{20}\Delta_{5,8,11,14}$)

**TABLE 11–1 Ratio of Saturated and Unsaturated Fatty Acids from Common Fats and Oils***

| Oil or Fat | Percentage of Total Fatty Acids by Weight | | |
|---|---|---|---|
| | *Saturated* | *Monounsaturated* | *Polyunsaturated* |
| Coconut oil | 93 | 6 | 1 |
| Corn oil | 14 | 29 | 57 |
| Cottonseed oil | 26 | 22 | 52 |
| Lard | 44 | 46 | 10 |
| Olive oil | 15 | 73 | 12 |
| Palm oil | 57 | 36 | 7 |
| Peanut oil | 21 | 49 | 30 |
| Safflower oil | 10 | 14 | 76 |
| Soybean oil | 14 | 24 | 62 |
| Sunflower oil | 11 | 19 | 70 |

**Saturated* means full complement of hydrogen (no C=C double bonds); *monounsaturated* means one C=C double bond per fatty acid molecule: *polyunsaturated* means two or more C=C double bonds per molecule of fatty acid. The chief unsaturated fatty acid is linoleic acid. Although derived from vegetable rather than animal fats, both coconut oil and peanut oil have been associated recently with hardening of the arteries when combined with a high cholesterol intake.

**Prostaglandins are synthesized from the essential fatty acids. Even in very small amounts, prostaglandins have powerful effects on the human body.**

The presence of any one of these in the diet permits the body to synthesize a very important group of compounds, the prostaglandins. The key compound here is linoleic acid, which the body cannot make from more saturated fatty acids. If linoleic acid is available, the body can make arachidonic acid and linolenic acid.

***Prostaglandins*** are a group of more than a dozen related compounds with potent effects on physiological activity such as blood pressure, relaxation and contraction of smooth muscle, gastric acid secretion, body temperature, food intake, and blood platelet aggregation. Their potential use as drugs is currently under widespread investigation. Two of the prostaglandins that have been characterized are prostaglandin $E_1$ (used to induce labor to terminate pregnancy) and prostaglandin $E_2$.

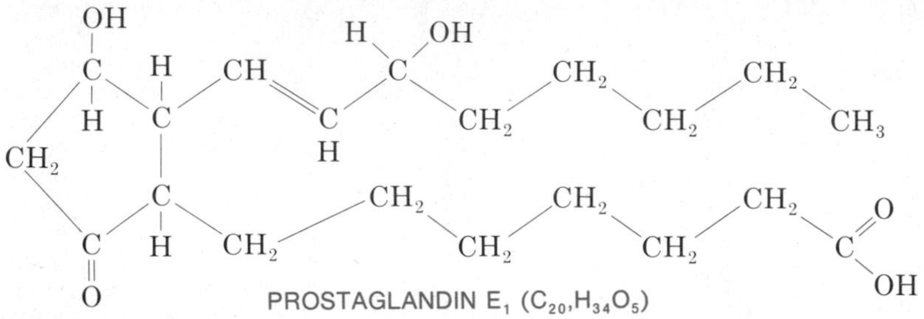

PROSTAGLANDIN $E_1$ ($C_{20}$,$H_{34}O_5$)

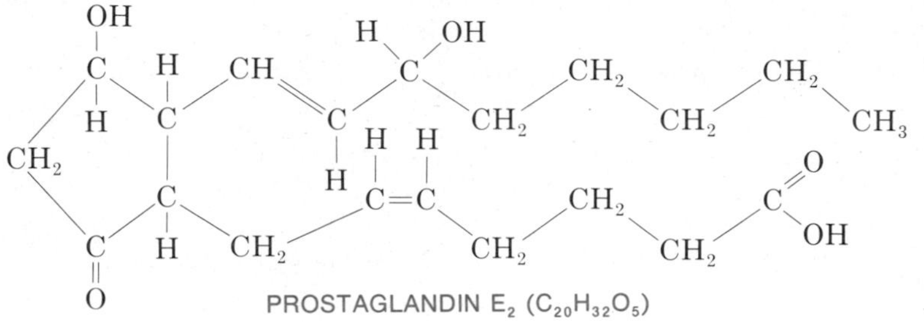

PROSTAGLANDIN $E_2$ ($C_{20}H_{32}O_5$)

Note that both of these prostaglandins contain exactly the same number of carbon atoms as arachidonic acid.

## Self-Test 11-A

1. Carbohydrates contain the elements ________, ________, and ________.
2. The complete hydrolysis of a polysaccharide yields ________.
3. When a molecule of sucrose is hydrolyzed, the products are one molecule each of the monosaccharides ________ and ________.
4. The sugar referred to as blood sugar, grape sugar, or dextrose is actually the compound ________.
5. Starch is a polymer built up out of ________ units.
6. Essential fatty acids are needed by the body to synthesize ________.

# Proteins, Amino Acids, and the Peptide Bond

Proteins occur in all the major regions of living cells. These compounds serve a wide variety of functions, including motion of the organism, defense mechanism against foreign substances, metabolic regulation of cellular processes, and cell structure. The close relationship between proteins and living organisms was first noted by the German chemist G. T. Mulder in 1835. He named these compounds proteins from the Greek *proteios,* meaning first, indicating this to be the starting point in the chemical understanding of life.

**Proteins are high molecular weight compounds made up of amino acid units.**

Proteins are macromolecules with molecular weights ranging from 5000 to several million. Like the polysaccharides, these macrostructures are composed of recurring units of molecular structure. The fundamental units in the case of proteins are ***amino acids.*** Proteins and amino acids are made primarily from four elements: carbon, oxygen, hydrogen, and nitrogen. Other elements occur in trace amounts; the one most often encountered is sulfur.

## Amino Acids

**Amino acids are compounds that generally have the structure**

$$R-\underset{NH_2}{\overset{H}{\overset{|}{\underset{|}{C}}}}-C\begin{matrix}\nearrow O \\ \searrow OH\end{matrix}$$

The complete hydrolysis of a typical protein yields a mixture of about 20 different amino acids. Some proteins lack one or more of these acids, others have small amounts of other amino acids characteristic of a given protein, but the 20 given in Table 11–2 are predominant. In a few instances, one amino acid will constitute a major fraction of a protein (the protein in silk, for example, is 44% glycine), but this is not common.

**There are about 20 common amino acids.**

As the name suggests, amino acids contain an amino group ($-NH_2$) and an acid (carboxyl) group ($-COOH$). In all of the amino acids listed in Table 11–2, the amine group and the acid group are bonded to the same carbon atom. Of these acids, 18 have the general formula

$$R-\underset{NH_2}{\overset{H}{\overset{|}{\underset{|}{C}}}}-C\begin{matrix}\nearrow O \\ \searrow OH\end{matrix}$$

where R is a characteristic group for each amino acid. The simplest amino acid is ***glycine,*** in which R is a hydrogen atom:

$$H-\underset{NH_2}{\overset{H}{\overset{|}{\underset{|}{C}}}}-C\begin{matrix}\nearrow O \\ \searrow OH\end{matrix}$$

**Essential amino acids are amino acids that the body needs but cannot make.**

**For good nutrition we require *all* of the essential amino acids in our daily diet, but the amount required does not exceed 1.5 g per day for any of them.**

The human body is capable of synthesizing some amino acids needed for protein structures, but it is unable to provide others necessary for normal growth and development. The latter are designated ***essential amino acids*** and must be ingested in the food supply. The ***nonessential amino acids*** are just as necessary for life as the essential amino acids but can be made by the body from other compounds. The essential amino acids are indicated in Table 11–2 by asterisks.

**TABLE 11–2 Common Amino Acids**

All of the amino acids except proline and hydroxyproline have the general formula

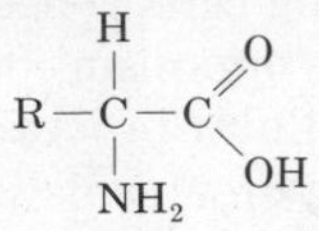

$$R-\underset{\displaystyle NH_2}{\overset{\displaystyle H}{\underset{|}{\overset{|}{C}}}}-C\genfrac{}{}{0pt}{}{\nearrow O}{\searrow OH}$$

in which R is the characteristic group for each acid. The R groups are as follows.

1. Glycine $-H$
2. Alanine $-CH_3$
3. Serine $-CH_2OH$
4. Cysteine $-CH_2SH$
5. Cystine $-CH_2-S-S-CH_2-$
*6. Threonine $-\underset{\displaystyle OH}{\underset{|}{CH}}-CH_3$
*7. Valine $CH_3-\underset{|}{CH}-CH_3$
*8. Leucine $-CH_2-\underset{\displaystyle CH_3}{\underset{|}{CH}}-CH_3$
*9. Isoleucine $-CH\genfrac{}{}{0pt}{}{\nearrow CH_3}{\searrow CH_2-CH_3}$
*10. Methionine $-CH_2-CH_2-S-CH_3$
11. Aspartic acid $-CH_2CO_2H$
12. Glutamic acid $-CH_2-CH_2-CO_2H$
*13. Lysine $-CH_2-CH_2-CH_2-CH_2-NH_2$
*14. Arginine $-CH_2-CH_2-CH_2-NH\overset{\displaystyle NH}{\overset{\|}{C}}NH_2$
*15. Phenylalanine $-CH_2-C_6H_5$ (benzene ring)
16. Tyrosine $-CH_2-C_6H_4-OH$ (benzene ring)
*17. Tryptophan $-CH_2-$ (indole ring, N–H)
*18. Histidine $-CH_2-$ (imidazole ring, N, N–H)

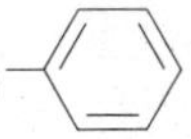

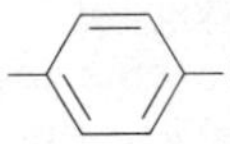

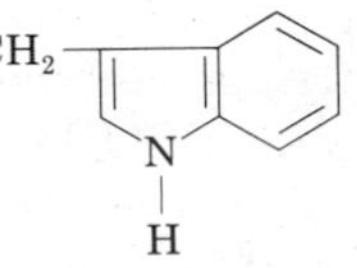

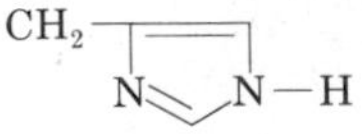

The structures for the other two are:

19. Proline

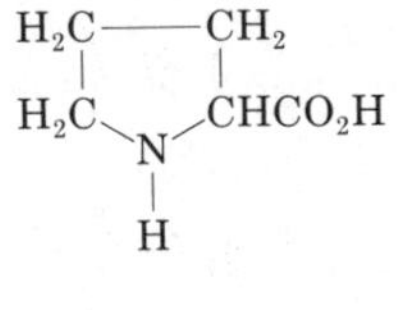

$$\begin{array}{ccc} H_2C & — & CH_2 \\ | & & | \\ H_2C & \diagdown N \diagup & CHCO_2H \\ & | & \\ & H & \end{array}$$

20. Hydroxyproline

$$\begin{array}{ccc} HOHC & — & CH_2 \\ | & & | \\ H_2C & \diagdown N \diagup & CHCO_2H \\ & | & \\ & H & \end{array}$$

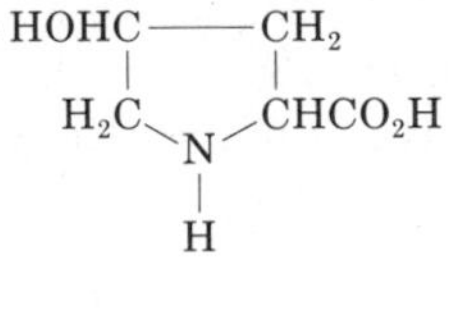

*Essential amino acids; arginine and histidine are essential for children but not essential for adults.

**The peptide bond**

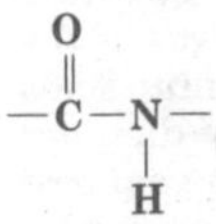

$$-\overset{\displaystyle O}{\overset{\|}{C}}-\underset{\displaystyle H}{\underset{|}{N}}-$$

**binds amino acid units together in proteins.**

## The Peptide Bond

Amino acid units are linked together in protein structures by peptide bonds. This same linkage is found in nylon 66, in which a carboxylic acid and an amine are condensed to form the polymer and the amide bond. As it applies to proteins, the peptide bond can be understood in terms of the reaction between two glycine molecules.

If the acid group of one glycine molecule reacts with the basic amine group of another, the two are joined through the peptide linkage, and one molecule of water is eliminated for each bond formed.

**Starch, glycogen, cellulose, and proteins are condensation polymers, Chapter 10.**

PEPTIDE BOND

$$H_2N-CH_2-C(=O)OH + H_2N-CH_2-C(=O)OH \longrightarrow H_2N-CH_2-\boxed{C(=O)-N(H)}-CH_2-C(=O)OH + HOH$$

GLYCINE GLYCINE GLYCYLGLYCINE

If this hypothetical reaction is carried out with two different amino acids, glycine and alanine, two different ***dipeptides*** are possible.

PEPTIDE BONDS

$$H_2N-CH_2-\boxed{C(=O)-N(H)}-CH(CH_3)-C(=O)OH \qquad H_2N-CH(CH_3)-\boxed{C(=O)-N(H)}-CH_2-C(=O)OH$$

GLYCYLALANINE ALANYLGLYCINE

**A very large number of different proteins can be prepared from a small number of different amino acids.**

Twenty-four ***tetra***peptides are possible if four amino acids (for example, glycine, Gly; alanine, Ala; serine, Ser; and cystine, Cys) are linked in all possible combinations. They are:

| | | | |
|---|---|---|---|
| Gly-Ala-Ser-Cys | Ala-Gly-Ser-Cys | Ser-Ala-Gly-Cys | Cys-Ala-Gly-Ser |
| Gly-Ala-Cys-Ser | Ala-Gly-Cys-Ser | Ser-Ala-Cys-Gly | Cys-Ala-Ser-Gly |
| Gly-Ser-Ala-Cys | Ala-Ser-Gly-Cys | Ser-Gly-Ala-Cys | Cys-Gly-Ala-Ser |
| Gly-Ser-Cys-Ala | Ala-Ser-Cys-Gly | Ser-Gly-Cys-Ala | Cys-Gly-Ser-Ala |
| Gly-Cys-Ser-Ala | Ala-Cys-Gly-Ser | Ser-Cys-Ala-Gly | Cys-Ser-Ala-Gly |
| Gly-Cys-Ala-Ser | Ala-Cys-Ser-Gly | Ser-Cys-Gly-Ala | Cys-Ser-Gly-Ala |

If 17 different amino acids are used, the sequences alone would make $3.56 \times 10^{14}$ uniquely different 17-unit molecules.* Although there are numerous protein structures in nature, these represent an extremely small fraction of the possible structures. Of all the many different proteins that could possibly be made from a set of amino acids, a living cell will make only a relatively small, select number.

Linus Pauling (1901– ) is a scientist of great versatility and accomplishment. His interests have included the determination of the molecular structures of crystals by X ray diffraction and theories of the chemical bond. His work led to the Nobel prize in 1954. For his fight against the nuclear danger confronting the world, he was awarded the 1963 Nobel peace prize.

## Protein Structures

The ***primary structure*** of a protein indicates only the sequence of amino acid units in the polypeptide chain. Since the single bonds in the chain allow free rotation around the bond, an almost infinite number of conformations is possible. Because of interactions, such as hydrogen bonds, between atoms in the same chain, certain conformations called ***secondary structures*** are favored. Linus Pauling, along with R. B. Corey, suggested the two secondary structures for polypeptides discussed below, the sheet structure and the helical structure.

*If the amino acids are all different, the number of arrangements is $n!$ (read $n$ factorial). For five different amino acids, the number of different arrangements is $5!$ (or $5 \times 4 \times 3 \times 2 \times 1 = 120$).

Polyglycine is a synthetic protein made entirely of the amino acid glycine. In polyglycine the hydrogen attached to the nitrogen atom and the oxygen bonded to the carbon are both well suited to engage in hydrogen bonds. In the two stable conformations of polyglycine, maximum advantage is taken of the hydrogen bonds available. In the sheet structure, the hydrogen bonds are between adjacent chains of the polypeptide; in the helical structure, hydrogen bonds occur between atoms within the same chain.

**The amino acids in a protein chain interact with each other via hydrogen bonds.**

Figure 11–8 illustrates a sheetlike structure in which several chains of the polypeptide are joined by hydrogen bonds. Note that all the oxygen and nitrogen atoms are involved in hydrogen bonds. Most of the properties of silk can be explained in terms of this type of structure for fibroin, the protein of silk.

Hydrogen bonds are possible within a single polypeptide chain if the secondary structure is helical (Fig. 11–9). Bond angles and bond lengths are such that the nitrogen atom forms hydrogen bonds with the oxygen atom in the third amino acid unit down the chain.

**A coiled spring is helical in structure.**

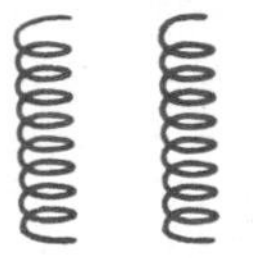

***Collagen*** is the principal fibrous protein in mammalian tissue. It has remarkable tensile strength, which makes it important in the structure of bones, tendons, teeth, and cartilage. Three polypeptide chains, each of which is twisted into a left-

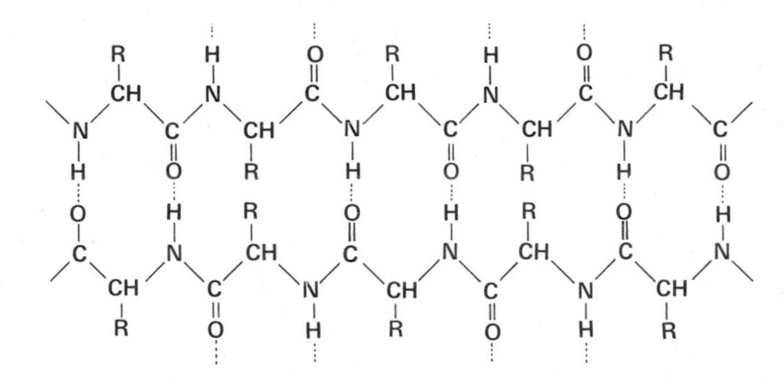

(a)

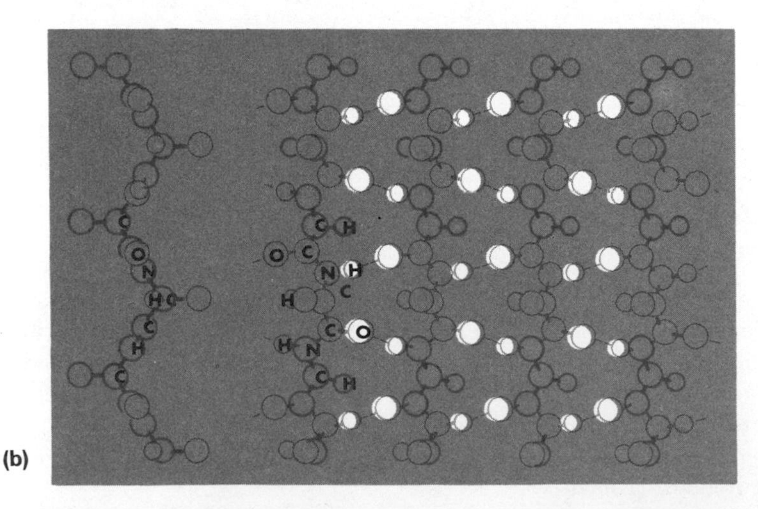

(b)

**Figure 11–8** Sheet structure for polypeptide. In (*a*) the two-dimensional drawing emphasizes that all of the oxygen and nitrogen atoms are involved in hydrogen bonds for the most stable structure. (*b*) Illustrates the bonds in perspective, showing that the sheet is not flat; rather, it is sometimes called a pleated sheet structure.

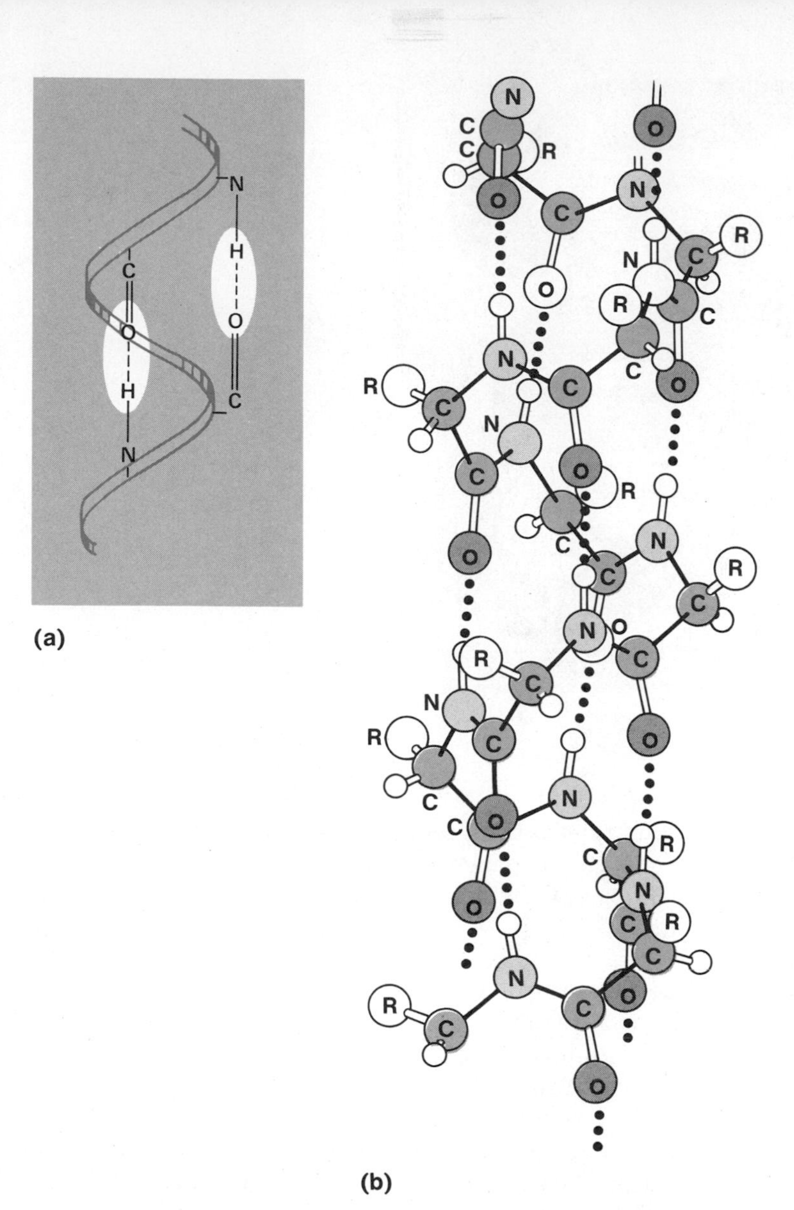


Figure 11–9 (*a*) Helix structure for a polypeptide in which each oxygen atom can be hydrogen bonded to a nitrogen atom in the third amino acid unit down the chain. (*b*) α-Helix structure of proteins. The sketch represents the actual position of the atoms and shows where intra-chain hydrogen bonds occur.

handed helix, are twisted into a right-handed super-helix to form an extremely strong fibril, as shown in Figure 11–10. A bundle of such fibrils forms the macroscopic protein.

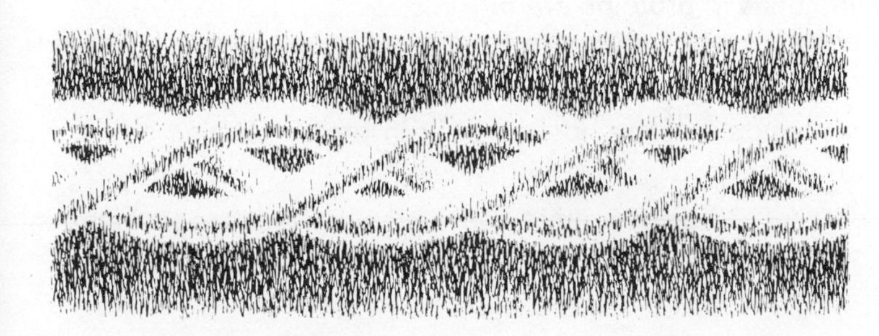

Figure 11–10 The imaginary structure of collagen.

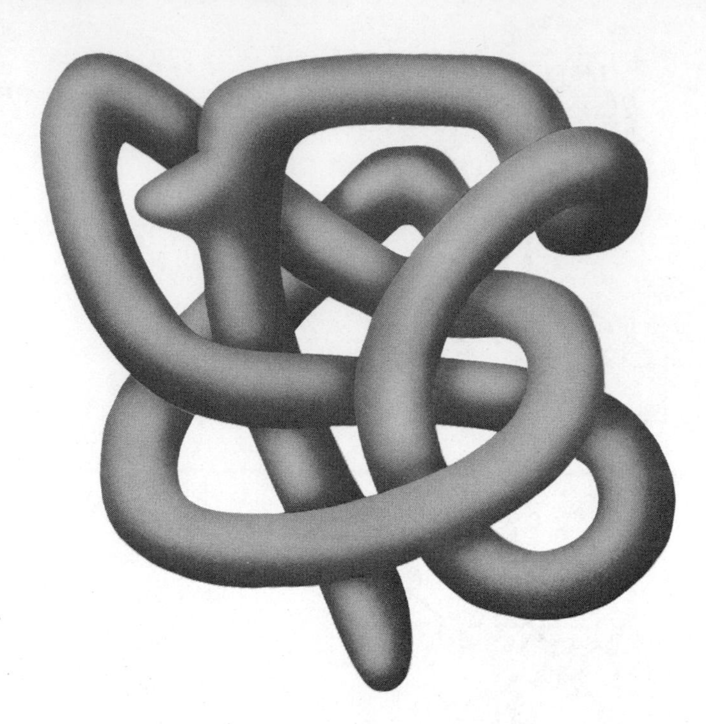

**Figure 11–11** Imaginary folded structure of the helix in a globular protein.

The structure of collagen illustrates a third level of protein structure, ***tertiary structure.*** The primary structure is the sequence of amino acids in the protein, the secondary structure is the helical form of the protein chain, and the tertiary structure is the twisted or folded form of the helix. Another tertiary structure is found in globular proteins. In these structures, the helix chain (secondary structure) is folded and twisted into a definite geometric pattern. This pattern may be held in place by one or more of several different kinds of chemical bonds, such as —S—S— bonds, depending on the particular functional groups in the amino acids involved (Table 11–2). Figure 11–11 illustrates the folded structure of a typical globular protein.

The ***quaternary structure*** of proteins refers to the degree of aggregation of protein units. Native hemoglobin (molecular weight of 68,000) must have its four polypeptide chains properly aggregated in order to form active hemoglobin. Insulin is also composed of subunits, properly arranged into its quaternary structure.

If hemoglobin, a globular protein, has an abnormal primary, secondary, tertiary, or quaternary structure because of a wrong amino acid in a given position, it may be unable to transfer oxygen in the blood. In sickle cell anemia, only one specific amino acid of the 146 in one of the hemoglobin chains is altered with respect to normal hemoglobin. Models of hemoglobin are shown in Figure 11–12.

## Self-Test 11-B

1. The fundamental building units in proteins are the ____________.
2. Amino acids that the body cannot synthesize from other molecules are called ____________.
3. The peptide linkage that bonds amino acids together in protein chains has the structure ____________.
4. The basic structure present in almost all of the amino acids can be represented as ____________.

(a)

**Figure 11–12** (*a*) The structure of heme. (*b*) Two views of a model of the hemoglobin structure. Two light-colored chains of protein, two dark-colored chains, and two heme "disks" in proper arrangement compose the quaternary structure of hemoglobin. M. F. Perutz received a Nobel Prize for determining this structure.

(b)

5. The formula for glycylglycine is: ______________

6. a. The primary structure of a protein refers to its ______________;
   b. the secondary structure refers to its ______________;
   c. its tertiary structure refers to ______________;
   d. and its quaternary structure refers to ______________.

7. a. If we have three different amino acids, we can make a total of ______________ different tripeptides from them if we can use an amino acid up to three times in any given tripeptide.
   b. If we can use each amino acid only once, there are still ______________ possible different tripeptides.

## Enzymes

**Enzymes are protein molecules that speed up chemical reactions.**

An important group of globular proteins is the ***enzymes,*** molecules that function as catalysts for reactions in living systems. Like other catalysts, a given enzyme increases the rate of a reaction without requiring an increase in temperature. As an example of a simple type of catalysis, consider the oxidation of glucose, which burns in air with some difficulty and is hard to light with a match. If cigarette ashes or other catalysts are placed on its surface, combustion can be initiated easily with a match. When the glucose burns, it liberates a large amount of energy, 688 kilocalories, or 688,000 calories, per mole.

$$\underset{\text{GLUCOSE}}{C_6H_{12}O_6} + \underset{\text{OXYGEN}}{6O_2} \longrightarrow \underset{\text{CARBON DIOXIDE}}{6CO_2} + \underset{\text{WATER}}{6H_2O} + 688\text{ kcal}$$

The energy required to get the reaction started is the ***activation energy;*** catalysts, in general, work by lowering the activation energy. If an enzyme can lower the activation energy to a point where the average kinetic energy of the molecules in a living cell (or in a laboratory system) is sufficient for reaction, then the reaction can proceed rapidly. Glucose is oxidized rapidly and efficiently at ordinary temperatures in the presence of the proper enzymes. To be sure, the oxidation of glucose in a living cell requires many enzymes and many steps, but enzymatic catalysis produces the same final result as combustion at elevated temperature, namely carbon dioxide, water, and 688 kilocalories of usable energy per mole of sugar. Figure 11–13 graphically illustrates the concepts of activation energy, the energy available from an energy-producing reaction, and the reduction of the activation energy by an enzyme.

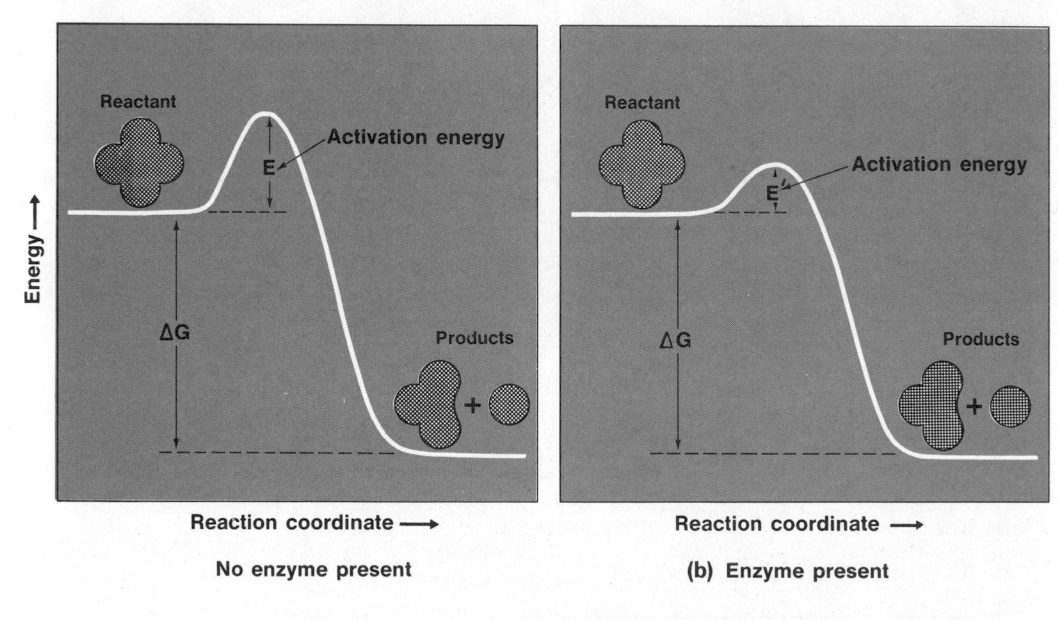


**Figure 11–13** Enzyme effect on activation energy. The vertical coordinate represents increasing energy, and the horizontal one represents time. For energy-producing reactions the reactant molecules are at a higher energy than the product molecules, as illustrated in (*a*). The difference between these energies is the net useful energy provided by the chemical reaction. The useful energy is tabulated as the free energy change of the reaction (ΔG). However, it is necessary for the reactant molecules to "get over" the energy barrier (acquire the activation energy, E) in going from reactants to products. Note that the activation energy is given back along with the free energy. The enzyme lowers the activation energy to E′, as illustrated in (*b*), while the free energy change remains the same. The net effect is to obtain the free energy of the reaction with a smaller expenditure of activation energy. Free energy is discussed in the next chapter.

Enzymes are remarkable catalysts because they are highly specific for a given reaction. Maltase, an enzyme, catalyzes the hydrolysis of maltose into two molecules of D-glucose. This is the only known function of maltase, and no other enzyme can substitute for it. The explanation for the specific activity of enzymes can be found in their molecular structures.

**Enzymes are specific catalytic molecules with a specific catalytic task.**

Enzymes are globular proteins with definite tertiary structures. The highly specific action of maltase can be explained if its globular structure accurately accommodates a maltose molecule at the point where the reaction occurs, the reactive site. When the two units come together, strain is placed on the bonds holding the two simple sugar units together. As a result, water is allowed to enter and hydrolysis occurs. Sucrose cannot be hydrolyzed by maltase because of the different structure involved. Another enzyme, sucrase, hydrolyzes sucrose effectively. Some enzymes, however, are less specific. The digestive enzyme trypsin, for example, acts predominantly on peptide bonds in proteins, but it will also catalyze the hydrolysis of some esters because of somewhat similar structure and polarity at the active site.

How can an enzyme lower the activation energy and be so specific for a given reaction? Just as a key can separate a padlock into two parts and subsequently remain unchanged, ready to unlock other identical locks, so the enzyme makes possible a molecular change (Fig. 11–14). With enough energy the lock could be separated without the key, and with enough energy the molecular alteration could occur without the enzyme. An enzyme cannot make an unnatural chemical reaction occur.

The tremendous speed of enzyme-catalyzed reactions requires more than just random collisions to fit "the key in the lock." For example, a molecule of $\beta$-amylase catalyzes the breaking of four thousand bonds between the $\alpha$-glucose units in amylose per second. Speed like this requires something to attract the key into the lock, such as electrically polar regions, partially charged groupings, or ionic sections on the enzyme and the ***substrate*** (the reactant molecule). These regions attract as well as guide the substrate to the proper position on the enzyme and thereby speed up the reaction. The electrically charged portions of the enzyme are believed to be the chemically ***active sites*** in the enzyme.

**Enzyme structure is the key to specific catalytic activity.**

Sometimes an enzyme is more than a globular protein; in such cases, the protein is not a catalyst by itself. In addition to the protein part of the enzyme there is also another chemical species called a ***coenzyme.*** The coenzyme, required for catalytic activity, may be an ion (e.g. $Co^{3+}$, $Fe^{3+}$, $Mg^{2+}$, or another of the essential minerals) or may be derived in part from a vitamin (as we shall see later in this section).

**Some enzymes require coenzymes: metal ions or vitamins.**

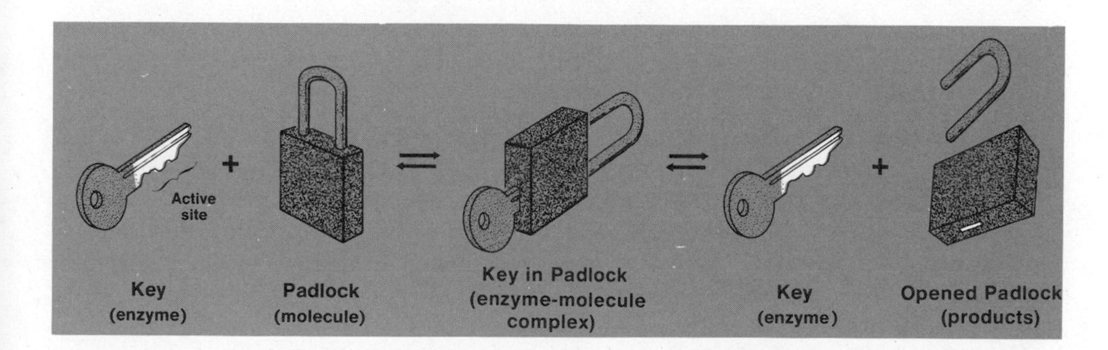


**Figure 11–14** Lock-and-key theory for enzymatic catalysis. Although it is generally agreed that this analogy is an oversimplification, it does make one very important point: the enzyme makes a difficult job easy by reducing the energy required to get the job started. It also suggests that the enzyme has a particular structure at an active site that will allow it to work only for certain molecules, similar to a key that fits the shape of a particular keyhole and a particular sequence of tumblers.

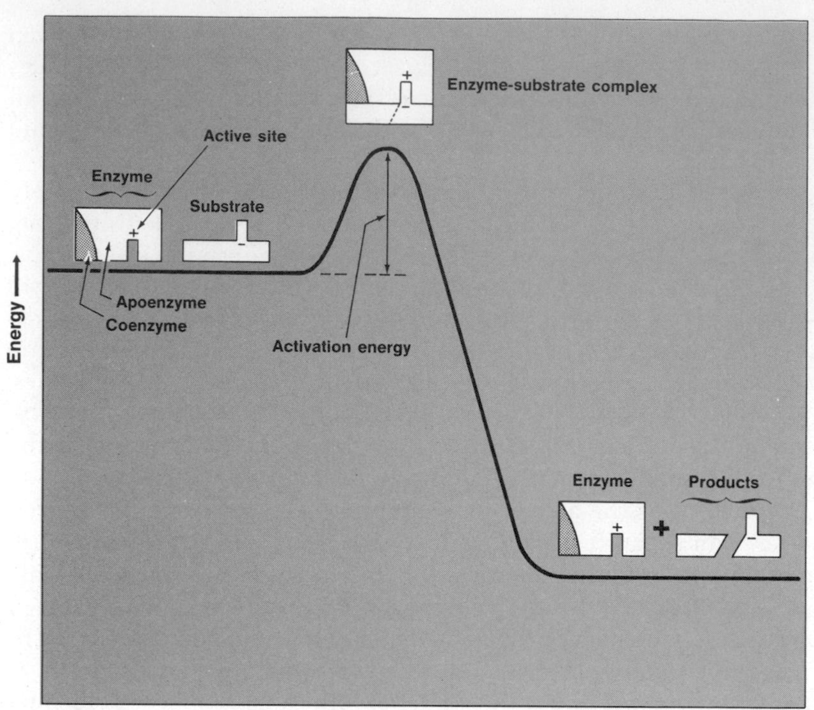


**Figure 11–15** A reaction utilizing an enzyme that requires a coenzyme. Here the essential feature is that the combination of enzyme plus coenzyme allows the reaction to proceed with a lower activation energy. Apoenzyme is the name given to the enzyme structure that combines with the coenzyme.

The protein part of such an enzyme is called the ***apoenzyme.*** The coenzyme alone does not have enzymatic activity; neither does the apoenzyme. Before the enzyme becomes active, the apoenzyme and the coenzyme must combine like the two keys required to open a bank lock-box. Neither your key nor the bank's key will open the box when used alone, but both together will (Fig. 11–15).

Let's use a simple chemical example to demonstrate the relationship between coenzyme, substrate, and apoenzyme. The peptide glycylglycine is hydrolyzed very slowly by water to give two molecules of glycine.

$$\underset{\text{GLYCYLGLYCINE}}{\mathrm{H_2N{-}CH_2{-}C(=O){-}NH{-}CH_2{-}C(=O)OH}} + H_2O \longrightarrow \underset{\text{GLYCINE}}{2\,\mathrm{CH_2(NH_2){-}C(=O)OH}}$$

The glycylglycine substrate forms an intermediate compound with an enzyme, and as a result the substrate is activated for further reaction. Activation can result from extensive hydrogen bonds, interaction with a metal ion in the enzyme, or a number of other processes. In this case the coordination of the glycylglycine to the positive charge of a cobalt ion ($Co^{2+}$), the coenzyme, makes it more susceptible to attack by the negative end of a water molecule (Fig. 11–16).

The names of most enzymes end in *-ase.*

Most of the names of enzymes end in ***-ase.*** There are a few exceptions, such as pepsin and trypsin, which are both digestive enzymes. Hydrolases promote the breakdown of foodstuffs and other substances by hydrolysis. Carbohydrases (such as maltase, lactase, sucrase, ptyalin, and amylase) help to effect hydrolysis of carbohydrates. Proteases (such as pepsin, trypsin, and chymotrypsin) hydrolyze large protein molecules into groups of smaller proteins. Lipases hydrolyze esters such as fats and oils.

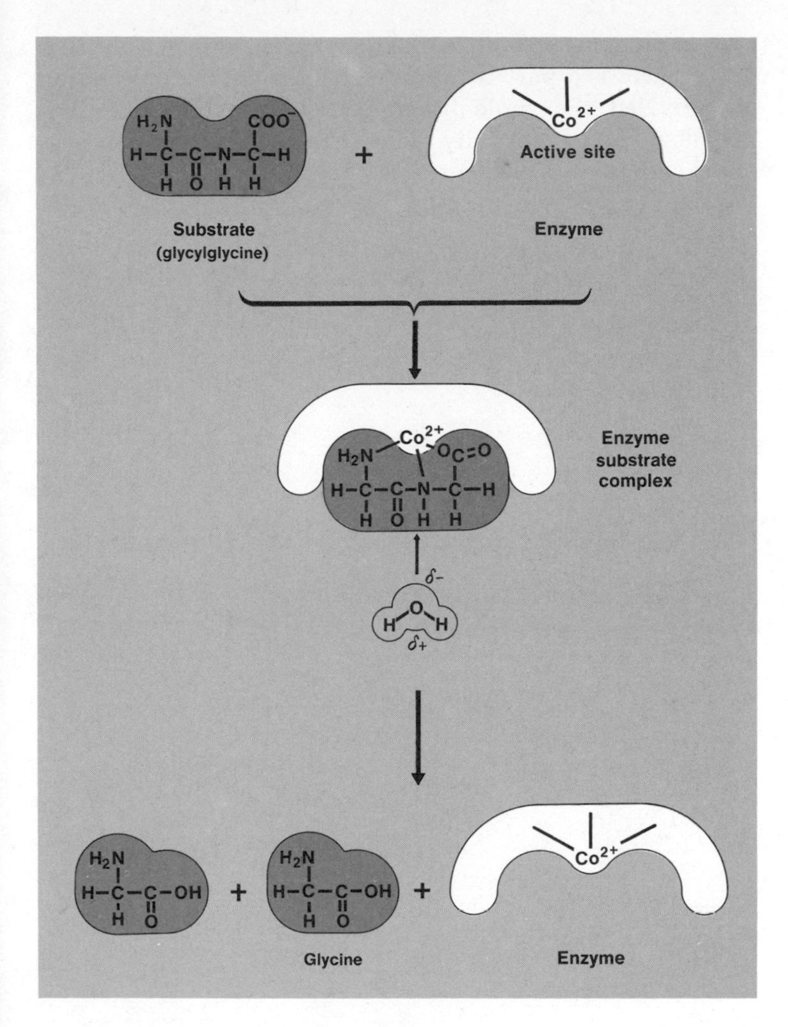


**Figure 11–16** Action of an enzyme. The substrate molecule is chemically bonded to the enzyme (glycylglycine dipeptidase). The negative oxygen and the nitrogen atoms of the substrate bond to the positive cobalt ion in the enzyme. The bonding of the substrate makes it more susceptible to attack by water. Hydrolysis occurs and the glycine molecules are released by the enzyme, which is then ready to play its catalytic role again.

Examples of the oxidizing enzymes, called oxidases, are catalase, which speeds up the conversion of hydrogen peroxide to water and oxygen, and dehydrogenases, which assist in the removal of hydrogen from molecules. There are other categories of enzymes, but these illustrate the wide variety of biochemical catalysts.

Besides being biochemical middlemen, speeding up and directing all the chemical reactions that go into the continuous breakdown and buildup of our cells (three million red blood cells are renewed in the human body every second), enzymes may be the answer to future food problems. Scientists have already developed a way to produce sugar (on a limited experimental basis) by bubbling carbon dioxide into water containing enzymes. Trash fish can be converted into palatable animal feed by using enzymes. Work is under way to convert oil spills into edible products for sea organisms. A little later in this text we shall encounter the use of enzymes as meat tenderizers.

**Enzymes are also important in commercial products.**

Many ***vitamins*** function as coenzymes. They not only trigger specific enzymes, they pitch in and help out. The B vitamins are found in every cell as coenzymes in various oxidative processes. For example, niacin (vitamin $B_3$) becomes part of an enzyme that prevents pellagra, at one time a common vitamin-deficiency disease in the United States. Doctors and biochemists now know that the body suffers from pellagra when it lacks sufficient niacin. Some foods such as yeast, liver, meats, fish, eggs, whole wheat, brown rice, and peanuts contain niacin. The body needs only

0.06 gram of niacin each day in order to prevent pellagra. This isn't much, but it is vital. The coenzyme of which niacin is a part is necessary to the energy production in the body. If energy is not provided, the whole process of renewing cells and building needed compounds slows down and eventually stops. The coenzyme involved is a large molecule, ***nicotinamide adenine dinucleotide (NAD⁺).***

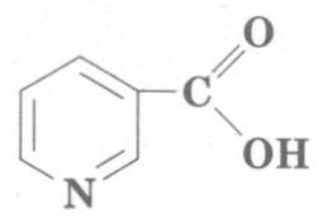

*Niacin*
*(Nicotinic acid)*

**Some of the bonds in this structure are shown abnormally long for clarity.**

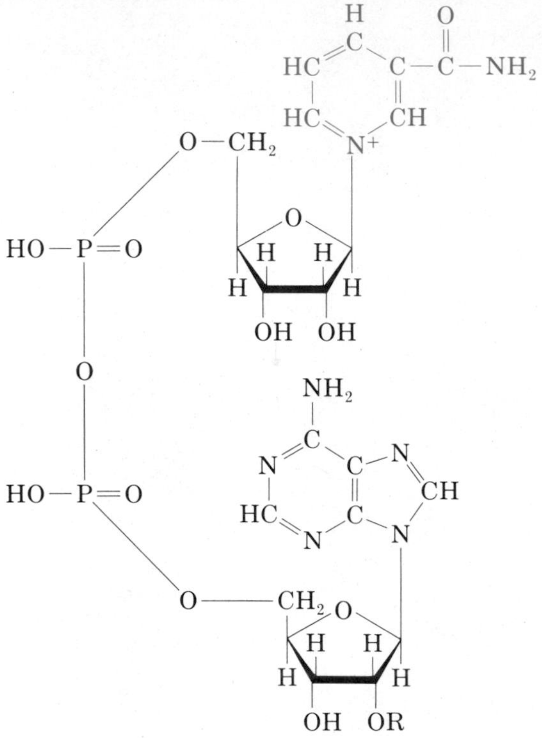

$NAD^+$, R = H
$NADP^+$, R = $PO(OH)_2$

Niacin is also an integral part of another oxidative enzyme, ***nicotinamide adenine dinucleotide phosphate (NADP⁺).*** In $NAD^+$, R = H; in $NADP^+$, R = $H_2PO_3$. The part of the structure printed in color comes from niacin and is the active part in oxidation-reduction. It accepts a hydrogen atom during biological oxidation-reduction processes.

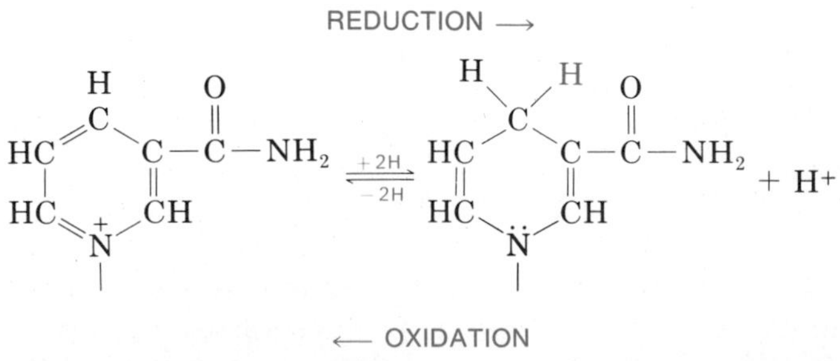

REDUCTION ⟶

+2H / −2H

+ $H^+$

⟵ OXIDATION

Biochemists often keep track of oxidation and reduction by the movement of hydrogen atoms, which can be considered from one of two viewpoints. When hydrogen combines with oxygen to form water

$$2H_2 + O_2 \longrightarrow 2H_2O$$

hydrogen is oxidized (it combines with oxygen, loses some control of its one electron per atom, and increases its positive charge), and oxygen is reduced (it combines with hydrogen, gains some control of additional electrons, and decreases its positive charge). Biochemists usually choose the viewpoint of oxygen, which, in this example, represents any substrate. If the substrate *adds* hydrogen atoms, the substrate is reduced; if the substrate *loses* hydrogen atoms, the substrate is oxidized.

Riboflavin (vitamin $B_2$) is a necessary part of another important hydrogen acceptor coenzyme, ***flavin adenine dinucleotide (FAD).***

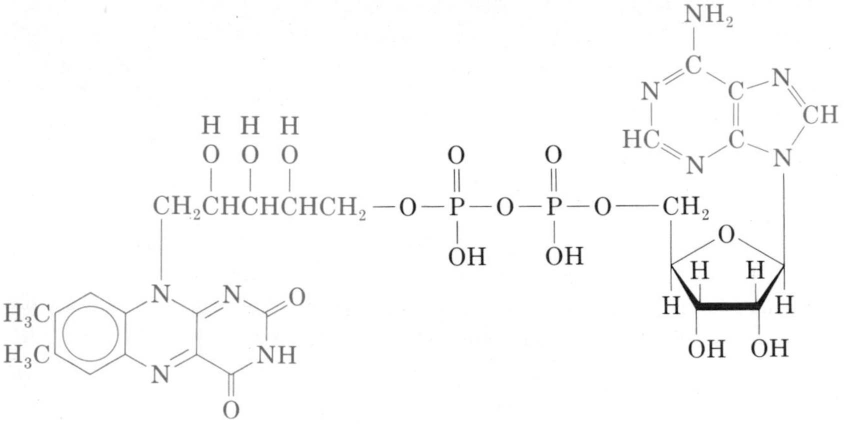

FLAVIN ADENINE DINUCLEOTIDE (FAD)
(The portion in color is riboflavin.)

FAD accepts hydrogen atoms in the following manner:

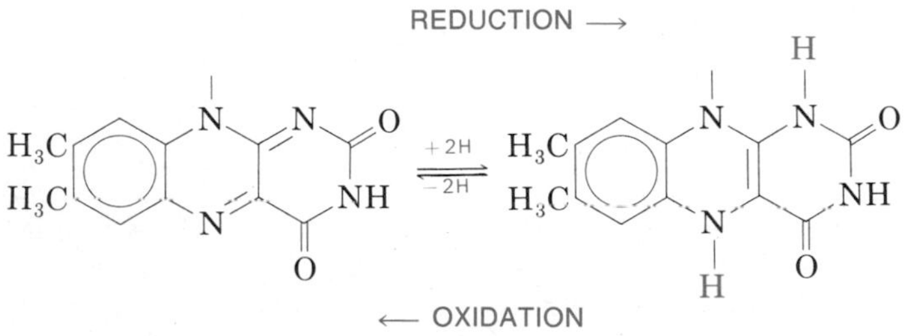

REDUCTION ⟶

+2H / −2H

⟵ OXIDATION

Glucose and glycogen are the principal sources of energy in the body. How $NAD^+$, $NADP^+$, and FAD fit into the oxidation of glucose and glycogen will be considered in the next chapter.

## Nucleic Acids

Like the polysaccharides and the polypeptides, the ***nucleic acids*** are polymeric substances with molecular weights up to several million. Nucleic acids are found in all living cells, with the exception of the red blood cells of mammals. The almost infinite variety of possible structures for nucleic acids allows information in coded form to be recorded in molecular structures in a somewhat similar fashion to the way a few language symbols can be used to convey the many ideas in this book. Such stored information is believed to control the inherited characteristics of the next generation as well as many of the ongoing life processes of the organism.

**The coded information of nucleic acids tells cells which molecules to synthesize.**

Hydrolysis of nucleic acids yields one of two simple sugars, phosphoric acid ($H_3PO_4$), and a group of nitrogen compounds that have basic (alkaline) properties.

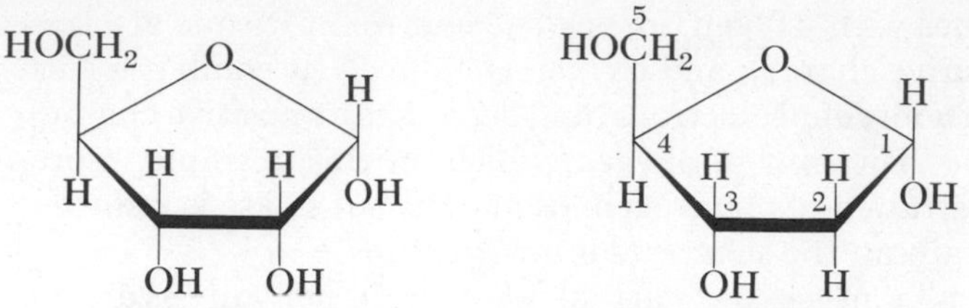

*α-D-ribose*

*α-2-deoxy-D-ribose*

**Figure 11–17** The structure of α-D-ribose and α-2-deoxy-D-ribose. In the IUPAC names given, α indicates the one of two ring forms possible; D distinguishes the isomers that rotate plane polarized light in opposite directions, and the 2 indicates the carbon to which no oxygen is attached in the second sugar.

The structures of the two sugars in nucleic acids are shown in Figure 11–17. The names and formulas of the basic nitrogen compounds are given in Figure 11–18.

The nucleic acids can be classified as those containing the sugar ***α-2-deoxy-D-ribose*** and those containing ***α-D-ribose.*** The former are called ***deoxyribonucleic acids (DNA)*** and the latter ***ribonucleic acids (RNA).*** DNA is found primarily in the nucleus of the cell, whereas RNA is found mainly in the cytoplasm outside the nucleus (Fig. 11–19).

**One nucleotide is joined to another by an ester-forming reaction.**

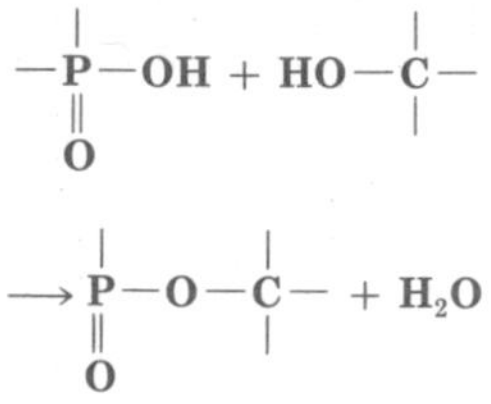

$$-\underset{\underset{O}{\|}}{\overset{|}{P}}-OH + HO-\overset{|}{\underset{|}{C}}- \longrightarrow \underset{\underset{O}{\|}}{\overset{|}{P}}-O-\overset{|}{\underset{|}{C}}- + H_2O$$

## Nucleotides

The repeating units of DNA or RNA are ***nucleotides.*** These substances contain a simple sugar unit, one of the nitrogenous bases, and one or two units of phosphoric acid. An example of a nucleotide structure is illustrated in Figure 11–20.

## Polynucleotides

In addition to the mononucleotides, partial hydrolysis of DNA or RNA yields oligonucleotides that have a few nucleotide units bonded as shown in Figure 11–21.

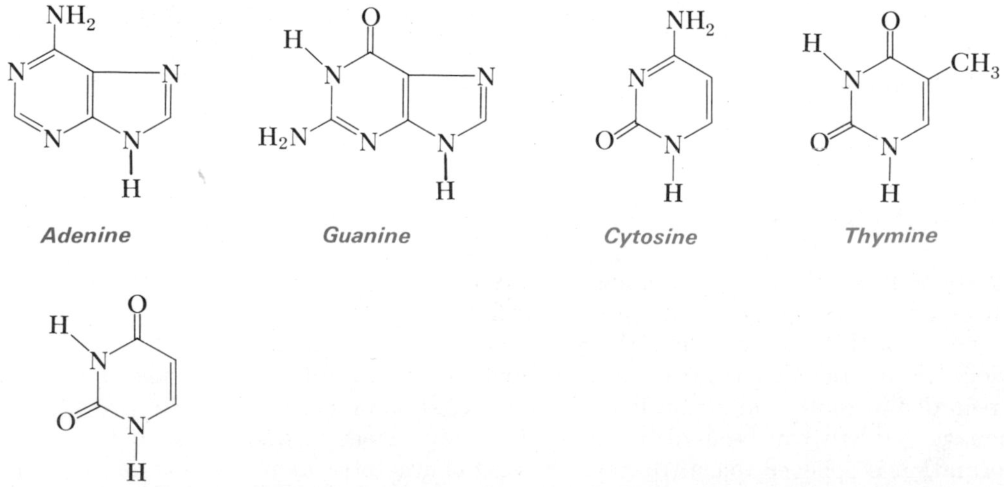

*Adenine* *Guanine* *Cytosine* *Thymine*

*Uracil*

**Figure 11–18** Some nitrogenous bases obtained from the hydrolysis of nucleic acids.

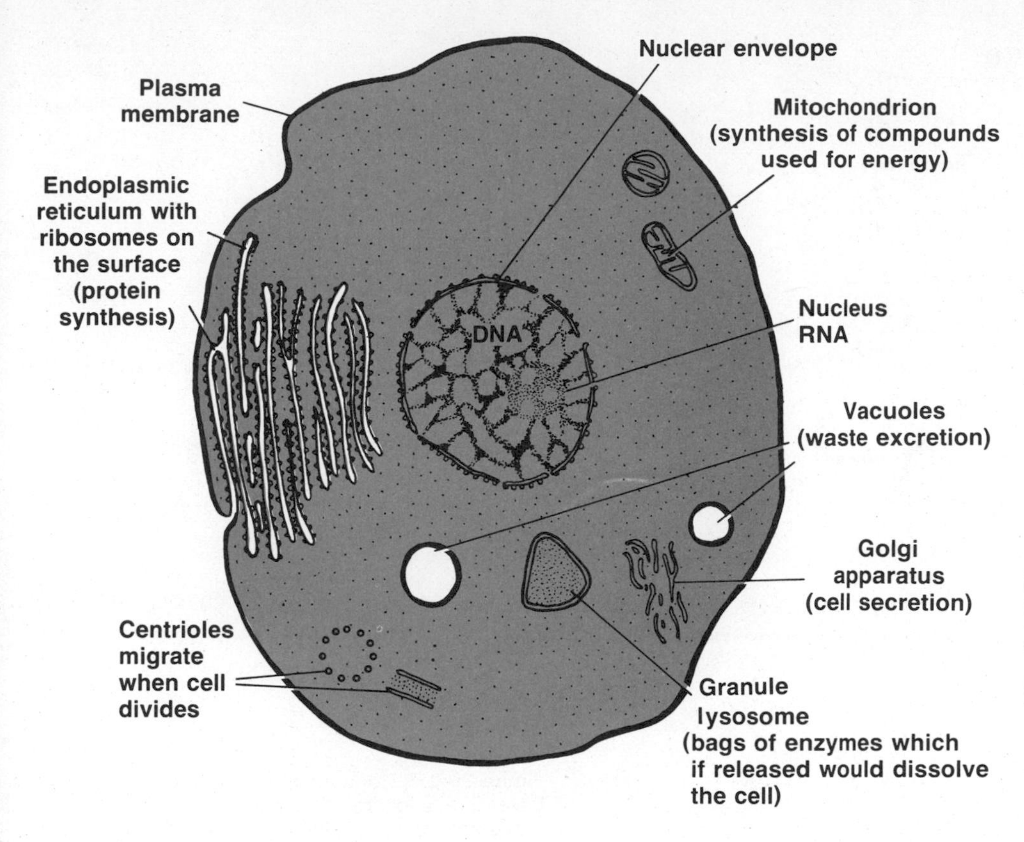


**Figure 11–19** Diagrammatic generalized cell to show the relationships between the various components of the cell. Cytoplasm is the material of the cell, exclusive of the nucleus. Many of the components shown are not visible through an ordinary optical microscope.

Obviously, a large number of oligonucleotides is possible when one considers the choice of base structures and the different sequence possibilities for the chain of nucleotides.

DNA and RNA are polynucleotides. The number of possible structures for these molecules, which have molecular weights as high as a few million, appears to be almost limitless. Since DNA is a major part of the chromosome material in the nucleus of a cell, it seems reasonable to assume that the organism's characteristics are coded in the DNA structure. Even if we assume that there are over two million different species of organisms and that each individual of each species requires a different DNA structure, there are ample combinations of nucleotides for each individual to be unique. It is now known that some kinds of RNA transfer the information coded in the DNA structure to the cytoplasmic region of the cell, where they control the thousands of reactions that occur.

**The inherited traits of an organism are controlled by DNA molecules.**

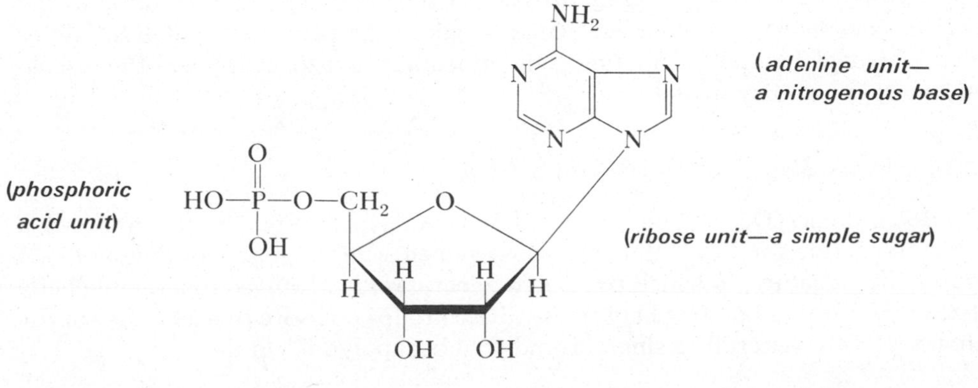

*(adenine unit—a nitrogenous base)*

*(phosphoric acid unit)*

*(ribose unit—a simple sugar)*

**Figure 11–20** A nucleotide. If other bases are substituted for adenine, a number of nucleotides are possible for each of the two sugars shown in Figure 11–17. There is ample evidence that the nucleotides found in both DNA and RNA have the general structure indicated for this nucleotide.

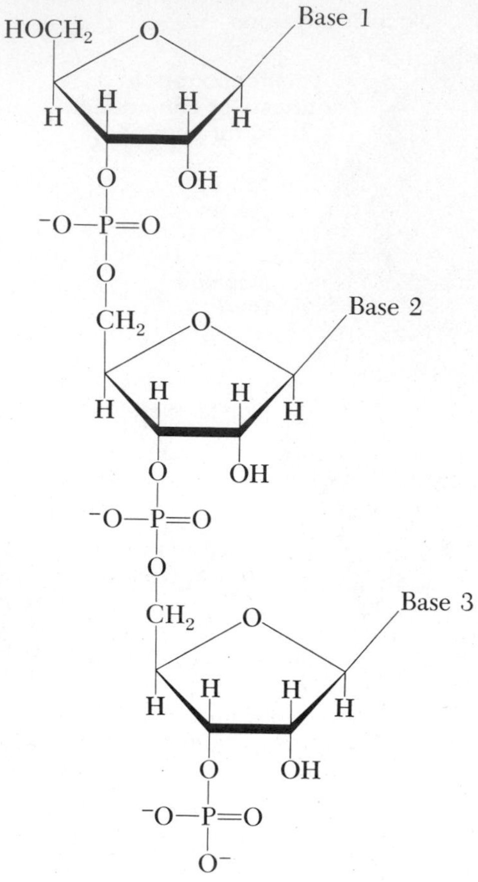

**Figure 11–21** Bonding structure of a trinucleotide. Bases 1, 2, and 3 represent any of the nitrogenous bases obtained in the hydrolysis of DNA and RNA. The primary structure of both DNA and RNA is an extension of this structure to produce molecular weights as high as a few million.

Three major types of RNA have been identified. They are messenger RNA (mRNA), transfer RNA (tRNA), and ribosomal RNA (rRNA). Each has a characteristic molecular weight and base composition. Messenger RNA's are generally the largest, with molecular weights between 25,000 and one million. They contain from 75 to 3000 mononucleotide units. Transfer RNA's have molecular weights in the range of 23,000 to 30,000 and contain 75 to 90 mononucleotide units. Ribosomal RNA's have molecular weights between those of mRNA's and tRNA's and make up as much as 80% of the total cell RNA.

Most RNA is found in the cytoplasm and ribosomes of the cell (Fig. 11–19), but in liver cells as much as 11% (largely mRNA) of the total cell RNA is found in the nucleus. Besides having different molecular weights, the three types of RNA differ in function. One difference in function is described in the next chapter in the discussion of natural protein synthesis.

## Secondary Structure of DNA and RNA

In 1953, James D. Watson and Francis H. C. Crick (Fig. 11–22) proposed a secondary structure for DNA that has since gained wide acceptance. Figure 11–24 illustrates the structure, in which two polynucleotides are arranged in a double helix stabilized by hydrogen bonding between the base groups opposite to each other in the two chains. RNA is generally a single strand of helical polynucleotide.

**Figure 11–22** Francis H. C. Crick (1916– ) (*right*) and James D. Watson (1928– ) (*left*), working in the Cavendish Laboratory at Cambridge, built scale models of the double helical structure of DNA based on the x-ray data of Maurice H. F. Wilkins. Knowing distances and angles between atoms, they compared the task to the working of a three-dimensional jigsaw puzzle. Watson, Crick, and Wilkins received the Nobel Prize in 1962 for their work relating to the structure of DNA.

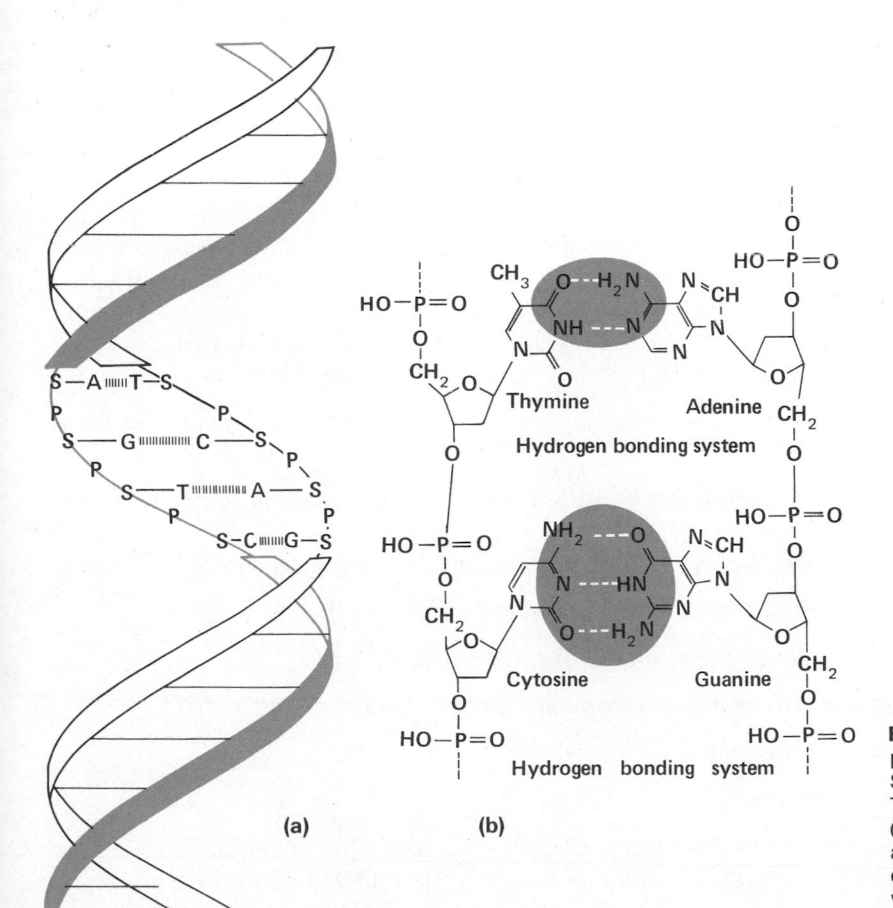


**Figure 11–23** (*a*) Double helix structure proposed by Watson and Crick for DNA. S-sugar, P-phosphate, A-adenine, T-thymine, G-guanine, C-cytosine. (*b*) Hydrogen bonds in the thymine-adenine and cytosine-guanine pairs stabilize the double helix. Adenine will also base pair with uracil in the mRNA since there is no thymine in mRNA.

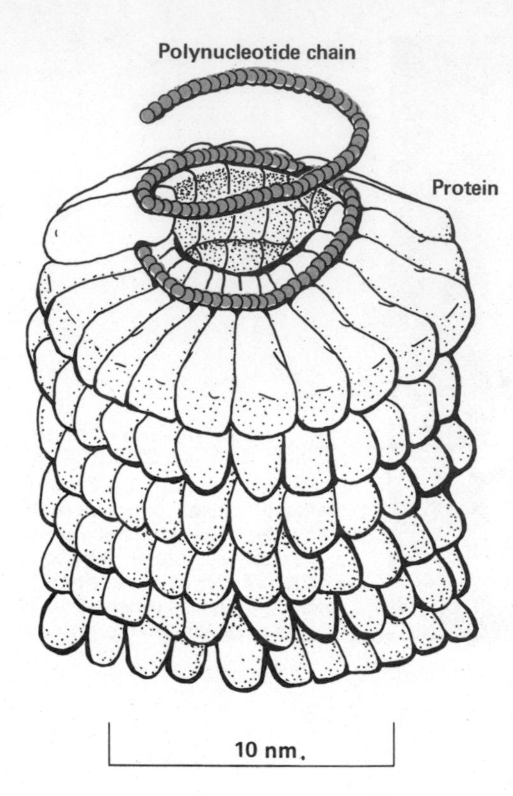


**Figure 11–24** The imagined structure of the tobacco mosaic virus. The polynucleotide chain is coiled, and there is one protein unit for each three nucleotide units. Part of the polynucleotide chain is exposed for clarity. This structure is based on x-ray studies.

### Virus Structure

A virus is a parasitic chemical complex that can reproduce only when it has invaded a host cell. It has the ability to disrupt the life processes of the host cell and order its cell contents to reproduce the virus structure. The isolated virus unit has neither the enzymes nor the smaller molecules necessary to reproduce itself alone.

A virus is a polynucleotide surrounded by a layer of protein. One virus that has been studied in detail is the tobacco mosaic virus, illustrated in Figure 11–24.

## Self-Test 11-C

1. The best term to describe the general function of enzymes is (catalyst/intermediate/oxidant).
2. In the lock-and-key analogy of enzyme activity, the enzyme functions as the ______________ while the substrate molecule serves as the ______________.
3. Pellagra can be prevented by intake of the vitamin named ______________.
4. The activation energy of many biological reactions is decreased if a(n) ______________ is present.
5. Apoenzyme + coenzyme $\longrightarrow$ ______________.
6. Riboflavin is a vitamin that is needed because it is part of an essential ______________.
7. The coenzyme nicotinamide adenine dinucleotide ($NAD^+$) cannot be made by the human body unless it has a supply of ______________.

8. That portion of the enzyme at which the reaction is catalyzed is called the ______________ ______________.

9. a. The sugar in RNA is ______________,

   b. whereas the one in DNA is ______________.

10. A nucleotide contains ______________, ______________, and ______________.

11. The secondary structure of DNA is in the shape of a(n) ______________ ______________.

12. A virus is a chemical compound that can reproduce itself. (True/False)

## Matching Set

| | | | |
|---|---|---|---|
| ________ | 1. D-glucose | **a** | made from essential fatty acids |
| ________ | 2. linoleic acid | **b** | polymer consisting of $\alpha$-D-glucose units |
| ________ | 3. methionine | **c** | proteins |
| ________ | 4. enzymes | **d** | sugar present in blood |
| ________ | 5. carbohydrate storage in animals | **e** | an essential amino acid |
| ________ | 6. starch | **f** | sucrase |
| ________ | 7. polypeptides | **g** | an unsaturated fatty acid |
| ________ | 8. prostaglandins | **h** | a polynucleotide |
| ________ | 9. DNA | **i** | biochemical catalysts |
| ________ | 10. fibrous protein | **j** | coenzymes |
| ________ | 11. cellulose | **k** | glycogen |
| ________ | 12. vitamins | **l** | collagen |
| ________ | 13. enzyme that splits sucrose into fructose and glucose | **m** | polymer consisting of $\beta$-D-glucose units |

## Questions

**1.** Show the structure of the product that would be obtained if two alanine molecules (Table 11–2) react to form a dipeptide.

**2.** Biochemistry is a special field within what branch of chemistry?

**3.** What is an essential amino acid?

**4.** The ketone structure of D-fructose has three asymmetric carbon atoms per molecule. How many isomers result from the asymmetric centers?

**5.** Name a polysaccharide that yields only D-glucose upon complete hydrolysis. Name a disaccharide that yields the same hydrolysis product.

**6.** What is the difference between the starch, amylopectin, and the "animal starch," glycogen?

**7.** What is the chief function of glycogen in animal tissue?

**8.** Explain the basic difference between starch, amylose, and cellulose.

**9.** Why does cotton, a cellulose material, absorb moisture so well in contrast to nylon 66?

**10.** What functional groups are always present in each molecule of an amino acid?

**11.** Give the name and formula for the simplest amino acid. What natural product has a high percentage of this amino acid?

**12.** The amino acid of silk is glycine. Name the peptide bond in silk.

**13.** If six different amino acids formed all the possible different tripeptides, how many would there be?

**14.** Prostaglandins belong to what group of biochemicals?

**15.** What is the meaning of the terms primary, secondary, and tertiary structures of proteins?

**16.** In a protein, what type of bond holds the helix structure in place?

**17.** Enzymes are what type of proteins?

**18.** What three molecular units are found in nucleotides?

**19.** Based on the structures in Figure 11–17, explain the meaning of the prefix *deoxy-* in deoxyribonucleic acid.

**20.** What is the acid of a nucleotide?

**21.** How many trinucleotides with the structure indicated in Figure 11–21 could be made with the nitrogenous bases listed in Figure 11–18?

**22.** What are the basic differences between DNA and RNA structures?

**23.** What stabilizing forces hold the double helix together in the secondary DNA structure proposed by Watson and Crick?

**24.** Consult a medical dictionary and determine the difference between atherosclerosis and arteriosclerosis.

**25.** What two molecular structures are present in viruses?

**26.** What important type of chemical can function as a coenzyme?

**27.** How many kilocalories of heat energy are liberated when 90 g of glucose, $C_6H_{12}O_6$, are burned to carbon dioxide and water?

**28.** Calculate the molecular weight of the nucleotide shown in Figure 11–20. Recall that the representation of the adenine unit omits two hydrogen atoms.

**29.** Calculate the molecular weight of the coenzyme $NAD^+$.

**30.** What is another name for niacin?

**31.** a. Which of the following biochemicals are polymers: starch, cellulose, glucose, fats, proteins, DNA, RNA? b. What are the monomer units for those which are polymers?

**32.** What is a chemical function of vitamins? Give some examples.

**33.** Why are carbohydrates considered "energy-rich"?

**34.** Why is it that humans cannot digest cellulose?

**35.** The molecular structures of enzymes (particularly apoenzymes) are most closely related to which structures: proteins, fats, carbohydrates, or polynucleic acids?

**36.** a. What element is necessarily present in proteins that is not present in either carbohydrates or fats? b. Name another element that is probably present in proteins but not present in either carbohydrates or fats.

# Biochemical Processes 12

How are carbohydrates, fats, and proteins used in our bodies? What happens to them in the mouth, in the stomach, and in individual cells? How is energy derived, received, stored, and used by the chemicals of our bodies? How are our custom-made proteins constructed? Why are we like other human beings and especially like our parents?

In this chapter, we shall examine some of the chemical reactions by which proteins, fats, and carbohydrates are disassembled, reassembled, serve their function, and eventually become water, carbon dioxide, and urea.

There are three aspects of biochemical systems that should be kept in mind in this study. First, the contents of a living cell are in a dynamic state; the molecules are constantly being synthesized and degraded. However, the healthy living cell is characterized by a "steady state" condition in which the rates of buildup and breakdown are nearly the same at any time. Second, biochemical processes are very general in that the same basic chemistry is employed by a wide variety of cells. For example, the same types of reactions utilized to obtain energy from a compound in human beings also occur in simple unicellular organisms. Third, biochemical systems are composed of literally thousands of different kinds of molecules. Out of such an apparent chaos comes an ordered array of reactions that support all life forms. In view of this complexity, it is evident that an elementary treatment of what is known about biochemical reactions will have to deal only with selected highlights.

**Similar biochemical reactions occur in a wide variety of living cells.**

## Biochemical Energy and ATP

**ATP is an energy-rich molecule that furnishes the energy required by many biochemical processes.**

Most useful biochemical processes require energy to drive the reactions toward the desired products. This energy comes from the ultimate energy source in our environment, the sun. How is the energy of sunlight stored in various compounds, and in turn supplied to feed the life processes? Energy is stored primarily by ***adenosine triphosphate,*** often written ATP (Fig. 12–1). The energy obtained by the oxidation of foods is mostly used to synthesize ATP. ATP is the immediate source of energy in muscular contraction, and the energy released by it allows many energy-requiring biochemical reactions to occur.

Each ATP molecule contains two so-called high-energy phosphate bonds. These are marked by wiggle bonds ($\sim$) in Figure 12–1. In the presence of a suitable catalyst, ATP will undergo a three-step hydrolysis. The hydrolysis of ATP to adenosine diphosphate (ADP) and phosphoric acid releases about 12 kilocalories per mole. The second hydrolysis of ADP to adenosine monophosphate (AMP) also produces about 12 kilo-

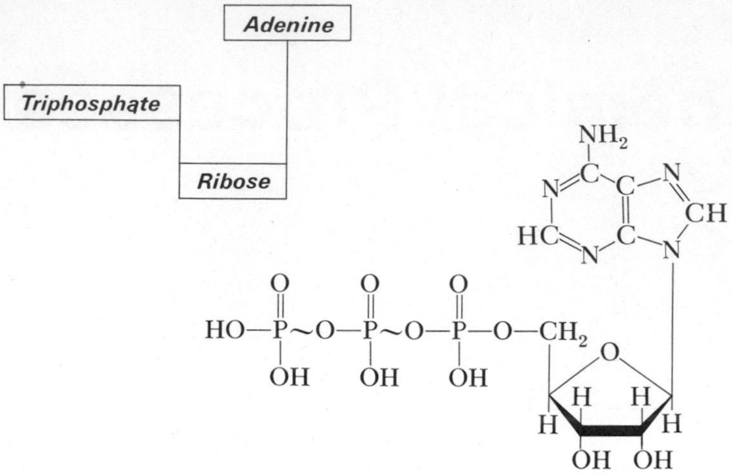


**Figure 12–1** Molecular structure of adenosine triphosphate (ATP). Note the similarity between ATP and the fundamental unit of a nucleic acid, the nucleotide (See Chapter 11).

calories of energy per mole. Finally, the hydrolysis of AMP to adenosine, which involves a low energy bond, releases only about 2.5 kilocalories per mole.

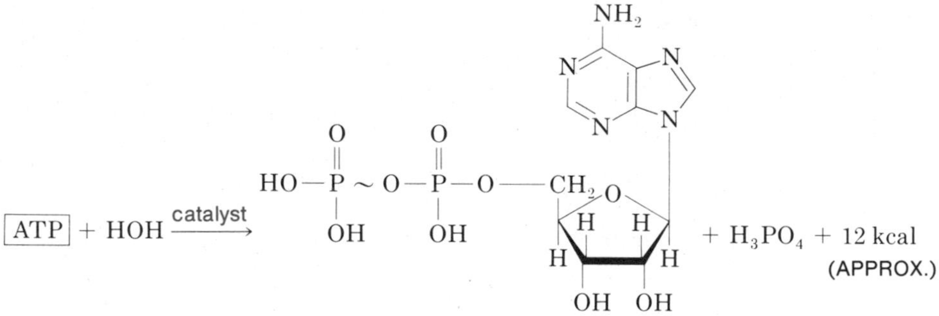


ADENOSINE DIPHOSPHATE
(ADP)

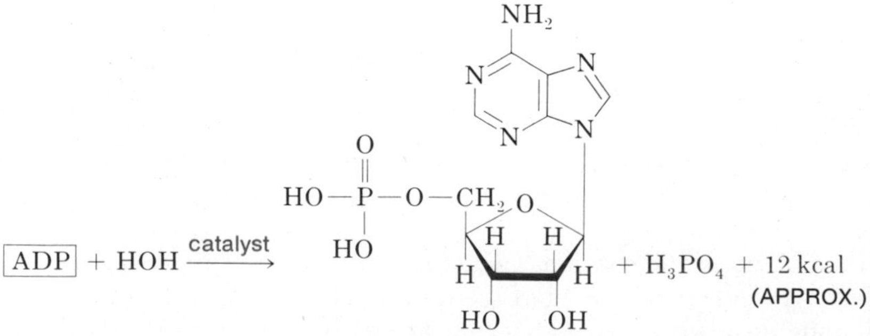


ADENOSINE MONOPHOSPHATE
(AMP)

$$\boxed{AMP} + HOH \xrightarrow{\text{catalyst}} \text{Adenosine} + H_3PO_4 + 2.5\ \text{kcal}$$

(APPROX.)

ADENOSINE

The energies for these three reactions are given only to show the relatively large amount of energy associated with the hydrolysis of the first two phosphate units in the ATP structure; however, not all of the energy released by chemical reactions is available to initiate other chemical changes. A more meaningful value is the energy that *is* available to initiate other chemical change; this is termed the change in ***free energy.*** For the three hydrolytic reactions of ATP, the approximate amounts of free energy released are:

**The change in free energy is generally represented by the symbol ΔG, and is the energy provided by a reaction to do work. The symbol G honors J. Willard Gibbs, who did much to establish the concept of free energy.**

ATP hydrolysis: 7.4 kcal per mole released
ADP hydrolysis: 6.8 kcal per mole released
AMP hydrolysis: 2.2 kcal per mole released

A visual conception of the release of free energy by hydrolysis of ATP and ADP is shown in Figure 12–2.

These free energy values indicate the amount of chemical potential energy in the phosphate bonds in the ATP molecule. Such energy-producing reactions provide the energy necessary for energy-requiring biochemical reactions to occur, and thus supply the energy necessary for the basic life processes.

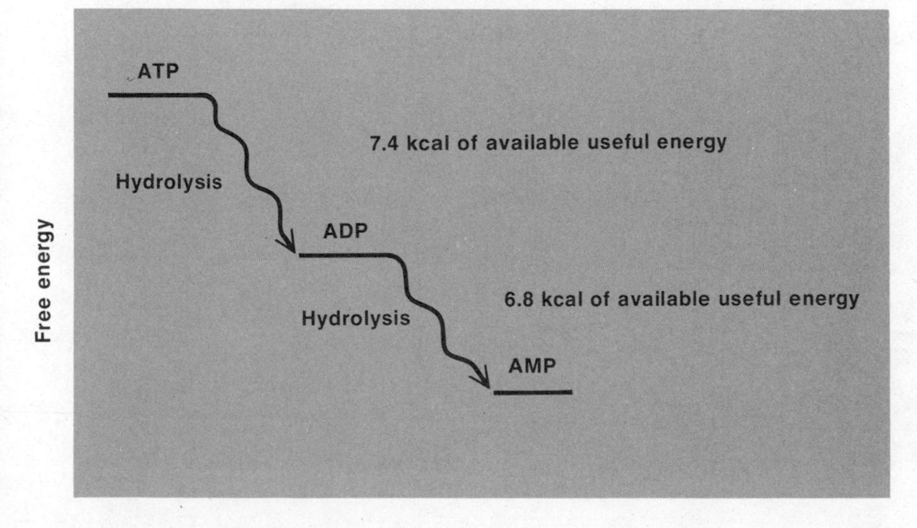


**Figure 12–2** Free energy furnished by successive hydrolyses of phosphate groups from ATP. This energy is available to bring about other useful chemical reactions.

## Photosynthesis

Photosynthesis is a very complex process that produces the relatively simple overall reaction in which carbon dioxide and water are converted into energy-rich carbohydrates by solar energy

$$\underset{\text{CARBON DIOXIDE}}{6CO_2} + \underset{\text{WATER}}{6H_2O} + \underset{\text{ENERGY (SUNLIGHT)}}{688\ \text{kcal}} \longrightarrow \underset{\text{GLUCOSE}}{C_6H_{12}O_6} + \underset{\text{OXYGEN}}{6O_2}$$

*Reduction:* gain of electrons or hydrogen.
*Oxidation:* loss of electrons or hydrogen.

In photosynthesis carbon dioxide is ***reduced*** to form a sugar

$$6CO_2 + 24H^+ + 24e^- \longrightarrow C_6H_{12}O_6 + 6H_2O$$

and water is ***oxidized*** to oxygen

$$12H_2O \longrightarrow 6O_2 + 24H^+ + 24e^-$$

**Photosynthesis involves a number of different steps and is a very complex process.**

Note that these two half reactions, the first reduction and the second oxidation, give the overall reaction when added together.

Not all the details of photosynthesis are fully understood. However, some aspects are presented here because it is the beginning of the energy flow through biochemical systems.

Photosynthesis is generally considered in terms of the ***light reaction,*** which can occur only in the presence of light energy, and the ***dark reaction,*** which can occur in the dark, but feeds on the high energy structures produced in the light reaction. Actually, both the light and dark reactions are a series of reactions, all occurring simultaneously in the green plant cell. The light reaction is unique to green plants, although the dark reaction is characteristic of both plant and some animal cells.

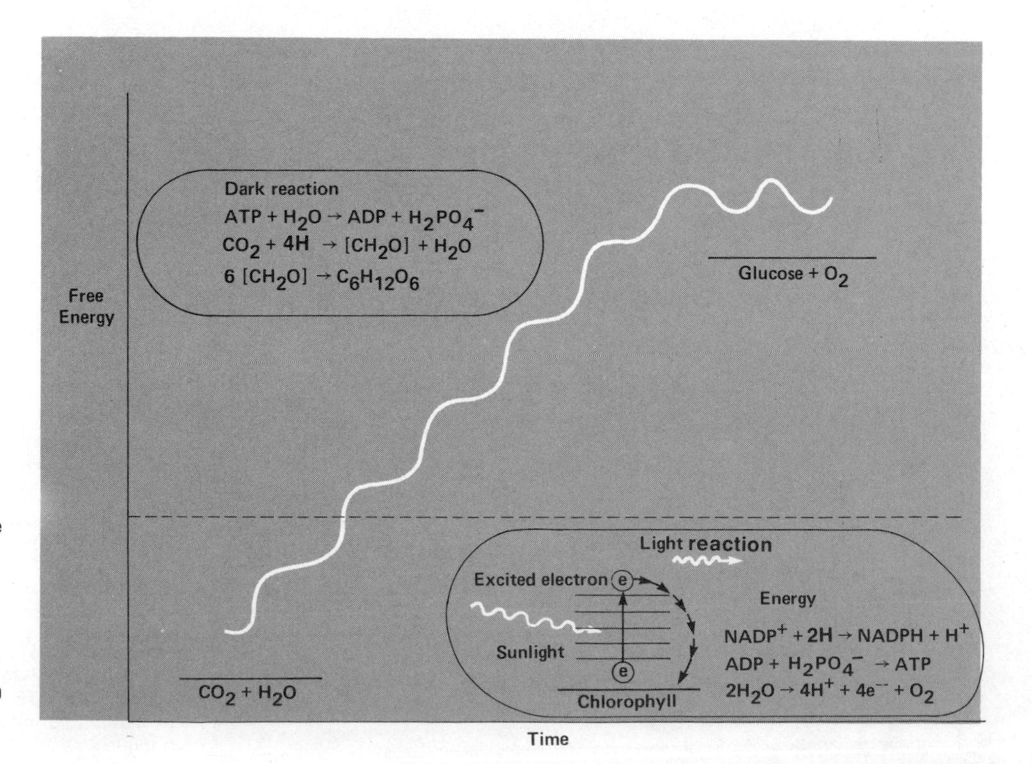


**Figure 12–3** Free energy within the chemical system is increased as carbon dioxide and water are converted into glucose and oxygen by photosynthesis. This results in stored, useful energy in glucose. A very simplified mechanism for photosynthesis is discussed in the text.

Photosynthesis is initiated by a quantum of light energy. The green plant contains certain pigments that readily absorb light in the visible region of the spectrum. The most important of these are the chlorophylls, ***chlorophyll a*** and ***chlorophyll b.*** Note that both chlorophylls are compounds of magnesium and both have complex ring systems. Such ring systems usually absorb light in the visible region of the spectrum; consequently, they are colored. For example, chlorophyll is green because it absorbs light in the violet region (about 400 nanometers) and the red region (about 650 nanometers) and allows the green light between those wavelengths to be reflected or transmitted.

**The energy of a photon is captured by chlorophyll by raising an electron to a higher energy state.**

When chlorophyll absorbs photons of light, electrons are raised to higher energy levels. As these electrons move back down to the ground state, very efficient subcellular components of the plant cell known as chloroplasts grab this energy and, through a series of steps that are not all completely known, store the energy as chemical potential energy. As shown in Figure 12–3, one of the chemicals used to store this energy is nicotinamide adenine dinucleotide phosphate ($NADP^+$).

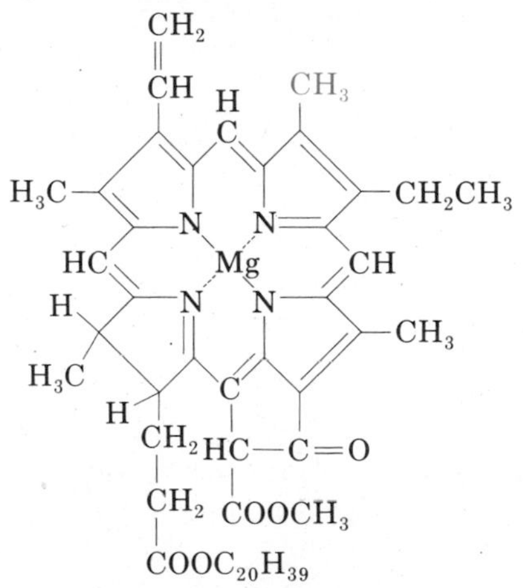

CHLOROPHYLL A

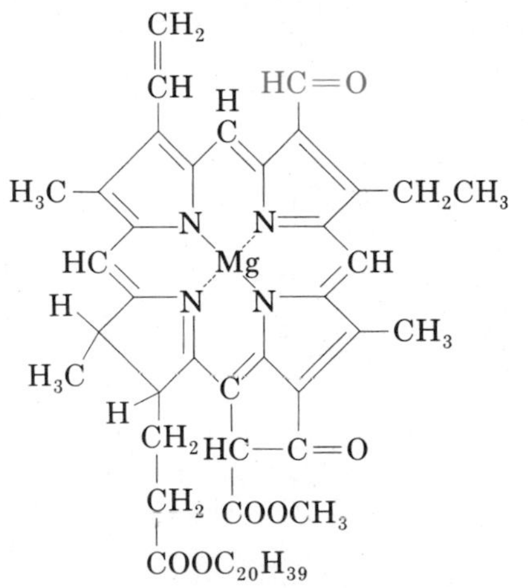

CHLOROPHYLL B

**The structure of $NADP^+$ is on p. 252.**

Energy is absorbed by $NADP^+$ in the process of being reduced by a hydrogen donor to $NADPH + H^+$.

$$NADP^+ + 2H + \text{energy} \longrightarrow NADPH + H^+$$

NADPH eventually transfers its precious energy to the universal storehouse of biochemical energy, adenosine triphosphate, ATP.

**Green plants produce oxygen by oxidizing water.**

Thus far little mention has been made of the electrochemical charge balance and the oxygen produced by photosynthesis, both of which are important parts of the light reaction. Water, in the presence of chloroplasts and light, is decomposed into oxygen, hydrogen ions, and electrons:

$$2H_2O \longrightarrow 4H^+ + 4e^- + O_2$$

The hydrogen ions and electrons are available for maintaining the balance of charge, and oxygen is liberated from the plant cell.

At this point we no longer need the energy of the sun. We have the energy necessary to run biochemical systems stored in the ATP structure. If the plant cell is given the minerals along with carbon dioxide and water, it, or subsequent living cells, can use the energy in ATP in the dark to provide energy for the complex biochemical reactions that take place.

## The Dark Reaction

The mechanisms of many reactions can be studied by the use of radioactive atoms known as *tracers*.

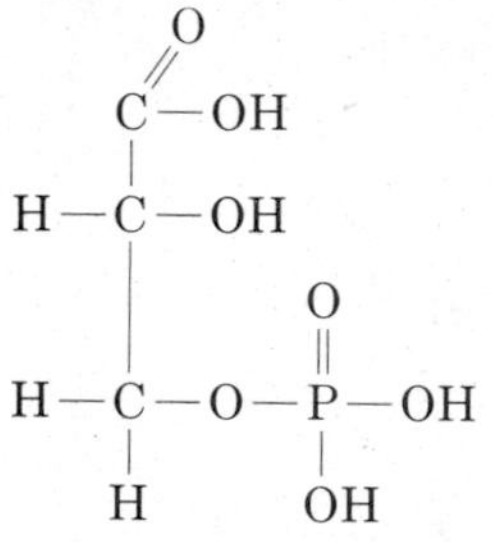

The dark reaction is responsible for the ultimate conversion of gaseous carbon dioxide to glucose. It was discovered by Melvin Calvin (1911– ; Nobel prize in 1961). Calvin studied the uptake of radioactive carbon in carbon dioxide by plant cell chloroplasts. He illuminated the plants for definite short periods of time and then analyzed the plant cells to determine which compounds contained the most radioactive carbon. As the time periods were reduced, more of the radioactive carbon was found in those compounds into which it had been initially incorporated and less in compounds that had been formed in subsequent reactions. For example, after only five seconds' illumination, radioactive carbon is found in the compound 3-phosphoglyceric acid:

```
     O
    //
   C—OH
   |
H—C—OH
   |          O
   |          ||
H—C—O—P—OH
   |          |
   H         OH
```

3-PHOSPHOGLYCERIC ACID

This compound is apparently formed in the initial reaction in which $CO_2$ from the air reacts with some molecule present in the plant. Calvin discovered that the key was the reaction of atmospheric $CO_2$ with ribulose 1,5-diphosphate to give two molecules of 3-phosphoglycerate:

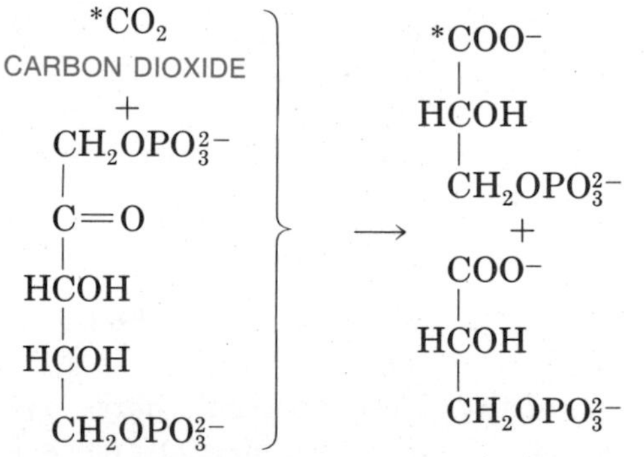

```
   *CO2                    *COO−
CARBON DIOXIDE              |
     +                     HCOH
  CH2OPO3 2−                |
   |                       CH2OPO3 2−
  C=O            ⟶           +
   |                       COO−
  HCOH                      |
   |                       HCOH
  HCOH                      |
   |                       CH2OPO3 2−
  CH2OPO3 2−
```

RIBULOSE 1,5-DIPHOSPHATE 3-PHOSPHOGLYCERATE

*C indicates radioactive carbon (Calvin's experiments)

The 3-phosphoglycerate is then transformed into other carbohydrates in reactions that produce glucose and regenerate ribulose 1,5-diphosphate (for further uptake of more atmospheric $CO_2$). The energy needed to carry out these reactions is furnished by the NADPH and ATP generated from the light reaction.

Figure 12–4 presents the cyclic character of the dark reaction of photosynthesis. Note that carbon dioxide enters the cycle at the upper left and that sugars are removed at the lower right. Since just one carbon atom enters the cycle per revolution, only one fructose 6-phosphate molecule out of seven is converted to glucose.

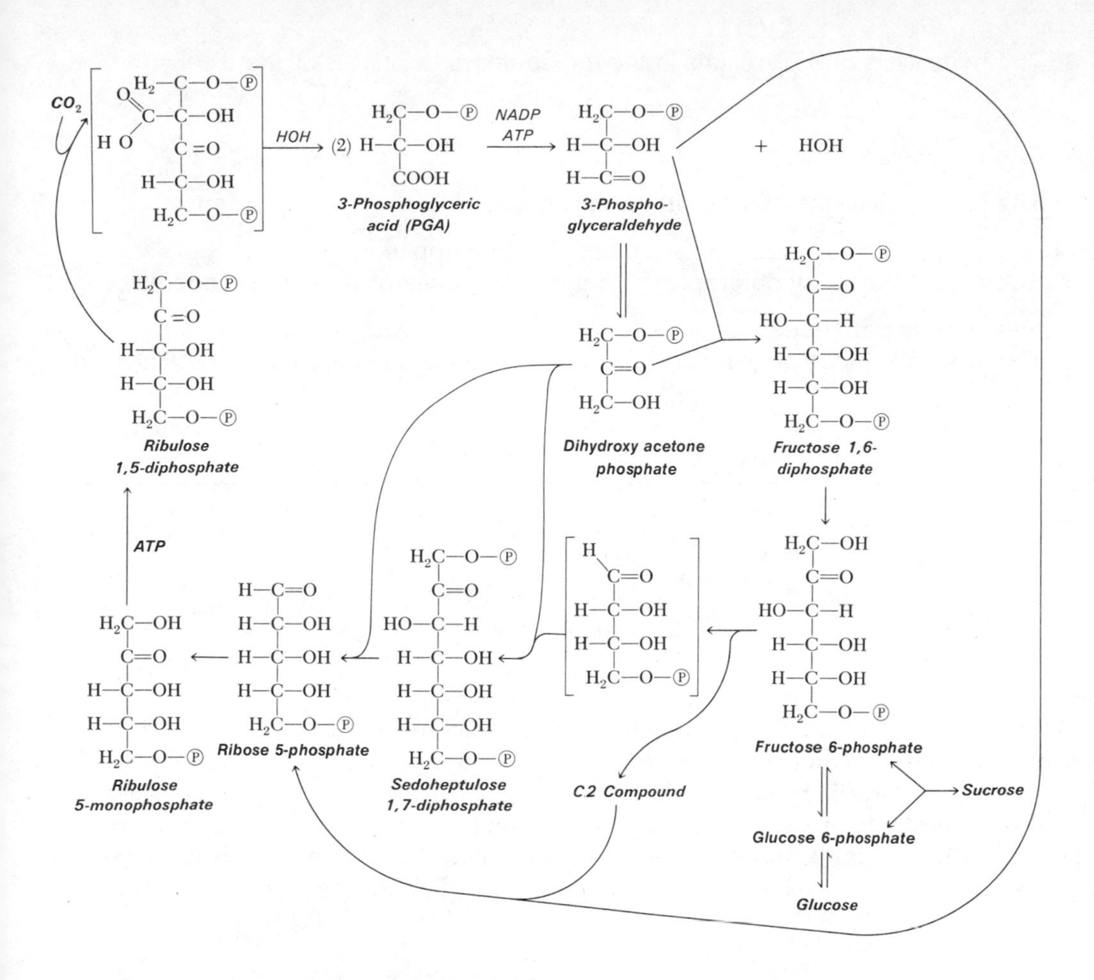


**Figure 12–4** Abbreviated version of the dark reaction of photosynthesis. The energy needed to carry out the dark reactions is furnished from the high energy compounds produced in the light reaction. Ⓟ indicates a phosphate group.

When the light and dark reactions are considered, one can readily see that the net equation for photosynthesis, given at the beginning of this section, is a tremendous simplification of the actual process. According to the latest research, the light reaction may be summarized as follows:

Oxygen and ATP are the products of the light reaction.

$$12H_2O + 12NADP^+ + 24ADP + 24H_3PO_4 \longrightarrow 24ATP + 12H^+ + 12NADPH + 6O_2 + 24H_2O$$

Water has been placed on both sides of the equation because water is part of the input of the reaction (each $NADP^+$ requires one $H_2O$ for reaction) and is part of the output of the reaction (the formation of each ATP splits out one $H_2O$). The coefficients of the equation are in the correct ratios to give $6O_2$, the production of which provides the electrons to form one glucose molecule ($C_6H_{12}O_6$) in the dark reaction. The dark reaction (carbon fixation) may be represented as follows:

Glucose is a product of the dark reaction.

$$6CO_2 + 18ATP + 12NADPH + 12H^+ + 12H_2O \longrightarrow \underset{\text{GLUCOSE}}{C_6H_{12}O_6} + 18ADP + 12NADP^+ + 18H_3PO_4$$

## Self-Test 12-A

1. The source of energy for photosynthesis is ______________.
2. Most of the energy obtained by food oxidation is used immediately to synthesize the molecule ______________.

3. The hydrolysis of ATP results in the molecules ____________ and ____________. The other "product" is ____________.
4. Energy available to do work is called ____________ energy.
5. The reactants in the photosynthesis process are ____________ and ____________; ____________ must also be supplied.
6. Energy absorbed by chloroplasts in the green cells of a plant is transferred by means of the molecule ____________ to ATP.
7. The first chemical product of the dark reaction, as determined by Calvin, is ____________.
8. The final product of the dark reaction is the substance ____________.

## Digestion

**Digestion is the hydrolysis of carbohydrates, fats, and proteins to provide small molecules that can be absorbed.**

From a chemical point of view, digestion is the breakdown of ingested foods through hydrolysis. The products of these hydrolytic reactions are relatively small molecules that can be absorbed through the intestinal walls into the body fluids where they are used for metabolic processes. The hydrolytic reactions of digestion are catalyzed by enzymes, there being a specific enzyme for each hydrolysis. The hydrolysis of carbohydrates ultimately yields simple sugars, proteins yield amino acids, and fats yield

**TABLE 12–1 Principal Digestive Enzymes**

| Enzyme | Source | Substrate | Products | Optimal pH |
|---|---|---|---|---|
| Ptyalin | salivary glands | starch | smaller carbohydrate polymers (minor physiological role) | 6–7 |
| Pepsin | chief cells of stomach | protein | polypeptides | 1.6–2.4 |
| Gastric lipase | stomach | fat | glycerides, fatty acids (minor physiological role) | – |
| Enterokinase | duodenal mucosa | trypsinogen | trypsin | – |
| Trypsin | exocrine pancreas | denatured proteins and polypeptides | small polypeptides (also activates chymotrypsinogen to chymotrypsin) | 8.0 |
| Chymotrypsin | exocrine pancreas | proteins and polypeptides | small polypeptides | 8.0 |
| Nucleases | exocrine pancreas | nucleic acids | nucleotides | – |
| Carboxypeptidases | exocrine pancreas | polypeptides | smaller polypeptides* | – |
| Pancreatic lipase | exocrine pancreas | fat | glycerides, fatty acids, glycerol | 8.0 |
| Pancreatic amylase | exocrine pancreas | starch | maltose units | 6.7–7.0 |
| Aminopeptidases | intestinal glands | polypeptides | smaller polypeptides† | 8.0 |
| Dipeptidase | intestinal glands | dipeptide | amino acids | – |
| Maltase | intestinal glands | maltose | hexoses | 5.0–7.0 |
| Lactase | intestinal glands | lactose | (glucose, galactose, and fructose) | 5.8–6.2 |
| Sucrase | intestinal glands | sucrose | | 5.0–7.0 |
| Nucleotidase | intestinal glands | nucleotides | nucleosides, phosphoric acid | – |
| Nucleosidase | intestinal glands | nucleosides | purine or pyrimidine base, pentose | – |
| Intestinal lipase | intestinal glands | fat | glycerides, fatty acids, and glycerol | 8.0 |

*Removal of C-terminal amino acid.
†Removal of N-terminal amino acid.

fatty acids. The activities of some of the more important enzymes are described in Table 12–1. Note that the pH of the stomach fluid is much lower than the pH of the intestines.

### Carbohydrate Digestion and Absorption

The principal forms of carbohydrates in our food are (1) high molecular weight polymers such as starch and glycogen, (2) disaccharides such as sucrose and lactose, and (3) simple sugars such as glucose and fructose. The first enzyme capable of acting on ingested food is furnished by the saliva and is named salivary amylase, or ptyalin. Its action on starch or glycogen produces limited amounts of the disaccharide maltose. Ptyalin is inactivated by the high acidity in the stomach. The stomach furnishes no enzymes that can catalyze the splitting of carbohydrate polymers. Indeed, the high acidity of the gastric juice would destroy or inactivate all such complex structures.

**Acidic solutions: pH below 7**
**Basic solutions: pH above 7**

When food passes from the stomach, the acidity is neutralized by a secretion of the pancreas. The enzymes secreted by the pancreas can split some of the polysaccharides to maltose and maltose to glucose, and can catalyze other hydrolytic reactions. The final result is a mixture of simple sugars such as glucose, fructose, and galactose. These simple sugars are then absorbed into the bloodstream, where the concentration of blood sugar is regulated by the hormone insulin. If the sugar level is too high, the simple sugars are converted into the polysaccharide glycogen in the liver; if the blood sugar level is too low, the stored glycogen is hydrolyzed to raise the level. Malfunctions of our biological systems can lead to too much blood sugar (hyperglycemia) or too little blood sugar (hypoglycemia); either condition, if sustained, indicates one type or another of diabetes.

**Human blood normally contains between 0.08 and 0.1% glucose.**

**Insulin is a protein.**

### Fat and Oil Digestion and Absorption

The term ***lipid*** denotes a group of compounds that includes fats, oils, and other substances whose solubility characteristics are similar to those of fats and oils. These compounds are not all structurally related, and we shall consider only the triglycerides (triesters of fatty acids and glycerol) in this chapter. A typical triglyceride is palmitooleostearin. Its structure and its hydrolytic products are shown in the following equation:

$$\begin{array}{l} H_2C-O-\overset{O}{\overset{\|}{C}}-(CH_2)_{14}CH_3 \\ \;|\\ HC-O-\overset{O}{\overset{\|}{C}}-(CH_2)_7CH=CH(CH_2)_7CH_3 \\ \;| \\ H_2C-O-\overset{O}{\overset{\|}{C}}-(CH_2)_{16}CH_3 \end{array} + \underset{\text{WATER}}{3HOH} \longrightarrow \begin{array}{l} \underset{\text{PALMITIC ACID}}{CH_3(CH_2)_{14}COOH} \\ \underset{\text{OLEIC ACID}}{CH_3(CH_2)_7CH=CH(CH_2)_7COOH} \\ \underset{\text{STEARIC ACID}}{CH_3(CH_2)_{16}COOH} \end{array} + \underset{\text{GLYCEROL}}{\begin{array}{l} HO-CH_2 \\ \quad\;\; | \\ HO-CH \\ \quad\;\; | \\ HO-CH_2 \end{array}}$$

PALMITOOLEOSTEARIN
(A TRIGLYCERIDE)

Digestion of fats and oils occurs primarily in the intestinal tract where bile salts secreted by the liver aid in the process. The enzyme that aids in the hydrolysis of the fatty acid esters is water-soluble, whereas the fats and oils are insoluble in water. Bile salts emulsify the oil, that is, they break up the oil into very tiny drops and prevent the drops from recombining readily. The tiny drops provide more surface area for the enzyme to attack so digestion can occur. The bile salts form an interface between the nonpolar oil and the polar water and make it possible for the oil to "dissolve" in water. For a molecule to be an emulsifier between polar and nonpolar molecules, the emul-

**Bile salts act much like detergent molecules.**

sifier must have characteristics of both. One of the principal bile salts is derived from glycocholic acid.

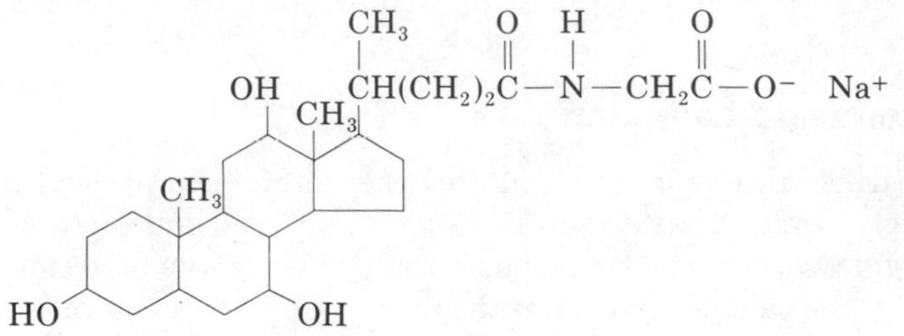

$$\text{(steroid nucleus with HO, OH, OH, } CH_3\text{, } CH_3\text{)}-\overset{CH_3}{\overset{|}{C}H}(CH_2)_2\overset{O}{\overset{\|}{C}}-\overset{H}{\overset{|}{N}}-CH_2\overset{O}{\overset{\|}{C}}-O^- \ Na^+$$

SODIUM SALT OF GLYCOCHOLIC ACID

Notice that the bulky hydrocarbon groups of this bile salt are compatible with oil or fat and that the —OH and ionic groups anchor to water molecules. The bile salts emulsify oil in a manner similar to the action of a soap or detergent during the cleaning process (Chapter 9).

## Protein Digestion and Absorption

The hydrolysis of proteins begins in the stomach and continues in the small intestine. Several different types of enzymes are known to be involved. These enzymatic systems must be controlled very carefully, for they have the potential of digesting the walls of the stomach and intestines. A number of these enzymes are secreted in an inactive form. For example, pepsin, which is secreted in the stomach, is first present in a form called pepsinogen. The molecular weight of pepsinogen is 42,600. In the presence of the acid of the stomach, pepsinogen is broken down to ***pepsin.*** The molecular weight of pepsin is 34,500. It is reasonable to believe that pepsinogen, pepsin, other enzymes, and the stomach acid normally have no effect on the stomach wall. However, pepsin would have considerable action on the stomach protein if it were formed under the mucous lining, a lining which is constantly sloughing off like the outer skin.

**The stomach is protected from protein-splitting enzymes by a mucous lining.**

Pepsin facilitates the breakdown of only about 10% of the bonds in a typical protein, leaving polypeptides with molecular weights from 600 up to 3000. In the small intestine, hydrolysis is completed to amino acids, which are absorbed through the intestinal wall.

Some protein enzymes are sold commercially. Meat tenderizers are proteases, materials that speed up partial digestion of meat. Enzymes are used as stain removers in detergents, although they may irritate the skin of some individuals. Related enzymes are also used to free the lens of the eye prior to cataract surgery.

## Enzymes and Heredity

Genetic effects are often observed in the pattern of enzymes produced by individuals or races. An example of this is found in "lactose intolerance," common in certain peoples of Asia (e.g., Chinese and Japanese) and Africa (many black tribes), whose diets have traditionally contained little milk after the age of weaning. While infants, such people manufacture the enzyme ***lactase*** that is necessary to digest lactose, a sugar occurring in all mammals' milk. As they grow older their bodies stop producing this enzyme because their diets normally contain no milk, and the ingestion of milk products containing lactose can lead to considerable discomfort in the form of stomach aches and diarrhea. People whose ancestral adult diets contained substantial amounts of milk or milk products (African tribes such as the Masai, Mongols, Caucasians, etc.)

continue to produce lactase as adults and can eat such foods and digest the lactose they contain. It is quite possible that this is only one of several similar cases in which a traditional tribal diet has altered the pattern of production of digestive enzymes.

## Glucose Metabolism

The biochemical process by which living cells obtain energy from glucose and similar compounds involves many chemical reactions. The initial sequence of reactions can follow two courses; one of these does not use oxygen (***anaerobic***) but the other one does (***aerobic***).

*Aerobic:* use oxygen. *Anaerobic:* use no oxygen.

When a muscle is used, glucose is anaerobically converted by a series of steps known as the ***Embden-Meyerhof pathway*** to lactic acid (Fig. 12–5). We are con-

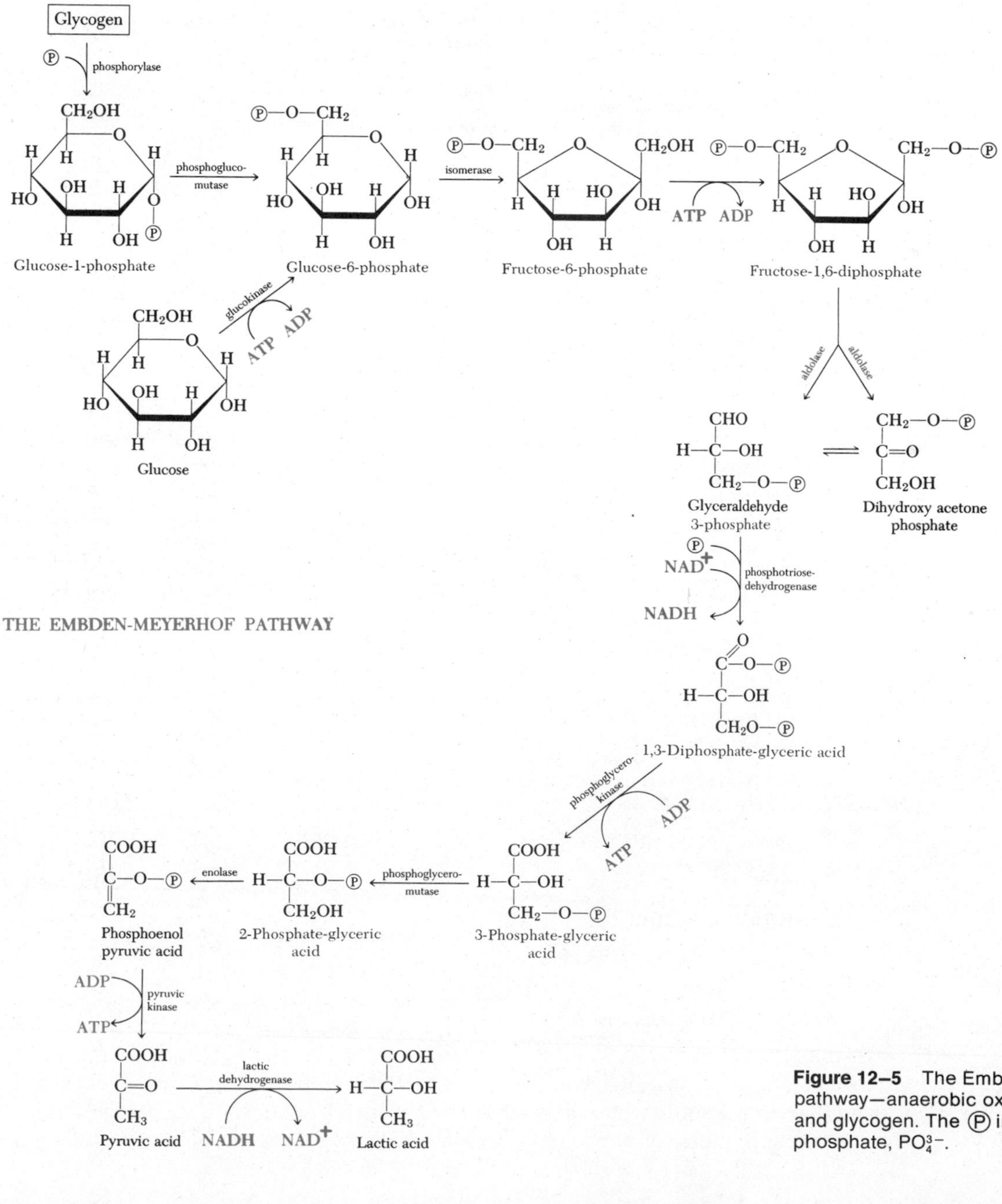

THE EMBDEN-MEYERHOF PATHWAY

**Figure 12–5** The Embden-Meyerhof pathway—anaerobic oxidation of glucose and glycogen. The Ⓟ indicates inorganic phosphate, $PO_4^{3-}$.

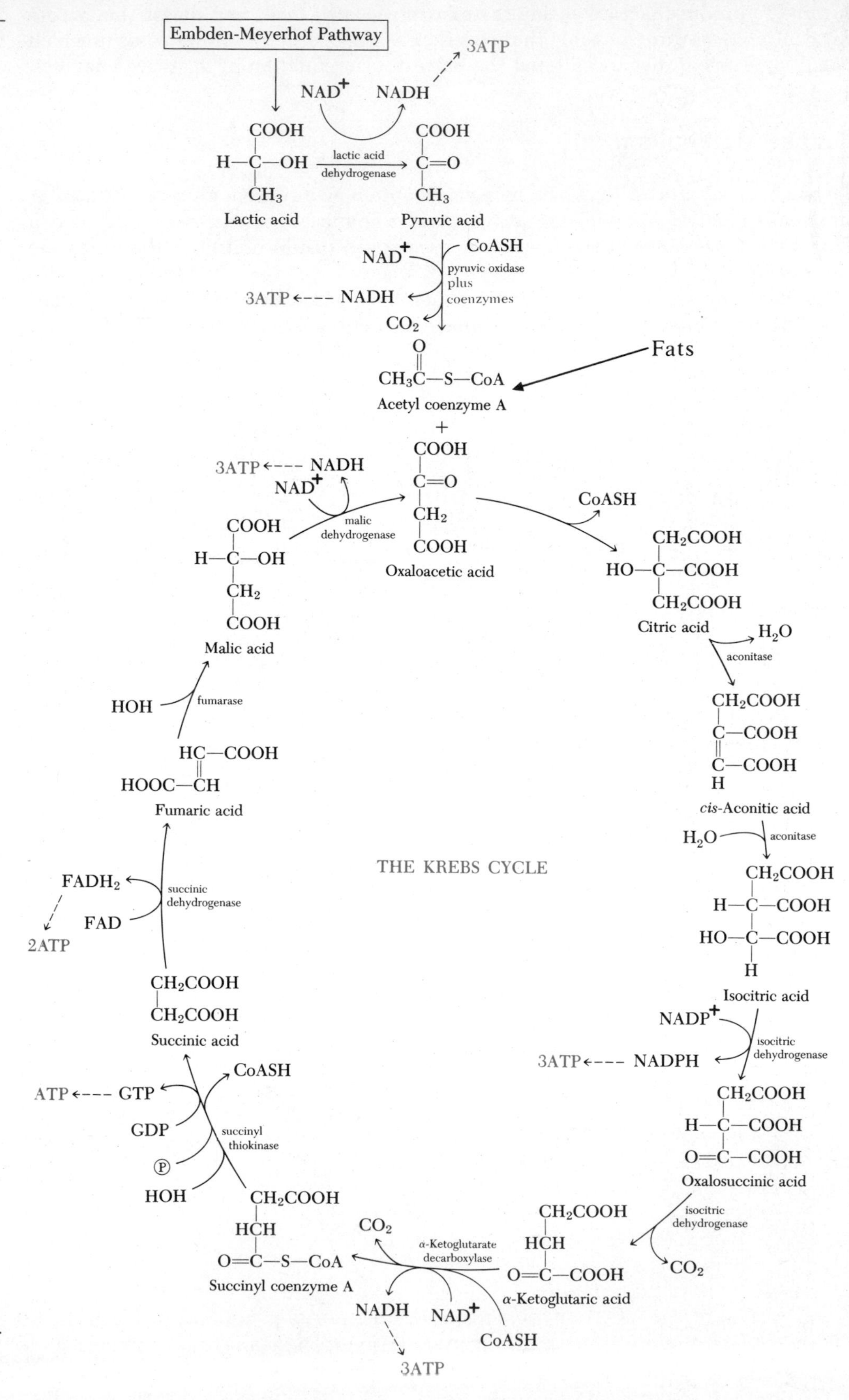


**Figure 12–6** The Krebs cycle. Derivatives of fats enter the cycle as acetylcoenzyme A. Derivatives of proteins enter at various points depending upon the specific amino acid.

cerned here only with the starting material, the energy flow, and the final products. The overall reaction can be represented by the equation:



$$\underset{\text{GLUCOSE}}{C_6H_{12}O_6} + 2ADP + 2H_3PO_4 \longrightarrow \underset{\text{LACTIC ACID}}{2CH_3-\overset{\displaystyle H}{\overset{|}{\underset{\underset{\displaystyle OH}{|}}{C}}}-\overset{\displaystyle O}{\overset{\|}{C}}-OH} + 2ATP + 2H_2O$$

If the muscle is used strenuously for a sufficiently long period of time, the lactic acid buildup will produce tiredness and a painful sensation in the muscles. The bloodstream eventually carries away the lactic acid, but time and oxygen are needed to convert the lactic acid to carbon dioxide and water, which are excreted. This is accomplished by the slower, aerobic (with air) process known as the ***Krebs cycle.*** As you look at Figure 12–6, imagine how a molecule of citric acid reacts with the enzyme aconitase. The citric acid has —OH and —H exchange sites which exchange and produce isocitric acid. The isocitric acid dissociates from aconitase and goes to the next enzyme, isocitric dehydrogenase, where two hydrogen atoms are removed by $NADP^+$, and so on around the cyclic pathway. Hydrogen atoms are removed, carboxyl groups are destroyed as carbon dioxide is formed, and hydrolysis occurs. Since two carbon atoms are fed into the cycle at oxaloacetic acid by acetyl coenzyme A, two carbon atoms in the form of two molecules of carbon dioxide ($CO_2$) must be eliminated before the cycle returns to oxaloacetic acid. This is exactly what happens. See whether you can find the two molecules of $CO_2$ formed in the Krebs cycle. Figure 12–7 will help.

**Muscular activity converts glucose to lactic acid, which produces fatigue in muscles as the lactic acid accumulates.**

If the Krebs cycle is aerobic, where does the oxygen enter the cycle? Oxygen is required to regenerate the coenzyme $NADP^+$ from NADPH. It is also required to remove the hydrogen atoms from NADH and $FADH_2$. There are about seven known steps involved in the removal of hydrogen from NADH and NADPH and six steps for $FADH_2$. The steps can be summarized by the following equations:

$$NADPH + 3ADP + 3H_3PO_4 + H^+ + \tfrac{1}{2}O_2 \longrightarrow NADP^+ + 3ATP + 4H_2O$$

$$NADH + 3ADP + 3H_3PO_4 + H^+ + \tfrac{1}{2}O_2 \longrightarrow NAD^+ + 3ATP + 4H_2O$$

$$FADH_2 + 2ADP + 2H_3PO_4 + \tfrac{1}{2}O_2 \longrightarrow FAD + 2ATP + 3H_2O$$

The $NADP^+$, $NAD^+$, and FAD are now ready to be fed back into the Krebs cycle, where the process is continued by extracting other hydrogen atoms.

When these equations are combined with the principal sequence of reactions for the Krebs cycle, we have an equation for the conversion of lactic acid into carbon dioxide and water aerobically.

$$\underset{\text{LACTIC ACID}}{C_3H_6O_3} + 18ADP + 18H_3PO_4 + 3O_2 \longrightarrow 3CO_2 + 21H_2O + 18ATP$$

The key product of this oxidation reaction is ATP. Energy derived from glucose (and lactic acid) is stored and transported in the phosphorus-oxygen bonds of ATP. This is the usable product of the Krebs cycle. Waste products are carbon dioxide and water. Since two lactic acid molecules are formed from one glucose molecule, 36 ATP's are formed per glucose molecule oxidized via the Krebs cycle. Two more ATP's are formed in the Embden-Meyerhof pathway, making a total of 38 ATP molecules formed in the complete oxidation of a glucose molecule. If we burn a mole of glucose to carbon dioxide and water outside the body, 688,000 calories of energy are released. This is the total energy available. Earlier in this chapter, the point was made that breaking a mole of P—O—P bonds and forming $H_2PO_4^-$ gives up about 7,400 calories of free energy

**The key products of glucose metabolism are ATP (energy), $CO_2$, and $H_2O$.**

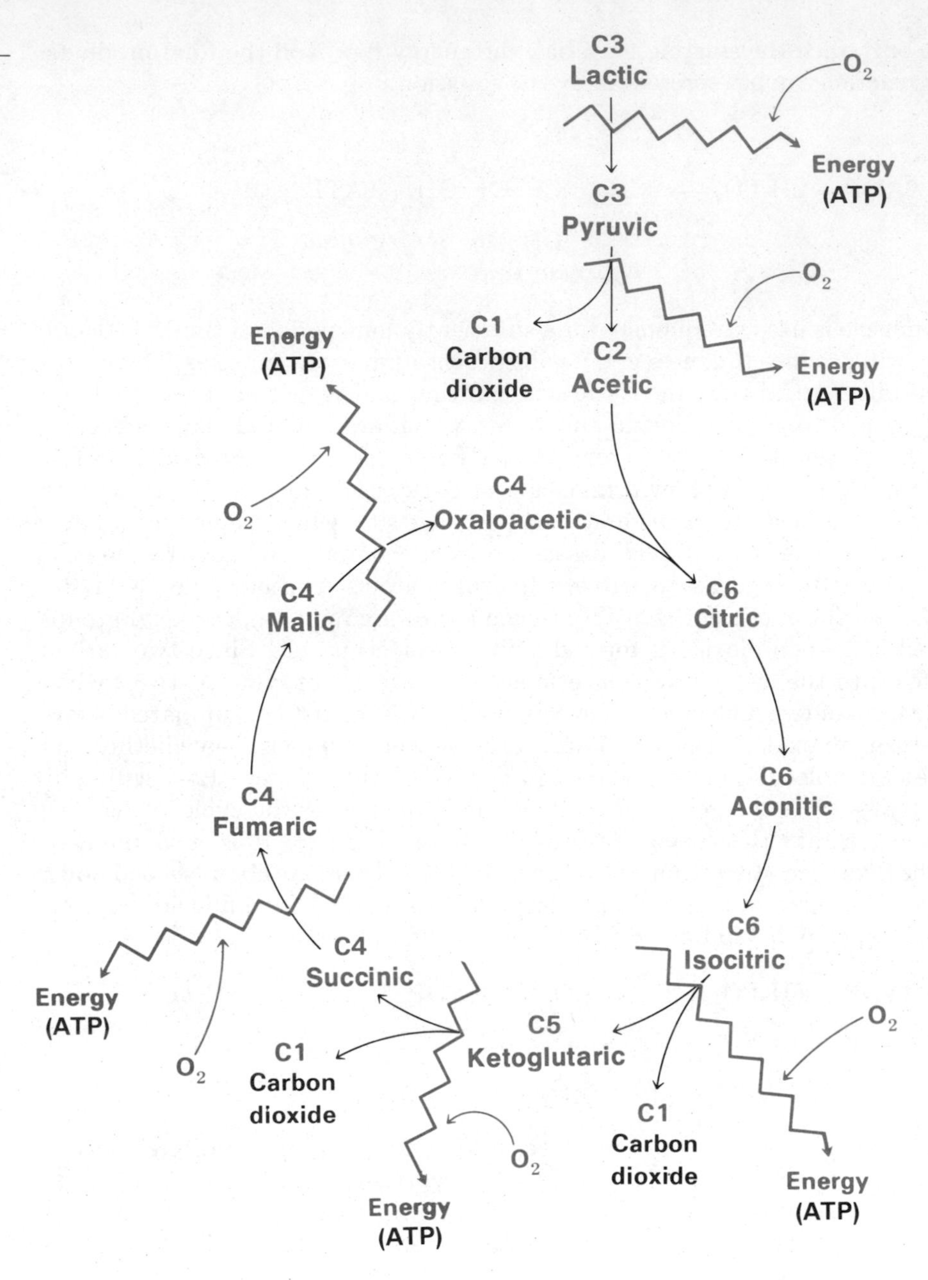


**Figure 12–7** A summary of the Krebs cycle showing the number of carbon atoms per molecule for the principal substances in the cycle.

(ΔG). Thirty-eight moles of ATP molecules would provide 281,200 calories of free energy.

$$38 \cancel{\text{moles ATP}} \times 7{,}400 \frac{\text{calories}}{\cancel{\text{mole ATP}}} = 281{,}200 \text{ calories}$$

This is an efficiency of 41% for obtaining stored energy from the total usable energy available.

$$\frac{281{,}200 \cancel{\text{calories}} \times 100\%}{688{,}000 \cancel{\text{calories}}} = 41\%$$

This figure is remarkable when you consider that the efficiency of the automobile engine is only about 20%, and real heat engines of any size seldom go above 35%.

When ATP converts back to ADP by losing a phosphate group, the energy is used to move muscles, such as those that provide the pumping action of the heart, move the diaphragm so we can breathe, or produce hundreds of other movements required for everyday life.

The energy is stored in ATP and released from ATP by means of ***coupled reactions.*** In Figures 12–6 and 12–7, curved arrows denote reactions in which sufficient free energy is provided by one reaction to drive another reaction. These are examples of coupled reactions. In fact, the first step in the oxidation of glucose is a specific example of a coupled reaction. The substitution of a phosphate group for a hydrogen on glucose is an unfavorable reaction, as far as energy is concerned.

**Coupled reactions allow the energy from one reaction to be used to run another reaction.**

(UNFAVORABLE—REQUIRES ENERGY)

$$\text{Glucose} + H_3PO_4 + \text{about 3,000 cal of free energy per mole of glucose} \rightleftharpoons \text{Glucose-6-}PO_4 + H_2O$$

If this reaction is coupled with the favorable hydrolysis of ATP,

(FAVORABLE—RELEASES ENERGY)

$$\text{ATP} + H_2O \rightleftharpoons \text{ADP} + H_3PO_4 + \text{about 7,400 cal of free energy per mole of ATP}$$

the net reaction is favorable:

$$\begin{array}{lll} \text{Glucose} + \cancel{H_3PO_4} + \text{3,000 cal of free energy} & \rightleftharpoons & \text{Glucose-6-}PO_4 + \cancel{H_2O} \\ \text{ATP} + \cancel{H_2O} & \rightleftharpoons & \text{ADP} + \cancel{H_3PO_4} + \text{7,400 cal of free energy} \\ \hline \text{Glucose} + \text{ATP} & \rightleftharpoons & \text{Glucose-6-}PO_4 + \text{ADP} + \text{4,400 cal of free energy, net} \end{array}$$

Even with this favorable free energy change, the enzyme glucokinase is required to hasten the process.

**Reactions that are favored energetically may still require a catalyst to proceed at a reasonable rate.**

Fats and proteins (as amino acids) can enter the Krebs cycle also. Amino acids can enter at several places after first being converted by a series of reactions into one of the compounds in the cycle. Each amino acid has its particular point of entry. For example, aspartic acid is converted into $\alpha$-ketoglutaric acid and enters at that point in the Krebs cycle.

Fatty acids go through a series of at least five reactions in which the hydrocarbon chain of the fatty acid is decreased by a two-carbon fragment. The fragment completes a molecule of acetyl coenzyme A, which is important in the Krebs cycle (Fig. 12–6). Palmitic acid, a 16-carbon fatty acid (see Chapter 7), would do seven turns around the cycle to form eight acetyl CoA molecules.

In conclusion, you now have seen some of the detailed chemistry involved in simply raising your arm, and you are now aware of what happens to some of the sugars and starches that disappear down the hatch. Of course, there is much more known than is presented here and there appears to be no end to what is left to be discovered.

## Self-Test 12-B

1. Digestion is the breakdown of foodstuffs by ______________.
2. Substances whose solubility characteristics are similar to those of fats and oils are termed ______________.
3. Bile salts act as (catalysts/emulsifying agents/enzymes).

4. The two products of the Embden-Meyerhof pathway are ______________ and ______________.

5. The end products of the Krebs energy cycle are ______________ and ______________ ______________.

6. Energy is stored and released from ATP by chemical reactions called ______________ ______________ reactions.

## Synthesis of Living Systems

**Our current knowledge is far from sufficient to allow the synthesis of a living cell.**

In the quest for a molecular understanding of living systems, theories must finally be put to the ultimate test of synthesis. If some of the complex biochemical substances can be synthesized and can successfully participate in the life processes, we can be reassured that we are on the right track. It should be emphasized that the biochemist is currently working at the molecular level, and as yet only a relatively few of the giant

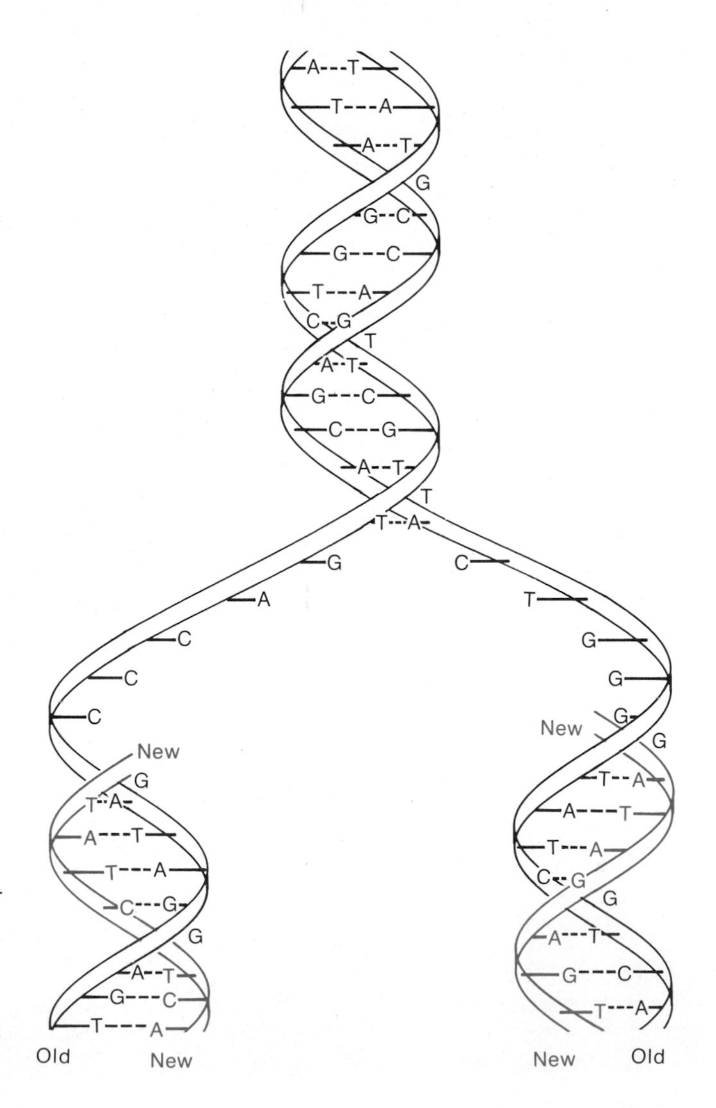


**Figure 12–8** Replication of DNA structure. When the double helix of DNA (*black*) unwinds, each half serves as a template on which to assemble subunits (*color*) from the cell environment.

molecules have been characterized. Most of the syntheses in a living cell are far too complex for our present methods to duplicate. However, in spite of the enormousness of this undertaking, remarkable strides have been made recently, and interest in current research in this area is intense.

## Replication of DNA

Almost all of the cells of one organism contain the same chromosome structure in their nuclei. This structure remains constant regardless of whether the cell is starving or has an ample supply of food materials. Each organism begins life as a single cell with this same chromosome structure; in sexual reproduction one half of this structure comes from each parent. These well-known biological facts, along with recent discoveries concerning polynucleotide structures, lead to the conclusion that the DNA structure is faithfully copied during normal cell division (***mitosis***—both strands) and only one half is copied in cell division producing reproductive cells (***meiosis***—only one strand).

**The DNA molecule is capable of causing the synthesis of its duplicate.**

**To make a replicate is to make a complement (something that fits) of the original.**

A prominent theory of DNA replication, based on verifiable experimental facts, suggests that the double helix of the DNA structure unwinds and each half of the structure serves as a template or pattern to reproduce the other half from the molecules in the cell environment (Fig. 12–8). Replication of the DNA occurs in the nucleus of the cell. (The components of a typical cell are illustrated in Fig. 11–19 on p. 255.)

**Replicates:**
**object—cast of object**
**screw—hole for screw**
**bolt—bolt hole**
**adenine—thymine**

## Natural Protein Synthesis

The proteins of the body are being replaced and resynthesized continuously from the amino acids available to the body. The amino acids and proteins in the body can be considered as constituents of a "nitrogen pool"; additions to and losses from the pool are shown in Figure 12–9.

The use of isotopically labeled amino acids has made possible studies of the average lifetimes of amino acids as constituents in proteins—that is, the time it takes the body to replace a protein in a tissue. For a process that must be extremely complex, replacement is very rapid. Only minutes after radioactive amino acids are injected into animals, radioactive protein can be found. Although all the proteins in the body are continually being replaced, the rates of replacement vary. Half of the proteins in the liver and plasma are replaced in *six days.* The time is longer for muscle proteins, about 180 days, and replacement of protein in other tissues, such as bone collagen, takes even longer.

**The proteins in the human body are continuously being replaced.**

Recall that each organism has its own kinds of proteins. The number of possible arrangements of 20 amino acid units is $2.43 \times 10^{18}$, yet proteins characteristic of a given organism can be synthesized by the organism in a matter of a few minutes.

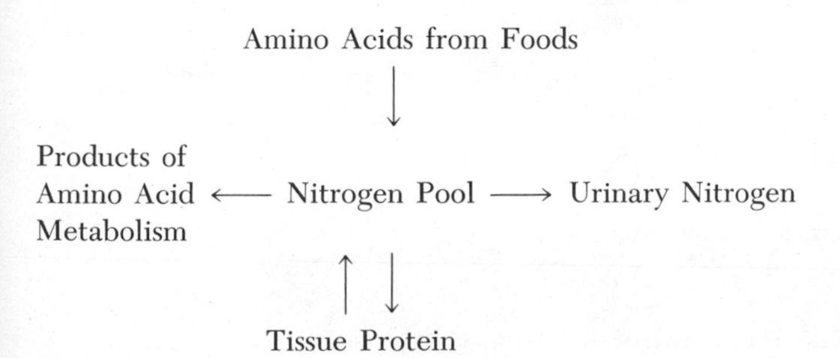


**Figure 12–9** The nitrogen pool.

The DNA in the cell nucleus holds the code for protein synthesis. Messenger RNA, like all forms of RNA, is synthesized in the cell nucleus. The sequence of bases in one strand of the chromosomal DNA serves as the template for monoribonucleotides to order themselves into a single strand of mRNA (Fig. 12–10). The bases of the mRNA strand complement those of the DNA strand. A pair of complementary bases is so structured that each one fits the other and forms one or more hydrogen bonds. Messenger RNA contains only the four bases adenine (A), guanine (G), cytosine (C), and uracil (U). DNA contains principally the four bases adenine (A), guanine (G), cytosine (C), and thymine (T). The base pairs are as follows:

**The DNA molecule tells the cell what kind of protein to synthesize.**

| DNA | mRNA |
|---|---|
| A | U |
| G | C |
| C | G |
| T | A |

This means that every place a DNA has an adenine base (A), the mRNA will transcribe a uracil base (U), and so on, provided the necessary enzymes and energy are present.

After transcription, mRNA passes from the nucleus of the cell to a ribosome, where it serves as the template for the sequential ordering of amino acids during protein synthesis. As its name implies, messenger RNA contains the sequence message, in the form of a three-base code, for ordering amino acids into proteins. Each of the thousands of different proteins synthesized by cells is coded by a specific mRNA or segment of an mRNA molecule.

**Transfer RNA carries specific amino acids to the messenger RNA.**

Transfer RNA's carry the specific amino acids to the messenger RNA. Each of the 20 amino acids found in proteins has at least one corresponding tRNA, and some

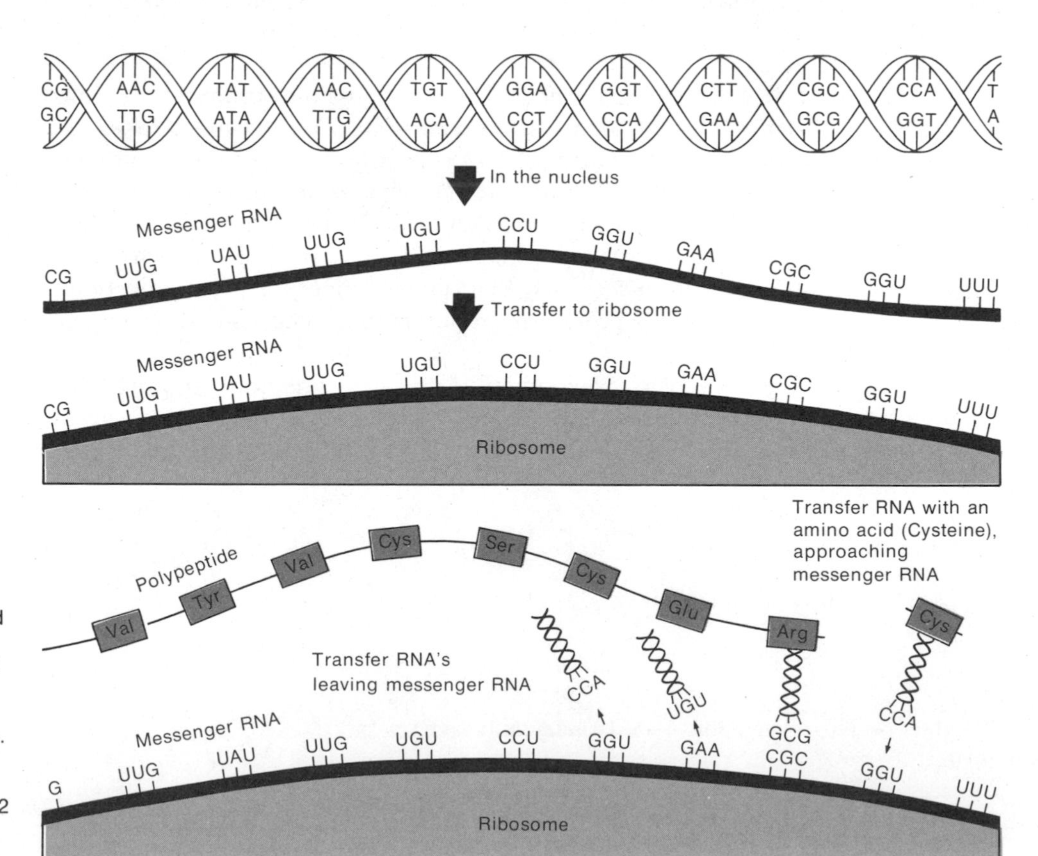


**Figure 12–10** A schematic illustration of the role of DNA and RNA in protein synthesis. A, C, G, T, and U are nitrogen bases characteristic of the individual nucleotides. See Figure 11–18 for structures of the bases, and Table 12–2 for abbreviations of the amino acids used.

**TABLE 12–2 Messenger RNA Codes for Amino Acids***

| Amino Acid | Shortened Notation Used for Amino Acids In Figs. 12–10 and 12–12 | Base Code on mRNA |
|---|---|---|
| Alanine | Ala | GCA, GCC, GCG, GCU |
| Arginine | Arg | AGA, AGG, CGA, CGG, CGC, CGU |
| Asparagine | Asp-$NH_2$ | AAC, AAU |
| Aspartic acid | Asp | GAC, GAU |
| Cysteine | Cys | UGC, UGU |
| Glutamic acid | Glu | GAA, GAG |
| Glutamine | Glu-$NH_2$ | CAG, CAA |
| Glycine | Gly | GGA, GGC, GGG, GGU |
| Histidine | His | CAC, CAU |
| Isoleucine | Ileu | AUA, AUC, AUU |
| Leucine | Leu | CUA, CUC, CUG, CUU, UUA, UUG |
| Lysine | Lys | AAA, AAG |
| Methionine | Met | AUG |
| Phenylalanine | Phe | UUU, UUC |
| Proline | Pro | CCA, CCC, CCG, CCU |
| Serine | Ser | AGC, AGU, UCA, UCG, UCC, UCU |
| Threonine | Thr | AGA, ACG, ACC, ACU |
| Tryptophan | Try | UGG |
| Tyrosine | Tyr | UAC, UAU |
| Valine | Val | GUA, GUG, GUC, GUU |

*In groups of three (called codons), bases of mRNA code the order of amino acids in a polypeptide chain. A, C, G, and U represent adenine, cytosine, guanine, and uracil, respectively. Some amino acids have more than one codon, and hence more than one transfer RNA can bring the amino acid to messenger RNA. The research on this coding was initiated by Nirenberg.

**The bases in groups of three are called codons.**

have multiple tRNA's (Table 12–2). For example, there are five distinctly different tRNA molecules specifically for the transfer of the amino acid leucine in cells of the bacterium *Escherichia coli.* At one end of a tRNA molecule is a trinucleotide base sequence (called the "***anticodon***") that fits a trinucleotide base sequence on mRNA (the "***codon***"). At the other end of a tRNA molecule is a specific base sequence of three terminal nucleotides—CCA—with a hydroxyl group on the sugar exposed on the terminal adenine nucleotide group. This hydroxyl group reacts with a specific amino acid by an esterification reaction with the aid of enzymes.

$$\underset{\text{tRNA}}{\text{(MONONUCLEOTIDES)}_{75-90}\text{CCA}-\text{OH}} + \underset{\text{AMINO ACID}}{\text{HO}\overset{\overset{\text{O}}{\|}}{\text{C}}\text{CH(NH}_2\text{)R}} \longrightarrow$$

$$\underset{\text{tRNA-AMINO ACID}}{\text{(MONONUCLEOTIDES)}_{75-90}\text{CCA}-\text{O}\overset{\overset{\text{O}}{\|}}{\text{C}}\text{CH(NH}_2\text{)R}} + H_2O$$

The P—O bonds of adenosine triphosphate (ATP) provide energy for the reaction between an amino acid and its transfer RNA. A molecule of ATP first activates an amino acid.

ATP + Amino acid ⟶ ATP—amino acid-activated species

The activated complex then reacts with a specific transfer RNA and forms the products shown in Figure 12–11.

The transfer RNA and its amino acid migrate to the ribosome, where the amino acid is used in the synthesis of a protein. The transfer RNA is then free to migrate back to the cell cytoplasm and repeat the process.

**The ribosome is the part of the cell in which protein synthesis takes place.**

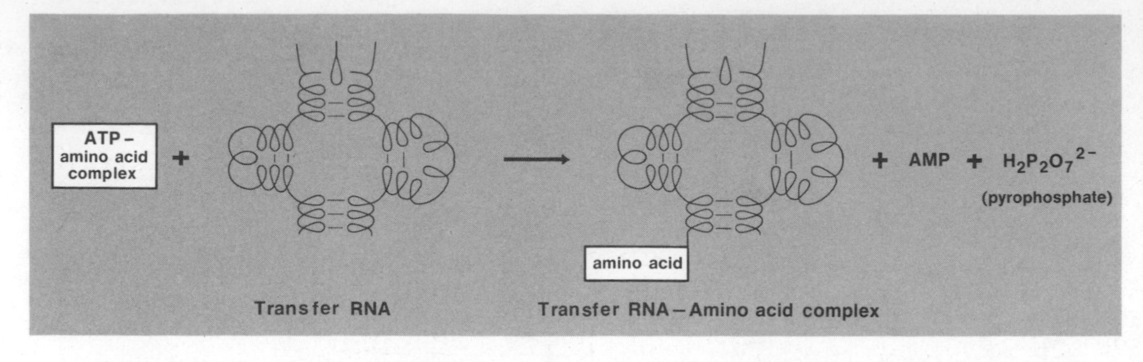


**Figure 12–11** Bonding of activated amino acid to transfer RNA. AMP is adenosine monophosphate.

Messenger RNA is used only once, or at most a few times, before it is depolymerized. While this may seem to be a terrible waste, it allows the cell to produce different proteins on very short notice. As conditions change, a different type of messenger RNA comes from the nucleus, a different protein is made, and the cell responds adequately to a changing environment.

## Synthetic Protein

**The proteins in living things are not a random sequence of amino acids.**

A mixture of amino acids can react to form polypeptide structures outside of a living cell. Only relatively simple catalytic agents are required, and the complex enzymatic environment of cytoplasm is not necessary. However, a mixture of only a few amino acids will result in a multitude of different protein structures. The synthesis of a particular sequence of amino acids presents many problems.

Methods too complex to be detailed here have been devised to construct protein structures with a desired sequence of amino acids. In order to obtain the desired bond between an amino acid and the peptide already constructed, it is necessary to block all the functional groups except those which are to undergo the peptide reaction. With the chemical blocking groups in place, the particular amino acid is added and the peptide bond is made. This is followed by removal of the blocking groups. To build a polypeptide with 20 amino acid units, this complicated process would have to be repeated 19 times. As the peptide chain grows, it becomes increasingly difficult to carry out the chemical operations without disturbing the bonds previously formed.

**Many large protein molecules have now been made in the laboratory.**

In spite of the difficulties, the customized synthesis of prescribed protein has progressed steadily. In 1953, Vincent du Vigneaud synthesized a hormone from nine amino acids. For his work he received a Nobel prize in 1955. Other, more lengthy molecules have been synthesized similarly by starting with the individual amino acids and adding them in proper sequence to make the desired protein. Notable hormone syntheses were those of $\beta$-corticotropin with 39 amino acid units, and insulin with 51 units. In 1969, two teams of researchers synthesized the first enzyme, ribonuclease (Fig. 12–12), to be assembled outside the living cell from individual amino acids. Ribonuclease contains 124 amino acid units. One team was led by Rockefeller University's Robert B. Merrifield and Bernd Gutte, and the other team at Merck Sharp & Dohme research laboratories was led by Robert G. Denkewalter and Ralph F. Hirschmann.

Merrifield's method has had wider application. In this automated technique, an insoluble solid support, polystyrene, acts as an anchor for the peptide chain during the synthesis (Fig. 12–13). The first amino acid is firmly bonded to a small polystyrene bead, and each of the other amino acids is then added one at a time in a stepwise man-

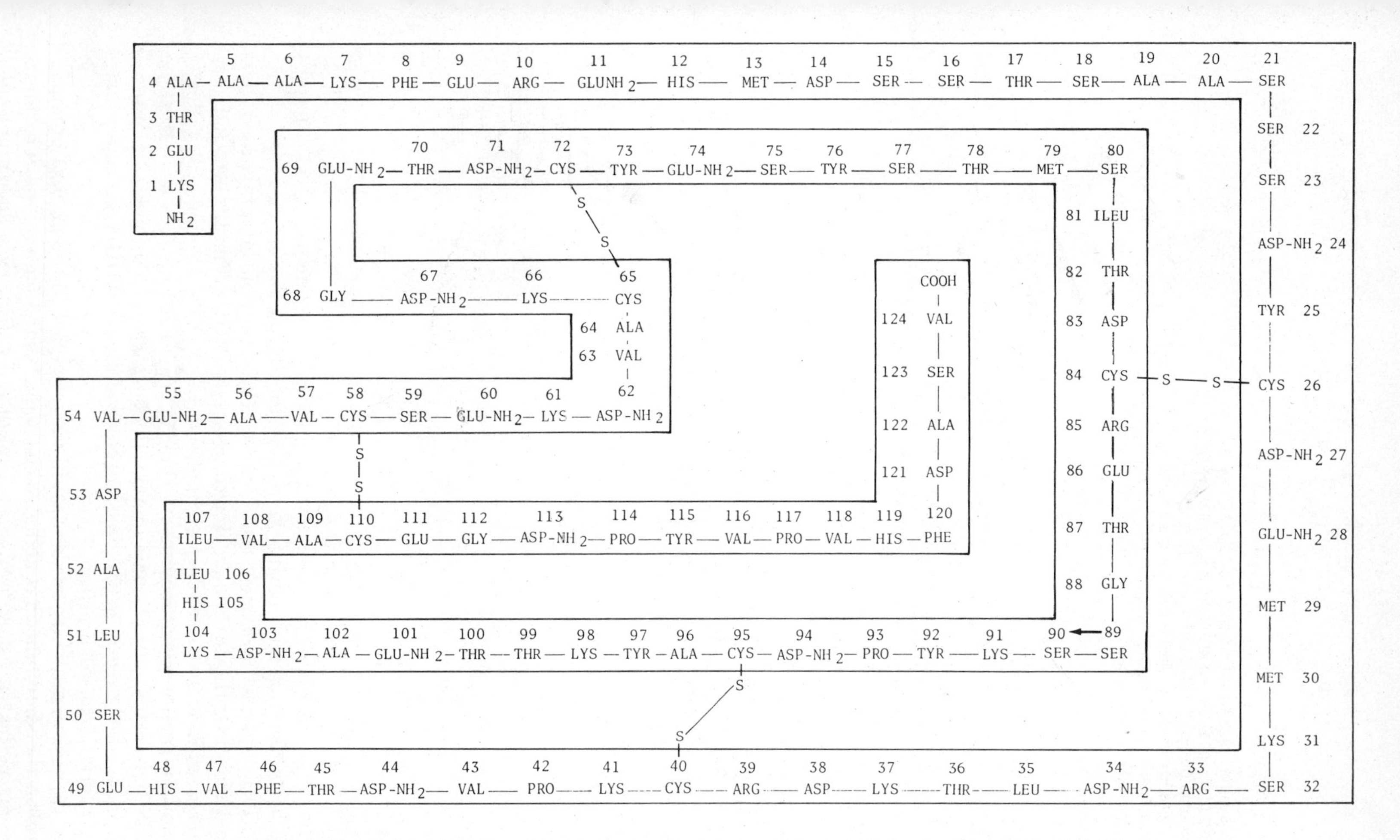


**Figure 12–12** The structure of ribonuclease A, an enzyme. This protein structure is composed of 124 amino acid units. The structure is partially explained by sulfur-sulfur bonds between some of the acid units—for example, between 40 and 95. Refer to Table 12–2 for the names of the amino acids abbreviated in this figure.

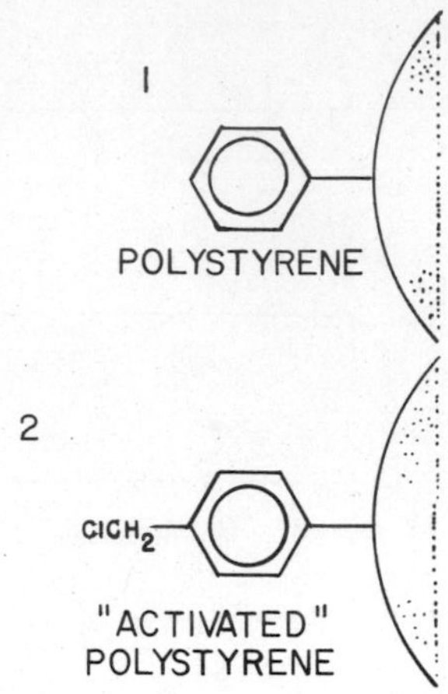

Figure 12–13 Solid phase method of synthesizing a protein is carried out stepwise from the carboxyl end toward the amino end of the peptide. An aromatic ring of the polystyrene (1) is activated by attaching a chloromethyl group (2). This first amino acid (black), protected by a butyloxycarbonyl (Boc) group (*black box*), is coupled to the site (3) by a benzyl ester bond and is then deprotected (4). Subsequent amino acid units are supplied in one of two activated forms; a second unit is shown in one of these forms, the nitrophenyl ester of the amino acids (5). The ester (*colored box*) is eliminated as the second unit couples to the first. Then the second unit is deprotected, leaving a dipeptide (6). These processes are repeated to lengthen the peptide chain.

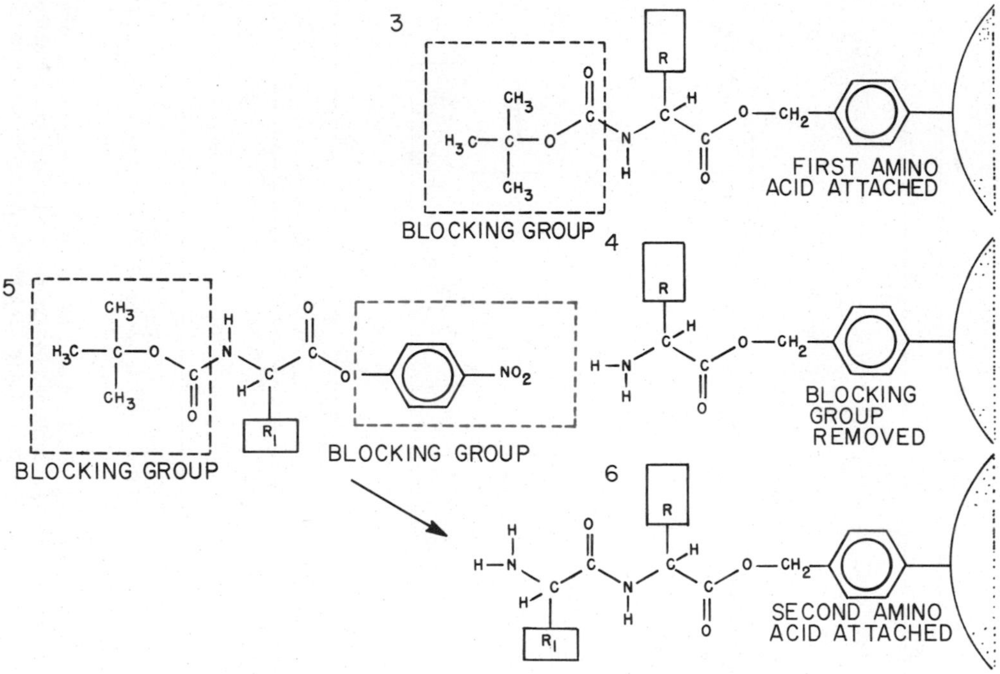

ner. The synthesis of ribonuclease required 369 chemical reactions and 11,931 steps in a continuous operation on a machine developed for this purpose.

In 1970, a group led by Choah Hao Li at the University of California used Merrifield's method to synthesize human growth hormone (HGH) (Fig. 12–14). HGH has 188 amino acid units and a molecular weight of about 21,500. This hormone is produced naturally by the front lobe of the pituitary gland, a pea-sized body located at the base of the brain. In humans, it has a dual function—it controls milk formation and it regulates many aspects of growth. In childhood, excess secretions of HGH can cause giantism; too little HGH causes dwarfism.

These synthetic proteins are identical in every respect to the natural products. They perform the same functions and give the same results on analysis. The limits of the present methods are unknown; at the time of this writing even longer molecules have been duplicated.

**Figure 12–14** Dr. Li (*center*) checks experiment with Dr. Richard Noble (*left*) and Dr. Donald Yanashiro (*right*). This group was the first to synthesize HGH polypeptide.

## Synthetic Nucleic Acids

Progress in the synthesis of polynucleotides has been difficult, principally because of the difficulties involved in determining the proper blocking groups. Progress, although slow, is being made.

In 1959, Arthur Kornberg synthesized a DNA-type polynucleotide, for which he received a Nobel prize. He used natural enzymes as templates to arrange the nucleotides in the order of a desired polynucleotide. His product was not biologically active. In 1965, Sol Spiegelman synthesized the polynucleotide portion of an RNA virus. This polynucleotide was biologically active and reproduced itself readily when introduced into living cells. In 1967, Mehran Goulian and Kornberg synthesized a fully infectious virus of the more complicated DNA type.

In 1970, Gobind Khorana synthesized a complete, double-stranded, 77-nucleotide gene. He, too, used natural enzymes to join previously synthesized, short, single-stranded polynucleotides into the double-stranded gene.

If scientists can construct DNA, can they then control the genetic code? Genes are the submicroscopic, theoretical bodies proposed by early geneticists to explain the transmission of characteristics from parents to progeny. It was thought that genes composed the chromosomes, which are large enough to be observed through the microscope as the central figures in cell division. It is now generally believed that DNA structures carry the message of the genes; hence, DNA contains the ***genetic code.*** If scientists can construct DNA, they could very well alter its structure and thereby control the genetic code. The ability to alter genes has led to a new field of science known as ***biogenetic engineering.*** This is a very active field of research that has developed (among other accomplishments) bacteria that can clean up an oil spill. In this case a patent was granted to the General Electric Co. for the production of life—a unique patent. Other bacteria have been produced that can synthesize protein, human growth hormone, and insulin. The method of producing bacteria for a particular function involves removing a gene from the bacterium, splicing in part of a gene from a human or other organism (that part that produces human insulin, for example), placing the spliced gene back into the bacterium, and letting the bacterium make millions

**Biogenetic engineering—alteration of genes for a desired purpose.**

of other insulin-producing bacteria. The process of splicing and recombining genes is referred to as ***recombinant DNA technology.*** The implications of gene splicing are tremendous—for both good and bad—and will demand responsible human decision-making for guidance toward the common good.

A ***mutation*** occurs whenever an individual characteristic appears that has not been inherited but is duly passed along as an inherited factor to the next generation. A mutation can readily be accounted for in terms of an alteration in the DNA genetic code; that is, some force alters the nucleotide structure in a reproductive cell. Some sources of energy, such as gamma radiation, are known to produce mutations. This is entirely reasonable because certain kinds of energy can disrupt some bonds, which can re-form in another sequence.

**A mutation results when there has been an alteration of the genetic code contained within the DNA molecule.**

If scientists can control the genetic code, can they control hereditary diseases such as sickle cell anemia, gout, some forms of diabetes, or mental retardation? If the understanding of detailed DNA structure and the enzymatic activity in building these structures continues to grow, it is reasonable to believe that some detailed relationships between structure and gross properties will emerge. If this happens, it may be possible to build compounds that, when introduced into living cells, can combat or block inherited characteristics.

## Self-Test 12-C

1. a. The basic code for the synthesis of protein is contained in the ____________ molecule.

   b. The synthesis of a protein is carried out when ____________ molecules bring up the required amino acids to messenger RNA.
2. When DNA replicates itself, each nitrogenous base in the chain is matched to another one via ____________ bonds.
3. The energy for DNA replication and natural protein synthesis is supplied by substances such as ____________.
4. What nitrogenous base complements (matches through hydrogen bonding) adenine (A)? ____________ cytosine (C)? ____________ guanine (G)? ____________ thymine (T)? ____________ uracil (U)? ____________
5. At this time, a gene has been synthesized from individual nucleotides in the laboratory without the aid of natural enzymes. (True/False)
6. Replication means the same as duplication. (True/False)

## Matching Set

| | | |
|---|---|---|
| ______ | 1. energy "cash" in the living cell | **a** ptyalin |
| ______ | 2. mutation | **b** ADP + energy |
| ______ | 3. enzyme that splits polysaccharides in the mouth | **c** Krebs cycle |
| ______ | 4. natural protein | **d** chlorophylls |
| ______ | 5. first oxidation product of glucose | **e** coupled reaction |

| | | |
|---|---|---|
| ______ | 6. occurs under aerobic (with air) conditions | **f** structure determined by DNA and RNA |
| ______ | 7. reaction utilizing ATP | **g** lactic acid |
| ______ | 8. product of ATP hydrolysis | **h** altered DNA |
| ______ | 9. molecule synthesized in a laboratory | **i** ATP |
| ______ | 10. molecules that absorb light energy | **j** insulin |
| ______ | 11. due to inadequate supply of lactase | **k** lactose intolerance |

## Questions

**1.** Write an equation for the digestion of:
a. starch to a disaccharide
b. a disaccharide to a simple sugar
c. a protein to amino acids
d. a triglyceride to fatty acids

**2.** What is the metal in chlorophyll?

**3.** If you were to "feed" radioactive carbon dioxide to a green plant, what would be the first radioactive carbon compound formed? Who made this discovery?

**4.** Give the structure of ATP and point out the region of the molecule that contains bonds that are hydrolyzed in coupled reactions.

**5.** What is meant by coupled reactions? Give an example. Why do the energetics of biochemical systems make coupled reactions necessary?

**6.** What type of compound first absorbs light energy in photosynthesis? Give an example.

**7.** In photosynthesis, why is it partially correct to say that light is an oxidizing agent?

**8.** What are the two major divisions of reactions in photosynthesis? Express in words what is accomplished in each.

**9.** What is the source of oxygen in photosynthesis?

**10.** What part of photosynthesis could take place in an animal cell?

**11.** Since chlorophyll loses electrons because of light, it must subsequently gain electrons from somewhere. Where do they come from?

**12.** What is the basic nature of the digestion processes for large molecules?

**13.** The chemical changes in the Krebs cycle can be classified as dehydrogenation (removal of hydrogens, a type of oxidation), dehydration (removal of water), hydrolysis (reaction with water in which water loses its molecular identity), decarboxylation (removal of —COOH group and formation of $CO_2$), and phosphorylation (adding a phosphate group, such as $H_2PO_4^-$). Beginning with citric acid and progressing around the cycle to oxaloacetic acid, determine the total number of each kind of chemical change.

**14.** What compound produces soreness in the muscles after a period of vigorous exercise?

**15.** Which compounds in the $CO_2$ fixation scheme in photosynthesis could enter directly the reactions of the Embden-Meyerhof pathway or the Krebs cycle?

**16.** If protein digestion is facilitated by enzymes, and these enzymes are produced in body organs made of proteins, explain why the enzymes do not cause rapid digestion of the organs themselves.

**17.** What is pyruvic acid? Why is it so important in extracting energy from sugars?

**18.** What happens if an amino acid is needed for protein synthesis and the amino acid can neither be made by the body nor obtained from the diet? Does the modern theory of protein synthesis include an explanation of the role of essential amino acids? (These are amino acids that cannot be manufactured by the human body; they must be obtained in the diet.)

**19.** Does a strand of DNA actually duplicate itself base for base in the formation of a strand of messenger RNA? Explain.

**20.** a. Describe the general method of synthesizing DNA ***in vitro*** (in a test tube) at present.
b. Why would the synthesis of a polynucleotide from the individual phosphoric acid, sugar, and nitrogenous bases be a breakthrough in controlling the genetic code?

**21.** What is a storehouse chemical for biochemical energy?

**22.** What are the end products in the digestion of carbohydrates? of fats? of proteins?

**23.** A mutation can be explained in terms of a change in which chemical in the cell?

**24.** The replication of DNA occurs in which part of the cell?

**25.** What is meant by a base pair in protein synthesis? What type of bonds holds base pairs together?

**26.** What is recombinant DNA?

**27.** Triphosphoric acid has the formula $H_5P_3O_{10}$. After drawing a structure for triphosphoric acid, look at the structure of ATP. Do you see any similarities?

**28.** When ATP or ADP is hydrolyzed, what types of bonds are broken and formed, and how many of each type of bond?

**29.** Draw a structure of glyceric acid.

**30.** The dark reaction of photosynthesis is a(n) (oxidation/reduction) reaction. The chemical that reduces the $CO_2$ is (NADPH/ATP).

**31.** How do living beings store and transfer energy?

**32.** When water reacts with ATP, is this an energy-releasing or an energy-requiring process?

**33.** What is the role of enzymes in digestion?

**34.** What is a purpose of ATP?

**35.** The importance of water in living systems is emphasized by incidences of hydrolysis. Cite examples.

**36.** In the structure of ATP, indicate the particular bond that yields the most energy when hydrolyzed. Why is this bond more energy rich than other P—O bonds?

**37.** Discuss some aspect of the feasibility of solving sociological and psychological problems via religion, DNA alteration, chemical suppressants, political pressures, and/or persuasive dialogue.

**38.** Check the recent issues of *Science* or other scientific news publications to find out about more recent work done on synthesis of proteins and polynucleotides.

# Science and Technology as a New Philosophy

13

This chapter is intended to be read and thoughtfully considered, perhaps discussed, but not to be taught. It is an interlude between fundamental chemistry and practical, applied chemistry. We hope that this chapter will raise some questions about relationships between you and your environment and cause you to think about a few values.

## A Philosophical and Historical Background

Direct observation and reobservation, not who did it, are the ultimate bases of truth in the sciences.

The various factors that have led to the systematic development of science and its associated technologies are both numerous and complex. Rather than attempt to develop an overall "complete" view, we shall discuss only a few of the empirical results. One of the key points seems to be the development of *printing techniques,* which enormously increased the availability of accumulated knowledge (see Fig. 13–1). Another seems to be the continuous *refusal to accept authority of a person or a group of people as the ultimate judge of truth* in certain areas; this indeed still seems to be one of the major differences between mathematics and the sciences on the one hand, and other areas of human knowledge. No one ever states, as proof of the validity of a piece of scientific information, that so-and-so said this and therefore it must be true. This skeptical attitude is to be found throughout all nations at all times; however, it was only in Western Europe that such a frame of mind came to be cultivated systematically and ultimately formed the basis of some intellectual organizations whose primary purpose was the search for and propagation of a kind of ***objective knowledge.***

The most important aspects of this type of knowledge are the ways in which it changes. With the passage of time, it becomes *more extensive, more accurate, more broadly disseminated, and,* most important of all, *more concisely summarized* in terms of generalizations or their equivalent: mathematical equations giving the quantitative relationships connecting a set of properties.

Technology: the manipulation of our environment.

Because the physical sciences deal with our universe, our world, and our environment, this growth of knowledge leads to an ever increasing understanding of the physical world and the ways it can be manipulated to obtain desired ends. This manipulation of our environment can attain a high degree of sophistication. Modern ***technology*** has two types of roots. One derives from the practical knowledge of craftsmen and frequently has been transmitted orally or by apprenticeships. The other is in the sciences and has been transmitted by publications of various sorts (Fig. 13–1)

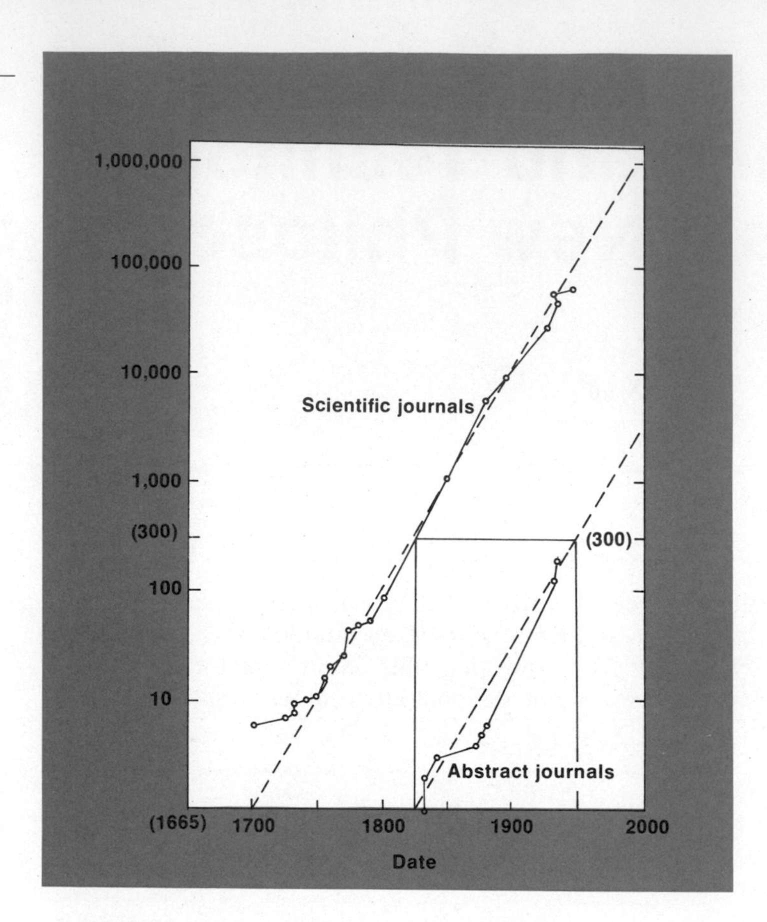


**Figure 13–1** Rate of increase in number of scientific journals since 1665.

and by formal education. In practice the dual roots tend to fuse and reinforce each other. The older method of oral transmission had its ups and downs, as is especially obvious in certain areas where knowledge of techniques has died out (e.g., the early Middle Ages in Europe as contrasted with the Roman Empire). The newer method of transmission by publication tends to make such knowledge available to all who can read and who have the required training for understanding (see Fig. 13–2).

As time has passed, the effectiveness of this new method of collecting and extending knowledge has become more apparent to the ordinary citizen. It has also become accepted by scholars in more disciplines and has led to the extension of this method to new areas of knowledge that previously had not been the subject of systematic study.

**We are in the midst of a knowledge explosion.**

At this point one might well ask for a definition of ***science*** or the scientific method. While it is difficult to give such definitions that are acceptable to all, a close approximation may be achieved. First of all, "science" consists of that which has been collected by the use of the ***scientific method.*** The term "scientific method" refers more to a mental attitude than to a procedure. The attitude is one that requires the strictest intellectual honesty in the collection of observable facts, and in the arrangement of these facts into a pattern that reveals the underlying basis of the *observed* behavior. The data normally must be collected under conditions that can be reproduced anywhere in the world, so that new data can be obtained to confirm or to refute the correctness of the suggested pattern (geologists and astronomers may sometimes be excused from these rigid requirements). The results obtained are thus independent of differences in language, culture, religion, or economic status in the various observers, and so represent a unique type of objective truth.

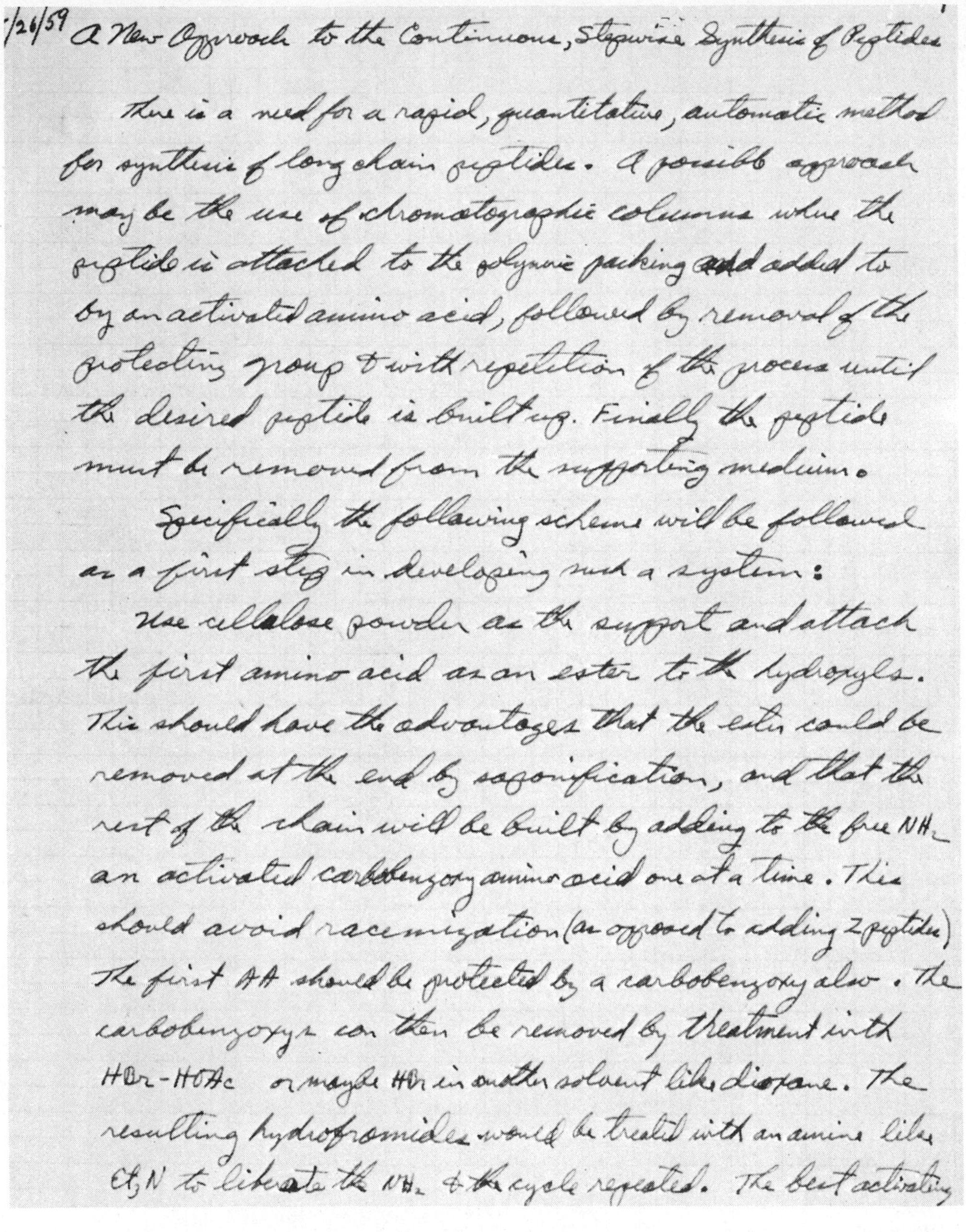

/26/59 A New Approach to the Continuous, Stepwise Synthesis of Peptides

There is a need for a rapid, quantitative, automatic method for synthesis of long chain peptides. A possible approach may be the use of chromatographic columns where the peptide is attached to the polymeric packing and added to by an activated amino acid, followed by removal of the protecting group & with repetition of the process until the desired peptide is built up. Finally the peptide must be removed from the supporting medium.

Specifically the following scheme will be followed as a first step in developing such a system:

Use cellulose powder as the support and attach the first amino acid as an ester to the hydroxyls. This should have the advantages that the ester could be removed at the end by saponification, and that the rest of the chain will be built by adding to the free $NH_2$ an activated carbobenzoxy amino acid one at a time. This should avoid racemization (as opposed to adding 2 peptides) The first AA should be protected by a carbobenzoxy also. The carbobenzoxys can then be removed by treatment with HBr-HOAc or maybe HBr in another solvent like dioxane. The resulting hydrobromides would be treated with an amine like $Et_3N$ to liberate the $NH_2$ & the cycle repeated. The best activating

**Figure 13–2** A page from the notes of Dr. Bruce Merrifield (Professor of Biochemistry at Rockefeller University), the designer of a protein-making machine. Progress in science is made through disciplined human thought, taking advantage of previous knowledge and useful theories. See Figure 12–13.

The methods of collecting observed, objective data can be grouped generally into about four classes: trial and error, planned research, accident, and "let's do some experiments and see what happens."*

Discovery by trial and error begins by having a problem to solve and doing various experiments in the hope that something desirable will emerge. The next set of experiments then depends on the results obtained. The discovery of the Edison battery by Thomas Edison's group is an example of discovery by trial and error. His group

*A. B. Garrett, "The Discovery Process and the Creative Mind," *Journal of Chemical Education,* Vol. 41, pp. 479–482 (1964).

performed more than 2000 experiments, each guided by the last, before settling on the composition of Edison's battery.

Discovery by planned research comes from plotting out specific experiments to test a well-defined hypothesis. The carcinogenic nature of some compounds is determined by progressing through a set pattern of experimental tests.

**Carcinogens are substances that cause cancer.**

Discovery by accident may be a misnomer. The investigator is usually actively involved in investigating nature through experimentation, but "accidentally" finds some phenomenon not originally imagined or conceived. Thus the "accident" has an element of serendipity, and is not seen unless the investigator is a trained observer. Pasteur's statement sums it up: "chance favors the prepared mind." The discovery of penicillin by Fleming was a type of "accidental" scientific discovery.

The method of "let's try some experiments and see what happens" lacks the formality of the other methods. Experimentation evolves with vague hypotheses or no specific hypotheses because no well-defined pathway is obvious. Joseph Priestley, who discovered oxygen in 1774, utilized this approach on occasion.

Regardless of the method of inquiry and discovery, there is often a period of incubation time between the discovery of a usable idea supported by evidence and the fruition of the idea through application of technology. Some examples of such incubation times are listed in Table 13–1.

Over the last 200 years, accumulated scientific and technological knowledge has been put to use on an extensive scale in Europe and in those areas of the world that have had the means and the will to follow the examples set by England, which underwent this "industrial revolution" first (see Fig. 13–3). The result has been the development of a society largely dependent upon and supported by a technology which is itself undergoing constant change. The first consequence of this technology has been to increase the *rate* at which things can be produced. This in turn has continually changed the occupational patterns of millions of human beings, and has brought forcefully to mind the persistence of *change* in our pattern of life.

**Technology increases the rate at which things can be produced.**

These changes have influenced profoundly the way in which people think about their material wants and the ways in which they can be satisfied. For example, there

**TABLE 13–1 How Long It Has Taken Some Fruitful Ideas to Reach Technological Realization**

| Innovation | Conception | Realization | Incubation Interval (years) |
|---|---|---|---|
| Antibiotics | 1910 | 1940 | 30 |
| Automatic transmission | 1930 | 1946 | 16 |
| Ballpoint pen | 1938 | 1945 | 7 |
| Cellophane | 1900 | 1912 | 12 |
| Dry soup mixes | 1943 | 1962 | 19 |
| Filter cigarettes | 1953 | 1955 | 2 |
| Heart pacemaker | 1928 | 1960 | 32 |
| Hybrid corn | 1908 | 1933 | 25 |
| Instant camera | 1945 | 1947 | 2 |
| Instant coffee | 1934 | 1956 | 22 |
| Liquid shampoo | 1950 | 1958 | 8 |
| Long-playing records | 1945 | 1948 | 3 |
| Nuclear energy | 1919 | 1965 | 46 |
| Nylon | 1927 | 1939 | 12 |
| Photography | 1782 | 1838 | 56 |
| Radar | 1907 | 1939 | 32 |
| Roll-on deodorant | 1948 | 1955 | 7 |
| Self-winding wristwatch | 1923 | 1939 | 16 |
| Video tape recorder | 1950 | 1956 | 6 |
| Xerox copying | 1935 | 1950 | 15 |
| Zipper | 1883 | 1913 | 30 |

**Figure 13–3** Name some discoveries that led to this change in transportation. How many of them are chemical in nature?

seems to be little argument with the statement, "If the number of human beings on the earth could be stabilized, a much higher standard of living could prevail over most of the earth." A statement such as this would have been greeted with widespread derision 500 years ago. While depending on technology, people today are beginning to doubt its ability to solve both personal and social problems on a long-range basis. It is obvious that confusion exists on this point since the cries about the curses of technology come from people who are highly dependent on it and who are constantly asking for even more help from technology.

## Technology: Its Triumphs and Problems

Almost as soon as the industrial revolution began in England, the public realized that technological progress brought with it a series of problems. The first to be noticed was the necessity for progress to be accompanied by changing patterns of employment.

**Technological unemployment results when people are displaced from their jobs by machines.**

It is obvious that if a machine makes as much thread as 100 men can make, the men are released to do other work. The 100 men, however, do not look on this as an advantage, especially if they are settled in their place of employment with their families. The new opportunities that result from such a machine are rarely of benefit directly to the men displaced. The wealth of their country is increased since there are now 100 men able to do other work. However, the initial reaction of the men in eighteenth century England was to riot and break up the machinery.

The increased use of fuels of all sorts, especially the introduction of coal and coke into metallurgical plants and then the use of coal to fuel steam engines, led to widespread problems with air pollution which were recognized and discussed over 200 years ago (Fig. 13–4).

The most important point of these results is the realization that technological progress is always obtained at some cost, and the cost may not be obvious at the outset.

**The amount of food that can be grown is related to the nitrogen content of the soil.**

An important technological development was recognized as necessary in 1890 by Sir William Crookes, who addressed the British Association for the Advancement of Science on the problem of the fixed nitrogen supply (that is, nitrogen in a chemical form that plants are capable of using in growth). At that time, scientists recognized that nitrogen compounds were necessary in fertilizers and that the world's future food supply would be determined by the amount of nitrogen compounds that could be made available for this purpose. The source of these nitrogen supplies was then limited to rapidly-depleting supplies of guano (bird droppings) in Peru and to sodium nitrate in Chile. It was realized that when these were exhausted, widespread famine would result unless an alternative supply could be developed. This problem was recognized first by English scientists as a potentially acute one, because by the 1890's

FUMIFUGIUM:

OR,

The Inconvenience of the AER,

AND

SMOAKE of LONDON

DISSIPATED.

TOGETHER

With ſome REMEDIES humbly propoſed

By J. E. Eſq;

To His Sacred MAJESTIE,

AND

To the PARLIAMENT now Aſſembled.

*Publiſhed by His Majeſties Command.*

Lucret. l. 5.

Carbonumque gravis vis, atque odor inſinuatur
Quam facile in cerebrum?——

LONDON:

Printed by W. GODBID, for GABRIEL BEDEL, and THOMAS COLLINS; and are to be ſold at their Shop at the Middle Temple Gate, neer Temple Bar. M.DC.LXI.

Re-printed for B. WHITE, at Horace's Head, in Fleet-ſtreet. MDCCLXXII.

**Figure 13–4** Title page from J. Evelyn, F.R.S., *The Smoake of London.* The Latin quotation is from the Roman poet Lucretius (97–53 B.C.). It may be translated, "How easily the heavy potency of carbons and odors sneaks into the brain!"

England had become very dependent upon imported food supplies. The Industrial Revolution allowed the population to grow rapidly, so that it soon outstripped the domestic food supply.

**The $N_2$ molecule is rather unreactive because of its strong triple bond ($N{\equiv}N$).**

Widespread interest in this problem led to research on a number of chemical reactions by which nitrogen can be obtained from the relatively inexhaustible supply present in the air. Air is 21% oxygen ($O_2$) and 79% nitrogen ($N_2$). The nitrogen in the air is present as the rather unreactive molecule $N_2$, and in this form it can be used as a source of other nitrogen compounds only by a few kinds of bacteria. Some is also transformed into NO by lightning, and when this is washed into the soil by rain it can be utilized by plants. Needless to say, the amount of nitrogen transformed into chemical compounds useful to plants by these processes is quite limited and cannot be increased easily.

Several chemical reactions were developed to form useful compounds from atmospheric nitrogen, but the best known and most widely used one has an ironic history. While England was interested in nitrogen for fertilizers, Germany was interested in nitrogen for explosives. The German General Staff realized that the British Navy could blockade German ports and cut them off from the sources of nitrogen

Figure 13–5 An early German ammonia plant.

compounds in South America. As a consequence, when a German chemist named Fritz Haber showed the potential of an industrial process in which nitrogen reacts with hydrogen in the presence of a suitable catalyst to form ammonia, the German General Staff was quite interested and furnished support through the German chemical industry for the study of the reaction and the development of industrial plants based on it. The first such plant was in operation by 1911, and by 1914 such plants were being built very rapidly.

**The Haber process transforms the nitrogen of the air into ammonia.**

When the First World War broke out in August of 1914, many people thought that a shortage of explosives based on nitrogen compounds would force the war to end within a year. Unfortunately, by this time the nitrogen-fixing industry in Germany was capable of supplying the needed compounds in large amounts. This process thus prolonged the war considerably and resulted in an enormous increase in mortality. Subsequently, the ammonia process has been used on a huge scale to prepare fertilizers and now is largely responsible for the fact that the earth can support a population of more than four billion (see Fig. 13–6). Ammonia production by this process exceeds 40,000 tons per day in the United States alone.

$$N_2\,(g) + 3H_2\,(g) \rightleftharpoons 2NH_3\,(g)$$

The same type of problem seems to arise from the development of many technological processes. The utilization of nuclear energy brings with it the ability to make nuclear explosives from one of the by-products. The development of rapid and convenient means of transportation, such as the automobile and the airplane, also brings forth new weapons of war and air pollution problems. We must learn to control our technology in such a manner as to maximize its benefits and minimize its disad-

**Ammonia is the number two commercial chemical.**

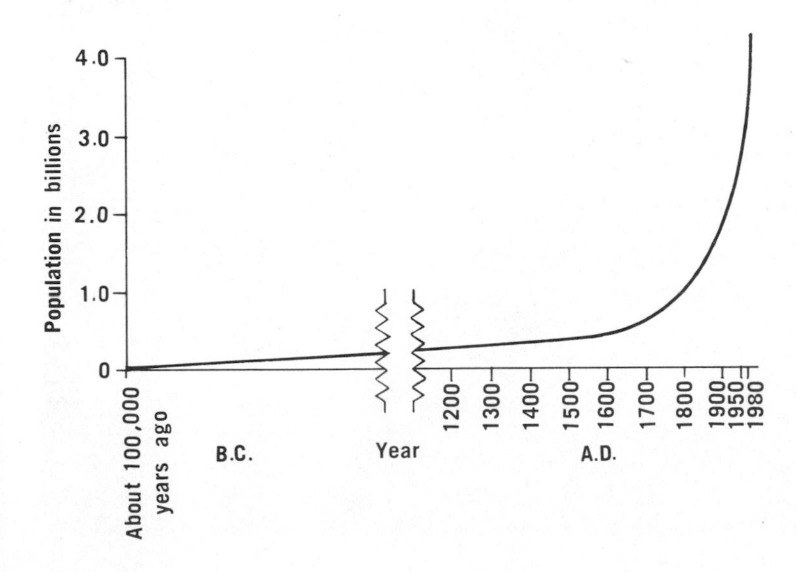


Figure 13–6 World population.

**Understanding nature does not necessarily lead to a better world.**

vantages. These problems arise with *all* technological developments, even the most primitive. The discovery of the techniques necessary to the manufacture of iron led first to the development of weapons (swords) by their discoverers, the Hittites, who then proceeded to conquer their neighbors and lead the first successful invasion of Egypt (ca. 1550 B.C.).

## Technology and the Human Environment

**Technology usually alters the environment drastically.**

The growth in large-scale technology has a large number of effects, both direct and indirect, on the human environment. The examination of a few of these shows just how complex these consequences can be.

An obvious case is the development of atomic energy. When the incredibly large amounts of energy released by nuclear reactions were first recognized, the development of such energy sources was given a top priority. After some nuclear reactors had been built and actually placed in operation, it was evident that their operation was accompanied by some serious potential risks to their human users. The first was the danger of some potential disaster such as the explosion of a boiler, with the consequent dispersal of radioactive material. The second danger was in the generation of radioactive products as the nuclear reaction proceeded. The uranium used in such reactors was transformed into a wide variety of fission products, which made the operation of the pile more difficult as time went on. The repurification of the uranium could be accomplished chemically, but what was to be done with the radioactive wastes generated in the process? Obviously they could not be dumped into a sewer, since many of them have relatively long half-lives. Present practice calls for solid radioactive wastes to be placed in vaults and buried at carefully chosen sites (Fig.

**Nuclear wastes constitute an unsolved problem. See Chapter 14.**

**Figure 13–7** Land burial trench at the Oak Ridge National Laboratory reservation. Each day's accumulation of waste containers is buried under three or more feet of earth.

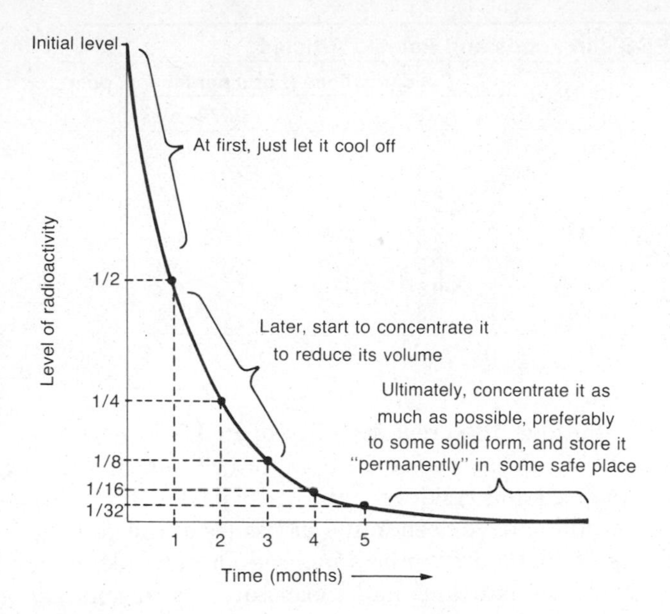


**Figure 13–8** Disposal of radioactive wastes, for a hypothetical waste product with a one-month half-life.

13–7). Liquid wastes are placed in underground storage tanks. The level of radioactivity of such wastes as a function of time is shown in Figure 13–8. The problem that faces us here is obtaining the benefits of the technology with minimum disruption of our own environment.

The same type of problem arises whenever we introduce a specific chemical compound into our environment to accomplish one single thing. The compound is often capable of a variety of actions and can lead to consequences undreamed of by those who introduce it. Many such examples can be found among drugs, industrial chemicals, and insecticides.

**Technological procedures have both obvious and not so obvious consequences.**

## Polychlorinated Biphenyls (PCB's)

Polychlorinated biphenyls were discovered in 1881, but their commercial importance was not recognized until about 40 years later. By 1935, the Monsanto Chemical Company was manufacturing PCB's in large quantities. There are 209 different molecular structures that are called PCB's. Most commercial PCB's are actually mixtures of many different chlorinated biphenyls. As a class of compounds, PCB's are very stable to heat, acids, and bases. Their stability and insulating ability led to their wide use as insulating fluids and coolants in electrical transformers, commercial capacitors, hydraulic systems, gas turbines, and vacuum pumps, and as fire retardants and plasticizers. More than 95% of all power capacitors contain PCB's and approximately 5% of the transformers are PCB-filled. Documented production of PCB's totaled about 1.7 billion pounds by 1977. It has been estimated that of this production about 750 million pounds still remain in use in transformers and other electrical devices (Table 13–2).

**TABLE 13–2 Fate of the PCB's**

| | Millions of Pounds |
|---|---|
| Remaining in use | 750 |
| Buried in landfills | 300 |
| Scattered in the environment | 150 |
| Exported | 150 |
| Destroyed | 50 |

**General structure of PCB's. Z is either an H atom or a Cl atom.**

**TABLE 13–3 FDA Tolerances for PCB's in Foods and Related Articles**

| | Concentrations (Parts per Million, ppm) | |
|---|---|---|
| | *1973* | *1979–present* |
| Paper, food-packaging materials intended for food | 10 | 10 |
| Animal feed components | 5.0 | 2.0 |
| Fish and shellfish (edible parts) | 5.0 | 5.0 |
| Poultry (based on fat only) | 5.0 | 3.0 |
| Milk and dairy products (based on fat only) | 2.5 | 1.5 |
| Feed for food-producing animals | 0.5 | 0.2 |
| Eggs | 0.5 | 0.3 |
| Infant and junior foods | 0.2 | 0.2 |

While very desirable in some situations, the great stability of PCB's is very undesirable in others. When the human body is exposed to PCB vapor or fluid, the PCB received by the body is poorly metabolized and resides for the most part in fatty tissues. As much as 29 parts of PCB per billion have been found in tissues of the general population. The known toxic effects of PCB's in humans include an acne-like skin eruption (chloracne), pigmentation of the skin and nails, excessive eye discharge, swelling of eyelids, and distinctive hair follicles. PCB's are known to cause cancer in test animals and are suspected of being carcinogenic in humans.

The ubiquitous nature of PCB's in the environment was first noticed by a Swedish scientist, Jensen, in 1966, who recorded their occurrence in fish and birds. By 1971 it was obvious that PCB contamination of wildlife was a widespread problem. The human toxicity of PCB's was demonstrated in 1968 when a group of Japanese developed symptoms of what was called "Yusho" poisoning as a result of consuming PCB-contaminated rice oil. By 1971, PCB usage began to be curtailed, and by 1977 its production stopped. The Toxic Substances Control Act (TSCA) of 1976 forbade the production of PCB's after 1979. The U.S. Food and Drug Administration (FDA) has now established tolerances of PCB's in food articles (Table 13–3)—mute testimony that we may be living with the presence of PCB's for many years to come.

**PCB's are still being manufactured in Germany, France, Spain, Italy, and some Eastern European countries.**

## Insecticides

It seems that we must always combat insects, since they and other biological pests still claim about 35% of our food supply (Table 13–4). If we could eliminate that loss, we could feed most of the world's hungry people. The most effective aid we now receive in this struggle against insects is from insecticides. Unfortunately, many insecticides are rather unspecific poisons.

Arsenic compounds were used on a large scale as insecticides, but a realization of their great toxicity for human beings has led to their replacement by other compounds equally effective but usually less dangerous. However, some of these com-

**TABLE 13–4 Estimated Annual Crop Loss Due to Pests and Disease (Percentage of Total U.S. Crop Destroyed)**

| Crop | Insects | Weeds | Disease |
|---|---|---|---|
| Corn | 12 | 10 | 12 |
| Rice | 4 | 17 | 7 |
| Wheat | 6 | 12 | 14 |
| Potatoes | 14 | 3 | 19 |
| Cotton | 19 | 8 | 12 |

pounds which are most effective in protecting crops are nevertheless toxic to humans and are even capable of causing death when improperly handled or ingested.

DDT is far more harmful to many species of wild animals than it is to human beings.

Even those insecticides that are not obviously harmful to us often have side effects that make them unattractive. DDT, for example, is effective for the control of a wide variety of insects and, as far as can be ascertained, is nontoxic to humans. Its widespread use from 1943 to 1965 virtually eliminated plague from the world population by controlling fleas and lice. The plague brought terrible devastation during the Middle Ages and persisted well into our century. DDT has also been responsible for huge increases in crop yields, has greatly improved the health of armies and refugees during the awful years of World War II, has improved the health of cattle, and has even eliminated the housefly from homes where it was used. Hence there were strong reasons for using DDT; it was not a willful or arbitrary decision.

Because of the widespread use of DDT, and the fact that it is accumulated in fish and animal fats, it is found to concentrate enormously in birds of prey (eagles, pelicans, and ospreys), whose diet is principally fish, and to disrupt their reproductive cycles (Fig. 13–9). A side effect such as this led to extensive agitation to replace DDT with other compounds that are equally effective as insecticides yet free from this particular side effect. It is possible, however, that the compounds used to replace DDT will have other side effects that will not become obvious until after they have been used for some time. It would seem that the use of any insecticide will have a considerable effect upon the bird population of an area, if only through the changes that it introduces into the food supply of the birds. The perplexing problem in this area and others is to know and implement the balance between benefits and subsequently determined drawbacks. Most chemical substances, like DDT, cannot be labeled "good"

**Figure 13–9** The concentration of DDT in birds of prey, such as hawks, results in poor egg shell formation along with other abnormalities in the reproductive cycle.

**TABLE 13–5 U.S. Department of Transportation Chemical Hazard Classes with Examples**

Chemicals exhibiting the properties of the hazard class must be shipped in proper containers and labeled to alert others to the hazards.

| Class | Examples |
|---|---|
| Class A, B, and C explosives | A, dynamite; B, gunpowder; C, firecrackers |
| Flammable compressed gases | liquefied natural gas |
| Nonflammable compressed gases | argon, nitrogen |
| Flammable liquids | gasoline |
| Combustible liquids | diesel fuel |
| Flammable solids | phosphorus |
| Oxidizers | oxygen |
| Class A and B poisons | A, cyanide; B, parathion |
| Radioactive materials | radioisotopes |
| Corrosive materials | sulfuric acid, lye |
| Miscellaneous hazardous materials | hazardous wastes |

or "bad." Ozone, to take another example, is necessary in our upper atmosphere to filter out most of the incoming ultraviolet light, but it is dangerous for us to breathe.

**Most technical problems require the efforts of several disciplines for a solution.**

Chemistry is not, however, the only science that can furnish technological means for the control of insects. There is a wide variety of biological processes that can be used for the same purposes. One is to furnish assistance to the biological species that destroy insect pests, such as parasitic insects. This can lead to problems with the thriving new species, too. Another method is the introduction of large numbers of sterile insects. When these mate with normal insects, no offspring are produced; the numbers of a type of undesirable insect can be reduced considerably by this process, at least temporarily. These examples are given to emphasize the fact that there are usually several different kinds of processes that can be used to solve any given practical problem. In the future these will be assessed, at least in part, on their ability to leave the environment undamaged as well as on their monetary cost.

## Hazardous Chemicals

Some commonly used chemicals exhibit properties of extreme reactivity, flammability, corrosiveness, or toxicity (Table 13–5). When produced and used in the

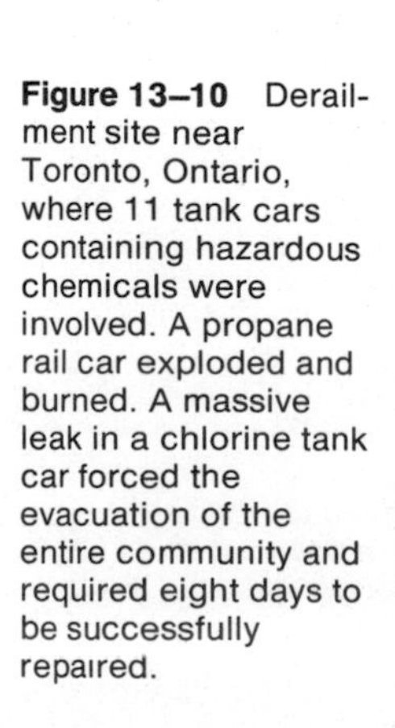

**Figure 13–10** Derailment site near Toronto, Ontario, where 11 tank cars containing hazardous chemicals were involved. A propane rail car exploded and burned. A massive leak in a chlorine tank car forced the evacuation of the entire community and required eight days to be successfully repaired.

chemical plant, they represent only a slight possibility of harming the public. But when they are transported from producer to consumer, they represent a greater hazard. Often the shipments are made in large quantities and routed through population centers. Large cities have responded to the apparent dangers by rerouting shipments around the major population density, but small towns and villages must still contend with the problems.

In 1978, sixteen persons were killed in Waverly, Tennessee, as a result of a train accident involving liquefied petroleum gas. In 1980, a quarter of a million people were evacuated from their homes and businesses near Toronto, Ontario, as a result of a chlorine railcar derailment in the suburbs. No deaths or injuries resulted from this accident. Numerous other transportation accidents involving hazardous chemicals have occurred during the past half century.

## Technological Development and Its Environmental Consequences

A good example of a highly desirable technological development that has consequences that are obviously not so desirable is seen in the development of fertilizers. Most of us agree that an abundant food supply is both necessary and good. Most would admit, too, that every crop harvested from a field removes essential nutrients from that field: nitrogen, phosphorus, potassium, and so forth. By replacing these lost elements with fertilizer, we can restore or enhance the amount of food we obtain from the field. The increased yields obtained with increased application of fertilizers have been well established and commonly held to be desirable. It is also well established that increased fertilization of a field increases the concentration of these essential nutrients in the rain water that runs off such land.

The change in the mineral concentration in the runoff water is capable of causing drastic changes in the rivers and lakes into which it drains (Fig. 13–11). By increasing the amount of minerals available, we have greatly stimulated the growth of surface algae in rivers and lakes, and as these grow they choke out other forms of life. The growth of such slimes makes the water much less capable of supporting its normal population of fish, and ultimately they die off. The algae also make the water

**Figure 13–11** This pond was fertilized with chemicals to produce an excessive growth of algae. The left photo shows the water covered with the plant growth; the right photo shows the decay that follows when the concentrated form of life cannot be sustained.

**Fertilizers increase the growth of both desirable and undesirable organisms.**

**What are the questions that should be asked before a new chemical is widely used?**

more difficult to purify for drinking purposes. After a time the increased food supply is paid for by a general disruption of the biology of the waterways that drain an area. This disruption has occurred in many areas in the United States where the movement of water through lakes and rivers is slow.

Other examples of the same kind of process can be seen in the development of energy sources, the use of antibiotics, weed killers, and, in fact, any kind of process that releases chemical compounds into the environment in sufficient quantities to cause an appreciable percentage increase in its concentration in the environment. This holds for the hydrocarbons released from automobile engines and for the carbon dioxide and sulfur dioxide released by electrical power plants. The changes that increased concentrations of various compounds cause cannot always be estimated on the basis of experiments covering only a short period of time, since some of the effects may be long-range ones that build up slowly.

## Is Technology Escaping Control?

It is quite easy to imagine a situation in which a technological process can introduce into our environment drastic and irreversible changes, which set a chain of events into action before we can stop them. This is especially easy to visualize in the case of

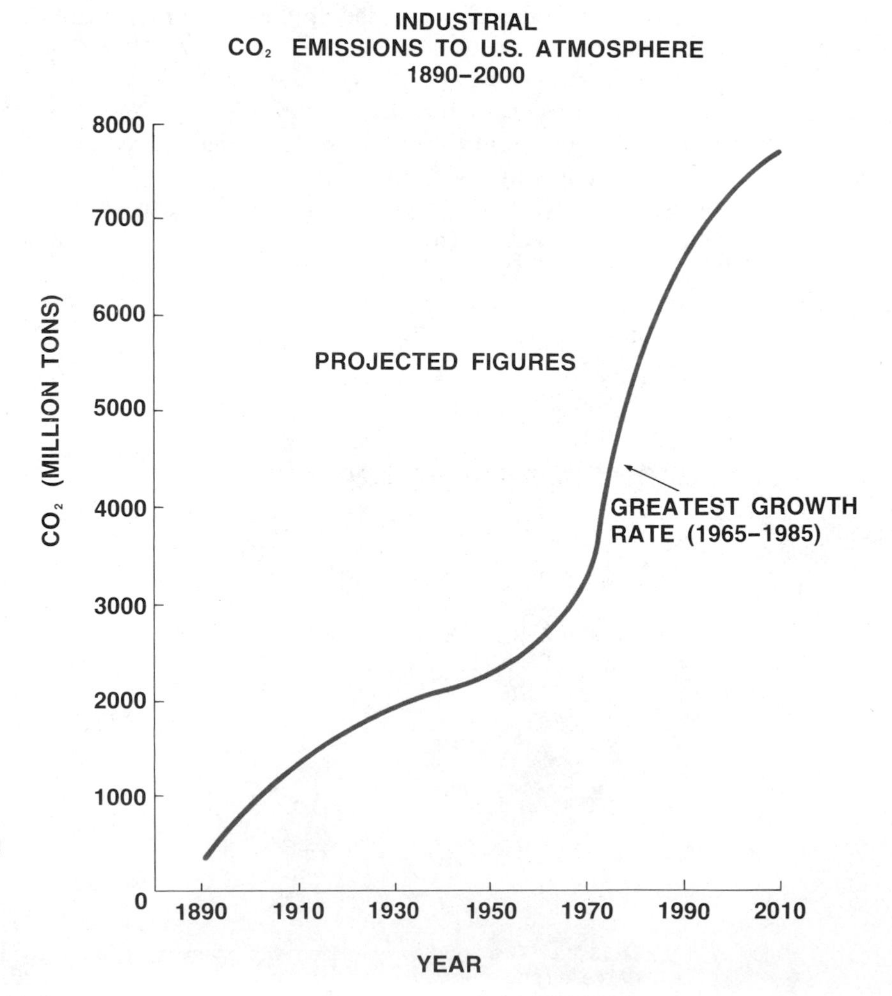


**Figure 13–12** United States' industrial emissions of $CO_2$ into the atmosphere in millions of tons per year.

changes that might be provided by the continuous increase in the carbon dioxide content of our atmosphere (Fig. 13–12). The increased emission of $CO_2$ is caused by the increased use of "fossil" fuels, such as coal and petroleum, to furnish power for energy generation and transportation. What will be the long-range effects, if any, from this steadily increasing amount of carbon dioxide in our atmosphere? At the present time, no one really knows.



There are other situations where the consequences of technology seem more obviously to be moving beyond our control. The ability to build nuclear weapons and power plants is now spreading quite rapidly, and soon may be well within the reach of all but the smallest nations. How are these to be controlled? What mechanisms can be developed to prevent humanity from destroying itself in an atomic holocaust triggered by some insignificant local dispute? This is obviously a very urgent problem!

**It is not presently clear that we will effectively control technology; we may destroy ourselves with it.**

The application of technology to the problems of warfare has already produced some frightening developments, including chemical and biological warfare agents whose discovery has closely followed the extension of human knowledge in these sciences. The same kind of knowledge that allows more effective drugs and insecticides to be synthesized facilitates the synthesis of more effective agents for gas warfare. The understanding of cellular behavior that allows us to produce new varieties of high-yielding grains also can be used to develop new techniques of biological warfare.

Obviously, a distinction must be made between scientific knowledge, which can be used for good or evil purposes, and the types of technology that are developed specifically for destructive purposes. The kind of emotional thinking that puts all technological developments under some kind of moral ban can only lead us rapidly

**Figure 13–13** Human thought versus chemicals. Will mankind ever completely control chemicals—or will they control us?

to a new Dark Age. Most of the materials of the world can be used for a variety of purposes, which are ethical to varying degrees. Selection must be made between good developments, the indifferent ones, and the bad ones. This has *always* been a problem and probably always will be. In such a situation, ignorance can be catastrophic; only by a study of such processes and an evaluation of their probable consequences can rational selections be made. In this case it makes more sense to ask if we are being given control over things beyond our understanding than to ask, "Is technology escaping control?" Technology is always *initiated* by human beings and is under some sort of actual or potential control by us. Whatever the minds of human beings devise, the minds of human beings can control. Only the will to do so is needed. Our worst enemies are ignorance and selfishness.

**Technology is always initiated by human beings and is therefore under our control in principle.**

There is still another manner in which fears arise over the developing course of technology and our ability to control it. The vast majority of people have no knowledge of the scientific principles upon which technology is based and accordingly must accept the opinions of others on technological matters. This dependence upon experts, in turn, arouses fears of these experts and the damage they can do either by error or by evil intent. This fear, in some people, is an unreasoning, blind, driving force that leads them to condemn all technology and, indeed, all thinkers and all knowledge. (How do you feel about someone smarter than you?) The only antidote to such fear is an understanding of technology based on study. Humanity's perennial enemy has been ignorance and the prejudice it generates. Human progress has always been the result of activities of that small percentage of people who accept neither the ignorance nor the popular prejudices of their fellow human beings.

In all fairness, it must be noted that the benefits of technology far outweigh its disadvantages, and also that its transformation of the human condition is still in its infancy. There is, even now, an enormous number of practical problems facing us that can be solved only by new scientific discoveries and the technological advances that they will make possible. It is the responsibility of all of us to learn about the sciences and to develop an understanding of the possible ways in which this knowledge may be developed to the advantage of the human race. In these learning processes, we shall discover an avenue to responsible participation in decision making, an understanding of technological advances, and a thorough enjoyment of our investigations.

**People should never hesitate to probe into technology to find how it affects their lives. This is necessary if they are to make good decisions on many basic questions.**

## Self-Test 13-A

1. The ultimate test of a scientific theory is its agreement with ______________.
2. Different workers, in different countries, who carry out a particular laboratory experiment in exactly the same way should get ______________ result.
3. Technology allows us to produce things at a greater ______________.
4. We now expect that over the period of our own lifetime our way of life will ______________.
5. Nuclear power plants produce competitively priced energy but dangerously ______________ by-products.
6. The principal justification for the use of chemical fertilizers and pesticides is that they allow us to produce more ______________.

## Subjective Test



Rate the following on a scale: *good, bad, indifferent.* Be prepared to state your reasons.

1. DDT ______________
2. Nuclear energy ______________
3. Coal as a fuel ______________
4. Petroleum as a fuel ______________
5. Fertilizers ______________
6. Birth control pills ______________
7. Plastic containers ______________
8. Synthetic foods ______________
9. Solar energy ______________
10. Genetic manipulation ______________

## Matching Set

| | | | |
|---|---|---|---|
| ______ | 1. transmission of scientific knowledge | **a** | appeal to authority |
| ______ | 2. allowed World War I to be prolonged | **b** | DDT |
| ______ | 3. source of the element $N_2$ | **c** | technology |
| ______ | 4. alternative to insecticides | **d** | nitrogen compounds |
| ______ | 5. unsatisfactory way to establish scientific truth | **e** | used to study nature |
| ______ | 6. insecticide harmful to wildlife | **f** | printed journals |
| ______ | 7. increases rate of change in life | **g** | Haber process (ammonia production) |
| ______ | 8. used in explosives | **h** | air |
| ______ | 9. needed to grow crops | **i** | observation |
| ______ | 10. scientific methods | **j** | biological controls |
| ______ | 11. basis of scientific truth | | |

## Topics for Themes

These are suggested topics for short themes in which you can express your views. The depth of research is to be determined by you and (or) your instructor. If a topic is to be researched, the *Readers' Guide* is a good place to start, and *Scientific American* (well-indexed!), *Chem Tech,* and *Environmental Science and Technology* are good places to continue. Rather than discuss bona fide solutions to a particular problem, it might be more interesting to state the questions that must be answered before you could vote informatively on the issue. A primary requirement for an informed citizenry is that we ask enough good questions to assure that all sides of a given issue have been covered. Asking investigative questions is our privilege and responsibility.

**1.** Contrast knowledge gained by scientific methods with knowledge in philosophy, religion, art and (or) some other disciplines.

**2.** What role should scientists play in deciding moral uses of their discoveries?

**3.** Would it be a good idea to limit the number of people on earth in order to have more material and energy per person?

**4.** Explore some ways in which technology has changed the occupational patterns of people.

**5.** Find examples in which a technological development at first was considered more good than bad, but with passage of time was looked upon as more bad than good.

**6.** Is technological "progress" always obtained at some cost? Explain.

**7.** Discuss ways in which side effects of chemicals can be determined prior to putting a product on the market.

**8.** Should every conceivable side effect of a product be tested for and eliminated before the product is brought on the market? Justify your answer.

**9.** Techniques have been developed to analyze for some elements when only a nanogram ($10^{-9}$ g) is present in the sample. What effects may this have on the consciousness of side effects of chemicals?

**10.** Modern society demands more energy with less pollution and less expenditure of money. From your point of view, is this a "pipe dream"? Discuss reasons for your answer.

**11.** To what degree should food chains be considered when eradication of insects via insecticides seems to be the only practical way to protect a food supply and health?

**12.** What should be done with radioactive wastes? What problems are involved?

**13.** Discuss the need to recycle chemicals in view of limited resources.

**14.** What can one do to make a better, safer, more livable world for everybody with regard to waste materials?

**15.** To what extent should political decisions influence scientific investigation?

**16.** Should scientists be morally responsible in their investigations of nature?

**17.** Resolved: Education is the solution to problems concerning energy, pollution, and the proper use and reuse of materials. Defend or deny.

**18.** Discuss one or more ways in which food production creates a pollution problem and wastes energy.

**19.** Cite a technological advance that was not given specifically in this chapter; state its benefits, and briefly discuss problems arising from it.

**20.** How should wastes be dealt with? In your discussion consider minimum use of energy, minimum pollution, and maximum recycling.

**21.** Discuss significant differences and similarities between science and technology.

**22.** Discuss an important practical problem now facing us in terms of how science and technology have helped to solve the problem and, in turn, have created other problems.

**23.** Describe evidence of technological advances that caused a decline of civilization. [Consider the article in the *Journal of Occupational Medicine,* Vol. 7, pp. 53–60 (1965).]

**24.** Technology is more beneficial than detrimental to humanity. Defend or deny.

**25.** Discuss areas or problems in which technology is most needed today.

**26.** Discuss the limitations of science.

**27.** Discuss the limitations of technology.

**28.** What factors should be considered before a chemical is banned for a particular use? Defend each factor.

**29.** Chemical dumps have developed into a national shame. Do you think that the users of chemical products should pay for the long-term disposition of associated waste materials when they purchase the product?

**30.** Do you think it possible or likely that chemicals will ultimately cause our destruction in a sea of chemical poison?

**31.** Debate the statement: "Scientific and technological knowledge is dangerous in the absence of human values."

**32.** How dangerous do you consider the possibility that scientists and others will "play God" in the control of new human life? Do you consider genetic control a threat to you?

**33.** Should we refrain from seeking certain types of new chemicals for fighting insects?

**34.** The term "chemophobia" has been coined to describe a certain public phenomenon. Can you explain?

**35.** Should we refrain from seeking certain types of new knowledge for fear of the consequences? Can people be trusted to understand and show proper restraint? Does nature have a trap for us waiting to be sprung?

**36.** Should a proliferation of fruit flies be eradicated by spraying malathion? What facts should be known and weighed before making this kind of decision?

**37.** From a chemical standpoint, what do we owe future generations?

**38.** Among the following, which are the three most pressing problems? Justify your choices. Current problems: conservation of metals, pollution, energy, production of food and feeding the world, controlling pests, storing nuclear wastes, transportation, biogenetic engineering.

**39.** What new technological advances would you most like to see?

**40.** What is the proper balance or equilibrium point between the use of petroleum as a fuel and the use of petroleum as the source of medicine, plastics, paints, and other consumer products?

**41.** Is there really a "miracle fuel" that is not currently popular but will solve our energy problems? Speculate on the prospects of such a fuel, and its existence or nonexistence.

**42.** Before you could vote informatively on the following items, what questions must be answered?
a. Clean Air Act
b. nuclear energy for electrical power plants
c. what to do with solid refuse

**43.** In other parts of the world it is more acceptable to buy fruit at the market that might contain a few worms. In the United States, insecticides have been used to produce commercial quantities of insect-free fruit. Do you think it wise to spray large sections of California or Florida to have absolute control over the medfly?

# 14 Energy And Our Society

Today we are more aware than ever before of the need for energy. We feel its importance through increased prices of electricity, natural gas, gasoline, and energy-produced and -transported consumer goods. In our highly industrialized and appliance-oriented society, large amounts of energy are now a daily necessity for our way of life (Fig. 14–1).

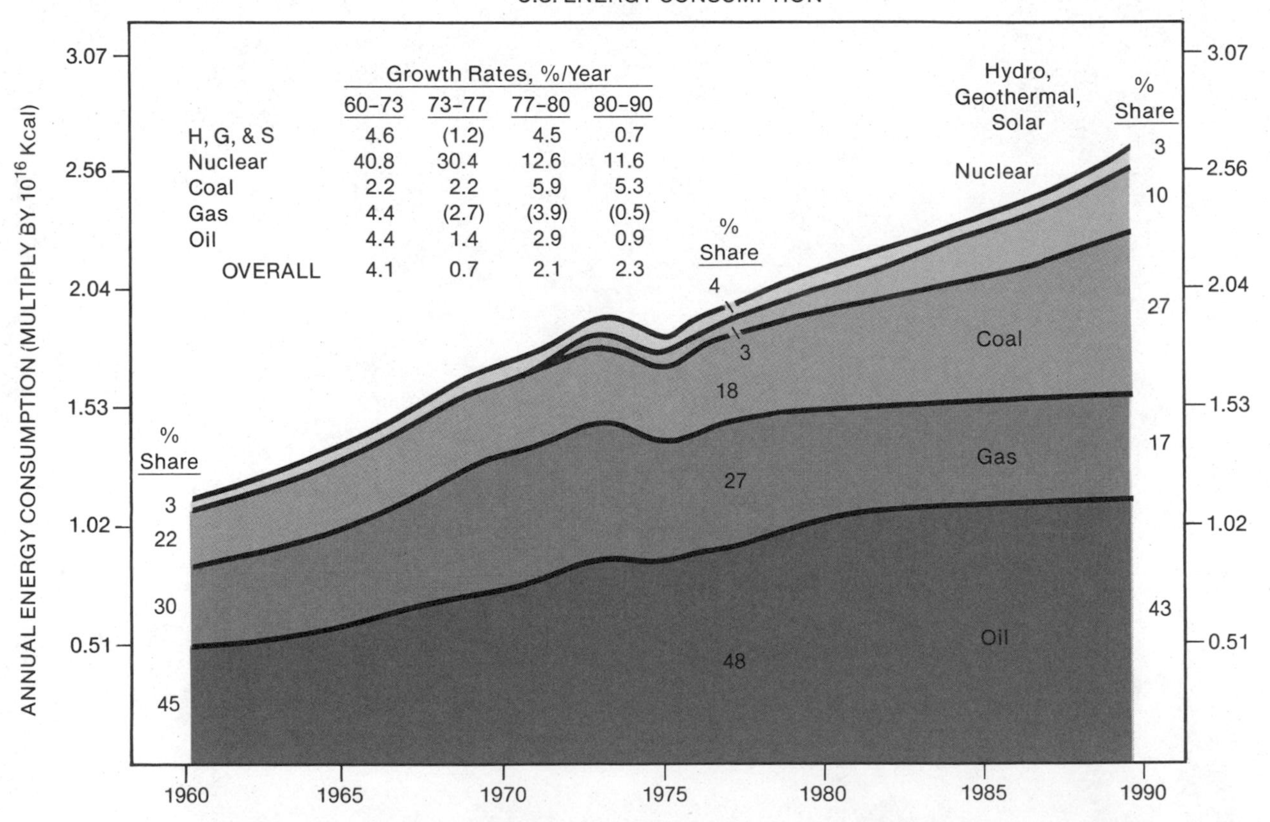


**Figure 14–1** Energy consumption in the United States—the immediate view. Note that the burning of fossil fuels furnishes nearly all of our present energy in spite of all the talk about hydroelectric and nuclear energy.

Our dependence upon the energy from coal, petroleum, and natural gas has been uncomfortably and abruptly realized recently, but the present shortages and high prices of energy have been on the horizon for several years. The power brownouts in the eastern United States in the late 1960's were a signal that something was amiss. In the early 1970's, gasoline shortages were caused by inadequate refining capacity, low return on investment capital, and a war in the Middle East. In 1973, the Organization of Petroleum Exporting Countries (OPEC) declared an embargo on oil shipments to all nations that supported Israel, demanding and getting political concessions. This made energy even more a matter of public concern; the price of oil was quadrupled, and has been raised a number of times since.

**Why an energy shortage?**

To put the increasing dependence upon energy into perspective, consider Figure 14–2. As late as 1940, 80% of the world's population was living at the primitive or advanced agricultural level. Consequently, little energy was being used. Today, about 25 to 30% of the present world population lives in a highly industrialized society and uses 80% of the world's energy. In the United States, 6% of the world's population uses 31% of the world's energy production in any given year. Through conservation of energy in the United States and increased worldwide consumption of energy, our relative position in the use of the world's production of energy is declining. The advanced development of other countries causes serious worldwide competition for the types of energy sources currently available, and the supplies of the types being used are becoming depleted.

**Who is using energy?**

Our primary purposes in this chapter are to examine our present sources of energy to see what chemicals are used and how energy is obtained from these chemicals.

**Purposes of this chapter.**

**Figure 14–2** Energy use per capita based on various types of societies.

However, we shall first discuss some of the fundamental principles that guide and limit the extraction and use of energy.

## Fundamental Principles of Energy

### Definitions of Energy and Power

**Energy is the ability to move matter.**

**Power is the rate of using energy.**

**Thermodynamics is the movement of energy.**

***Energy*** is the ability to move matter, that is, to do work. Some comparisons among units of energy and sources of energy are shown in Table 14–1. Types of energy include chemical, light, heat, sound, electrical, nuclear, and mechanical (a wound clock spring or moving car). All of these types of energy can move things. A closely related term that is often confused with energy is ***power,*** which is the rate at which energy is used. Power has the units of energy used per time, such as calories per second or joules per second (watts).

### Law of Conservation of Energy

**First statement of the first law of thermodynamics: "A force [translated: energy] once in existence cannot be annihilated." Julius Robert Mayer, a ship's doctor, 1840.**

**The law of conservation of energy has been extended by the discovery of Albert Einstein, who showed the interrelationship between mass and energy by $E = mc^2$.**

**The more general law is stated: The total amount of matter and energy in the universe is constant.**

Also known as the first law of thermodynamics, this law asserts that energy is neither lost nor gained in all energy processes. In heating a beaker of water with a burner, all of the energy given off by the flame can be accounted for. For example, the heat, light, and sound of the flame heat the beaker, stand, air, and water. No energy is lost or gained in the transformation of chemical energy to heat energy. Furthermore, the transformation is quantitative in that a certain amount of gas burned produces a certain amount of energy (see Table 14–1). In the changing of one kind of energy into another, the exchange rate is definite, reliable, and reproducible.

This law also implies that the total amount of energy in the universe is constant. The energy is being transformed regularly from one kind to another, but the total remains the same. Of interest to Earth travellers is the fact that the sun and the energy stored in the chemicals on earth are what we have to use—that is all! No creation of energy is ongoing.

One last implication of the law applicable to this study is the limitation the law puts on perpetual motion. The law recognizes that a machine, by running, cannot produce enough energy to run itself, much less the creation of enough energy to be used elsewhere. At a minimum, the machine would have to create enough energy to move its parts and to overcome friction. Since this is a creation process, and not a transformation, the law says this is impossible.

**TABLE 14–1 A Handy Chart of Energy Units***

| Cubic Feet of Natural Gas | Barrels of Oil | Tons of Bituminous Coal | British Thermal Units (BTU) | Kilowatt Hours of Electricity | Joules | Kilo-calories† |
|---|---|---|---|---|---|---|
| 1 | 0.00018 | 0.00004 | 1000 | 0.293 | $1.055 \times 10^6$ | 252 |
| 1000 | 0.18 | 0.04 | $1 \times 10^6$ | 293 | $1.055 \times 10^9$ | $0.25 \times 10^6$ |
| 5556 | 1 | 0.22 | $5.6 \times 10^6$ | 1628 | $5.9 \times 10^9$ | $1.40 \times 10^6$ |
| 25,000 | 4.50 | 1 | $25 \times 10^6$ | 7326 | $26.4 \times 10^9$ | $6.30 \times 10^6$ |
| $1 \times 10^6$ | 180 | 40 | $1 \times 10^9$ | 293,000 | $1.055 \times 10^{12}$ | $0.25 \times 10^9$ |
| $3.41 \times 10^6$ | 614 | 137 | $3.41 \times 10^9$ | $1 \times 10^6$ | $3.6 \times 10^{12}$ | $0.86 \times 10^9$ |
| $1 \times 10^9$ | 180,000 | 40,000 | $1 \times 10^{12}$ | $293 \times 10^6$ | $1.055 \times 10^{15}$ | $0.25 \times 10^{12}$ |
| $1 \times 10^{12}$ | $180 \times 10^6$ | $40 \times 10^6$ | $1 \times 10^{15}$ | $293 \times 10^9$ | $1.055 \times 10^{18}$ | $0.25 \times 10^{15}$ |

*Based on normal fuel heating values. $10^6$ = 1 million, $10^9$ = 1 billion, $10^{12}$ = 1 trillion, $10^{15}$ = quadrillion (quad).
†A food Calorie = 1000 calories = 1.000 kilocalorie.

## Energy Is Conserved in Quantity but Not in Quality

This is but one of the many ways to state the second law of thermodynamics; another statement pertinent to this discussion will be given presently. But, first, what does it mean that energy is conserved in quantity but not in quality? Perhaps two of the many available examples will clarify the concept. Consider as a first example the commonly known facts that coal, petroleum, and wood, along with air, have energy stored in their chemical structures, and that some of this energy is released during burning. It is also well known that the main products of the burning process, carbon dioxide and water, will not burn and release more energy. In the burning process, both matter and energy are conserved, which is required by the laws of conservation of matter and energy, respectively. However, the reactants and their stored energy are more useful in energetic terms than the products and their spent energy.

**Usable energy is not conserved.**

As a second example, consider an electrical motor. The electricity that runs the motor is more useful than the heat that comes from the warm motor. Again, energy is conserved in the process of running an electrical motor, yet the usable energy is not conserved.

**No matter how we try, we can never convert all of the stored energy in a system into usable energy.**

In concept, the energy relationships in the second law of thermodynamics can be compared to the relationships among gross income, deductions, and net pay (or realizable income) in a paycheck. A certain amount of energy is available for the process considered; this is analogous to gross income. Some of the energy is not usable due to frictional losses, electrical shorts and drains, retention of some energy in the chemical products, or some other factor affecting efficiency; the energy that is not usable is represented in the analogy by the paycheck deductions. Finally, some of the energy is usable and is represented by net pay in the analogy. Both the energy (and the money) are accounted for as required by the first law of thermodynamics.

In all processes, then, some energy is wasted—not lost—by conversion into energy that is not usable in the process. The wasted (or not usable) energy is represented by ***entropy*** (Greek, meaning disorder). Another statement of the second law of thermodynamics is based on entropy: In all natural processes, entropy is increased. Taken to its extreme, this means that the entropy of the whole universe is increasing at the expense of stars running down in usable energy at a tremendous rate. This is not a reason for worry because the universe is so vast that enough usable energy is there for all conceivable purposes for many billions of years. However, a source of usable energy that is not limitless is the so-called fossil fuels (coal, petroleum, and natural gas), which when gone are not easily restored. It would take eons of time for photosynthesis to regenerate the material for new fossil fuel deposits.

***Entropy* means disorder, and measures nonuseful energy.**

Why, then, is the energy used to increase entropy not usable energy? The derivation of the word entropy, meaning disorder, explains. The ultimate fate of any change in energy is a form of heat energy caused by the random, disordered motion of molecules. Have you ever thought about what happens to the light energy coming from a light bulb, or what happens to the electrical energy once it is used to run an electrical motor, or what happens to the sometimes large amounts of energy that result from an explosion? All forms of energy, including sound, are converted eventually into random molecular motion—the molecules move faster and (or) further apart. Molecules moving in all directions are not as useful in bringing about controlled change as are electrons, photons of light, or molecules when they are moving from one point to another in organized fashion. The type of energy is important if it is to be useful (moving molecules cannot run an electrical motor; moving electrons can), but for all types of usable energy the usefulness also comes from the organized, nonrandom movement, for example, of electrons from a generator or battery to the motor, of light from its source outward, and of molecules streaming from a gas jet. Useful energy involves organized flow.

**The end of the line for energy is the random motion of molecules.**

Let us summarize this brief encounter with the second law of thermodynamics by describing the energy coming from a burning match. Some of the energy is usable to ignite other objects, or to heat an object, or to provide light. This is the directional, organized energy. All the while that the usable energy is being used, some of the total energy emanating simply heats molecules in the vicinity and increases the entropy of the molecules. Eventually, all of the heat and light coming from the match will become increased random motion of the molecules.

When fossil fuels are gone, then what?

What does the second law of thermodynamics mean to the informed citizen? Simply stated, when usable energy-rich chemicals such as coal and petroleum are consumed, the usable energy is lost to us.

## The Efficiency of Energy Use Is Low

Efficiency is $\frac{\text{used energy}}{\text{available energy}}$.

Primary source of energy: one transformation on site (e.g., chemical ⟶ heat via combustion). Secondary source: usually more than one transformation plus long-distance transport (e.g., chemical ⟶ heat via combustion ⟶ steam ⟶ mechanical ⟶ electricity).

In every energy process, the efficiency of the use of the energy for doing work is less than 100%—usually far less. Automobiles are about 22–25% efficient, that is, about 80% of the useful energy available to do work is lost and not applied to turning the wheels. The human body is about 45% efficient. Photosynthesis is 2–10% efficient; steam turbines for producing electricity are about 38% efficient; heating homes with electricity is about 38% efficient, whereas heating homes with natural gas is about 70% efficient. The efficiency is usually greater when using a ***primary source*** of energy on site (burning gas) than when using a ***secondary source*** (electricity). For example, it takes about 10,000 BTU's to produce one kilowatt-hour (kwh) of electricity. If this one kwh is then used for heating, only 4,000 BTU's of heat are produced. Natural gas burned on site would be more efficient than if burned in a steam generator plant to produce the electricity.

## Energy Not Lost Is Energy Gained

Energy can be transported through wires (electricity), stored in chemicals (batteries), and carried through the air (radio waves). On the other hand, energy can be prevented from moving by means of insulators. The insulation of houses has popularized the ***R-value*** for heat insulators. The R-value (the resistance) is inversely proportional to the conductivity of heat through a slab of material. A common unit of R-value is ($ft^2$)(°F)(hr/BTU). The typical recommendation of an R-value of 30 for the ceilings of single-family dwellings means that an average square foot of such a ceiling would lose heat by conduction at a rate of (1/30) BTU/hr for every 1°F difference in temperature. The higher the R-value, the less BTU's escape per hour per square foot of ceiling. Some R-values for one inch slabs of the material (in units of $ft^2$ °F hr/BTU) are air, 5.9; polyurethane foam, 5.9; rock wool, 3.3; fiberglass 3.0; white pine, 1.3; and window glass, 0.14. Dry, still air has an insulating value (R-value) as great as almost any building material. In fact, many commercial materials owe their heat insulating ability to entrapped, isolated pockets of air.

The R-factor for heat insulators.

## Some Materials Have a Higher Energy Cost Than Other Materials

It costs less energy to make a ton of bricks than a ton of aluminum.

It costs more in energy terms to produce a ton of some substances than to produce the same amount of other substances (Table 14–2). Certain applications now using plastics or metals might more efficiently use ceramics or brick in order to conserve energy. Of course, other factors such as labor costs also influence the economic decisions involved.

Now let's consider the chemicals that provide energy, and see how energy is obtained from them.

**TABLE 14–2 Energy Requirements to Produce Some Common Products**

| Product | Millions of BTU/Ton |
|---|---|
| Titanium | 482 |
| Aluminum | 244 |
| Copper | 112 |
| Polyethylene | 100 |
| Polystyrene | 64 |
| Polyvinylchloride | 49 |
| Plate glass | 25 |
| Steel slabs | 24 |
| Paper | 22 |
| Portland cement | 8 |
| Brick | 4 |

## Fossil Fuels

**The world fossil fuel usage is doubling every 10 years.**

**Fossil fuels store the energy of the sun.**

As indicated in Figures 14–1 and 14–3, nearly all of the present energy needs in the United States are being supplied by fossil fuels: petroleum, natural gas, and coal. This is also true on a worldwide basis. Although the chemistry and geology of fossil fuel formation are not thoroughly understood, it is generally agreed that buried plant material formed these fuels over millions of years. So far as we know, no coal and oil are being formed underground today.

When fossil fuels are burned, chemical energy is released. The amount of energy varies with the type of fuel (Table 14–3). In each case, the complete combustion products are carbon dioxide and water.

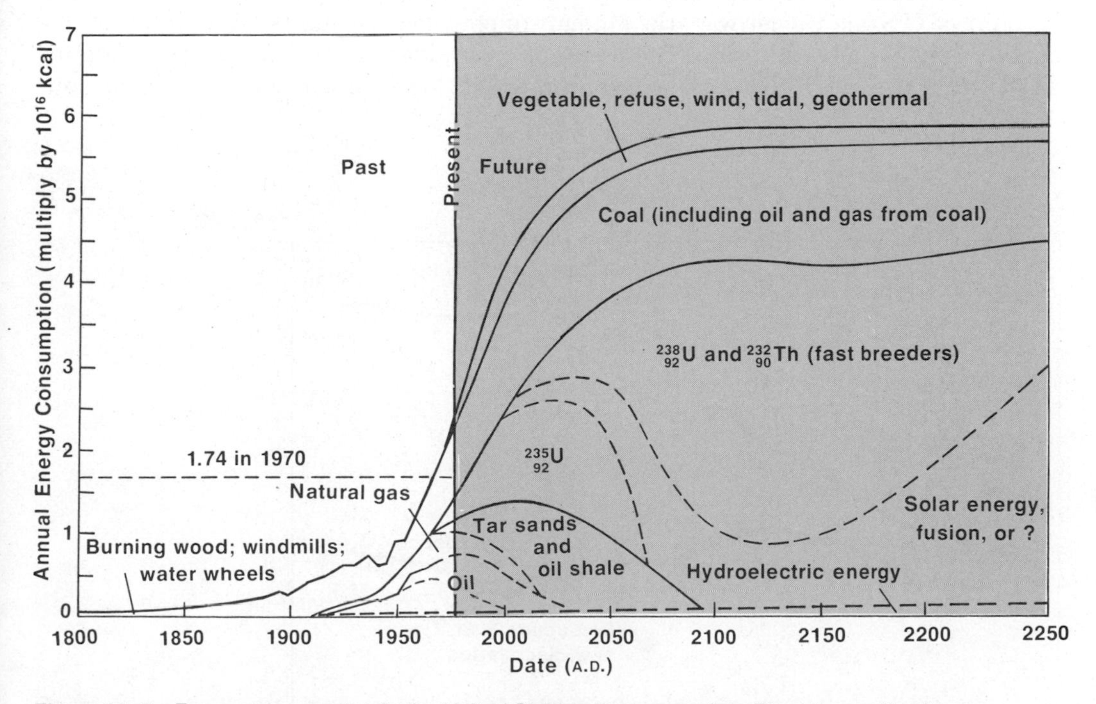


**Figure 14–3** Energy consumption in the United States—the longer view. The future estimates are based on 400,000 kilocalories per day per person after 2000 A.D., a zero population growth by 2100 A.D., and a total population of 400 million.

**TABLE 14–3 Energy Content of Fossil Fuels**

| Fuel | Reaction Equation | Heat Energy Evolved* |
|---|---|---|
| Coal | $C + O_2 \longrightarrow CO_2$ | 94 kcal/mole; 7.8 kcal/g |
| Natural gas** | $CH_4 + 2O_2 \longrightarrow CO_2 + 2H_2O$ | 211 kcal/mole; 13.2 kcal/g |
| Petroleum† | $2C_8H_{18} + 25O_2 \longrightarrow 16CO_2 + 18H_2O$ | 1,303 kcal/mole; 11.4 kcal/g |

*Numbers are for the quantity of *fuel* (C, $CH_4$, or $C_8H_{18}$) specified.
**$CH_4$, methane, is the principal constituent in natural gas (up to 97%).
†$C_8H_{18}$, octane, is only one of many hydrocarbons present in petroleum.

## Petroleum

**One barrel of petroleum contains 42 gallons.**

Vast deposits of petroleum were first discovered in the United States (Pennsylvania) in 1859 and in the Middle East (Iran) in 1908. Since that time petroleum has found wide use as an energy source, first as kerosene for lighting, then as gasoline and aircraft fuel for transportation, and more recently as fuel oil to produce electricity. All of this has resulted in the use of over 5.4 billion barrels of oil per year in the United States in recent years.

Figure 14–4 illustrates the overall U.S. oil supply. There were an estimated 200 billion barrels of oil to be recovered in the United States (including Alaska) when oil production first began. Half of this amount has now been produced (see Figure 14–5). Numbers of this type are often quoted, and a few qualifications should be noted. ***"Recoverable oil"*** is based on 30% recovery, i.e., pumping 30% from the ground and leaving 70% behind, since to recover more would cost more and hence raise the cost of the average barrel of oil. It stands to reason that as supplies decrease and costs go up, it will prove economically feasible to "recover" more of the oil.

**The definition of recoverable oil is based on economics.**

A second estimate of U.S. reserves of oil can be based on as yet undiscovered oil under the continental shelf. This estimate is 400 billion barrels, thus doubling the amount of U.S. oil. Either way, the amount of petroleum is limited. Figure 14–6 presents the situation worldwide. The assumption underlying the projection in Figure 14–6 is that the rest of the world will continue in its lower rate of energy consumption.

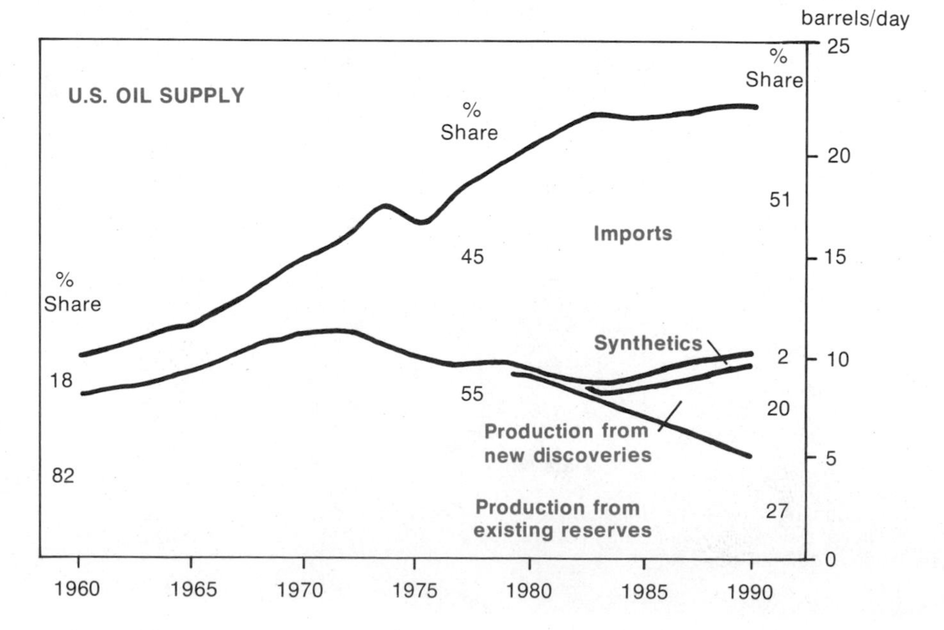


**Figure 14–4** The U.S. oil supply, since 1970, has come increasingly from imported oil. Our demand for petroleum has continued to increase despite greater gasoline mileage from automobiles and somewhat lessened driving.

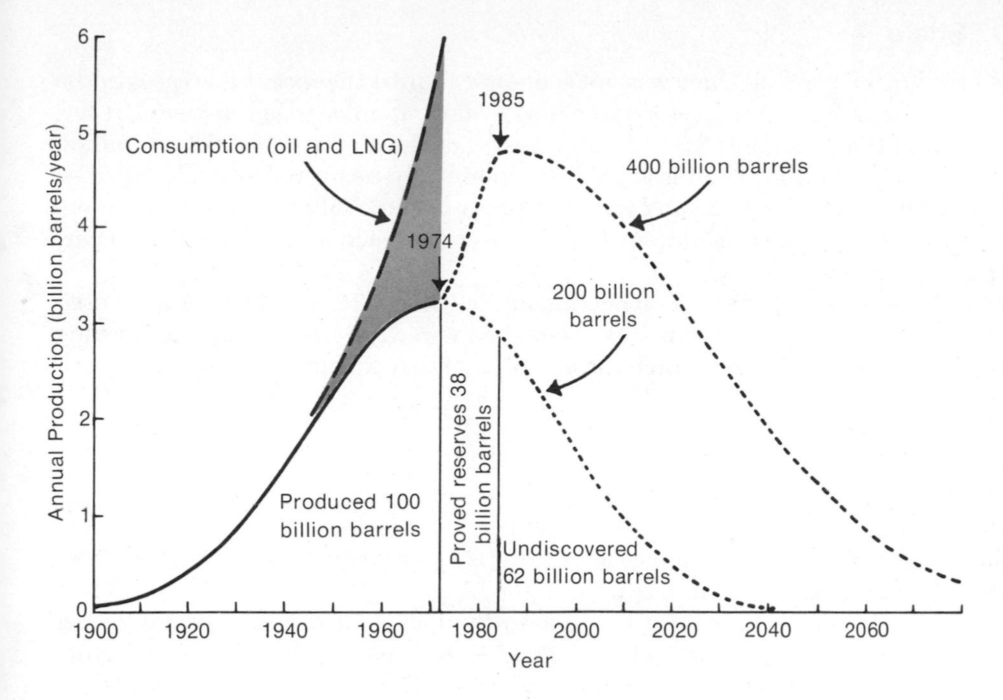


**Figure 14–5** History of oil production in the United States (including Alaska). The dashed curve represents oil imports. The lower curve is based on known oil reserves, and the higher dotted curve is based on estimates of what is likely to be found with large deposits expected under the continental shelf. Regardless of how much is actually present, it is a finite amount and apparently will be consumed in a relatively short period of time. (LNG = natural gas liquefied).

This is not a feasible assumption. Based on the present trends, a reasonable projection is that 80% of the world's petroleum will be used in a 60- to 70-year period, ending at about the year 2025. Unless drastic changes are made at some point in the relatively near future, perhaps as early as the year 2000, petroleum will become an increasingly rare commodity.

**Oil may become more valuable as a chemical raw material than as an energy source.**

As indicated earlier in this text, petroleum is the base for a large petrochemical industry whose products include plastics, fibers, paints, and medicines. A real chemical consideration that must come into focus in the future as the petroleum supply decreases is whether we want to burn petroleum for its energy content or save it as a starting material for petrochemical products.

The petrochemical industry may very well be required to produce basic foodstuffs in the future. Edible fats were produced in Germany during World War II. Glycerol is now made from petroleum on a commercial scale, and we know how to make sugar from oil.

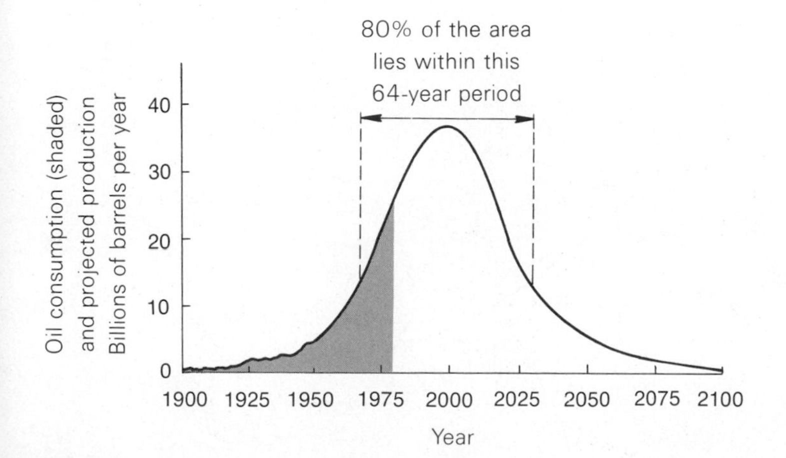


**Figure 14–6** World oil use: past and projected.

## Oil Shale

Extracting oil from oil shale was not economical until the recent increases in the cost of crude oil prompted at least a dozen major oil companies to get involved. It has been estimated that the United States has reserves of shale oil that total 1,800 billion barrels and that 600 billion barrels could be economically recovered at current prices. Occidental Petroleum Corporation expects to produce shale oil near Grand Junction, Colorado by 1986, and the production of this mine could reach as much as 200,000 barrels per day.

The oil comes out of the rock on heating in the absence of air. The technique used to get the oil to the surface involves: (1) blasting the rock to break it up, (2) burning a part of the oil underground to force the oil out of the rock, and (3) pumping the collected oil to the surface.

## Coal

**The largest supply of fossil fuel is in the form of coal.**

Unlike petroleum, where discoveries are being made almost daily, geologists believe that all the world's coal supplies have now been discovered. ***Minable coal*** is defined as 50% of all coal that is in a seam at least 12 inches thick and within 4000 feet of the surface. Estimates have been made as to the length of time this minable coal will supply our energy needs. Figure 14–7 indicates that coal will be available as a fuel for a much longer period of time than oil. If new mining techniques are developed, then more of the deposited coal might be termed minable.

Two of coal's major drawbacks are that it is a relatively dirty fuel and difficult to handle. Coupled with the atmospheric pollution caused by sulfur-containing coal, these drawbacks were prime reasons for a major shift from coal to petroleum in electrical generating plants, particularly in the industrialized nations, late in the 1960's. The dangerous and unhealthful character of deep coal mining and the environmental disruption caused by strip mining contributed to the shift.

### Coal Gasification

**Gasification can make coal cleaner to burn and easier to handle from supplier to user.**

Coal can be converted into a relatively clean-burning fuel by a process known as ***gasification*** (Fig. 14–8). In this process, coal is made to react with either a limited supply of hot air or steam. In the reaction of coal with air, the product is a gaseous mixture known as ***"power gas."***

$$\text{Coal} + \text{Air} \longrightarrow \underbrace{CO\,(g) + H_2\,(g) + N_2\,(g)}_{\text{POWER GAS}}$$

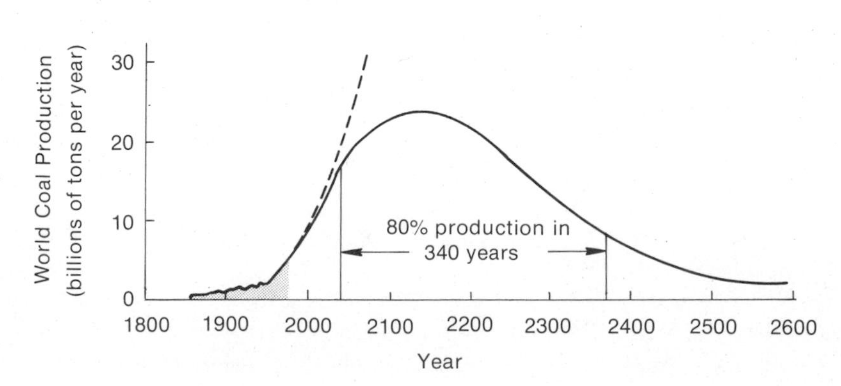


**Figure 14–7** The coal mined to date (*shaded area*) represents only a small fraction of the minable coal. The rate of increase in coal consumption (*dashed line*) is 3.56% per year. It is obvious that such an exponential rise cannot continue long after the year 2000. At the present usage (held constant) coal would last for many hundreds of years.

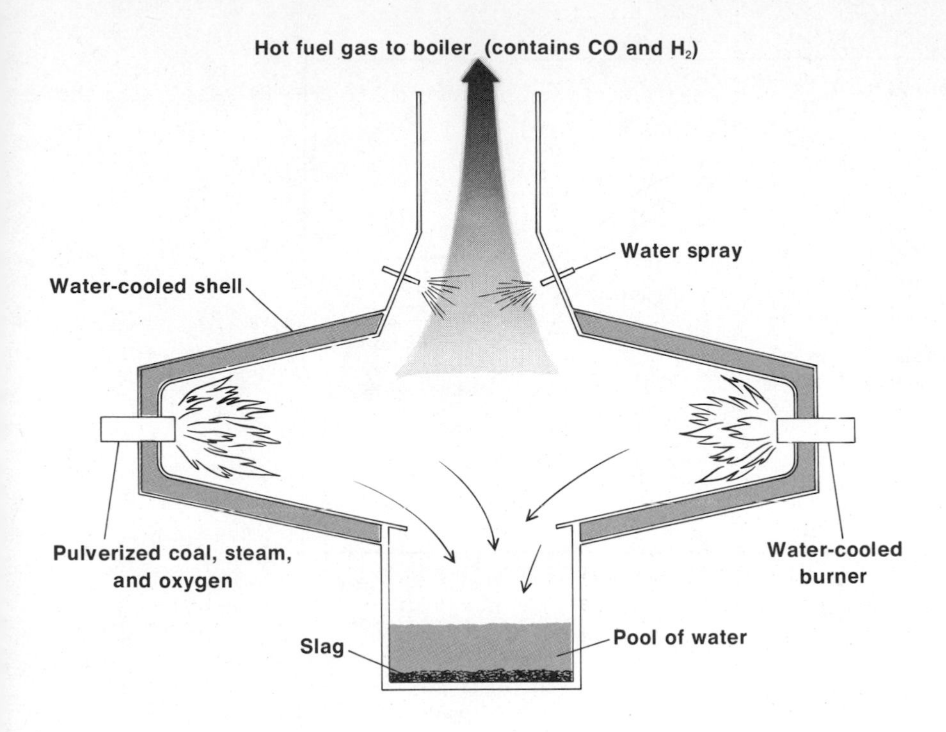


**Figure 14–8** Schematic drawing of coal gasifier. A relatively cool combustion of powdered coal in a limited supply of oxygen produces a mixture of carbon monoxide and hydrogen along with other gases. The mineral content in the coal collects in the slag.

Power gas contains up to 50% nitrogen by volume and is consequently a relatively poor fuel. In fact, power gas of this composition has only one sixth the heat content of methane.

If the coal is allowed to react with high-temperature steam, a mixture of carbon monoxide and hydrogen (which contains no nitrogen) known as ***synthesis gas*** or ***coal gas*** is obtained.

$$\underset{\text{COAL}}{C} + \underset{\text{STEAM}}{H_2O\,(g)} \longrightarrow \underset{\text{SYNTHESIS GAS OR COAL GAS}}{CO\,(g) + H_2\,(g)} - 31\ \text{kcal/mole C}$$

**This reaction is endothermic.**

In either mixture, the CO and $H_2$ are burned in the air to produce heat. The heat produced is about one third that of an equal volume of methane (natural gas).

$$2CO + O_2 \longrightarrow 2CO_2 + 135.3\ \text{kcal}\ (67.6\ \text{kcal/mole CO})$$

$$2H_2 + O_2 \longrightarrow 2H_2O + 115.6\ \text{kcal}\ (57.8\ \text{kcal/mole}\ H_2)$$

**These reactions are exothermic.**

Other opportunities exist in the area of coal modification, since liquid fuels can also be obtained from coal. Knowledgeable estimates indicate that we will not be able to rely on coal modification to supply our energy needs on an extensive scale before 1990.

## Natural Gas

Natural gas approaches an ideal fuel. It burns with a high heat output (Table 14–3), with little or no residue, and is easily transported. The problem with natural gas is its limited supply. Federal price regulations begun in 1954 held down prices, stimulated demand, and decreased incentives for exploration of new gas deposits. These price regulations have been removed. It remains to be seen whether new gas will become available. Certainly the present gas will be higher priced.

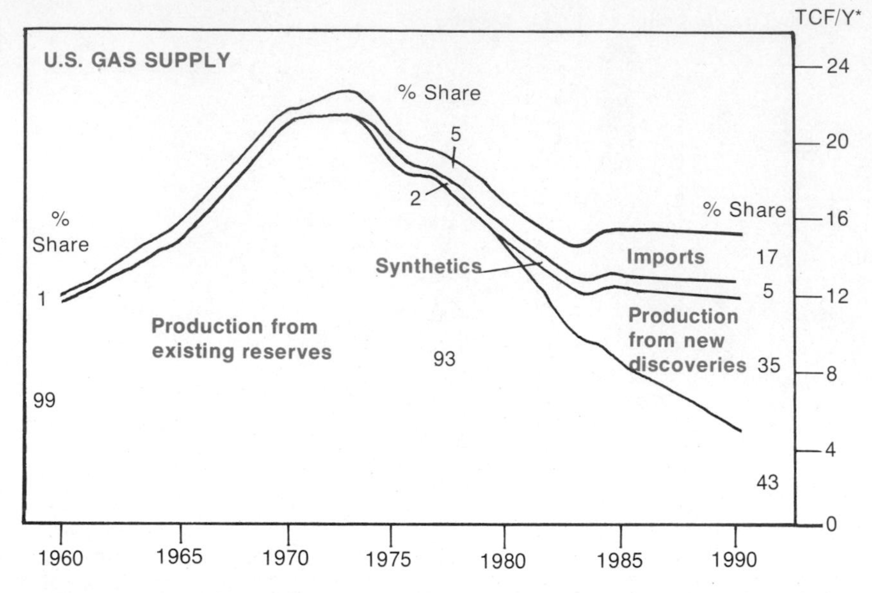


**Figure 14–9** The U.S. gas supply has been mostly from existing reserves within the U.S., unlike the oil situation in recent years. Natural gas consumption may never again equal that of the mid-1970's.

In 1973, United States gas production peaked at 22 trillion ($10^{12}$) cubic feet per year. Even if significant new deposits of natural gas are discovered in such locations as the outer continental shelf, the North American natural gas deposits are about 50% depleted. Figure 14–9 illustrates the U.S. gas supply. Importation of natural gas is complicated by difficulties encountered in condensing natural gas to a liquid for ocean transit and the danger of transporting and storing such concentrated and very volatile fuel.

## Electricity Production

When electricity is discussed, electrical power is usually mentioned. While energy is the ability to do work, power is the *rate* of doing work (Table 14–4).

Most consumers purchase electrical energy (kwh) and not power, although appliances and other electrical devices are rated according to power. A 100-watt light bulb operating at 100 volts would draw 1 ampere of current; in one hour it would use 100 watt-hours or 0.1 kwh of energy.

**When a power unit is multiplied by a time unit, the result is an equivalent energy unit. energy units = watts × time i.e., 1 kilowatt-hour = $10^3$ w × 1 hr = $\frac{10^3 \text{ joules}}{\text{second}} \times \frac{3600 \text{ sec}}{\text{hr}} \times 1 \text{ hr} = 3.6 \times 10^6$ joules.**

Ultimately, the cost of electricity (except hydroelectricity) must be related to the cost of the fuel used to produce the electricity. In 1978, about 30% of all the energy con-

**TABLE 14–4 Power Units**

| |
|---|
| 1 watt (w) = 1 joule per second |
| 1 kilowatt (kw) = $10^3$ w |
| 1 megawatt (Mw) = $10^3$ kw = $10^6$ w |
| 1 watt = 1 volt × 1 ampere |

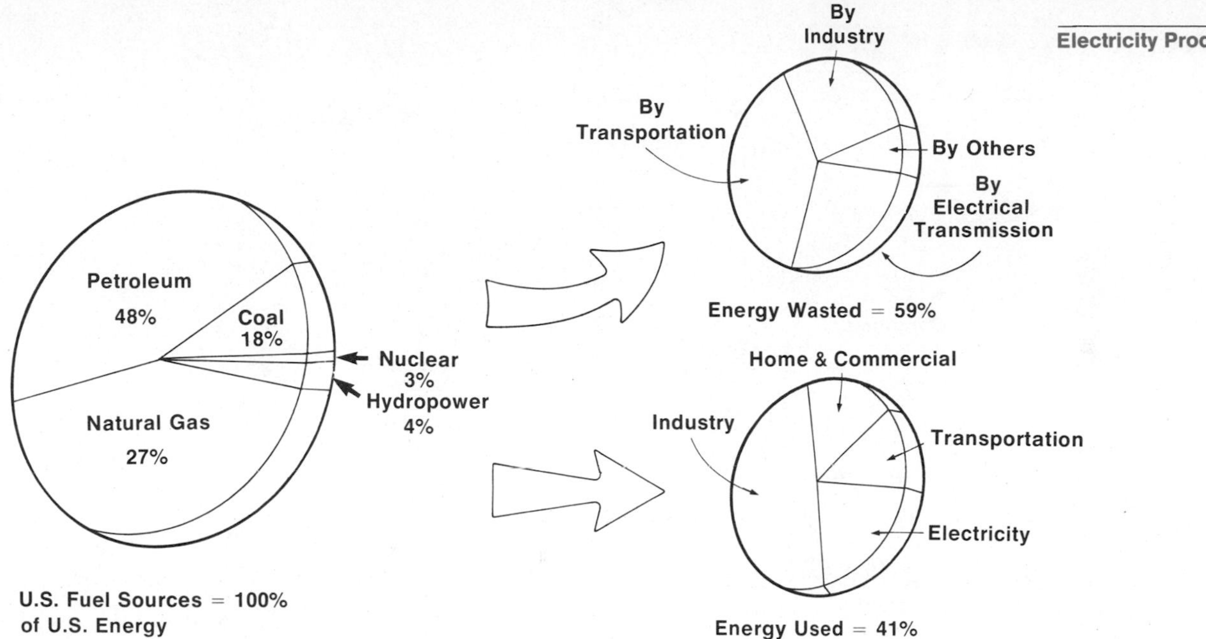


**Figure 14–10** The annual fuel "pie" in the United States. Of the fuel sources left, this energy is more than half (59%) wasted. Only 41% of the fuel we pay for is actually used. The situation is partly due to the laws of thermodynamics and partly due to wasteful practices, sloppy planning, and poor design of fuel-using devices.

sumed in the United States was used to produce electricity. This accounted for 22% of all the fossil fuels consumed in this country. However, only 30% of the energy contained in those fuels is available at the point where the electricity is used (Fig. 14–10).

**Electrical generating plants yield about one third of the fuel energy in the form of electrical energy.**

Part of the energy loss in electricity production is illustrated in a schematic diagram of a large modern fossil-fuel generating plant (Fig. 14–11). For a 1,000-megawatt coal-burning plant, one hour of operation might look like this:

| | |
|---|---|
| Coal consumed | 696 tons producing 2.270 billion kilocalories |
| Smokestack heat loss | 0.227 billion kilocalories |
| Heat loss in plant | 0.106 billion kilocalories |
| Heat loss in evaporator to cool condenser | 1.080 billion kilocalories |
| Electrical energy delivered to power lines | 0.857 billion kilocalories |
| Percentage of energy delivered as electricity before transmission losses | $\frac{0.857}{2.27} \times 100\% = 37.8\%$ |

There is a further energy loss in the power lines and the transformers, which lowers the useful output of the plant to 30% of the energy consumed. This is the ***efficiency*** figure for the overall operation. It is important to note that we pay for 300 kcal of heat energy in the form of coal or fuel oil but receive less than 100 kcal of energy in the form of electricity. Obviously it requires much less fuel to heat homes with the fuel itself than with electricity made from the fuel.

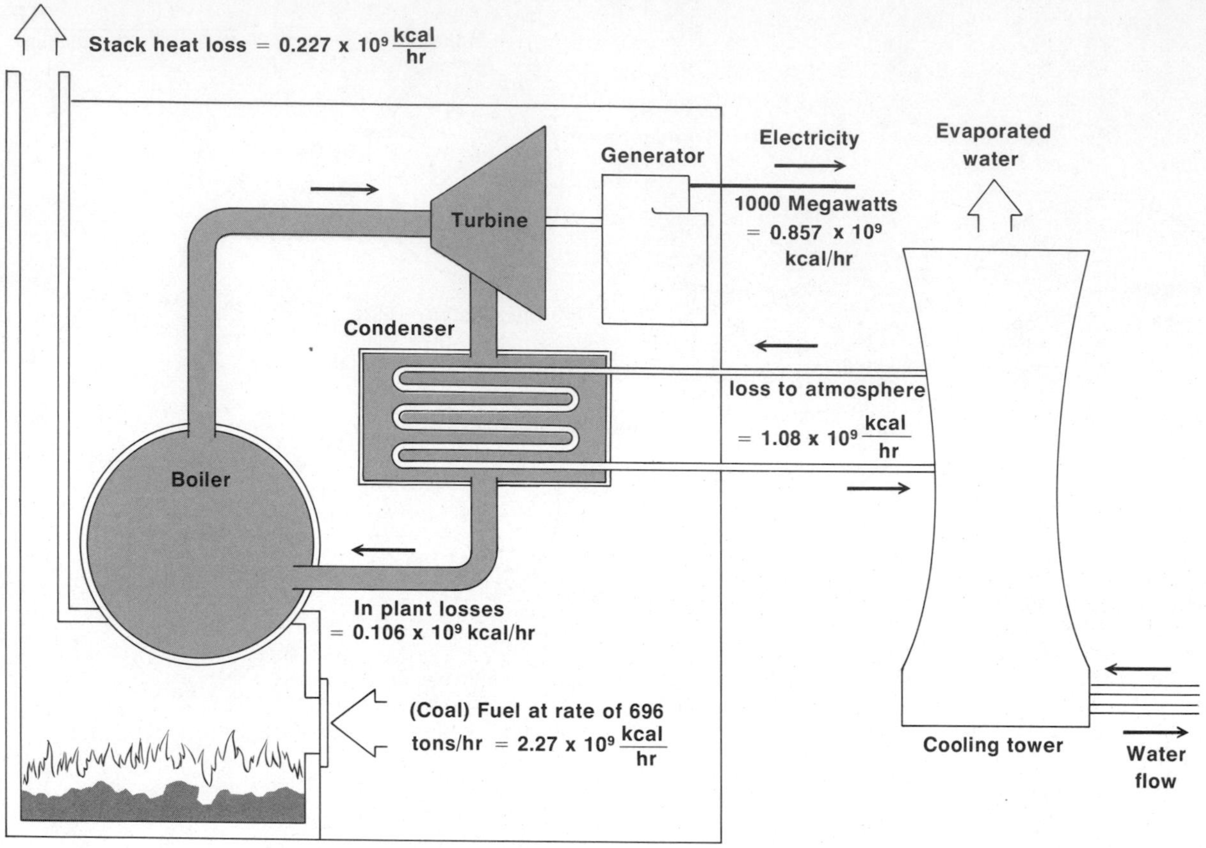

Stack heat loss = $0.227 \times 10^9 \frac{kcal}{hr}$

Generator

Turbine

Electricity

1000 Megawatts = $0.857 \times 10^9$ kcal/hr

Evaporated water

Condenser

loss to atmosphere = $1.08 \times 10^9 \frac{kcal}{hr}$

Boiler

In plant losses = $0.106 \times 10^9$ kcal/hr

(Coal) Fuel at rate of 696 tons/hr = $2.27 \times 10^9 \frac{kcal}{hr}$

Cooling tower

Water flow

**Figure 14–11** The heat balance of a 1000-megawatt coal burning electrical generating plant. Note that the 969 tons of coal burned per hour furnish $2.27 \times 10^9$ kcal of heat energy, but only $0.857 \times 10^9$ kcal of energy, or 38%, is converted to electricity. Note also the large amounts of heat energy lost to the cooling water and atmosphere.

## Self-Test 14-A

1. In 1980, approximately what percentage of the total U.S. energy consumption was furnished by coal? See Figure 14–1. ____________
2. Which furnishes the most heat energy per gram—coal, petroleum, or natural gas? ____________
3. How many gallons of oil are there in one barrel? ____________
4. Using the recovery rate of 30% for oil in place, approximately how large a percentage of the oil in the United States has been thus far pumped from the earth (Fig. 14–5)? ____________
5. Is the composition of "power gas" obtained from coal gasification $CO$, $H_2$, $N_2$ or $CO$, $H_2$?
6. The typical efficiency of an electrical generating plant is about (100, 50, 33, 10) percent.
7. Examples of fossil fuels are ____________, ____________, and ____________.
8. Natural gas and petroleum react with ____________ to produce $CO_2$ and ____________.

9. All combustions of fossil fuels give off energy (True/False).
10. Energy is the ability to do ____________.
11. One type of energy (for example, light) is always transformed into another type of energy (for example, heat) (1) quantitatively, (2) not quantitatively, or (3) sometimes quantitatively, sometimes not quantitatively.
12. The ultimate fate of all types of energy is an increase in ____________.
13. Although the quantity of energy is conserved, the ____________ of energy is not conserved.
14. Three units of energy are ____________, ____________, and ____________.
15. Two units of power are ____________ and ____________.
16. Which costs more energy to produce, a ton of aluminum or a ton of brick? ____________ Which costs more money to buy? ____________

## Nuclear Energy

Few issues have captured the awe, imagination, and scrutiny of mankind to quite the extent that nuclear energy has in the past four decades. Nuclear energy has been acclaimed, on the one hand, as the source of all of our energy needs, and accused, on the other hand, of being our eventual destroyer. Recently, nuclear power plants and neutron bombs have received much public attention. Part of the interest in nuclear power is the tremendous amount of energy achieved from a relatively small amount of fuel. The vast amounts of energy are released when heavy atomic nuclei split, the ***fission*** process, and when small atomic nuclei combine to make heavier nuclei, the ***fusion*** process. Consider the energy contrast between combustion of a fossil fuel and a nuclear fusion reaction. When one mole ($6.02 \times 10^{23}$ molecules, or 16 g) of methane is burned, over 200 kcal of heat are liberated:

**Methane is the major component of natural gas.**

$$CH_4 + 2O_2 \longrightarrow CO_2 + 2H_2O + 211 \text{ kilocalories (kcal/mole } CH_4)$$

**Energy changes associated with nuclear events may be many thousands of times larger than those associated with chemical events.**

In contrast, a lithium nucleus can be made to react with a hydrogen nucleus to form two helium nuclei in a nuclear reaction. The energy released per mole of lithium in this reaction is 23,000,000 kcal. This means that 7 g of lithium and 1 g of hydrogen produce 100,000 times more energy through fusion of nuclei than 16 g of methane and 64 g of oxygen produce by electron exchange.

$$^{7}_{3}Li + ^{1}_{1}H \longrightarrow 2^{4}_{2}He + 23{,}000{,}000 \text{ kcal/mole of } ^{7}_{3}Li$$

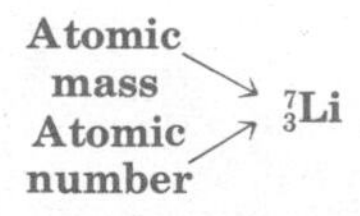


**Note in nuclear reactions that the sum of the atomic numbers on the left-hand side of the equation equals the sum of the atomic numbers on the right-hand side of the equation. Likewise, for the atomic masses.**

Realizing that nuclear changes could involve giant amounts of energy relative to chemical changes for a given amount of matter, Otto Hahn, Fritz Strassman, Lise Meitner, and Otto Frisch discovered in 1938 that $^{235}_{92}U$ is fissionable. Subsequently the dream of controlled nuclear energy became a reality, followed by the bomb and nuclear power plants. In the 1950's it was hoped that nuclear energy would soon relieve the shortage of fossil fuels. To date this has not been accomplished, although the production of nuclear energy has grown very rapidly in recent years.

### Fission Reactions

Fission can occur when a thermal neutron (with a kinetic energy about the same as that of a gaseous molecule at ordinary temperatures) enters certain heavy nuclei with an odd number of neutrons ($^{235}_{92}U$, $^{233}_{92}U$, $^{239}_{94}Pu$). The splitting of the heavy nucleus

Fission is the breakup of heavy nuclei.

produces two smaller nuclei, two or more neutrons (an average of 2.5 neutrons for $^{235}_{92}U$), and much energy. Typical nuclear fission reactions are written as follows:

$^{1}_{0}n$ represents a neutron.

$$^{235}_{92}U + ^{1}_{0}n \longrightarrow ^{141}_{56}Ba + ^{92}_{36}Kr + 3^{1}_{0}n + \text{energy}$$

$$^{235}_{92}U + ^{1}_{0}n \longrightarrow ^{103}_{42}Mo + ^{131}_{50}Sn + 2^{1}_{0}n + \text{energy}$$

Note that the same nucleus may split in more than one way. The fission products, such as $^{141}_{56}Ba$ and $^{92}_{36}Kr$, emit beta particles ($_{-1}^{0}e$) and gamma rays ($^{0}_{0}\gamma$) until stable isotopes are reached.

$_{-1}^{0}e$ or $_{-1}^{0}\beta$ represents a beta particle.

$$^{141}_{56}Ba \longrightarrow _{-1}^{0}e + ^{0}_{0}\gamma + ^{141}_{57}La$$

$$^{92}_{36}Kr \longrightarrow _{-1}^{0}e + ^{0}_{0}\gamma + ^{92}_{37}Rb$$

The products of these reactions emit beta particles, as do their products. After several such steps, stable isotopes are reached: $^{141}_{59}Pr$ and $^{90}_{40}Zr$, respectively.

A low-energy neutron will disrupt some large nuclei.

The neutrons emitted can cause the fission of other heavy atoms if they are slowed down by a moderator, such as graphite. For example, the three neutrons emitted in the first reaction above could produce fission in three more uranium atoms, the nine neutrons emitted by those nuclei could produce nine more fissions, the 27 neutrons from these fissions could produce 81 neutrons, the 81 neutrons could produce 243, the 243 neutrons could produce 729, and so on. This process is called a "***chain reaction***" (Fig. 14–12), and it occurs at a maximum rate when the uranium sample is large enough for most of the neutrons emitted to be captured by other nuclei before passing out of the sample. Sufficient sample in a certain volume to sustain a chain reaction is termed the ***critical mass.***

In the atomic bomb the critical mass is kept separated into several smaller subcritical masses until detonation, at which time the masses are driven together by an implosive device. It is then that the tremendous energy is liberated and everything in the immediate vicinity is heated to temperatures of 5 to 10 million degrees. The sudden expansion of hot gases literally explodes everything nearby and scatters the radioactive fission fragments over a wide area. In addition to the movement of gases, there is the tremendous vaporizing heat that makes the atomic bomb so devastating.

There is no danger of an atomic explosion in the uranium mineral deposits in the earth for two reasons. First, uranium is not found pure in nature—it is found only in compounds, which in turn are mixed with other compounds. Second, less than 1% of the uranium found in nature is fissionable $^{235}_{92}U$. The other 99% is $^{238}_{92}U$, which is not fissionable by thermal neutrons. In order to make nuclear bombs or nuclear fuel for electrical generation, a purification enrichment process must be carried out on the

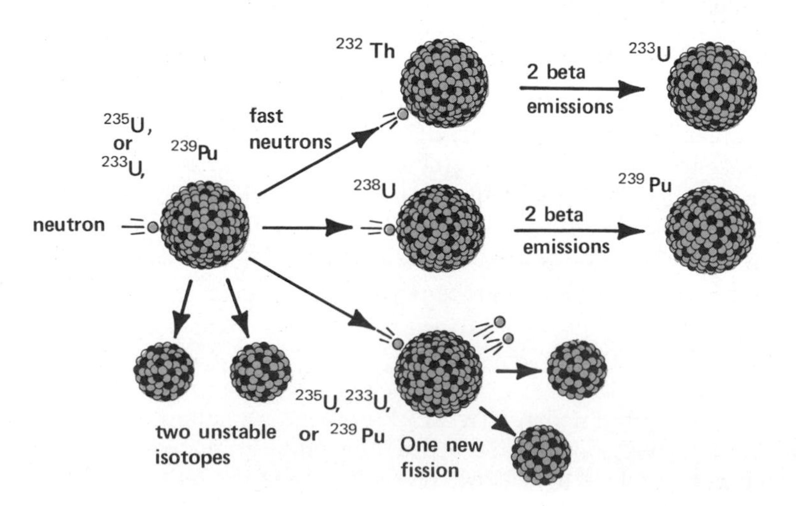


**Figure 14–12** A chain reaction. A thermal neutron collides with a fissionable nucleus and the resulting reaction produces three additional neutrons. These neutrons can either convert non-fissionable nuclei such as $^{232}_{90}Th$, to fissionable ones or cause additional fission reactions. If enough fissionable nuclei are present, a chain reaction will be sustained.

uranium isotopes, thus increasing the relative amount of $^{235}_{92}U$ atoms in a sample. Ordinary isotopic uranium such as that found in ores is only 0.711% $^{235}_{92}U$.

It is interesting to note that fission products can be found in the Gabon Republic of West Africa, which indicate that a uranium ore deposit "went critical" about 150,000 years ago. At that time the natural uranium-235 content would have been higher than it is now.

**Separation of uranium isotopes had to precede the control of atomic energy.**

## Mass Defect—The Ultimate Nuclear Energy Source

What is the source of the tremendous energy of the fission process? It ultimately comes from the conversion of mass into energy, according to Einstein's famous equation, $E = mc^2$, where E is energy that results from the loss of an amount of mass, m, and $c^2$ is the speed of light squared. If separate neutrons, electrons, and protons are combined to form any particular atom, there is a loss of mass called the ***mass defect.*** For example, the calculated mass of one $^{4}_{2}He$ atom from the masses of the constituent particles is 4.032982 amu:

**amu = atomic mass unit**

**$6.02 \times 10^{23} \frac{\text{amu}}{\text{gram}}$**

**or**

**$\frac{\text{1 mole amu}}{\text{gram}}$**

$$2 \times 1.007826 = 2.015652 \text{ amu, mass of two protons and two electrons}$$
$$2 \times 1.008665 = \underline{2.017330} \text{ amu, mass of two neutrons}$$
$$\text{total} = 4.032982 \text{ amu, calculated mass of one } ^{4}_{2}He \text{ atom}$$

Since the measured mass of a $^{4}_{2}He$ atom is 4.002604 amu, the mass defect is 0.030378 amu:

**The mass that is lost leaves in the form of energy: $E = mc^2$.**

$$\begin{array}{r} 4.032982 \text{ amu} \\ -\underline{4.002604} \text{ amu} \\ 0.030378 \text{ amu, mass defect} \end{array}$$

Because the atom is more stable than the separated neutrons, protons, and electrons, the atom is in a lower energy state. Hence, the 0.030378 amu lost per atom would be released in the form of energy if the $^{4}_{2}He$ atom were made from separate protons, electrons, and neutrons. The energy equivalent of the mass defect is called the ***binding energy.*** The binding energy is analogous to the earlier concept of bond energy, in that both are a measure of the energy necessary to separate the package (nucleus or molecule) into its parts.

Atoms with atomic numbers between 30 and 63 have a greater mass defect per nuclear particle than very light elements or very heavy ones, as shown in Figure 14–13. This means the most stable nuclei are found in the atomic number range from 30 to 63.

**Separated nuclear particles have more mass than when combined in a nucleus.**

Because of the relative stabilities, it is in the intermediate range of atomic numbers that most of the products of nuclear fission are found. Therefore, when fission occurs and smaller, more stable nuclei result, these nuclei will contain less mass per nuclear particle. In the process, mass must be changed into energy. This energy gives the fission process its tremendous energy. It takes only about 1 kg of $^{235}_{92}U$ or $^{239}_{94}Pu$ undergoing fission to be equivalent to the energy released by 20,000 tons (20 kilotons) of ordinary explosives like TNT. The energy content in matter is further dramatized when it is realized that the atomic fragments from the 1 kg of nuclear fuel weigh 999 g, so only one tenth of one percent of the mass is actually converted to energy. The fission bombs dropped on Japan during World War II contained approximately this much fissionable material.

**Intermediate - sized nuclei tend to have the greatest nuclear stability.**

## Controlled Nuclear Energy

The fission of a $^{235}_{92}U$ nucleus by a slow-moving neutron to produce smaller nuclei, extra neutrons, and large amounts of energy suggested to Enrico Fermi and others that

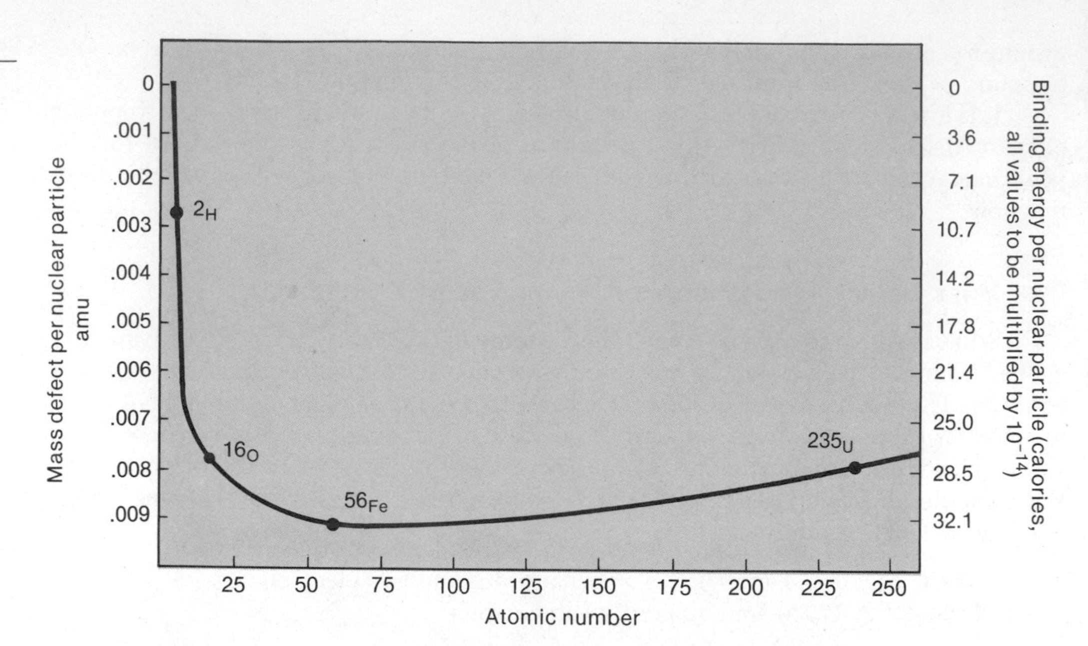


**Figure 14–13** Mass defect for different nuclear masses. The most stable nuclei center around $^{56}_{26}Fe$, which has the largest mass defect per nuclear particle.

the reaction could proceed at a moderate rate if the number of neutrons could be controlled. If a neutron control could be found, the concentration of neutrons could be maintained at a level sufficient to keep the fission process going but not high enough to allow an uncontrolled explosion. It would then be possible to drain the heat away from such a reactor on a continuing basis to do useful work. In 1942, Fermi, working at the University of Chicago, was successful in building the first atomic reactor, called an ***atomic pile.***

Atomic pile:

1. carefully diluted fissionable material;
2. moderator to control fission reaction;
3. coolant to control heat;
4. shielding to limit radiation.

The first one was piled together at the University of Chicago in 1942.

An atomic reactor has a number of essential components. The charge material (fuel) must be fissionable or contain significant concentrations of a fissionable isotope such as $^{235}_{92}U$, $^{239}_{94}Pu$, or $^{233}_{92}U$. Ordinary uranium, which is mostly the nonfissionable $^{238}_{92}U$, cannot be used since it has a small concentration of the $^{235}_{92}U$ isotope. A moderator is required to slow the speed of the neutrons produced in the reactions without absorbing them. Graphite, water, and other substances have been used successfully as moderators. A substance that will absorb neutrons, such as cadmium or boron steel, is present in order to have a fine control over the neutron concentration. Shielding, to protect the workers from dangerous radiation, is an absolute necessity. Shielding tends to make reactors heavy and bulky installations. A heat-transfer fluid provides a large and even flow of heat away from the reaction center.

Once the heat is produced in a nuclear reactor and safety measures are employed to protect against radiation, conventional technology allows this energy to be used to generate electricity, to power ships, or to operate any device that uses heat energy. A system for the nuclear production of electricity is illustrated in Figure 14–14.

What are the fuel requirements in nuclear fission energy production? In a typical fission event such as

1 million electron volts (Mev) = 3.827 × $10^{-14}$ calorie.

$$^{1}_{0}n + ^{235}_{92}U \longrightarrow ^{93}_{37}Rb + ^{141}_{55}Cs + 2^{1}_{0}n + 200 \text{ Mev}$$

the energy release, 200 Mev, is equivalent to $7.7 \times 10^{-12}$ calorie per atom of $^{235}_{92}U$, or $4.64 \times 10^{9}$ kcal/mole. Since 1 g of pure $^{235}_{92}U$ contains $2.56 \times 10^{21}$ atoms, the total energy release for 1 g of uranium-235 undergoing fission would be

$$1\,\cancel{g} \times 2.56 \times 10^{21}\,\frac{\cancel{\text{atoms}}}{1\,\cancel{g}} \times 7.7 \times 10^{-12}\,\frac{\text{cal}}{\cancel{\text{atom}}} = 2.0 \times 10^{10} \text{ cal}$$

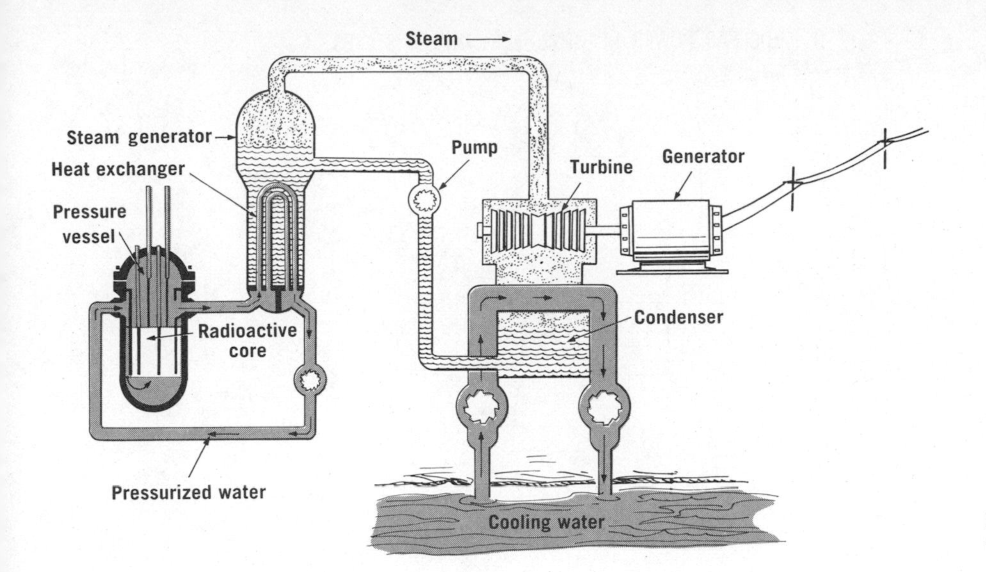


**Figure 14–14** Schematic illustration of a nuclear power plant.

This is the amount of energy that would be released if 5.95 tons of coal were burned, or if 13.7 barrels of oil were burned to produce heat to power a boiler. This means that about 3 kilograms of $^{235}_{92}U$ fuel per day would be required for a 1,000-megawatt electric generator. The fuel used, however, is not pure $^{235}_{92}U$, but ***enriched*** uranium containing up to 3% $^{235}_{92}U$.

**When one gram of $^{235}U$ undergoes fission, it provides the same energy as burning about six tons of coal.**

Nuclear energy is mostly used for electricity production. In 1965, when nuclear energy usage began, and until about 1974, more energy was produced in the United States from burning firewood than from nuclear energy. In 1977, 3% of the U.S. energy supply came from nuclear sources; by 1990, almost 10% of our energy need is expected to be met by nuclear sources (Fig. 14–15). There appears to be enough uranium ore available to meet the electrical generating needs until the last decade of this century; by then, however, the cheap ore will have run out and dramatic increases in electrical rates will be necessary, brought on by fuel costs alone.

It is possible to convert the nonfissionable $^{238}_{92}U$ and $^{232}_{90}Th$ into fissionable fuels by using a ***breeder reactor.*** In such a reactor, a blanket of nonfissionable material is placed outside the fissioning $^{235}_{92}U$ fuel (Fig. 14–16), which serves as the source of neutrons in the breeder reactions. The two breeder reaction sequences are the following:

**$^{235}_{92}U$ is in very short supply.**

$$^{238}_{92}U + {}^{1}_{0}n \longrightarrow {}^{239}_{92}U \xrightarrow{\beta} {}^{239}_{93}Np \xrightarrow{\beta} {}^{239}_{94}Pu$$

$$^{232}_{90}Th + {}^{1}_{0}n \longrightarrow {}^{233}_{90}Th \xrightarrow{\beta} {}^{233}_{91}Pa \xrightarrow{\beta} {}^{233}_{92}U$$

**High temperature and pressure are required for a fusion reaction to occur.**

The products of the breeder reactions, $^{233}_{92}U$ and $^{239}_{94}Pu$, are both fissionable, and neither is found in the earth's crust.

Breeder reactors present many technological problems, not the least of which is the potential of a disaster caused by mishandling of the $^{239}_{94}Pu$ isotope, which is extremely toxic and can also be fabricated into a fission bomb. Nevertheless, the expected benefit from the breeder program is massive amounts of energy. For example, if all the uranium used for electrical generation were used in breeder reactors, instead of running out of uranium fuel in several decades, the breeder fuels would supply the United States' electrical requirements for about 2600 years, assuming 1970 electricity-use levels were maintained, something euphemistically called ***Zero Energy Growth!*** If a breeder program cannot start in earnest soon enough, then the $^{235}_{92}U$ isotope will be used up only to generate electricity and another, as yet undiscovered, source of neutrons will be needed to convert the $^{238}_{92}U$ and $^{232}_{90}Th$ into fissionable fuels.

**$^{239}_{94}Pu$ is toxic from a radiation as well as from a chemical point of view.**

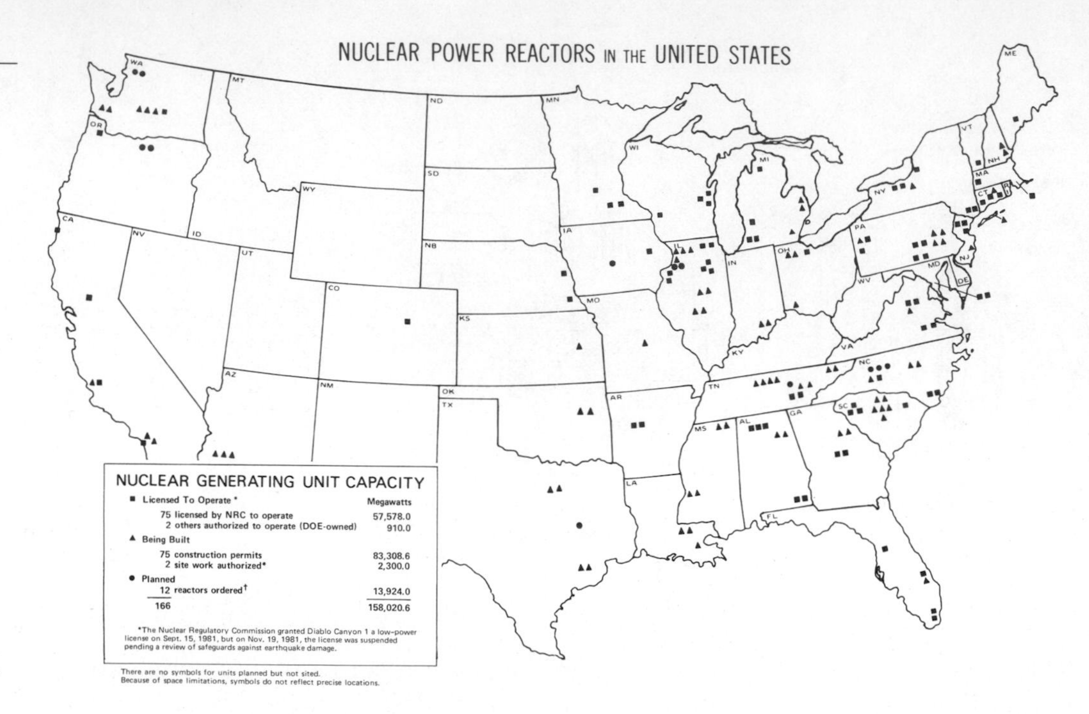


**Figure 14–15** Location of nuclear power reactors.

**The half-life of a radioactive isotope is the time required for one half of the isotope to react.**

Radioactive isotopes are produced in all nuclear fission reactors. The radiation levels produced by these isotopes are dangerous to life, and some of the half-lives involved extend into millions of years. We are, then, in the production of nuclear energy, building up large amounts of very dangerous materials. Serious objections have been raised to every suggested method of radioactive waste disposal. The radioactive waste is presently being held in underground depots and surface tanks. It is argued by some that the exploitation of fission energy should await adequate methods for the disposal of these radioactive wastes.

## Radiation Damage

We are constantly bombarded by radiation from a number of sources; this radiation includes cosmic rays, medical X rays, radioactive fallout from countries that do

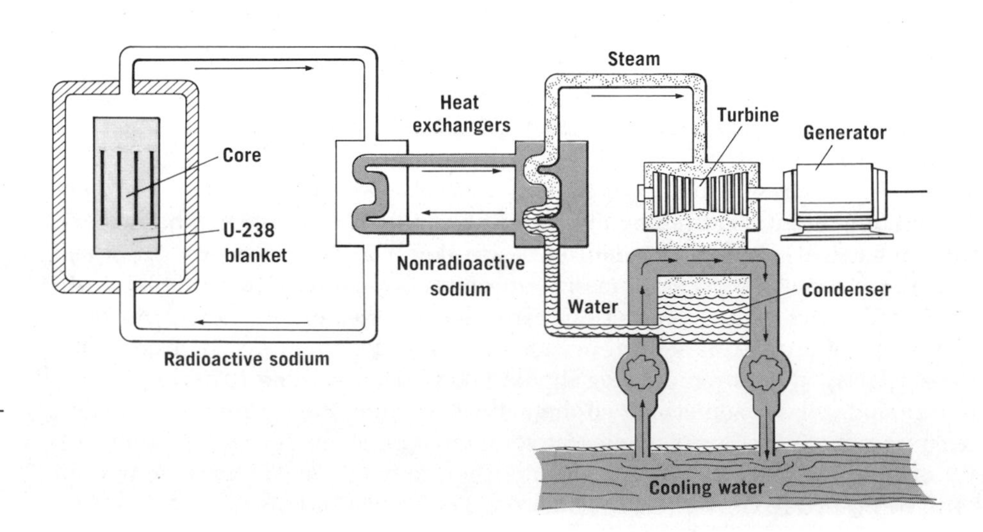


**Figure 14–16** Schematic diagram of a fast breeder reactor and steam-turbine power generator.

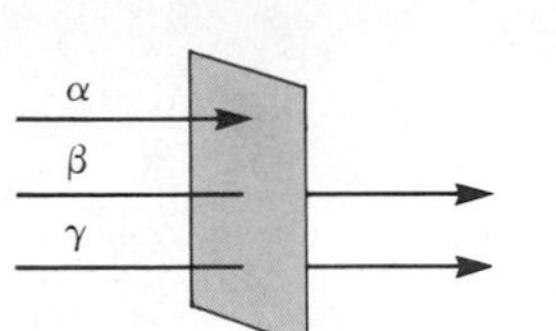


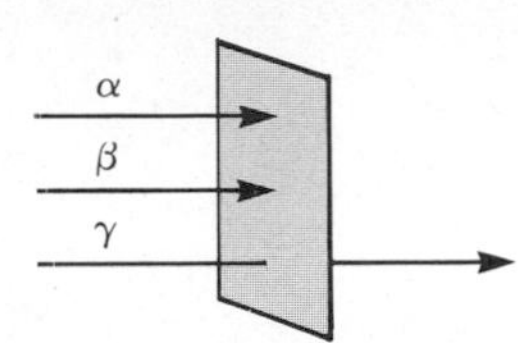


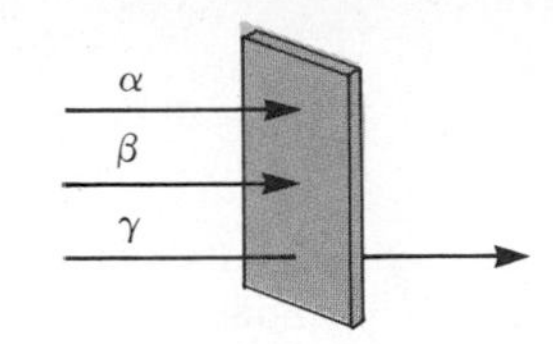


An alpha particle $\left({}^{4}_{2}He\right)$ and gamma ray emitter:

$$^{220}_{88}Ra \rightarrow {}^{4}_{2}He + {}^{0}_{0}\gamma + {}^{216}_{86}Rn$$

A beta particle $\left({}^{0}_{-1}e\right)$ and gamma ray emitter:

$$^{131}_{53}I \rightarrow {}^{0}_{-1}e + {}^{0}_{0}\gamma + {}^{131}_{54}Xe$$

**Figure 14–17.** Penetrating ability of alpha (α), beta (β), and gamma (γ) radiation. Gamma rays even penetrate an 8-mm lead sheet. Skin will stop alpha rays but not beta rays.

nuclear testing, and naturally occurring radioisotopes, which are widespread. Fortunately, most radiation damage is too slight to be noticed immediately, although its very presence should be regarded as one of the hazards of everyday life.

**There is a normal background radiation from natural causes.**

A radioisotope is an atom that will disintegrate, producing one of three types of radiation, alpha (α), beta (β), or gamma (γ) (Fig. 14–17). The resulting atom may be stable or another radioisotope. In a sample of radioactive matter large enough to measure, there will be many disintegrations over a given time if the half-life is short, or few disintegrations over the same time interval if the half-life is long. A ***half-life*** represents the period of time required for half of the radioactive material originally present to undergo transmutation. Normal background radiation on the human body is two or three disintegrations per second.

**Transmutation is the changing of one element into another by a nuclear event.**

There are two principal factors that render a radioactive substance dangerous: (1) the number of disintegrations per second and (2) the type (or energy) of the radiation produced. In addition, radiation can be very damaging if the radioactive substance is of a chemical nature such that it can be incorporated into a food chain or otherwise enter a living organism.

**Marie Curie (1867–1934) was awarded Nobel Prizes in 1903 and 1911 for her work on radioactivity; her daughter, Irène Joliot (1897–1956) received one in 1935.**

Radioactive disintegrations are measured in ***curies*** (Ci; one Ci is 37 billion disintegrations per second). A more suitable unit is the microcurie (μCi), which is 37,000 disintegrations per second. One curie of a radioisotope is a potent sample if the energy per quantum is large enough to cause a biochemical change.

The unit ***roentgen*** is used to measure the intensity of X rays or gamma rays. One roentgen is the quantity of X or gamma radiation delivered to 0.001293 gram of air such that the ions produced in the air carry $3.34 \times 10^{-10}$ coulomb of charge. A single dental X ray represents about one roentgen.

**Wilhelm Roentgen discovered X rays in 1895 and was awarded the Nobel prize for this work in 1901.**

The three types of natural radioactive emissions differ in their penetrating ability (Fig. 14–17), with gamma rays being by far the most penetrating.

Damage by radiation is due to ionization caused by the fast-moving particles colliding with matter, and by the excitation of matter by gamma and X rays that in turn produces ionization. Neutrons are produced in nuclear explosions, in nuclear reactions, and by background cosmic radiation. A neutron does not itself produce ionization but instead imparts its kinetic energy to atoms, which in turn may ionize or break away from the atom to which they are bonded. Neutrons render many engineering materials such as plastics and metals structurally weak over long periods of time because of the decay caused by breaking chemical bonds.

**Neutrons can damage metals, causing structural failure. This is a severe problem in nuclear reactors.**

Biological tissue is easily harmed by radiation. A flow of high-energy particles may cause destruction of a vital enzyme, hormone, or chromosome needed for life of a cell. The radiation may also produce free radicals, which poison the cell. In general, those cells that divide most rapidly are those most easily harmed by radiation (Fig. 14–18).

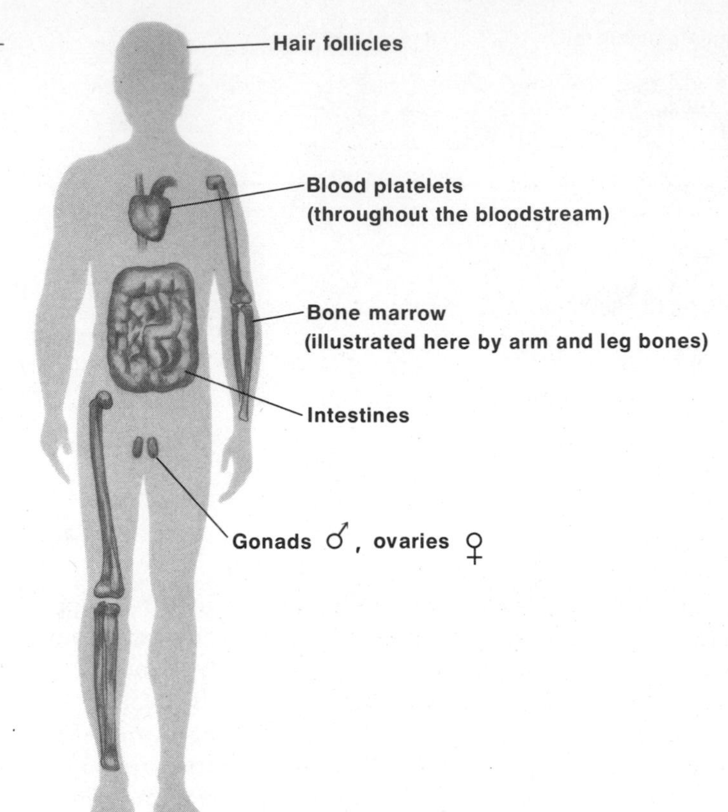


**Figure 14–18** The fast-dividing cells within the body are the ones most harmed by radiation. These include cells in bone marrow, white cells, platelets of the blood, those lining the gastrointestinal tract, hair follicles, and gonads. In addition, the lymphocytes (cells producing the immune responses) are easily killed by radiation.

One *rad* is roughly the energy absorbed by tissue exposed to one roentgen of gamma rays.

1 joule = $10^7$ ergs
1 calorie = $4.184 \times 10^7$ ergs

Whole-body radiation effects are divided into ***somatic effects,*** which are confined to the population exposed, and ***genetic effects,*** which are passed on to subsequent generations. A unit of measurement of radiation density is helpful in measuring the effect of radiation on tissue. The ***rad*** is defined as 100 ergs of energy imposed on a gram of tissue. Whole-body doses of radiation of up to 150 rads produce scarcely any symptoms, whereas doses of 700 rads produce death. Intermediate doses produce vomiting, diarrhea, fatigue, and loss of hair. Often the somatic effects are delayed. Perhaps the best studied of the delayed effects are the incidences of cancer related to exposure of radiation. It has been estimated that 11% of all leukemia cases and about 10% of all forms of cancer are attributable to background radiations. Certainly an individual who is exposed to a higher than normal level of radiation over a considerable length of time has an increased chance of getting cancer. The alteration of normal cells to cancerous cells caused by radiation is undoubtedly a series of changes, since in almost all cases the onset of cancer lags behind the exposure to radiation by an induction period of 5 to 20 years.

The genetic effects of radiation are the result of radiation damage to the germ cells of the testes (sperm) or the ovaries (egg cells). Ionization caused by radiation passing through a germ cell may break a DNA strand or cause it to be altered in some other way. Replication of this altered DNA means transmission of a new message to successive generations, a ***mutation.*** Every type of laboratory animal upon which radiation damage experiments have been performed has shown increased incidence of mutation. Therefore, the necessity of protecting the population of childbearing age from radiation should be apparent. Theoretically, at least, one photon or one high-energy particle can ionize a chromosomal DNA structure and produce a genetic effect that will be carried for generations.

## Radiation Chemistry—The Good Side

The chemical effects induced by the interaction of ionizing radiation (e.g., alpha particles, high-energy electrons, gamma rays, X rays, etc.) with matter have increasingly important roles in industrial, pollution, and public health matters.

Radiotherapy along with chemotherapy and surgery are the three principal approaches to the treatment of cancer. Radiotherapy involves the selective killing of localized tumor cells by focused beams of radiation.

The radiation of certain foods can reduce or eliminate biological or physiological agents that would cause spoilage. It can destroy insects, retard mold growth, and eliminate disease-causing and spoilage-causing insects. Foods susceptible to effective radiation include flour, potatoes, strawberries, poultry, and other precooked meats. The elimination of microorganisms renders such foods stable indefinitely, even when stored at room temperature.

Paint is cured by radiation. Foamed plastics are radiated to produce proper linkages. Disposable materials in hospitals are frequently sterilized by gamma radiation. Oil and gas pipelines are protected from corrosion by radiation processes. Smokestack gases radiated even in the presence of high concentrations of particulates form substances readily removed by electrostatic precipitators. The radiation of sewage destroys pathogenic (disease-causing) bacteria.

## Fusion Reactions

When very light nuclei, such as H, He, and Li, are combined or ***fused*** to form an element of higher atomic number, energy must be given off consistent with the greater stability of the elements in this intermediate atomic number range (Fig. 14–13). This energy, which comes from a decrease in mass, is the source of the energy released by the sun and by hydrogen bombs. Typical examples of fusion reactions are the following:

**Fusion is the combination of very light nuclei.**

$$4\,{}^{1}_{1}\mathrm{H} \longrightarrow {}^{4}_{2}\mathrm{He} + 2\,{}^{\;0}_{+1}\mathrm{e} + 26.7\ \mathrm{Mev}\ \text{for four}\ {}^{1}_{1}\mathrm{H}\ \text{fused}$$

$${}^{2}_{1}\mathrm{H} + {}^{2}_{1}\mathrm{H} \longrightarrow {}^{3}_{2}\mathrm{He} + {}^{1}_{0}\mathrm{n} + 3.2\ \mathrm{Mev}$$

$${}^{2}_{1}\mathrm{H} + {}^{2}_{1}\mathrm{H} \longrightarrow {}^{3}_{1}\mathrm{H} + {}^{1}_{1}\mathrm{H} + 4.0\ \mathrm{Mev}$$

$${}^{3}_{1}\mathrm{H} + {}^{2}_{1}\mathrm{H} \longrightarrow {}^{4}_{2}\mathrm{He} + {}^{1}_{0}\mathrm{n} + 17.6\ \mathrm{Mev}$$

**${}^{2}_{1}\mathrm{H}$ = Deuterium**
**${}^{3}_{1}\mathrm{H}$ = Tritium**
**${}^{\;0}_{+1}\mathrm{e}$ = Positron**

**1 million electron volts (Mev) = 3.827 $\times 10^{-14}$ calorie.**

The net reaction for the last three reactions given here is

$$5\,{}^{2}_{1}\mathrm{H} \longrightarrow {}^{4}_{2}\mathrm{He} + {}^{3}_{2}\mathrm{He} + {}^{1}_{1}\mathrm{H} + 2\,{}^{1}_{0}\mathrm{n} + 24.8\ \mathrm{Mev}\ \text{for five}\ {}^{2}_{1}\mathrm{H}\ \text{fused}$$

or, for every five ${}^{2}_{1}\mathrm{H}$ fused, 24.8 Mev of energy are released.

Deuterium is a relatively abundant isotope—out of 6,500 atoms of hydrogen in sea water, for example, one is a deuterium atom. What this means is that the oceans are a potential source of fantastic amounts of deuterium. There are $1.03 \times 10^{22}$ atoms of deuterium in a single liter of sea water. In a single cubic kilometer of sea water, therefore, there would be enough deuterium atoms with enough potential energy to equal the burning of 1360 billion barrels of crude oil, and this is approximately the total amount of oil originally present in this planet.

**Materials for fusion reactions are available in enormous amounts.**

Fusion reactions take place rapidly only when the temperature is of the order of 100 million degrees or more. At these high temperatures atoms do not exist as such; instead, there is a ***plasma*** consisting of unbound nuclei and electrons. In this plasma nuclei merge or combine. In order to achieve the high temperatures required for the fusion reaction of the hydrogen bomb, a fission bomb (atomic bomb) is first set off.

One type of hydrogen bomb depends on the production of tritium (${}^{3}_{1}\mathrm{H}$) in the bomb. In this type lithium deuteride (${}^{6}_{3}\mathrm{Li}{}^{2}_{1}\mathrm{H}$, a solid salt) is placed around an ordinary

$^{235}_{92}U$ or $^{239}_{94}Pu$ fission bomb. The fission is set off in the usual way. A $^{6}_{3}Li$ nucleus absorbs one of the neutrons produced and splits into tritium, $^{3}_{1}H$, and helium, $^{4}_{2}He$.

$$^{6}_{3}Li + ^{1}_{0}n \longrightarrow ^{3}_{1}H + ^{4}_{2}He$$

The temperature reached by the fission of $^{235}_{92}U$ or $^{239}_{94}Pu$ is sufficiently high to bring about the fusion of tritium and deuterium:

$$^{3}_{1}H + ^{2}_{1}H \longrightarrow ^{4}_{2}He + ^{1}_{0}n + 17.6 \text{ Mev}$$

A 20-megaton bomb usually contains about 300 pounds of lithium deuteride, as well as a considerable amount of plutonium and uranium.

The enhanced radiation weapon (ERW) commonly known as the ***neutron bomb*** is a modified hydrogen bomb (Fig. 14–19). The weapon is not a bomb in the sense that it is delivered aboard a bomber and then dropped over a target. Rather it is small enough to be fired from field artillery. One bomb would deliver a destructive force of roughly one kiloton, which is about one twentieth the destructive power of the 20-kiloton fission bomb dropped on Hiroshima, Japan in August of 1945.

The major modification of a hydrogen bomb into a neutron bomb requires removal of the U-238 shield. In the hydrogen bomb, the shield enables a larger buildup of pressure and hence a larger explosion. The shield reduces the velocity of the neutrons and is a source of harmful radioactivity. With the shield absent, as in neutron bombs, 80% of the energy released is in the form of high-speed neutrons. The neutrons readily penetrate iron and steel, so armored tanks provide no protection from the effects of the neutrons. Since there is minimal heat and minimal blast from the neutron bomb, little destruction is done to inanimate matter. However, people and animals within a 1,000 foot radius would be paralyzed within five minutes and dead in two days. Other living beings within a 1,000–2,000 foot radius would be dead in four to six days, although those in concrete shelters would be safe. There is virtually no residual radiation or fallout.

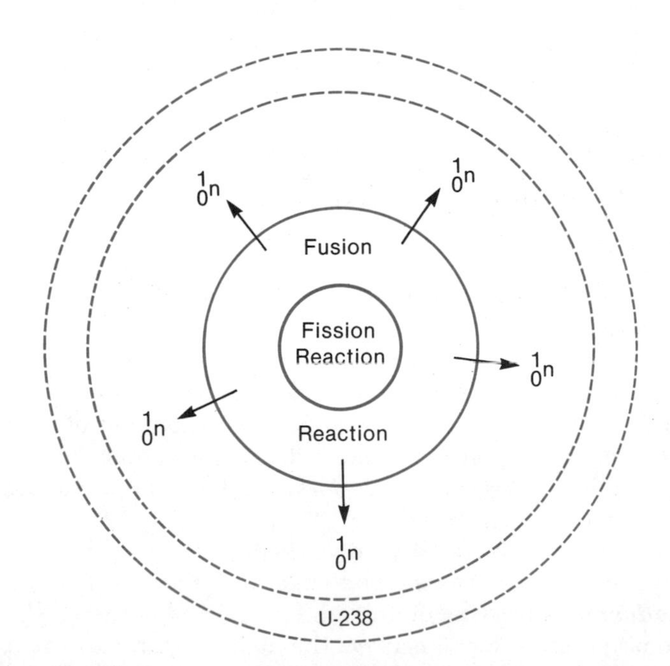


**Figure 14–19** The general arrangement within a neutron bomb with the U-238 shield of the hydrogen bomb shown in its place. The neutron bomb (or enhanced radiation weapon [ERW]) does not have a U-238 shield.

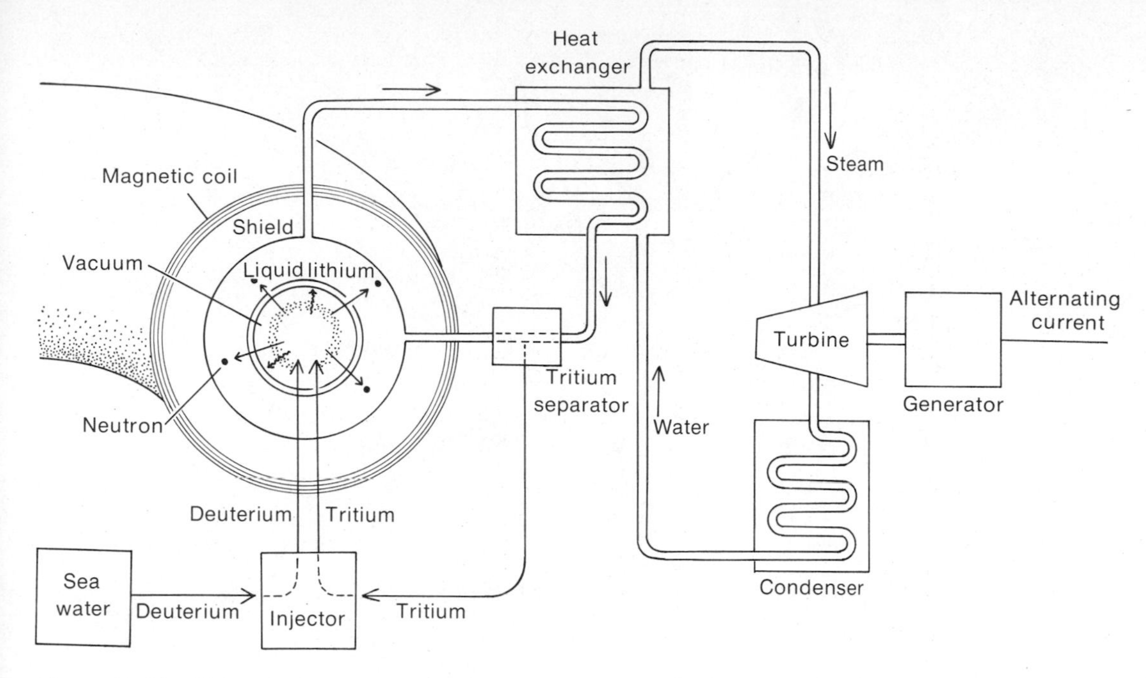


**Figure 14–20** Thermal energy conversion from nuclear fusion. Such a scheme, based on the deuterium-tritium fuel cycle, relies on the energy of highly energetic neutrons. A liquid lithium shield absorbs these neutrons and is heated. The lithium then exchanges this heat with water to generate steam.

## Controlled Fusion

**Containment is one of the biggest problems in developing controlled fusion.**

There are three critical requirements for controlled fusion. First, the temperature must be high enough for ignition to occur. For the deuterium-tritium combination given earlier, a temperature of about $10^8$–$10^9$° is needed. Second, the plasma must be confined for a long enough time to release a net output of energy. Third, the energy must be recoverable in some usable form.

As yet the fusion reactions have not been "controlled." No physical container can contain the plasma without cooling it below the critical fusion temperature. Magnetic "bottles," enclosures in space bounded by a magnetic field, have confined the plasma, but not for long enough periods of time. Recent developments suggest that these "bottles," with further development, may hold the plasma long enough for the fusion reaction to occur.

Thermal energy conversion (Fig. 14–20) could be used to take the power from a deuterium-tritium–fueled fusion reaction. Liquid lithium would be used to absorb kinetic energy of fast-moving neutrons and then exchange this heat with water to drive a steam turbine, thus producing electricity. This system is actually a breeder reactor, as some of the lithium is converted into fuel—tritium, ${}^3_1H$—by neutron absorption.

A newer confinement method is based on a laser system that simultaneously strikes tiny hollow glass spheres called ***microballoons*** that enclose the fuel, consisting of equal parts of deuterium and tritium gas, at high pressures (Fig. 14–21).

There is hope that controlled fusion will be demonstrated during the next decade, but it appears that fusion will not furnish any significant fraction of the world's energy needs before the turn of the century.

**Controlled fusion might end many of the world's energy problems.**

Controlled fusion energy should result in a rather limited production of dangerous radioactivity. The lighter elements involved can be radioactive enough to be

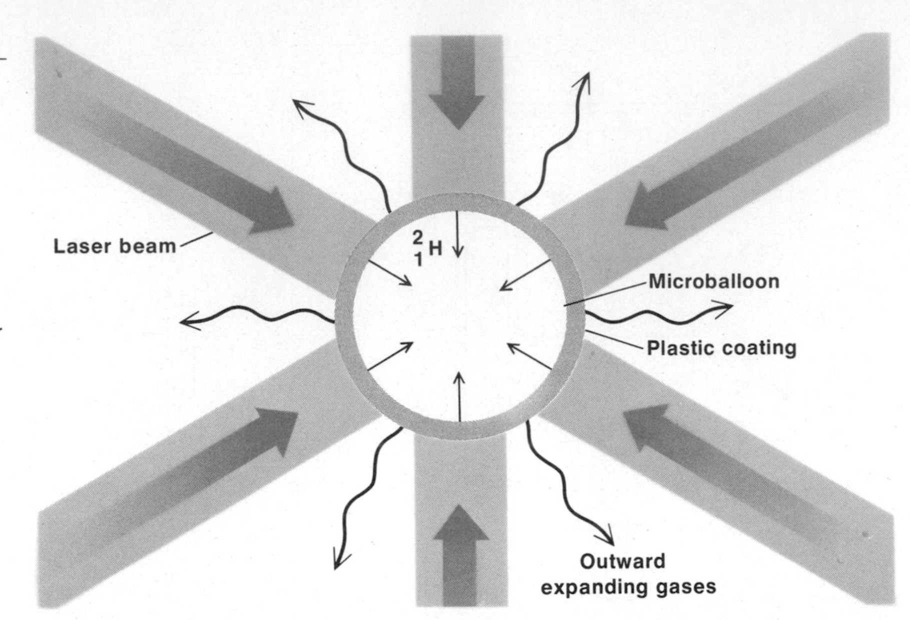


**Figure 14–21** Focused laser light strikes the microballoon filled with deuterium and tritium, causing a plastic outer layer to burn off (or *ablate*). The outwardly expanding gases from the plastic material drive the glass sphere and its fuel contents inward. The high density and high temperatures produced might result in fusion.

a serious hazard, but only for a short period of time; the half-lives of these isotopes are short. Storage and then return to the environment would be quite satisfactory.

## Solar Energy

**Solar energy is transmitted nuclear energy.**

Earth's ultimate source of energy is the sun. The sun provides energy for photosynthesis, which gives us food to power our bodies, wood to burn, and the starting materials for coal and petroleum. In addition, the sun provides the energy for the water cycle, which is the cause of our weather.

Although the earth receives only about three ten-millionths (0.0000003) of the total energy emitted from the sun, this amount of energy is enormous—about $2 \times 10^{15}$

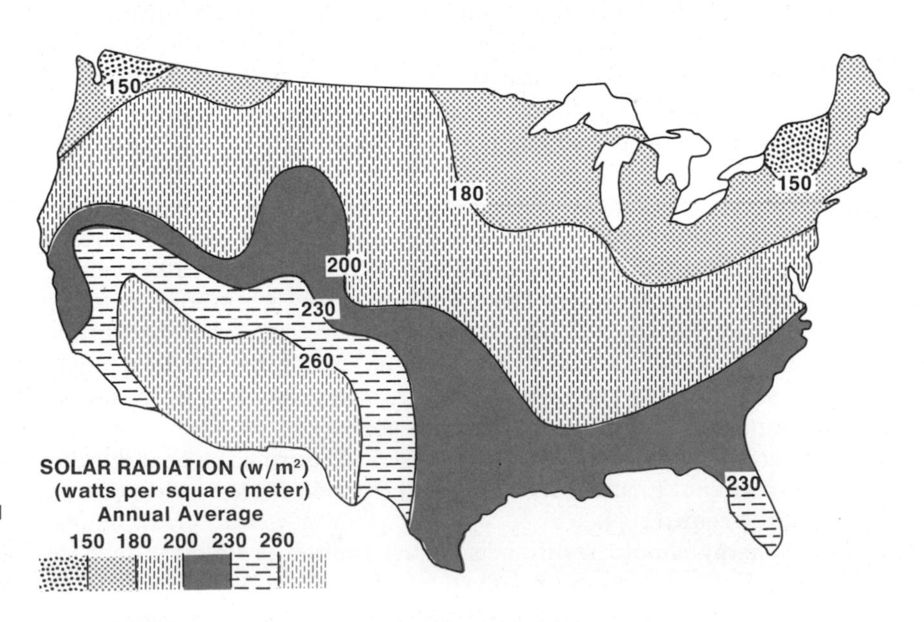


**Figure 14–22** Annual rate of delivery of solar radiation in the United States.

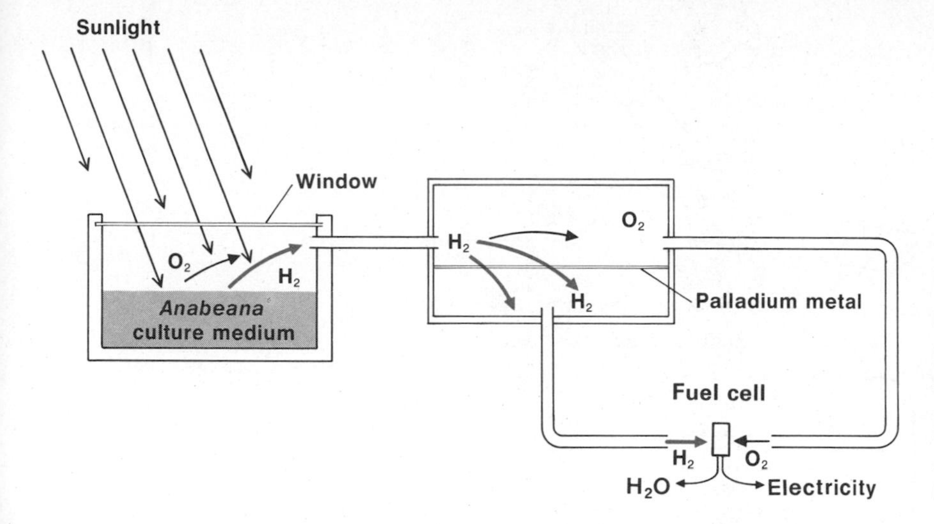


**Figure 14–23** Schematic diagram of an electricity-producing photosynthesis process. $H_2$ and $O_2$ produced by the *Anabeana* are separated by palladium metal which is permeable toward $H_2$ but not $O_2$. The $H_2$ and $O_2$ are then combined in the fuel cell to produce electricity.

kcal/min, or 2.0 cal/cm$^2$/min. Owing to reradiation from the atmosphere, and the absorption and scattering of radiant energy by molecules in the lower portions of the atmosphere, the amount of radiation actually reaching the surface of this planet is about 1 cal/cm$^2$/min. The actual value depends on location, season, and weather conditions. Even this is a large amount of energy. For example, the roof of an average-sized house will receive about $10^8$ calories/day when the radiation level is 1 cal/cm$^2$/min. This is equivalent to the heat energy derived from burning about 32 pounds of coal per day. This is also equal to 120 kilowatt-hours per day of electrical energy—more than enough to heat an average American home on a winter day.

Some catalysts reported in late 1982 for the decomposition of water in the presence of sunlight are indium phosphide (InP), phosphorus-doped silicon coated with Pt or Ni, and p-type iron oxide semiconductor. All are presently too expensive and (or) inefficient (12%, 12%, and 0.05%, respectively) to compete with hydrogen produced by the reaction of coal with steam.
$C + H_2O \rightarrow CO + H_2$

Only about 1% of the solar radiation used by plants in photosynthesis ends up as stored chemical energy such as foodstuffs. After fossil fuels are depleted, could photosynthesis be considered as a viable source of energy outside of food materials? Much of the energy used in photosynthesis is for "fixing" carbon in the form of cellulose and carbohydrates. Recently it has been shown that certain blue-green algae, *Anabaena cylindrica,* can convert sunlight and water into hydrogen and oxygen. A colony of such algae, coupled with a fuel cell utilizing hydrogen and oxygen, could be a source of electricity during sunlight hours (Fig. 14–23).

Solar energy, used efficiently, could solve many energy problems. For example, the world's present energy requirements per year are about $10^{18}$ kcal. An area of desert of about 28,000 square miles with little cloud cover or dust, near the equator, would receive about $10^{17}$ kcal/year of solar radiation. Such deserts exist in northern Chile, and could, if the need were great enough to justify the costs, supply a large portion of the world's energy needs.

The solar energy could heat water to steam, which in turn could generate electricity (Fig. 14–24), which could electrolyze water to hydrogen and oxygen. The hydrogen could be piped to where the energy is needed and then converted to electricity. Such an arrangement would give rise to a ***hydrogen economy,*** one that has many advantages over present energy sources such as fossil fuels.

Hydrogen can be burned in most devices that now burn natural gas.

Some hydrogen-powered buses and cars are now operating on an experimental basis.

Another approach to the direct utilization of solar energy is the ***solar battery,*** known as a photovoltaic device. The solar battery converts energy from the sun into electron flow. Solar batteries are about 13–14% efficient and are capable of generating electrical power from sunlight at the rate of at least 90 watts per square yard of illuminated surface. They are now used in space flight applications and communication satellites, and in Israel, India, Pakistan, South Africa, and Azerbaijan SSR to obtain electrical power.

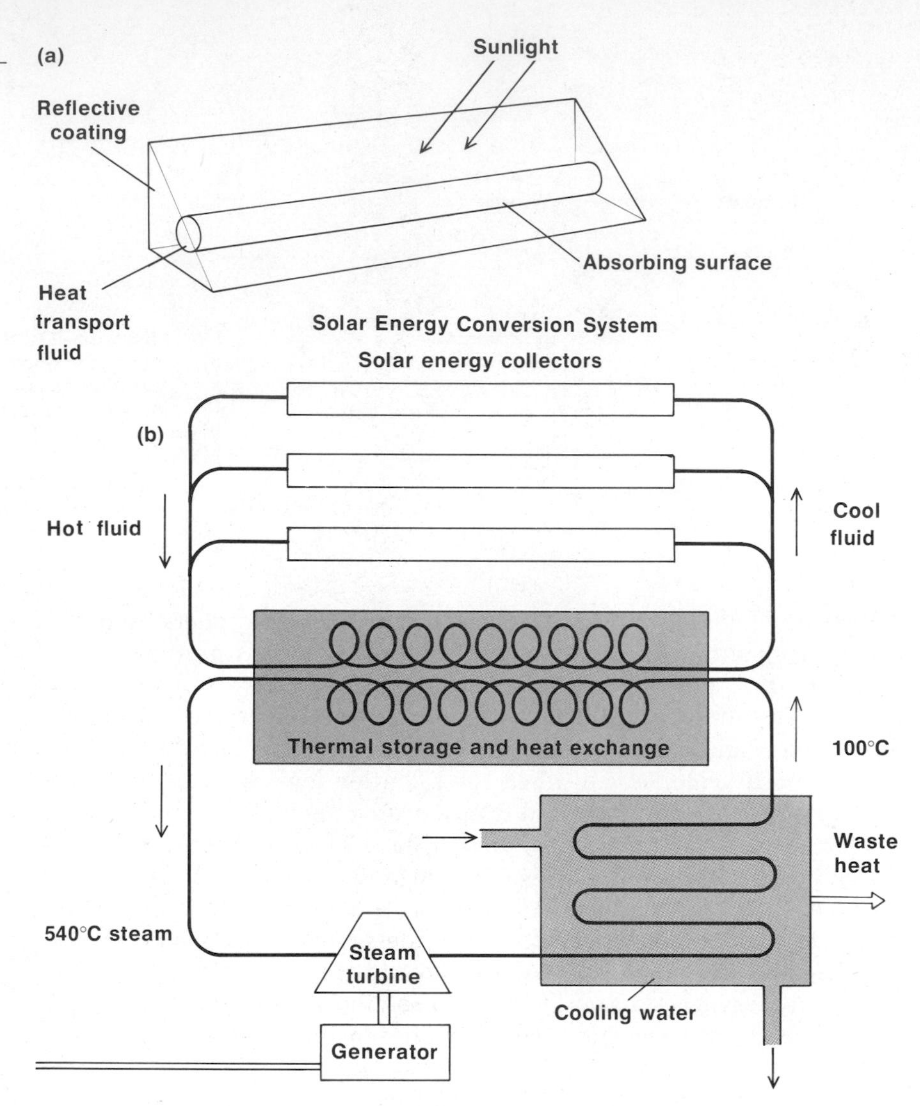


**Figure 14–24** (*a*) Solar energy collector. (*b*) Solar energy conversion system.

**The goal is to capture and use the solar energy "on the run" without upsetting the energy flow into (light side) and away from (dark side) the earth.**

One type of solar battery consists of two layers of almost pure silicon (Fig. 14–25). The lower, thicker layer contains a trace of arsenic and the upper, thinner layer a trace of boron. Silicon has four valence electrons and forms a tetrahedral, diamond-like, crystalline structure. Each silicon atom is covalently bonded to four other silicon atoms. Arsenic has five valence electrons. When arsenic atoms are included in the silicon structure, only four of the five valence electrons of arsenic are used for bonding with four silicon atoms; one electron is relatively free to roam (Fig. 14–26). Boron has three valence electrons. When boron atoms are included in the silicon structure, there is a deficiency of one electron around the boron atom; this creates "holes" in the boron-enriched layer. Even without sunlight, the "extra" electrons of arsenic diffuse into the holes in the boron. The driving force is the strong tendency to pair electrons. Externally applied potentials as high as 1000 volts are incapable of reversing the flow. The negative charge built up in the boron layer would hinder the flow of electrons into that layer and eventually stop the flow. The opposing factors—repulsion between free electrons and the drive to pair electrons—finally bring about an equilibrium.

When sunlight strikes the boron layer, the equilibrium is disturbed. If the wafer is connected as an ordinary battery would be, electrons flow from the arsenic layer

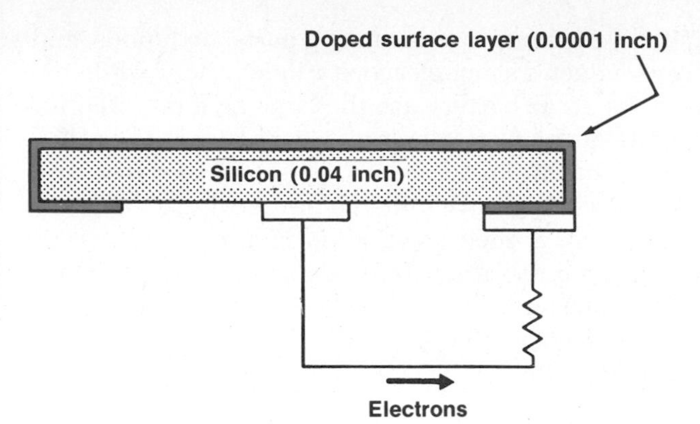


Figure 14–25 Silicon photodiode (Bell solar battery).

through the circuit and back into the boron layer (Fig. 14–25). The fact that electrons enter the circuit from the arsenic layer can be explained if sunlight unpairs electrons in the boron layer, and the freed electrons are repelled to the arsenic layer and into the circuit. To complete the circuit, electrons enter the boron layer, where there are holes for electrons. If the sun is still shining on the cell, unpaired electrons will continue to be repelled to the positive arsenic layer. Since there are no places for the incoming electrons in the structure of the arsenic layer, the electrons go into the external circuit, where there is less opposition to their flow. An analogy to the operation of the solar cell would be the benches on which people rest in a crowded amusement park (the boron layer). The flow is toward the benches, but they are too hot to sit on. The flow is continued by those who do not know (or forget) that the benches are too hot.

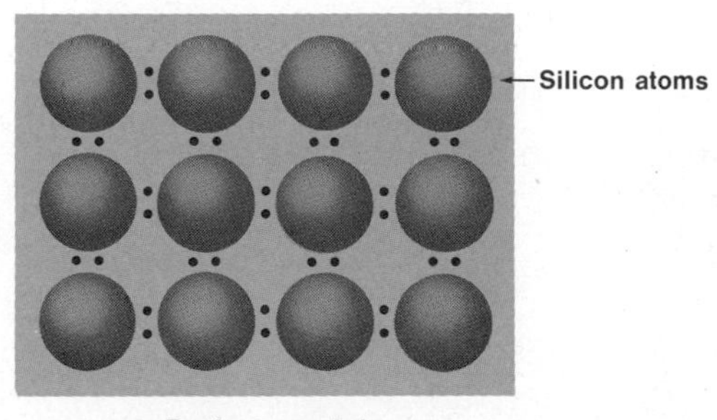


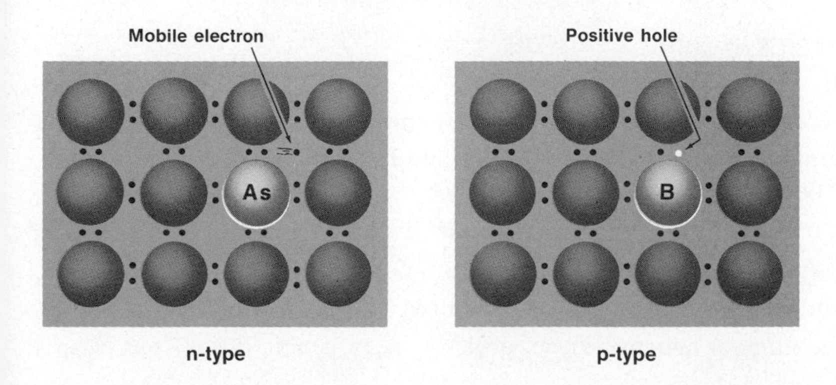


Figure 14–26 Schematic drawing of semiconductor crystal layers derived from silicon.

Solar batteries are still very expensive.

The advantage of the solar battery is that it has no moving parts, no liquids, and no corrosive chemicals—it just keeps on generating electricity indefinitely while exposed to sunlight. The drawbacks of the solar battery are the large area required for large amounts of power, the high costs of the pure materials, and the fact that they work only when the sun is shining. Since the first practical use of solar batteries in 1955 to power eight rural telephones in Georgia, they have undergone a great deal of development, and much more is expected because of their great potential use.

In theory, a number of huge energy sources are potentially available for our use. As yet we have not learned to control them sufficiently well to have cheap energy for all. However, progress is being made. It is evident that both chemical sources of energy and the fission of heavy nuclei are limited in their ability to supply effectively our growing energy needs, and we must develop a wide variety of other energy sources.

## Self-Test 14-B

1. The splitting of an unstable nucleus to produce energy is termed (fission/fusion).
2. A sufficiently large sample to sustain a chain reaction in a fission process is called the ______________ ______________.
3. Of the two major isotopes of uranium, which is fissionable, uranium-238 or uranium-235? ______________
4. When light nuclei combine to form heavy nuclei and energy, the process is (fission/fusion).
5. The major problem with obtaining fusion energy is (containment of reactants at high temperature/enough fuel/costs).
6. When and if fusion is used to produce energy in a breeder reaction, the fuel produced will be ______________.
7. When fission is used to produce energy in a breeder reaction, the fuel produced will be ______________ and ______________.
8. Which is the more penetrating type of radiation? (alpha/beta) ______________
9. The curie is a unit related to (numbers of disintegrations/intensity). ______________
10. Complete the nuclear reactions:

    a. ${}^{85}_{34}Se \longrightarrow {}^{0}_{-1}e +$ ______________

    b. ${}^{246}_{96}Cm \longrightarrow {}^{4}_{2}He +$ ______________

## Matching Set

| | | |
|---|---|---|
| ______ | 1. user of 35% of world's energy | **a** 1859 |
| ______ | 2. fossil fuels | **b** $CO_2$ and $H_2O$ |
| ______ | 3. combustion products of fossil fuels | **c** uranium-235 (${}^{235}_{92}U$) |
| ______ | 4. minable coal | **d** mass defect |
| ______ | 5. synthesis gas | **e** plutonium-239 (${}^{239}_{94}Pu$) |
| ______ | 6. fissionable isotope | **f** United States |
| ______ | 7. basis of nuclear energy | **g** ${}^{2}_{1}H$ |

| | | | |
|---|---|---|---|
| ______ | 8. product of a fission breeder reactor | h | sea water |
| ______ | 9. deuterium | i | 700 rads |
| ______ | 10. date of petroleum discovery in United States | j | within 4,000 feet of surface |
| ______ | 11. tritium | k | microballoons |
| ______ | 12. source of deuterium | l | coal, petroleum, natural gas |
| ______ | 13. used to confine fusion fuel | m | photosynthesis |
| ______ | 14. one use of solar radiation | n | CO and $H_2$ |
| ______ | 15. approximate efficiency of a solar battery | o | 10–14% |
| ______ | 16. nuclear radiation death | p | $^3_1H$ |

## Questions

**1.** What is your attitude toward using up the fossil fuels within a few decades? Do we owe future generations a supply of these resources? Would you agree to give up air-conditioning, private cars, and power tools, to mention a few examples, and to limit heating and cooking if necessary to share these fuels with your grandchildren?

**2.** Which theoretically yields the greatest energy per mole?
a. the burning of gasoline
b. the fission of uranium-235
c. the fusion of hydrogen

**3.** Which is the more efficient use of energy: burning coal in a house to heat it or heating the house electrically with energy produced in a coal-burning power plant?

**4.** Give three examples of systems that contain chemical energy that can be used as a source of heat energy.

**5.** Is the electrical energy where you live produced by burning fossil fuels? If not, what is the energy source? Are there pollution problems associated with the generation of the electrical power?

**6.** What produces the tremendous energy of a fission reaction?

**7.** What was the original source of energy that is tied up in fossil fuels?

**8.** Why is it difficult to fuse two $^{52}_{24}Cr$ nuclei? Explain.

**9.** What is meant by a chain reaction?

**10.** What major problem is associated with harnessing the energy from a fusion reaction?

**11.** Suggest several ways solar energy might be harnessed.

**12.** Name two sources of energy not specifically mentioned in this chapter.

**13.** Explain how useful energy might be obtained from garbage.

**14.** Which is more fundamental—a supply of energy or a supply of food? Explain.

**15.** The energy consumption of the United States in 1970 was $2 \times 10^{13}$ kilowatt-hours. What is this amount of energy expressed in kilocalories? In BTU's?

**16.** Assume the world population to be 4.5 billion and calculate the earth's energy needs if everyone used as much energy as is used in the United States.

**17.** What is meant by "delayed somatic effect"? Give an example.

**18.** Define energy.

**19.** List three so-called fossil fuels. Why are they called fossil fuels?

**20.** Which fuel—coal, petroleum, or natural gas—burns naturally with the least amount of pollution?

**21.** Describe what is meant by coal gasification. How is it accomplished?

**22.** Do you pay for electricity as electrical power (kilowatts) or electrical energy (kilowatt-hours)? What is the difference?

**23.** How do mass defect and binding energy relate? How do they arise?

**24.** How do the neutron bomb and the hydrogen bomb differ? How are they alike?

**25.** What is meant by an insulator R-value of 30?

**26.** What are some constructive applications of radioactive radiation?

**27.** What is the principal element in the photoelectric diode? What two elements are used in trace quantities in a photocell?

**28.** What are two dangerous properties of plutonium-239?

**29.** Is it the fission or the fusion atomic reaction that is controlled in an atomic reactor?

**30.** What gases are contained in both coal gas and power gas?

**31.** Which is the more efficient transport of energy: gas through pipes or electricity through wires?

**32.** If solar energy is so clean, why are we so slow in moving to its use?

**33.** Do you think the United States should go forward with the use of the breeder reactor? Why? Are other nations now ahead of the United States in breeder reactor use?

**34.** How much has the price of oil increased during your lifetime?

**35.** If the mass of a proton is 1.007275 amu, the mass of an electron is 0.000551 amu, and the mass of a neutron is 1.008665 amu, what is (a) the calculated mass of one atom, and (b) the mass defect when the measured masses are as given below?

(1) $^{7}_{3}Li$, 7.01601 amu
(2) $^{9}_{4}Be$, 9.01219 amu
(3) $^{18}_{8}O$, 17.99916 amu

**36.** Is the energy crisis a crisis of quality or quantity of energy?

**37.** Is electricity a primary or secondary source of energy? Explain your answer.

# Toxic Substances in Our Environment

15

Toxic substances are materials that upset the incredibly complex system of chemical reactions occurring in the human body. Sometimes toxic substances cause mere discomfort; sometimes they cause illness, sometimes disability or even death. Toxic symptoms can be caused by very small amounts of extremely toxic materials (an example is sodium cyanide) or larger amounts of a less toxic substance. The term ***toxic substances*** usually is limited to materials that are dangerous in small amounts. However, as most of us know, ill effects can be caused by excessive intake of substances normally considered harmless (eating too much candy, for example). Fortunately, in most cases the human body is capable of recognizing "foreign" chemicals and ridding itself of them. In this chapter, we shall focus on the chemical mechanisms by which toxic substances act.

**A large enough dose of any compound can result in poisoning.**

## Dose

Lethal doses of toxic substances are customarily expressed in milligrams (mg) of substance per kilogram (kg) of body weight of the subject. For example, the cyanide ion ($CN^-$) is generally fatal to human beings in a dose of 1 mg of $CN^-$ per kg of body weight. For a 200-pound (90.7 kg) person, about one tenth of a gram of cyanide is a lethal dose. Examples of somewhat less toxic substances and the range of lethal doses for human beings are the following:

**"Dosis sola facit venenum"—the dose makes the poison.**

| | |
|---|---|
| Morphine | 1–50 mg per kg |
| Aspirin | 50–500 mg per kg |
| Methyl alcohol | 500–5,000 mg per kg |
| Ethyl alcohol | 5,000–15,000 mg per kg |

A quantitative measure of toxicity is obtained by introducing various dosages of substances to be tested into laboratory animals (such as rats). That dosage that is found to be lethal in 50% of a large number of the animals under controlled conditions is called the $LD_{50}$ (lethal dosage—50%) and is reported in milligrams of poison per kilogram of body weight. Thus, if a statistical analysis of data on a large population of rats showed that a dosage of 1 mg per kg was lethal to 50% of the population tested, the $LD_{50}$ for this poison would be 1 mg per kg. Obviously, metabolic variations

**Metabolism (from the Greek, *metaballein*, to change or alter) is the sum of all the physical and chemical changes by which living organisms are produced and maintained.**

**TABLE 15–1 Approximate Comparison of $LD_{50}$ Values with Lethal Doses for Human Adults**

| Oral $LD_{50}$ for Any Animal (mg/kg) | Probable Lethal Oral Dose for Human Adult |
|---|---|
| Less than 5 | a few drops |
| 5 to 50 | "a pinch" to 1 teaspoonful |
| 50 to 500 | 1 teaspoonful to 2 tablespoonfuls |
| 500 to 5,000 | 1 ounce to 1 pint (1 pound) |
| 5,000 to 15,000 | 1 pint to 1 quart (2 pounds) |

and other differences between species will produce different $LD_{50}$ values for a given poison in different kinds of animals. For this reason such data cannot be extrapolated to human beings with any assurance, but it is safe to assume that a substance with a low $LD_{50}$ value for several animal species will also be quite toxic to humans (Table 15–1).

Toxic substances can be classified according to the way in which they disrupt the chemistry of the body. Some of the modes of action of toxic substances can be described as ***corrosive, metabolic, neurotoxic, mutagenic, teratogenic,*** and ***carcinogenic,*** and these will serve as the bases of our discussion.

## Corrosive Poisons

Toxic substances that actually destroy tissues are corrosive poisons. Examples include strong acids and alkalies, and many oxidants such as those found in laundry products, which can destroy tissues. Sulfuric acid (found in auto batteries) and hydrochloric acid (also called muriatic acid; used for cleaning purposes) are very dangerous corrosive poisons. So is sodium hydroxide, used in clearing clogged drains. Death has resulted from the swallowing of 1 ounce of concentrated (98%) sulfuric acid, and much smaller amounts can cause extensive damage and severe pain.

Concentrated mineral acids such as sulfuric acid act by first dehydrating cellular structures. The cell dies because its protein structures are destroyed by the acid-catalyzed hydrolysis of the peptide bonds.

$$\mathrm{R-\overset{\overset{\displaystyle O}{\|}}{C}-\overset{\overset{\displaystyle H}{|}}{N}-R' + H_2O \xrightarrow[\text{from acid}]{H^+} R-\overset{\overset{\displaystyle O}{\|}}{C}-OH + H-\overset{\overset{\displaystyle H}{|}}{N}-R'}$$

PEPTIDE LINK (IN PROTEIN) — CARBOXYL END OF SMALLER PEPTIDE OR AMINO ACID — AMINE END OF SMALLER PEPTIDE OR AMINO ACID

**Strong acids and bases destroy cell protoplasm.**

In the early stages of this process there will be a large proportion of larger fragments present. Subsequently, as more bonds are broken, smaller and smaller fragments result, leading to the ultimate disintegration of the tissue.

Some poisons act by undergoing chemical reaction in the body to produce corrosive poisons. Phosgene, the deadly gas used during World War I, is an example. When inhaled, it is hydrolyzed in the lungs to hydrochloric acid, which causes pulmonary edema (a collection of fluid in the lungs) due to the dehydrating effect of the strong acid on tissues. The victim dies of suffocation because oxygen cannot be absorbed effectively by the flooded and damaged tissues.

$$\underset{\text{PHOSGENE}}{Cl-\overset{\overset{O}{\|}}{C}-Cl} + H_2O \longrightarrow \underset{\text{HYDROCHLORIC ACID}}{2HCl} + \underset{\text{CARBON DIOXIDE}}{CO_2}$$

Chemical "warfare gases," such as phosgene, were outlawed by an international conference in 1925.

Sodium hydroxide, NaOH (caustic soda—a component of drain cleaners), is a very strongly alkaline, or basic, substance that can be just as corrosive to tissue as strong acids. The hydroxide ion also catalyzes the splitting of peptide linkages:

$$R-\overset{\overset{O}{\|}}{C}-\overset{\overset{H}{|}}{N}-R' + H_2O \xrightarrow[\text{base}]{OH^-} R-\overset{\overset{O}{\|}}{C}-OH + H-\overset{\overset{H}{|}}{N}-R'$$

Both acids and bases, as well as other types of corrosive poisons, continue their action until they are consumed in chemical reactions.

Some corrosive poisons destroy tissue by oxidizing it. This is characteristic of substances such as ozone, nitrogen dioxide, and possibly iodine, which destroy enzymes by oxidizing their functional groups. Specific groups, such as the $-SH$ and $-S-S-$ groups in the enzyme, are believed to be converted by oxidation to nonfunctioning groups; alternatively, the oxidizing agents may break chemical bonds in the enzyme, leading to its inactivation.

A summary of some common corrosive poisons is presented in Table 15–2.

**TABLE 15–2 Some Corrosive Poisons**

| Substance | Formula | Toxic Action | Possible Contact |
|---|---|---|---|
| Hydrochloric acid | $HCl$ | acid hydrolysis | tile and concrete floor cleaner; concentrated acid used to adjust acidity of swimming pools |
| Sulfuric acid | $H_2SO_4$ | acid hydrolysis, dehydrates tissue—oxidizes tissue | auto batteries |
| Phosgene | $ClCOCl$ | acid hydrolysis | combustion of chlorine-containing plastics (PVC or Saran) |
| Sodium hydroxide | $NaOH$ | base hydrolysis | caustic soda, drain cleaners |
| Trisodium phosphate | $Na_3PO_4$ | base hydrolysis | detergents, household cleaners |
| Sodium perborate | $NaBO_3 \cdot 4H_2O$ | base hydrolysis—oxidizing agent | laundry detergents, denture cleaners |
| Ozone | $O_3$ | oxidizing agent | air, electric motors |
| Nitrogen dioxide | $NO_2$ | oxidizing agent | polluted air, automobile exhaust |
| Iodine | $I_2$ | oxidizing agent | antiseptic |
| Hypochlorite ion | $OCl^-$ | oxidizing agent | bleach |
| Peroxide ion | $O_2^{2-}$ | oxidizing agent | bleach, antiseptic |
| Oxalic acid | $H_2C_2O_4$ | reducing agent, precipitates $Ca^{2+}$ | bleach, ink eradicator, leather tanning, rhubarb, spinach, tea |
| Sulfite ion | $SO_3^{2-}$ | reducing agent | bleach |

## Metabolic Poisons

Metabolic poisons are more subtle than the tissue-destroying corrosive poisons. In fact, many of them do their work without actually indicating their presence until it is too late. Metabolic poisons can cause illness or death by interfering with a vital biochemical mechanism to such an extent that it ceases to function or is prevented from functioning efficiently.

### Carbon Monoxide

The interference of carbon monoxide with extracellular oxygen transport is one of the best understood processes of metabolic poisoning. As early as 1895, it was noted that carbon monoxide deprives body cells of oxygen (asphyxiation), but it was much later before it was known that carbon monoxide, like oxygen, combines with hemoglobin:

$O_2$ + hemoglobin ⟶ oxyhemoglobin

CO + hemoglobin ⟶ carboxyhemoglobin

Laboratory tests show that carbon monoxide reacts with hemoglobin to give a compound (carboxyhemoglobin) that is 140 times more stable than the compound of hemoglobin and oxygen (oxyhemoglobin) (see Fig. 15–1). Since hemoglobin is so effectively tied up by carbon monoxide, it cannot perform its vital function of transporting oxygen.

An organic material that undergoes incomplete combustion will always liberate carbon monoxide. Sources include auto exhausts, smoldering leaves, lighted cigars or cigarettes, and charcoal burners. In the United States alone, combustion sources of all types dump about two hundred million tons of carbon monoxide per year into the atmosphere.

***ppm*—parts per million—a measure expressing concentration. 50 ppm CO means 50 ml CO for every million ml of air.**

While the best estimates of the maximum global background level of carbon monoxide are of the order of 0.1 ppm, the background concentration in cities is higher. In heavy traffic, sustained levels of 100 or more ppm are common; for offstreet sites

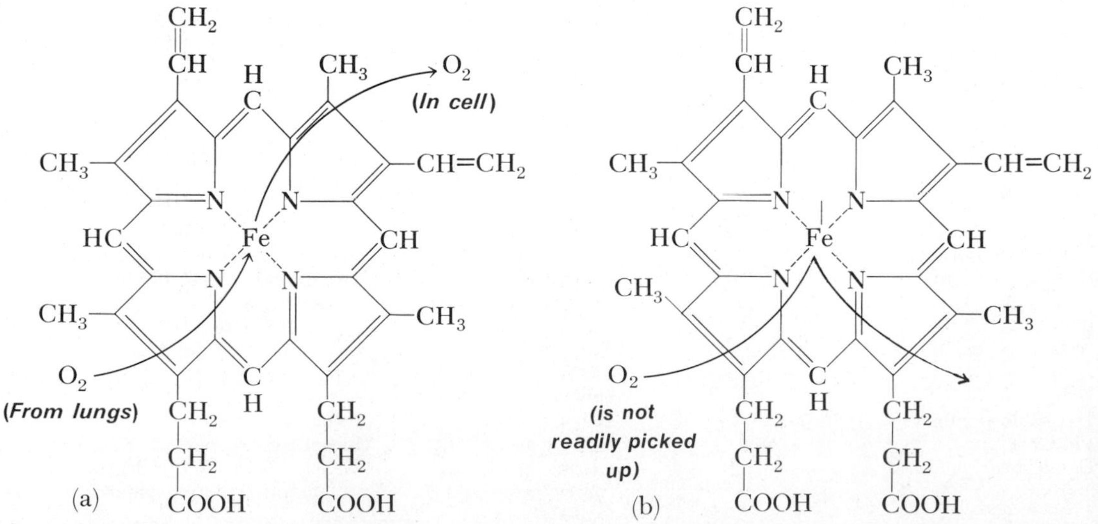

**Figure 15–1** Structure of the heme portion of hemoglobin. (*a*) Normal acceptance and release of oxygen. (*b*) Oxygen blocked by carbon monoxide.

**Figure 15–2** A healthy adult can tolerate 100 ppm carbon monoxide in air without suffering ill effect. A one-hour exposure to 1000 ppm causes a mild headache and a reddish coloration of the skin develops. A one-hour exposure to 1300 ppm turns the skin cherry red and a throbbing headache develops. One-hour exposure to concentrations greater than (>) 2000 ppm will likely cause death.

an average of about 7 ppm is typical for large cities. A concentration of 30 ppm for eight hours is sufficient to cause headache and nausea. Breathing an atmosphere that is 0.1% (1,000 ppm) carbon monoxide for four hours converts approximately 60% of the hemoglobin of an average adult to carboxyhemoglobin (Table 15–3), and death is likely to result (Fig. 15–2).

**To convert ppm to percent, divide by 10,000.**

Since both the carbon monoxide and oxygen reactions with hemoglobin involve easily reversed reactions, the concentrations, as well as relative strengths of bonds, affect the direction of the reaction. In air that contains 0.1% CO, oxygen molecules outnumber CO molecules 200 to 1. The larger concentration of oxygen helps to counteract the greater combining power of CO with hemoglobin. Consequently, if a carbon monoxide victim is exposed to fresh air or, still better, pure oxygen (provided he is still breathing), the carboxyhemoglobin (HbCO) is gradually decomposed, owing to the greater concentration of oxygen:

**Air is 21% $O_2$ by volume; in 1,000,000 "air molecules" there would be 210,000 $O_2$ molecules.**

$$HbCO + O_2 \rightleftarrows HbO_2 + CO$$

REACTION SHIFTED TO RIGHT BECAUSE OF GREATER CONCENTRATION OF OXYGEN

Although carbon monoxide is not a cumulative poison, permanent damage can occur if certain vital cells (e.g., brain cells) are deprived of oxygen for more than a few minutes.

Individuals differ in their tolerance of carbon monoxide, but generally those with anemia or an otherwise low reserve of hemoglobin (e.g., children) are more susceptible. No one is helped by carbon monoxide, and smokers suffer chronically from its effects. It is a subtle poison, since it is odorless and tasteless.

**TABLE 15–3 Concentration of CO in Atmosphere versus Percentage of Hemoglobin (Hb) Saturated***

| CO concentration in air | 0.01% (100 ppm) | 0.02% (200 ppm) | 0.10% (1,000 ppm) | 1.0% (10,000 ppm) |
|---|---|---|---|---|
| Percentage of hemoglobin molecules saturated with CO† | 17 | 20 | 60 | 90 |

*A few hours of breathing time is assumed.
†Normal human blood contains up to 5% of the hemoglobin as carboxyhemoglobin (HbCO).

## Cyanide

The cyanide ion, $CN^-$, is the toxic agent in cyanide salts such as sodium cyanide used in electroplating. Since the cyanide is a relatively strong base, it reacts easily with many acids (weak and strong) to form volatile hydrogen cyanide, HCN:

$$\underset{\text{ACETIC ACID}}{CH_3COOH} + \underset{\text{SODIUM CYANIDE}}{Na^+CN^-} \rightleftharpoons \underset{\text{HYDROGEN CYANIDE}}{HCN} + \underset{\text{SODIUM ACETATE}}{Na^+CH_3COO^-}$$

Since HCN boils at a relatively low temperature (26°C), it is a gas at temperatures slightly above room temperature. It is often used as a fumigant in storage bins and holds of ships because it is toxic to most forms of life and, in gaseous form, can penetrate into tiny openings, even into insect eggs.

Natural sources of cyanide ions include the seeds of the cherry, plum, peach, apple, and apricot fruits. Hydrogen cyanide is produced by hydrolysis of certain compounds, such as amygdalin, contained in the seeds:

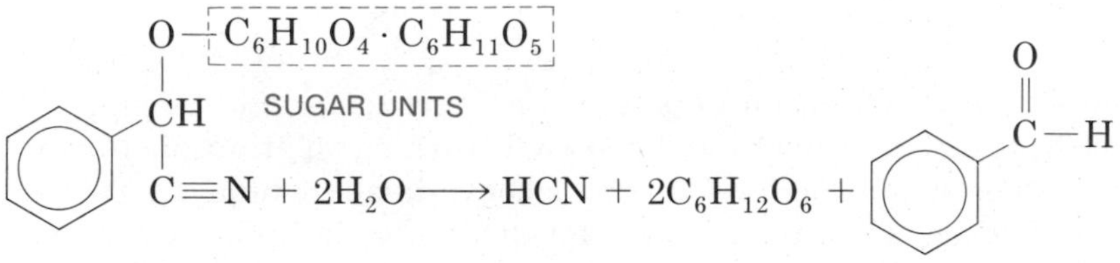

$$\underset{\text{AMYGDALIN}}{C_6H_5\text{—}CH(O\text{—}\underbrace{C_6H_{10}O_4 \cdot C_6H_{11}O_5}_{\text{SUGAR UNITS}})\text{—}C{\equiv}N} + 2H_2O \longrightarrow \underset{\text{HYDROGEN CYANIDE}}{HCN} + \underset{\text{GLUCOSE}}{2C_6H_{12}O_6} + \underset{\text{BENZALDEHYDE}}{C_6H_5\text{—}CHO}$$

The cyanide is not toxic as long as it is tied up in the amygdalin, but presumably if enough apple or peach seeds were hydrolyzed in warm acid, sufficient HCN would result to cause considerable danger. There are a few recorded instances of humans poisoned by eating large numbers of apple seeds. Amygdalin is not confined to the seeds; amounts as high as 66 mg per 100 g have been reported in peach leaves.

The cyanide ion is one of the most rapidly working poisons. Lethal doses taken orally act in minutes. Cyanide poisons by asphyxiation, as does carbon monoxide, but the mechanism of cyanide poisoning is different (Fig. 15–3). Instead of preventing the cells from getting oxygen, cyanide interferes with oxidative enzymes, such as cytochrome oxidase. Oxidases are enzymes containing a metal, usually iron or copper. They catalyze the oxidation of substances such as glucose:

**A metabolite is any substance in a metabolic process.**

$$\text{Metabolite } (H)_2 + \tfrac{1}{2}O_2 \xrightarrow{\text{oxidase}} \text{Oxidized metabolite} + H_2O + \text{energy}$$

**$Fe^{2+} \longrightarrow Fe^{3+} + e^-$ Oxidation**

The iron atom in cytochrome oxidase is oxidized from $Fe^{2+}$ to $Fe^{3+}$ to provide electrons for the reduction of $O_2$. The iron regains electrons from other steps in the pro-

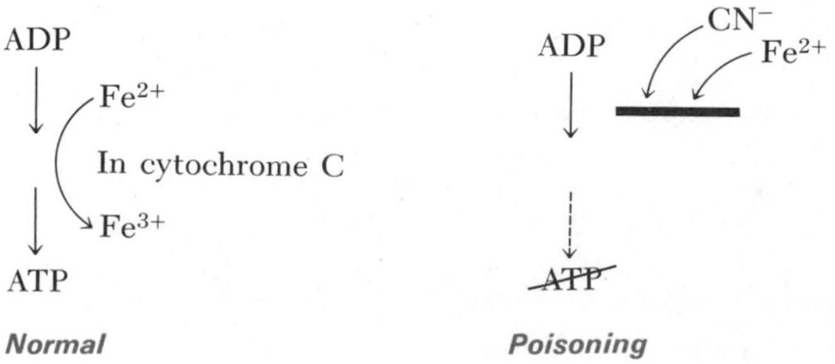

**Figure 15–3** The mechanism of cyanide ($CN^-$) poisoning. Cyanide binds tightly to the enzyme cytochrome C, an iron compound, thus blocking the vital ADP-ATP reaction in cells.

cess. The cyanide ion forms stable cyanide complexes with the metal ion of the oxidase and renders the enzyme incapable of reducing oxygen or oxidizing the metabolite.



$$\text{Cytochrome oxidase (Fe)} + CN^- \longrightarrow \text{Cytochrome oxidase} \underbrace{\text{(Fe)} \cdots CN^-}_{\text{complex}}$$

In essence the electrons of iron are "frozen"—they cannot participate in the oxidation-reduction processes. Plenty of oxygen gets to the cells, but the mechanism by which the oxygen is used in the support of life is stopped. Hence the cell dies, and if this occurs fast enough in the vital centers, the victim dies.

**The body can rid itself of a large number of toxic substances if the dose is small enough and sufficient time is allowed.**

The body has a mechanism for ridding itself slowly of cyanide ions. The cyanide-oxidative enzyme reaction is reversible, and other enzymes such as rhodanase, found in almost all cells, can convert cyanide ions to relatively harmless thiocyanate ions. For example,

$$CN^- + \underset{\text{THIOSULFATE}}{S_2O_3^{2-}} \xrightarrow{\text{rhodanase}} \underset{\text{THIOCYANATE}}{SCN^-} + SO_3^{2-}$$

This mechanism is not as effective in protecting a cyanide-poisoning victim as it might appear, since there is only a limited amount of thiosulfate available in the body at a given time.

### Fluoroacetic Acid

Nature has used the synthesis of fluoroacetic acid as part of a defense mechanism for certain plants. Native to South Africa, the *gilfbaar* plant contains lethal quantities of fluoroacetic acid. Cattle that eat these leaves usually sicken and die.

$$F-\underset{H}{\overset{H}{\underset{|}{\overset{|}{C}}}}-C\begin{matrix}\nearrow O \\ \searrow O-H\end{matrix}$$

FLUOROACETIC ACID

Sodium fluoroacetate, the sodium salt of this acid (Compound 1080), is a potent rodenticide (rat poison). Because it is odorless and tasteless it is especially dangerous, and its sale in this country is strictly regulated by law.

Fluoroacetate is toxic because it enters the Krebs cycle, producing fluorocitric acid, which in turn blocks the Krebs cycle (Fig. 15–4). The C—F linkage apparently ties up the enzyme aconitase, thus preventing it from converting citrate to isocitrate.

**Some poisons can act by mimicking other compounds.**

In this instance, the poison is similar enough to the normal substrate to compete effectively for the active sites on the enzyme. If a poison has sufficient affinity for the active site on the enzyme, it blocks the normal function of the enzyme. The blocking of the Krebs cycle by fluorocitrate is a typical example of this affinity. If fluoroacetate is not present in excessive amounts, its action can be reversed simply by increasing the concentration of available citrate.

### Heavy Metals

Heavy metals are perhaps the most common of all the metabolic poisons. These include such common elements as lead and mercury, as well as many less common ones such as cadmium, chromium, and thallium. In this group we should also include the infamous poison, arsenic, which is really not a metal but is metal-like in many of its properties, including its toxic action.

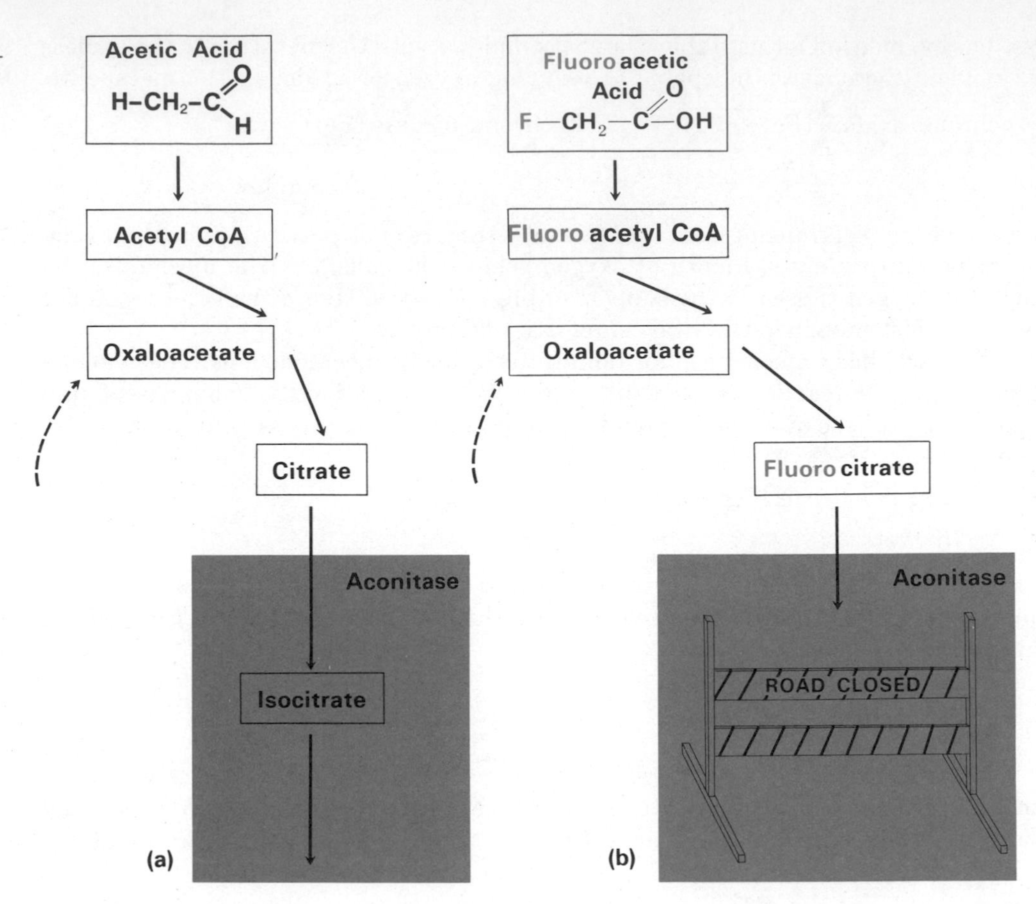


**Figure 15–4** Fluoroacetic acid is converted into fluorocitrate, which then forms a stable bond with the enzyme aconitase. This blocks the normal Krebs cycle, a portion of which is shown.

**The effects of cumulative poisons add up.**

Arsenic, a classic homicidal poison, occurs naturally in small amounts in many foods. Shrimp, for example, contain about 19 ppm arsenic, while corn may contain 0.4 ppm arsenic. Some agricultural insecticides contain arsenic (Table 15–4), and so some arsenic is observed in very small amounts on some fruits and vegetables. The Federal Food and Drug Administration (FDA) has set a limit of 0.15 mg of arsenic per pound of food, and this amount apparently causes no harm. Several drugs, such as arsphenamine, which has found some use in treating syphilis, contain covalently bonded arsenic. In its ionic forms, arsenic is much more toxic.

**Most heavy metals are cumulative poisons.**

Arsenic and heavy metals owe their toxicity primarily to their ability to react with and inhibit sulfhydryl (—SH) enzyme systems, such as those involved in the production of cellular energy. For example, glutathione (a tripeptide of glutamic acid, cysteine, and glycine) occurs in most tissues; its behavior with metals illus-

**TABLE 15–4 Some Arsenic-Containing Insecticides**

| Name | Formula |
|---|---|
| Lead arsenate | $Pb_3(AsO_4)_2$ |
| Monosodium methanearsenate | $CH_3—As(=O)(OH)—O^-Na^+$ |
| Paris green (copper acetoarsenite) | $3CuO \cdot 3As_2O_3 \cdot Cu(C_2H_3O_2)_2$ |

$$2\ \text{Glutathione} + \text{Metal ion}\ (M^{2+}) \longrightarrow M\ (\text{Glutathione})_2 + 2H^+$$

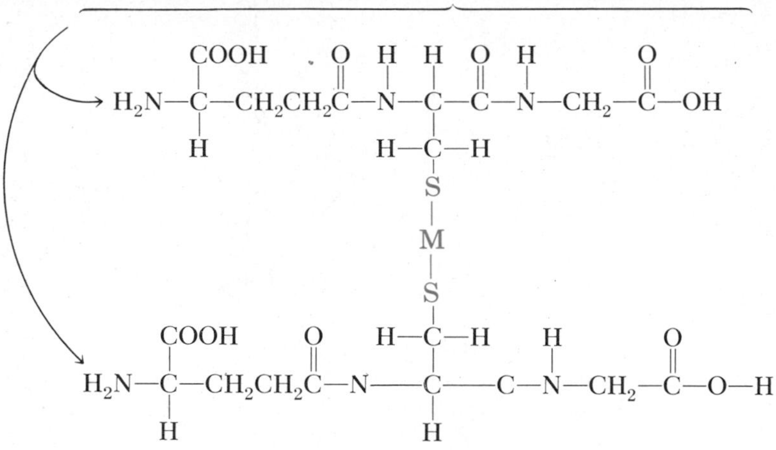

*Glutathione-Metal Complex*

**Figure 15–5** Glutathione reaction with a metal (M).

trates the interaction of a metal with sulfhydryl groups. The metal replaces the hydrogen on two sulfhydryl groups on adjacent molecules (Fig. 15–5), and the strong bond that results effectively eliminates the two glutathione molecules from further reaction. Glutathione is involved in maintaining healthy red blood cells.

The typical forms of toxic arsenic compounds are inorganic ions such as arsenate ($AsO_4^{3-}$) and arsenite ($AsO_3^{3-}$). The reaction of an arsenite ion with sulfhydryl groups results in a complex in which the arsenic unites with two sulfhydryl groups, which may be on two different molecules of protein or on the same molecule:

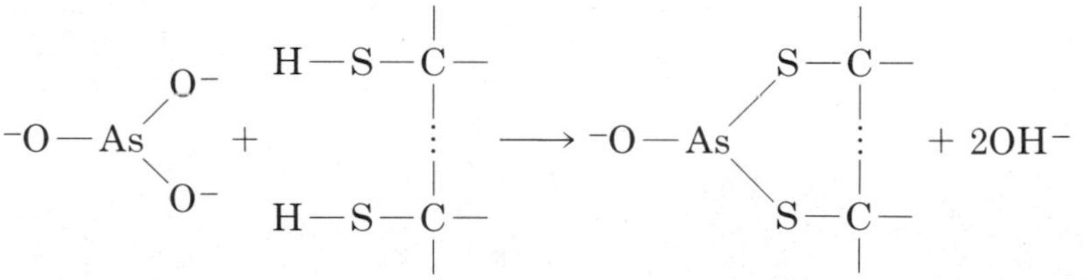

ARSENITE SULFHYDRYL GROUPS ARSENIC COMPLEX

The problem of developing a compound to counteract *Lewisite,* an arsenic-containing poison gas used in World War I, led to an understanding of how arsenic acts as a poison and subsequently to the development of an antidote. Once it was understood that Lewisite poisoned people by the reaction of arsenic with protein sulfhydryl groups, British scientists set out to find a suitable compound that contained highly reactive sulfhydryl groups that could compete with sulfhydryl groups in the natural substrate for the arsenic, and thus render the poison ineffective. Out of this research came a compound now known as British Anti-Lewisite (BAL).

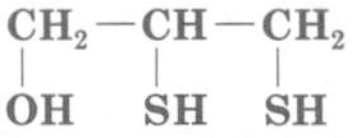

BAL
British Anti-Lewisite

The BAL, which bonds to the metal at several sites, is called a ***chelating agent*** (Greek, *chela,* claw), a term applied to a reacting agent that envelops a species such as a metal ion. BAL is one of a large number of compounds that can act as chelating agents for metals (Fig. 15–6).

**A chelating agent encases an atom or ion like a crab or an octopus surrounds a bit of food.**

With the arsenic or heavy metal ion tied up, the sulfhydryl groups in vital enzymes are freed and can resume their normal functions. BAL is a standard thera-

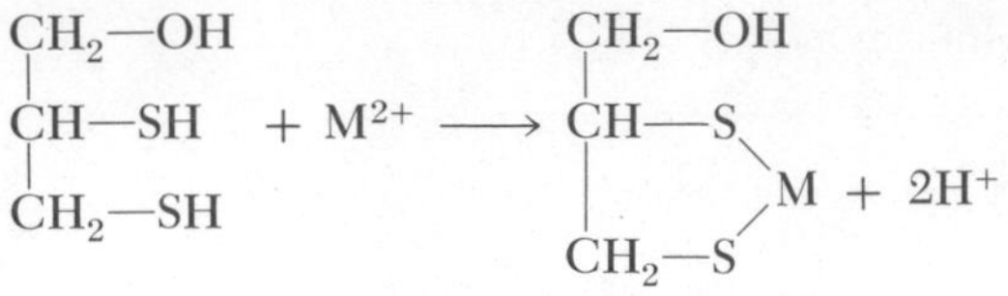

$$
\begin{array}{l} CH_2{-}OH \\ | \\ CH{-}SH \\ | \\ CH_2{-}SH \end{array} + M^{2+} \longrightarrow \begin{array}{l} CH_2{-}OH \\ | \\ CH{-}S \\ | \qquad\quad \rangle M \\ CH_2{-}S \end{array} + 2H^+
$$

*BAL* *Heavy metal ion* *Chelated metal ion*

**Figure 15–6** BAL chelation of arsenic or a heavy metal ion such as lead.

A vivid description of the psychic changes produced by mercury poisoning can be found in the Mad Hatter, a character in Lewis Carroll's *Alice in Wonderland.* The fur felt industry once used mercury (II) nitrate, $Hg(NO_3)_2$, to stiffen the felt. Chronic mercury poisoning accounted for the Mad Hatter's odd behavior; it also gave the workers in hat factories symptoms known as "hatter's shakes."

Amalgam: any mixture or alloy of metals in which mercury is a constituent.

peutic item in a hospital's poison emergency center and is used routinely to treat heavy metal poisoning.

Mercury deserves some special attention because it has a rather peculiar fascination for some people, especially children, who love to touch it (Fig. 15–7). It is poisonous and, to make matters worse, mercury and its salts accumulate in the body. This means the body has no quick means of ridding itself of this element and there tends to be a buildup of the toxic effects leading to ***chronic*** poisoning.

Although mercury is rather unreactive compared with other metals, it is quite volatile and easily absorbed through the skin. In the body the metal atoms are oxidized to $Hg_2^{2+}$ [mercury (I) ion] and $Hg^{2+}$ [mercury (II) ion]. Compounds of both $Hg_2^{2+}$ and $Hg^{2+}$ are known to be toxic.

Today mercury poisoning is a potential hazard to those working with or near this metal or its salts, such as dentists (who use it in making amalgams for fillings), various medical and scientific laboratory personnel (who routinely use mercury compounds or mercury pressure gauges), and some agricultural workers (who employ mercury salts as fungicides).

Mercury can also be a hazard when it is present in food. It is generally believed that mercury enters the food chain through small organisms that feed at the bottom of bodies of water that contain mercury from industrial waste or mercury minerals in the sediment. These in turn are food for bottom-feeding fish. Game fish in turn

**Figure 15–7** The first step to coating a coin with mercury. Children love to coat coins with metallic mercury—a very dangerous practice since mercury is easily inhaled and also passes through the skin.

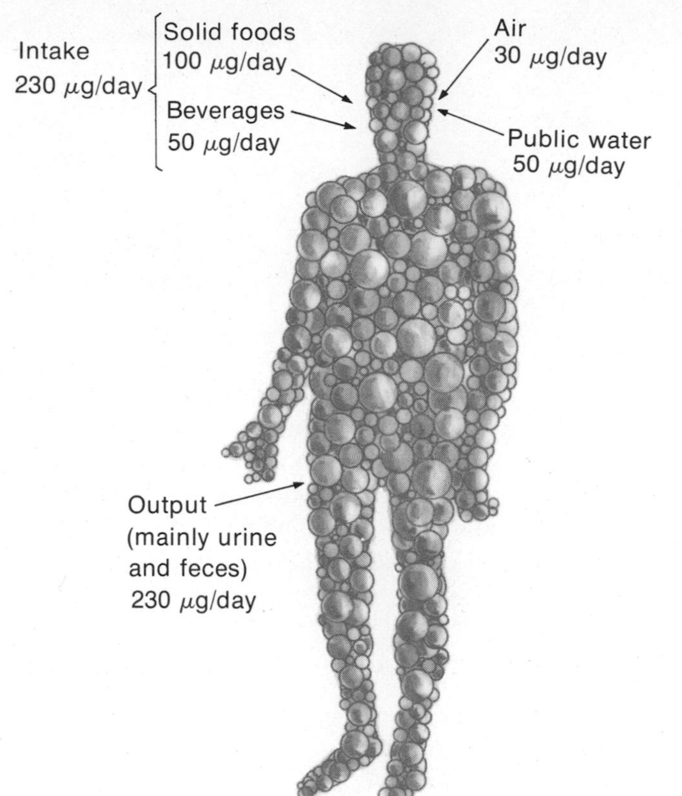


**Figure 15–8** Lead equilibrium in man. Figures chosen for intake are probable upper limits.

eat these fish and accumulate the largest concentration of mercury, the accumulation of poison building up as the food chain progresses.

Lead is another widely encountered heavy-metal poison. The body's method of handling lead provides an interesting example of a "metal equilibrium" (Fig. 15–8). Lead often occurs in foods (100–300 μg per kg), beverages (20–30 μg per liter), public water supplies (100 μg per liter, from lead-sealed pipes), and even air (2.5 μg per cubic meter, from lead compounds in auto exhausts). With this many sources and contacts per day, it is obvious that the body must be able to rid itself of this poison; otherwise everyone would have died long ago of lead poisoning! The average person can excrete about two milligrams of lead a day through the kidneys and intestinal tract; the daily intake is normally less than this. However, if intake exceeds this amount, accumulation and storage result. In the body lead resides not only in soft tissues but is also deposited in bone. In the bones lead acts on the bone marrow, while in tissues it behaves like other heavy-metal poisons, such as mercury and arsenic. Lead, like mercury and arsenic, can also affect the central nervous system.

**1 μg (microgram) = $10^{-6}$ g**

**1 mg (milligram) = $10^{-3}$ g = 1,000 μg**

Lead salts, unless they are very insoluble, are always toxic, and their toxicity is directly related to the salt's solubility. One common covalent lead compound, tetraethyllead, $Pb(C_2H_5)_4$, until recently a component of most gasolines, is different from most other metal compounds in that it is readily absorbed through the skin. Even metallic lead can be absorbed through the skin; cases of lead poisoning have resulted from repeated handling of lead foil, bullets, and other lead objects.

One of the truly tragic aspects of lead poisoning is the fact that even though lead-pigmented paints have not been used in this country for interior painting during the past 30 years, children are still poisoned by lead from old paint. Health experts estimate that up to 225,000 children become ill from lead poisoning each year, with many experiencing mental retardation or other neurological problems. The rea-

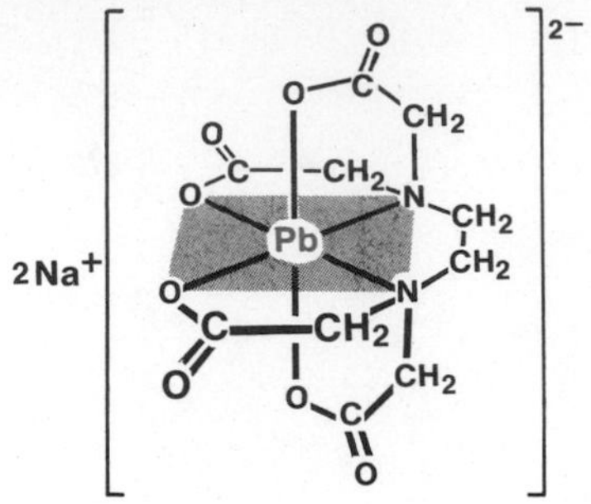

**Figure 15–9** The structure of the chelate formed when the anion of EDTA envelops a lead (II) ion.

son for this is two-fold. Lead-based paints still cover the walls of many older dwellings. Coupled with this is the fact that many children in poverty-stricken areas are ill-fed and anemic. These children develop a peculiar appetite trait called ***pica,*** and among the items that satisfy their cravings are pieces of flaking paint, which may contain lead. In 1969, about 200 children in the United States alone died of lead poisoning and untold thousands suffered permanent damage.

Toxicologists have discovered an effective chelating agent to remove lead from the human body—ethylenediaminetetraacetic acid, also called EDTA (Fig. 15–9).

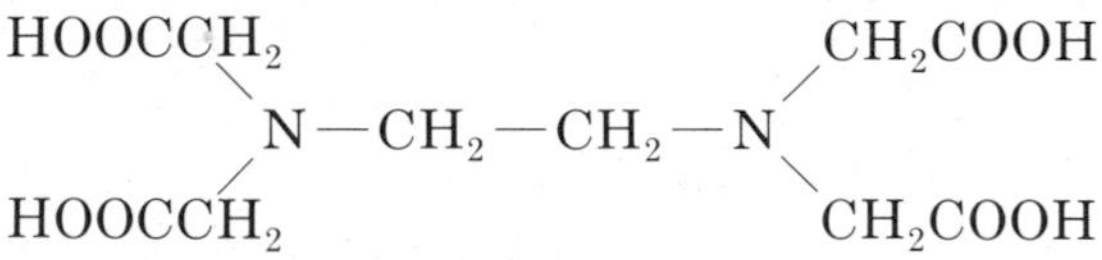

$$\begin{array}{lcl} HOOCCH_2 & & CH_2COOH \\ & N-CH_2-CH_2-N & \\ HOOCCH_2 & & CH_2COOH \end{array}$$

EDTA
(ETHYLENEDIAMINETETRAACETIC ACID)

The calcium disodium salt of EDTA is used in the treatment of lead poisoning because EDTA by itself would remove too much of the blood serum's calcium. In solution EDTA has a greater tendency to complex with lead ($Pb^{2+}$) than with calcium ($Ca^{2+}$). As a result, the calcium is released and the lead is tied up in the complex:

$$[CaEDTA]^{2-} + Pb^{2+} \longrightarrow [PbEDTA]^{2-} + Ca^{2+}$$

The lead chelate is then excreted in the urine.

## Self-Test 15-A

1. Corrosive poisons such as sulfuric acid destroy tissue by ______________ followed by ______________ of proteins.
2. Corrosive poisons, such as ozone, nitrogen dioxide, and iodine, destroy tissue by ______________ it.
3. Carbon monoxide poisons by forming a strong bond with iron in ______________ and thus preventing the transport of ______________ from the lungs to the cells throughout the body.
4. CO is a cumulative poison. (True/False)
5. The cyanide ion has the formula ______________. It poisons by complexing with iron in the enzyme ______________ ______________, thus preventing the use of ______________ in the oxidative processes in the cells.

6. Give an example of a metabolic poison that is toxic because its structure is so similar to a useful substance that it can mimic the useful substance. ____________
7. BAL is an antidote for ____________. BAL is effective because its sulfhydryl (—SH) groups ____________ arsenic and heavy metals and render them ineffective toward enzymes.
8. Mercury is a cumulative poison. (True/False)

## Neurotoxins

Some metabolic poisons are known to limit their action to the nervous system. These include poisons such as strychnine and curare (a South American Indian dart poison), as well as the dreaded nerve gases developed for chemical warfare. The exact modes of action of most neurotoxins are not known for certain, but investigations have discovered the action of a few.

**Investigations of the actions of neurotoxins have provided insight into how the nervous system works.**

$$CH_3\overset{O}{\overset{\|}{C}}OCH_2CH_2\overset{CH_3}{\underset{CH_3}{\overset{|}{\underset{|}{N^+}}}}CH_3,\ OH^-$$

*Acetylcholine*

A nerve impulse or stimulus is transmitted along a nerve fiber by electrical impulses. The nerve fiber connects with either another nerve fiber or with some other cell (such as a gland, or cardiac, smooth, or skeletal muscle) capable of being stimulated by the nerve impulse (Fig. 15–10). Neurotoxins often act at the point where

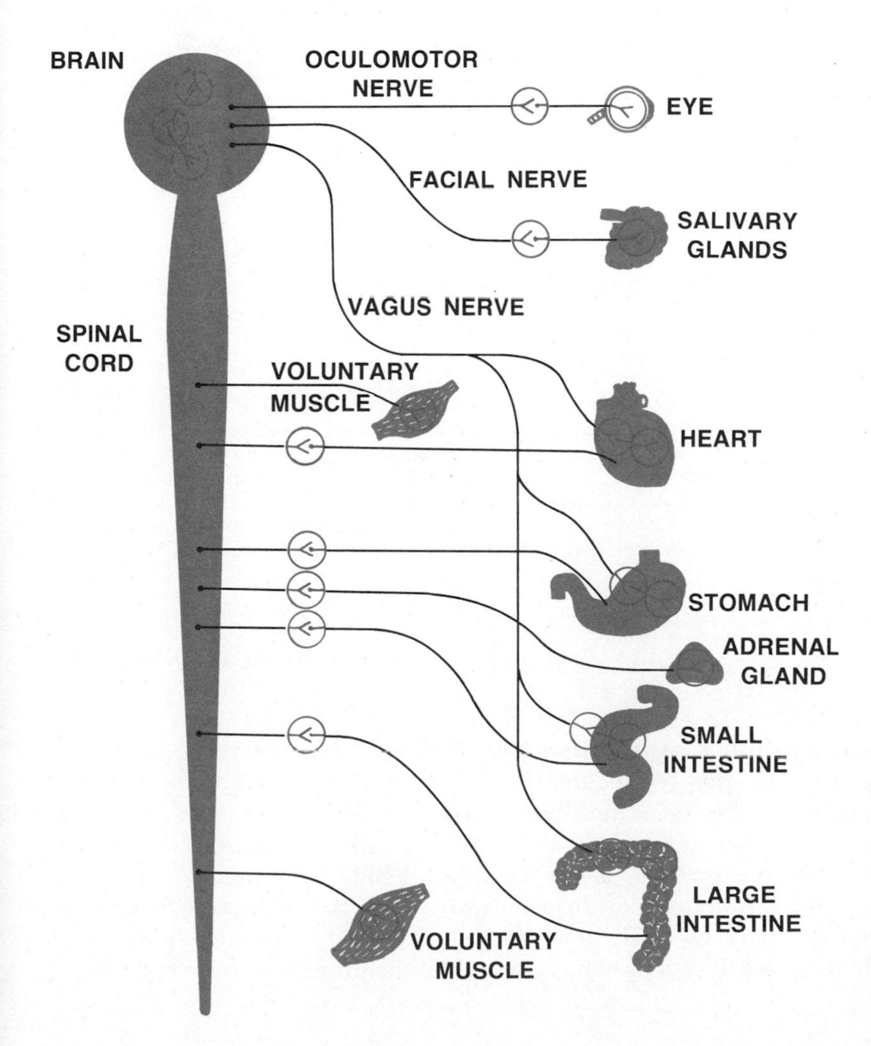


**Figure 15–10** "Cholinergic" nerves, which transmit impulses by means of acetylcholine, include nerves controlling both voluntary and involuntary activities. Exceptions are parts of the "sympathetic" nervous system that utilize norepinephrine instead of acetylcholine. Sites of acetylcholine secretion are circled in color; poisons that disrupt the acetylcholine cycle can interrupt the body's communications at any of these points. The role of acetylcholine in the brain is uncertain, as is indicated by the broken circles.

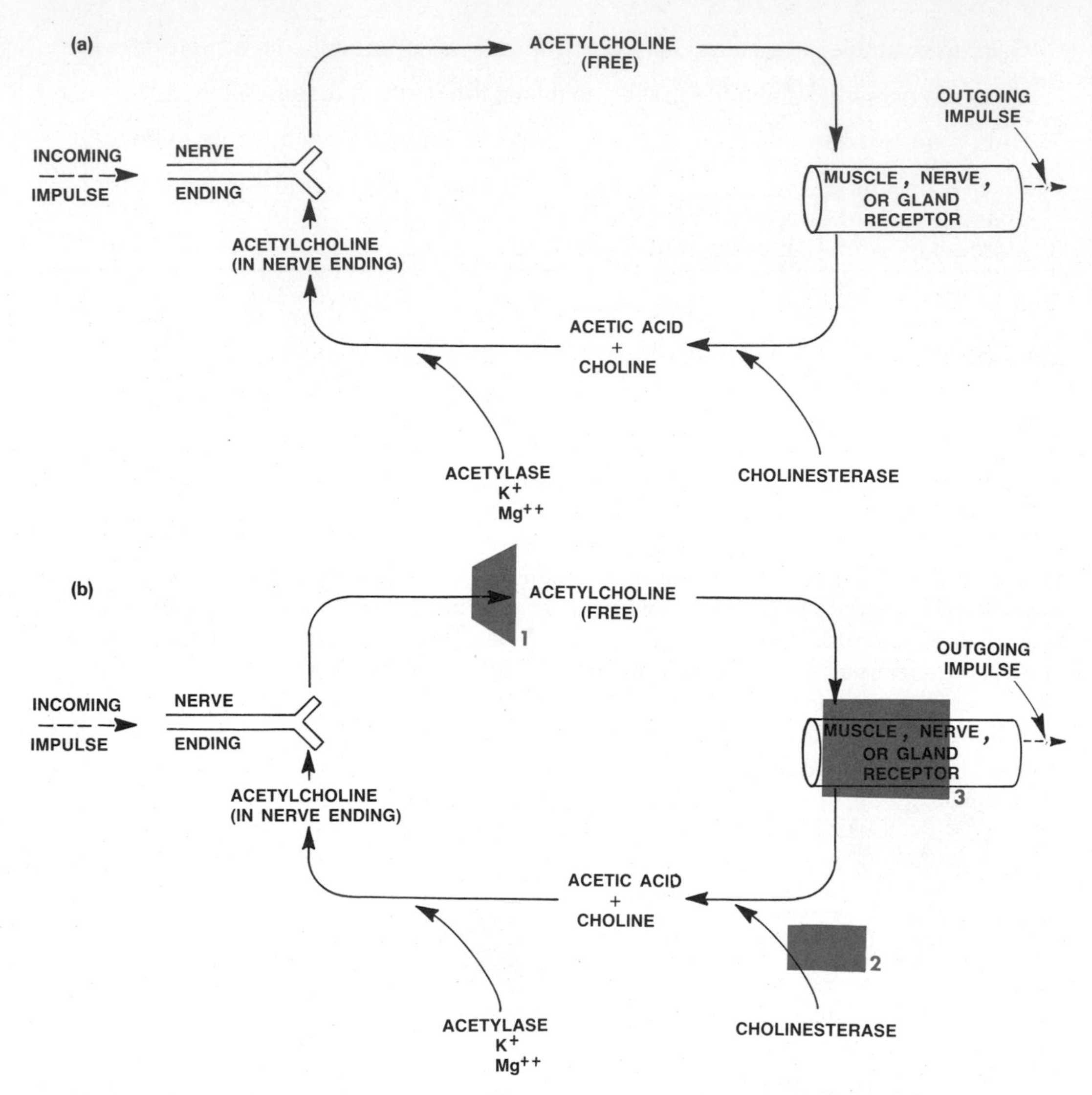


**Figure 15–11** The acetylcholine cycle, a fundamental mechanism in nerve impulse transmission, is affected by many poisons. An impulse reaching a nerve ending in the normal cycle (*a*) liberates acetylcholine, which then stimulates a receptor. To enable the receptor to receive further impulses, the enzyme *cholinesterase* breaks down acetylcholine into acetic acid and choline; other enzymes resynthesize these into more acetylcholine. (*b*) Botulinus and dinoflagellate toxins inhibit the synthesis, or the release, of acetylcholine (*1*). The "anticholinesterase" poisons inactivate cholinesterase, and therefore prevent the breakdown of acetylcholine (*2*). Curare and atropine desensitize the receptor to the chemical stimulus (*3*).

**Permeability: the ability of a membrane to let chemicals pass through it.**

**$10^{-6}$ of a mole of acetylcholine is $6 \times 10^{17}$ molecules.**

two nerve fibers come together, called a ***synapse.*** When the impulse reaches the end of certain nerves, a small quantity of ***acetylcholine*** is liberated. This activates a receptor on an adjacent nerve or organ. The acetylcholine is thought to activate a nerve ending by changing the permeability of the nerve cell membrane. The method of increasing membrane permeability is not clear, but it may be related to an ability to dissociate fat-protein complexes or to penetrate the surface films of fats. Such effects can be brought about by as little as $10^{-6}$ mole of acetylcholine, which could alter the permeability of a cell so ions can cross the cell membrane more freely.

To enable the receptor to receive further electrical impulses, the enzyme ***cholinesterase*** breaks down acetylcholine into acetic acid and choline (Fig. 15–11):

$$\underset{\text{ACETYLCHOLINE}}{CH_3\overset{\displaystyle O}{\overset{\|}{C}}OCH_2CH_2\overset{\displaystyle CH_3}{\overset{|}{\underset{\displaystyle CH_3}{\underset{|}{N^+}}}}-CH_3,\ OH^-} + \underset{\text{WATER}}{H_2O} \xrightarrow{\text{cholinesterase}} \underset{\text{ACETIC ACID}}{CH_3\overset{\displaystyle O}{\overset{\|}{C}}OH} + \underset{\text{CHOLINE}}{HOCH_2CH_2\overset{\displaystyle CH_3}{\overset{|}{\underset{\displaystyle CH_3}{\underset{|}{N^+}}}}-CH_3,\ OH^-}$$

In the presence of potassium and magnesium ions, other enzymes such as acetylase resynthesize new acetylcholine from the acetic acid and the choline within the incoming nerve ending:

$$\text{Acetic acid} + \text{Choline} \xrightarrow{\text{acetylase}} \text{Acetylcholine} + H_2O$$

The new acetylcholine is available for transmitting another impulse across the gap.

Neurotoxins can affect the transmission of nerve impulses at nerve endings in a variety of ways. The ***anticholinesterase poisons*** prevent the breakdown of acetylcholine by deactivating cholinesterase. These poisons are usually structurally analogous to acetylcholine, so they bond to the enzyme cholinesterase and deactivate it (Fig. 15–12). The cholinesterase molecules bound by the poison are held so effectively that the restoration of proper nerve function must await the manufacture of new cholinesterase. In the meantime, the excess acetylcholine overstimulates nerves, glands, and muscles, producing irregular heart rhythms, convulsions, and death. Many of the organic phosphates that are widely used as insecticides are metabolized in the body to produce anticholinesterase poisons. For this reason, they should be treated with extreme care. Some poisonous mushrooms also contain an anticholinesterase poison. Figure 15–13 contains the structures of some anticholinesterase poisons.

Neurotoxins such as ***atropine*** and ***curare*** are able to occupy the receptor sites on nerve endings of organs that are normally occupied by the impulse-carrying acetylcholine. When atropine or curare occupies the receptor site, no stimulus is transmitted to the organ. Acetylcholine in excess causes a slowing of the heartbeat, a decrease in blood pressure, and excessive saliva, whereas atropine and curare produce excessive thirst and dryness of the mouth and throat, a rapid heartbeat, and an increase in blood pressure. The normal responses to acetylcholine activation are absent, and the opposite responses occur when there is sufficient atropine present to block the receptor sites.

**Curare, used by South American Indians in poison darts, was brought to Europe by Sir Walter Raleigh in 1595. It was purified in 1865, and its structure was determined in 1935.**

Neurotoxins of this kind can be extremely useful in medicine. For example, atropine is used to dilate the pupil of the eye to facilitate examination of its interior. Applied to the skin, atropine sulfate and other atropine salts relieve pain by deactivating sensory nerve endings on the skin. Atropine is also used as an antidote for anticholinesterase poisons. Curare has long been used as a muscle relaxant.

A well-known, natural organic compound that blocks receptor sites in a manner similar to that of curare and atropine is ***nicotine.*** This powerful poison causes stimulation and then depression of the central nervous system. The probable lethal dose for a 70-kilogram person is less than 0.3 g. It is interesting to note that pure nicotine was first extracted from tobacco and its toxic action observed *after* tobacco use was established as an acceptable habit.

Natural or synthetic ***morphine*** is the most effective pain reliever known. It is widely used to relieve short-term acute pain resulting from surgery, fractures, burns, and so on, as well as to reduce suffering in the later stages of terminal illnesses such as cancer. The manufacture and distribution of narcotic drugs are stringently controlled by the federal government through laws designed to keep these products avail-

**Morphine is the most effective pain killer known.**

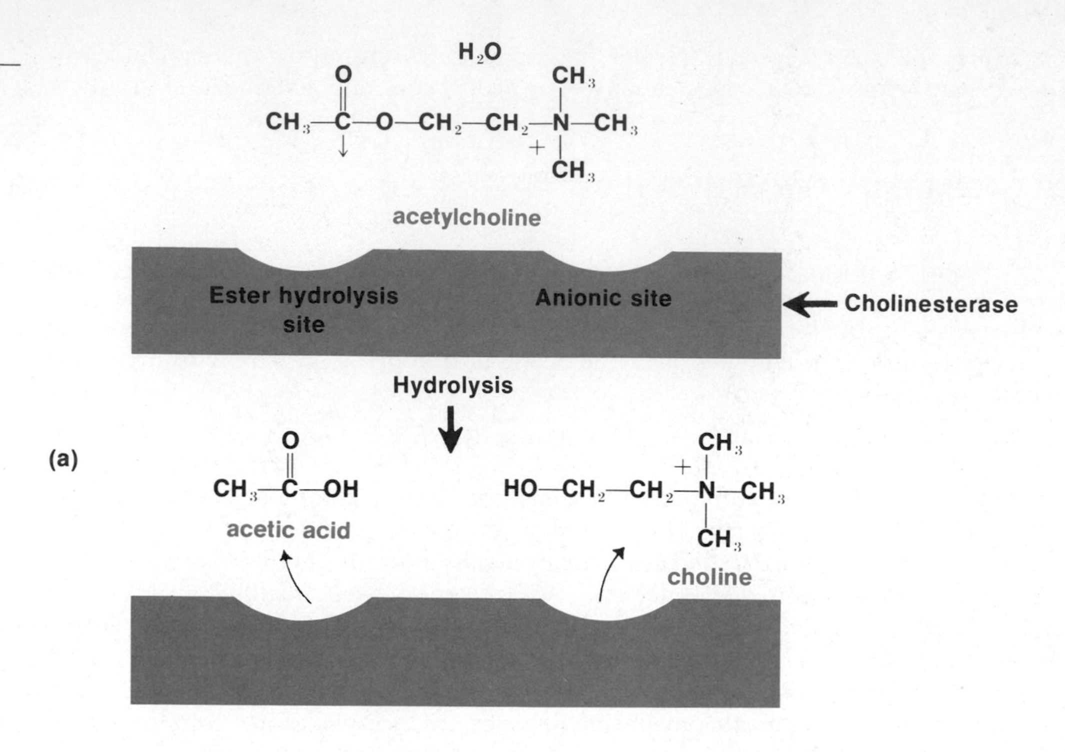


Diisopropylfluorophosphate
(DFP, nerve gas)
Strong bonding
acetylcholine
(b)

**Figure 15–12** (*a*) The mechanism of cholinesterase breakdown of acetylcholine. (*b*) The tie-up of cholinesterase by an anticholinesterase poison like the nerve gas DFP blocks the normal hydrolysis of acetylcholine since the acetylcholine cannot bind to the enzyme.

*Meperidine*

able for legitimate medical use. Under federal law, some preparations containing small amounts of narcotic drugs may be sold without a prescription (for example, cough mixtures containing codeine), but not many.

In spite of stringent controls, drugs like ***morphine, heroin, meperidine,*** and ***methadone*** are abused and illicitly used. Heroin is prepared from morphine, which is derived from sap in the opium poppy. It takes about 10 pounds of opium to prepare one pound of morphine. Morphine reacts with acetic anhydride in a one-to-one reaction to form heroin. Street-grade heroin is only 9–10% pure.

***Meperidine*** and ***methadone*** are products of chemical laboratories rather than of poppy fields. Meperidine was claimed to be nonaddictive when first produced.

| Name | Structure | $LD_{50}$ (rat;oral), mg/kg | Use |
|---|---|---|---|
| Sarin | 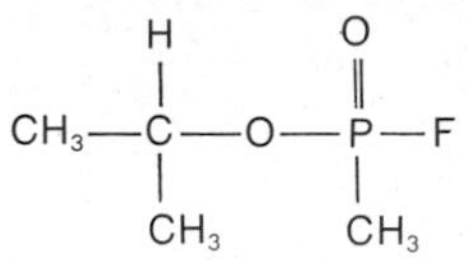 $CH_3-CH(CH_3)-O-P(=O)(CH_3)-F$ | 0.55 | World War II nerve gas |
| Tabun | 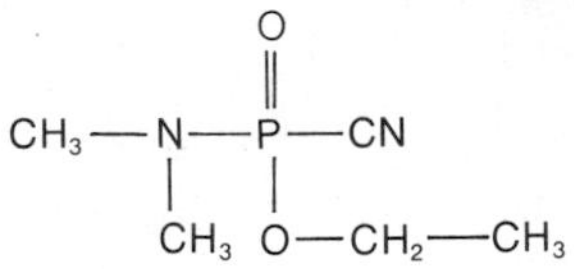 $CH_3-N(CH_3)-P(=O)(O-CH_2-CH_3)-CN$ | 3.7 | World War II nerve gas |
| Parathion | 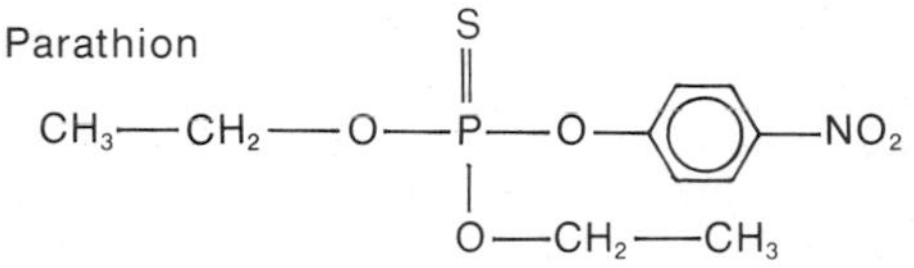 $CH_3-CH_2-O-P(=S)(O-CH_2-CH_3)-O-C_6H_4-NO_2$ | 20 | Insecticide |
| Paraoxon | 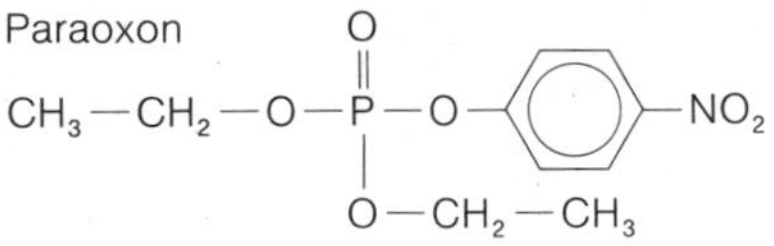 $CH_3-CH_2-O-P(=O)(O-CH_2-CH_3)-O-C_6H_4-NO_2$ | 1.8 | Insecticide |
| Malathion | 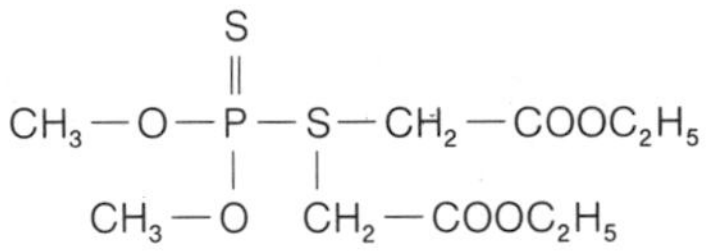 $CH_3-O-P(=S)(CH_3-O)-S-CH(CH_2-COOC_2H_5)-COOC_2H_5$ | 885 | Insectide |
| Carbyl (Sevin) | 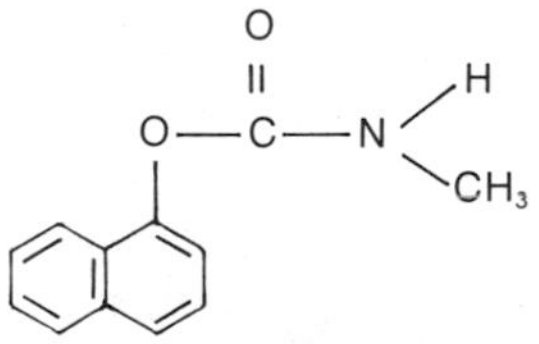 $C_{10}H_7-O-C(=O)-N(H)-CH_3$ | 400 | Insecticide |

**Figure 15–13** Some anticholinesterase poisons. In animals, parathion is converted into paraoxon in the liver. Carbyl and malathion do not bind to cholinesterase as strongly. Malathion was the insecticide used in California in July 1981 to eradicate Medflies.

Experience, however, proved otherwise (as it did with morphine and heroin). A major difference between methadone and morphine and heroin is that when methadone is taken orally, under medical supervision, it prevents withdrawal symptoms for approximately 24 hours.

Some other natural products that affect the central nervous system and can be neurotoxic in comparatively small amounts are listed in Table 15–5.

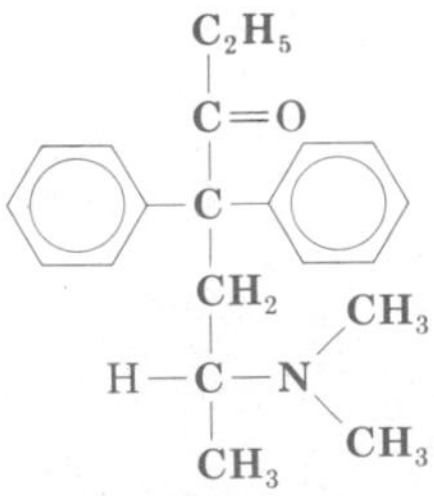

$C_2H_5-C(=O)-C(C_6H_5)_2-CH_2-CH(CH_3)-N(CH_3)_2$

*Methadone*

## Mutagens

Mutagens are chemicals capable of altering the genes and chromosomes sufficiently to cause abnormalities in offspring. Chemically, mutagens alter the structures of DNA and RNA, which compose the genes (and, in turn, the chromosomes) that transmit the traits of parent to offspring. Mature sex, or germinal, cells of humans normally have 23 chromosomes; body, or somatic, cells have 23 *pairs* of chromosomes.

**A mutagen is a chemical that can change the hereditary pattern of a cell.**

Although many chemicals are under suspicion because of their mutagenic effects on laboratory animals, it should be emphasized that no one has yet shown conclusively

**TABLE 15–5 Alkaloid Neurotoxins That Compete with Acetylcholine for the Receptor Site***

| Name | Normal Contact | Lethal Dose (for a 70-kg Human) | Formula |
|---|---|---|---|
| Atropine | dilation of pupil of the eye | 0.1g | 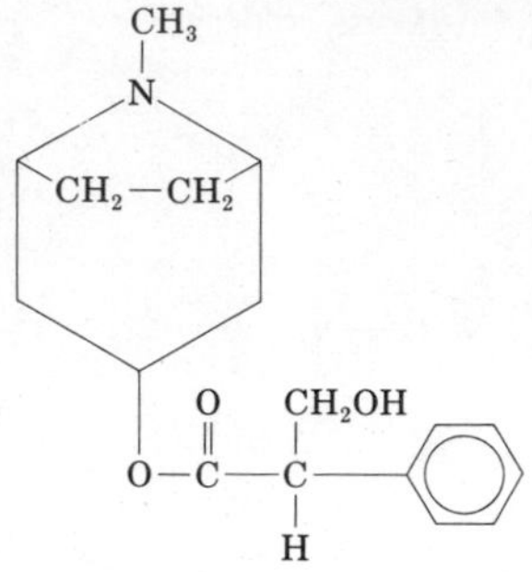 |
| Curare | muscle relaxant | 20 mg | 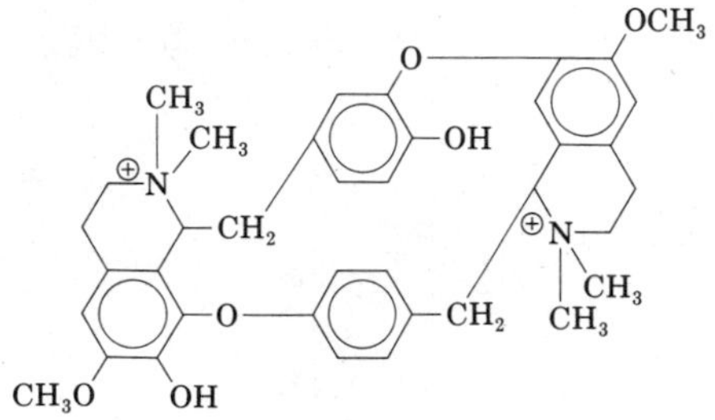 |
| Nicotine | tobacco, insecticide | 75 mg | 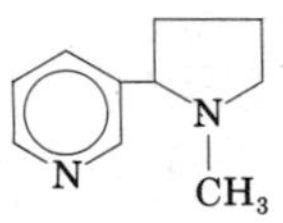 |
| Caffeine | coffee, tea, cola drinks | 13.4 g (one cup of coffee contains about 40 mg caffeine) | 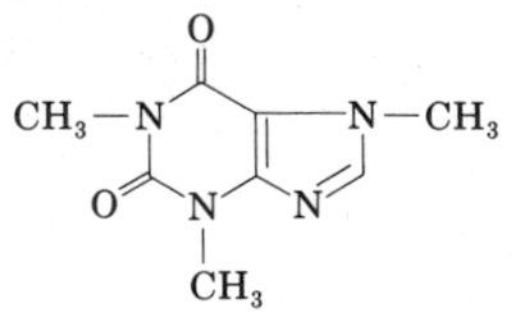 |
| Morphine | opium—pain killer | 100 mg | 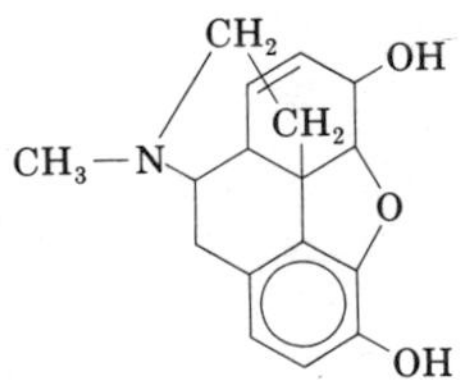 |
| Codeine | opium—pain killer | 0.3 g | 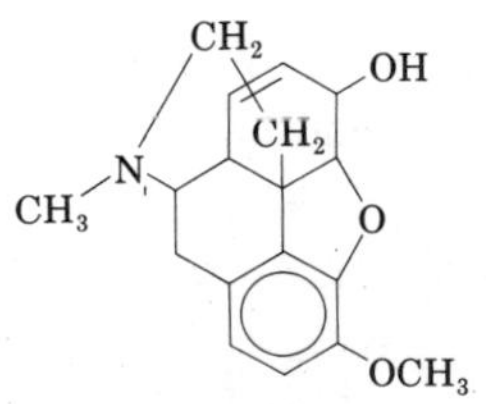 |
| Cocaine | leaves of *Erythroxylon coca* in South America | 1 g | 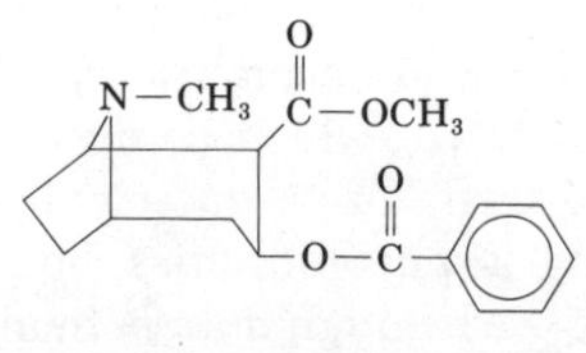 |

****Alkaloid*** is broadly defined as a physiologically active compound found in plants and containing amino nitrogen atoms (consequently it has basic properties). The nitrogen atom, or atoms, are frequently found as part of rings.

that any chemical induces mutations in human germinal cells. Part of the difficulty of determining the effects of mutagenic chemicals in humans is the extreme rarity of mutation. A specific genetic disorder may occur as infrequently as only once in 10,000 to 100,000 births. Therefore, to obtain meaningful statistical data, a carefully controlled study of the entire population of the United States would be required. In addition, the very long time between generations presents great difficulties, and there is also the problem of tracing a medical disorder to a single specific chemical out of the tens of thousands of chemicals with which we come in contact.

If there is no direct evidence for specific mutagenic effects in human beings, why, then, the interest in the subject? The possibility of a deranged, deformed human race is frightening; the chance for an improved human body is hopeful; and the evidence for chemical mutation in plants and lower animals is established. A wide variety of chemicals are known to alter chromosomes and to produce mutations in rats, worms, bacteria, fruit flies, and other plants and animals. Some of these are listed in Table 15–6.

Experimental work on the chemical basis of the mutagenic effects of nitrous acid ($HNO_2$) has been very revealing. Repeated studies have shown that nitrous acid is a potent mutagen in bacteria, viruses, molds, and other organisms. In 1953, at Columbia University, Dr. Stephen Zamenhof demonstrated experimentally that nitrous acid attacks DNA. Specifically, nitrous acid reacts with the adenine, guanine, and cytosine bases of DNA by removing the amino group of each of these compounds. The eliminated group is replaced by an oxygen atom (Fig. 15–14). The changed bases may garble a part of DNA's genetic message, and in the next replication of DNA the new base may not form a base pair with the proper nucleotide base.

**TABLE 15–6 Mutagenic Substances as Indicated by Experimental Studies on Plants and Animals**

| Substance | Experimental Results |
|---|---|
| Aflatoxin (from mold, *Aspergillus flavus*) | mutations in bacteria, viruses, fungi, parasitic wasps, human cell cultures, mice |
| Benzo($\alpha$)pyrene (from cigarette and coal smoke) | mutations in mice |
| Caffeine | chromosome changes in bacteria, fungi, onion root tips, fruit flies, human tissue cultures |
| Captan (a fungicide) | mutagenic in bacteria and molds; chromosome breaks in rats and human tissue cultures |
| Chloroprene | mutagenic in male sex cells; results in spontaneous abortions |
| Dimethyl sulfate (used extensively in chemical industry to methylate amines, phenols, and other compounds) | methylates DNA base guanine; potent mutagen in bacteria, viruses, fungi, higher plants, fruit flies |
| LSD (lysergic acid diethylamide) | chromosome breaks in somatic cells of rats, mice, hamsters, white blood cells of humans and monkeys |
| Maleic hydrazide (plant growth inhibitor; trade names Slo-Gro, MH-30) | chromosome breaks in many plants and in cultured mouse cells |
| Mustard gas (dichlorodiethyl sulfide) | mutations in fruit flies |
| Nitrous acid ($HNO_2$) | mutations in bacteria, viruses, fungi |
| Ozone ($O_3$) | chromosome breaks in root cells of broadleaf plants |
| Solvents in glue (glue sniffing) (toluene, acetone, hexane, cyclohexane, ethyl acetate) | 4% more human white blood cells showed breaks and abnormalities (6% versus 2% normal) |
| TEM (triethylenemelamine) (anticancer drug, insect chemosterilants) | mutagenic in fruit flies, mice |

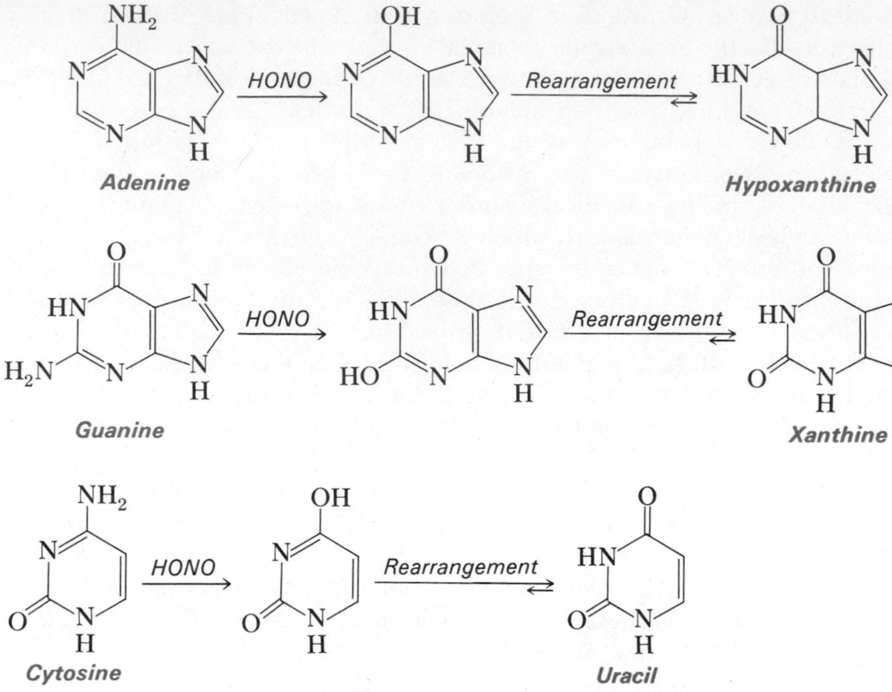

**Figure 15–14** Reaction of nitrous acid (HONO) with nitrogenous bases of DNA. Nitrogen gas ($N_2$) and water are also products of each reaction.

For example, adenine (A) typically forms a base pair with thymine (T) (Fig. 15–15). However, when adenine is changed to hypoxanthine, the new compound forms a base pair with cytosine (C). In the second replication, the cytosine forms its usual base pair with guanine (G). Thus, where an adenine-thymine (A-T) base pair existed originally, a guanine-cytosine (G-C) pair now exists. The result is an alteration in the DNA's genetic coding, so that a different protein is formed later.

**Sodium nitrite produces nitrous acid in the stomach.**

Do all of these findings mean that nitrous acid is mutagenic in humans? Not necessarily. We do know that ***sodium nitrite*** has been widely used as a preservative, color enhancer, or color fixative in meat and fish products for at least the past 30 years. It is currently used in such foods as frankfurters, bacon, smoked ham, deviled ham, bologna, Vienna sausage, smoked salmon, and smoked shad. The sodium nitrite is converted to nitrous acid by hydrochloric acid in the human stomach:

$$NaNO_2 + HCl \longrightarrow HNO_2 + NaCl$$

The Food and Drug Administration (FDA) now considers the mutagenic effects of nitrous acid in lower organisms sufficiently ominous to suggest strongly that the use of sodium nitrite in foods be severely curtailed, and a complete ban of this use of sodium nitrite is being considered. A number of European countries already restrict the use of sodium nitrite in foods. The concern is that this compound, after being converted in the body to nitrous acid, may cause mutation in somatic cells (and possibly in germinal cells) and thus could possibly produce cancer in the human stomach. Other scientists doubt that nitrous acid is present in germinal cells and, therefore, seriously question whether this compound could be a cause of genetically produced birth defects in man. The uncertainty of extrapolating results obtained in animal studies to human beings hovers over the mutagenic substances.

Thus far research has concentrated on the action of chemicals in causing mutations in bacterial viruses, molds, fruit flies, mice, rats, human white blood cells, and so on. Perhaps in the next 10 to 20 years it will be demonstrated that these chemicals

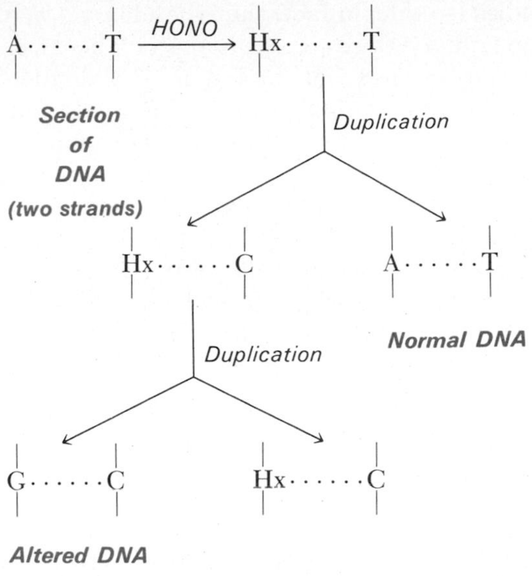

A······T —HONO→ Hx······T

Section of DNA (two strands)

Duplication

Hx······C    A······T

Normal DNA

Duplication

G······C    Hx······C

Altered DNA

**Figure 15–15** Alteration of DNA genetic code by base pairing nitrous acid-converted nitrogenous bases. The bases are adenine (A), cytosine (C), guanine (G), hypoxanthine (Hx), and thymine (T).

can produce transmissible alteration of chromosomes in human germinal cells. Meanwhile, many scientists are pressing for a more vigorous research effort to expand our knowledge of chemically induced mutations and of their potentially harmful effects. One intriguing theory that will surely invoke experimental examination is the belief that some compounds cause cancer because they are first and foremost mutagenic. The supporting evidence at present is still extremely inconclusive.

## Teratogens

The effects of chemicals on human reproduction are a frightening aspect of toxicity. The study of birth defects produced by chemical agents is the discipline of ***teratology.*** The word root *terat* comes from the French word meaning "monster." There are three known classes of teratogens: radiation, viral agents, and chemical substances.

Birth defects occur in 2–3% of all births. About 25% of these occur from genetic causes, some possibly due to contact with mutagens, and 5–10% are the result of teratogens. The remaining 60% or so result from unknown causes.

In the development of the newborn there are three basic periods during which the fetus is at risk. For a period of about 17 days between conception and implantation in the uterine wall, a chemical "insult" will result in cell death. The rapidly multiplying cells often recover, but if a lethal dose is administered, death of the organism occurs followed by abortion or reabsorption. During the critical embryonic stage (18 to 55 days) organogenesis, or development of the organs, occurs. At this time the fetus is extremely sensitive to teratogens. During the fetal period (56 days to term) the fetus is less sensitive. Contact with teratogens results in reduction of cell size and number. This is manifested in growth retardation and failure of vital organs to reach maturity.

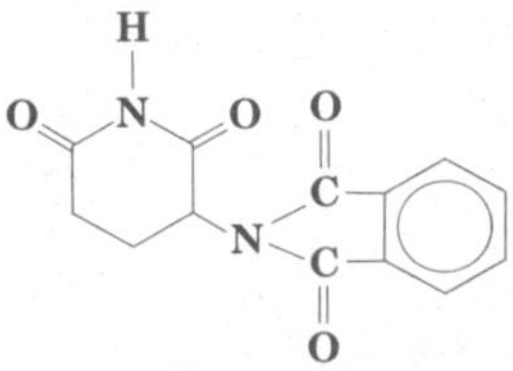

The horrible thalidomide disaster in 1961 focused worldwide attention on chemically induced birth defects. Thalidomide, a tranquilizer and sleeping pill, caused gross deformities (flipperlike arms, shortened arms, no arms or legs, and other defects) in children whose mothers used this drug during the first two months of pregnancy. The use of this drug resulted in more than 4000 surviving malformed babies in West Germany, more than 1000 in Great Britain, and about 20 in the United States. With shattering impact, this incident demonstrated that a compound can appear to be

*Thalidomide (A teratogen)*

remarkably safe on the basis of animal studies (so safe, in fact, that thalidomide was sold in West Germany without prescription) and yet cause catastrophic effects in humans. While the tragedy focused attention on chemical mutagens, thalidomide presumably does not cause genetic damage in the germinal cells, and is really not mutagenic. Rather, thalidomide, when taken by a woman during early pregnancy, causes direct injury to the developing embryo.

Any chemical substance that can cross the placenta is a potential teratogen, and any activity resulting in the uptake into the mother's blood of these chemicals might prove a dangerous act for the health and well-being of the fetus. Smoking a cigarette results in higher-than-normal blood levels of such substances as carbon monoxide, hydrogen cyanide, cadmium, nicotine, and benzo($\alpha$)pyrene. Of course, many of these substances are present in polluted air as well. Table 15–7 lists a number of chemical substances known to be teratogenic in humans and laboratory animals.

## Carcinogens

Cancer of the epithelial tissue—*carcinoma.*

Cancer of the connective tissue—*sarcoma.*

Carcinogens are chemicals that cause cancer. ***Cancer*** is an abnormal growth condition in an organism that manifests itself in at least three ways. The rate of cell growth (that is, the rate of cellular multiplication) in cancerous tissue differs from the rate in normal tissue. Cancerous cells may divide more rapidly or more slowly than normal cells. Cancerous cells spread to other tissues; they know no bounds. Normal liver cells divide and remain a part of the liver. Cancerous liver cells may leave the liver and be found, for example, in the lung. Most cancer cells show partial or complete loss of specialized functions. Although located in the liver, cancer cells no longer perform the functions of the liver.

BENZO($\alpha$)PYRENE

Attempts to determine the cause of cancer have evolved from early studies in which the disease was linked to a person's occupation. It was first noticed in 1775 that persons employed as chimney sweeps in England had a higher rate of skin cancer than the general population. It was not until 1933 that ***benzo($\alpha$)pyrene,*** $C_{20}H_{12}$ (a 5-ringed

**TABLE 15–7 Teratogenic Substances**

| Substances | Species | Effects on Fetus |
|---|---|---|
| *Metals* | | |
| Arsenic | mice | increase in males born with |
| | hamsters | eye defects, renal damage |
| Cadmium | mice | abortions |
| | rats | abortions |
| Cobalt | chickens | eye, lower limb defects |
| Gallium | hamsters | spinal defects |
| Lead | humans | low birth weights, brain damage, |
| | rats | stillbirth, early and late deaths |
| | chickens | |
| Lithium | primates | heart defects |
| Mercury | humans | Minamata disease (Japan) |
| | mice | fetal death, cleft palate |
| | rats | brain damage |
| Thallium | chickens | growth retardation, abortions |
| Zinc | hamsters | abortions |
| *Organic Compounds* | | |
| DES (diethylstilbestrol) | humans | uterine anomalies |
| Caffeine (15 cups per day equivalent) | rats | skeletal defects, growth retardation |
| PCB's (polychlorinated biphenyls) | chickens | central nervous system and eye defects |
| | humans | growth retardations, stillbirths |

aromatic hydrocarbon), was isolated from coal dust and shown to be metabolized in the body to produce one or more carcinogens. In 1895, the German physician Rehn noted three cases of bladder cancer, not in a random population, but in employees of a factory that manufactured dye intermediates in the Rhine Valley. Rehn attributed these cancers to his patients' occupation. These and other cases confirmed that at times as many as 30 years passed between the time of the initial employment and the occurrence of bladder cancer. The principal product of these factories was aniline. Although aniline was first thought to be the carcinogenic agent, it was later shown to be noncarcinogenic. It was not until 1937 that continuous long-term treatment with ***2-naphthylamine,*** one of the suspected dye intermediates, in dosages of up to 0.5 g per day produced bladder cancer in dogs. Since then other dye intermediates have been shown to be carcinogenic.

$NH_2$ ANILINE $NH_2$ 2-NAPHTHYLAMINE

A vast amount of research effort has verified the carcinogenic behavior of a large number of diverse chemicals. Some of these are listed in Table 15–8. This research has led to the formulation of a few generalizations concerning the relationship between chemicals and cancer. For example, carcinogenic effects on lower animals are

**TABLE 15–8 Chemicals Carcinogenic for Humans**

| Compound | Formula | Use or Source | Site Affected | Confirming Animal Tests* |
|---|---|---|---|---|
| *Inorganic Compounds* | | | | |
| Arsenic (and compounds) | As | insecticides, alloys | skin, lung liver | − |
| Asbestos | $Mg_6(Si_4O_{11})(OH)_6$ | brake linings, insulation | respiratory tract | + |
| Beryllium | Be | alloy with copper | bone, lung | + |
| Cadmium | Cd | metal plating | kidney, lung | + |
| Chromium | Cr | metal plating | lung | + |
| *Organic Compounds* | | | | |
| Benzene | | solvent, chemical intermediate in syntheses | blood (leukemia) | + |
| Acrylonitrite | $CH_2{=}CH(CN)$ | monomer | colon, lung | + |
| Aflatoxins B₁ (shown) | $OCH_3$ | mold or peanuts | liver | + |
| Carbon tetrachloride | $CCl_4$ | solvent | liver | + |
| Diethylstilbestrol | HO–C₆H₄–C($C_2H_5$)=C($C_2H_5$)–C₆H₄–OH | hormone | female genital tract | + |
| Benzo($\alpha$)pyrene | | cigarette and other smoke | skin, lung | + |
| Benzidine | $H_2N$–C₆H₄–C₆H₄–$NH_2$ | dye manufacture, rubber compounding | bladder | + |
| Ethylene oxide | $CH_2{-}CH_2$ (–O–) | chemical intermediate used to make ethylene glycol, surfactants | gastrointestinal tract | ± |
| Soots, tars, and mineral oils | | roofing tar, chimney soot, oils of hydrocarbon nature | skin, lung, bladder | + |
| Vinyl chloride | $CH_2{=}CHCl$ | monomer for making PVC | liver, brain, lung, lymphatic system | + |

*For animal tests, (+) means positive supporting data, (−) means a lack of supporting data, (±) means conflicting data.

commonly extrapolated to humans. The mouse has come to be the classic animal for studies of carcinogenicity. Strains of inbred mice and rats have been developed that are genetically uniform and show a standard response.

Some carcinogens are relatively nontoxic in a single, large dose, but may be quite toxic, often increasingly so, when administered continuously. Thus, much patience, time, and money must be expended in carcinogenic studies. The development of a sarcoma in humans, from the activation of the first cell to the clinical manifestation of the cancer, takes from 20 to 30 years. With life expectancy of an average person in the United States now set at about 70 years, it is not surprising that the number of deaths due to cancer is increasing.

Cancer does not occur with the same frequency in all parts of the world. Breast cancer occurs with lower frequency in Japan than in the United States or Europe. Cancer of the stomach, especially in males, is more common in Japan than in the United States. Cancer of the liver is not widespread in the Western Hemisphere but accounts for a high proportion of the cancers among the Bantu in Africa and in certain populations in the Far East. The widely publicized incidence of lung cancer is higher in the industrialized world and is increasing at an appreciable rate.

**Cancer spreads from one tissue to another via *metastases*.**

Some compounds cause cancer at the point of contact. Other compounds cause cancer in an area remote from the point of contact. The liver, the site at which most toxic chemicals are removed from the blood, is particularly susceptible to such compounds. Since the original compound does not cause cancer on contact, some other compound made from it must be the cause of cancer. For example, it appears that the substitution of an $>$NOH group for an $>$NH group in an aromatic amine derivative produces at least one of the active intermediates for carcinogenic amines. If R denotes a two- or three-ring aromatic system, then the process can be represented as follows:

**An abnormal growth is classified as cancerous or malignant when examination shows it is invading neighboring tissue. A growth is benign if it is localized at its original site.**

$$\underset{\text{INACTIVE ON CONTACT}}{\mathrm{RN(H)COCH_3}} \longrightarrow \underset{\text{ACTIVE ON CONTACT}}{\mathrm{RN(OH)COCH_3}} \longrightarrow \underset{\text{OTHER UNKNOWN INTERMEDIATES}}{\mathrm{RX? \longrightarrow RY?}} \xrightarrow{\text{tissue}} \text{Tumor cell}$$

As indicated by the variety of chemicals in Table 15–8, many molecular structures produce cancer whereas closely related ones do not. The 2-naphthylamine mentioned earlier is carcinogenic, but repeated testing gives negative results for 1-naphthylamine.

[structures: $NH_2$ on naphthalene ring]

1-NAPHTHYLAMINE (NONCARCINOGENIC) — 2-NAPHTHYLAMINE (CARCINOGENIC)

**Smoking is thought to play both an initiation and promotion role in cancer causation.**

For some types of cancer there are distinct stages that ultimately result in cancer. These may be identified as the *initiation period,* the *development* or *promotion period,* and the *progression period.* A single, minute dose of a carcinogenic polynuclear aromatic hydrocarbon, such as benzo($\alpha$)pyrene, applied to the skin of mice produces the permanent change of a normal cell to a tumor cell. This is the initiation step. No noticeable reaction occurs unless further treatment is made. If the area is painted repeatedly with noncarcinogenic croton oil, even up to one year later, carcinomas appear (Fig. 15–16). This is the development period. Additional fundamental alterations in the nature of the cells occur during the progression pe-

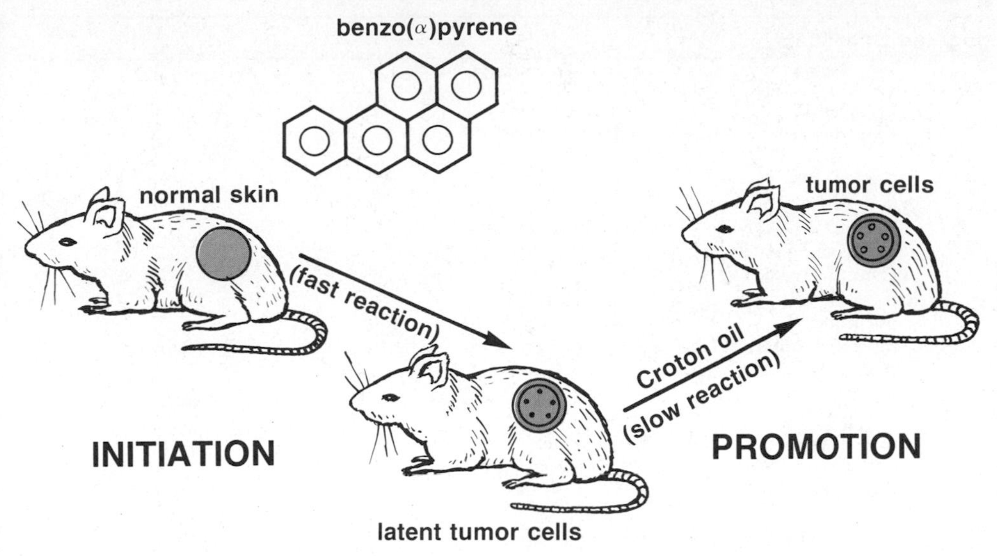


**Figure 15–16** A second chemical can promote tumor growth in mice after an initiation period. Treatment of mice with croton oil produces no tumor nor does treatment with small quantities of benzo (α) pyrene alone. Croton oil is an irritant oil similar to castor oil. Both are derived from plants.

riod. If there is no initiator, there are no tumors. If there is initiator but no promoter, there are no tumors. If the initiator is followed by repeated doses of promoter, tumors appear. This seems to indicate that cancer cannot be contracted from chemicals unless repeated doses are administered or applied.

Almost all chemical carcinogens have an induction period.

Just how do these toxic substances work? Cancer might be caused if the carcinogen combines with growth control proteins, rendering them inactive. During the normal growth process the cells divide and the organism grows to a point and stops. Cancer is abnormal in that cells continue to divide and portions of the organism continue to grow. One or more proteins are thought to be present in each cell with the specific duty of preventing replication of DNA and cell division. Virtually all of the carcinogens bind firmly to proteins, but so do some similar compounds that are noncarcinogenic. The specific growth proteins involved are not yet known for any of the carcinogens, despite considerable efforts to find them.

Smoking is associated with over 20% of all cancers; asbestos with between 3 and 18%.

Another theory suggests that carcinogens react with and alter nucleic acids so the proteins ultimately formed on the messenger RNA are sufficiently different to alter the cell's function and growth rate. The carcinogen may be included in the DNA or RNA strands by covalent bonding or it may be entangled in the helix and held by weak van der Waals attractions. The carcinogenic compounds nitrosodimethylamine and mustard gas have been shown to react with nucleic acids.

While researchers collect data in their laboratories and speculate on the theoretical structural causes of cancer, we can studiously avoid compounds known to cause cancer in man. It has been proposed that as much as 80% of all human cancer has its origin in carcinogenic chemicals.

## Hallucinogens

All of the hallucinogenic drugs are toxic.

Hallucinogens can produce temporary changes in perception, thought, and mood. These substances include ***mescaline, lysergic acid diethylamide (LSD), tetrahydrocannabinol*** (the active component of marihuana), and a broadening field of more than 50 other substances. They are included in this chapter as a separate section because they can be toxic in comparatively small doses.

Several characteristics of some of the more famous hallucinogens are given in Table 15–9. Each of these substances is capable of disturbing the mind and produc-

**TABLE 15–9 Some Hallucinogenic Chemicals**

| Chemical | Aliases | Notes | Active Compound | Formula |
|---|---|---|---|---|
| LSD | acid<br>crackers<br>the chief<br>the hawk | very powerful hallucinogen | D-lysergic acid diethylamide | 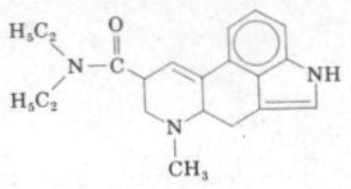 |
| Mescaline | peyote | extracted from mescal buttons from peyote cactus in South and Central America | 3,4,5-trimethoxy-phenethylamine | 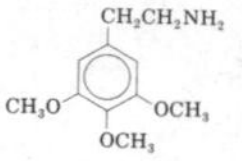 |
| Marihuana | grass<br>pot<br>reefers<br>locoweed<br>hash<br>Mary Jane | leaves of the 5-leaf-per-frond marihuana plant (*Cannabis sativa*) | 3,4-*trans*-tetra-hydrocannabinol | 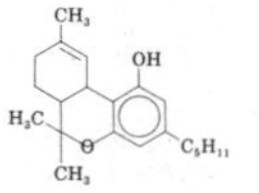 |
| Amphetamines | pep pills<br>speed<br>bennies<br>ups | eighteen times more active than mescaline | benzedrine (for example) | 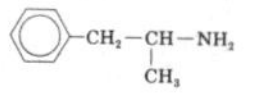 |
| Phencyclidine | PCP<br>hog<br>elephant<br>peace pills<br>angel dust | powerful analgesic-anesthetic, used in veterinary medicine | phencyclidine | 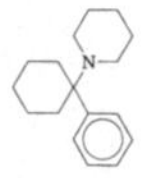 |
| Methaqualone | quaalude<br>ludes | hypnotic | methaqualone | |

ing bizarre and even colored interpretations of visual and other external stimuli. Mescaline is one of the oldest known hallucinogens, having been isolated from the peyote plant in 1896 by Heffter. Indeed, as early as 1560 the Mexican Indians who ate or drank the peyote were described by Spanish explorers as experiencing "terrible or ludicrous visions; the inebriation lasting for two or three days and then disappearing."

Although known for nearly 5,000 years, marihuana is one of the least understood of all natural drugs. Very early in China's history it was used as a medicine, and in the United States it had early use as an analgesic and a poultice for corns. Marihuana has little acceptable medical use in this country at present, although it is being studied experimentally for the treatment of glaucoma.

The body does not become dependent on the continuing use of marihuana as it does with heroin or other narcotics. However, reliable scientific data are not available with regard to chronic toxicity resulting from long-term use of the drug. It is known to cause dryness of the mouth, leading to dental problems.

Today there are many substances that produce hallucinogenic experiences; the most powerful one known is LSD. Our brief discussion will be restricted to this compound.

## Lysergic Acid Diethylamide (LSD)

Literally hundreds of scientists are doing research on the effects of hallucinogens, including LSD. They are investigating the influence of these drugs on nerve and brain function, the possibility of chromosome alteration, tissue damage, psychological changes, and a host of other physiological and psychological effects. Scientists are not near a consensus on the toxic effects that LSD can cause. Collecting valid data is very difficult; many users of LSD overestimate the quantity of the drug they have used, its purity is often in question, and some take other drugs in addition. Even if the purity of the LSD is known, the results are difficult to interpret. For example, a study in which mice were given 0.05–1.0 microgram of LSD on the seventh day of pregnancy showed a 5% incidence of badly deformed mouse embryos. Another study, probably equally valid and meticulous, showed no unusual fetal damage to rat embryos when 1.5–300 micrograms of LSD were administered during the fourth or fifth day of pregnancy. Babies with depressed skulls were aborted from two women who had taken LSD during the early weeks of pregnancy. Two other women, also LSD users, delivered full-term, apparently normal babies.

**LSD has been linked with birth defects.**

There are, however, some dangers that are well documented. These drugs destroy one's sense of judgment. Such things as height, heat, or even a moving truck may seem to hold no danger for the person under the influence of a hallucinogen. A dose of 50–200 micrograms will take the user on a "trip" for approximately eight to sixteen hours. Excessive or prolonged use of LSD can cause a person to "freak out" and possibly to sustain permanent brain damage. After one "trip," a user can experience another "trip" some time later, unexpectedly, without taking any more of the drug. The debate over LSD's dangers is continuing and is not likely to be resolved soon.

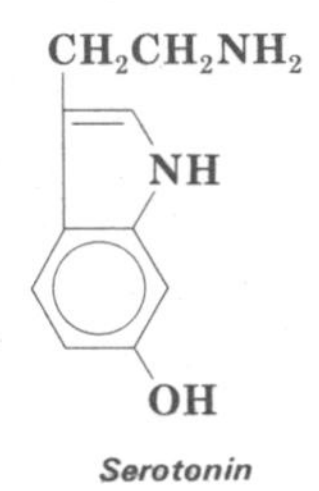

*Serotonin*

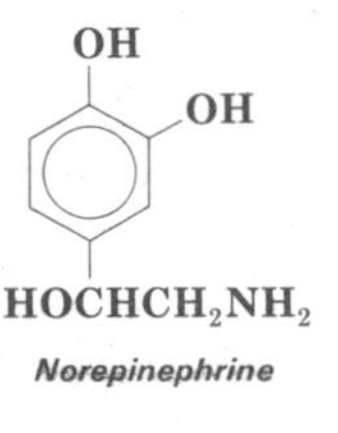

*Norepinephrine*

Out of the darkness of LSD-induced suicides and brain damage has come some new understanding of how the brain works. It is an interesting coincidence that soon after LSD was found localized in areas of the brain responsible for eliciting a human being's deep-seated emotional reactions, the compounds were discovered that are probably responsible for transmitting the impulse across synapses in the brain. Serotonin and norepinephrine are thought to act in a manner similar to acetylcholine, which was discussed earlier. These compounds carry the message from the end of one neuron to the end of another across the synapse. Somewhat later Dilworth Wooley discovered that LSD blocks serotonin action (Fig. 15–17). This has led to an interesting theory to explain the LSD trip.

The theory of the hallucinogenic mechanism is, roughly, as follows. The LSD releases in some chemical way those emotional experiences that are generally hidden away, or chemically stored in the lower midbrain and the brain stem. Serotonin or noreprinephrine, or both, inhibit the escape of these experiences into consciousness, as they turn off certain excitations. LSD interacts with serotonin or norepinephrine

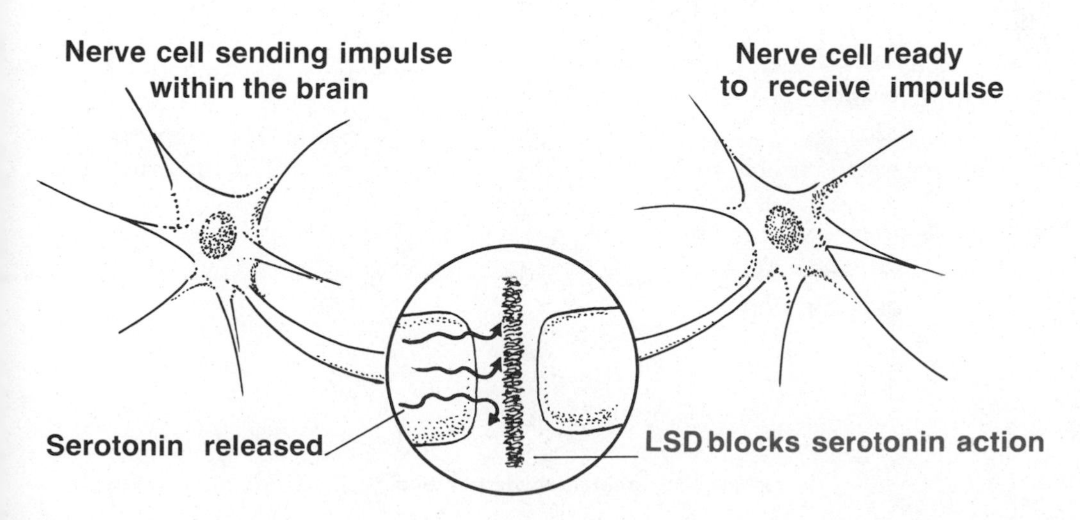


**Figure 15–17** LSD blocks the flow of serotonin from one brain nerve cell to another.

at the synapse and nullifies their blocking effect. Thus, these stored, previously rejected thoughts are allowed to enter the conscious part of the brain where the imaginary trip occurs.

This theory is built on the experimental fact that LSD blocks the action of serotonin, but there are other theories as well as some unexplained phenomena involved. Psychiatry is fraught with theories of why LSD causes an uplifting experience in some people and a frightening one in others. There are theories about dosage, purity of the compound, psychiatric state of the tripper, and conditions of administration. If there is anything close to a consensus, it is that the more controlled and relaxed the surroundings, the "better" the trip will be; even this, however, is debatable. Many trips, even under medical supervision, have not been satisfactory.

## Alcohols

Alcohols have some well-known toxic effects, but a complete chemical explanation has eluded scientists so far.

Methyl alcohol (methanol or wood alcohol) is highly poisonous, and unlike the other simple alcohols, it is, in effect, a cumulative poison in human beings. It has a specific toxic effect on the optic nerve, causing blindness with large doses. After its

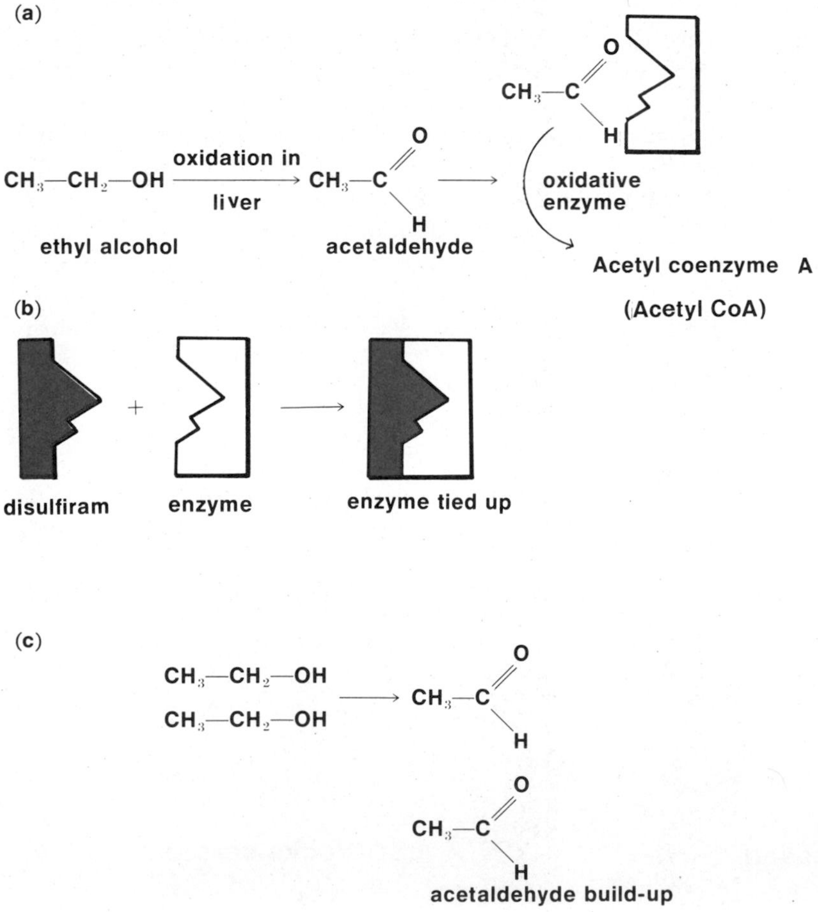

(a)

$$CH_3-CH_2-OH \xrightarrow[\text{liver}]{\text{oxidation in}} CH_3-CHO \longrightarrow \text{oxidative enzyme} \longrightarrow \text{Acetyl coenzyme A (Acetyl CoA)}$$

ethyl alcohol — acetaldehyde

(b)

disulfiram + enzyme → enzyme tied up

(c)

$$2\,CH_3-CH_2-OH \longrightarrow 2\,CH_3-CHO$$

acetaldehyde build-up

**Figure 15–18** The blocking of ethyl alcohol oxidation by disulfiram (Antabuse). When the normal oxidative process (*a*) is stopped by blockage of the oxidative enzyme (*b*), acetaldehyde builds up in the body (*c*).

rapid uptake by the body, oxidation occurs in which the alcohol is first converted into formaldehyde and then to formic acid, which is eliminated in the urine.

$$\underset{\text{METHYL ALCOHOL}}{CH_3OH} \xrightarrow{\text{oxidation}} \underset{\text{FORMALDEHYDE}}{H-\overset{\overset{\displaystyle O}{\|}}{C}-H} \xrightarrow{\text{oxidation}} \underset{\text{FORMIC ACID}}{H\overset{\overset{\displaystyle O}{\|}}{C}OH}$$

This is a slow process and, for this reason, daily exposure to methyl alcohol can cause an extremely dangerous buildup of the alcohol in the body. The toxic effect on the optic nerve is thought to be caused by the oxidative products.

Ethyl alcohol (ethanol, grain alcohol) is found in alcoholic beverages, yet it is toxic like the other simple alcohols and is quantitatively absorbed by the gastrointestinal tract. About 58% of a dose is absorbed in 30 minutes, 88% in 1 hour, and 93% in 90 minutes. Over 90% of the ethyl alcohol is then slowly oxidized to carbon dioxide and water, mainly in the liver.

The intoxicated person's staggering gait, stupor, and nausea are caused by the presence of acetaldehyde, but the chemical reactions involved are not fully understood. The compound disulfiram (Antabuse) is sometimes given as a treatment for chronic alcoholism because it blocks the oxidative steps beyond acetaldehyde (Fig. 15–18). The accumulation of acetaldehyde causes nausea, vomiting, blurred vision, and confusion. This is supposed to encourage the partaker to avoid this severe sickness by avoiding alcohol. Interestingly, this drug was discovered by two researchers who took a dose for another purpose and got violently ill that evening at a cocktail party.

$$(CH_3CH_2)_2N-C(=S)-S-S-C(=S)-N(CH_2CH_3)_2$$

*Disulfiram*

## Self-Test 15-B

1. Substrates that poison the nervous system are called ______________.
2. Most neurotoxins affect chemical reactions that occur in the opening between two nerve cells. These openings are called ______________.
3. The electrical impulse is carried across a synapse by the chemical ______________.
4. Mutagens alter the structures of ______________ or ______________.
5. If a substance is mutagenic in test animals, particularly dogs, it must necessarily be mutagenic in human beings. (True/False)
6. The first occupation definitely linked to cancer was ______________.
7. An active component of marihuana is ______________.
8. The nausea and stupor of drunkenness from ethyl alcohol are caused by ______________ and not by the alcohol itself.
9. Two dangers associated with smoking are ______________ and ______________.
10. A chemical that can cross the placenta and harm the fetus is called a ______________

## Matching Set

| | | |
|---|---|---|
| ______ | 1. metabolic poison | **a** lysergic acid diethylamide |
| ______ | 2. metabolism | **b** cyanide ion |
| ______ | 3. corrosive poison | **c** cancer in connective tissue |
| ______ | 4. neurotoxin | **d** benzo($\alpha$)pyrene metabolite |

| | | | |
|---|---|---|---|
| ______ | 5. mutagen | **e** | cancerous growths of lung tissue located in liver |
| ______ | 6. carcinogen | **f** | use of chemicals in the body |
| ______ | 7. carcinoma | **g** | cancer of skin |
| ______ | 8. metastases | **h** | sodium hydroxide (caustic soda) |
| ______ | 9. sarcoma | **i** | atropine |
| ______ | 10. chelating agent | **j** | alters DNA |
| ______ | 11. hallucinogen | **k** | EDTA |

## Questions

**1.** Give an example of a toxic substance that is toxic as a result of
a. binding to an oxygen-carrying molecule
b. disguising itself as another compound
c. attack on an enzyme
d. hydrolysis

**2.** True/False. Explain each answer concisely.
a. Lead is a corrosive poison.
b. Carbon monoxide and cyanide poison in the same way
c. There are no known chemical compounds that cannot be toxic under some circumstances.

**3.** The application of a single, minute dose of a fused-ring hydrocarbon such as benzo($\alpha$)pyrene fails to produce a tumor in mice. Does this mean this compound is definitely noncarcinogenic? Give a reason for your answer.

**4.** Describe the chemical mechanism by which the following substances show their toxic effects.
a. fluoroacetic acid
b. phosgene
c. curare

**5.** Should any laws and regulations be placed on the use of any of the following? Justify your answers.
a. LSD
b. marihuana
c. ethanol

**6.** Discuss some of the pros and cons of testing toxic substances on animals.

**7.** Give chemical reactions in words for
a. action of NaOH on tissue
b. action of carbon monoxide in blood
c. reaction of EDTA and lead ion

**8.** What questions do you think need to be answered before the action of ethyl alcohol is understood?

**9.** Assume a normal diet has the quantity of lead in a given quantity of food stated in the text. What would a person's total food intake of lead be per day?

**10.** What is the meaning of the symbolism $LD_{50}$?

**11.** Write chemical equations for:
a. the hydrolysis of acetylcholine
b. acid hydrolysis of a protein having a glycine-glycine primary structure

**12.** Describe how the corrosive poisons lye (NaOH), $NO_2$, and the hypochlorite ion ($OCl^-$) destroy tissue.

**13.** What are some common sources of carbon monoxide?

**14.** If a relatively small amount of carbon monoxide is inhaled, are the chemical reactions reversible, or is carbon monoxide a cumulative poison?

**15.** What poisons can be rendered ineffective by wrapping a large molecule around them?

**16.** What is the cause of pica?

**17.** What is a common structural feature of alkaloids?

**18.** Give two examples of each: (a) corrosive poison, (b) metabolic poison, (c) neurotoxic poison, (d) mutagenic poison, (e) carcinogenic poison.

**19.** Phosgene hydrolyzes in the lungs to produce what acid?

**20.** What concentration of carbon monoxide in the air is likely to cause death in one hour?

**21.** Is it possible that one molecule of a mutagen could cause a problem to human life?

**22.** Two new chemicals are prepared in the lab. One is a relatively simple acid with corrosive properties. The other chemical has carcinogenic properties. Which chemical's toxic property is more likely to be discovered?

**23.** An old laboratory chemical is discovered to have a new property. It reacts with amine groups to produce —OH groups. Could this chemical be a possible mutagen?

**24.** If a poisoning victim has pinpoint pupils and is salivating excessively, what type of poison was the probable cause of these symptoms?

**25.** If your drinking water contains numerous toxic substances, why are you not normally harmed by drinking it?

# Water: Its Use and Misuse

Water is certainly one of the most important of all chemical compounds. Its intimate participation in life processes marks it as indispensable to all known organisms. The rapidly expanding scope of human life, with its associated industrial development, is increasing the demand for water supplies so quickly that future needs may easily be underestimated. The time has passed when we can take an abundant water supply for granted.

Water pollution arises from two main sources. First, water, being an unusually good solvent, readily dissolves materials from the soil. Second, and far more serious, are the waste materials of our culture; mercury and raw sewage effluents are but two examples.

**Water—the universal solvent.**

It is important to realize at the outset that the technical know-how is available to solve water pollution problems, and much of this know-how is patterned after natural purification processes. Considerable progress has been made in improving the quality of the water in our lakes and streams and along our shores during the past decade. However, our drinking water is still a matter of considerable concern. Many chemicals are in our waters as a result of our technologies and, at this point, we simply do not know the level of concentrations that pose a threat to human life. It should be emphasized that we know how to purify water to any level of purity we desire. Pure water thus becomes a question of costs.

## Water Reuse

It is estimated that an average of 4,350 billion gallons of rain (and snow) fall on the contiguous United States each day. Of this amount, 3,100 billion gallons return to the atmosphere by evaporation and transpiration. The discharge to the sea and underground reserves amounts to 800 billion gallons daily, leaving 450 billion gallons of surface water each day for domestic or commercial use. The 48 contiguous states withdrew from natural sources 40 billion gallons per day in 1900 and 325 billion gallons per day in 1960, and it is estimated that the demand will be at least 900 billion gallons per day by the year 2000. It is evident that the reuse of surface water is the only way that human needs for water can be met in the future. Much progress can be, and some has been, made in the reuse of water. For example, a Dow Chemical plant in Pittsburg, California used 20,000 gallons per minute (gpm) prior to its water reuse study. After its reuse plans were activated, the same plant required only 1000 gpm of fresh water. The use that we make of the water we withdraw from nature is given in Table 16–1.

TABLE 16–1 Uses of Water in the United States

| | Percentage of Total Use |
|---|---|
| Agriculture | 48 |
| Steam and power generation | 26 |
| Manufacturng and minerals production | 17 |
| Domestic, commercial, and public | 9 |
| | 100% |

Reuse of water may be approaching a critical point in some areas. Within a 16-mile radius of Paris, eight million people receive water of which 70% has been previously used, mostly in industrial processes.

Since the reuse of water is necessary, and since the purification of used water before reuse is often necessary, the question arises as to where the purification should occur. Purification immediately prior to reuse actually may lead to greater chemical problems because of the complex mixtures that result. Purification immediately after use is likely to involve only one type of problem characteristic of that use. How pure must the purified water be? There are no general agreements on standards in this area. Present analytical techniques are so good that we are approaching the identification of single molecules. The water engineers say, "Tell us what is safe and we'll meet it; it is simply a matter of costs."

## Water Purification in Nature

Rain water in clean air is very pure.

Volatile: goes easily into the gaseous state.

Water is a natural resource which, within limitations, is continuously renewed. The familiar water cycle (Fig. 16–1) offers a number of opportunities for nature to purify its water. The worldwide ***distillation*** process results in rain water containing only traces of nonvolatile impurities, along with gases dissolved from the air. ***Crystallization*** of ice from ocean saltwater results in relatively pure water in the form of icebergs. ***Aeration*** of groundwater as it trickles over rock surfaces, as in a rapidly running brook, allows volatile impurities to be released to the air. ***Sedimentation*** of solid particles occurs in slow-moving streams and lakes. ***Filtration*** of water through sand rids the water of suspended matter such as silt and algae. Next, and of very great importance, are the ***oxidation processes.*** Essentially all naturally occurring organic materials—plant and animal tissue, as well as their waste materials—are changed through a complicated series of oxidation steps in surface waters to simple substances common to the environment. Finally, another process used by nature is ***dilution.*** Most, if not all, pollutants found in nature are rendered harmless if reduced below certain levels of concentration by dilution with pure water.

Pure water:
Chemist: "Pure $H_2O$—no other substance."
Parent: "Nothing harmful to human beings."
Game and Fish Commission: "Nothing harmful to animals."
Sunday boater: "Pleasing to the eye and nose, no debris."
Ecologist: "Natural mixture containing necessary nutrients."

Before the advent of the exploding human population and the industrial revolution, natural purification processes were quite adequate to provide ample water of very high purity in all but desert regions. Nature's purification processes can be thought of as massive but somewhat delicate. In many instances the activities of humans push the natural purification processes beyond their limit, and polluted water accumulates.

A simple example of nature's inability to handle increased pollution comes from dragging gravel from stream beds. This excavation leaves large amounts of suspended matter in the water. For miles downstream from a source of this pollutant, aquatic life is destroyed. Eventually, the solid matter settles, and normal life can be found again in the stream.

Only about 1% of groundwater supplies in the United States are now considered unsafe.

A more complex example, and one for which there is not nearly so much reason to hope for the eventual solution by natural purification, is the degradability of organic

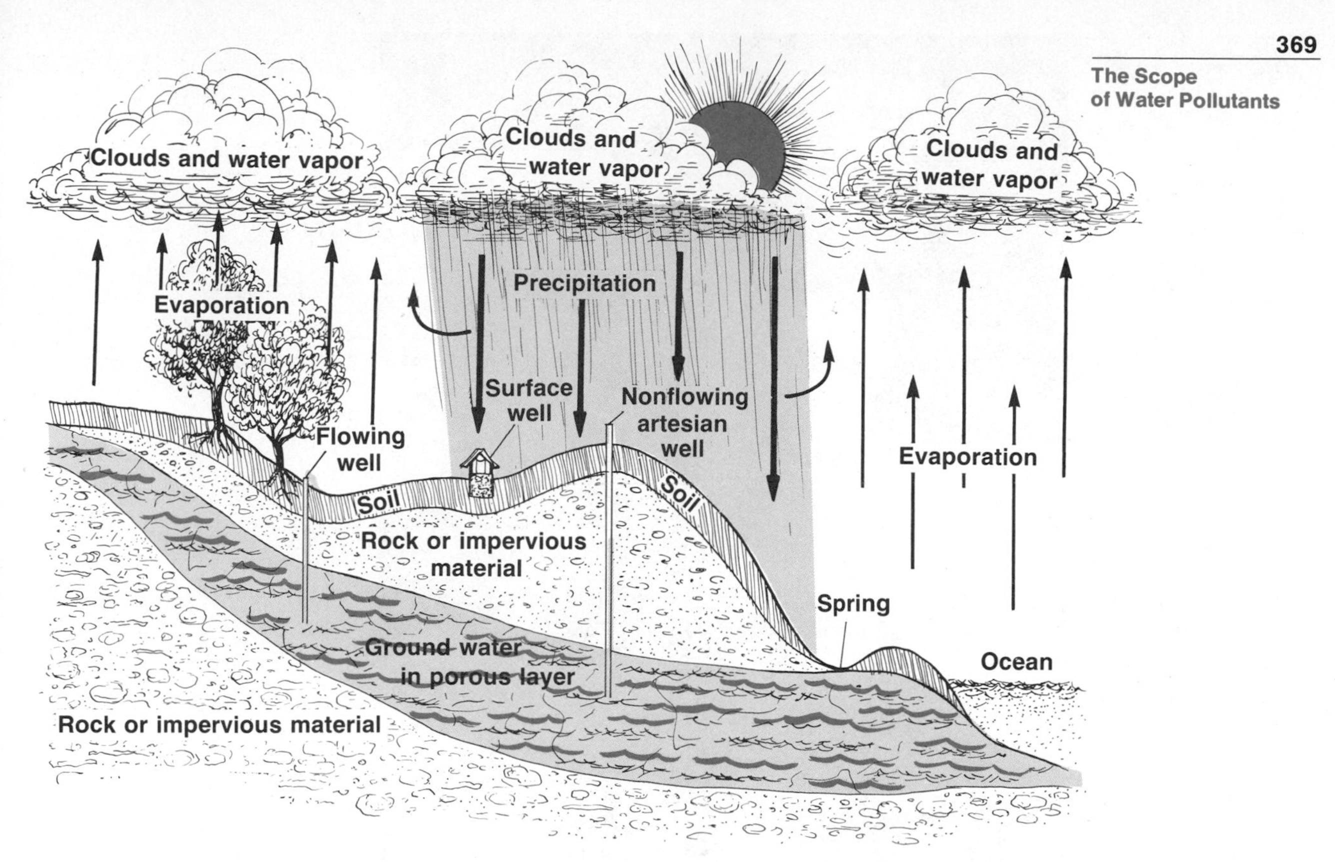


**Figure 16–1** The water cycle in nature.

materials. A ***biodegradable*** substance is composed of molecules that are broken down to simpler ones in the natural environment. For example, cellulose suspended in water will eventually be converted to carbon dioxide and water. Some organic compounds, notably some of those synthetically produced, are not easily biodegradable; these substances simply stay in the natural waters or are absorbed by life forms and remain intact for long periods of time. An example is DDT.

Some synthetic molecules are not biodegradable and therefore are very persistent in natural waters.

Even nature's pure rain water is in jeopardy. If the acidic air pollutants, such as the oxides of sulfur, are concentrated enough, the absorbing rain water will become acidic enough to harm life forms and mar metal and stone structures. The government of Canada has complained to Washington because of acid rains arising from the industrial Northeast. In areas in which heavy concentrations of automobile fumes collect, poisonous lead compounds have been found in rain water in concentrations many times higher than the 0.01 ppm generally allowed in drinking water. The concentration of the lead can be correlated with the concentration of exhaust fumes in the air. Fortunately, lead does not long remain in water, since it generally forms insoluble compounds.

Acid rain in Pasadena, California (1976–77) contained:
1. acid (pH = 4.06)
2. $NH_4^+$
3. $K^+$
4. $Ca^{2+}$
5. $Mg^{2+}$
6. $Cl^-$
7. $NO_3^-$
8. $SO_4^{2-}$

(2–8: Concentrations in $10^{-5}$ to $10^{-6}$ molar range.)

## The Scope of Water Pollutants

When natural purification processes cannot cope with materials added to water, pollution results.

There was a time when polluted water could be thought of in terms of dissolved minerals, natural silt, and contaminants associated with the natural wastes of animals

**Figure 16–2** Foam in natural waters. A high concentration of detergent materials dissolved in water causes foam because of the lowered surface tension of the water. If the water is badly polluted, the foam may exist miles from the pollution source.

and humans. As our use of water has increased, the pollution has become more diversified. The U.S. Public Health Service now classifies water pollutants into the eight broad categories listed in Table 16–2.

*Eutrophication* is the enrichment of a body of water with nutrients such that growth of organisms in the water makes it unfit for human purposes.

## Biochemical Oxygen Demand

The way in which organic materials are oxidized in the natural purification of water deserves special attention. The process opposes ***eutrophication.*** Even in the

**TABLE 16–2 Classes of Water Pollutants, with Some Examples**

| | |
|---|---|
| 1. Oxygen-demanding wastes | plant and animal material |
| 2. Infectious agents | bacteria and viruses |
| 3. Plant nutrients | fertilizers, such as nitrates and phosphates |
| 4. Organic chemicals | pesticides, such as DDT, detergent molecules |
| 5. Other minerals and chemicals | acids from coal mine drainage, inorganic chemicals such as iron from steel plants |
| 6. Sediment from land erosion | clay silt on stream bed may reduce or even destroy life forms living at the solid-liquid interface |
| 7. Radioactive substances | waste products from mining and processing of radioactive material, radioactive isotopes after use |
| 8. Heat from industry | cooling water used in steam generation of electricity |

natural state, living organisms found in natural waters are constantly discharging organic debris into the water. To change this organic material into simple inorganic molecules (such as $CO_2$ and $H_2O$) requires oxygen. The amount of oxygen required to oxidize a given amount of organic material is called the ***biochemical oxygen demand (BOD).*** The oxygen is required by microorganisms, such as many forms of bacteria, to metabolize the organic matter that constitutes their food. Ultimately, given near normal conditions and enough time, the microorganisms will convert huge quantities of organic matter into the following end products:

Organic carbon $\longrightarrow CO_2$
Organic hydrogen $\longrightarrow H_2O$
Organic oxygen $\longrightarrow H_2O$
Organic nitrogen $\longrightarrow NO_3^-$

One way to determine the amount of organic pollution is to determine how much oxygen a given sample of polluted water will require. For example, a known volume of the polluted water is diluted with a known volume of standardized sodium chloride solution of known oxygen content. This mixture is then held at 20° C for five days in a closed bottle. At the end of this time the amount of oxygen that has been consumed is taken to be the biochemical oxygen demand.

**A quantitative relationship exists between oxygen needs and organic pollutants to be destroyed. This is BOD.**

**A standardized solution is one of known concentration.**

Highly polluted water often has a high concentration of organic material, with resultant large biochemical oxygen demand (Fig. 16–3). In extreme cases, more oxygen is required than is available from the environment, and putrefaction results. Fish and other freshwater aquatic life can no longer survive. The aerobic bacteria (those that require oxygen for the decomposition process) die. As a result of the death of these organisms, even more lifeless organic matter results and the BOD soars. Nature, however, has a back-up system for such conditions. A whole new set of microorganisms (anaerobic bacteria) takes over; these organisms take oxygen from oxygen-containing compounds to convert organic matter to $CO_2$ and water. Organic nitrogen is converted to elemental nitrogen by these bacteria. Given enough time, enough oxygen may become available, and aerobic oxidation will then return.

**Fish cannot live in water that has less than 0.004 g $O_2$ per liter (4 ppm).**

| Temperature °C | Solubility $g_{O_2}$/liter $H_2O$ |
|---|---|
| 0 | 0.0141 |
| 10 | 0.0109 |
| 20 | 0.0092 |
| 25 | 0.0083 |
| 30 | 0.0077 |
| 35 | 0.0070 |
| 40 | 0.0065 |

**These data are for water in contact with air at 760 mm mercury pressure.**

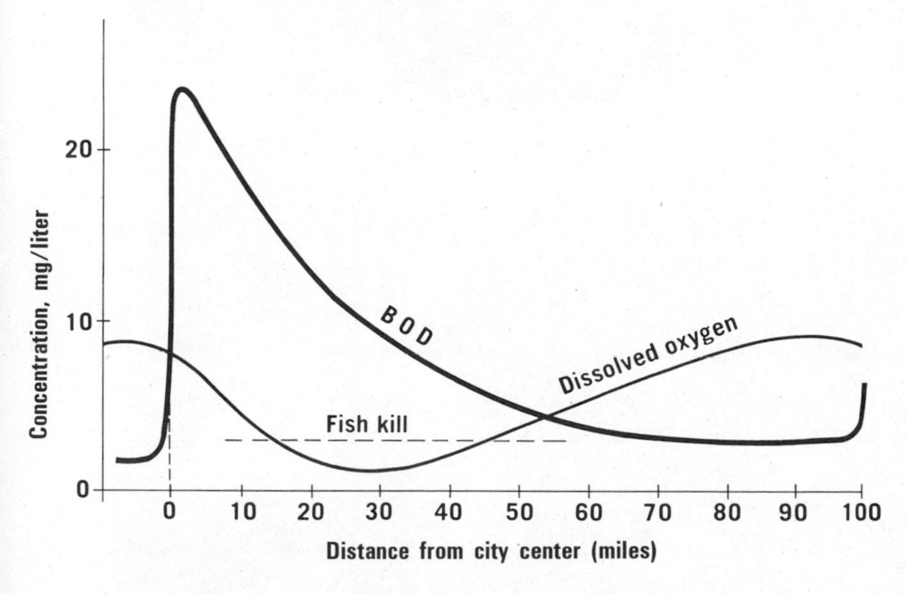


**Figure 16–3** Graph showing oxygen content and oxidizable nutrients (BOD) as a result of sewage introduced by a city. The results are approximated on the basis of a river flow of 750 gallons per second. Note that it takes 70 miles for the stream to recover from a BOD of .023 gram oxygen per liter.

A stream containing 10 parts per million (ppm) by weight (just 0.001%) of an organic material, the formula of which can be represented by $C_6H_{10}O_5$, will contain 0.01 g of this material per liter. The calculation used to obtain this is:

$$?g = 1 \text{ liter of water}$$

$$?g = 1 \text{ liter} \times \frac{1000 \text{ ml}}{1 \text{ liter}} \times \frac{1 \text{ g}}{\text{ml}} = 1000 \text{ g}$$

0.001% of this is the pollutant:
0.001% of 1000 g = (0.00001)(1000 g) = 0.010 g

**Characteristic BOD Levels**

| | $g_{O_2}$/liter |
|---|---|
| Untreated municipal sewage | 0.1–0.4 |
| Runoff from barnyards and feed lots | 0.1–10 |
| Food processing wastes | 0.1–10 |

To transform this pollutant to $CO_2$ and $H_2O$, the bacteria present use oxygen as described by the equation:

$$C_6H_{10}O_5 + 6O_2 \longrightarrow 6CO_2 + 5H_2O$$

RELATIVE WEIGHT 162    RELATIVE WEIGHT 192

The 0.010 g of pollutant requires 0.012 g of dissolved oxygen.

$$\frac{?\text{g oxygen}}{\text{liter}} = \frac{0.010 \text{ g pollutant}}{\text{liter}} \times \frac{192 \text{ g oxygen}}{162 \text{ g pollutant}} = \frac{0.012 \text{ g oxygen}}{\text{liter}}$$

**High concentration of organic pollutants**
↓
**Low oxygen concentration**
↓
**Dead organisms**
↓
**Higher concentration of organic pollutants**
↓
**Lower oxygen concentration**
↓
**Anaerobic conditions**

At 68° F (20° C), the solubility of oxygen in water under normal atmospheric conditions is 0.0092 g of oxygen per liter.

Since the BOD (0.012 g per liter) is greater than the equilibrium concentration of dissolved oxygen (0.0092 g per liter), as the bacteria utilize the dissolved oxygen in this stream, the oxygen concentration of the water will soon drop too low to sustain any form of fish life. Life forms can survive in water where the BOD exceeds the dissolved oxygen if the water is flowing vigorously in a shallow stream (this facilitates the absorption of more oxygen from the air via aeration).

**Figure 16–4** Fish kills can be caused by the lack of a substance necessary for life, such as oxygen, or by the presence of toxic materials that interfere with the life processes. A heavy concentration of organic matter in a stream may depress the oxygen concentration below that required to support fish life.

BOD values can be greatly reduced by treating industrial wastes and sewage with oxygen and/or ozone. Numerous commercial clean-up operations now being developed and used employ this type of "burning" of the organic wastes. Another benefit of treating waste water with oxygen is that some of the nonbiodegradable material becomes biodegradable as a result of partial oxidation.

## Thermal Pollution

Thermal pollution results when water is used for cooling purposes and in the process has its own temperature raised. Water has a high ***heat capacity*** (the heat required to raise a unit weight of water one degree) and a high ***heat of vaporization*** (the heat required to change a unit weight of liquid water to gaseous water); this combination makes water an ideal cooling fluid for thermal power stations, nuclear energy generators, and industrial plants.

**Water will absorb large amounts of heat before it is vaporized.**

The most obvious result of thermal pollution is to make the water less efficient for further cooling applications. Far more important, however, are the biological and biochemical implications.

**The solution of oxygen in water is facilitated by:**
1. **exposed surface area of water;**
2. **low temperature;**
3. **low concentration of oxygen in the water.**

The oxygen content of water in contact with air is dependent on the temperature of the water, since more oxygen can dissolve in a quantity of cold water than in the same quantity of warm water. Also, the *rate* at which water dissolves oxygen is directly proportional to the difference between the actual concentration of oxygen present and the equilibrium value. This is extremely fortunate, since it means the rate of solution of oxygen increases sharply as the oxygen is consumed. Since a larger surface area will allow quicker absorption of oxygen, the rate of absorption of oxygen from the air is much greater in a shallow, cold mountain stream than in a deep lake behind a dam in a warm river.

**Aquatic life is very sensitive to temperature. Lethal temperatures for various species of fish in Wisconsin and Minnesota:**

| | |
|---|---|
| **trout** | **77°F** |
| **white sucker** | **84–85°F** |
| **walleye** | **86°F** |
| **yellow perch** | **84–88°F** |
| **fathead minnow** | **93°F** |

The increased temperature of the Thames River decreased the oxygen content by 4% over what it otherwise would have been. However, the biochemical results of this factor alone could not be determined, since the river was anaerobic in 1950 as a result of other pollutants. Even though more is to be learned about thermal pollution, two conclusions seem obvious: (1) thermal pollution aggravates the problem of oxygen supply; and (2) a significant rise in the temperature of a stream can drastically change or even destroy entire biological populations.

Solutions to thermal pollution involve cooling the water in evaporation towers or storage lakes before returning it to the natural body of water. Cycling the water for reuse after cooling has obvious advantages. Table 16–3 shows a quantitative forecast of the thermal pollution problem in the United States.

**TABLE 16–3 A Projection of the Thermal Pollution Problem in the United States***

| Year | Per Capita Electrical Energy Consumption in U.S. (48 States, Actual or Projected) | Cooling Water Needs (Billion Gallons per Day) |
|---|---|---|
| 1950 | 2,000 kwh | ~30 |
| 1968 | 6,500 kwh | ~100 |
| 1980 | 11,500 kwh | ~200 |
| 2000 | 24,000 kwh | 450† |

*Notes: a. Surface runoff is 1,250 billion gallons per day.
b. Generation of electrical energy has been doubling every 10 years.
c. 500 new power stations will be needed by the year 2000, at the present growth rate.
d. kwh is kilowatt-hour, an energy unit equivalent to 860 kilocalories.

†Projected.

## Fertilizers

Fertilizers contain such chemicals as ammonium nitrate, $NH_4NO_3$, and "superphosphate," $CaH_4(PO_4)_2 \cdot CaSO_4$. These substances are soluble enough to be carried away by the surface runoff during and after a hard rain. Contamination of water by fertilizers leads to undesirable effects. In a large measure, this contamination results from the phosphate ($PO_4^{3-}$) and nitrate ($NO_3^-$) present in fertilizers. It is generally believed that phosphate and nitrate encourage the growth of large amounts of algae. It is certain that nitrate in sufficient concentration is toxic to most higher organisms.

**Limnology—the study of fresh-water systems.**

***Limnology*** is the study of physical, biochemical, and biological aspects of fresh-water systems. One of the conclusions of limnology is that massive but relatively delicate balances exist in such systems. The growth of a limited kind and quantity of algae is good for a lake; such growth is a source of needed oxygen via photosynthesis:

$$\underset{\text{CARBON DIOXIDE}}{6CO_2} + \underset{\text{WATER}}{6H_2O} \xrightarrow[\text{catalysts}]{\text{sunlight}} \underset{\text{SUGAR}}{C_6H_{12}O_6} + \underset{\text{OXYGEN}}{6O_2}$$

**Leached—dissolved by rain water and carried into streams or lakes.**

However, in waters that contain excessive amounts of phosphate and nitrate leached from fertilized farmland, algal growths are sometimes so massive that they tend to choke out desirable life forms. Such "blooms" of algae may lead to eutrophication. In this state, the dead algae actually give rise to an oxygen-deficient environment, especially at lower depths. This is because the decaying algae quickly consume the available oxygen produced by the algae; the BOD rises sharply.

## Detergent Builders

Detergents contain ***detergent builders,*** which are additives that make the surfactant (detergent) molecules more efficient and effective. In some detergents, as much as 40% of the total weight is in the form of detergent builders. A common detergent builder is sodium tripolyphosphate.

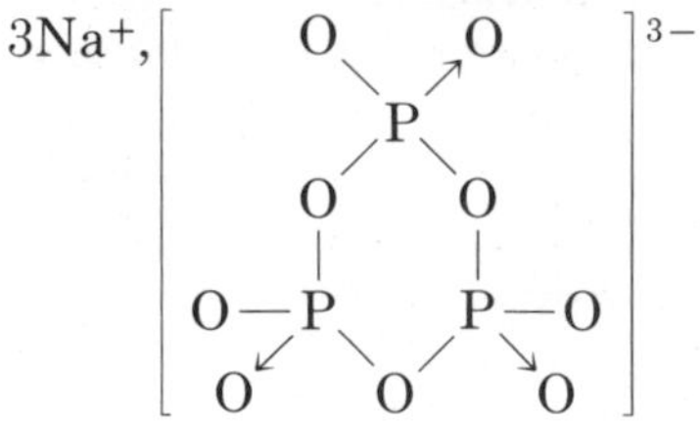

```
3Na+, [      O     O       ]3−
      [       \   ↗        ]
      [         P          ]
      [       /   \        ]
      [      O     O       ]
      [      |     |       ]
      [  O — P     P — O   ]
      [     ↙  \  /  ↘     ]
      [    O    O     O    ]
```

**Detergents and detergent builders both may present water pollution problems.**

Many polyphosphate structures exist, and a commercial preparation is likely to contain several. At least one positive action of detergent builders is to tie up metal ions such as $Fe^{3+}$.

$$Fe^{3+} + 2PO_4^{3-} \longrightarrow \underset{\text{IRON-PHOSPHATE COMPLEX}}{Fe(PO_4)_2^{3-}}$$

Because of the use of phosphorus in detergent builders, more than 500 million pounds of this element end up in our waste waters each year in the form of phosphate. A typical detergent contains about 9% phosphorus by weight.

By 1979 six states had passed laws banning the retail sale of phosphate detergents. The Environmental Protection Agency reported in 1978 that the phosphate concentration was greatly reduced in Lake Michigan along the shore of Indiana after that state banned phosphate detergents.

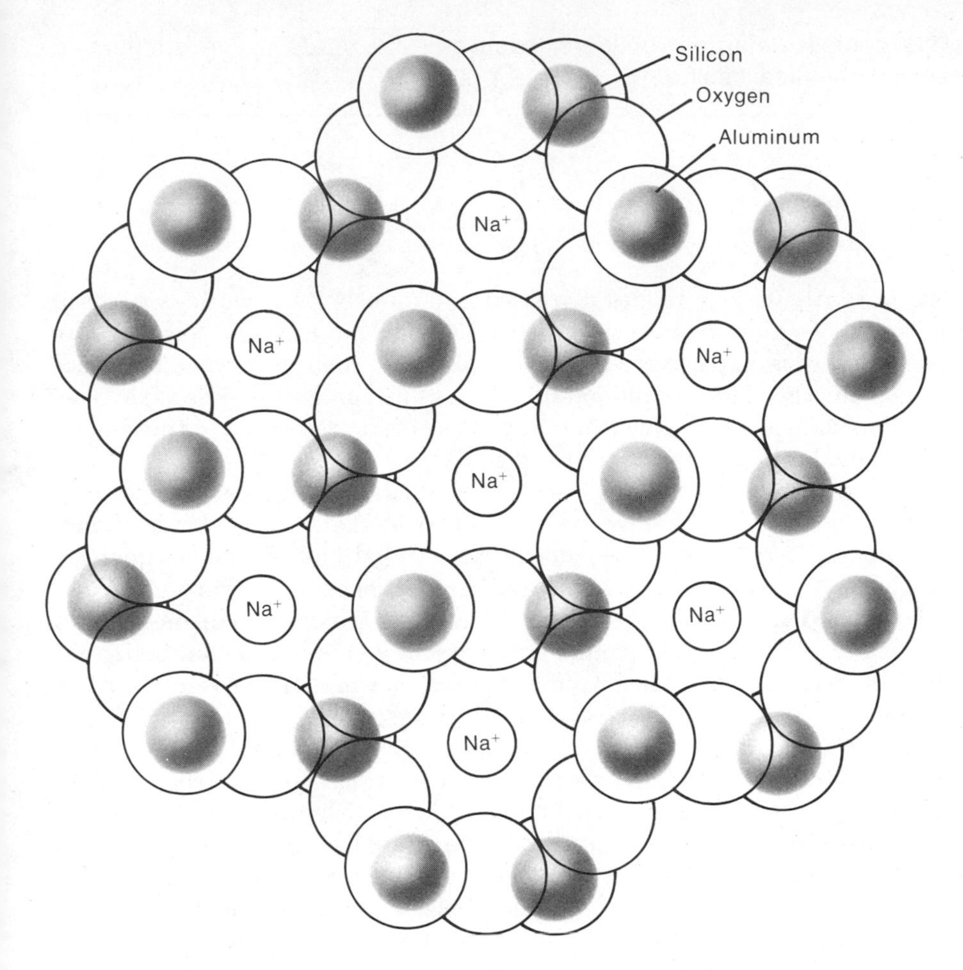


**Figure 16–5** The structure of natural zeolite. The aluminum, oxygen, and silicon atoms form a macromolecular structure that is negatively charged. The positive sodium ions are loosely held in the holes within the lattice. Water, containing other metal ions, will exchange some of these ions for sodium ions when passed through the zeolite.

No substitute with qualities equal to those of phosphates has yet been found. Sodium citrate is somewhat effective and is biodegradable. Synthetic zeolites are currently being introduced and show promise. Natural zeolites have been used for many years to remove metal ions from hard water (Fig. 16–5).

## Self-Test 16-A

1. What liquid is sometimes called the universal solvent? ______________
2. Which process does nature not use in purifying natural water: distillation, crystallization, sedimentation, chlorination, or aeration? ______________
3. BOD stands for ______________.
4. BOD is the amount (grams) of ______________ per ______________ required to oxidize a given amount of organic material.
5. ______________ is the study of physical, biochemical, and biological aspects of freshwater systems.

6. Detergents contain detergent builders, such as ______________ in addition to the detergent chemical itself.

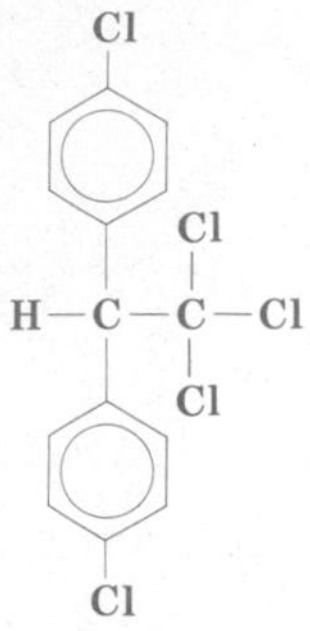

DDT

**Half-life is the time required for half of the substance to disappear (see Chapter 14).**

**Do we really have to decide between pests and eagles?**

**City water treatment generally does not remove DDT.**

## Pesticides

The use of synthetic insecticides increased enormously on a world-wide basis after World War II. As a result, insecticides such as DDT have found their way into lakes and rivers. There is a great variety of pesticides, and their use frequently leads to severe damage to other forms of animal life, such as fish and birds. The toxic reactions and peculiar biological side effects of many of the pesticides were not thoroughly studied or understood prior to their widespread use.

A good case in point is ***DDT.*** This insecticide, which had not been shown to be toxic to humans in doses up to those received by factory workers involved in its manufacture (400 times the average exposure), does have peculiar biological consequences. The structure of DDT is such that it is *not* metabolized (broken down) very rapidly by animals; it is deposited and stored in the fatty tissues. The biological half-life of DDT is about eight years; that is, it takes about eight years for an animal to metabolize one half of an amount it assimilates. The enzymes are just not present for rapid breakup of this molecule. If ingestion continues at a steady rate, it is evident that DDT will build up within the animal over a period of time. For many animals this is not a problem, but for some predators, such as eagles and ospreys, which feed on other animals and fish, the consequences are disastrous. The DDT in the fish eaten by such birds is concentrated in the bird's body, which attempts to metabolize the large amounts by an alteration in its normal metabolic pattern. This alteration involves the use of compounds that normally regulate the calcium metabolism of the bird and are vital to its ability to lay eggs with thick shells. When these compounds are diverted to their new use, they are chemically modified and are no longer available for the egg-making process. As a consequence, the eggs the bird does lay are easily damaged, and the survival rate decreases drastically. This process has led to the nearly complete extinction of eagles and ospreys in some parts of the United States where formerly they were numerous.

Dieldrin and aldrin 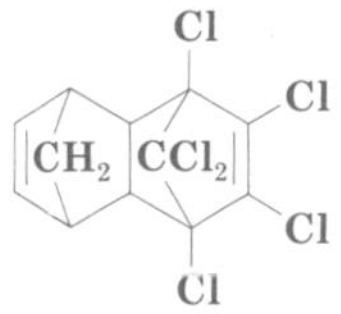 were banned by the courts from their major uses (such as on corn) in 1974.

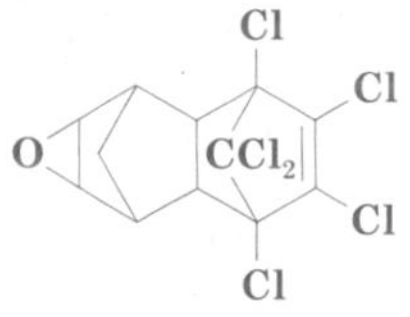

Since DDT is not readily biodegradable, there is a resultant buildup of this substance in natural waters. However, it is not an irreversible process; the Environmental Protection Agency reported a 90% reduction of DDT in Lake Michigan fish by 1978 as a result of the ban on the use of this insecticide. DDT and other insecticides such as ***dieldrin*** and ***heptachlor*** are referred to as ***persistent pesticides.*** Substitutions of other substances with biodegradable structures are now made where possible. The compound ***chlordan*** is an example of just such a substitution. It is interesting to note that the structural differences between heptachlor (persistent) and chlordan (short-lived) are relatively slight (look at the chlorine atom on the lower five-membered ring).

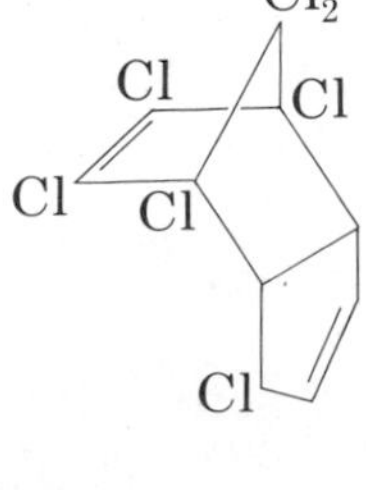

HEPTACHLOR 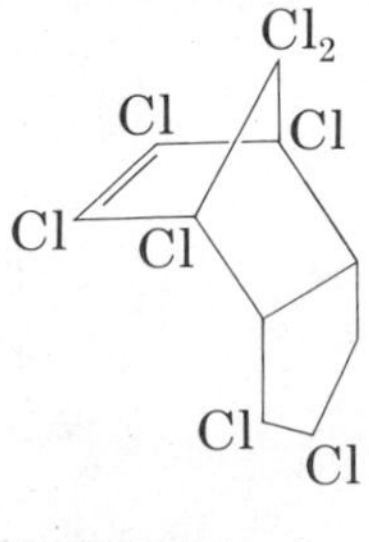 CHLORDAN

There are many other insecticides that are actually much more toxic to humans than is DDT. These include inorganic materials based on arsenic compounds, as well as a wide variety of phosphorus derivatives based on structures of the type

**Heptachlor and chlordan were banned for most garden and home use in December, 1975.**

$$\begin{matrix} R \\ R' \end{matrix} \!\!> \overset{\overset{\displaystyle Z}{\|}}{P} - X$$

where Z is oxygen or sulfur, R and R′ are alkyl, alkoxy, alkylthio or amide groups, and X is a group that can be split easily from the phosphorus. Insecticides of this type include ***parathion,*** which is effective against a large number of insects but is also *very* poisonous to human beings. These compounds are anticholinesterase poisons. One of their most important properties, however, is that they are readily hydrolyzed to less toxic substances that are not residual poisons.

$(C_2H_5O)_2P(=S)-O-C_6H_4-NO_2$

*Parathion*

$(C_2H_5O)_2P(=O)-S-CH(COOC_2H_5)-CH_2-COOC_2H_5$

*Malathion*

Perhaps the most publicized case of a poorly handled insecticide was the manufacture of ***kepone*** ($C_{10}Cl_{10}O$, a complex molecule containing several fused rings) in a converted filling station. The resulting contamination of the James River was extensive, as was the personal suffering of the workers. The dredging costs to clean the river are estimated by the EPA to be one billion dollars, taking one dredge 120 years to complete the job. Total clean-up costs are estimated at $7.2 billion. Obviously, there are no immediate plans to clean up this chemical spill.

**The goal of the insecticide quest; a selectively toxic chemical that is quickly biodegradable.**

The choice of solutions to our problems with pesticides is not an easy one. The use of insecticides introduces them into our environment and our water supplies. A refusal to use insecticides means that we must tolerate malaria, plague, sleeping sickness, and consumption of a large part of our food supply by insects. It is obvious that continuing research is needed on new methods and materials for the control of insect populations.

## Industrial Wastes

Industrial wastes can be an especially vexing sort of pollution problem because often they are either not removed or are removed very slowly by naturally occurring purification processes, and are generally not removed at all by a typical municipal water treatment plant. Table 16–4 lists some of the industrial pollutants that result from products important to us.

Chemicals buried in dumps have in some cases infiltrated groundwater or surface water supplies. The Love Canal chemical landfill subsequently caused near panic conditions in Niagara Falls, New York because of polluted ground- and surface waters. The water in a private well near Toone, Tennessee was shown to contain carbon tetrachloride, chloroform, benzene, heptachlor, and chlordan along with other chemicals.

**TABLE 16–4 Important Industrial Products and Consequent Hazardous Wastes**

| The Products We Use | The Potentially Hazardous Waste They Generate |
|---|---|
| Plastics | organic chlorine compounds |
| Pesticides | organic chlorine compounds, organic phosphate compounds |
| Medicines | organic solvents and residues, heavy metals like mercury and zinc, for example |
| Paints | heavy metals, pigments, solvents, organic residues |
| Oil, gasoline, and other petroleum products | oils, phenols, and other organic compounds, heavy metals, ammonia, salts, acids, alkalies |
| Metals | heavy metals, fluorides, cyanides, acids, and alkaline cleaners, solvents, pigments, abrasives, plating salts, oils, phenols |
| Leather | heavy metals, organic solvents |
| Textiles | heavy metals, dyes, organic chlorine compounds, organic solvents |

**TABLE 16–5 Categories of Industrial Materials Producing Pollutants and Related Toxic Pollutant Classes—Environmental Protection Agency**

| | 34 Industrial Categories | |
|---|---|---|
| adhesives | plastics processing | pesticides |
| leather tanning and finishing | porcelain enamel | pharmaceuticals |
| soaps and detergents | gum and wood chemicals | plastic and synthetic materials |
| aluminum forming | paint and ink | rubber |
| battery manufacturing | printing and publishing | auto and other laundries |
| coil coating | pulp and paper | mechanical products |
| copper forming | textile mills | electric and electronic components |
| electroplating | timber | explosives manufacturing |
| foundries | coal mining | inorganic chemicals |
| iron and steel | ore mining | |
| nonferrous metals | petroleum refining | |
| photographic supplies | steam electric | |
| | organic chemicals | |
| | **65 Toxic Pollutant Classes** | |
| acenapthene | DDT and metabolites | nitrobenzene |
| acrolein | dichlorobenzenes | nitrophenols |
| acrylonitrile | dichlorobenzidine | nitrosamines |
| aldrin/dieldrin | dichloroethylenes | pentachlorophenol |
| antimony and compounds | 2,4-dimethylphenol | phenol |
| arsenic and compounds | dinitrotoluene | phthalate esters |
| asbestos | diphenylhydrazine | polychlorinated biphenyls (PCB's) |
| benzene | endosulfan and metabolites | polynuclear aromatic hydrocarbons |
| benzidine | endrin and metabolites | selenium and compounds |
| beryllium and compounds | ethylbenzene | silver and compounds |
| cadmium and compounds | fluoranthene | 2,3,7,8-tetrachlorodibenzo-*p*-dioxin (TCDD) |
| carbon tetrachloride | haloethers | tetrachloroethylene |
| chlordane | halomethanes | thallium and compounds |
| chlorinated benzenes | heptachlor and metabolites | toluene |
| chlorinated ethanes | hexachlorobutadiene | toxaphene |
| chloralkyl ethers | hexachlorocyclopentadiene | trichloroethylene |
| chlorinated phenols | hexachlorocyclohexane | vinyl chloride |
| chloroform | isophorone | zinc and compounds |
| 2-chlorophenol | lead and compounds | |
| chromium and compounds | mercury and compounds | |
| copper and compounds | napthalene | |
| cyanides | nickel and compounds | |

The well, which "smells like shoe polish," is $1\frac{1}{4}$ miles from a chemical dump that was closed six years prior to the testing of the well. The chemical dump is estimated to contain 16.5 million pounds of chemical wastes from a pesticide plant.

The Environmental Protection Agency has designated 34 industrial categories that produce polluted waste water and 65 classes of pollutants (Table 16–5). Under the law it is now illegal for these pollutants to be discharged into the U.S. waterways without a permit.

Relatively few legal limits have been set for inorganic chemicals in municipal drinking water supplies. The federal Safe Drinking Water Act (SDWA) has set maximum contaminant levels for arsenic, barium, cadmium, chromium, lead, mercury, nitrate, selenium, silver, and fluoride.

Industrial organic wastes have been referred to as "toxic time bombs" because of their latent ability to cause cancer. The new industrial organic chemicals produced since World War II number 50,000, and 70% of these are deemed harmful to human health. Yet, most of these wastes have been dumped into or onto the earth. The amounts of these chemicals that are dumped add up to billions of pounds each year. The Safe Drinking Water Act regulates only a handful of organic wastes in our municipal water supplies: six pesticides and the trihalomethanes (see **Chlorination,** at the end of this chapter.) The legal effort at the federal level is obviously aimed at the point of pollution rather than at the clean-up job required in the use of polluted resources.

Synthetic organic chemicals have been identified in water withdrawn from underground aquifers located close to urban and industrial areas. While most groundwater is still quite pure for domestic purposes, such findings raise important questions as to how the organics got into the groundwater and how long they will remain in such systems. It is known that such organics are removed by dispersion, absorption or adsorption, and biodegradation. However, these processes are slower in groundwater and some synthetic organics may persist for decades in such systems. The technology necessary to remove industrial wastes from water before it is returned to natural waters is available now. The limiting factors are cost (major cost is energy) and doing it (by will or by governmental enforcement). The control of industrial pollution is made easier by national and even international standards for all parties to observe. For example, steel mills have been and, to a lesser extent, still are notorious polluters of water. Some companies would find it very difficult to compete if they had to clean up all of their wastes while other companies did not. Concerted action, whether voluntary or enforced, is necessary.

**Chemical methods capable of eliminating industrial water pollution are known for most pollutants.**

### Silt

Silt, which is finely divided solid material, is picked up and carried along by flowing water until the velocity of the water is reduced as it enters a lake, dam area, or the sea. The silt is stabilized by two factors: (1) the kinetic energy of the moving water keeps at least some of the silt moving against gravity, and (2) charges build up on the silt particles, causing them to repel each other (Fig. 16–6). In the quiet water of a lake, silt tends to settle out. The saltwater of the sea is very effective in removing silt; hence, the great river deltas at the river mouths. The ions in saltwater neutralize the electrostatic charge on the particles and then collisions between them result in coagulation. The larger particles settle readily.

**A sticky precipitate will settle silt.**

Silt is undesirable in municipal water because it is esthetically unappealing and also because the silt particles can be carriers of chemicals or microorganisms, or both, that are harmful to humans.

## Water Purification: Classical and Modern Processes

The outhouses of some rural dwellers had their counterparts in city cesspools. The terrible job of cleaning led to the development of cesspools that could be flush-cleaned with water, followed by a connecting series of such pools that could be flushed from time to time. City sewer systems with no holding of the wastes were the next step.

**Cesspools were an early and crude form of the modern activated sludge process.**

Since there were not enough pure wells and springs to serve the growing population, water purification techniques were developed. The classical method, which is now termed ***primary water treatment,*** involved settling and filtration (Fig. 16–7).

**Sewage is still 99.9% water!**

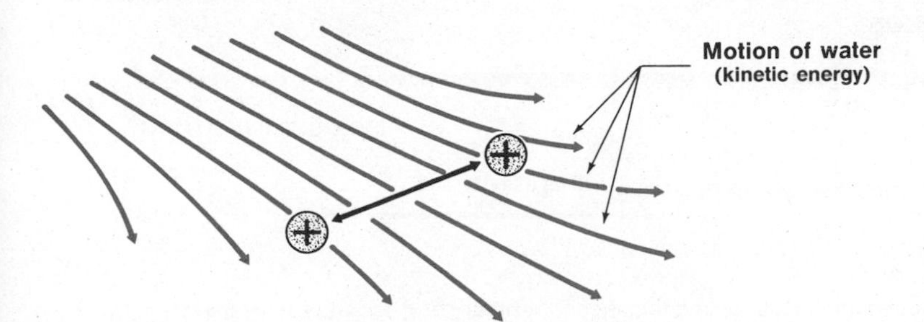


**Figure 16–6** Silt particles are stabilized by kinetic energy of the water and repulsive electrostatic forces.

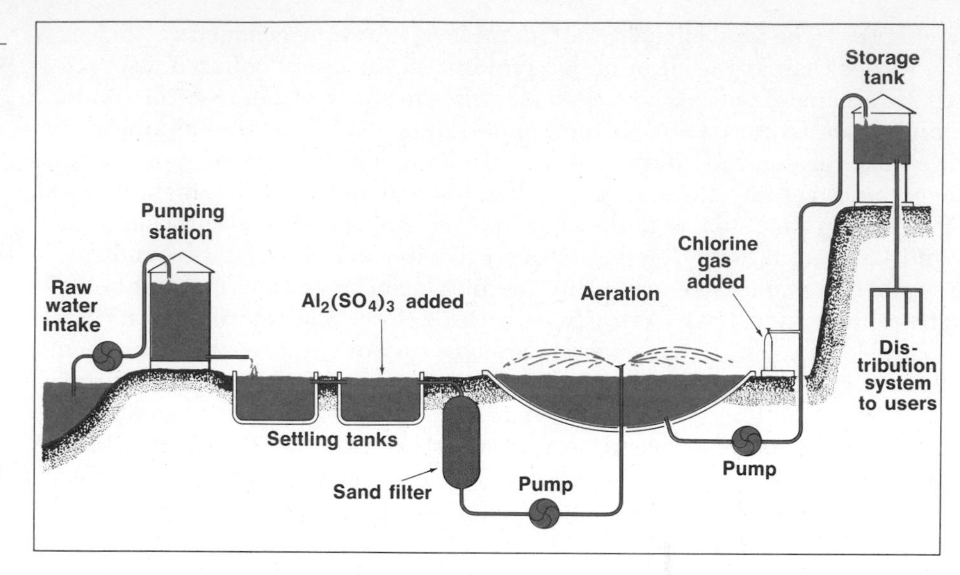


**Figure 16–7** Primary water purification removes particles that will settle or can be filtered, usually by sandbed filters. Aeration adds oxygen and gets rid of foul gases, and chlorination kills microbes. These are more recent additions. The system outlined here is almost as common as communities in the civilized world.

**Americans spend about $350 million a year on bottled water. Buyer beware: a very wide variety of standards exist for bottled water.**

If the intake water is polluted enough with biological wastes, the primary treatment, even with chlorination, cannot render the water safe. To be sure, enough chlorine or other oxidizing agents could be added to kill all life forms, but the result would be water loaded with a wide variety of noxious chemicals, especially chlorinated organics, many of which are suspected carcinogens. Some way had to be found to coagulate and separate out the organic material that passed through the primary filters.

***Secondary water treatment*** revives the old cesspool idea under a more controlled set of conditions and acts only on the material that will not settle or cannot be filtered (Fig. 16–8). It operates in an oxygen-rich environment (aerobic), whereas the cesspool operates in an oxygen-poor environment (anaerobic). The results are the same: the organic molecules that will not settle are consumed by organisms; the resulting sludge will settle. Bacteria and even protozoa are introduced into the oxygen-rich environment for this purpose. Two techniques, the trickle filter (Fig. 16–9) and the activated sludge method (Fig. 16–10), have been widely used in secondary water treatment.

**Primary, secondary, and tertiary water treatment methods can be used in both the purification of water to be consumed and the preparation of sewage to be sent back into a stream.**

Primary and secondary water treatment systems will not remove dissolved inorganic materials such as poisonous metal ions or even residual amounts of organic materials. These materials are removed by a variety of ***tertiary water treatments.***

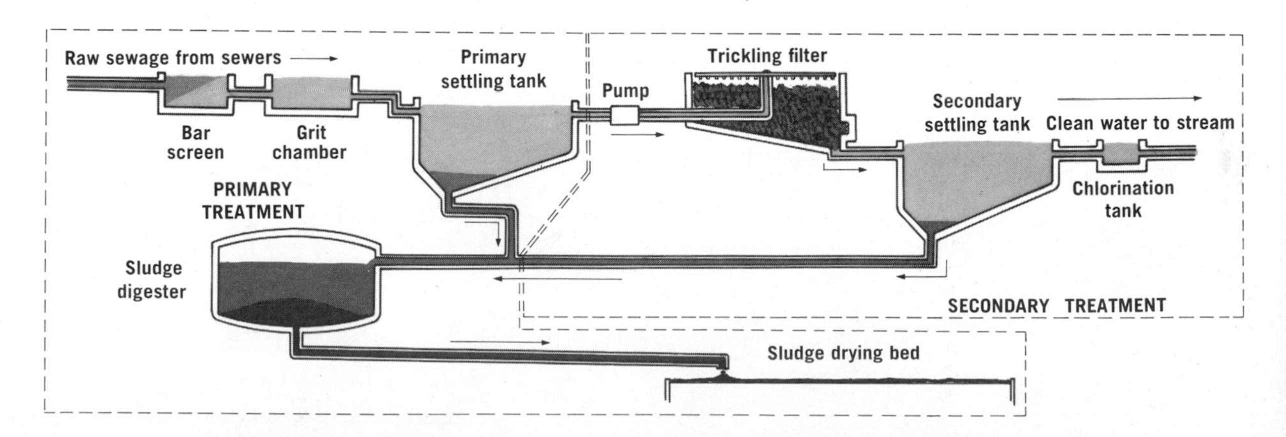


**Figure 16–8** Sewage plant schematic, showing facilities for primary and secondary treatment.

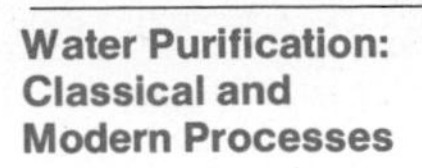

**Figure 16–9** Picture of a trickle filter with a section removed to show construction details. Rotating pipes discharge the water over a bed of stones. As a result, the organic molecules are "eaten" by microorganisms.

Three technologies are now being used for the removal of toxic materials from wastewater; these are carbon adsorption, activated sludge, and steam stripping. Carbon black has been used for many years for absorbing vapors and solute materials from liquid streams. Many toxic organic materials can be removed with activated or baked carbon powder. The drawbacks are the large amounts of energy needed to prepare the activated carbon and the unpredictability of the effectiveness of this technique for new applications. Activated sludge is a hurry-up version of natural stream purification. Bacteria and other microorganisms degrade the water pollutants in the sludge medium. Steam stripping, Figure 16–11, involves the removal of volatile organic pollutants from wastewater through steam distillation.

## Fresh Water from the Sea

**The sea is a largely untapped source of water.**

As we look for new supplies of water, attention is inevitably drawn to the sea, as well as to the vast supplies of brackish water frequently found by drilling in arid lands. Sea water contains 3.5% salts; brackish waters have smaller salt concentrations. If we can convert such water to potable (drinkable) water inexpensively, we have a

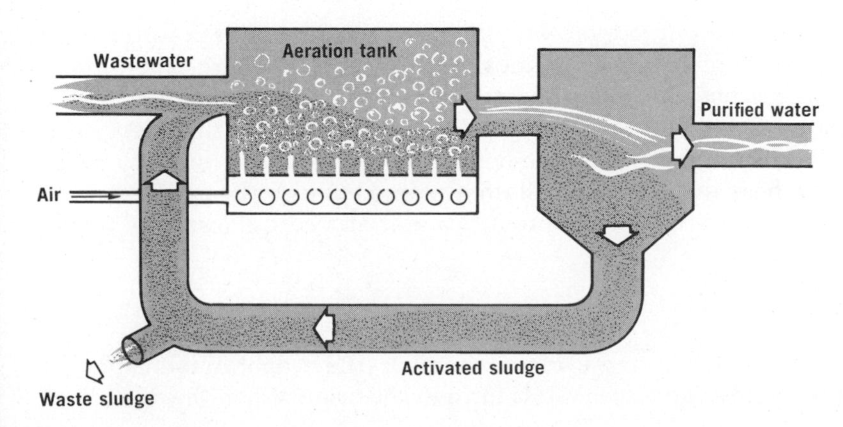


**Figure 16–10** The activated sludge process provides a closed-system environment in which organic molecules can be consumed by organisms that will readily settle.

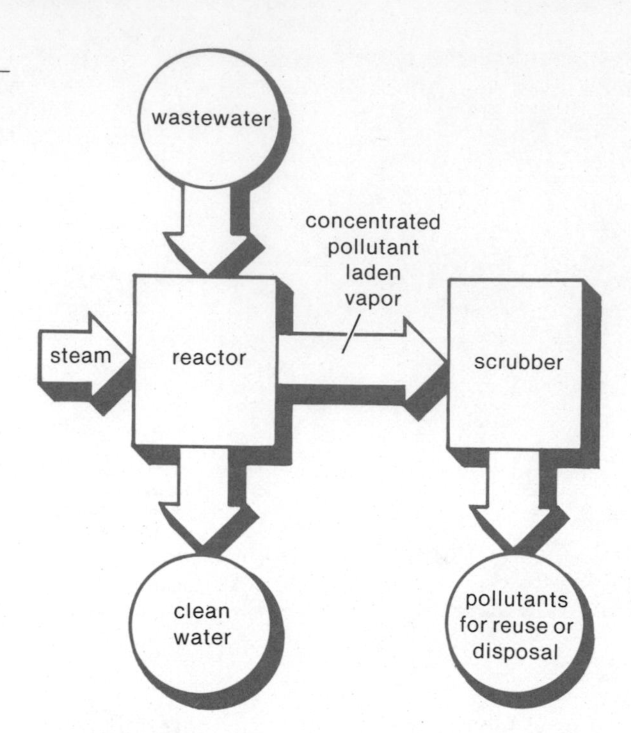


**Figure 16–11** Steam stripping. Volatile pollutants are extracted from waste water by steam.

potentially valuable process. Thus far, the cost of purification of saltwater is high in comparison with the cost of water supplied by a typical European or American municipal water system. However, one can expect the cost of freshwater purification to increase if pollution continues to increase, and the cost of purifying saltwater to decrease as technology is improved.

The purification of saltwater is at present approached in two general ways. The first way involves separation of the water from the salt by evaporation or by freezing. A second approach is to use the influence of electrical charge, chemical attraction, or selective membranes to cause the ions in the saltwater to move out of the stream of water flow. If this process continues long enough, pure water results. Changing the state of water is expensive because of the large amount of energy that must move either into or out of the water. It takes 540 cal to vaporize 1 g of water at its boiling point (100° C), and 80 cal must be removed to freeze each gram of water at its freezing point (0° C). The removal of ions from water is expensive because it is relatively slow for large flows of water and involves costly electric power.

## Distillation

**Because water needs much heat for vaporization, distillation requires a great deal of energy and is expensive.**

Distillation processes for the purpose of obtaining fresh water from sea water have been developed to a considerable state of refinement. In view of the large amounts of heat required, various methods have been devised to preheat the water with "waste heat." Since the steam has to lose 540 cal per gram in order to liquefy, this energy can be used to heat up the raw water to near the boiling point. Solar and nuclear energy have also been used to heat the water. Distillation plants are now the most common method in use to refine sea water and can produce water at a cost of less than $1 per 1000 gallons.

## Freezing

**The ice in icebergs can be melted and drunk by humans.**

When cold sea water is sprayed into a vacuum chamber, the evaporation of some of the water cools the remainder, and ice crystals form in the brine. When the crystals of ice form, they tend to exclude the salt ions. Any solid separating from a liquid will tend to take only the molecules or ions that fit into the particular solid pattern. Recall that this generalization is the basis for purification by recrystallization. Even though the separation of salt and water is not complete in one step, the ice has less salt than the same weight of liquid solution. The ice crystals can be collected on a filter, washed with a small amount of fresh water, and then melted to obtain purified water. The process is repeated until the desired degree of purity is achieved. Plants have been built and successfully operated that produce as much as 250,000 gallons of pure water per day by this purification technique.

## Ion Exchange

In the process called ion exchange, brackish water or sea water is first passed through a cation exchange resin to replace the cations with $H^+$, and then through an anion exchange resin to replace the anions with $OH^-$; these two ions then neutralize each other.

Modern ion-exchange resins are high molecular weight polymers containing firmly bonded functional groups that can exchange one ion for another as an ionic solution is passed over the giant molecules. ***Cation exchangers,*** or positive-ion exchangers, swap their hydrogen ions for the positive ions present in solution. They are usually organic derivatives of sulfuric acid, and their action can be depicted as:

**The chloride ion is shown on both sides of this reaction to indicate that it is not removed.**

$$\text{Polymer-SO}_3\text{H} + \underbrace{\text{Na}^+ + \text{Cl}^-}_{\text{BRACKISH WATER}} \longrightarrow \text{Polymer-SO}_3^-\text{Na}^+ + \underbrace{\text{H}^+ + \text{Cl}^-}_{\text{FROM CATION EXCHANGER}}$$

**Ion exchange resins can be regenerated.**

When the polymer is saturated with cations, it can be regenerated by treatment with strong acid, which reverses the above reaction. Because they generate weakly acidic solutions, cation exchange resins themselves do not do a complete job. When they are followed by ***anion exchange resins,*** they provide a good route to very pure water (Fig. 16–12). An anion exchange resin can replace the anions in solution by hydroxide ions. These resins are again high molecular weight polymers, but now the poly-

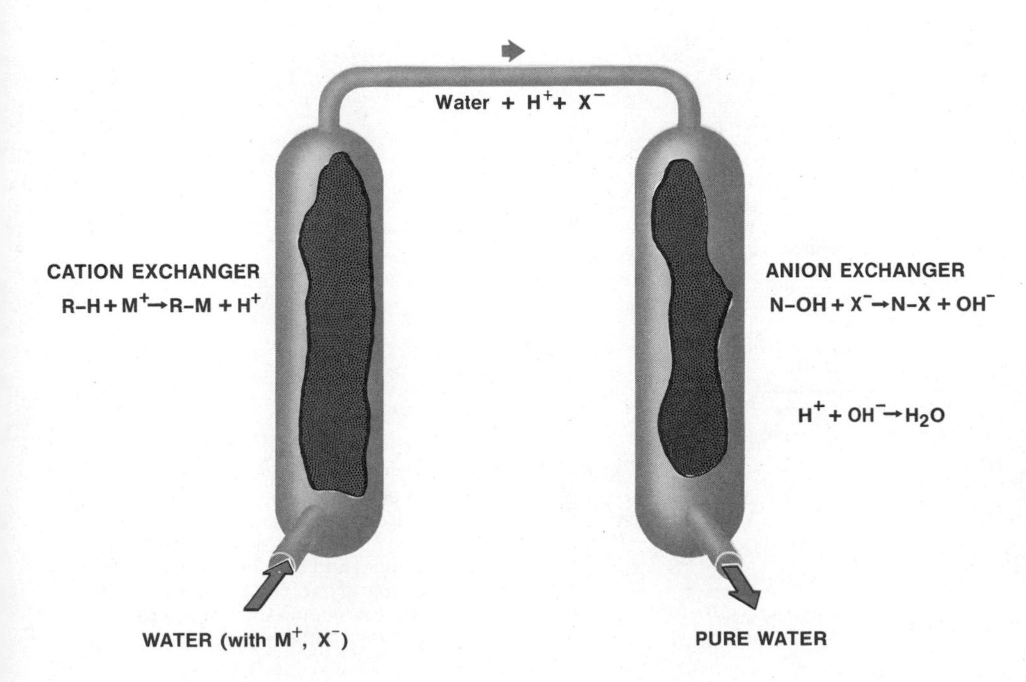


**Figure 16–12** Ion exchange purification. The tank on the left contains a cation exchange resin that replaces metal ions in solution with hydrogen ions. The tank on the right contains an anion exchange resin that replaces nonmetal ions in solution with hydroxide ions. One might well ask why this kind of process cannot be used to turn seawater into pure water. The answer is that the process can be used, but its cost is much *greater* than that of distillation. It is best suited to the removal of small amounts of salts from water that is already quite pure.

mer contains a nitrogen atom bonded to the polymer and three other groups. The reaction is:

$$\text{Polymer}-\overset{|}{\underset{|}{N}}{}^{+}OH^{-} + \underset{\text{FROM CATION EXCHANGER}}{H^{+} + Cl^{-}} \longrightarrow \text{Polymer}-\overset{|}{\underset{|}{N}}{}^{+}Cl^{-} + \underset{\text{PURE WATER FROM CATION AND ANION EXCHANGERS}}{H_2O}$$

When the anion exchange resin has exchanged all its hydroxide, it can be regenerated by treatment with a strong solution of sodium hydroxide:

$$\text{Polymer}-\overset{|}{\underset{|}{N}}{}^{+}Cl^{-} + Na^{+} + OH^{-} \longrightarrow \text{Polymer}-\overset{|}{\underset{|}{N}}{}^{+}OH^{-} + Na^{+} + Cl^{-}$$

Unfortunately, the amount of sea water that can be purified by a given amount of ion-exchange resin is quite small, and the resulting water is relatively costly.

## Electrodialysis

**In electrodialysis, ions are attracted to oppositely charged electrodes on the other side of a membrane.**

**Home treatment devices on the market:**
**1. disinfection: chlorine, ultraviolet light, ozone, iodine, bromine, silver salt**
**2. filtration**
**3. adsorption**
**4. deionization**
**5. reverse osmosis**

We have learned that in the electrolysis of saltwater, positive ions migrate toward the negative electrode and negative ions move toward the positive electrode. If an electrolysis cell is divided into three compartments by semipermeable membranes, one permeable to positive ions and the other permeable to negative ions (Fig. 16–13), the resulting ionic separation is called ***electrodialysis.*** Dialysis is the passage of selected species in solution through membranes while other species are excluded. Electrodialysis is the special case in which the passage of ions is influenced by an electrical field.

In Figure 16–13, note that positive ions (cations) can move to the left out of the center compartment but cannot move through the negative ion (anion) membrane

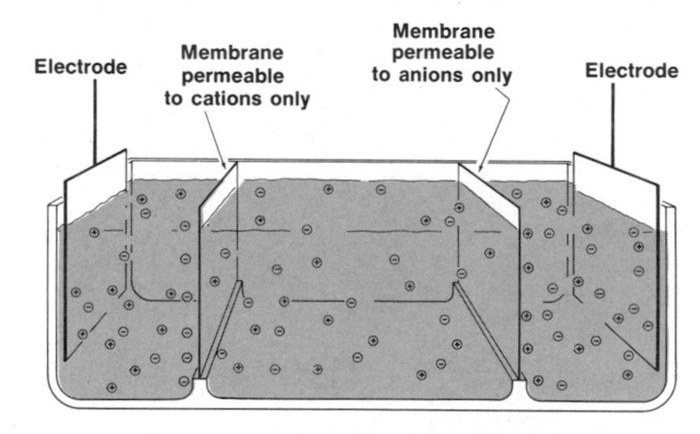


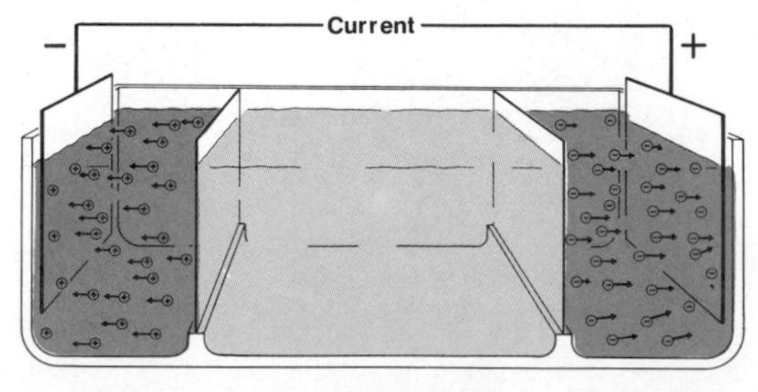


**Figure 16–13** The essential features of the electrodialysis process. Each compartment of the cell contains brackish water. Application of an electrical potential across the cell causes the cations to move from the center into the left compartment and the anions to move into the right compartment. The salt content of the water in the center compartment is thus reduced, and raised in the end compartments.

from the right compartment to the center one. In a similar way, anions move out of the center compartment to the right. Thus, the ionic concentration of the water in the central compartment is reduced. If the process is continued long enough, the water in the central compartment loses most of its salt content.



## Reverse Osmosis

***Osmosis*** is the process whereby water moves through a semipermeable membrane from a region of relatively pure water into a region containing a concentrated solution. For example, water will move through a living cell membrane into the cell's protoplasm, causing the cell to become turgid. The resulting pressure inside the cell is often very high. If enough pressure is brought to bear on the solution inside the membrane (in other words, if a pressure greater than the osmotic pressure is applied in reverse), the water can actually be made to flow from the concentrated solution inside to the region outside (Fig. 16–14). This is called ***reverse osmosis.***

Reverse osmosis is a promising process. Costs are becoming more and more competitive. Membranes used thus far are mostly cellulose acetate and work very well for saltwater. Much effort is being made in developing new films for a wide variety of applications of reverse osmosis.

It has been demonstrated that membrane processes are effective in the removal of polychlorinated biphenyls (PCB's), phenol, chromium, lead, and silver. Both nitrates and phosphates are effectively removed from municipal water supplies by reverse osmosis. It is probable that membrane processes can be developed to remove any dissolved species from water.

### Softening of Hard Water

**Hard water contains metal ions that react with soaps and give precipitates.**

The presence of $Ca^{2+}$, $Mg^{2+}$, $Fe^{3+}$, or $Mn^{2+}$ will impart "hardness" to waters. Hardness in water is objectionable because (1) it causes precipitates (scale) to form

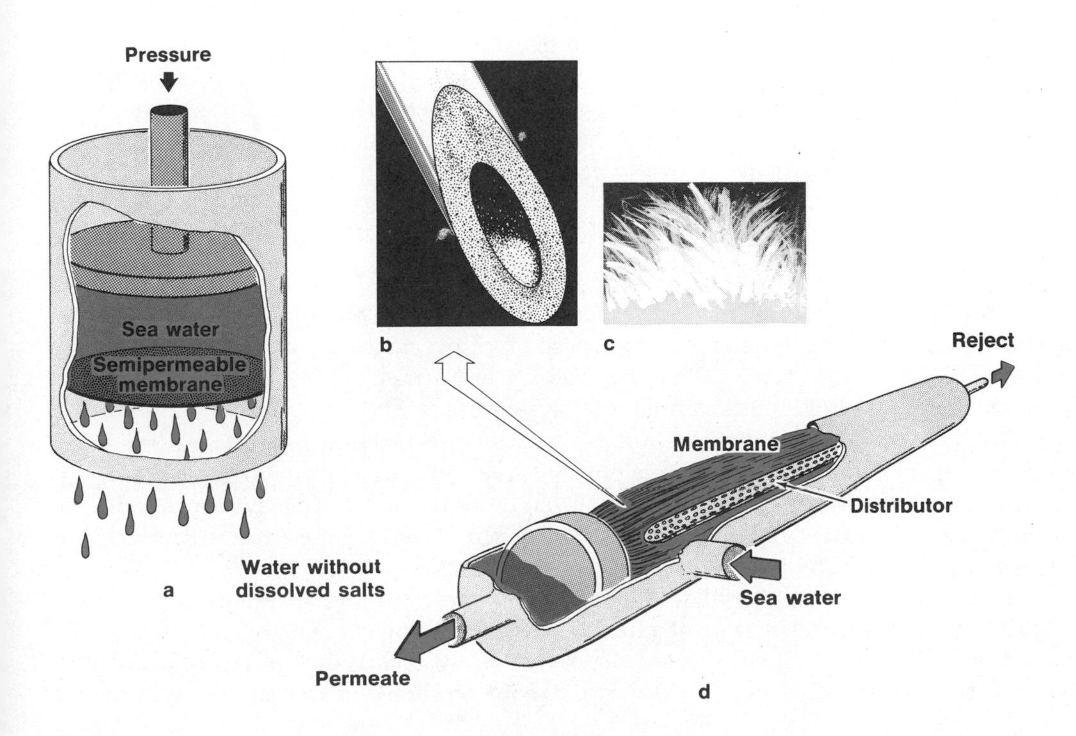


**Figure 16–14**
Reverse osmosis.
(*a*) Mechanical pressure forces water against osmotic pressure to region of pure water.
(*b*) Enlargement of individual membrane.
(*c*) Mass of many membranes.
(*d*) Industrial unit; feed water (salt) that passes through membranes collects at the left end (permeate). The more concentrated salt solution flows out to the right as the reject.

in boilers and hot water systems, (2) it causes soaps to form insoluble curds (this reaction does not occur with some synthetic detergents), and (3) it can impart a disagreeable taste to the water.

Hardness consisting of calcium or magnesium, present as their bicarbonates, is produced when water containing dissolved carbon dioxide trickles through limestone or dolomite:

$$\underset{\text{LIMESTONE}}{CaCO_3} + CO_2 + H_2O \longrightarrow Ca^{2+} + 2HCO_3^-$$

$$\underset{\text{DOLOMITE}}{CaCO_3 \cdot MgCO_3} + 2CO_2 + 2H_2O \longrightarrow Ca^{2+} + Mg^{2+} + 4HCO_3^-$$

**Water softeners that act like ion-exchange resins are used to make soft water. They remove the hard water ions, $Ca^{2+}$, $Mg^{2+}$, and $Fe^{3+}$, and put $Na^+$ ions in the water in exchange.**

Such "hard water" can be softened by removing these compounds. The principal methods for achieving this are (1) the lime-soda process and (2) ion-exchange processes.

The lime-soda process is based on the fact that calcium carbonate ($CaCO_3$) is much less soluble than calcium bicarbonate [$Ca(HCO_3)_2$] and that magnesium hydroxide is much less soluble than magnesium bicarbonate. The raw materials added to the water in this process are hydrated lime [$Ca(OH)_2$] and soda ($Na_2CO_3$). In the system, several reactions take place, which can be summarized:

$$HCO_3^- + OH^- \longrightarrow CO_3^{2-} + H_2O$$

$$Ca^{2+} + CO_3^{2-} \longrightarrow CaCO_3\downarrow$$

$$Mg^{2+} + 2OH^- \longrightarrow Mg(OH)_2\downarrow$$

**Soft water—**
**less than 65 mg of metal ion per gallon**
**Slightly hard—**
**65–228 mg**
**Moderately hard—**
**228–455 mg**
**Hard—**
**455–682 mg**
**Very hard—**
**above 682 mg**

(The downward arrow indicates that the chemical falls out as a precipitate.)

The overall result of the lime-soda process is to precipitate all the calcium and magnesium ions and to leave sodium ions as replacements.

Iron present as $Fe^{2+}$ and manganese present as $Mn^{2+}$ can be removed from water by oxidizing them with air (aeration) to higher oxidation states. If the pH of the water is 7 or above (either naturally or by adding lime), the insoluble compounds $Fe(OH)_3$ and $MnO_2(H_2O)_x$ are produced.

The desire for and achievement of soft water for domestic use has sparked a rather heated health debate during the past two decades. Soft water is usually acidic and contains $Na^+$ ions in the place of di- and trivalent metal ions. An increased intake of sodium is known to be related to heart disease. Also, the acidic soft water is more likely to attack metallic pipes, resulting in the solution of dangerous ions such as $Pb^{2+}$. The safe course would be to drink naturally hard water and wash in soft water.

## Chlorination

Chlorine is introduced into water as the gaseous free element ($Cl_2$), and it acts as a powerful oxidizing agent for the purpose of killing bacteria that remain in water after preliminary purification. The principal water-borne diseases spread by bacteria include cholera, typhoid, paratyphoid, and dysentery.

Most city water supplies are not bacteria-free. Surviving bacteria will usually produce counts numbering in the tens of thousands, but only rarely do these surviving bacteria cause disease. The most common bacterial disease borne by water today is giardiasis, a gastrointestinal disorder. Most often this disease comes from surface water but, on occasion, it can be traced to city water systems.

Chlorination of industrial wastes and city water supplies presents a potential threat because of the reaction of chlorine with residual concentrations of organic compounds that are formed and are not removed by traditional purification procedures. City water supplies have been shown to contain such chloroorganics as dibromo-

**Figure 16–15** This apparatus adds chlorine in sufficient amounts to meet health standards (1 ppm residual) for a 60-million-gallons per day water treatment plant.

dichloromethane, trichloroethylene, bromoform, bromodichloromethane, chlorobenzene, chloroform, and dibromochloromethane in concentrations of a few parts per million or less. A number of these chemicals in the same concentration range have been shown to be mutagenic to salmonella bacteria. Studies show an increased risk of 50 to 100% in rectal, colon, and bladder cancers in individuals who drink chlorinated water. According to the Environmental Protection Agency, mutagenic or carcinogenic chemicals have been found in fourteen major river basins in the United States. It is estimated that more than 500 water systems in the United States exceed EPA's maximum for THM's, 0.1 ppm. If chlorination turns out to do more harm than good, which is not established at this point, other oxidizing agents such as ozone, chlorine dioxide, and bromine chloride have shown promise. However, potential problems with these chemicals are yet to be explored.

**The trihalomethanes (THM's) are chloroform, $CHCl_3$, bromodichloromethane, $CHBrCl_2$, dibromochloromethane, $CHBr_2Cl$, and bromoform, $CHBr_3$.**

**92% of the U.S. water systems serve fewer than 10,000 people. Water quality is in the hands of many people.**

## Self-Test 16-B

1. Pick the pesticide with the shortest half-life in nature: DDT, dieldrin, heptachlor, chlordan. ______________
2. Heavy metal ions in more than trace concentrations are usually ______________ to life forms.
3. Silt is often stabilized in suspension by ______________ forces.
4. Select the ions that may cause water to be hard: sodium, calcium, magnesium, potassium. ______________
5. The element ______________ is added to water to kill microbes in water.
6. Electrodialysis involves ion migration in a(n) ______________ field.

7. Ion-exchange systems for removing salts from water involve ________ (how many) types of resins.
8. Ice formed from impure water will probably be (more pure than, as pure as, less pure than) the original water. ________
9. Primary water treatment involves ________ and ________ of particles.
10. Tertiary water treatment removes ________ ions and trace amounts of ________.

## Matching Set

| | | | |
|---|---|---|---|
| ____ | 1. sedimentation | a | a measure of organic material in water |
| ____ | 2. biodegradable | b | widely used as a detergent builder |
| ____ | 3. detergent | c | results mostly from water used to cool a process |
| ____ | 4. BOD | d | caused by metal ions such as $Ca^{2+}$ and $Mg^{2+}$ in solution |
| ____ | 5. thermal pollution | e | primary purification process |
| ____ | 6. phosphate | f | persistent pesticide |
| ____ | 7. activated sludge process | g | secondary purification process |
| ____ | 8. reverse osmosis | h | naturally reducible to simpler compounds |
| ____ | 9. water hardness | i | soap substitute |
| ____ | 10. DDT | j | requires high pressure |

## Questions

**1.** If four fifths of the earth is covered with water, why is there a problem with water supply for humans?

**2.** Which can dissolve more oxygen to support marine or aquatic life, cold or warm water?

**3.** Find out what industrial wastes are produced in your community. Are you satisfied with the way these wastes are handled? Explain.

**4.** Name three ways to obtain fresh water from sea water. Name one way that does not involve a change in state.

**5.** What are some processes that *decrease* the amount of dissolved oxygen in a stream? What are some processes that *increase* the amount of dissolved oxygen in a stream? Which ones are most readily subject to human control?

**6.** Explain why each of the following introduces a pollution problem when its wastes are emptied into a stream:

a. a chlorine-producing plant
b. a steel mill
c. an electricity-generating plant burning oil or coal
d. an agricultural area that is intensively cultivated

**7.** An old rule of thumb is, "Water purifies itself by running two miles from the source of incoming waste." What processes are active in purifying the water? Is this adage foolproof? Explain.

**8.** What is a detergent builder? What property should it have to increase the cleansing power of the detergent? What property must it have to avoid environmental problems?

**9.** What are some consequences of requiring by law that all water put into our national waters be clinically and chemically pure $H_2O$? Do you think this is the proper solution to water pollution? If not, what is the proper compromise?

**10.** What are some ecological consequences of thermal pollution?

**11.** From your experience, add one additional example for each of the classes of pollutants listed in Table 16–2.

**12.** What pertinent facts would you try to gather if it were your responsibility to vote on a bill to regulate water pollution?

**13.** At what point should pollutants be removed from used water? Who should be responsible for this removal? Would you distinguish between industrial wastes and household wastes?

**14.** The most abundant elements in organic compounds are carbon, hydrogen, oxygen, and nitrogen. What are the oxidation products for these elements in the decomposition that occurs in nature?

**15.** Relate molecular structure to biodegradability for detergent molecules.

**16.** Classify water pollutants into as few major groups as you can. Describe some effects of each group and a removal process.

**17.** From your study of biochemistry, Chapters 11 and 12, explain why there is no question concerning the biodegradability of the proposed detergent builder, sodium citrate.

**18.** In your judgment, what are the most serious pollution problems? Be ready to defend your points in class discussion.

**19.** What is natural osmosis? Explain the significance of the word "reverse" in reverse osmosis.

# 17 Air Pollution

What was in the air you just breathed? Has acid rain fallen on you lately? Chances are harmful chemicals were in the air and in the rain; what are they, how did they get there, what do they do once in the air, and how can they be removed? These questions will form the bases of our study of the chemistry of air pollution.

Prior to 1960 there was little concern about air pollution. Most smoke, carbon monoxide, sulfur dioxide, nitrogen oxides, and organic vapors were emitted into the air with little apparent thought of the harmful nature of these pollutants as long as they were scattered into the atmosphere and away from human smell and sight (Fig. 17–1). Earth inhabitants acted as though the atmosphere were infinite. But not so. Ninety-nine percent of the estimated 5,500 trillion tons of gases that compose the atmosphere is below an altitude of 19 miles. Sufficient oxygen to sustain life extends upward to only about 4 miles above sea level, and most of our weather takes place within an average altitude of just 7 miles. In this limited region, pollutants collect and react; they do not escape earth and venture into outer space.

**Figure 17–1** Coppertown Basin (Ducktown), Tennessee, as photographed in 1943. Copper ore (principally copper sulfide, $Cu_2S$) had been mined and smelted in this area since 1847. In the early years, large quantities of sulfur dioxide, a by-product, were discharged directly into the atmosphere and killed all vegetation for miles around the smelter. Today the sulfur is reclaimed in the exhaust stacks to make sulfuric acid, but the denuded soil remains a monument to the misuse of the atmosphere.

**Figure 17–2** Region of Donora, Pennsylvania. In late October, 1948, Donora had a five-day siege of extreme air pollution. Before rain cleansed the air, more than 800 domestic animals died and 43% of the population, or 5910 people, became ill. Eighteen died; the normal rate was two deaths every five days.

By the early 1960's, air pollution achieved notoriety as air pollution disasters began to occur more often. Some of the more devastating disasters were:

October 27–31, 1948, Donora, Pennsylvania, 18 excess deaths (Fig. 17–2)
November 26–December 1, 1948, London, 700–800 excess deaths
December 5–9, 1952, London, 4,000 excess deaths
January 3–6, 1956, London, 1,000 excess deaths
December 5–10, 1962, London, 700 excess deaths
December 7–10, 1962, Osaka, Japan, 60 excess deaths
January 29–February 12, 1963, New York, 200–400 excess deaths
February 27–March 10, 1964, New York, 168 excess deaths

Increased pollution of the air was brought on by increased urbanization, industrialization, and transportation via the automobile and airplane. More public and governmental concern about air pollution led to the federal Clean Air Act of 1970 and additional state regulations. As a result, controls were placed on emissions from automobiles, industries, and power companies. The effects of the controls can be seen in Fig. 17–3.

By 1979, actual air pollutant emissions from the major human-controlled sources were at the levels shown in Table 17–1. Volcanic action, forest fires, dust storms, and even growing plants are natural sources of air pollutants, but these contribute very little compared with the sources that we control.

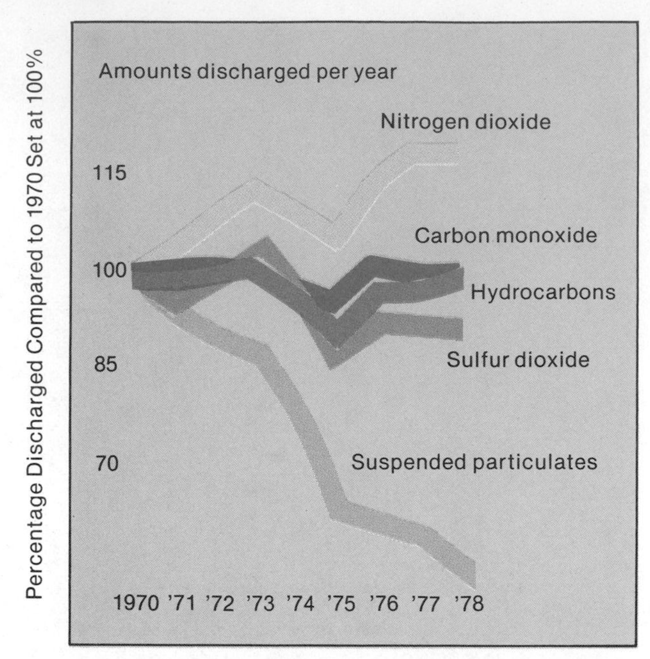


**Figure 17–3** Changes in discharges of air pollutants since 1970, the birth date of the Clean Air Act. Although the total tonnage of particulates discharged has declined, suspended fine particulates may be on the rise.

The energy crisis of the 1970's and 1980's caused a relaxation of some air quality standards (Table 17–2).

**A quadrillion is $10^{15}$.**

Although pollution occurs in relatively small amounts compared with oxygen, nitrogen, and carbon dioxide, about 200 quadrillion *pollutant* molecules are inhaled per breath on a *clear* day in Los Angeles, where the average breath would contain:

| | |
|---|---|
| Carbon monoxide | 175,000,000,000,000,000 molecules |
| Hydrocarbons | 10,000,000,000,000,000 |
| Peroxides | 5,000,000,000,000,000 |
| Nitrogen oxides | 4,000,000,000,000,000 |
| Lower aldehydes | 3,500,000,000,000,000 |
| Ozone | 3,000,000,000,000,000 |
| Sulfur dioxide | 2,500,000,000,000,000 |

On a smoggy day, the numbers increase by a factor of five or more. The air you are now inhaling could contain pollutant molecules in comparable amounts, give or take a few quadrillion.

**Air pollutants tend to attack the site where they first enter the body, i.e., the lungs.**

The long-range effects of air pollution on materials and the health of plants, animals, and human beings are beginning to emerge. The lung cancer rate in large

**TABLE 17–1 Air Pollutant Emissions in the United States in 1979 (Millions of Tons per Year)**

| | Totals | % of Totals | Carbon Monoxide | Sulfur Oxides | Hydro-carbons | Nitrogen Oxides | Particulates |
|---|---|---|---|---|---|---|---|
| Transportation | 104.1 | 55 | 81.9 | 0.9 | 9.7 | 10.1 | 1.5 |
| Fuel combustion in stationary sources | 40.2 | 21 | 2.1 | 21.6 | 0.2 | 13.5 | 2.8 |
| Industry | 30.6 | 16 | 6.9 | 4.5 | 13.6 | 0.9 | 4.7 |
| Solid waste disposal | 4.1 | 2 | 2.7 | 0 | 0.9 | 0.1 | 0.4 |
| Miscellaneous | 10.6 | 6 | 6.8 | 0 | 2.6 | 0.2 | 1.0 |
| Totals: | 189.6 | 100 | 100.4 | 27.0 | 27.0 | 24.8 | 10.4 |

**TABLE 17–2 Federal Air Quality Standards, July 1, 1979**

| | Concentration |
|---|---|
| *Sulfur Dioxide* | |
| Arithmetic mean (annual) | 0.03 ppm |
| 24-hour concentration not to be exceeded more than once per year | 0.14 ppm |
| *Suspended Particulates* | |
| Geometric mean (annual) | 75 $\mu g/m^3$* |
| 24-hour concentration not to be exceeded more than once per year | 260 $\mu g/m^3$ |
| *Carbon Monoxide* | |
| 8-hour concentration not to be exceeded more than once per year | 9 ppm |
| 1-hour concentration not to be exceeded more than once per year | 40 ppm |
| *Ozone* | |
| 1-hour concentration not to be exceeded more than once per year | 0.12 ppm |
| *Hydrocarbons* | |
| 3-hour concentration not to be exceeded more than once per year | 0.24 ppm |
| *Nitrogen Dioxide* | |
| Arithmetic mean (annual) | 0.05 ppm |
| *Lead* | |
| Arithmetic mean (annual) | 1.5 $\mu g/m^3$ |

*$1\ \mu g = 10^{-6}$ g

metropolitan areas is twice as great as the rate in rural areas, even after full allowance is made for differences in cigarette smoking habits. The incidence of the serious pulmonary disease emphysema shot up eight-fold during the decade of the sixties. However, only after we comprehend the short- and long-range effects of air pollution can we evaluate wisely its relative importance.

A serious study of the material in this chapter, combined with a study of the chapter on energy (Chapter 14), provides a sound basis for the decisions society must make concerning energy, pollution, and health, and its willingness to pay for them.

**Aerosol particles are intermediate in size between small molecules and easily visible small particles.**

## Do Air Pollutants Solo or Aggregate?

**1 nanometer = $10^{-9}$ meter.**

Pollutants may exist and react as single, isolated molecules, ions, or atoms. More often, because of the polar nature of pollutants such as $SO_2$ and $NO_2$, the pollutants are attracted into water droplets and form ***aerosols*** or onto larger particles called ***particulates.***

**_Adsorption_ is the attachment of particles to a surface.**

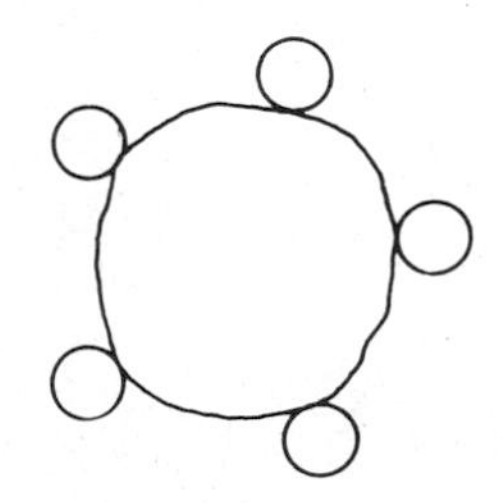

Aerosols range upward from a diameter of 1 nanometer to about 10,000 nanometers and may contain as many as a trillion atoms, ions, or small molecules per particle. They are small enough to remain suspended in the atmosphere for long periods of time. Smoke, dust, clouds, fog, mist, and sprays are typical aerosols. Since they are small, many aerosol particles can exist in a small volume of gas. Because of their vast combined surface area, aerosol particulates have enormous capacities to *adsorb* and concentrate gases on the surfaces of the particles. At other times, liquid aerosols *absorb* air pollutants, thereby concentrating them and providing a water medium in which reactions can occur readily. Thanks to the concentration and reaction effects, aerosols can be more devastating than isolated air pollutant molecules.

**_Absorption_ is pulling particles inside.**

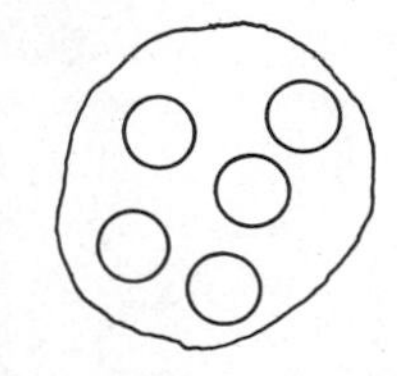

Particulates are generally large enough to be seen as individual particles. They range in size from 1 to 10 microns in diameter. Millions of tons of soot, dust, and smoke particulates are deposited into the atmosphere of the United States each year. Average suspended particulate concentrations in the United States range from 0.00001 gram per cubic meter ($g/m^3$) of air in remote rural areas to about six times that value in urban locations. In heavily polluted areas, concentrations up to 0.002 $g/m^3$, or 200 times the usual value, have been measured.

**1 micron = $10^{-6}$ meter.**

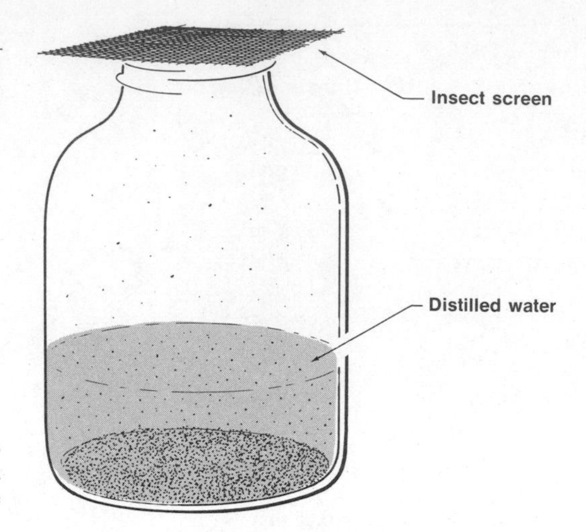


**Figure 17–4** A dust sample collector for use in the backyard. The open jar containing water is exposed to the air for a known interval. Fifteen days is sufficient exposure time. The water is then evaporated and the residue weighed. Balances sensitive to 0.001 g should be used; the area ($cm^2$) of the opening must be known.

**Particulate effects depend heavily on the chemical nature of the particle.**

**Major contributors to the amount of atmospheric particulates are volcanic eruptions by: Krakatoa, Indonesia, 1883; Mt. Katmai, Alaska, 1912; Hekla, Iceland, 1947; Mt. Spurr, Alaska, 1953; Bezymyannaya, U.S.S.R., 1956; Mt. St. Helens, Washington, 1980.**

Particulates can cause damage in several ways. As small solid particles, particulates may cause damage by abrasive action, fouling and shorting electrical contacts and switches, and by blocking breathing passages.

Some particulates are intrinsically toxic to human beings and animals. The toxic effects of lead and arsenic were described in Chapter 15. Lead components are emitted from automobiles that use leaded gasoline. Arsenic compounds are used as insecticides to dust growing plants. Particulates containing fluorides, commonly emitted from aluminum-producing and fertilizer factories, have caused weakening of bones and loss of mobility in animals that have eaten plants covered with the dust. Asbestos particulates have been shown to be carcinogenic.

**Figure 17–5** The General Electric CF6-6D engines that power this McDonnel Douglas DC-10 were designed to eliminate exhaust smoke and reduce by one half the noise level at takeoff. American Airlines advertises that it spent in a 10-year period an amount equal to 43% of its profits on noise and air pollution control systems.

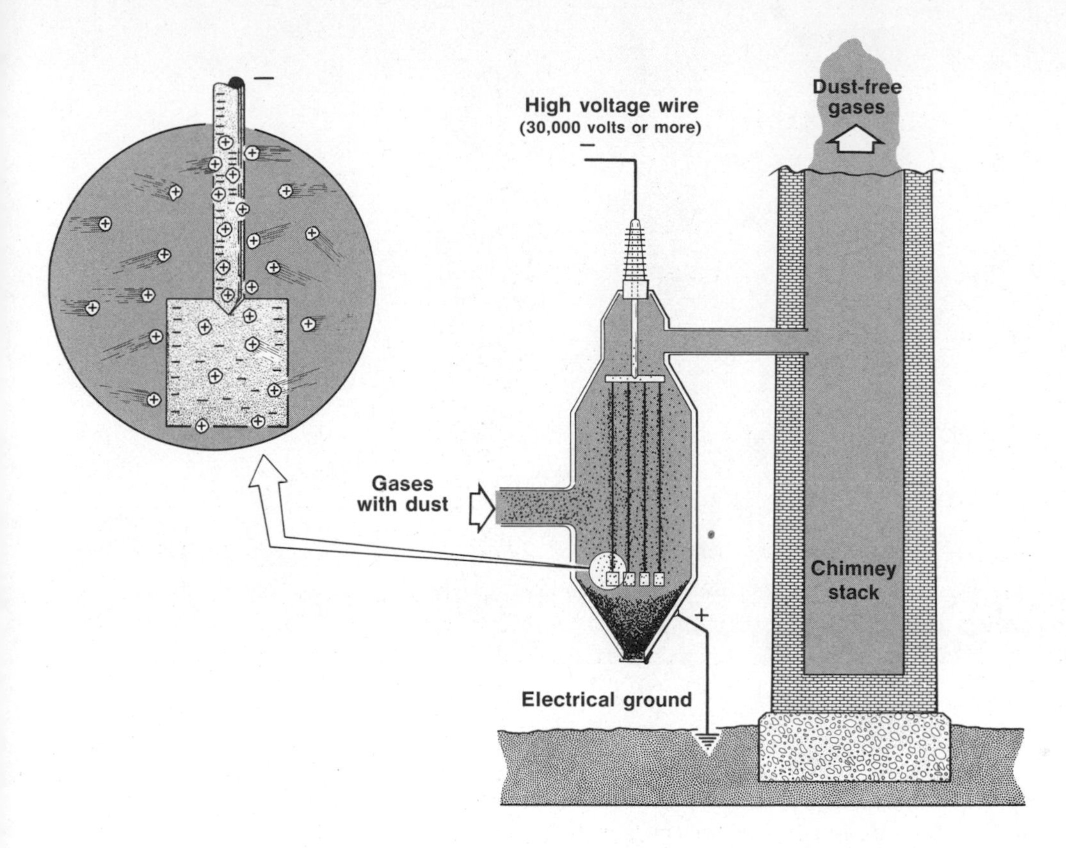


**Figure 17–6** The Cottrell electrostatic precipitator.

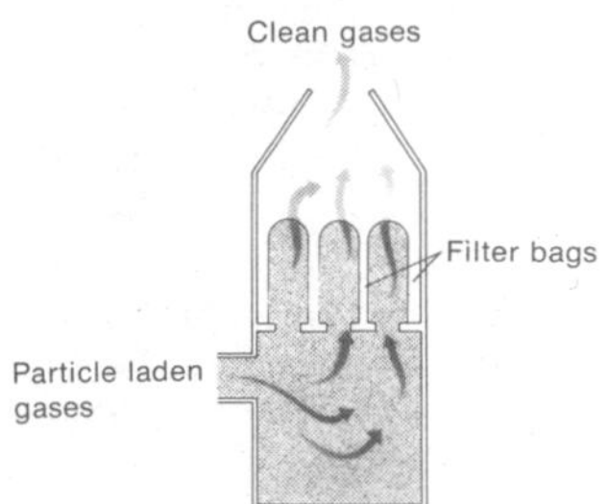


Removing particulates by filtration

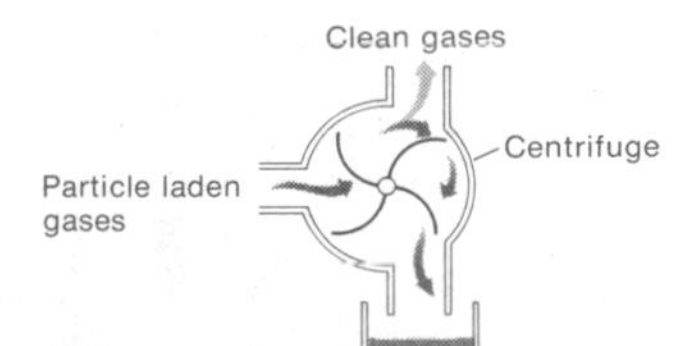
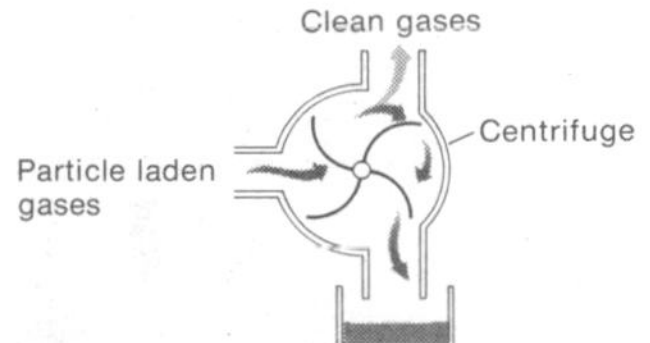


Removing particulates by centrifugal separation

Like aerosols, particulates can adsorb and concentrate air pollutants. Sulfur dioxide, nitrogen oxides, hydrocarbons, and carbon monoxide do their greatest damage when concentrated on the surface of particulates or aerosols.

Particulates in the atmosphere can cool the earth by partially shielding the earth from the sun. Large volcanic eruptions such as that from Mt. St. Helens in 1980 have a small effect in cooling the earth.

Particulates and aerosols are removed naturally from the atmosphere by gravitational settling and by rain and snow. They can be prevented from entering the atmosphere by treating industrial emissions by one or more of a variety of physical methods such as filtration, centrifugal separation, spraying, and electrostatic precipitation. A method often used is electrostatic precipitation, which is better than 98% effective in removing aerosols and dust particulates even smaller than 1 micron from exhaust gases of industrial plants. A diagram of a Cottrell electrostatic precipitator is shown in Figure 17–6. The central wire is connected to a source of direct current at high voltage (about 50,000 volts). As dust or aerosols pass through the strong electrical field, the particles attract ions that have been formed in the field, become strongly charged, and are attracted to the electrodes. The collected solid grows larger and heavier and falls to the bottom, where it is collected.

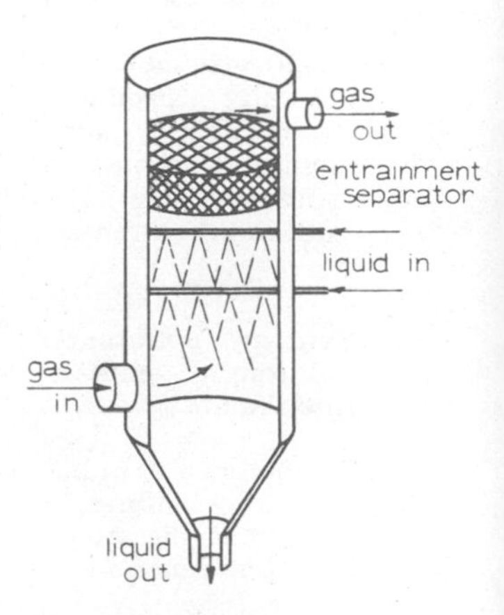


Removing particulates and aerosols by scrubbing. Schematic drawing of a spray collector, or scrubber.

## A Major Air Pollutant—Sulfur Dioxide

Sulfur dioxide is produced by burning sulfur or sulfur-containing substances in air:

$$S + O_2 \longrightarrow SO_2 \text{ (gas)}$$

Figure 17–7 Smog over New York City. A heavy haze hangs over Manhattan Island, viewed from the roof of the RCA Building. The Empire State Building is barely visible in the background.

**The nation's power plants accounted for 80% of the atmospheric $SO_2$ in 1976.**

Most of the atmospheric $SO_2$ comes from electrical power plants, smelting plants (which treat sulfide ores), and sulfuric acid plants. Almost 50% of the nation's total $SO_2$ output is isolated in the seven industrialized states of New York, Pennsylvania, Michigan, Illinois, Indiana, Ohio, and Kentucky.

$CH_3-SH$
*Methyl mercaptan*

When coal or petroleum is burned as in electrical power plants, the sources of sulfur are elemental sulfur (S), iron pyrite ($FeS_2$), and organic compounds such as mercaptans (compounds containing —SH groups). The average sulfur content of all coal mined in the United States is about 2.0%. Much of the petroleum used in the eastern United States is Caribbean residual fuel oil, which has an average sulfur content of 2.6%. If coal and petroleum containing up to 5% sulfur are the fuel for a 1,000-megawatt electrical power plant, about 600 tons of $SO_2$ are produced each day. More than 25 million tons of $SO_2$ were put into the air in 1978.

**A large modern power station (e.g., a station of 2,000 megawatts capacity) annually produces about the same amount of $SO_2$ as an industrial city of a million inhabitants.**

Once in the air, what happens to the primary pollutant, $SO_2$? It can be oxidized to a secondary pollutant, $SO_3$, in the presence of oxygen, sunlight, and water vapor:

$$2SO_2 + O_2 \longrightarrow 2SO_3$$

Either $SO_2$ or $SO_3$ can be an active participant in the formation of industrial ***smog.*** The hazy mixture of smoke, fog, air, and other chemicals was first called smog in 1911 by Dr. Harold de Voeux in his report on a London air pollution episode that caused the deaths of 1,150 people. Two general kinds of smog have been identified. Sulfur dioxide mixed with soot, fly ash, smoke, and partially oxidized organic compounds produces ***industrial smog,*** which is typical of London and other industrialized areas. A second type of smog, typical of Los Angeles, is called ***photochemical smog*** because light initiates the process. Photochemical smog is practically free of $SO_2$, and it will be described in the next section on nitrogen oxides.

**Primary pollutants: pollutants emitted into the air.**

**Secondary pollutants: pollutants formed in the air by chemical reaction.**

For either type of smog to exist, certain physical conditions must be present. There must be minimal horizontal and vertical movement of the air. Horizontal movement is restricted if the land area is bowl-shaped (surrounded by mountains or cliffs, for example). Vertical movement is diminished by a ***thermal inversion.*** Normally, warmer air is nearer the warm earth and cooler air is higher up. The warmer, less

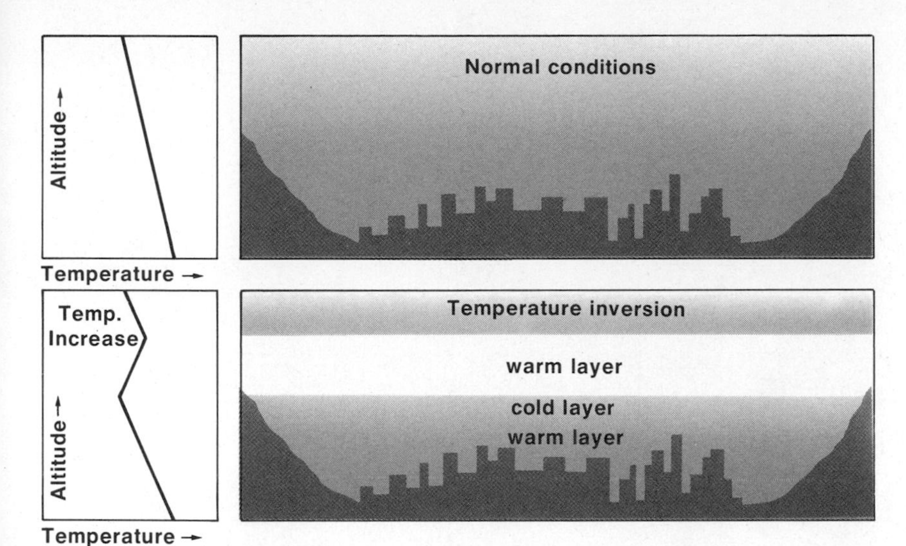


**Figure 17–8** A diagram of a temperature inversion over a city. Warm air over a polluted air mass effectively acts as a lid, holding the polluted air over the city until the atmospheric conditions change. The line on the left of the diagram indicates the relative-air temperature.

dense air rises and transports most of the pollutants to the upper air where the pollutants are dispersed. In a thermal inversion, the warmer air is above a mass of cooler, more dense air (Fig. 17–8). Vertical movement is shut off, and stagnation occurs.

Industrial smog is thought to be caused by $SO_2$. It is known through laboratory experiments that $SO_2$ increases the formation and stabilization of aerosols, particularly in the presence of hydrocarbons, nitrogen oxides, air, and sunlight. For example, a mixture of 3 ppm unsaturated hydrocarbon, 1 ppm $NO_2$, and 0.5 ppm $SO_2$ at 50% relative humidity forms aerosols. A major product is sulfuric acid, which is particularly devastating to breathing passages, vegetation, metals, and rock construction materials.

**ppm = number of molecules of pollutant per 1,000,000 molecules of air.**

**To change from percent to ppm multiply by 10,000.**

**1 ppm is the same as 1 inch in 16 miles, 1 minute in 2 years, 1 cent in $10,000.**

$$2SO_2 + O_2 \longrightarrow 2SO_3$$

$$SO_3 + H_2O \longrightarrow H_2SO_4$$

A third activity of $SO_2$ in the air is its active participation in the formation of ***acid rain,*** a much bantered-about term these days. A source of acid is the $H_2SO_4$ formed in smog. You will recall that $H_2SO_4$ is a strong acid.

**Relative humidity is a measure of the amount of water vapor air contains compared with the maximum amount it can contain.**

$$\underset{\text{STRONG ACID}}{H_2SO_4} \longrightarrow H^+ + HSO_4^-$$

Sulfurous acid, a weak acid, can be formed in the presence of $SO_2$ and $H_2O$.

$$H_2O + SO_2 \rightleftharpoons \underset{\text{WEAK ACID}}{H_2SO_3} \rightleftharpoons H^+ + HSO_3^-$$

Sulfuric, sulfurous, and other acids (such as those formed from nitrogen oxides and also ozone) can be leached from the air by rain or snow and produce precipitation with a pH below 7. Recent sampling in Florida shows an average rain-acidity range from pH 5.6 in the south to 4.6 in the panhandle, with the lowest sample at 3.7. In a recent study by Gilbert S. Raynor at Brookhaven National Laboratory, covering four years and 2,500 samples, it was concluded that thunderstorms, cold fronts, and squall lines produce higher $H^+$ concentrations (lower pH) than warm fronts. These aggravated weather systems may produce higher concentrations because they tap parts of the atmosphere that are not tapped by warm-front clouds. He also noted

**The pH range of acid rain is shown in the figure on page 398.**

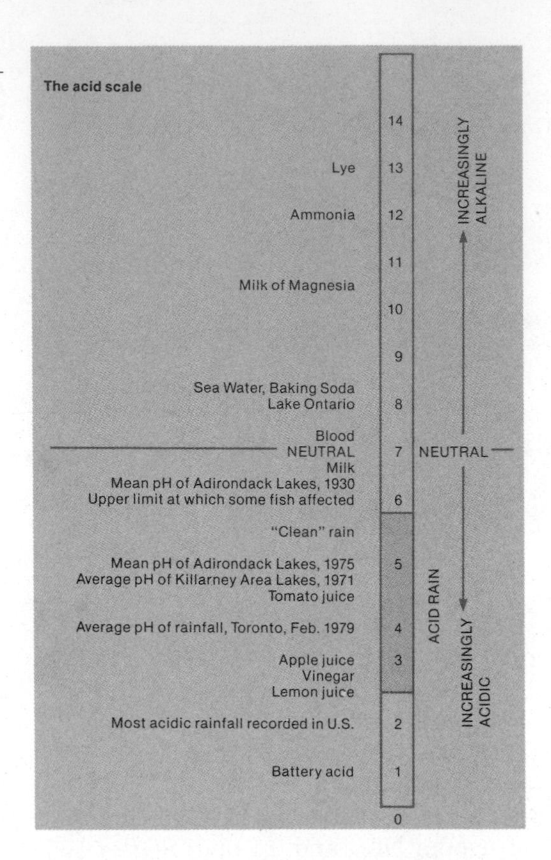


that the greatest total deposition of $H^+$ and sulfate occurs in the summer when the total rainfall is the least, perhaps because smaller droplet size, associated with slower rain, is more effective in scavenging material from the sky.

Two controversies associated with acid rain center around sample taking (whether a plastic bucket accepts from the air like a plant leaf) and completeness of extraction of $SO_2$, for example, by rain. Some contend that as much as 50% (more likely 10–30%) comes to earth in some way other than by precipitation. In any event, to Samuel Johnson's statement (July 6, 1773), "Rain is good for vegetables and for the animals who eat those vegetables and for the animals who eat those animals," we could now add, "as long as it is not acid rain."

**Sulfurous acid, $H_2SO_3$, is a weak acid; sulfuric acid, $H_2SO_4$, is a strong acid. Both can make the water in streams and lakes too acidic for fish.**

Once $SO_2$ (or as $SO_3$, $H_2SO_3$, or $H_2SO_4$) is on earth, then what?

When dissolved in rivers, lakes, and streams, the acidity can increase considerably. If the pH varies much, aquatic life suffers. Salmon, for example, cannot survive if the pH is as low as 5.5. The lower limit of tolerance for most organisms is a pH of 4.0. Several years ago, certain sections of the Netherlands had precipitation with a pH less than 4.

Sulfur dioxide and its attendant forms can corrode, damage vegetation, affect breathing, and decay building stones, in particular marble and limestone. Both stones are forms of calcium carbonate ($CaCO_3$), which reacts readily with acid ($H^+$) and with $SO_2$ and $H_2O$.

**Sulfur dioxide in the air is harmful to people, animals, plants, and buildings.**

$$CaCO_3 + 2H^+ \longrightarrow Ca^{2+} + H_2O + CO_2$$

$$CaCO_3 + SO_2 + 2H_2O \longrightarrow \underset{\text{(Soluble)}}{CaSO_3 \cdot 2H_2O} + CO_2$$

An alarming example is the disintegration of marble statues and buildings on the Acropolis in Athens, Greece. As all coatings have failed to protect the marble adequately, the only known solution is to bring the prized objects into air-conditioned museums protected from $SO_2$ and other corroding chemicals.

How can we have less $SO_2$ in the usable air and still maintain productivity? Three methods are now in use to accomplish the goal: use low-sulfur fuels, trap the $SO_2$ before it gets to the atmosphere, and (or) dilute the $SO_2$ high in the sky.

If low-sulfur fuels are used, there should be less $SO_2$ formed. This obvious conclusion was confirmed by a six-year study in Bayonne, New Jersey. The sulfur in the fuel was decreased from 1% to an average of 0.25% over the six-year span. The four-fold reduction of sulfur in the fuel correlated well with a four-fold reduction in the average $SO_2$ concentration in the air (200 micrograms per cubic meter to 50 micrograms per cubic meter).

What, then, is the problem? Why not use only low-sulfur fuels?

Most low-sulfur coals are mined far from the major metropolitan areas where they are most needed for power generation. The cleansing of sulfur from closer, high-sulfur coal is costly and incomplete. One method is to pulverize the coal to the consistency of talcum powder and remove the pyrite ($FeS_2$) by magnetic separation. Technology is available to decrease the sulfur content of fuel oil to 0.5%, but this process, too, is costly. It involves the formation of hydrogen sulfide ($H_2S$) by bubbling hydrogen through the oil in the presence of metallic catalysts, such as a platinum-palladium catalyst.

Several efficient methods are available to trap $SO_2$. In one method, limestone is heated to produce lime. The lime reacts with $SO_2$ to form calcium sulfite, a solid particulate, which can be removed from an exhaust stack by an electrostatic precipitator.

$$\underset{\text{LIMESTONE}}{CaCO_3} \xrightarrow{\text{heat}} \underset{\text{LIME}}{CaO} + CO_2$$

$$CaO + SO_2 \longrightarrow \underset{\text{CALCIUM SULFITE}}{CaSO_3\ (\text{solid})}$$

Another trapping method involves the passage of $SO_2$ through molten sodium carbonate. Solid sodium sulfite is formed.

$$SO_2 + \underset{\text{SODIUM CARBONATE}}{Na_2CO_3} \xrightarrow{800°C} \underset{\text{SODIUM SULFITE}}{Na_2SO_3} + CO_2$$

**TABLE 17–3 Physiological and Corrosive Effects of $SO_2$**

| $SO_2$ Exposure (ppm) | Duration | Effect | Comment |
|---|---|---|---|
| 0.03–0.12 | annual average | corrosion | moist temperate climate with particulate pollution |
| 0.3 | 8 hr | vegetation damage (bleached spots, suppression of growth, leaf drop, and low yield) | laboratory experiment; other environmental factors optimal. Field studies are consistent but dose is difficult to estimate |
| 0.47 | <1 hr | odor threshold (50% of subjects detect) | may be higher for many persons or when other methods are used |
| 0.2 | daily average | respiratory symptoms | community exposure exceeding 0.2 ppm more than 3% of the time |
| >0.05 | long-term average | respiratory symptoms | with particulates $> 100\ \mu g/m^3$ |
| 0.2 | daily average | respiratory symptoms | with particulates |
| 0.9 | hourly average | respiratory symptoms | with particulates |
| >0.05 | monthly average | respiratory symptoms, including impairment of lung function in children | with particulates |

**Figure 17–9** World's largest chimney (as of 1972) standing 1250 feet high (as tall as the Empire State Building) was built at a cost of $5.5 million for the Copper Cliff smelter in the Sudbury District of Ontario, Canada.

**Tall stacks contribute to the formation of acid rain.**

The less desirable method of dissipating $SO_2$ is by tall stacks (Fig. 17–9). Although tall stacks emit $SO_2$ into the upper atmosphere away from the immediate vicinity and give $SO_2$ a chance to dilute itself on the way down, the fact remains that $SO_2$ will come down, and the longer it stays up the greater chance it has to become sulfuric acid. Nevertheless, a 10-year study in Great Britain showed that although $SO_2$ emissions from power plants increased by 35%, the construction of tall stacks decreased the ground level concentrations of $SO_2$ by as much as 30%. The question is, who got the $SO_2$? In this case, Britain's solution was others' pollution.

## Major Air Pollutants—Nitrogen Oxides

There are eight known oxides of nitrogen, two of which are recognized as important components of the atmosphere: dinitrogen oxide ($N_2O$) and nitrogen dioxide ($NO_2$).

Most of the nitrogen oxides emitted are in the form of NO, a colorless reactive gas. In a combustion process involving air, some of the atmospheric nitrogen reacts with oxygen to produce NO:

$$N_2 + O_2 + \text{heat} \longrightarrow 2NO$$

**About 97% of the nitrogen oxides in the atmosphere are naturally produced and only 3% result from human activity.**

Nitrogen oxide is formed in this manner during electrical storms. Since the formation of nitrogen oxide requires heat, it follows that a higher combustion temperature would produce relatively more NO. This will be an important point to consider later in our discussion of the automobile and its nitrogen oxide emissions, since one way to achieve greater burning efficiency of fuels in automobile engines is to operate them at higher temperatures.

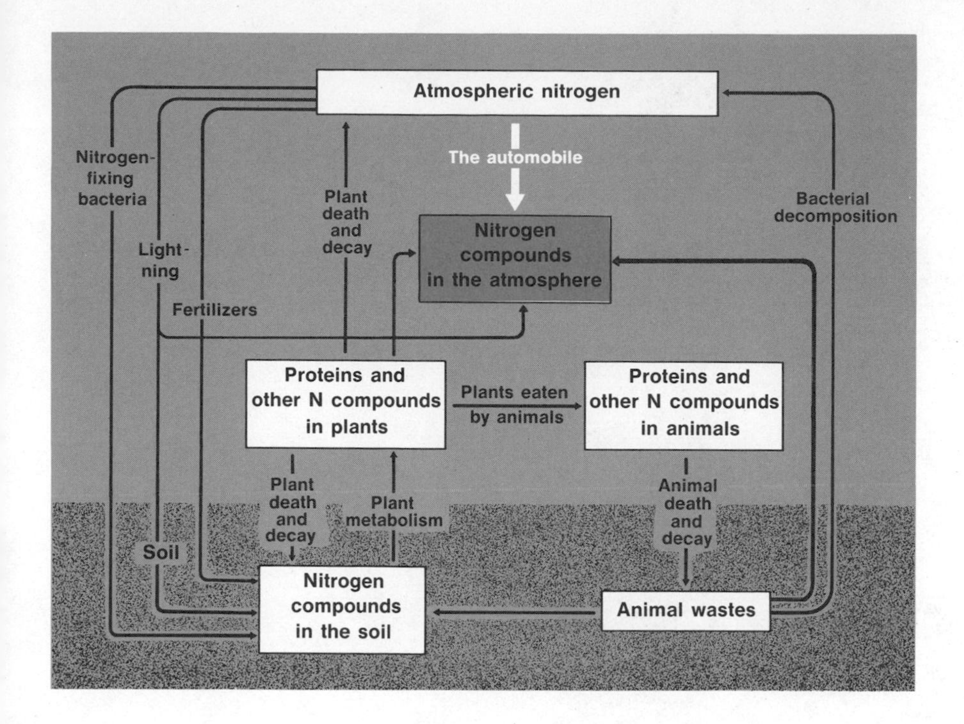


**Figure 17-10** The nitrogen cycle.

In the atmosphere NO reacts rapidly with atmospheric oxygen to produce $NO_2$:

$$2NO + O_2 \longrightarrow \underset{\text{NITROGEN DIOXIDE}}{2NO_2}$$

**Fixed nitrogen is nitrogen chemically bonded to another element.**

Normally the atmospheric concentration of $NO_2$ is a few parts per billion (ppb) or less.

Fixed nitrogen (nitrogen oxides are one type) is necessary to perpetuate nature's cycle (Fig. 17–10). In this respect, nitrogen oxides are useful. However, too large a quantity of nitrogen oxides in the air can lead to photochemical smog and bronchial problems. In these respects and many others, nitrogen oxides are harmful.

***Photochemical smog*** is thought to be initiated by the absorption of a quantum of ultraviolet light by a molecule of brown, choking nitrogen dioxide, $NO_2$ (Fig. 17–11).

$$\begin{aligned}
NO_2 + \text{light} &\longrightarrow NO + O\cdot \\
O\cdot + O_2 + M &\longrightarrow O_3 + M \\
O_3 + NO &\longrightarrow NO_2 + O_2 \\
O\cdot + Hc &\longrightarrow HcO\cdot \\
HcO\cdot + O_2 &\longrightarrow HcO_3\cdot \\
HcO_3\cdot + Hc &\longrightarrow RCHO \text{ or } R\overset{\overset{O}{\|}}{C}R \\
HcO_3\cdot + NO &\longrightarrow HcO_2\cdot + NO_2 \\
HcO_3\cdot + O_2 &\longrightarrow O_3 + HcO_2\cdot \\
HcO_3\cdot + NO_2 &\longrightarrow \underset{\text{(PAN)}}{R{-}\overset{\overset{O}{\|}}{C}{-}O{-}O{-}N{\genfrac{}{}{0pt}{}{\nearrow O}{\searrow O}}} + \text{other products}
\end{aligned}$$

**Figure 17-11** Simplified reaction scheme for photochemical smog. Ultraviolet light initiates the process to produce oxygen atoms. Hc is hydrocarbon (unsaturated or aromatic); M is a third body to absorb the energy released from forming the ozone. Among many possibilities, M could be a $N_2$ molecule, an $O_2$ molecule, or a solid particle. A species with a dot, as $HcO\cdot$, is a chemical radical. R is a saturated hydrocarbon group; PAN is peroxy acyl nitrate, a very reactive secondary pollutant that causes eyes to water and lungs to hurt.

**Figure 17–12**
Photochemical smog (a brown haze) enveloping the city of Los Angeles.

The energy absorbed causes a nitrogen-oxygen bond to break. The products are two very reactive species, nitrogen oxide and atomic oxygen, having one and two unpaired valence electrons respectively.

UNPAIRED VALENCE ELECTRONS

$$:\dot{N}::\ddot{\underset{\cdot\cdot}{O}} \qquad \cdot\ddot{\underset{\cdot}{O}}:$$

In chainlike fashion, these reactive species react with hydrocarbons and oxygen to form a variety of other reactive species with unpaired electrons. In turn, the most commonly encountered secondary pollutants in photochemical smog are formed. The hydrocarbons come mostly from unburned petroleum products such as gasoline.

This scheme accounting for all of the major secondary pollutants in photochemical smog was uncovered by brilliant detective work in 1951 when the smog was reproduced in the laboratory. It was found that only 0.2 ppm of nitrogen oxides and 1 ppm of reactive hydrocarbons are sufficient to initiate these reactions. Sulfur dioxide may be present in photochemical smog, but $SO_2$ is not a necessary ingredient.

Measurements made in several U.S. cities reveal that the amount of airborne aerosols increased ten-fold during the 1960's. Despite the stricter controls effected during the last decade, Los Angeles did not show a downward trend in the occurrence and amount of smog, which is present about 300 days a year. A comparison of the levels of pollutants on a clear day and on a smoggy day is presented in Table 17–4.

**TABLE 17–4 Approximate Concentrations of Pollutants in Los Angeles on a Clear Day (Visibility 7 Miles) and on a Smoggy Day (Visibility 1 Mile)**

| | Concentration (ppm) | | |
|---|---|---|---|
| **Pollutant** | *Clear Day* | *Smoggy Day* | *Increase* |
| Carbon monoxide | 3.5 | 23.0 | ×6.5 |
| Hydrocarbons | 0.2 | 1.1 | ×5.5 |
| Peroxides | 0.1 | 0.5 | ×5.0 |
| Oxides of nitrogen | 0.08 | 0.4 | ×5.0 |
| Lower aldehydes | 0.07 | 0.4 | ×6.0 |
| Ozone | 0.06 | 0.3 | ×5.0 |
| Sulfur dioxide | 0.05 | 0.3 | ×6.0 |

If $NO_2$ does not react photochemically, it can react with water vapor in the air to form nitric and nitrous acids:

**Nitrogen oxides can make "acid rain."**

$$2NO_2 + H_2O \longrightarrow \underset{\text{NITRIC ACID}}{HNO_3} + \underset{\text{NITROUS ACID}}{HNO_2}$$

In addition, nitrogen dioxide and oxygen yield nitric acid:

$$4NO_2 + 2H_2O + O_2 \longrightarrow 4HNO_3$$

These acids in turn can react with ammonia or metallic particles in the atmosphere to produce nitrate or nitrite salts. For example,

**Nitrates are important components of fertilizers.**

$$\underset{\text{AMMONIA}}{NH_3} + HNO_3 \longrightarrow \underset{\text{AMMONIUM NITRATE (A SALT)}}{NH_4NO_3}$$

The acids or the salts, or both, ultimately form aerosols, which eventually settle from the air or dissolve in raindrops. Nitrogen dioxide, then, is a primary cause of haze in urban or industrial atmospheres because of its participation in the process of aerosol formation. Normally nitrogen dioxide has a lifetime of about three days in the atmosphere.

At present, the emission of nitrogen oxides by human activities and our machines is only a minor part of that emitted by natural processes (Fig. 17–10).

In laboratory studies, nitrogen dioxide in concentrations of 25–250 ppm inhibits plant growth and causes defoliation. The growth of tomato and bean seedlings is inhibited by 0.3–0.5 ppm $NO_2$ applied continuously for 10 to 20 days.

**In 1979, haze near Abbeville, Louisiana, had two times more ozone and particulates and four times more $(NH_4)_2SO_4$ than normal.**

In a concentration of 3 ppm for 1 hour, nitrogen dioxide causes bronchioconstriction in humans, and short exposures at high levels (150–220 ppm) produce changes in the lungs that produce fatal results. A seemingly harmless exposure one day can cause death a few days later.

## Self-Test 17-A

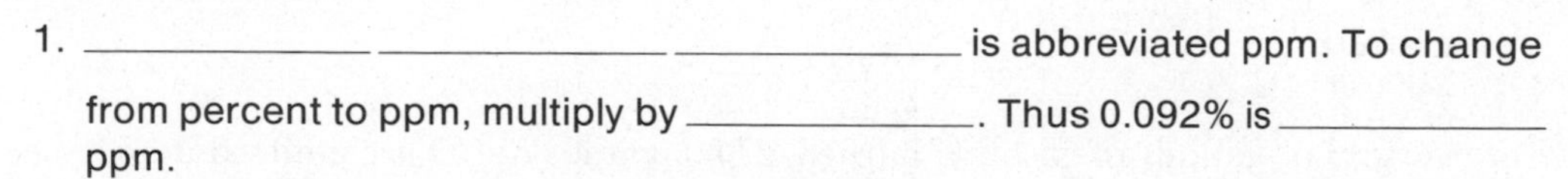
1. ______ ______ ______ is abbreviated ppm. To change from percent to ppm, multiply by ______. Thus 0.092% is ______ ppm.

2. Three substances normally considered primary pollutants are ______________, ______________, and ______________.
3. Two secondary pollutants in photochemical smog are ______________ and ______________. Two secondary pollutants in industrial smog are ______________ and ______________.
4. Particles that can remain suspended in air for long periods of time and are intermediate in size between individual molecules and particulates are called ______________.
5. Much of the effect of aerosols is due to their large ______________.
6. A device that imparts charges to dust and aerosol particles so they can be attracted out of a gaseous effluent stream is known as a(n) ______________.
7. The initial step in the formation of photochemical smog occurs when ultraviolet light decomposes a molecule of ______________ ______________.
8. During a thermal inversion, a (warm/cool) mass of air is above a mass of (warm/cool) air.
9. For initiation, industrial smog requires the substance ______________, and photochemical smog requires the substance ______________.
10. When sulfur is burned in air, the major product is ______________.
11. The major source of sulfur dioxide is the burning of ______________.
12. How does the production of electrical power rank as a producer of sulfur dioxide? ______________
13. Fixed nitrogen is nitrogen combined with ______________.
14. The reaction of nitrogen with oxygen is (endothermic/exothermic).
15. In all combustion processes in air, some nitrogen ______________ are formed.
16. The artificial emission of nitrogen oxides has greatly disrupted the nitrogen cycle. (True/False)
17. Nitrogen oxides remain in the atmosphere for an indefinite length of time. (True/False)
18. Acid rain is formed by dissolving ______________ and(or) ______________ in rain water.

---

## A Major Air Pollutant—Carbon Monoxide

**The toxic effects of carbon monoxide are discussed in Chapter 15.**

Carbon monoxide (CO) is the most abundant and widely distributed air pollutant found in the atmosphere. It is produced in combustion processes when carbon or some carbon-containing compound is burned in an insufficient amount of oxygen:

$$2C + \underset{\text{LIMITED SUPPLY}}{O_2} \longrightarrow 2CO$$

For every 1,000 gallons of gasoline burned, 2,300 pounds of CO are emitted if no emission controls are placed on the engine.

The highest concentration of CO is around heavy traffic areas where levels of 50 ppm or more are encountered, although 7 ppm or less is normal. In the countryside, the CO level is close to the global level of about 0.1 ppm. A bit of a mystery is that the low global level does not seem to be changing in spite of huge amounts of CO being dumped into the atmosphere (about $10^{14}$ grams per year in the United States, or about 1,000 pounds per person per year). Although polar CO will dissolve readily in water droplets, and CO will react slowly with oxygen to form $CO_2$, the bulk of its disappearing act is probably accounted for in the natural carbon cycle.

**At least ten times more CO enters the atmosphere from natural sources than from all industrial and automotive sources combined. Of the about 3.8 billion tons of CO emitted per year, about 0.3 billion come from human sources and about 3 billion from the oxidation of methane produced by decaying organic matter.**

## Major Air Pollutants—Certain Hydrocarbons

As we have seen in Chapter 8, hydrocarbons come in all shapes and sizes, beginning with methane, $CH_4$, and continuing to molecules containing hundreds of carbon atoms. Some have all single bonds, some have double bonds, and a few have triple bonds. Some are aromatic hydrocarbons and some are nonaromatic. Literally hundreds of these hydrocarbons and their oxygen, sulfur, nitrogen, and halogen derivatives find their way into the atmosphere.

Trees and other plants silently release turpentine, pine oil, and thousands of other hydrocarbons into the air. Bacterial decomposition of organic matter emits very large amounts of marsh gas, principally methane. We contribute our share of 15% (of the total global emissions; a greater quantity in urban areas) through incomplete incineration, leakage of industrial solvents, unburned fuel from the automobile, evaporation of gasoline from tanks, incomplete combustion of coal and wood, and petroleum processing, transfer, and use. An estimated 163,000 pounds of polynuclear aromatic hydrocarbons pass from the air into Lake Superior each year.

**A commercial synthetic fuel plant, if not controlled, could emit as much as 10,000 kg of polynuclear aromatic hydrocarbons per day.**

In Chapter 15 we saw that polynuclear aromatic hydrocarbons such as benzo($\alpha$)pyrene are capable of causing cancer in mice and in humans. In the late 1950's, the U.S. Public Health Service, Division of Air Pollution, surveyed 103 urban and 28 nonurban areas of the United States and found that the air in all of the 103 urban areas contained benzo($\alpha$)pyrene (BaP). Concentrations ranged from 0.11 to 61 micrograms per 1,000 cubic meters of air, with the average concentration being 6.6 micrograms. In 1967, the estimated annual emission of BaP in the United States was 422 tons from burning coal, oil, and gas, 20 tons from refuse burning, 19 tons from industries (petroleum catalytic cracking, asphalt road mix, and the like), and 21 tons from motor vehicles. British researchers report that lung cancer in nonsmokers closely parallels the 10-times-greater amount of BaP in city air than in rural air; there is 9 times more lung cancer in cities than in rural areas. A resident of a large town may inhale 0.20 g BaP a year. If he or she is a heavy smoker (two packs a day without filters), add another 0.15 g for a total of 0.35 g. This is about 40,000 times the amount of BaP necessary to produce cancer in a mouse. Coal smoke contains about 300 ppm BaP. Every million tons of coal burned in England in 1958 produced smoke laden with 750 tons of BaP. Many authorities attribute England's high lung cancer rate today to this enormous production of BaP.

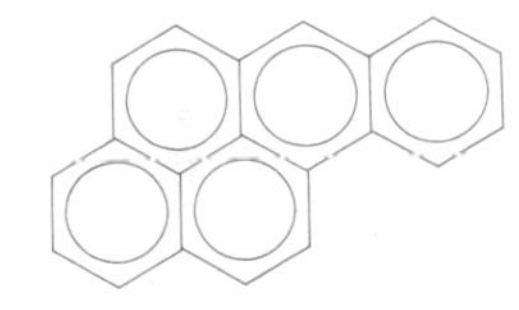

**Benzo($\alpha$)pyrene, a carcinogenic polynuclear aromatic hydrocarbon found in smoke.**

**In situ air pollution: cigarette smoking.**

Other polynuclear aromatics have also shown carcinogenic activity. Particulates from the atmosphere around Los Angeles, London, Newcastle, Liverpool, and eight other urban sites were extracted with organic solvents. The extracts produced cancer in mice.

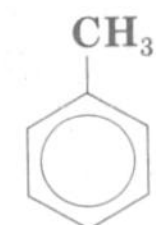


*Toluene*

In the section on smog we discussed the role of unsaturated and aromatic hydrocarbons in photochemical smog formation. In a study made in Los Angeles in 1970, an average of 0.106 ppm (maximum of 0.33 ppm) aromatics was found in that city's atmosphere. (About 38% of the total was toluene and 40% the more reactive dialkyl-

**Alkyl: hydrocarbon group such as ethyl, $-C_2H_5$, and octyl, $-C_8H_{17}$.**

and trialkylbenzenes.) These compounds are about as reactive as propylene and higher molecular weight unsaturated hydrocarbons in causing smog formation. The automobile is responsible for emitting most of these hydrocarbons to the atmosphere. In fact, the automobile without pollution controls emits more than 200 different hydrocarbons and hydrocarbon derivatives.

## Ozone—A Secondary Pollutant and a Sunscreen

Ozone is a pungent-smelling gas that can be detected by the human nose at concentrations as low as 0.02 ppm. Sparking electrical appliances, lightning, and even silent electrical discharges convert oxygen into ozone, which is a more reactive form of oxygen.

$$\text{Energy} + 3O_2 \longrightarrow \underset{\text{OZONE}}{2O_3}$$

Pure oxygen can be breathed for weeks by humans and animals without apparent injurious effects. Several studies have shown that concentrations of 0.3–1.0 ppm ozone, well within the recorded range of photochemical oxidant levels, after 15 minutes to 2 hours cause marked respiratory irritation accompanied by choking, coughing, and severe fatigue. For these reasons, outdoor recreation classes in Los Angeles public schools are cancelled on days when the ozone level reaches 0.35 ppm. Ozone at these levels for 1 hour depresses the body temperature, perhaps by an impairment of the brain center that regulates body temperature or by opening the pores of the skin. These levels (0.2–0.5 ppm) cause a considerable decrease in night vision in addition to other effects on vision.

Ozone attacks mercury and silver, which are not affected by molecular oxygen at room temperature. A typical reaction might be

$$6Ag + O_3 \longrightarrow 3Ag_2O$$

but

$$Ag + O_2 \longrightarrow \textit{No reaction}$$

Even with all the electrical sparks from lightning, electrical motors and such, very little ozone is emitted into the air. It decomposes into molecular oxygen or reacts with other molecules too quickly to leave its source.

**Atmosphere: troposphere—sea level to about 7 miles up ($N_2$, $O_2$, $H_2O$, $CO_2$); stratosphere—7 to about 50 miles up ($N_2$, $O_2$, ozone).**

Ozone is found in the lower troposphere only as a secondary pollutant; that is, it is formed from other substances, as in photochemical smog (Fig. 17–11). When sunlight impinges on automobile exhaust fumes, a considerable amount of ozone is produced. The stratosphere contains about 10 ppm of ozone in a layer that has the important function of filtering out some of the sun's ultraviolet light and providing an effective shield against radiation damage to living things.

### Halogenated Hydrocarbons and the Ozone Layer

Most pollutants are adsorbed on surfaces or react with other pollutants in the troposphere, and eventually wash out in the rain. One group of unreactive pollutants, the halogenated hydrocarbons, does not react quickly enough to be consumed in the troposphere, so these eventually mix with air in the stratosphere. The common halogenated hydrocarbon pollutants are listed in Table 17–5.

**TABLE 17–5 The Major Halogenated Hydrocarbons**

| Name | Formula | Uses | World Production (1973, Millions of kg) | Emission into Atmosphere |
|---|---|---|---|---|
| Fluorocarbon-11 | $CCl_3F$ | aerosol propellant | 302 | 89% |
| Fluorocarbon-12 | $CCl_2F_2$ | propellant, refrigerant | 441 | 76% |
| Carbon tetrachloride | $CCl_4$ | making fluorocarbons | 941 | 4% |
| Methyl chloroform | $CH_3CCl_3$ | metal cleaning | 405 | 98% |
| Perchloroethylene | $C_2Cl_4$ | dry cleaning | 743 | 83% |

**The United States accounts for about half the emissions of each of these chemicals. Released, they become atmospheric pollutants.**

In the stratosphere, where the ultraviolet radiation from the sun is most intense, bonds of the halogenated hydrocarbons are broken and species with unpaired electrons are formed. The most common reactive species produced is the chlorine atom (Cl·), as the breakdown of fluorocarbon-11 illustrates.

$$F-\underset{Cl}{\overset{Cl}{\underset{|}{\overset{|}{C}}}}-Cl + \text{light} \longrightarrow F-\underset{Cl}{\overset{Cl}{\underset{|}{\overset{|}{C}}}}\cdot + Cl\cdot$$

The reactive chlorine atom then combines with ozone ($O_3$), which is present in high concentration in the stratosphere.

**Why is Cl· reactive?**

$$O_3 + Cl\cdot \longrightarrow ClO\cdot + O_2$$

Since many oxygen atoms are available in the upper atmosphere as participants in the production of ozone,

$$O_2 + \text{light} \longrightarrow \dot{O}\cdot + \dot{O}\cdot$$

$$\dot{O}\cdot + O_2 \longrightarrow O_3$$

the ClO· species can react with an oxygen atom to release the Cl· radical

$$ClO\cdot + \dot{O}\cdot \longrightarrow O_2 + Cl\cdot$$

to react with another ozone molecule.

The net effect of these reactions is the destruction of ozone. The total number of ozone molecules destroyed by a single chlorine atom can run into the thousands. Eventually the chlorine atom reacts with a water molecule to form HCl, which mixes into the troposphere and washes out in rain.

In 1974, M. J. Molina and F. S. Roland of the University of California predicted that the increasing use of halogenated hydrocarbons could seriously deplete the stratospheric ozone, with a corresponding increase in ultraviolet rays reaching the earth's surface. Since living things are sensitive to ultraviolet rays (for instance, they increase skin cancer in humans), the depletion theory was judged to suggest a serious threat to our safety and health. Later research has shown that this theory is probably correct in its predictions if halogenated hydrocarbon emissions go unchecked. The EPA has now limited the emissions of these compounds.

## Carbon Dioxide—An Air Pollutant . . . Or Is It?

How can carbon dioxide be considered a pollutant when it is a natural product of respiration and a required reactant for photosynthesis? Carbon dioxide is not a pollutant per se; its increasing amount is. Between 1900 and 1970, the global concentration of $CO_2$ increased from 296 ppm to 318 ppm, an increase of 7.4%. Look at Figure 13–12 on p. 300 for the pretty curve of an ugly trend in industrial emissions of $CO_2$. It is the only atmospheric substance whose global concentration is known to be rising.

**$CO_2$ is a necessary ingredient for photosynthesis.**

The increased amount of $CO_2$ comes primarily from electrical power plants, internal combustion engines, and the manufacture of cement. But there are numerous sources: home heating, trash burning, forest fires, and bacterial oxidation of soil humus, to mention a few. Carbon dioxide has been entering the atmosphere faster than oceans and growing plants can remove it.

**Carbon dioxide in the atmosphere produces a greenhouse effect.**

A substantial increase of $CO_2$ in the air can cause two detrimental effects: $CO_2$ can increase the temperature of the atmosphere, and it can increase the acidity of the oceans.

Carbon dioxide increases the temperature of the atmosphere by the ***greenhouse effect*** (Fig. 17–13). Carbon dioxide, like the glass or plastic of a greenhouse, lets the shorter wavelengths of light through but absorbs the longer wavelengths as they are emitted from the surface of the earth. The carbon dioxide (and the glass), in turn, emits the absorbed energy as heat radiation; some returns to earth and some energizes other molecules. The net effect is an increase in the temperature of the lower atmosphere.

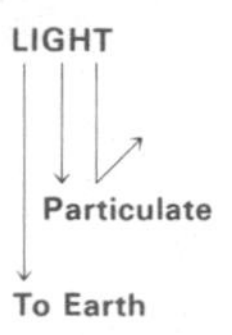


Particulates and aerosols counteract the warming effect of $CO_2$. They decrease the amount of solar energy reaching the earth by scattering incoming sunlight of all wavelengths. The net effect is a cooling of the lower atmosphere. Calculations indicate that a 25% increase in aerosols would counteract a 100% increase in $CO_2$.

**Another chance for "acid rain."**

Carbon dioxide can make surface waters acidic by reacting with water to form carbonic acid, $H_2CO_3$, a weak acid.

$$CO_2 + H_2O \rightleftharpoons H_2CO_3 \rightleftharpoons H^+ + HCO_3^-$$

Only a small part of the $CO_2$ in an ocean affects its acidity. A major consumer of $CO_2$ (through photosynthesis) is the huge amount of phytoplankton (small plants) in the oceans.

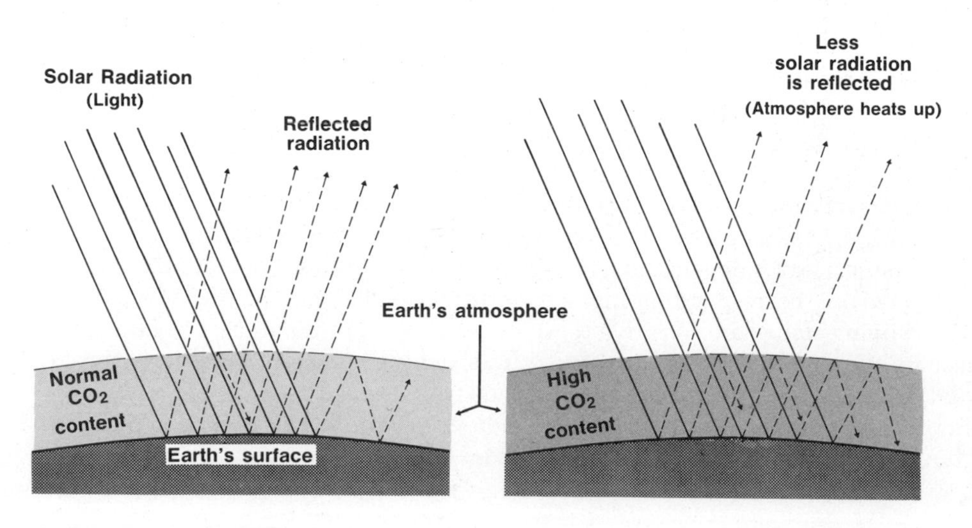


**Figure 17–13** The greenhouse effect. Owing to a balance of incoming and outgoing energy in the earth's atmosphere, the mean temperature of the earth is 14.4° C (58° F). Carbon dioxide permits the passage of visible radiation from the sun to the earth but traps some of the heat radiation attempting to leave the earth.

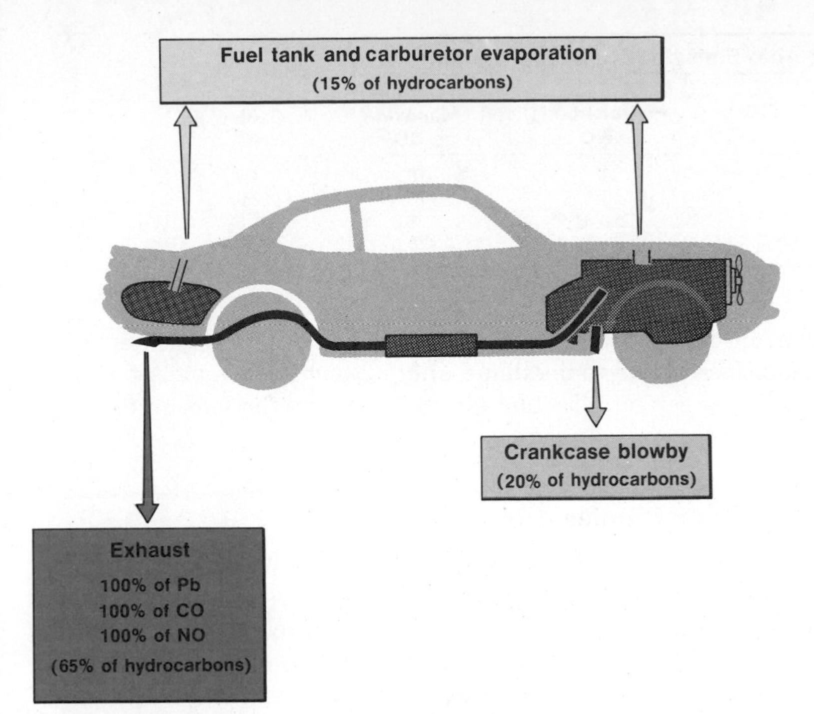


**Figure 17–14** Sources of automobile pollutant emissions. The percentages show the relative amounts of the various types of pollutants from the three major automobile sources for automobiles that have no pollution controls.

## The Automobile—A Special Case of Air Pollution

The automobile is a special case of air pollution simply because there are so many of them—more than 100 million now travel the U.S. roadways. Collectively, automobiles are *the* major source of air pollution (see Table 17–1).

**Automobiles are the major emitters of air pollutants in the United States.**

Air pollutants may enter the atmosphere from three major locations in the automobile (Fig. 17–14): from the exhaust, from the crankcase blowby (gases that escape around the piston rings), and by evaporation from the fuel tank and the carburetor.

**The composition and properties of gasoline are discussed in Chapter 9.**

The mixture of pollutants emitted by a given automobile depends on the composition of the gasoline and the effectiveness (or lack) of pollution controls. Generally when gasoline is burned, the products are CO, $CO_2$, $H_2O$, a variety of nitrogen oxides (principally NO and a little $NO_2$), newly formed hydrocarbons, and some unburned, original hydrocarbons (Fig. 17–15). If tetraethyllead is present in the gaso-

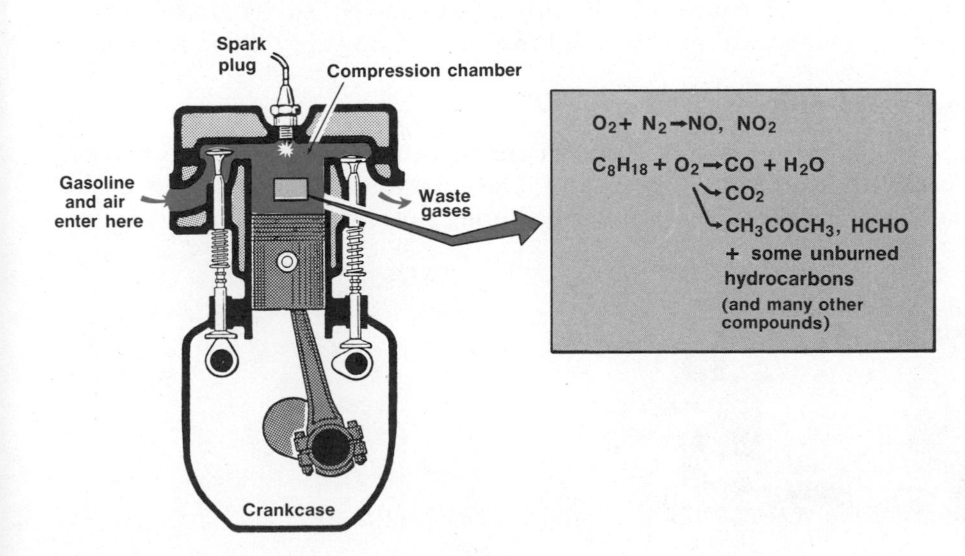


**Figure 17–15** Diagram of combustion chamber and some of the reactions that occur in it.

**TABLE 17–6 Levels of Auto Emissions in Grams per Mile**

| | Hydrocarbons, HC | Carbon Monoxide, CO | Nitrogen Oxides, $NO_x$ |
|---|---|---|---|
| Prior to control | 11 | 80 | 4.0 |
| EPA standards for 1977 model cars | 1.5 | 15.0 | 2.0 |
| California standards, 1978 | 0.41 | 9.0 | 1.5 |
| EPA standards for 1982 model cars | 0.41 | 7.0 | 1.0 |

line, additional products are $PbBrCl$, $PbBr_2$, and $PbCl_2$, which are formed when lead reacts with gasoline additives 1,2-dibromoethane and 1,2-dichloroethane. If pollution controls are absent or imperfect, gasoline evaporates from the carburetor and the gas tank.

## Ways to Control Pollutant Emissions

As an impetus, both state and federal governments have passed legislation that regulates automobile emissions and establishes standards. Governmental standards for hydrocarbons, CO, and nitrogen oxides are summarized in Table 17–6. Standards are being met by mechanical changes and changes in fuel. Fuel modifications (discussed in Chapter 9) are more effective because they apply to all cars, old or new.

Mechanical changes in cars for the purpose of decreasing air pollutant emissions include the following devices and adjustments.

### Positive Crankcase Ventilation Valve (PCV)

On all cars since 1963, fresh air (drawn by a vacuum system) sweeps the crankcase blowby gases through the PCV valve and connecting hose into the reaction cylinder of the engine for a second chance at combustion.

### Control Valves on Gas Tank Caps

Air can flow into the gas tank, but gasoline vapors cannot flow out into the air.

### Catalytic Converters (Fig. 17–16)

On most car models produced beginning in 1976, a platinum-based catalyst in the exhaust system (Fig. 17–17) converts hydrocarbons, CO, and NO to $CO_2$, $H_2O$, and $N_2$. Since the catalyst is inactivated by lead, lead-free gasoline must be used in these cars.

### Afterburner Technique

Air is injected into the hot gases as they exit through the exhaust valves. A portion of the gas completes the combustion process. The energy emitted contributes to the problem of dissipation of energy without contributing any power to the engine.

**Figure 17–16** Cutaway view of catalytic exhaust muffler showing catalyst pellets.

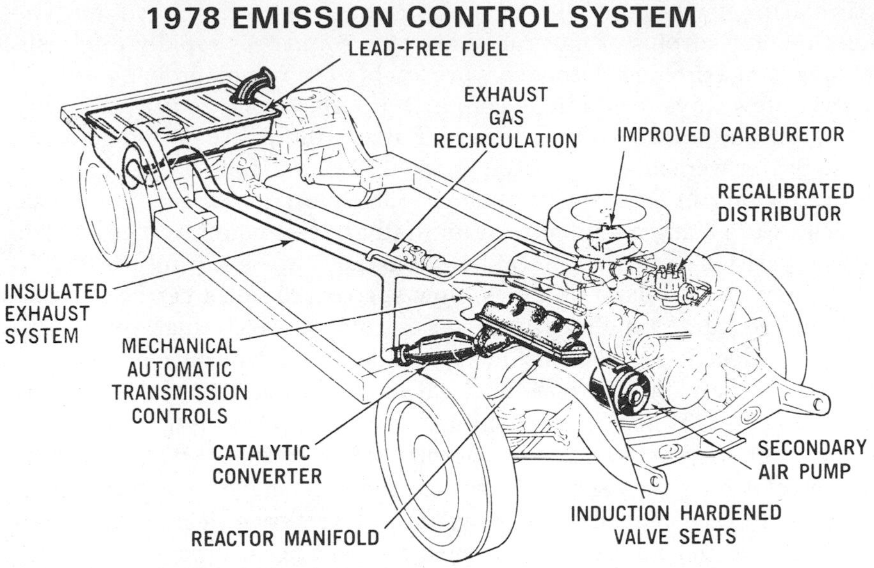

1978 EMISSION CONTROL SYSTEM
LEAD-FREE FUEL
EXHAUST GAS RECIRCULATION
IMPROVED CARBURETOR
RECALIBRATED DISTRIBUTOR
INSULATED EXHAUST SYSTEM
MECHANICAL AUTOMATIC TRANSMISSION CONTROLS
CATALYTIC CONVERTER
SECONDARY AIR PUMP
INDUCTION HARDENED VALVE SEATS
REACTOR MANIFOLD

**Figure 17–17** Automotive hardware for pollution control.

## Adjustment of Air-Fuel Ratio

This adjustment is an interesting trade-off type of chemical problem. For example, the *complete* combustion of one mole of octane or an isomer of octane requires 12.5 moles of oxygen.

**Isomers of octane are the major components of gasoline.**

$C_8H_{18} + 12.5O_2 \longrightarrow 8CO_2 + 9H_2O + \text{heat}$

Experimentally the maximum power is obtained when the oxygen-fuel molar ratio is 12.5:1, but this ratio does not give the lowest emissions of CO and hydrocarbons. In the split instant of combustion, not all of the fuel is burned and only part of the carbon goes all the way to $CO_2$; some stops at CO. A higher ratio of oxygen to fuel (15.1:1) produces about eight times less CO (than a ratio of 12.5:1) and slightly fewer hydrocarbons. Therefore, the 15.1:1 oxygen-fuel ratio gives a more efficient use of the fuel—more complete burning. The extra heat from the more complete burning when put with the extra oxygen (and *nitrogen*) produces more nitrogen oxides. (Reactions of nitrogen and oxygen to form nitrogen oxides require heat; they are endothermic.) It's a case of trying to have your cake and eat it, too. To get less CO and hydrocarbons, you get more nitrogen oxides—and less power. The reduced power comes from energy being used to form the nitrogen oxides and to heat the extra air. The problem of nitrogen oxides is partially offset by recycling part of the exhaust. The combustion temperature is reduced and so is the amount of nitrogen oxides that form. Recirculation of about 15% of the exhaust reduces the nitrogen oxides by 80%, accompanied by a 16% cut in power output and a 15% decrease in fuel economy. Changes in timing can increase power and economy, but the amount of nitrogen oxides increases. Fuel injected specifically at the spark plug gap produces low hydrocarbon, CO, and nitrogen oxide emissions, but then particulate emissions become a problem.

**Endothermic:**
$N_2 + O_2 + 4.32\ \text{kcal} \longrightarrow 2NO$

**Exothermic:**
$C + O_2 \longrightarrow CO_2 + 94.1\ \text{kcal}$

**The formation of nitrogen oxides deducts heat energy that could be used for expansion. This reduces power.**

### New Kinds of Cars?

Several thousand automobiles in this country are now burning natural gas instead of gasoline. Natural gas, which is mostly methane, $CH_4$, burns cleanly to $CO_2$

and water. A regulating device positioned over the carburetor costs about $350. The problem here is that our surplus of natural gas in this country is rapidly diminishing, a potential long-range problem. Other cars are adapted to run on propane.

One of the very few ways to eliminate combustion processes is to use electricity in some way to power our transportation devices. Early attempts to develop electric cars in the United States were made in Boston in 1888, 28 years after the development of the first storage battery in 1860. By 1912, about 6,000 electric passenger and 4000 commercial vehicles were being manufactured annually in this country. The electric car lost out in competition with the internal combustion engine during the 1920's. The problems of short range (about 20 miles), low speeds (20 miles per hour maximum), 8 to 12 hours recharge time for the batteries, and relatively high price combined to eliminate the electric car from the race for leader in transportation.

**At the turn of the century, a Baker Electric Coupe cost $2,600 and a Borland Electric Delux could be bought for $5,500.**

At present, more than 100,000 rider-type, battery-powered, materials-handling vehicles are operating in U.S. plants and warehouses where it is vital to avoid air pollution from internal combustion engines. Despite this positive start, attempts to employ electric vehicles for street and highway use generally have been commercial failures. The energy storage capacities of conventional batteries (lead-acid, nickel-iron, and silver-zinc) are too limited or expensive to provide an acceptable energy source for electric passenger cars (Fig. 17–18). New battery designs may prove competitive for short-range urban travel. Nickel-zinc, nickel-iron, zinc-fluoride, and aluminum-air batteries are all being studied seriously as possible successors to the lead-acid batteries now in use.

**In 1981 at Chattanooga, Tennessee, TVA completed a million-dollar test facility for electric cars. Spokesmen for TVA expect mass production of electric cars to begin in 1985.**

Other developments currently either coming on the market or being seriously considered include the stratified charge engine and the turbine-generator engine. These relatively new inventions are reported to be improvements in transportation, fuel economy, and pollution control.

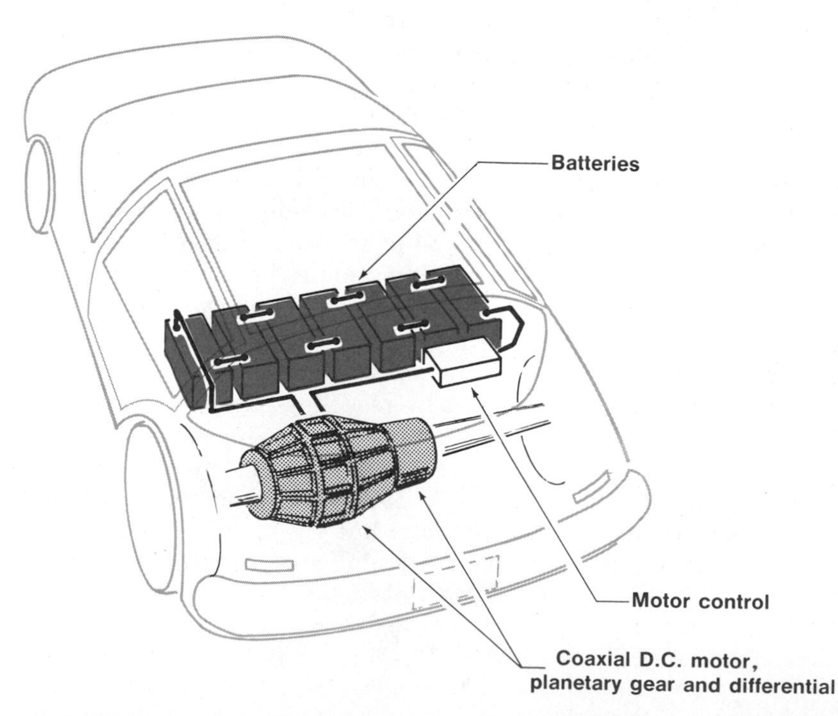


**Figure 17–18** Experimental electric car. Powered by conventional lead-acid batteries, it has a range of 50 miles. Lighter, more energetic batteries would increase its range significantly. Present-day batteries account for 30% of the total weight of the car.

## Self-Test 17-B

1. Carbon monoxide is always formed when hydrocarbons are burned in a (limited/plentiful) amount of air.
2. In the human body, carbon monoxide is a formidable competitor with oxygen for ______________.
3. The amount of carbon monoxide is definitely increasing in the atmosphere. (True/False)
4. A polynuclear aromatic hydrocarbon commonly found in the air and in cigarette smoke that is known to cause cancer is ______________.
5. A secondary pollutant around human beings but an upper atmosphere screen for ultraviolet energy is the substance ______________.
6. Carbon monoxide, CO, is very soluble in water because both substances are (polar/nonpolar).
7. How many carbon atoms are in the most plentiful hydrocarbon molecule in gasoline? ______________
8. What is the source of nitrogen used to form oxides of nitrogen in auto exhaust? ______________
9. What element found in "ethyl" gasoline poisons the catalysts in a catalytic muffler? ______________
10. In the process of adding more air to burn the residual CO and hydrocarbons, are more nitrogen oxides formed? ______________
11. If all of the fuel is burned in a regular automobile engine, maximum power is produced. (True/False)
12. Injecting air into the hot exhaust gases reduces the amount of ______________ and ______________ but increases the amount of ______________.
13. A metal catalyst that is capable of helping to convert CO and unburned hydrocarbons into $CO_2$ and $H_2O$ is ______________ (specific metal).
14. Two destructive effects of $CO_2$ on the environment are that it ______________ the temperature of the atmosphere and ______________ the acidity of the oceans.

## What Does the Future Hold?

There will undoubtedly be an abatement of air pollution in the future; the sheer pressures of population increase will demand it. But life also will undoubtedly have to be different. Perhaps the first major change will be the disappearance of the automobile from the city, followed by a gradual modification of the power plant of the automobile until it is relatively nonpolluting. One interesting effect of the lower legal speed limit on our nation's highways has been the reduction of $NO_x$ emissions due to the lower operating temperatures of the auto engines.

Most pollution exists because we demand the benefits of a technology that, for the most part, has given little consideration to the long-range effects of its products. When industry, automobile manufacturers, or power plant operators add equipment to stop noxious waste products from getting into the air, the costs are added to the already considerable manufacturing expense without adding one cent to the

**In 1981, the estimated cost to bring 147 plants in the Chicago area under the level of 250 micrograms of $NO_2$ per liter was $130 million.**

market value of the product being made. The cost to the consumer, however, will go up and will be reflected in the increased price of consumer goods, automobiles, and electrical power rates. This is a high price, of course, but what is the value of clean air?

At this time, it is uncertain whether the knowledge of the harmful, long-range effects of air pollution will bring people to the point where they are willing to give up the immediate activities that give rise to the pollution. Not many seem willing to use less energy or to give up the automobile.

**The cost of air pollution control in the United States is about $19.3 billion annually.**

In addition to existing studies and regulations, other problems in air pollution will be addressed, such as indoor air pollution, regulation of all hazardous chemicals, a standard for the transient fine particles, and a significant reduction in $SO_2$ to control acid rain.

There will be increased litigation to bring industry into conformance with the Clean Air Act. In 1981, of the estimated 6,500 major sources of industrial pollution, 2,400 had never complied or agreed to comply with the Clean Air Act.

In the final analysis, we all pollute the atmosphere. Much of the pollution is due to the misapplication of chemical techniques, yet the eradication of most forms of air pollution is within the capabilities of chemical technology. At present, there is some awareness of the problems, but the trade-offs among pollution, energy, and inflation are far from being settled. The process will be very slow—it is up to us.

## Matching Set

(Use each choice no more than once)

| | | | |
|---|---|---|---|
| ______ | 1. smog | **a** | heat absorbing |
| ______ | 2. primary pollutant | **b** | mixture of fog, $SO_2$, and (or) hydrocarbons and $NO_2$ |
| ______ | 3. secondary pollutant | **c** | parts per million |
| ______ | 4. aerosol | **d** | disease of lungs |
| ______ | 5. micron | **e** | hydrocarbon with double bond |
| ______ | 6. ppm | **f** | eradication |
| ______ | 7. unsaturated hydrocarbon | **g** | CO |
| ______ | 8. photochemical | **h** | destroys stratospheric ozone |
| ______ | 9. abatement | **i** | process of decreasing |
| ______ | 10. emphysema | **j** | intermediate in size between individual molecules and particulates |
| ______ | 11. brown gas | **k** | benzo($\alpha$)pyrene |
| ______ | 12. endothermic | **l** | peroxy acyl nitrate |
| ______ | 13. polynuclear hydrocarbon | **m** | chemical reaction energized by light |
| ______ | 14. chlorofluorocarbon | **n** | $10^{-6}$ meter |
| | | **o** | $NO_2$ |

## Questions

**1.** The formation of photochemical smog involves very reactive chemical species. What structural feature makes these species reactive?

**2.** Write a balanced chemical equation for the burning of iron pyrite ($FeS_2$) in coal to sulfur dioxide and $Fe_2O_3$, iron (III) oxide.

**3.** What conditions are necessary for thermal inversion?

**4.** What are the basic chemical differences between industrial smog and photochemical smog?

**5.** What are the major sources of the following pollutants?

a. carbon monoxide
b. sulfur dioxide
c. nitrogen oxides
d. ozone

**6.** What is a photochemical reaction? Give an example.

**7.** If air pollutants rise from the earth into the atmosphere, why do they not continue on into space?

**8.** Why is it so difficult to avoid the formation of either carbon monoxide or nitrogen oxides in the combustion process in an automobile engine?

**9.** What effects does weather have on local air pollution problems? on regional air pollution problems?

**10.** Knowing the chemistry of photochemical smog formation, list some ways to prevent its occurrence.

**11.** What part do aerosols play in the formation of smogs?

**12.** Describe the antipollution devices on your car.

**13.** Describe an air pollution problem in your community. How can this problem be solved?

**14.** Discuss the merits of abatement versus eradication of air pollution.

**15.** Of the following air pollutants—particulates, sulfur dioxide, carbon monoxide, ozone, and nitrogen dioxide—

a. which is normally a secondary pollutant?
b. which is emitted almost exclusively by human-controlled sources?
c. which can be removed from emissions by centrifugal separators?
d. which two react with water to form acids?

**16.** Define:

a. micron
b. ppm

**17.** What compound gives "ethyl" gasoline its name?

**18.** What two products are produced by the perfect combustion of hydrocarbons in gasoline?

**19.** Which are more effective in producing smog, hydrocarbons with all single bonds or with some double bonds?

**20.** Why is natural gas a good substitute for oil and coal as far as air pollution is concerned? What problem is related to its substitution?

**21.** What is the approximate concentration of air pollutants in the atmosphere during smoggy conditions?

**22.** Describe an effective way of eliminating carbon monoxide from the exhaust of an automobile.

**23.** What is the chief source of air pollution in the United States?

**24.** Why would $CO_2$, $SO_2$, and CO be more soluble in water than $N_2$ and $O_2$?

**25.** What is the molecular weight of BaP? See p. 405.

**26.** Sketch these generalized (no actual data required) graphs:

a. y axis: $SO_2$ emissions; x axis: concentration of S in coal
b. y axis: CO emissions; x axis: amount of oxygen present
c. y axis: CO emissions; x axis: gasoline consumed
d. y axis: CO emissions; x axis: controls applied

**27.** Why are our desires for cheap energy and a clean environment in conflict?

**28.** The Cottrell precipitator makes use of what property of particulate particles?

**29.** Why does a temperature inversion tend to act as a lid over polluted air?

**30.** Does sulfur dioxide have a long or short atmospheric life?

**31.** What brown oxide of nitrogen is a necessary component of photochemical smog?

**32.** Should human beings make a major effort to alter their activities in order to stop the increase of carbon dioxide in the atmosphere? What would be some of the costs to achieve this goal?

**33.** Why are small particulates in the atmosphere dangerous to human beings?

**34.** What is the effect on the catalytic ingredients in an automobile muffler if leaded fuel is burned in its engine?

**35.** Which substance, naturally found in the atmosphere, seems to be increasing in concentration over the years?

**36.** If nature emits more than 90% of the particulates (volcanic eruptions), nitrogen oxides (lightning), and carbon monoxide (decaying organic matter), why the concern about air pollution caused by people?

# 18 Consumer Chemistry

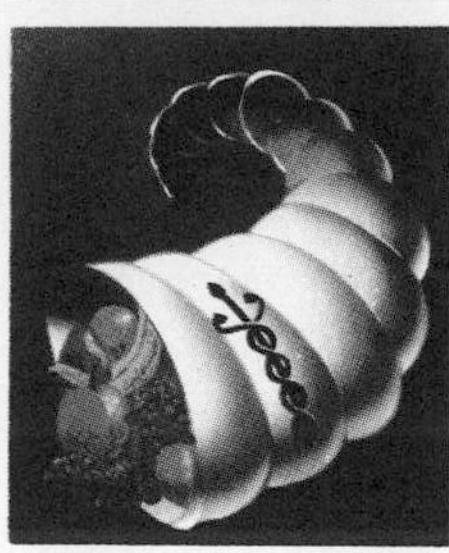

A major portion of the household budget is spent on consumer products. We buy chemical products to feed, cleanse, disinfect, wax, deodorize, paint, remove spots, relieve pain, cure disease, fertilize, preserve, destroy, and do hundreds of other things for us. You already know about the undesirable qualities of certain chemicals as pollutants (Chapters 16 and 17) and as toxic substances (Chapter 15). You are also familiar with many of the chemical products that make our lives more pleasant, healthy, and convenient, provided we use them in the proper amounts and for their intended purposes. You are aware that some, if not all, consumer products ultimately become pollutants, and it is a question of values as to how we use them in our society.

**Consideration of consumer products must involve disposal as well as preparation and use.**

To reap the full benefit of the products we purchase, we should know the types of raw materials in the products, the ways in which products perform their jobs, and the precautions to be taken when using the products. Many formulations under different brand names are essentially the same and cost about the same to produce. Competitive products vie for sales through the cleverness of the name, the attractiveness of the container, and the effectiveness of advertisements as well as the performance of the product. The selling costs are generally added to the price without, of course, increasing the true value of the product. When you have some knowledge of the chemistry of the product, the small print on the label becomes important. A comparison of lists of ingredients sometimes uncovers better formulations, and even harmful substances. In this way, chemical knowledge can then protect the consumer.

Most consumer products are mixtures, but their properties are not necessarily blends of individual properties of the components. In some formulations, the individual properties of certain chemicals are reinforced and enhanced by other chemicals in the mixture. This is called ***synergism,*** defined as the cooperative action of discrete agents such that the total effect is greater than the sum of the effects of each used alone. Citric acid has very little antioxidant effect on foods; butylated hydroxyanisole (BHA) has considerable effect on preventing oxidation of foods. When these two substances are used together in foods, the antioxidative powers of the BHA are increased several-fold. The citric acid is said to have synergistic action on BHA.

**Synergism—cooperative action so that the total effect is greater than the sum of the parts.**

A closely related term is ***potentiation.*** Potentiators do not have a particular effect themselves but exaggerate the effect of other chemicals. The 5′-nucleotides, for example, have no taste, but they enhance the flavor of meat or the effectiveness of salt.

We cannot discuss all consumer products in this chapter, but a few topics will be explained at the molecular level. You can, if you wish, extend this approach to understand the products you use and make wise choices in your selection of them.

## Preservation of Foods

Foods generally lose their usefulness and appeal a short time after harvest. Bacterial decomposition and oxidation are the prime reasons steps must be taken to lengthen the time that a foodstuff remains edible. Any process that prevents the growth of microorganisms and/or retards oxidation is generally an effective preservative process for food. Perhaps the oldest technique is the drying of grains, fruits, fish, and meat. Water is necessary for the growth and metabolism of microorganisms, and it is also important in oxidation. Dryness thus thwarts both the oxidation of food and the microorganisms that feed on it.

**Dry foods tend to be stable.**

Chemicals may also be added as preservatives. Salted meat, and fruit preserved in a concentrated sugar solution, are protected from microorganisms. The abundance of sodium chloride or sucrose in the immediate environment of the microorganisms forms a ***hypertonic*** condition in which water flows by ***osmosis*** from the microorganism to its environment. The salt and sucrose have the same effect on the microorganism as does dryness. Both dehydrate the microorganism.

**A hypertonic solution is more concentrated than solutions in its immediate environment.**

**Osmosis is the flow of water from a more dilute solution through a membrane into a more concentrated solution.**

The canning process for preserving food was developed around 1810, and involves first heating the food to kill all bacteria and then sealing it in bottles or cans to prevent access of other microorganisms and oxygen. Some canned meat has been successfully preserved for over a century. Newer techniques for the preservation of food include vacuum freezing, pasteurization, cold storage, irradiation, and chemical preservation.

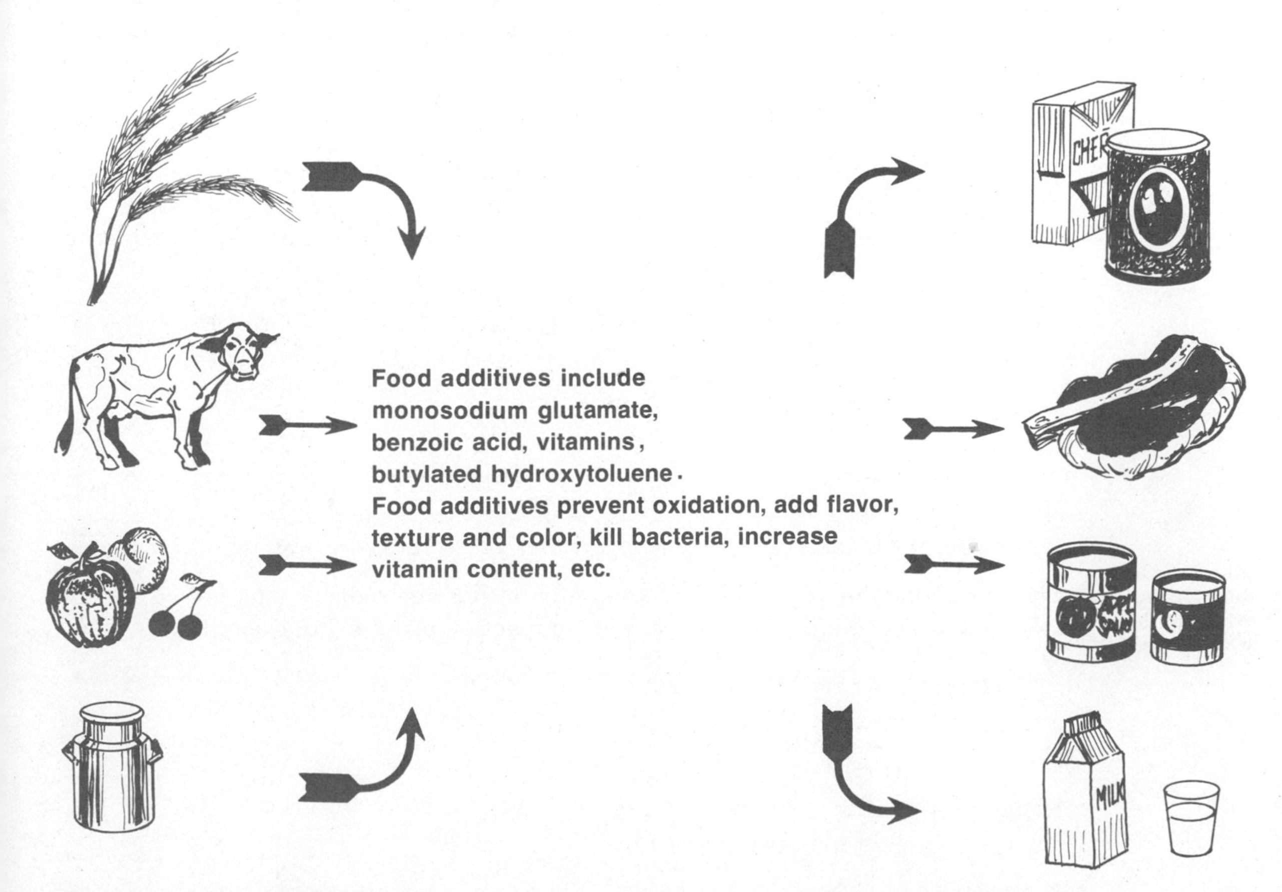

Food additives include
monosodium glutamate,
benzoic acid, vitamins,
butylated hydroxytoluene.
Food additives prevent oxidation, add flavor,
texture and color, kill bacteria, increase
vitamin content, etc.
MILK


**Figure 18–1** Between the harvested and the consumer-ready food one often finds the addition of a large variety of food additives.

## Antimicrobial Preservatives

Food spoilage caused by microorganisms is a result of the excretion of toxins. A preservative is effective if it prevents multiplication of the microbes during the shelf-life of the product. Sterilization by heat or radiation, or inactivation by freezing, is often undesirable since the quality of the food is impaired. Chemical agents seldom achieve sterile conditions but can preserve foods for considerable lengths of time.

**A preservative must interfere with microbes but be harmless to the human system—a delicate balance.**

Antimicrobial preservatives are widely used in a large variety of foods. For example, in the United States sodium benzoate is permitted in nonalcoholic beverages and in some fruit juices, fountain syrups, margarines, pickles, relishes, olives, salads, pie fillings, jams, jellies, and preserves. Sodium propionate is legal in bread, chocolate products, cheese, pie crust, and fillings. Depending on the food, the weight of the preservative permitted ranges up to a maximum of 0.1% for sodium benzoate and 0.3% for sodium propionate.

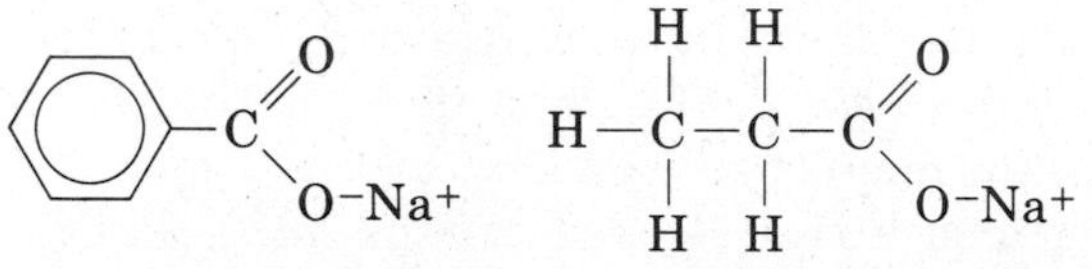

$$C_6H_5{-}C(=O){-}O^-Na^+ \qquad H{-}CH_2{-}CH_2{-}C(=O){-}O^-Na^+$$

SODIUM BENZOATE    SODIUM PROPIONATE

Postulated mechanisms for the action of food preservatives may be grouped into three categories: (1) interference with the permeability of cell membranes of the microbes in foodstuffs, so the bacteria die of starvation; (2) interference with bacterial genetic mechanisms so the reproduction processes are hindered; and (3) interference with intracellular enzyme activity so that metabolic processes such as the Krebs cycle cease.

## Atmospheric Oxidation

Microbial activity results in oxidative decay of food, but it is not the only means of oxidizing food. The direct action of oxygen in the air, ***atmospheric oxidation,*** is the chief factor in destroying fats and fatty portions of foods. Chemically, oxygen reacts with the fat to form a hydroperoxide (R—OOH).

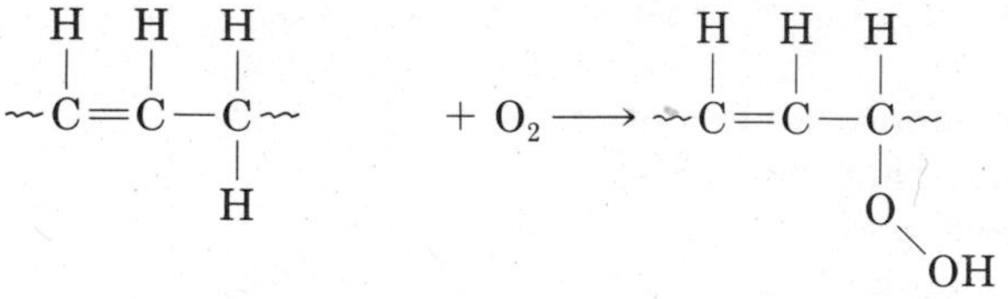

$$\sim CH{=}CH{-}CH_2\sim \; + O_2 \longrightarrow \; \sim CH{=}CH{-}CH(O{-}OH)\sim$$

PORTION OF AN UNSATURATED FAT MOLECULE    HYDROPEROXIDE

**Free radicals usually have short lives; in rare cases they have a stable structure.**

The mechanism involves the formation of reactive free radicals (species with one or more unpaired valence electrons) in a chain reaction process. For example:

$$\underset{\text{FAT}}{RH} + O_2 \longrightarrow \underbrace{R\cdot + HO_2\cdot}_{\text{FREE RADICALS}}$$

$$R\cdot + O_2 \longrightarrow ROO\cdot$$

$$ROO\cdot + RH \longrightarrow \underset{\text{HYDROPEROXIDE}}{ROOH} + R\cdot$$

Foods kept wrapped, cold, and dry are relatively free of oxidation. An antioxidant added to the food can also hinder oxidation. Antioxidants most commonly used in edible products contain various combinations of butylated hydroxyanisole (BHA), butylated hydroxytoluene (BHT), or propyl gallate:

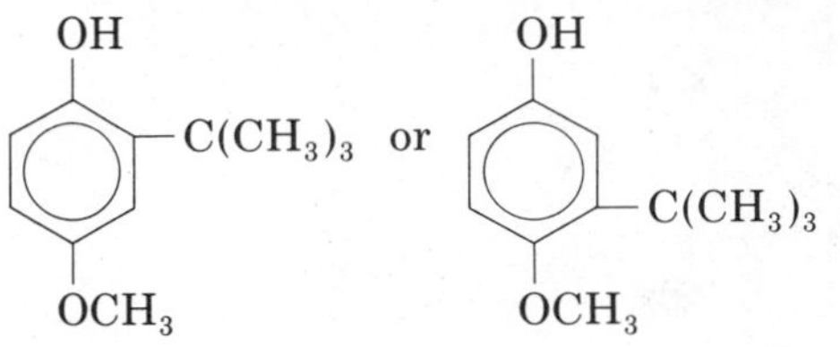

OH, $C(CH_3)_3$, $OCH_3$ or OH, $C(CH_3)_3$, $OCH_3$

BHA

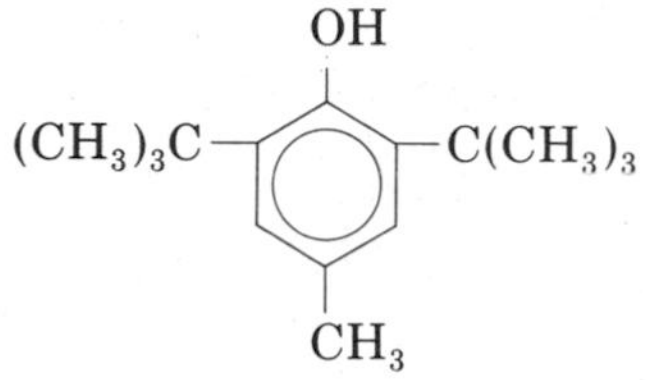

OH, $(CH_3)_3C$, $C(CH_3)_3$, $CH_3$

BHT

OH, HO, OH, $C(=O)-O-CH_2-CH_2-CH_3$

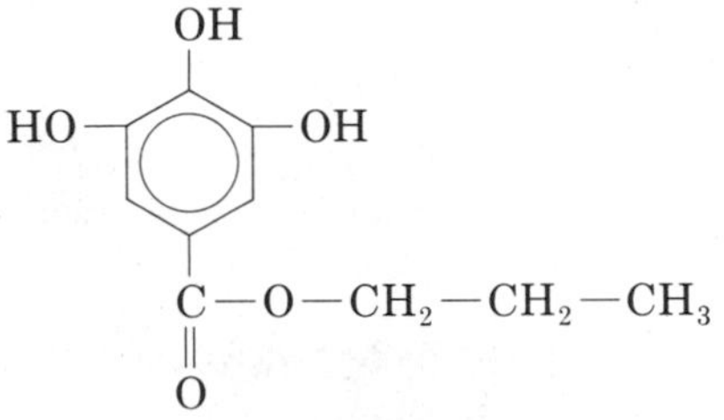

PROPYL GALLATE

To prevent the oxidation of fats, the antioxidant can donate the hydrogen atom in the —OH group to the free radicals and stop the chain reactions. The bulky aromatic radicals formed are relatively stable and unreactive; they add the unpaired electrons to their supply of delocalized electrons:

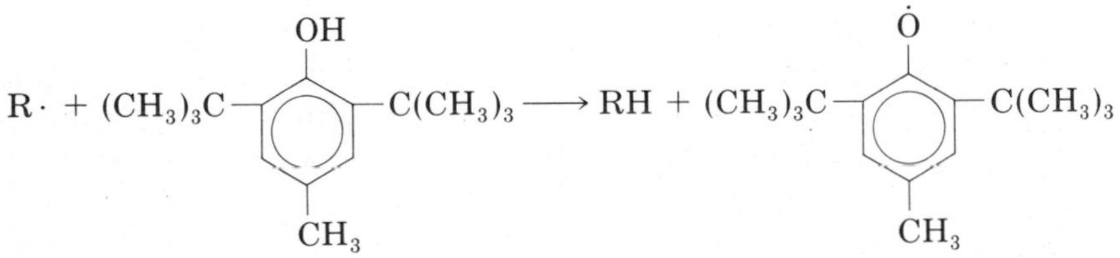

$$R\cdot + (CH_3)_3C\text{-}C_6H_2(OH)(CH_3)\text{-}C(CH_3)_3 \longrightarrow RH + (CH_3)_3C\text{-}C_6H_2(\dot{O})(CH_3)\text{-}C(CH_3)_3$$

ANTIOXIDANT

STABLE FREE RADICAL WITH LITTLE TENDENCY TO REACT

If antioxidants are not present, the hydroperoxy group will attack a double bond. This reaction leads to a complex mixture of volatile aldehydes, ketones, and acids, which cause the odor and taste of rancid fat.

## Sequestrants

Metals get into food from the soil and from machinery during harvesting and processing. Copper, iron, and nickel, and their ions, catalyze the oxidation of fats. However, a molecule of citric acid bonds with the metal ion, thereby rendering it ineffective as a catalyst. With the competitor metal ions tied up, antioxidants such as BHA and BHT can accomplish their task much more effectively.

To sequester means "to withdraw from use." The sequestering ability of EDTA accounts for its use in treating heavy metal poisoning (Chapter 15).

Citric acid belongs to a class of food additives known as ***sequestrants***. For the most part sequestrants react with trace metals in foods, tying them up in complexes so the metals will not catalyze the decomposition or oxidation of food. Sequestrants such as sodium and calcium salts of EDTA (ethylenediaminetetraacetic acid) are permitted in beverages, cooked crab meat, salad dressing, shortening, lard, soup, cheese, vegetable oils, pudding mixes, vinegar, confectioneries, margarine, and other

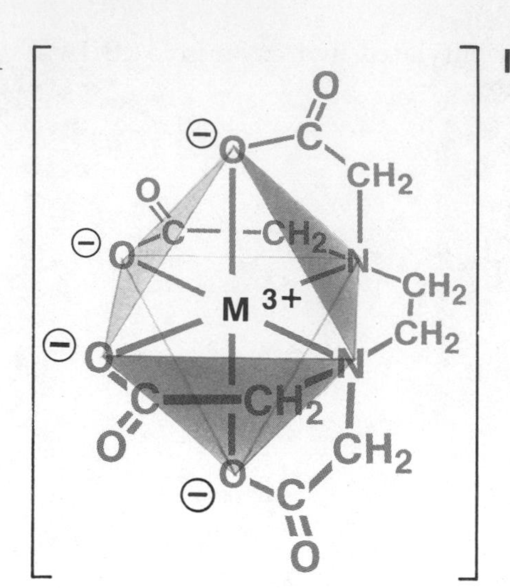


**Figure 18–2** The structural formula for the metal chelate of ethylenediaminetetraacetic acid (EDTA).

foods. The amount ranges from 0.0025 to 0.15%. The structural formula of EDTA bonded to a metal ion is shown in Figure 18–2.

## Flavor in Foods

Flavors result from a complex mixture of volatile chemicals. Since we have only four tastes (sweet, sour, salt, bitter), much of the sensation of taste in food is smell. For example, the flavor of coffee is determined largely by its aroma, and this in turn is due to a very complex mixture of over 100 compounds, mostly volatile oils.

Most flavor additives originally came from plants. The plants are crushed, and the compound is extracted with various solvents such as ethanol or carbon tetrachloride. Sometimes a single compound is extracted; more often, a mixture of several compounds occurs in the residue. By repeated efforts, relatively pure oils are obtained. Oils of wintergreen, peppermint, orange, lemon, and ginger, among others, are still

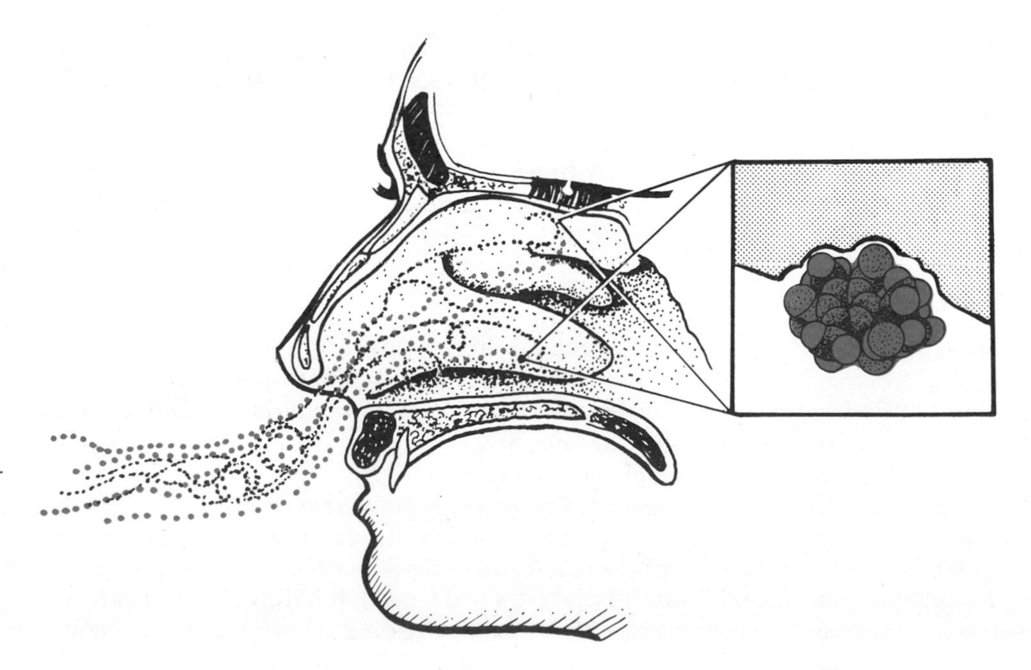

**Figure 18–3** A stereochemical interpretation of the sensation of smell. The substance fits a cavity in the back of the oral cavity. If the atoms are properly spaced, they sensitize nerve endings that transmit impulses to the brain. The brain identifies these sensations as a particular smell. A complete explanation of smell is certainly more involved than the simple idea presented here.

obtained this way. These oils, alone or in combination, are then added to foods to obtain the desired flavor. Gradually, analyses of the oils and flavor components of plants have revealed the active compounds responsible for the flavor. Today, the synthetic preparation of the same flavors actively competes with natural sources.

The Food and Drug Administration has recently banned some of the naturally occurring flavoring agents that were formerly used, including safrole, the primary root beer flavor, found in the root of the sassafras tree.

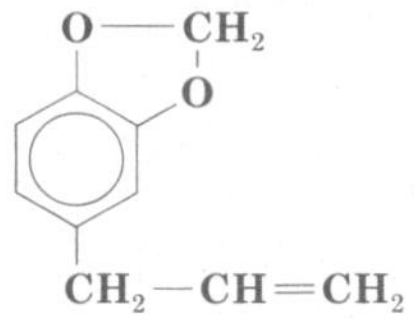

$CH_2-CH=CH_2$

*Safrole*

## Flavor Enhancers

Flavor enhancers have little or no taste of their own but amplify the flavors of other substances. They exert synergistic and potentiation effects. Potentiators were first used in meats and fish. Now they are also used to intensify the flavor and cover unwanted flavors in vegetables, bread, cakes, fruits, nuts, and beverages. Three common flavor enhancers are ***monosodium glutamate (MSG), 5′-nucleotides*** (similar to inosinic acid; Chapter 11), and ***maltol.***

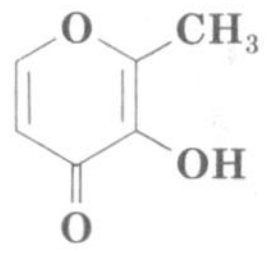

*Maltol*
*(from pine needles)*

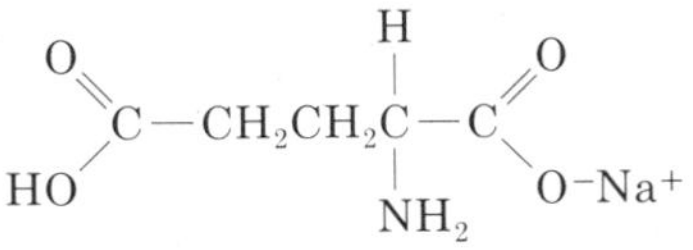

MONOSODIUM GLUTAMATE

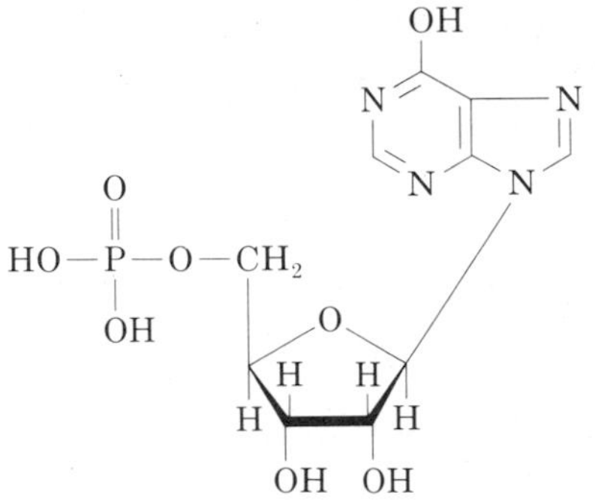

INOSINIC ACID
(a 5′-nucleotide)

**MSG is a natural constituent of many foods, such as tomatoes and mushrooms.**

Brain damage is caused when MSG is injected in very high dosages under the skin in 10-day-old mice. When these laboratory results were reported, considerable discussion ensued concerning the merits of MSG. National investigative councils have suggested that it be removed from baby foods since infants do not seem to appreciate enhanced flavor. However, in the absence of hard evidence that MSG is harmful in the amounts used in regular food, no recommendations were made relative to its use.

**In some people, MSG causes the so-called "Chinese restaurant syndrome," an unpleasant reaction that includes headaches, sweating, and other symptoms usually occurring after an MSG-rich Chinese meal. Tomatoes and strawberries affect some individuals in the same way.**

## Self-Test 18-A

1. Butylated hydroxyanisole plus citric acid in food is an example of (synergism/potentiation) ______________.
2. Name two of the oldest means for preserving food. ______________ ______________
3. Antimicrobial preservatives make foods sterile. (True/False)
4. Flavors result from (volatile/nonvolatile) compounds. ______________
5. Antioxidants are (more/less) easily oxidized than the food into which they are placed. ______________
6. Citric acid is an example of a(n) ______________. Such compounds tie up metals in stable complexes.

7. A flavor in a food can usually be traced to a single compound. (True/False)
8. Monosodium glutamate is a(n) ____________ ____________.
9. Salt is effective in preserving foods because it kills microorganisms by ________ ____________ them.
10. BHT serves as an antioxidant by destroying ____________.

## Sweetness

Insulin is a hormone that regulates glucose metabolism.

Sweetness is characteristic of a wide range of compounds, many of which are completely unrelated to sugars. Lead acetate, $Pb(CH_3COO)_2$, is sweet but poisonous. A number of ***artificial sweeteners*** are allowed in foods. These are primarily used for special diets such as those of diabetics. Artificial sweeteners have no known metabolic use in the body and do not require insulin.

### *Saccharin*

The most common artificial sweetener is saccharin.

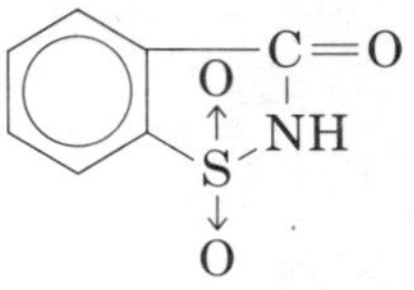

SACCHARIN

Saccharin is a synthetic chemical.

Saccharin is about 300 times sweeter than ordinary sugar (sucrose). When ingested, saccharin passes through the body unchanged. It therefore has no food value other than to render an otherwise bland mixture more tasty. Saccharin has a somewhat bitter aftertaste, which renders it unpleasant to some users. Glycine, the simplest amino acid, which is also sweet-tasting, is often added to counteract this bitter taste.

Laboratory studies have shown that high doses of saccharin cause cancer in mice. After months of consideration, the Institute of Medicine of the National Academy of Science joined the Food and Drug Administration in its 1978 statement that saccharin should be banned in U.S. foods.

### *Cyclamates*

The sweetness of cyclamate salts was discovered accidentally. Sodium cyclamate is about 30 times sweeter than sugar. Because a cyclamate does not have the aftertaste of saccharin, and because it could be used in cooked or baked products, it rapidly replaced saccharin in a wide variety of dietary products.

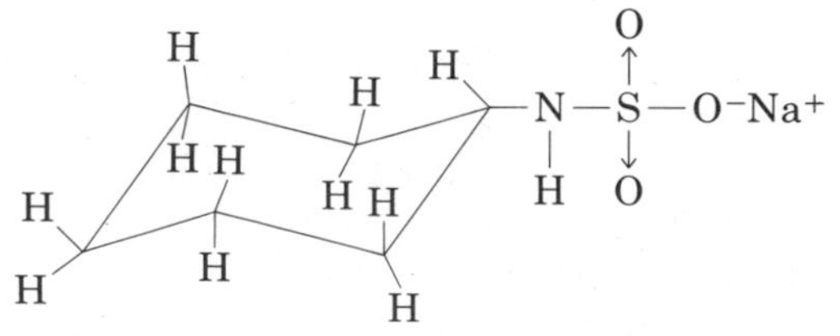

SODIUM CYCLAMATE

In 1969, experimental evidence indicated that cyclamates, under certain conditions, caused cancer of the bladder in laboratory mice. When this evidence was coupled with the fact that 20% of all cyclamate users metabolized cyclamate to a metabolite that caused chromosome breakage in rat cells, the Food and Drug Administration banned cyclamates from foods.

### *Aspartame*

Aspartame, a new entry into the sweetener market, was approved by the FDA in 1974 and subsequently withdrawn by its maker, G. D. Searle Co., when toxicity questions were raised. With these questions resolved, aspartame again received the approval of the FDA in 1981. Aspartame is about 180 times sweeter than table sugar (sucrose). The caloric value of aspartame is similar to that of proteins. It is metabolized in the body as a peptide. The caloric intake of consumers using the product is greatly reduced, since much smaller amounts are needed to produce the same sweetening effect. Aspartame does not have the bitter aftertaste associated with other artificial sweeteners. Aspartame has the drawback of being destroyed by heat; also, it is unstable in soft drinks.

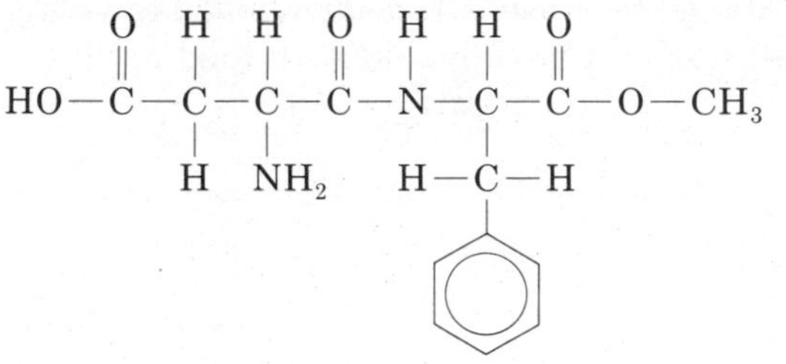

```
    O   H   H   O   H   H   O
    ‖   |   |   ‖   |   |   ‖
HO—C—C—C—C—N—C—C—O—CH3
        |   |           |
        H  NH2      H—C—H
                        |
                     (C6H5)
```

ASPARTAME

At present, there is no generally acceptable artificial sweetener for foods.

## Food and Esthetic Appeal

### Food Colors

Food colors are generally large organic molecules having several double bonds and aromatic rings. These conjugated structures can absorb only certain wavelengths of light and pass the rest; this behavior gives the substance its characteristic color. β-Carotene is an orange-red substance that occurs in a variety of plants, the carrot in particular, and is commonly used as a food color. It has a conjugated system of delocalized electrons. Most of the other food colors have similar conjugated systems.

**Colored organic substances often are *conjugated* molecules, having alternating double and single bonds in the carbon chain or ring.**

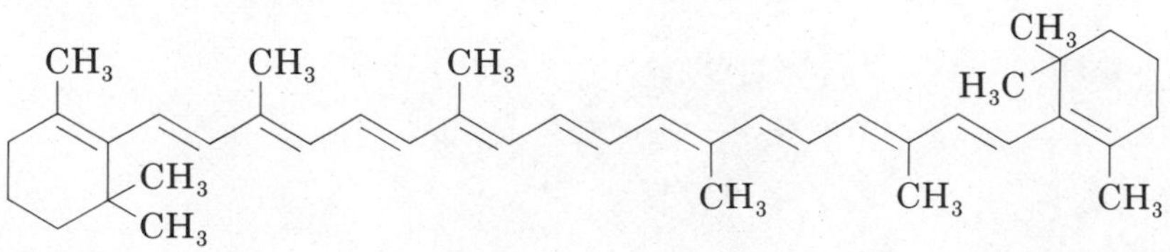

(Structure: CH3, CH3, CH3, CH3, CH3, CH3, CH3, H3C, CH3, CH3 substituents on a conjugated polyene chain joining two six-membered rings)

β-CAROTENE

**β-Carotene is a *precursor* of vitamin A. (It is changed into vitamin A in the body.)**

### pH Control

Weak organic acids are added to such foods as cheese, beverages, and dressings to give a mild acidic taste. They often mask undesirable aftertastes. Weak acids and acid salts, such as tartaric acid and potassium acid tartrate, react with bicarbonate to form $CO_2$ in the baking process.

Some acid additives control the pH of food during the various stages of processing as well as in the finished product. In addition to single substances, there are several combinations of substances that will adjust and then maintain a desired pH. These mixtures are called ***buffers.*** An example of one type of buffer is potassium acid tartrate, $KHC_4H_4O_6$.

**Buffer solutions resist change in acidity and basicity; pH remains constant.**

Adjustment of the pH of a fruit juice is allowed by the FDA. If the pH of the fruit is too high, it is permissible to add acid (called an ***acidulant***). Citric acid and lactic acid are the most common acidulants used since they are believed to impart good flavor; but phosphoric, tartaric, and malic acids are also used. These acids are often added at the end of the cooking time to prevent extensive hydrolysis of the sugar. In the making of jelly they are sometimes mixed with the hot product immediately after pouring. To raise the pH of a fruit that is unusually acid, buffer salts such as sodium citrate or sodium potassium tartrate are used.

**Small amounts of certain acids are allowed to be added to some foods.**

The versatile acidulants also function as preservatives to prevent growth of microorganisms, as synergists and antioxidants to prevent rancidity and browning, as viscosity modifiers in dough, and as melting point modifiers in such food products as cheese spreads and hard candy.

## Anticaking Agents

Anticaking agents are added to hygroscopic foods—in amounts of 1% or less—to prevent caking in humid weather. Table salt (sodium chloride) is particularly subject to caking unless an anticaking agent is present. The additive (magnesium silicate, for example) incorporates water into its structure as water of hydration and does not appear wet as sodium chloride does when it absorbs water physically on the surface of its crystals. As a result, the anticaking agent keeps the surface of sodium chloride crystals dry and prevents crystal surfaces from codissolving, which would join the crystals together.

**Hygroscopic substances absorb moisture from the air.**

## Stabilizers and Thickeners

Stabilizers and thickeners improve the texture and blends of foods. The action of carrageenin (a polymer from edible seaweed) is shown in Figure 18–4. Most of this group of food additives are polysaccharides (Chapter 11) having numerous hydroxyl groups as a part of their structure. The hydroxyl groups form hydrogen bonds with water to prevent the segregation of water from the less polar fats in the food, and to provide a more even blend of the water and oils throughout the food. Stabilizers and thickeners are particularly effective in icings, frozen desserts, salad dressing, whipped cream, confectioneries, and cheeses.

**Stabilizers and thickeners are types of emulsifying agents.**

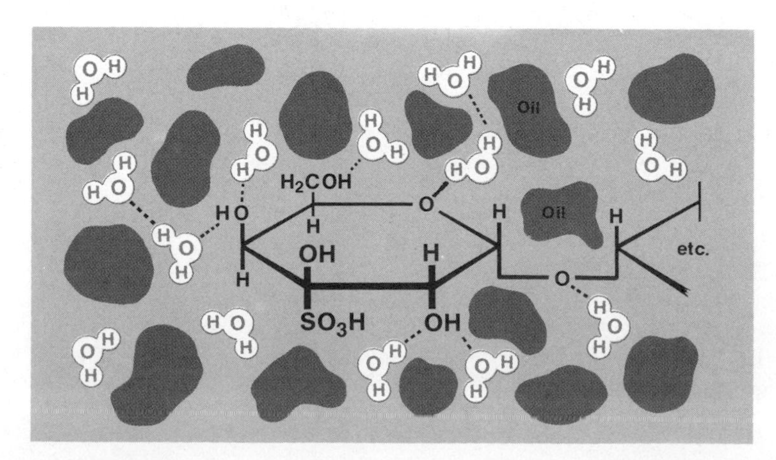


**Figure 18–4** The action of carrageenin to stabilize an emulsion of water and oil in salad dressing. An active part of carrageenin is a polysaccharide, a portion of which is shown above. The carrageenin hydrogen-bonds to the water, which keeps it dispersed. The oil, not being very cohesive, disperses throughout the structure of the polysaccharide. Gelatin (a protein) undergoes similar action in absorbing and distributing water to prevent ice crystals in ice cream.

## Surface Active Agents

Surface active agents are similar to stabilizers, thickeners, and detergents in their chemical action. They cause two or more normally incompatible (nonpolar and polar) chemicals to disperse in each other. If the chemicals are liquids, the surface active agent is called an ***emulsifier.*** If the surface active agent has a sufficient supply of hydroxyl groups, such as cholic acid has, the groups form hydrogen bonds to water. Cholic acid and its associated group of water molecules are distributed throughout dried egg yolk in a manner quite similar to that of carrageenin and water in salad dressing.

Cholic Acid

Some surface active agents have both hydroxyl groups and a relatively long nonpolar hydrocarbon end. Examples are diglycerides of fatty acids, polysorbate 80, and sorbitan monostearate. The hydroxyl groups on one end of the molecule are anchored via hydrogen bonds in the water, and the nonpolar end is held by the non-polar oils or other substances in the food. This provides tiny islands of water held to oil. These islands are distributed evenly throughout the food.

**Hydrogen bonding plays a major role in stabilizers, thickeners, surface active agents, and humectants.**

## Polyhydric Alcohols

Polyhydric alcohols are allowed in foods as humectants, sweetness controllers, dietary agents, and softening agents. Their chemical action is based on their multiplicity of hydroxyl groups that hydrogen-bond to water. This holds water in the food, softens it, and keeps it from drying out. Tobacco is also kept moist by the addition of polyhydric alcohols such as glycerol. An added feature of polyhydric alcohols is their sweetness. Two particularly effective alcohols added to sweeten sugarless chewing gum are mannitol and sorbitol. Compare the structures of these alcohols with the structure of glucose presented in Chapter 11. It is not surprising that the similar structures of these polyhydric alcohols and the isomers of glucose produce a similar taste sensation.

| D-*Sorbitol* | D-*Mannitol* |
|---|---|
| $CH_2OH$ | $CH_2OH$ |
| H—C—OH | HO—C—H |
| HO—C—H | HO—C—H |
| H—C—OH | H—C—OH |
| H—C—OH | H—C—OH |
| $CH_2OH$ | $CH_2OH$ |

### Kitchen Chemistry

## Leavened Bread

Sometimes cooking causes a chemical reaction that releases carbon dioxide gas. The carbon dioxide causes breads and pastries to rise. Yeast has been used since ancient times to make bread rise, and remains of bread made with yeast have been found in Egyptian tombs and the ruins of Pompeii. The metabolic processes of the yeast furnish gaseous carbon dioxide, which creates bubbles in the bread and makes it rise:

**Leavened bread is as old as recorded history.**

$$\underset{\text{GLUCOSE}}{C_6H_{12}O_6} \xrightarrow[\text{from yeast}]{\text{zymase}} \underset{\text{(GAS)}}{2CO_2} + \underset{\text{ETHANOL}}{2C_2H_5OH}$$

When the bread is baked, the $CO_2$ expands even more to produce a light, airy loaf.

Carbon dioxide can be generated in cooking by other processes. For example, baking soda (which is simply sodium bicarbonate, $NaHCO_3$, a base) can react with acidic ingredients in a batter to produce $CO_2$.

$$NaHCO_3 + H^+ \longrightarrow Na^+ + H_2O + CO_2 \text{ (gas)}$$

Baking powders contain sodium bicarbonate and an added acid salt or a salt that hydrolyzes to produce an acid. Some of the compounds used for this purpose are potassium hydrogen tartrate, $KHC_4H_4O_6$, calcium dihydrogen phosphate monohydrate, $Ca(H_2PO_4)_2 \cdot H_2O$, and sodium acid pyrophosphate, $Na_2H_2P_2O_7$. The reactions of these white, powdery salts with sodium bicarbonate are similar, although the compounds all have somewhat different appearances. For example:

$$KHC_4H_4O_6 + NaHCO_3 \xrightarrow{\text{water}} KNaC_4H_4O_6 + H_2O + CO_2 \text{ (gas)}$$

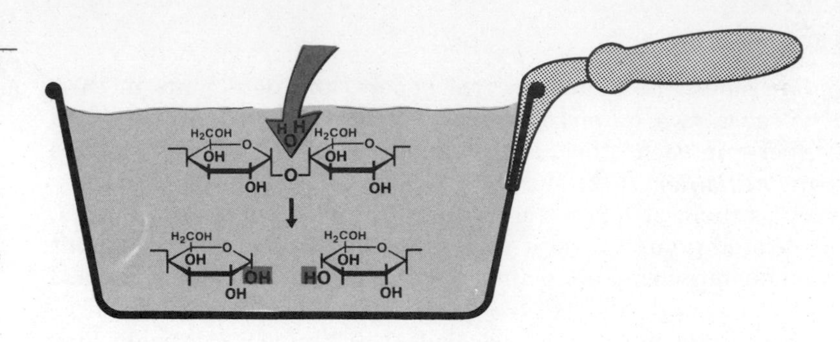


**Figure 18–5** The hydrolysis of starch during the cooking of foods such as potatoes and rice.

## Cooking and Precooking—"Preliminary Digestion"

The cooking process involves the partial breakdown of proteins or carbohydrates by means of heat and hydrolysis (Fig. 18–5). The polymers that must be degraded if cooking is to be effective are the carbohydrate cellular wall materials in vegetables and the collagen or connective tissues in meats. Both types of polymers are subject to hydrolysis in hot water or moist heat. In either case, only partial depolymerization is required.

**Cooking starts the digestive process, although it is rarely needed for this purpose. Since it may destroy nutrients, it is done mostly for esthetic reasons.**

In recent years several precooking additives have become popular; the ***meat tenderizers*** are a good example. These are simply enzymes that catalyze the breaking of peptide bonds in proteins via hydrolysis at room temperature. As a consequence, the same degree of "cooking" can be obtained in a much shorter heating time. Meat tenderizers are usually plant products such as papain, a proteolytic (protein-splitting) enzyme from the unripe fruit of the papaw tree. Papain has considerable effect on connective tissue, mainly collagen and elastin, and shows some action on muscle fiber proteins. On the other hand, microbial protease enzymes (from bacteria, fungi, or both) have considerable action on muscle fibers. A typical formulation for the surface treatment of cuts of beef contains 2% commercial papain or 5% fungal protease, 15% dextrose, 2% monosodium glutamate (MSG), and salt.

**TABLE 18–1 The Functions of Vitamins***

| Vitamin | Biochemical Function | Deficiency Effects |
|---|---|---|
| A | regeneration of rhodopsin (visual purple) | excessive light sensitivity; night blindness; increased susceptibility to infection |
| $B_1$ (Thiamine) | nerve activity; carbohydrate metabolism | beriberi; serious nervous disorders; muscular atrophy; serious circulatory changes |
| $B_2$ (Riboflavin) | coenzyme; affects sight | sores on the lips; bloodshot and burning eyes; excessive light sensitivity |
| $B_3$ (Nicotinamide) | metabolism of ATP | stunted growth; pellagra |
| $B_5$ (Pantothenic acid) | growth factor; component of coenzyme A | retarded growth |
| $B_6$ (Pyridoxine) | coenzyme; metabolism of fatty acids | retarded growth; anemia; leukocytosis; insomnia; lesions about eyes, nose, mouth |
| $B_7$ (Biotin) | growth factor; affects scaly and greasy skin; $CO_2$ fixation; coenzyme | scaly and greasy skin |
| $B_9$ (Folic acid) | coenzyme; tyrosine metabolism | anemia |
| $B_{12}$ (Cobalamin) | growth factor; involved in synthesis of DNA, hemoglobin | degeneration of the spinal cord, pernicious anemia |
| C (Ascorbic acid) | coenzyme; reducing agent; cholesterol metabolism | hemorrhages; lesions in the mouth; muscular degeneration; sterility; scurvy |
| D | Ca and P metabolism | abnormal development of bones and teeth; rickets |
| E ($\alpha$-Tocopherol) | antioxidant; cofactor between cytochromes *b* and *c* | sterility |
| K | synthesis of prothrombin, coenzyme Q | hemorrhages; slow clotting of blood |

*The vitamins participate in biochemical changes concerned with the utilization of foodstuffs.

## Vitamins and Minerals—Nutrient Supplements

A type of food additive that has received considerable attention is the nutrient supplement. The addition of iodide (as KI) to common table salt to prevent goiter is one example. Many food products now contain added vitamins: vitamin D is added to milk, the B vitamins are added to wheat flour, vitamin C is added to certain beverages, and minerals are added to many cereal products. Since these compounds are needed by the body, they fall into a somewhat different category from the other nonessential additives. The purpose of the various vitamins are given in Table 18–1.

**TABLE 18–2 A Partial List of Food Additives Generally Recognized as Safe***

**Food Colors**

Annatto (yellow)
Carbon black
Carotene (yellow-orange)
Cochineal (red)
Food dyes and colors
Titanium dioxide (white)

**Acids, Alkalies, and Buffers**

Acetates: Ca, K, Na
Acetic acid
Calcium lactate
Citrates: Ca, K, Na
Citric acid
Fumaric acid
Lactic acid
Phosphates, CaH, $Ca_3$, $Na_2$, $Na_3$, NaAl
Potassium acid tartrate
Sorbic acid
Tartaric acid

**Surface Active Agents**

Cholic acid
Glycerides: mono- and diglycerides of fatty acids
Polyoxyethylene (20) sorbitan monopalmitate
Sorbitan monostearate

**Polyhydric Alcohols**

Glycerol
Sorbitol
Mannitol
Propylene glycol

**Preservatives**

Benzoic acid
  Na benzoate
Methylparaben
Oxytetracycline
Propylparaben
Propionic acid
  Ca propionate
  Na propionate
Sorbic acid
  Ca sorbate
  K sorbate
  Na sorbate
Sulfites, $Na^+$, $K^+$

**Antioxidants**

Ascorbic acid
  Ca ascorbate
  Na ascorbate
Butylated hydroxyanisole (BHA)
Butylated hydroxytoluene (BHT)
Lecithin
Propyl gallate
Sulfur dioxide and sulfites
Trihydroxybutyrophenone (THBP)

**Flavor Enhancers**

Monosodium glutamate (MSG)
5′-Nucleotides
Maltol

**Sequestrants**

Citrate esters: isopropyl, stearyl
Citric acid
EDTA, $Ca^{2+}$ and $Na^+$ salts
Pyrophosphate, $Na^+$
Sorbitol
Tartaric acid
  Na tartrate

**Stabilizers and Thickeners**

Agar-agar
Algins: $NH_4^+$, $Ca^{2+}$, $K^+$, $Na^+$
Carrageenin
Gum acacia
Gum tragacanth
Sodium carboxymethyl cellulose

**Flavorings**

Acetanisole (slight haylike)
Allyl caproate (pineapple)
Amyl acetate (banana)
Amyl butyrate (pearlike)
Bornyl acetate (piney, camphor)
Carvone (spearmint)
Cinnamaldehyde (cinnamon)
Citral (lemon)
Ethyl cinnamate (spicy)
Ethyl formate (rum)
Ethyl propionate (fruity)
Ethyl vanillin (vanilla)
Eucalyptus oil (bittersweet)
Eugenol (spice, clove)
Geraniol (rose)
Geranyl acetate (geranium)
Ginger oil (ginger)
Linalool (light floral)
Menthol (peppermint)
Methyl anthranilate (grape)
Methyl salicylate (wintergreen)
Orange oil (orange)
Peppermint oil (peppermint) (menthol)
Pimenta leaf oil (allspice) (eugenol, cineole)
Vanillin (vanilla)
Wintergreen oil (wintergreen) (methyl salicylate)

*For precise and authoritative information on levels of use permitted in specific applications, the regulations of the U.S. Food and Drug Administration and the Meat Inspection Division of the U.S. Department of Agriculture should be consulted.

**If past history is a guide, at least some of these compounds will be taken off the GRAS list in the future.**

## The GRAS List

**The GRAS list is a noble effort—but at the present time it is not foolproof.**

The Food and Drug Administration lists about 600 chemical substances ***"generally recognized as safe" (GRAS)*** for their intended use. A small portion of this list is given in Table 18–2. It must be emphasized that an additive on the GRAS list is safe *only if it is used in the amounts and in the foods specified.* The GRAS list was published in several installments in 1959 and 1960. It was compiled by asking experts in nutrition, toxicology, and related fields to give their opinions about the safety of using various materials in foods. Since its publication, few substances have been added to the GRAS list and some, such as the cyclamates, saccharin, and Red Dye no. 2, have been removed.

It is evident, in view of the more than 2,500 known food additives, that many more chemicals than those that appear on the GRAS list are approved (or at least, not banned) for use as food additives by the FDA. It is quite expensive to introduce a new food additive with the approval of the FDA. Allied Chemical Corporation began research in 1964 on a new synthetic food color, Allura Red AC. It was approved by the FDA and went on the market in 1972. The cost for introducing this product was $500,000, and about half of this amount was spent on safety testing.

## Self-Test 18-B

1. Which has the sweetest taste when an equal amount of each is tasted: table sugar, aspartame, or saccharin? ______________
2. What molecular characteristic do most food colors have? ______________
3. What acids are added to foods to lower the pH? ____________ and ____________.
4. The gas released by leavening agents is ______________.
5. Monosodium glutamate (MSG) is used in meat as a ______________.
6. Cooking ______________ some chemical bonds.
7. The GRAS list is the FDA list of food additives that are ______________.
8. Hydrogen bonding generally plays a very important role in the action of surface ______________ agents.
9. In order to produce $CO_2$ in bread, sodium bicarbonate reacts with a(n) ________ ______________.

---

## Matching Set

| | | | |
|---|---|---|---|
| ______ | 1. β-carotene | **a** | nutrient supplement in food |
| ______ | 2. monosodium glutamate | **b** | food color |
| ______ | 3. copper, nickel, and iron | **c** | flavor enhancer |
| ______ | 4. sodium benzoate | **d** | catalyze oxidation of fats |
| ______ | 5. potentiator | **e** | antimicrobial preservative |
| ______ | 6. mineral | **f** | exaggerates some chemical effects |

## Part II—Medicines

The average life expectancy in the United States has risen from age 49 in 1900 to age 73 in 1976. It is expected to go beyond the age of 79 by the year 2000. The major

**TABLE 18–3 Widely Prescribed Drugs**

| Generic Name* | Medical Use |
|---|---|
| Tetracycline HCl | antibiotic |
| Ampicillin | antimicrobial (kills bacteria but is not derived from a plant, as is an antibiotic) |
| Phenobarbital | sedative, hypnotic, anticonvulsant |
| Thyroid | increases rates of metabolism |
| Prednisone | antiinflammatory, antiallergic agent similar to cortisone |
| Digoxin | decreases the rate of the heartbeat but increases the force of the heartbeat, similar to digitalis |
| Meprobamate | tranquilizer |
| Erythromycin | antimicrobial |
| Penicillin G potassium | antibiotic |
| Nitroglycerin | dilates the blood vessels of the heart |
| Penicillin VK | antibiotic |
| Quinidine sulfate | slows the heartbeat, also used for malaria and hiccups |
| Paregoric | a tincture (alcoholic solution) of camphorated opium used as an analgesic |
| Reserpine | tranquilizer |
| Nicotinic acid (niacin) | dilates blood vessels; also, essential B vitamin with antipellagra activity |

*The generic name for a drug is its generally accepted chemical name rather than a specific brand name. For example, the generic drug tetracycline HCl is sold under such brand names as Achromycin V, Ambracyn, Artomycin, Diacycline, Quatrex, and others. Medical doctors can prescribe either the generic name or a brand name. If the generic name is used, the prescription is often cheaper, particularly if the drug is not protected by patents and can be manufactured and marketed competitively by several companies. Only about 12% of the prescriptions written in 1974 used generic names.

**The generic name for a drug is its widely accepted chemical name.**

contributing factor is the widespread use of a large assortment of new medicinal compounds.

**Medicines rise and fall in popularity.**

The contents of the medicine cabinet have changed drastically in the past few decades. A survey of physicians shortly before World War I revealed the ten most essential drugs (or drug groups) to be ether, opium and its derivatives, digitalis, diphtheria antitoxin, smallpox vaccine, mercury, alcohol, iodine, quinine, and iron. When another survey was made at the end of World War II, at the top of the list were sulfonamides, aspirin, antibiotics, blood plasma and its substitutes, anesthetics and opium derivatives, digitalis, antitoxins and vaccines, hormones, vitamins, and liver extract. Today there is an even wider array of medicinal chemicals, but drugs for reducing fever, relieving pain, and fighting infection still head the list in all areas of medical practice.

**Many new drugs are tested each year, but few ever reach the marketplace.**

## Antacids

**The contents of the stomach are highly acidic.**

The walls of a human stomach contain thousands of cells that secrete hydrochloric acid, the main purposes of which are to suppress growth of bacteria and to aid in the hydrolysis (digestion) of certain foodstuffs. Normally, the stomach's inner lining is not harmed by the presence of this hydrochloric acid, since the mucosa, the inner lining of the stomach, is replaced at the rate of about a half million cells per minute. When too much food is eaten the stomach often responds with an outpouring of acid, which lowers the pH to a point where discomfort is felt.

**If the reduction of acidity is too great, the stomach responds by secreting an excess of acid. This is "acid rebound."**

Antacids are compounds used to decrease the amount of hydrochloric acid in the stomach. The normal pH of the stomach ranges from 1.2 to 3.0. Some alkaline compounds used for antacid purposes and their modes of action are given in Table 18–4.

## Analgesics

**Analgesics relieve pain, but they are harmful in large doses.**

Analgesics are pain killers. Most people need these compounds at one time or another. When we have a headache we take aspirin. When we have a tooth filled or extracted, the dentist uses novocaine. Intense suffering requires a strong pain killer, such as codeine or morphine. While these compounds are immensely useful, they are nevertheless dangerous if taken or used improperly. They can even become killers if taken in overdose.

**TABLE 18–4 The Chemistry of Some Antacids**

| Compound | Reaction in Stomach | Comments |
|---|---|---|
| Magnesium oxide $MgO$ | $MgO + 2H^+ \longrightarrow Mg^{2+} + H_2O$ | $MgO$ is white and tasteless |
| Milk of magnesia $Mg(OH)_2$ in water | $Mg(OH)_2 + 2H^+ \longrightarrow Mg^{2+} + 2H_2O$ | the water suspension has an unpleasant chalky consistency |
| Calcium carbonate $CaCO_3$ | $CaCO_3 + 2H^+ \longrightarrow Ca^{2+} + H_2O + CO_2$ | calcium carbonate is purified limestone |
| Sodium bicarbonate $NaHCO_3$ | $NaHCO_3 + H^+ \longrightarrow Na^+ + H_2O + CO_2$ | baking soda, like $CaCO_3$, produces $CO_2$ gas in the stomach |
| Aluminum hydroxide $Al(OH)_3$ | $Al(OH)_3 + 3H^+ \longrightarrow Al^{3+} + 3H_2O$ | $Al(OH)_3$ is a clear gel |
| Dihydroxyaluminum sodium carbonate $NaAl(OH)_2CO_3$ | $NaAl(OH)_2CO_3 + 4H^+ \longrightarrow Na^+ + Al^{3+} + 3H_2O + CO_2$ | sold as Rolaids, will not ordinarily cause pH to go above 5 |
| Sodium citrate $Na_3C_6H_5O_7 \cdot 2H_2O$ | $Na_3C_6H_5O_7 \cdot 2H_2O + 3H^+ \longrightarrow 3Na^+ + H_3C_6H_5O_7 + 2H_2O$ | mild |

Early societies may well have used opium. Although not all opium derivatives have therapeutic value, most of them are efficient pain killers. Their chief disadvantage lies in their addictive properties.

Opium is obtained from the opium poppy by scratching the seed pod with a sharp instrument. From this scratch flows a sticky mass that contains about 20 different compounds called ***alkaloids*** (organic nitrogenous bases containing basic nitrogen atoms). About 10% of this mass is the alkaloid ***morphine,*** which is primarily responsible for opium's effects.

**Morphine and its derivatives are addictive drugs.**

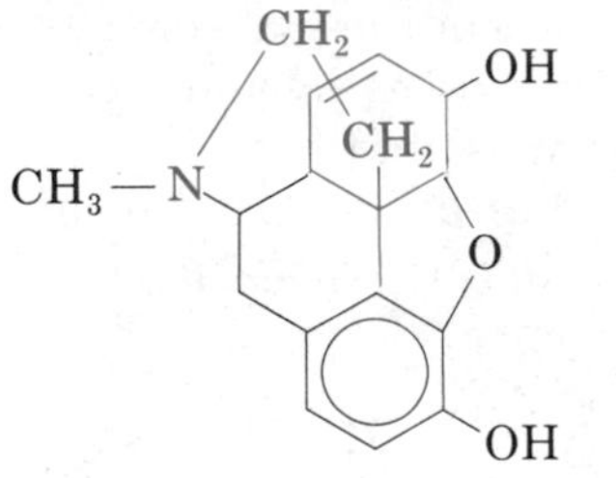

MORPHINE

**Analgesics may or may not be habit-forming.**

Two derivatives of morphine are of interest. One of these is ***codeine,*** a methyl ether of morphine, which is less addictive than morphine and is about as powerful an analgesic. The other compound is ***heroin,*** the diacetate ester of morphine. Heroin is much more addictive than morphine and for that reason finds no medical uses in the United States.

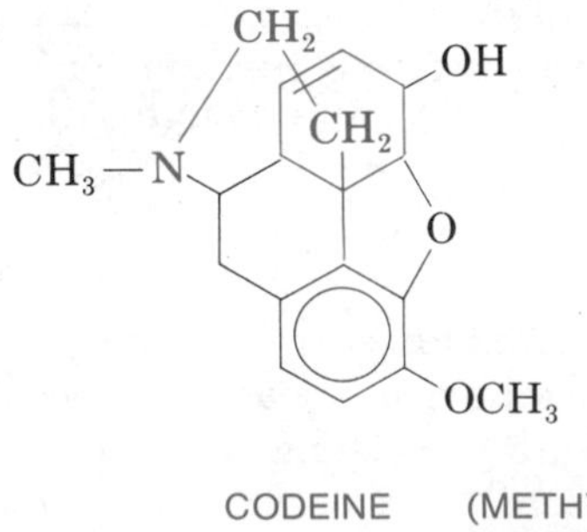

CODEINE (METHYL ETHER)

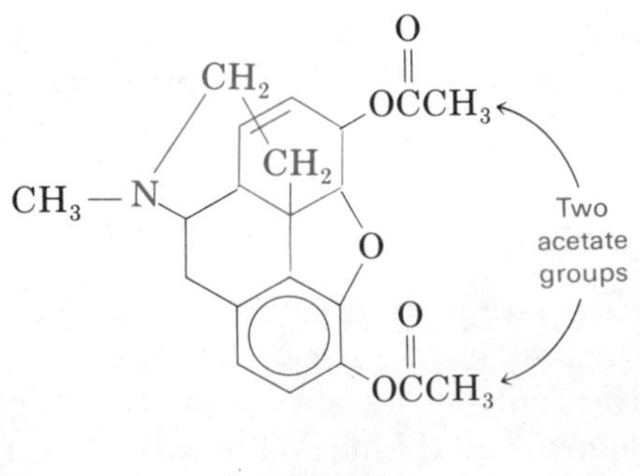

Two acetate groups

HEROIN

One of the most effective substitutes for morphine is ***meperidine,*** first reported in 1931, and now sold as Demerol. It is less addictive than morphine. Two other relatively strong pain relievers used today are ***pentazocine*** (Talwin) and ***propoxyphene*** (Darvon). Talwin is slightly addictive, while Darvon has not been shown to be. However, Darvon has been much abused; there were approximately 600 Darvon-related deaths in 1977. Critics argue it to be more dangerous and less effective in killing pain than other available opiates. Note that in the structures of these compounds there is a strong resemblance to the morphine structure.

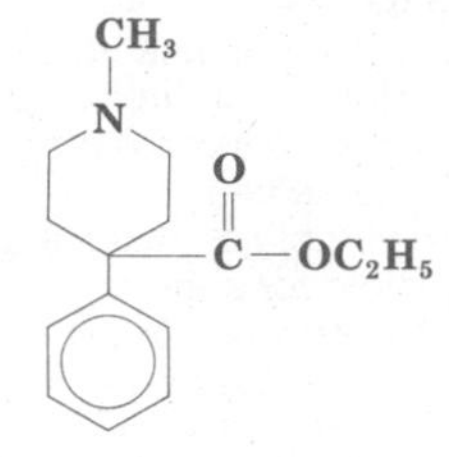

Considerable progress has been made in understanding the drug action of the opiates. Solomon Snyder, along with coworkers, discovered in 1973 that the brain and spinal cord contain specific bonding sites into which the opiate molecules fit as a key fits a lock. John Hughes and Hans Kosterlitz followed in 1975 with the discovery that vertebrates produce their own opiates, which they named ***enkephalins*** (from Greek, meaning in the head). Individuals with a high tolerance for pain produce more enkephalins and consequently tie up more receptor sites; hence there is less pain. A dose of heroin would temporarily bond to a high percentage (or all) of the sites and there would be little or no pain. Continued use of heroin would cause the body to reduce or cease its production of enkephalins. If the use of the narcotic is stopped, the receptor sites are empty and withdrawal symptoms appear.

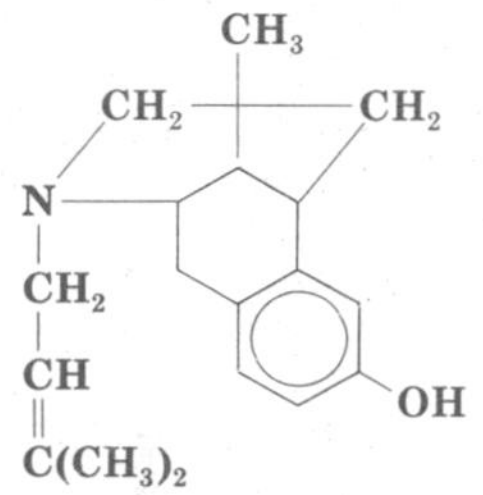

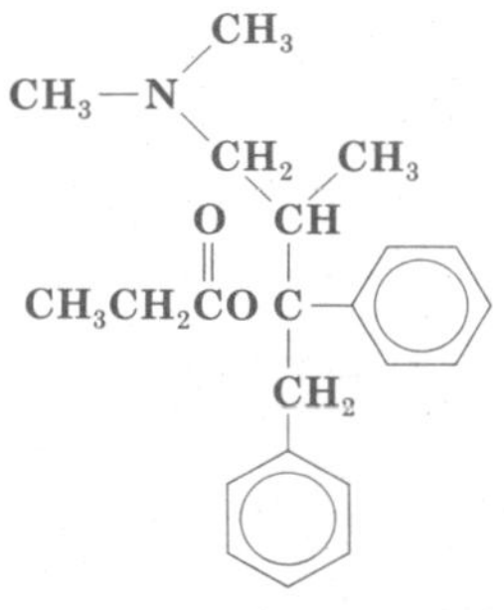

Many of the ***local analgesics,*** or local anesthetics, are nitrogen compounds, like the alkaloids (Table 18–5). Local analgesics include the naturally occurring ***cocaine,*** derived from the leaves of the coca plant of South America, and the familiar ***novocaine*** (procaine). All of these drugs act by some blockage of the nerves that transmit pain. Acetylcholine appears to be the "opener of the gate" for sodium ($Na^+$) and potassium ($K^+$) ions to flow into a nerve cell (Fig. 18–6).

There are times when milder, general analgesics are required, and few compounds work as well for as many people as ***aspirin.*** Not only is aspirin an analgesic but it is also an antipyretic, a fever reducer.

The synthesis of aspirin was outlined in Chapter 9. Each year, about 40 million pounds are manufactured in the United States.

$CH_3$ — N (ring) — C(=O)—$OC_2H_5$

MEPERIDINE (DEMEROL)

$CH_3$, $CH_2$, $CH_2$, N, $CH_2$, CH, $C(CH_3)_2$, OH

PENTAZOCINE (TALWIN)

$CH_3$, $CH_3$—N, $CH_2$, $CH_3$, O, CH, $CH_3CH_2CO$ C, $CH_2$

PROPOXYPHENE (DARVON)

**TABLE 18–5 Some Local Analgesics**

| | | |
|---|---|---|
| Cocaine | 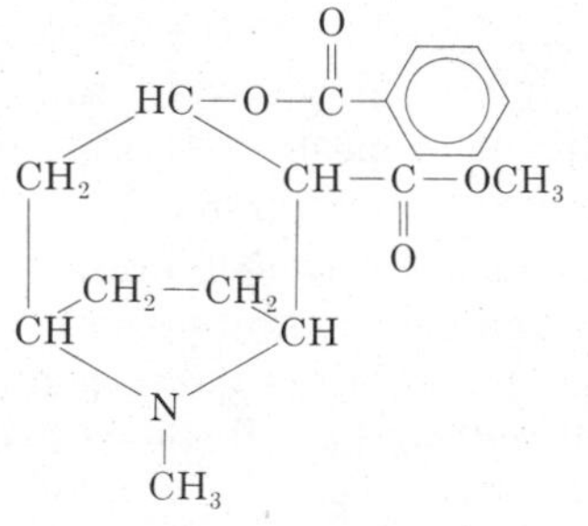 O; HC—O—C(=O)—(phenyl); $CH_2$; CH—C(=O)—$OCH_3$; $CH_2$—$CH_2$; CH; CH; N; $CH_3$ | probably the first local analgesic used |
| Procaine (Novocaine) | 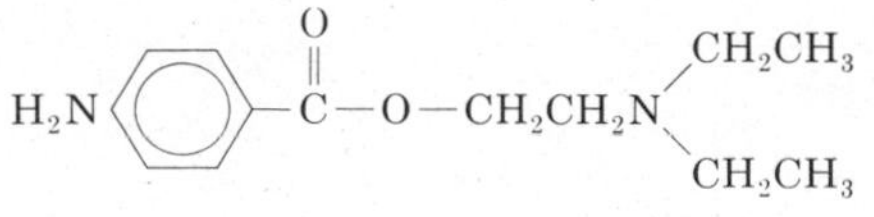 $H_2N$—(phenyl)—C(=O)—O—$CH_2CH_2N$($CH_2CH_3$)($CH_2CH_3$) | often used in dental work |
| Lidocaine (Xylocaine) | 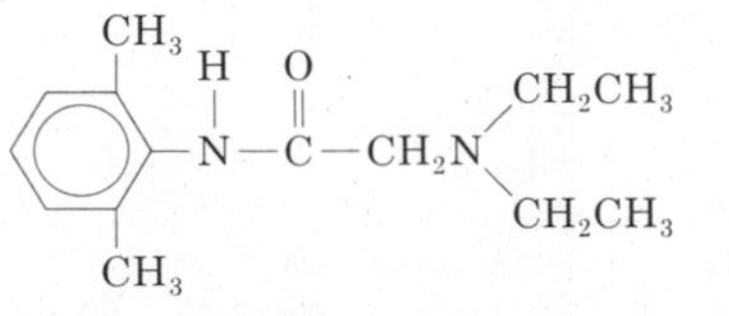 ($CH_3$)(phenyl)($CH_3$)—N(H)—C(=O)—$CH_2N$($CH_2CH_3$)($CH_2CH_3$) | more potent than procaine, can be applied to the skin |

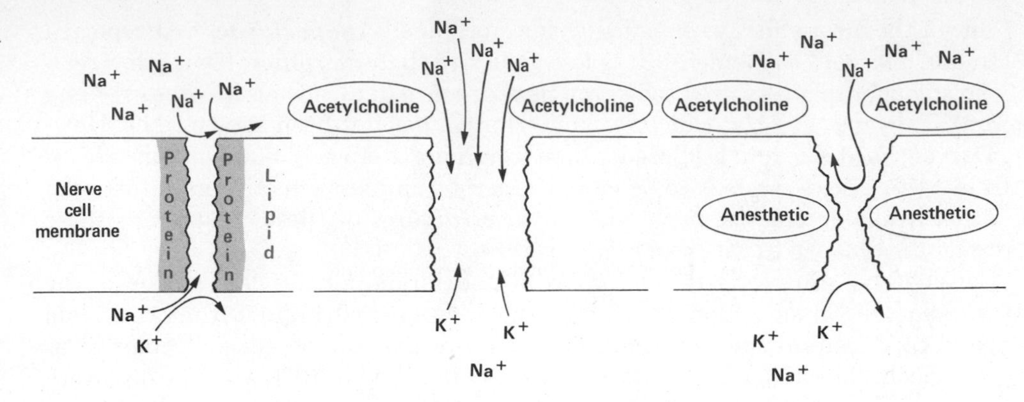


**Figure 18–6** Action of acetylcholine and anesthetics in depolarizing the membrane of a nerve cell. Acetylcholine makes it possible for sodium and potassium ions to neutralize the negative charge associated with a nerve impulse so another impulse can be transmitted. Anesthetics block the action of acetylcholine and do not allow repetitive impulses to travel along the nerve.

The greatest danger presented by aspirin is that of stomach bleeding, caused when an undissolved aspirin tablet lies on the stomach wall. As the aspirin molecules pass through the fatty layer of the mucosa, they appear to injure the cells, causing small hemorrhages. The blood loss for most individuals taking two 5-grain tablets is between 0.5 ml and 2 ml. Some people are more susceptible. Early aspirin tablets were not particularly fast-dissolving, which aggravated this problem greatly. Today, aspirin tablets are formulated to disintegrate quickly, although crushing the tablet in a little water might not be a bad idea.

**Many aspirin tablets contain starch to hasten their disintegration in the stomach.**

An alternative for pain-sufferers who have trouble with aspirin may be acetaminophen (Tylenol). Although known for about as long as aspirin (a century), acetaminophen is a relative newcomer to the market. Like aspirin, acetaminophen is both an analgesic and an antipyretic.

$C_6H_4(COOH)(O{-}CO{-}CH_3)$

*Acetylsalicylic acid (aspirin)*

$CH_3C(=O){-}N(H){-}C_6H_4{-}OH$

Acetaminophen (Tylenol)

## Antiseptics and Disinfectants

An antiseptic is a compound that prevents the growth of microorganisms. It now has the legal meaning ***"germicide,"*** or a compound that *kills* microorganisms. A disinfectant is a compound that destroys pathogenic bacteria or microorganisms, but usually not bacterial spores. Disinfectants are generally poisonous, and therefore suitable only for external use as on the skin or a wound.

**Pathogenic bacteria cause many illnesses.**

Some common germicides are listed in Table 18–6. Some of these, such as the halogens, sodium hypochlorite, hydrogen peroxide, and potassium permanganate, are effective because of their oxidizing properties. This is a general property and allows them to oxidize any kind of cell, including human cells. For this reason they are used mostly as disinfectants in destroying the germs on nonliving objects. Phenol (carbolic acid) is readily absorbed by cells and is a general poison. The quaternary ammonium compounds are surface active agents, and their bactericidal effect seems to be related to their ability to weaken the cell wall so the cell contents cannot be contained.

$R_4N^+\ Cl^-$

A quaternary ammonium chloride

**TABLE 18–6 Some of the More Common Antiseptics**

| | |
|---|---|
| Iodine | Phenols |
| Sodium hypochlorite | Mercurochrome |
| Potassium permanganate | Metaphen |
| Hydrogen peroxide | Merthiolate |
| Iodophors | Pine oil |
| Ethanol | Soap |
| Quaternary ammonium compounds | Hexylresorcinol |
| Chloramine-T | Mercuric chloride |

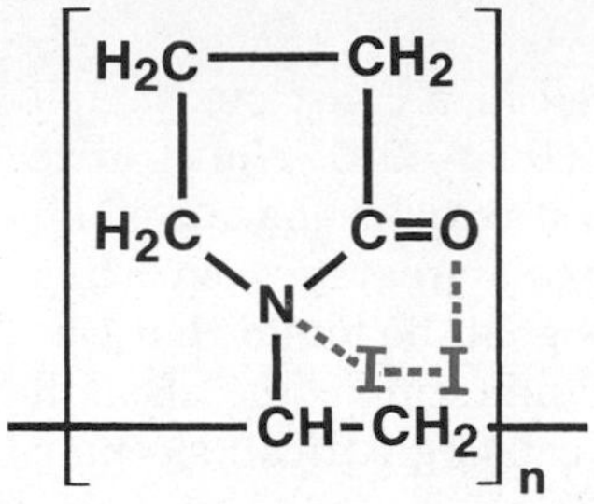

**Figure 18–7** An iodophor: complex of polyvinylpyrrolidone (PVP) and iodine.

One of the newer developments solves the problem of applying antiseptics to children. You may remember the sting of "iodine" when applied to a scratch or a wound. Old-fashioned iodine is a solution of iodine ($I_2$) in alcohol with a little potassium iodide (KI) to increase the solubility of the iodine. The alcohol causes most of the pain. Now there are polymers, such as polyvinylpyrrolidone, that complex iodine molecules (Fig. 18–7); the products (iodophors) are soluble in water. The resultant solution is a very efficient and painless disinfectant. The iodophors are active ingredients in a popular mouthwash and in restaurant glassware disinfectants.

**A tincture is an alcoholic solution. Tincture of iodine is a solution of water, iodine, and potassium iodide in ethanol.**

Because such compounds are generally toxic to living matter, it is necessary to utilize only dilute solutions and then only on the skin. Although they help to prevent the spread of disease, they are practically useless in its treatment because they act nonspecifically against all cells with which they come in contact. They are to be distinguished from antibiotics, which act more selectively against infecting bacteria than against "host" cells within the human body.

## Antimicrobial Medicines

In our time, the quest for drugs to wipe out disease due to microorganisms has been virtually fulfilled by the ***antibiotics*** (Table 18–7). In the original sense, an antibiotic is a substance such as penicillin, produced by a microorganism, that inhibits the growth of another organism. It has become common practice to include synthetic chemicals such as the sulfa drugs in a discussion of antibiotics.

Since the antibiotics are so efficient, they were the first of what came to be called "miracle" drugs. Their job generally is to aid the white blood cells by stopping bacteria from multiplying. When a person falls victim to or is killed by a disease, it means that the invading bacteria have multiplied faster than the white blood cells could devour them, and that the bacterial toxins increased more rapidly than the antibodies could neutralize them. The action of the white blood cells and antibodies plus an antibiotic is generally enough to repulse an attack of the disease germs.

**An antibody is a specific protein produced to protect the organism from harmful invading molecules.**

**TABLE 18–7 Deaths per 100,000 Americans Due to Different Causes**

| | 1900 | 1977 |
|---|---|---|
| Infectious diseases | 500 | 40 |
| Influenza and pneumonia | 210 | 23 |
| Diphtheria | 40 | fewer than 1 |
| Typhoid and paratyphoid | 30 | fewer than 1 |
| Whooping cough | 10 | fewer than 1 |
| Gastrointestinal problems | 150 | fewer than 10 |

## The Sulfa Drugs

Sulfa drugs represent a group of compounds discovered in a conscious search for antibiotics. In 1904, the German chemist Paul Ehrlich (1854–1915; Nobel prize in 1908) realized that infectious diseases could be conquered if toxic chemicals could be found that attacked parasitic organisms within the body to a greater extent than they did host cells. Ehrlich achieved some success toward his goal; he found that certain dyes that were used to stain bacteria for microscopic examination could also kill the bacteria. This led to the use of dyes against organisms causing African sleeping sickness and arsenic compounds against those causing syphilis.

**Large doses of sulfa drugs are required compared with the doses of "true antibiotics."**

In 1935, after experimenting with several drugs, Gerhard Domagk, a pathologist in the I. G. Farbenindustrie Laboratories in Germany, found that Prontosil, a dye, was somewhat effective against bacterial infection in mice. Prontosil can be changed to ***sulfanilamide,*** which is very effective.

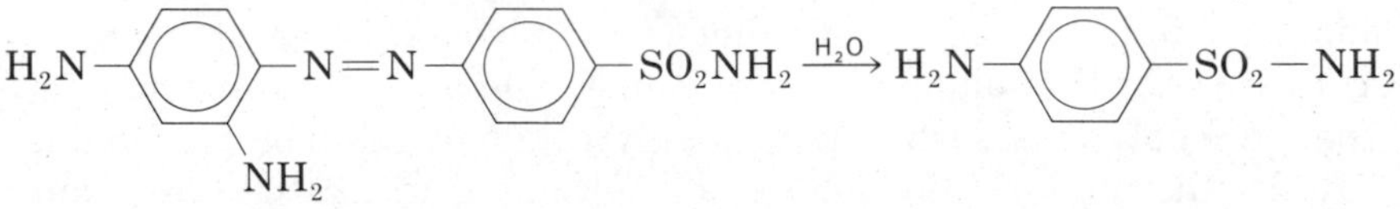

$$H_2N-C_6H_3(NH_2)-N=N-C_6H_4-SO_2NH_2 \xrightarrow{H_2O} H_2N-C_6H_4-SO_2-NH_2$$

PRONTOSIL SULFANILAMIDE

This discovery led to the synthesis and testing of a large number of related compounds in the search for drugs that are more effective or less toxic to the infected experimental animal. By 1964, more than 5,000 sulfa drugs had been prepared and tested.

**A sulfa drug mimics an essential compound.**

Sulfa drugs inhibit bacteria by preventing the synthesis of folic acid, a vitamin essential to their growth. The drugs' ability to do this apparently lies in their structural similarity to a key ingredient in the folic acid synthesis, *para*-aminobenzoic acid.

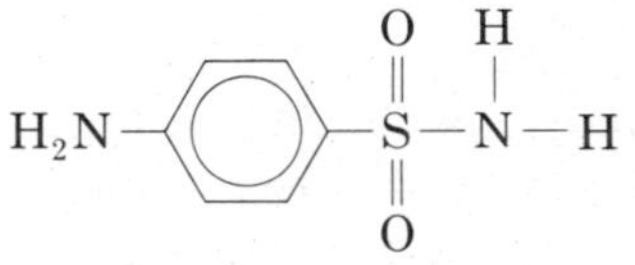

$$H_2N-C_6H_4-S(=O)_2-NH-H \qquad H_2N-C_6H_4-C(=O)OH$$

SULFANILAMIDE, A TYPICAL SULFA DRUG *p*-AMINOBENZOIC ACID

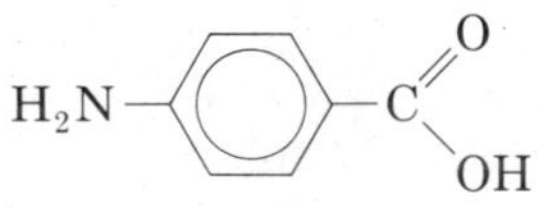

The close structural similarity of sulfanilamide and *p*-aminobenzoic acid permits sulfanilamide to be incorporated into the enzymatic reaction sequence instead of *p*-aminobenzoic acid. By bonding tightly, sulfanilamide shuts off the production of the essential folic acid, and the bacteria die of vitamin deficiency. In humans and the higher animals, *p*-aminobenzoic acid is not necessary for folic acid synthesis, so sulfa drugs have no effect on this mechanism.

### The Penicillins

Penicillin was discovered in 1928 by Alexander Fleming, a bacteriologist at the University of London, who was working with cultures of *Straphylococcus aureus,* a germ that causes boils and some other types of infections. In order to examine the cultures with a microscope, he had to remove the cover of the culture plate for a while. One day as he started work he noticed that the culture was contaminated by a blue-green mold. For some distance around the mold growth, the bacterial colonies were being destroyed. Upon further investigation, Fleming found that the broth in which this mold had grown also had an inhibitory or lethal effect against many pathogenic

A

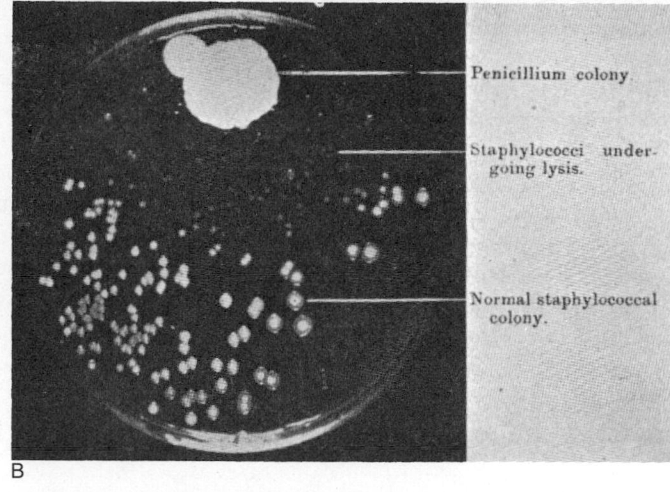


B

**Figure 18–8** *A,* Sir Alexander Fleming. *B,* A culture-plate showing the dissolution of staphylococcal colonies in the neighborhood of a colony of penicillium.

(disease-causing) bacteria. The mold was later identified as *Penicillium notatum* (the spores sprout and branch out in pencil shapes, hence the name).

***Penicillin,*** the name given to the antibacterial substance produced by the mold, apparently has no toxic effect on animal cells, and its activity is selective. The structure of penicillin (see margin) has now been determined. Many different penicillins exist, differing in the structure of the R group. Penicillin G is the most widely used in medicine. Several antibiotics, such as penicillin and bacitracin, are known to prevent cell-wall synthesis in bacteria, as shown in Figure 18–9.

R = $C_6H_5$–$CH_2$–; R–C(=O)–NH–CH–(C=O)–N–CH–S–C($CH_3$)($CH_3$)–CH(–N)–C(=O)–OH

*Penicillin G*

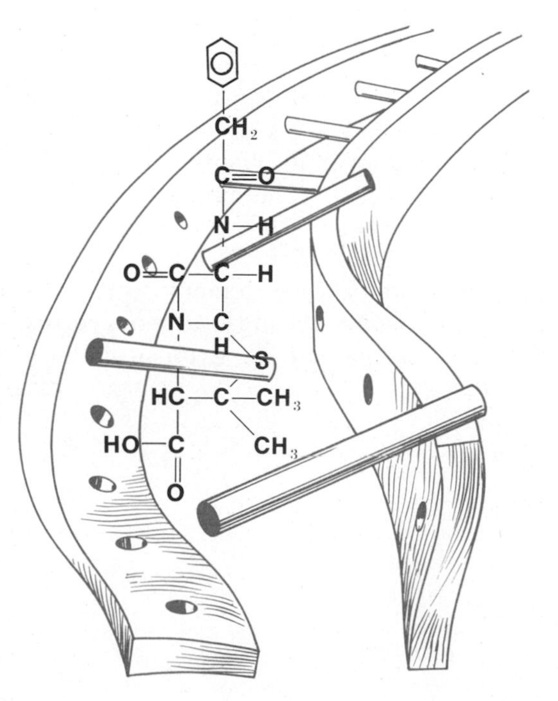


**Figure 18–9** Penicillin kills bacteria by interfering with the formation of cross links in the cell wall.

## Streptomycin and the Tetracyclines

In 1937, following collaboration with René Dubos, Selman Waksman isolated a compound from a soil organism, *Streptomyces griseus,* which came to be known as ***streptomycin*** and was released to physicians in 1947. This compound was quite successful in controlling certain types of bacteria but later had to be withdrawn because of its adverse side effects.

**Streptomycin has many undesirable side effects.**

In 1945, B. M. Duggan discovered that a gold-colored fungus, *Streptomyces aureofaciens,* produced a new type of antibiotic, ***aureomycin,*** the first of the ***tetracyclines.*** Research then stepped up to a fever pitch. Pfizer Laboratories tested 116,000 different soil samples before they discovered the next antibiotic, which they named ***terramycin.***

**Tetracyclines get their names from their four-ring structures.**

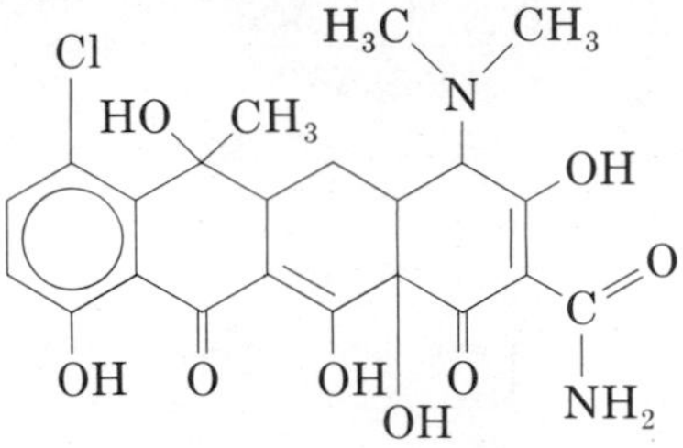

AUREOMYCIN

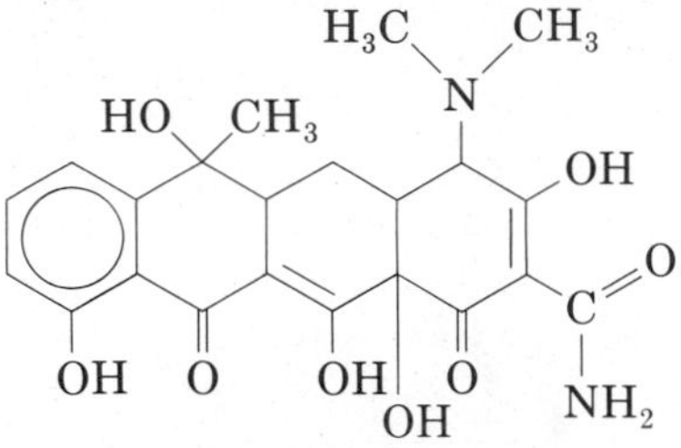

TERRAMYCIN

Compounds of the tetracycline family are so named because of their four-ring structure. One side effect of taking these drugs is diarrhea caused by the killing of the patient's intestinal flora (the bacteria normally residing in the intestines).

### The Steroid Drugs

A large and important class of naturally occurring compounds is derived from the tetracyclic structure given below.

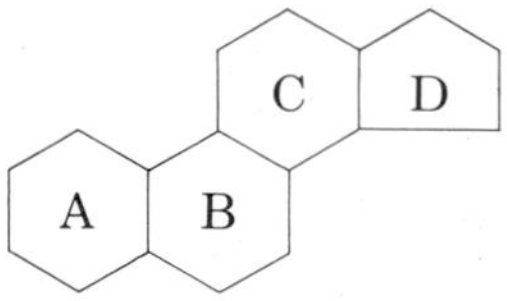

These compounds are known as ***steroids,*** and they occur in all plants and animals. The most abundant animal steroid is cholesterol, $C_{27}H_{46}O$. The human body synthesizes cholesterol and also readily absorbs dietary cholesterol through the intestinal wall. It is associated with gallstones and hardening of the arteries.

Biochemical alteration or degradation of cholesterol leads to many steroids of great importance in human biochemistry. When ***cortisone,*** one of the adrenal cortex hormones, is applied topically or injected into a diseased joint, it acts as an antiinflammatory agent and is of great use in treating arthritis.

**Cortisone is a "powerful drug" having a major effect on biological systems.**

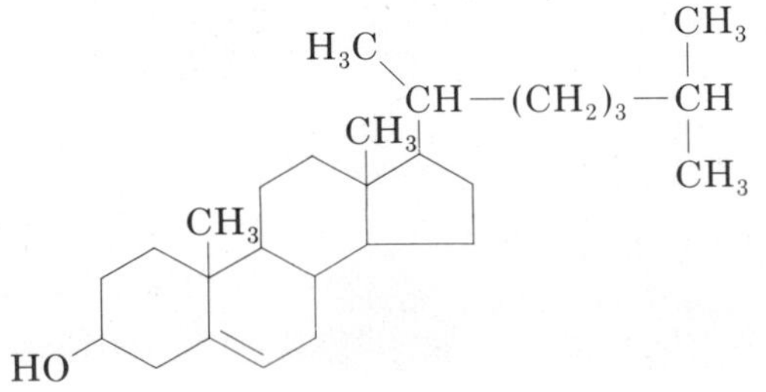

CHOLESTEROL

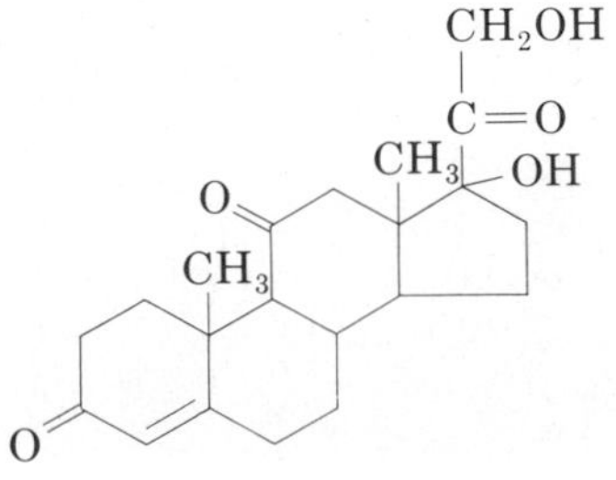

CORTISONE

Structurally related to cholesterol and cortisone are the sex hormones. One female sex hormone, ***progesterone,*** differs only slightly from an important male hormone, ***testosterone.***

$CH_3$ $C{=}O$ $CH_3$ $CH_3$ O ♀ $CH_3$ OH $CH_3$ O ♂

PROGESTERONE TESTOSTERONE

Other female hormones are estradiol and estrone, called ***estrogens.*** The estrogens differ from the other steroids discussed earlier in that they contain an aromatic A ring (in color).

O $CH_3$ HO OH $CH_3$ HO

ESTRONE ESTRADIOL

## Birth Control Pills

One of the most revolutionary medical developments of the 1960's was the worldwide introduction and use of "The Pill." More than ten million women in the United States use birth control pills. The basic feature of oral contraceptives for women is their chemical ability to simulate the hormonal processes resulting from pregnancy and, in so doing, prevent ovulation. Ovulation, the production of eggs by the ovary, ceases at the onset of pregnancy because of hormonal changes (Fig. 18–10). This same result can be produced by the administration of a variety of steroids, some of which are effective when taken orally, although the mechanism of their action and their long-term effects are not known in detail.

The active ingredients of the pill are the hormones progesterone and estrogen, or their derivatives.

### Allergens and Antihistamines

A person may have an unpleasant physiological response to poison ivy, pollen, mold, food, cosmetics, penicillin, aspirin, and even cold, heat, and ultraviolet light. In the United States about 5,000 people die yearly from bronchial asthma, at least 30 from the stings of bees, wasps, hornets, and other insects, and about 300 from ordinary doses of penicillin. The reason: ***allergy.*** About one person in ten suffers from some form of allergy; more than 16 million Americans suffer from hay fever.

**An allergy is a physiological response such as sneezing, runny nose, coughing, dermatitis, etc., to the introduction of a foreign substance. This foreign substance is called an *allergen.***

An allergy is an adverse response to a foreign substance or to a physical condition that produces no obvious ill effects in *most* other organisms, including humans. An ***allergen*** (the substance that initiates the allergic reaction) is, in most cases, a highly complex substance—usually a protein. Some are polysaccharides or compounds

**Figure 18–10** Some of the control processes in the female reproductive cycle. The pituitary gland, located in the brain area, sends out hormones that cause the ovary to release other hormones. Estrone and estradiol initiate the deterioration of the old wall of the uterus and the formation of a new wall. If the ovum is fertilized, it attaches to the uterine wall, and estrogens and progesterone are released to prevent further ovulation. The Pill serves the same function as the estrogens and progesterone in that it prevents the ovulation process.

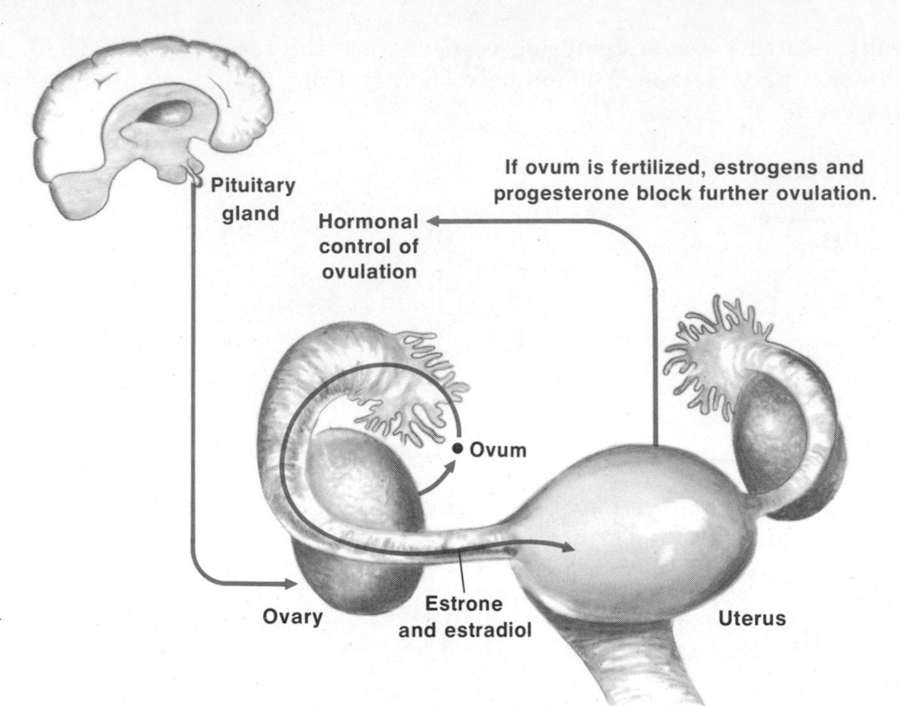


Most allergens are high molecular weight substances.

formed by combining a protein and a polysaccharide. Usually allergens have a molecular weight of 10,000 or more.

The principal allergen of ragweed pollen, a major allergy-producer, has been isolated and is named ragweed antigen E. It is a protein with a molecular weight of about 38,000; it represents only about 0.5% of the solids in ragweed pollen, but contributes about 90% of the pollen's allergenic activity. A mere $1 \times 10^{-12}$ g of antigen E injected into an allergic person is enough to induce a response.

$10^{-12}$ is 0.000000000001

The allergens come in contact with special cells in the nose and breathing passages to which a particular type of antibody is attached, the IgE antibody, which has a molecular weight of about 196,000. Allergic individuals have 6 to 14 times more IgE in their blood serum than nonallergic people. The IgE is formed in the nose, bronchial tubes, and gastrointestinal tract, and binds firmly to specific cells, called ***mast cells,*** in these regions.

Antibodies are high molecular weight proteins, called immunoglobulins, which attack foreign proteins such as allergens.

Antigen E from ragweed reacts with the IgE antibody attached to the mast cells, forming antigen-antibody complexes. The formation of these antigen-antibody complexes leads to the release of so-called "allergy mediators" from special granules in the mast cells (Fig. 18–11). The most potent of these mediators found so far is ***histamine.*** Although it is widely distributed in the body, it is especially concentrated in the 250–300 granules of the mast cells. Histamine accounts for many, if not most, of the symptoms of hay fever, bronchial asthma, and other allergies.

Histamine causes runny noses, red eyes, and other hay fever symptoms.

$H_2NCH_2CH_2$–(imidazole ring: N, N–H)

HISTAMINE

The chemical mediators such as histamine must be released from the cell to cause the symptoms of allergy. The release mechanism is an energy-requiring pro-

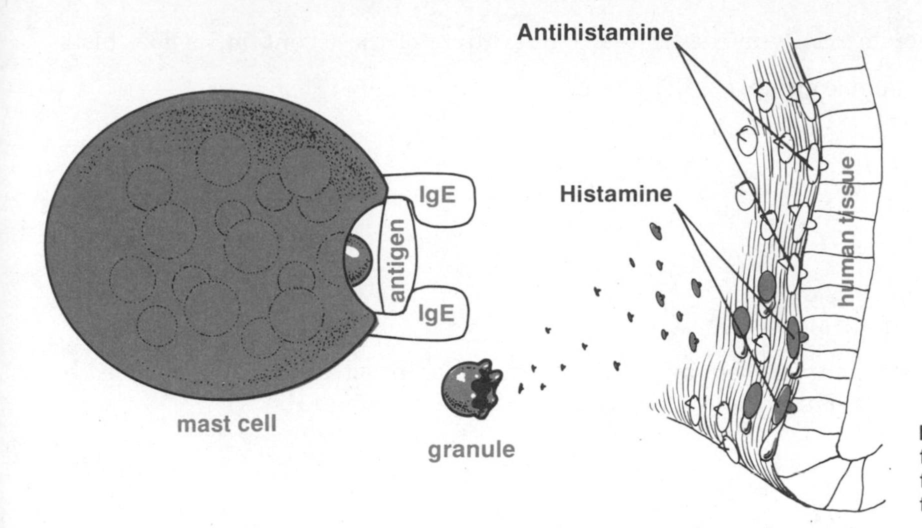


**Figure 18–11** A postulated mechanism for the cause of and relief from hay fever. The details are described in the text.

cess in which the granules may move to the outer edge of the living cell and, without leaving the cell, discharge their contents of histamine through a temporary gap in the cell membrane. This sends histamine on its way to produce the toxic effects of hay fever.

Treatment consists of three procedures: avoidance (Fig. 18–12), desensitization, and drug therapy. Desensitization therapy is costly and inconvenient, since 20 or more injections are required to achieve what is usually a partial cure. One chemical idea of desensitization is to inject a blocking antibody that preferentially reacts with the allergen so that it cannot react with the IgE allergy-sensitizing antibody. This breaks the chain of events leading to the release of histamine or other allergy-producing mediators. Many small injections, spaced in time, are required to build up a sufficient level of the blocking antibody.

Epinephrine (adrenalin), steroids, and antihistamines are effective drugs in treating allergies. The first two are particularly effective in treating bronchial asthma, whereas the ***antihistamines,*** introduced commercially in the United States in 1945, are the most widely used drugs for treating allergies. More than 50 antihistamines

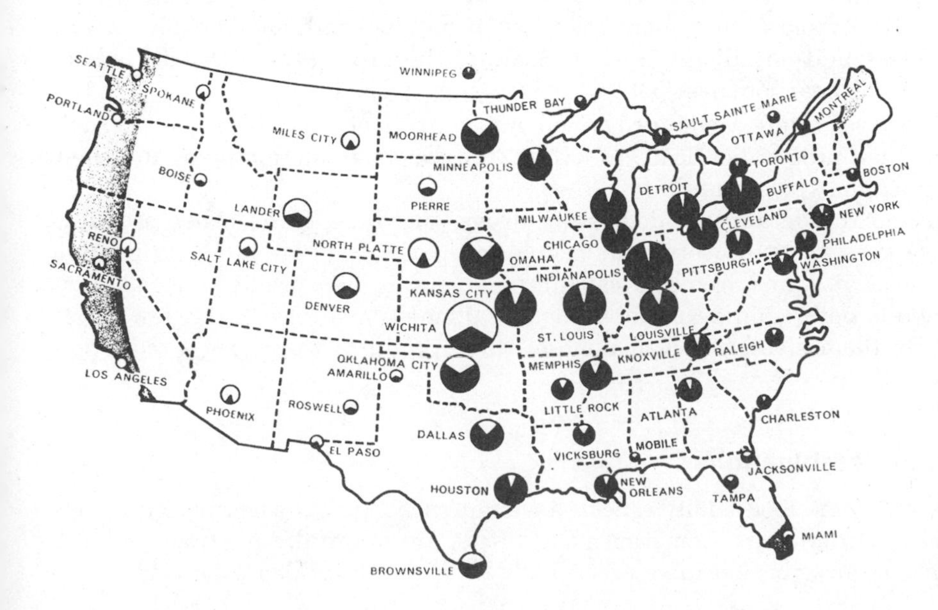


**Figure 18–12** In this Abbott Laboratories map, the size of each circle represents the amount of all late-summer and fall pollens found in the air in each city. Dark portions show amount of ragweed pollen. Shaded areas are regions of low pollen count.

are offered commercially in the United States. Many of these contain, as does histamine, an ethylamine group, ($-CH_2CH_2N\langle$):

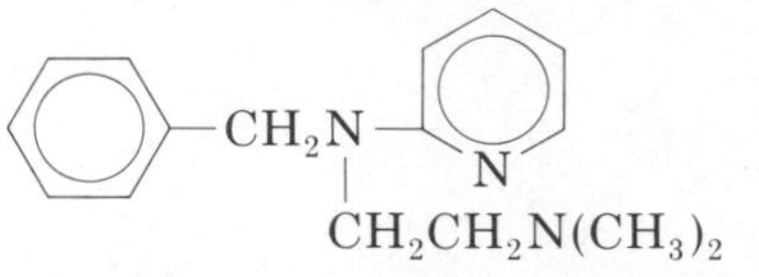

$CH_2N$ — N
$CH_2CH_2N(CH_3)_2$

PYRIBENZAMINE
(AN IMPORTANT ANTIHISTAMINE)

These drugs act competitively by occupying the receptor sites normally occupied by histamine on cells. This, in effect, blocks the action of histamine.

## Diet Pills

**A food Calorie is actually a kilocalorie, or 1,000 calories.**

**Many reducing diets are not nutritionally adequate.**

The desired rate of weight loss for most overweight people ranges from about one to two pounds per week. For most women, diets ranging between 1,200 and 1,500 food Calories per day will bring about a satisfactory weight loss. For men, the range is between 1,500 and 2,000 Calories per day. An intake of less than 1,200 Calories is generally not advised.

A variety of special, rather unusual diets have been reported to be effective in weight reduction. These diets have been known under such names as the grapefruit diet, the water diet, the 10-day diet, and others. The problem with the vast majority of these diets is that they are often made up of only a few foods, so that they are not nutritionally adequate. Five basic types of nutrients are essential: protein, fat, carbohydrate, vitamins, and certain minerals. They must be included in the diet for normal health. If the diet is lacking in any one category, the body suffers. Therefore, a balanced diet plus water is required, even if the amounts are reduced.

Many individuals have difficulty staying on a diet, and they must be highly motivated in order to maintain a caloric level low enough to bring about a loss of weight. Appetite-depressing drugs, known as ***anorexigenic drugs,*** have been used by some physicians to curb the appetite. More than two billion diet pills are distributed per year. The ingredients include amphetamines (which suppress the appetite), digitalis (which affects the heart), various diuretics (which increase the amount of water excreted), thyroid (which increases rates of metabolism), and prednisone (an antiinflammatory and antiallergic agent). Some of the drugs are potentially addictive; others tax the heart or cause potassium loss from the body.

**Diet pills should be used under medical supervision.**

By themselves anorexigenic drugs will not control obesity. Reliance on such drugs, rather than proper diet, leads to failure in a weight-reduction plan, and their unsupervised use may be harmful.

In addition to the appetite-depressing drugs, there are many other products that are advertised to help overweight people lose weight. These products include methyl cellulose, vitamins, iron and calcium compounds, benzocaine (a local anesthetic), dextrose, potassium *p*-aminobenzoate, caffeine, flavorings, and a few other substances. By themselves, however, none of these products will cause a reduction in weight.

## Drugs in Combinations

Drugs, like some food additives, can have enhanced effects when placed in certain chemical environments. Sometimes the effects are harmful, sometimes helpful. Take the case of an aging business executive who took an antidepressant and then ate a meal that included aged cheese and wine. The antidepressant is an inhibitor of monoamine oxidase, an enzyme that helps to control blood pressure. Both the aged cheese he ate and the wine he drank contained pressor amines, which raise blood pressure. Without the controlling effect of the enzyme, these amines skyrocketed his blood pressure and caused a stroke. Neither the amines nor the antidepressant alone would have been likely to cause the stroke, but the combination did.

**Pressor amines tend to increase blood pressure.**

Likewise, people who take digitalis for heart trouble and for reducing the sodium level in the blood should take aspirin only under medical supervision. Aspirin can cause a 50% reduction in salt excretion for three or four hours after it is taken.

Alcohol increases the action of many antihistamines, tranquilizers, and drugs such as reserpine (for lowering blood pressure) and scopolamine (contained in many over-the-counter nerve and sleeping preparations), making such combinations extremely dangerous. Staying away from dangerous alcohol-drug combinations is not as easy as it may seem. Many people fail to realize that a large number of over-the-counter preparations—such as liquid cough syrup and tonics—contain appreciable amounts of alcohol.

Not all drug combinations are bad. Doctors have been highly successful in prolonging the lives of leukemia and other cancer victims with combinations of drugs that individually could not do the job. Resistant kidney disease has also responded to drug combinations in cases in which single drugs were ineffective.

Perhaps the best advice is to take medicine only when you are seriously ill, making sure that a physician knows what you are taking.

## Self-Test 18-C

1. What acid is secreted by the walls of the stomach? ____________
2. Is an analgesic a pain killer or a germ fighter? ____________
3. Select the natural compound from which the other is made: morphine, heroin. ____________
4. A molecule of aspirin has two functional groups. What are they? ____________ and ____________
5. An antipyretic reduces ____________.
6. Penicillin was discovered by ____________ ____________.
7. Allergens are complex proteins with a usual molecular weight of 10,000 or more. (True/False)
8. Two groups of medicines that act by mimicking other compounds are ____________ ____________, which mimic *p*-aminobenzoic acid, and ____________, which mimic histamine.

## Matching Set

| | | |
|---|---|---|
| ______ | 1. histamine | **a** interferes with cell wall structure |
| ______ | 2. cortisone | **b** antiseptic |
| ______ | 3. tetracycline | **c** source of morphine |
| ______ | 4. penicillin | **d** analgesic |
| ______ | 5. sulfa drug | **e** antacid |
| ______ | 6. iodine | **f** a steroid |

| | | | |
|---|---|---|---|
| _______ | 7. procaine (novocaine) | **g** | causes symptoms of hay fever |
| _______ | 8. opium poppy | **h** | antibiotic containing a four-ring structure |
| _______ | 9. dihydroxyaluminum sodium carbonate | **i** | sulfanilamide |

## Part III—Beauty Aids

**Cosmetic products require careful testing before they are used on humans.**

People find many reasons for applying various chemical preparations ***(cosmetics)*** to their skin and hair. We wish to be clean, beautiful, healthy, and pleasing to others. Considerable progress has been made in producing chemical products that color hair and skin, disinfect our body surfaces, and suppress unwanted body odor. The purpose here is not to enable you to make face creams or hair sprays, but to give you a better understanding of the basic chemistry involved. In fact, amateur chemical preparations should be avoided, since toxic reactions are often encountered when impure or otherwise harmful concoctions are used without proper testing.

### Skin and Hair

The skin, hair, and nails are protein structures. Skin (Fig. 18–13), like the other organs of the body, is not uniform tissue. The outermost layer, called the ***stratum***

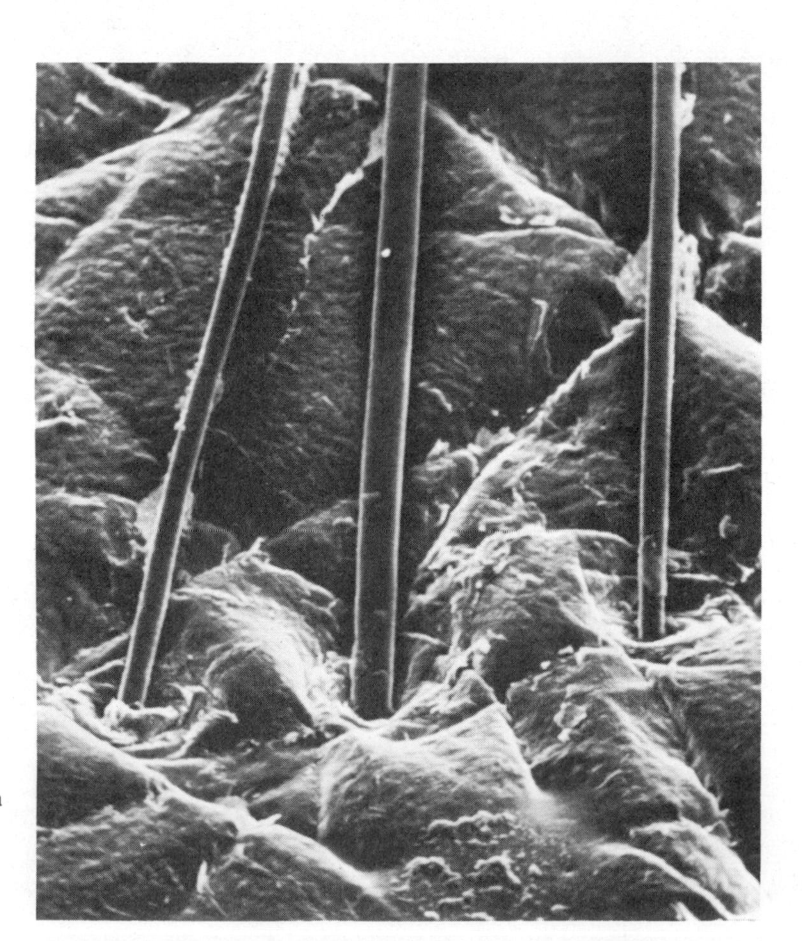

**Figure 18–13** Replica of the surface of human forearm skin, showing three hairs emerging from the skin (× 225).

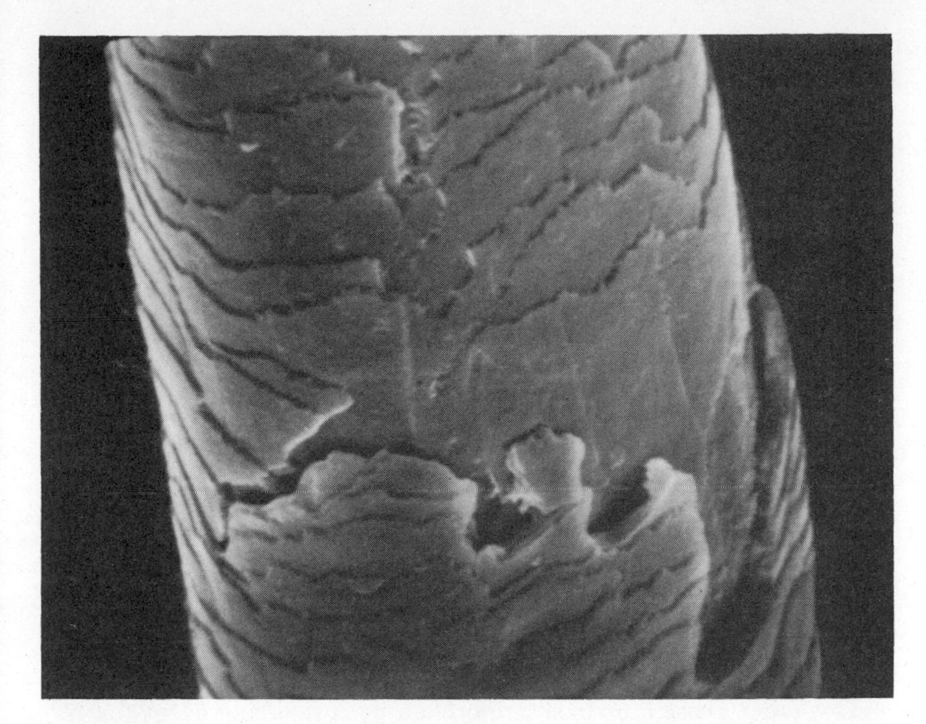

**Figure 18–14** Electron micrograph of human hair. Note the layers of keratinized cells

***corneum,*** or ***corneal layer,*** is where most cosmetic preparations for the skin act. The corneal layer is composed principally of dead cells with low moisture content and a slightly acidic surface pH of about 4. The principal protein of the corneal layer is ***keratin,*** which is composed of about 22 different amino acids. Its structure renders it insoluble in, but slightly permeable to, water. In order to control the moisture content of the corneal layer so that it does not dry out and slough off too quickly, moisturizers are added to the skin.

**Keratin is skin protein.**

Hair (Fig. 18–14) is composed principally of keratin. An important difference between hair keratin and other proteins is its high content of the amino acid ***cystine.*** About 16–18% of hair protein is cystine, but only 2.3–3.8% of the keratin in corneal cells is cystine. This amino acid plays an important role in the structure of hair.

$$\begin{array}{l}
\qquad\quad\;\; NH_2 \\
\qquad\quad\;\;\; | \\
HOOC-CH-CH_2-S \\
\qquad\qquad\qquad\qquad\; | \\
HOOC-CH-CH_2-S \\
\qquad\quad\;\;\; | \\
\qquad\quad\;\; NH_2
\end{array}$$

*Cystine*

The toughness of both skin and hair is due to the bridges between different protein chains, such as hydrogen bonds and $-S-S-$ linkages, called ***disulfide*** linkages.

$$\begin{array}{l}
\quad\;\; | \qquad\qquad\qquad\qquad\quad | \\
O=C \qquad\qquad\qquad\qquad N-H \\
\quad\;\; | \qquad\qquad\qquad\qquad\quad | \\
H-CCH_2-S-S-CH_2C-H \\
\quad\;\; | \qquad\qquad \uparrow \qquad\qquad | \\
H-N \qquad\qquad\qquad\qquad C=O \\
\quad\;\; | \qquad\qquad\qquad\qquad\quad |
\end{array}$$

*Disulfide linkage*

Another type of bridge between two protein chains, which is important in keratin as well as in all proteins, is the ***ionic bond.*** Consider the interaction between a lysine $-NH_2$ group and a carboxylic $-COOH$ group of glutamic acid on a neighboring protein chain. At pH 4.1, the presence of an $-NH_3^+$ group and a $-COO^-$

**The structures of protein tissues are due in part to disulfide crosslinks and to ionic bonds between "molecules."**

group is most favorable for keratin. If the two charges approach closely, an ionic bond is formed.

$$\text{HCCH}_2\text{CH}_2\text{CH}_2\text{CH}_2\text{NH}_2 + \text{HOOCCH}_2\text{CH}_2\text{CH} \xrightarrow{\text{at pH 4.1}}$$

LYSINE GLUTAMIC ACID

$$\text{HCCH}_2\text{CH}_2\text{CH}_2\text{CH}_2\text{NH}_3^+ \;{}^-\text{OOCCH}_2\text{CH}_2\text{CH}$$

IONIC BOND

As the pH rises above 4, keratin will swell and become soft as these crosslinks are broken. This is an important aspect of hair chemistry.

## Changing the Shape of Hair

When hair is wet, it can be stretched to one and a half times its dry length because water (pH 7) weakens some of the ionic bonds and causes swelling of the keratin. Imagine the disulfide crosslinks remaining between two protein chains in hair as in Figure 18–15(*a*). Winding the hair on rollers causes tension to develop at the crosslinks (*b*). In "cold" waving, these crosslinks are broken by a reducing agent (*c*), relaxing the tension. Then, an oxidizing agent regenerates the crosslinks (*d*) and the hair holds the shape of the roller. The chemical reactions in simplified form are shown in Figure 18–16.

$$\text{HS}-\text{CH}_2-\text{C}(=\text{O})-\text{OH}$$

THIOGLYCOLIC ACID

The most commonly used reducing agent is thioglycolic acid. The common oxidizing agents used include hydrogen peroxide, perborates ($NaBO_2 \cdot H_2O_2 \cdot 3H_2O$), and sodium or potassium bromate ($KBrO_3$). A typical neutralizer solution contains one or more of the oxidizing agents dissolved in water. The presence of water and strong base in the oxidizing solution also helps to break and re-form hydrogen bonds between adjacent protein molecules. However, too-frequent use of strong base causes hair to become brittle and lifeless.

**Hair can be straightened by the same solutions. It is simply "neutralized" (or oxidized) while straight (no rolling up).**

Various additives are present in both the oxidizing and the reducing solutions in order to control pH, odor, and color, and for general ease of application. A typical waving lotion contains 5.7% thioglycolic acid, 2.0% ammonia, and 92.3% water.

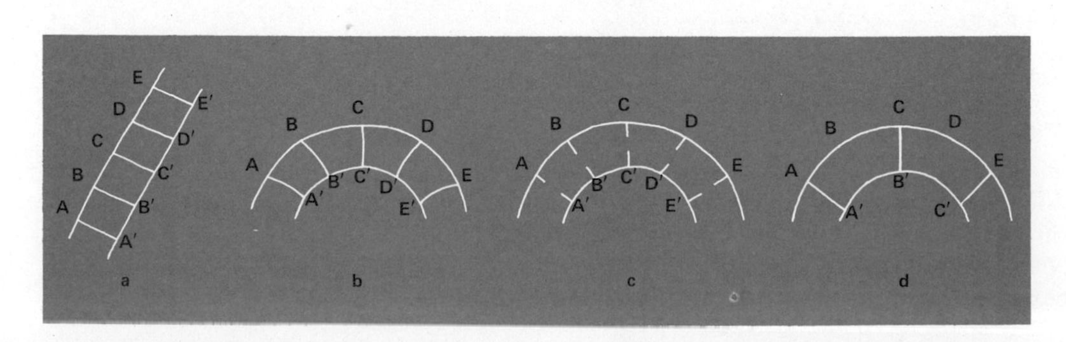


**Figure 18–15** A schematic diagram of a permanent wave.

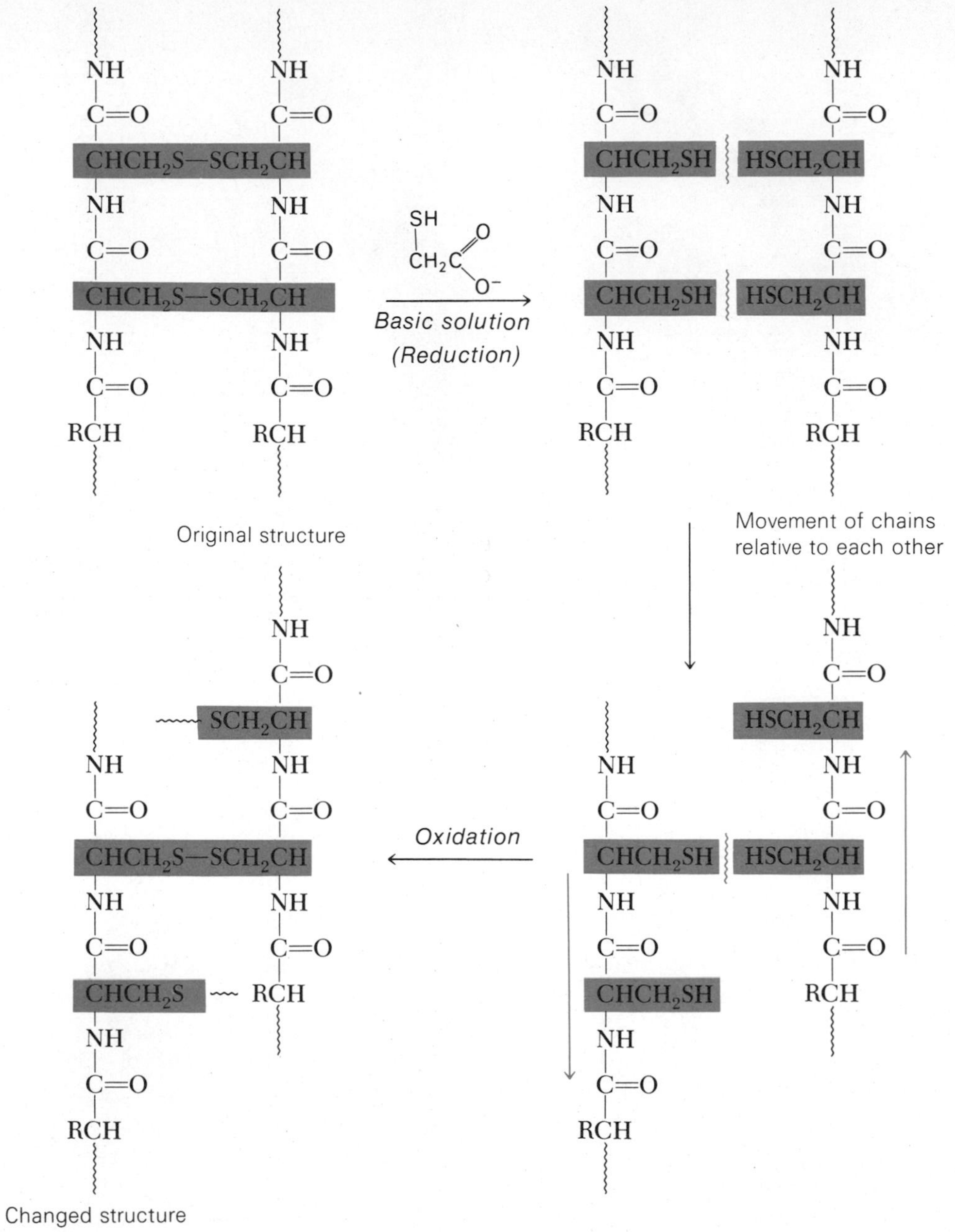

**Figure 18–16** Structural changes that occur in hair during a permanent wave.

## Hair Coloring and Bleaches

Hair contains two pigments: brown-black melanin and an iron-containing red pigment. The relative amounts of each actually determine the color of the hair. In deep black hair melanin predominates and in light-blond, the iron pigment predominates. The depth of the color depends upon the size of the pigment granules.

**Melanin—black.**

Formulations for dyeing hair vary from temporary coloring (removable by shampoo), which is usually achieved by means of a water-soluble dye that acts on the surface of the hair, to semipermanent dyes, which penetrate the hair fibers to a great extent (Fig. 18–17). These often consist of cobalt or chromium complexes of dyes dissolved in an organic solvent. Permanent dyes are generally "oxidation" dyes.

**Iron pigment—red.**

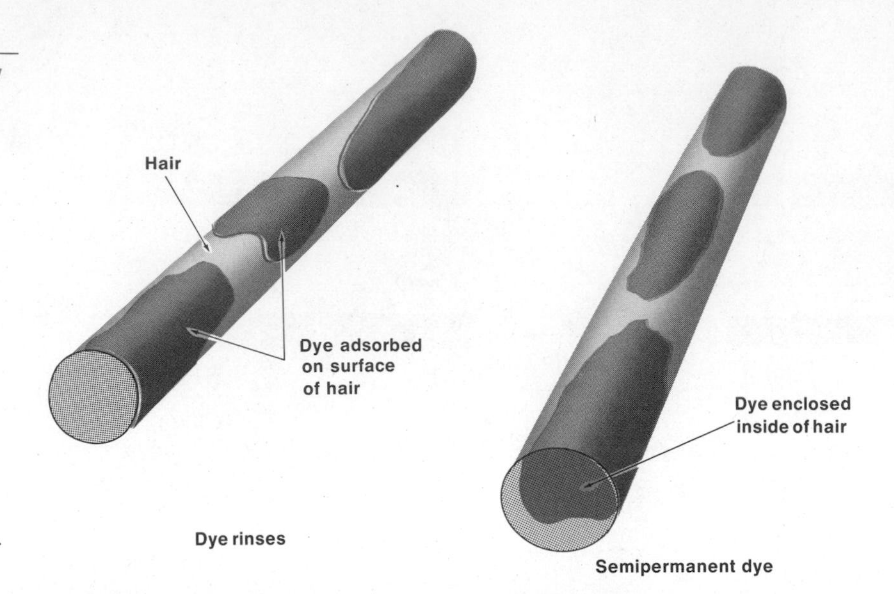


**Figure 18–17** Methods of dyeing hair.

They penetrate the hair, and then are oxidized to give a colored product that is permanently attached to the hair by chemical bonds or that is much less soluble than the reactant molecule. Permanent hair dyes generally are derivatives of phenylenediamine.

$$H_2N-C_6H_4-NH_2$$

*p*-PHENYLENEDIAMINE

Some of the hair dyes are suspected of being carcinogenic.

Phenylenediamine dyes hair black. A blond dye can be formulated with *p*-aminodiphenylaminesulfonic acid

$$NH_2-C_6H_4-NH-C_6H_4-SO_3H$$

or *p*-phenylenediaminesulfonic acid.

Just about any shade of hair color can be prepared by varying the modifying groups on certain basic dye structures.

$$H_2N-C_6H_3(SO_3H)-NH_2$$

The active compounds are applied in an aqueous soap or detergent solution containing ammonia to make the solution basic. The dye material is then oxidized by hydrogen peroxide to develop the desired color. The amines are oxidized to nitro compounds.

$$\underset{\text{AMINE}}{-NH_2} + 3H_2O_2 \xrightarrow{\text{oxidation}} \underset{\text{NITRO COMPOUND}}{-NO_2} + 4H_2O$$

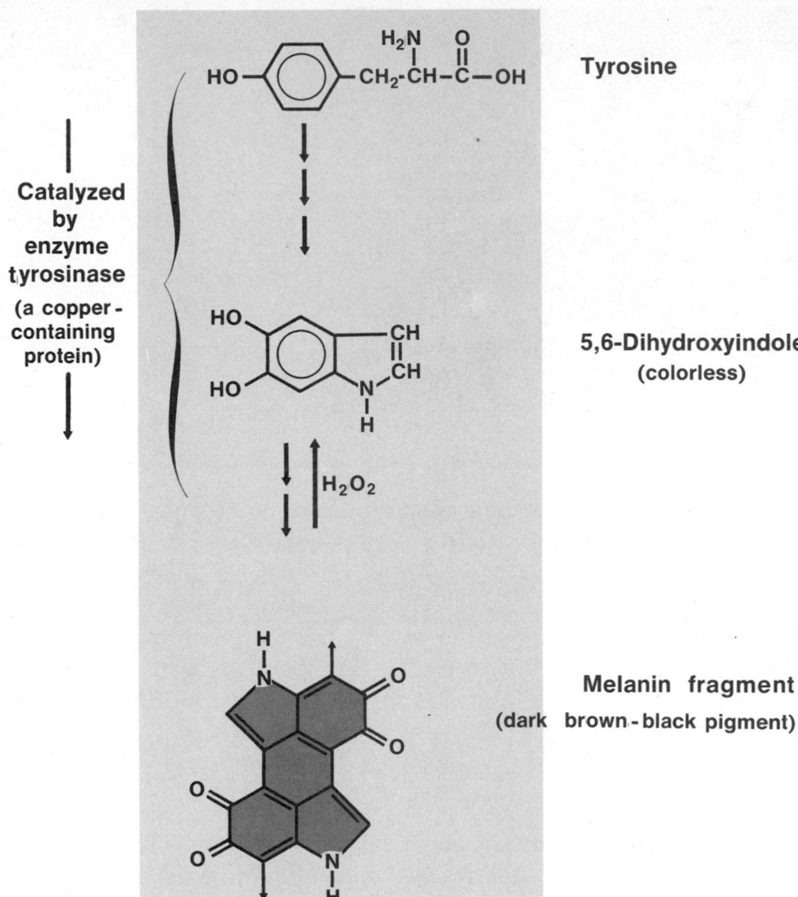


**Figure 18–18** Bleaching of the hair by hydrogen peroxide. There are several chemical intermediates between the amino acid—tyrosine—and the hair pigment—melanin, which is partly protein. Hydrogen peroxide oxidizes melanin back to colorless compounds, which are stable in the absence of tyrosinase (found only in the hair roots). Melanin is a high molecular weight polymeric material of unknown structure. The structure shown here is only a segment of the total structure.

Hair can be bleached by a more concentrated solution of hydrogen peroxide, which destroys the hair pigments by oxidation. The solutions are made basic with ammonia to enhance the oxidizing power of the peroxide. Parts of the chemical process are given in Figure 18–18. This drastic treatment of hair does more than just change the color. It may destroy sufficient structure to render the hair brittle and coarse.

## Hair Sprays

**Hair spray coats the hair with a plastic film.**

Hair sprays are essentially solutions of a resin in a volatile solvent whose purpose, when sprayed on hair, is to furnish a film with sufficient strength to hold the hair in place after the solvent has evaporated. A common resin in hair sprays is the addition polymer, polyvinylpyrrolidone (PVP).

```
CH2—CH2           CH2—CH2            CH2—CH2
|    |            |    |             |    |
CH2  C=O          CH2  C=O           CH2  C=O
  \ /               \ /                \ /
   N       H      /  N       H   \      N       H
   |       |     (   |       |    )     |       |
 —C————————C—————(———C———————C————)—————C———————C—
   |       |     (   |       |    )     |       |
   H       H      \  H       H   / n    H       H
```

POLYVINYLPYRROLIDONE (PVP)

Figure 18–19 Film of hair spray. Hair spray was allowed to dry on white surface and was then pulled up to reveal film.

**There are several dangers in breathing the vapor of hair sprays, such as the danger of possible carcinogens acting on delicate lung tissue and the danger of asphyxiation by the plastic coating lining the lungs.**

The resin is blended in hair spray formulations with a plasticizer, a water repellent, and a solvent-propellant mixture. The plasticizer makes the plastic more pliable. The resin concentration of hair sprays is of the order of 4%. Other additives, such as silicone oils, are often put into hair sprays to give the hair a sheen.

Since PVP tends to pick up moisture, other less hygroscopic polymers are beginning to replace PVP in hair sprays. For example, significantly better moisture control is obtained with a copolymer made from a 60/40 ratio of vinylpyrrolidone and vinyl acetate.

## Depilatories

The purpose of a depilatory is to remove hair chemically. Since skin is sensitive to the same kind of chemical attack as hair, such preparations should be used with caution and, even then, some attack on the skin is almost unavoidable. Because of this, the interval between applications of a depilatory should be of the order of a week or so. It should never be used on skin that is infected or that has a rash, and should not be followed by application of a deodorant with its *astringent* (contracting) action. If the sweat pores are closed by the deodorant, the caustic chemicals are retained and can do considerable harm. If the sweat pores are open, the body fluids will dilute and wash the caustic chemicals to the outside of the body.

The chemicals used as depilatories include sodium sulfide, calcium sulfide, strontium sulfide (water-soluble sulfides), and calcium thioglycolate [$Ca(HSCH_2COO)_2$], the calcium salt of the compound used to break S—S bonds between protein chains in permanent waving. A typical cream depilatory contains calcium thioglycolate (7.5%), calcium carbonate (filler, 20%), calcium hydroxide (provides basic solution, 1.5%), cetyl alcohol ([$CH_3(CH_2)_{15}OH$], skin conditioner, 6%), sodium lauryl sulfate (detergent, 0.5%), and water (64.5%).

The water-soluble sulfides are all strong bases in water, as indicated by the hydrolysis of the sulfide ion:

$$\underset{\text{SULFIDE}}{S^{2-}} + H_2O \longrightarrow HS^- + \underset{\text{HYDROXIDE}}{OH^-}$$

For example, a 0.1 M solution of $Na_2S$ has a pH of about 13, a strongly basic solution. The compounds act chemically on the hair to disrupt bonds in the protein chains and cause it to disintegrate by hydrolyzing to soluble amino acids and small peptides, which may be removed.

**0.1 M means 0.1 mole of $Na_2S$ (7.81 g) dissolved in sufficient water to make one liter of solution.**

The area on which a depilatory has been used should be washed with soap and water, dried, and then treated with small amounts of talcum powder.

## Killing Germs



Harmful microorganisms (germs) are always with us in large numbers. Fortunately, there are so many nonpathogenic microorganisms among them that there is little room left for the pathogenic kinds. Depending on body location, healthy skin may host as many as a million microorganisms per square centimeter.

Things we do to our skin, such as shaving, can upset the balance between harmful and harmless bacteria and promote infection by pathogenic microorganisms. ***Disinfectants*** are chemicals used to kill these organisms before they can overcome the skin's defenses. Most commonly used are the alcohols. These are the only disinfectants used in many aftershave preparations. Maximum effectiveness of ethanol is reached at a concentration of 70%, while isopropyl (rubbing) alcohol is most effective at a concentration of 50%. These alcohols kill germs apparently by hydrogen bonding with water, which dehydrates the cellular structure of the germ.

**Alcohols dehydrate microbes.**

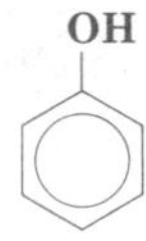


PHENOL

One widely used disinfectant is phenol; its aqueous solution is known as carbolic acid. An —OH group attached to the benzene ring is slightly acidic. The disinfectant action of phenol was discovered in 1867 by Sir Joseph Lister, who introduced it into surgery. It appears that phenol kills bacteria by denaturing their cellular proteins. Today, about one third of all toilet soaps sold contain some derivative of phenol.

**To denature a protein is to break down its structure.**

## Deodorants

The 2,000,000 sweat glands on the body surface are primarily used to regulate body temperature via the cooling effect produced by the evaporation of the water they secrete. This evaporation of water leaves solid constituents, mostly sodium chloride, as well as smaller amounts of proteins and other organic compounds. Body odor results largely from amines and hydrolysis products of fatty oils (fatty acids, acrolein, etc.) emitted from the body, and from bacterial growth within the residue from sweat glands. Sweating is both normal and necessary for the proper functioning of the human body; sweat itself is quite odorless, but the bacterial decomposition products are not.

**Body odor is promoted by bacterial action.**

There are three kinds of deodorants: those that directly "dry up" perspiration or act as astringents, those that have an odor to mask the odor of sweat, and those that remove odorous compounds by chemical reaction. Among those that have astringent action are hydrated aluminum sulfate, hydrated aluminum chloride ($AlCl_3 \cdot 6H_2O$), aluminum chlorohydrate [actually aluminum hydroxychloride, $Al_2(OH)_5Cl \cdot 2H_2O$ or $Al(OH)_2Cl$ or $Al_6(OH)_{15}Cl_3$], and alcohols. Those compounds that act as deodorizing agents include zinc peroxide, essential oils and perfumes, and a variety of mild antiseptics. Zinc peroxide removes odorous compounds by oxidizing the amines and fatty acid compounds. The essential oils and perfumes absorb or otherwise mask the odors, and the antiseptics are generally oxidizing or reducing agents that kill the odor-causing bacteria.

**An astringent closes the pores, thus stopping the flow of perspiration.**

## Skin Preparations for Health and Beauty

To remain healthy, the moisture content of skin must stay near 10%. If it is higher, microorganisms grow too easily; if lower, the corneal layer flakes off. Washing skin removes fats that help retain the right amount of moisture. If dry skin is treated with a fat after washing, it will be protected until enough natural fats have been regenerated.

**Skin with a low fat content tends to be dry.**

Lanolin is an excellent skin softener ***(emollient)*** and is a component of many cosmetics. It is a complex mixture of esters from hydrated wool fat. The esters are derived from 33 different alcohols of high molecular weight and 37 fatty acids. Cholesterol, a common alcohol in lanolin, is found both free and in esters. Cholesterol appears to give fat mixtures the property of absorbing water. This is one factor that

**Lanolin is grease from wool.**

makes lanolin an excellent emollient. With its high proportion of free alcohols, particularly cholesterol, and hydroxyacid esters, lanolin has the structural groups (—OH) to hydrogen-bond water (to keep the skin moist) and to anchor within the skin (the fatty acid and ester hydrocarbon structures; see Fig. 18–20).

### *Creams*

Colloids are intermediate in size between small molecules and clumps of molecules sufficiently large to precipitate.

Creams are generally emulsions of either an oil-in-water type or a water-in-oil type. An ***emulsion*** is simply a colloidal suspension of one liquid in another. The oil-in-water emulsion has tiny droplets of an oily or waxy nature dispersed throughout a water solution (homogenized milk is an example). The water-in-oil emulsion has tiny droplets of a water solution dispersed throughout an oil (natural petroleum and melted butter are examples). An oil-in-water emulsion can be washed off the hands with tap water, while a water-in-oil gives the hands a greasy, water-repellent surface.

Creams add oil or fat content to surface skin.

Cold cream originally was a suspension of rose water in a mixture of almond oil and beeswax. Subsequently, other ingredients were added to get a more stable emulsion. An example of a modified cold cream composition is: almond oil, 35%; beeswax, 12%; lanolin, 15%; spermaceti (from whale oil), 8%; and strong rose water, 30%. Other oils can be substituted for some or all of the almond oil. Lanolin stabilizes the emulsion.

Vanishing cream is a suspension of stearic acid in water, to which a stabilizer has been added to prevent the ingredients from separating. The stabilizer may be a soap, such as potassium stearate. These creams do not actually vanish; they merely spread as a smooth, thin covering over the skin.

Creams of various sorts may be used as the base for other cosmetic preparations; other ingredients are added to give desired properties to the creams. As an example, hydrated aluminum chloride can be added to prepare a cream deodorant.

### *Lipstick*

The skin on our lips is covered by a thin corneal layer that is free of fat and consequently dries out easily. A normal moisture content is maintained from the mouth. In addition to being a beauty aid, lipstick can be helpful under harsh conditions that tend to dry lip tissue.

A *lake* is a coloring agent made up of an organic dye adhering to an inorganic substance called a mordant. Some lakes are also approved as food colors.

Lipstick consists of a solution or suspension of coloring agents in a mixture of high molecular weight hydrocarbons or their derivatives, or both. The material must be soft enough to produce an even application when pressed on the lips, yet the film must not be too easily removed, nor may the coloring matter run. Lipstick is perfumed to give it a pleasant odor. The color usually comes from a dye or "lake" from the eosin group of dyes. A ***lake*** is a precipitate of a metal ion ($Fe^{3+}$, $Ni^{2+}$, $Co^{3+}$) with an organic dye. The metal ion enhances the color or changes the color of the dye.

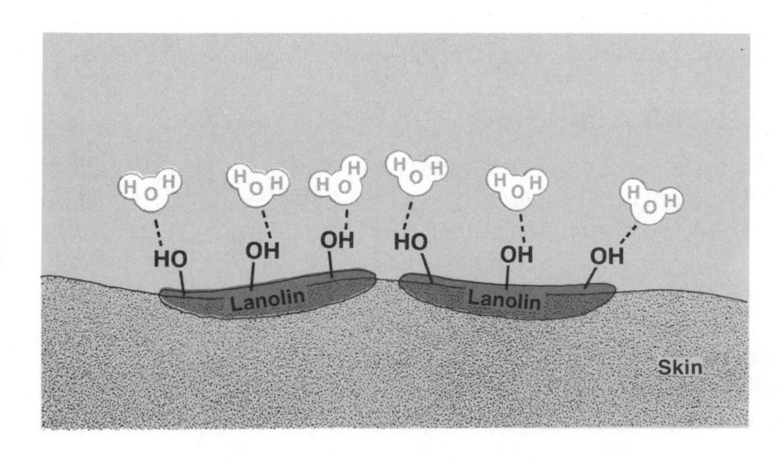


**Figure 18–20** The hydroxyl groups of lanolin form hydrogen bonds with water and keep the skin moist. The fat parts of the molecule are "soluble" in the protein and fat layers of the skin.

Two suitable dyes, used in admixture and with their lakes, are dibromofluorescein (yellow-red) and tetrabromofluorescein (purple):

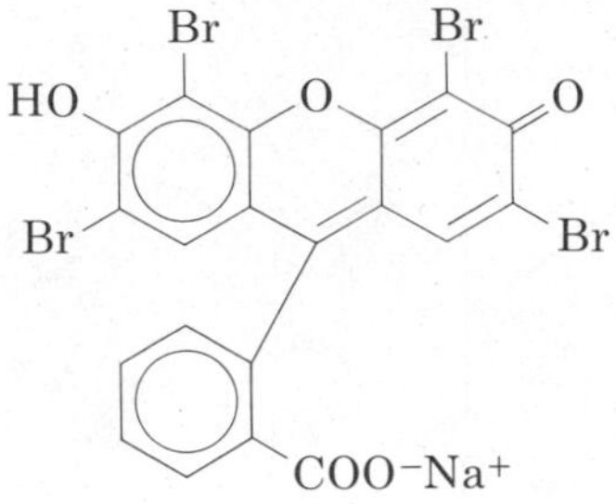

TETRABROMOFLUORESCEIN (Eosin)
(SODIUM SALT)

The ingredients in a typical formulation of lipstick include:

| | | |
|---|---|---|
| Dye | furnishes color | 4–8% |
| Castor oil, paraffins or fats | dissolves dye | 50% |
| Lanolin | emollient | 25% |
| Carnauba wax, Beeswax | makes stick stiff by raising the melting point | 18% |
| Perfume | imparts pleasant odor | small amount |

Carnauba wax and beeswax are high molecular weight esters.

***Suntan Lotions***

One of the agents most harmful to skin is the short wavelength (ultraviolet) light from the sun. It is considered desirable to exclude the shorter, more harmful wavelengths, while transmitting enough less energetic, longer wavelength ultraviolet to permit gradual tanning.

**Ultraviolet radiation tans skin.**

The variety of suntan products ranges from lotions, which selectively filter out the higher energy ultraviolet rays of the sun, to preparations that essentially dye light-colored skin a tan color.

The lotions that filter out the ultraviolet rays are more accurately described as sunscreens, and their ingredients are often mixed with other materials, to give a lotion that both screens and tans. A common ingredient in preparations used to *prevent* sunburn is *p*-aminobenzoic acid.

**Sunbathers refer to this compound as PABA.**

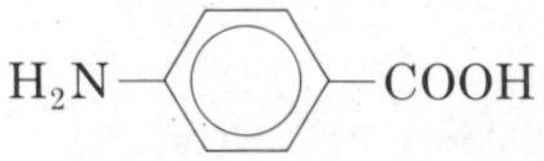

$H_2N-C_6H_4-COOH$

*p*-AMINOBENZOIC ACID
(PABA)

Like most aromatic compounds, it absorbs strongly in the ultraviolet region of the spectrum (Fig. 18–21).

In tanning, the skin is stimulated to increase its production of the pigment ***melanin.*** At the same time, the skin thickens and becomes more resistant to deep burning.

**Increased melanin protects sensitive lower layers of skin.**

Preparations for the relief of sunburn pain are solutions of local anesthetics such as benzocaine.

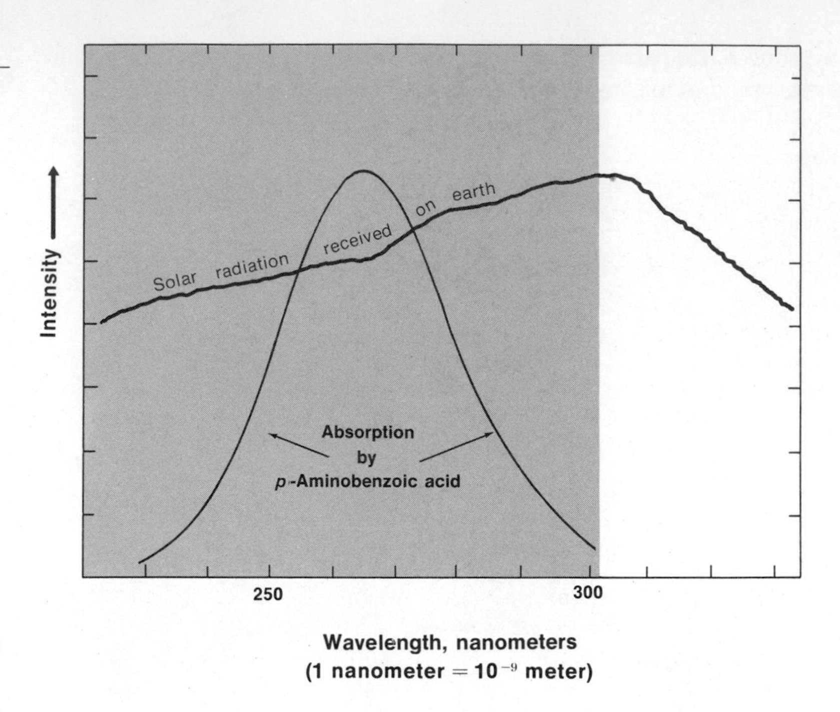


**Figure 18–21** Absorption spectrum of *p*-aminobenzoic acid and its relationship to solar ultraviolet radiation received on earth. Maximum absorption occurs at 265 nanometers, although it absorbs at other wavelengths as shown. The maximum of the deep-burning ultraviolet radiation received on earth is at about 308 nanometers.

### *Face Powder*

Face powder is used to give the skin a smooth appearance by covering up any oil secretions, which would otherwise give it a shiny look. The powder must have some hiding ability, but if it is too opaque it will look too obvious. A powder that has the proper appearance, sticking properties, absorbence for oily skin secretions, and spreading ability usually requires several ingredients. A typical formula is:

| | | |
|---|---|---|
| Talc | absorbent | 56% |
| Precipitated chalk | absorbent | 10% |
| Zinc oxide | astringent | 20% |
| Zinc stearate | binder | 6% |
| Perfume, dye | odor and color | trace |

### *Perfume*

**Perfumes are complex mixtures of odorous compounds.**

A typical perfume has at least three components of somewhat different volatility and molecular weight. (Recall that lower molecular weight compounds are generally more volatile.) The first, called the ***top note,*** is the most volatile and is the most obvious odor when the perfume is first applied. The second, called the ***middle note,*** is less volatile and is generally a flower extract (violet, lilac, etc.). The last, or ***end note,*** is least volatile, and is usually a resin or waxy polymer.

Most perfumes contain many components, and chemically they are often complex mixtures. As the analysis of natural perfume materials has progressed, the use of pure synthetic organic compounds to duplicate specific odors has become very common. An example is the isolation of civetone (see structure, below), a cyclic ketone from civet, a secretion of the civet cat of Ethiopia. It is highly valued for perfumes.

Civetone is now available in a synthetic form. It is prepared by forming 8-hexadecene-1,16-dicarboxylic acid into a ring. The thorium ion ($Th^{4+}$) catalyzes the closure of the ring.

$$\begin{matrix} HC-(CH_2)_7COOH \\ \| \\ HC-(CH_2)_7COOH \end{matrix} \xrightarrow[\Delta]{Th^{4+}} \begin{matrix} HC-(CH_2)_7 \\ \| \\ HC-(CH_2)_7 \end{matrix}\!\!>C{=}O + CO_2 + H_2O$$

CIVETONE

Other compounds used in perfumes include high molecular weight alcohols and esters. An example is geraniol (bp 230°C), a principal component of Turkish geranium oil.

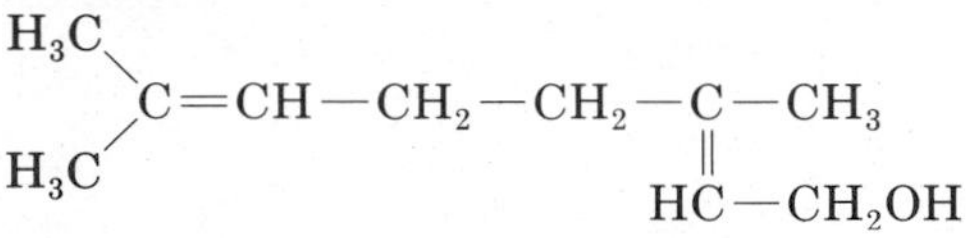

$$\begin{array}{l} H_3C \diagdown \\ \quad C{=}CH{-}CH_2{-}CH_2{-}\underset{\displaystyle \underset{\displaystyle HC{-}CH_2OH}{\|}}{C}{-}CH_3 \\ H_3C \diagup \end{array}$$

GERANIOL

Esters of this alcohol are used to make synthetic rose aromas for perfumes. For example, the ester formed by reaction between geraniol and formic acid has a rose odor.

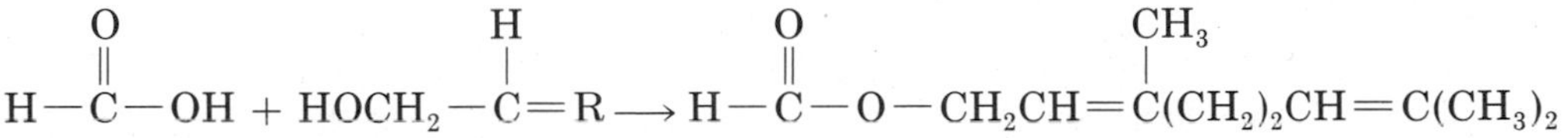

$$\underset{\text{FORMIC ACID}}{H{-}\overset{\overset{\displaystyle O}{\|}}{C}{-}OH} + \underset{\text{GERANIOL}}{HOCH_2{-}\overset{\overset{\displaystyle H}{|}}{C}{=}R} \longrightarrow \underset{\text{GERANYL FORMATE}}{H{-}\overset{\overset{\displaystyle O}{\|}}{C}{-}O{-}CH_2CH{=}\overset{\overset{\displaystyle CH_3}{|}}{C}(CH_2)_2CH{=}C(CH_3)_2}$$

Typical perfumes are 10–25% perfume essence and 75–90% alcohol. Perfumes are added to most cosmetics to give the products desirable odors; they also mask the natural odors of other constituents. They are often mildly bactericidal and antiseptic.

**Ethyl alcohol is a major constituent of most perfumes.**

### *Eye Makeup*

There are several types of eye makeup: eyebrow pencils, mascaras for eyelashes, and shading, among others. Eyebrow pencils are very much like lipstick, but they contain a different coloring matter. The coloring matter is a pigment such as lampblack; the other ingredients include fats, oils, petrolatum, and lanolin, blended to give the desired melting point, which may be raised by the addition of beeswax or paraffin. Petrolatum is a semisolid mixture of hydrocarbons (saturated, $C_{16}H_{34}$ to $C_{32}H_{66}$; and unsaturated, $C_{16}H_{32}$; etc.; melting point, 34–54°C). Brown pencils are made by using iron oxide pigments in place of lampblack.

Mascara is used to darken eyelashes and give them a longer appearance. The same colors as in eyebrow pencils are used, as well as other mineral coloring matters such as chromic oxide (dark green) and ultramarine (blue pigment of variable composition; probably a double silicate of sodium and aluminum silicate with some sodium sulfide). The coloring matter is suspended in a foundation that is a mixture of a soap, oils, fats, and waxes. The mascara may be water-soluble or water-resistant, depending upon the composition of the foundation. A typical formula consists of about 40% wax (beeswax, carnauba wax, and paraffin, adjusted for hardness), 50% soap, 5% lanolin and 5% coloring matter.

In the discussion of depilatories, the danger of the sulfide ion was emphasized. Through hydrolysis, the sulfide ion can produce a hydroxide ion, which functions as a corrosive poison. If very much mascara containing sodium sulfide gets into the eye, it can cause blindness.

$$S^{2-} + HOH \longrightarrow HS^- + OH^-$$

Eye shadow is a formulation of a coloring matter suspended in an oily-fatty-waxy base. A formula that has been used for this purpose is 60% petroleum jelly, 6% lanolin, 10% fats and waxes (approximately equal amounts of cocoa butter, beeswax, and spermaceti), and the balance zinc oxide (white) plus tinting or coloring dyes. Cocoa butter (melting point, 30–35°C) is composed of glycerides of stearic, palmitic, and lauric acids. It is obtained from natural products by compression of cacao seed. Spermaceti (melting point, 45°C) is chiefly cetyl palmitate, $CH_3(CH_2)_{14}COO(CH_2)_{15}CH_3$. It is taken from the solid fat from the head of the sperm whale.

**Titanium dioxide, a white powder, is also used as a base for many eye makeup preparations.**

### *Nail Polish*

Nail polish is essentially a lacquer or varnish. It can be made of nitrocellulose, a plasticizer, a resin, a solvent, and perhaps a dye. The nitrocellulose can be replaced

**Nail polish and hair sprays are formulated very much alike.**

by another polymer molecule that possesses similar qualities. The evaporation of the solvent leaves a film of nitrocellulose, plasticizer, resin, and dye. The nitrocellulose furnishes the shiny film; the plasticizer is added to make the film less brittle; and the resin is added to make the film adhere to the nail better and prevent flaking. Perfumes are added to cover the odor of the other constituents. A typical formulation is:

| | |
|---|---|
| Nitrocellulose | 15% |
| Acetone (solvent) | 45% |
| Amyl acetate (solvent) | 30% |
| Butyl stearate (plasticizer) | 5% |
| Ester gum (resin) | 5% |

Perfumes and colors are added as needed.

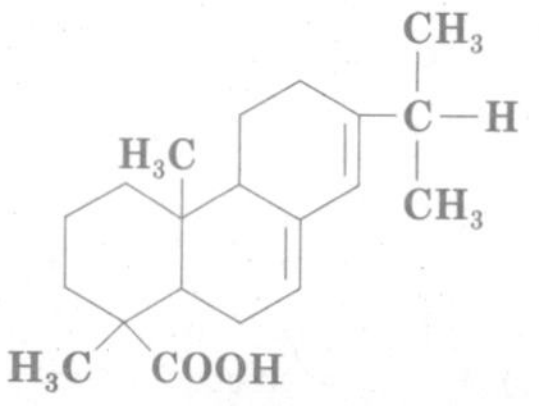

Ester gum is a combination of esters—mainly glyceryl, methyl, and ethyl esters of rosin. Rosin is the resin remaining after distilling turpentine from pine exudate. It is 80–90% abietic acid. Rosin is slightly toxic to mucous membranes and slightly irritating to the skin. Its sticky nature is well known. The ester gum is prepared by heating rosin and the alcohol under pressure until the esterification occurs. The gums are soluble in nonpolar solvents.

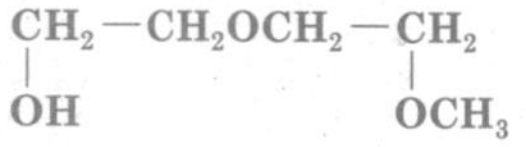

$CH_3$, $C-H$, $CH_3$, $H_3C$, $H_3C$, $COOH$

*Abietic acid*

$$\underset{OH}{CH_2}-CH_2OCH_2-\underset{OCH_3}{CH_2}$$

*Diethylene glycol monomethyl ether*

Nail polish removers are simply solvents that dissolve the film left by the nail polish. They consist largely of acetone or ethyl acetate, or both, to which small amounts of butyl stearate and diethylene glycol monomethyl ether have been added to reduce the drying effect of the solvent. However, some formulations contain combinations of amyl acetate, butyl acetate, ethyl acetate, benzene, olive oil, lanolin, and alcohol. Both nail polish and nail polish removers are very flammable, and care should be taken never to use them in the presence of open flames or lighted cigarettes.

Cuticle softeners are primarily wetting agents and alkalies used to soften skin around the fingernails so it can be shaped as desired. The use of alkali to soften and swell protein is well known. A typical cuticle softener contains potassium hydroxide (3%), glycerol (12%), and water (85%). It may also contain sodium carbonate (an alkali), triethanol amine (a detergent, used as wetting agent), and trisodium phosphate (an alkali).

## Self-Test 18-D

1. The surface pH of the human body is about __________.
2. Hair keratin contains considerable amounts of the sulfur-containing amino acid __________.
3. Two types of bonds that hold strands of protein in place include __________ and __________.
4. Wet hair swells and stretches because of the breaking of __________ bonds.
5. The peroxide oxidation of melanin produces a (colored/colorless) product.
6. A plasticizer causes a plastic material to be more __________.
7. What is the purpose of talc in powders? __________
8. Zinc peroxide acts in deodorants to oxidize amines and __________.
9. Commercial lanolin comes from what animal? __________
10. Vanishing cream is completely absorbed through the pores of the skin. (True/False)

## Matching Set

| | | | |
|---|---|---|---|
| ______ | 1. corneal layer | a | bonds two protein chains |
| ______ | 2. keratin | b | reducing agent in waving lotion |
| ______ | 3. sulfide | c | hair bleach |
| ______ | 4. isopropyl alcohol | d | protein in skin |
| ______ | 5. hydrogen peroxide | e | resin in hair spray |
| ______ | 6. hydrated aluminum chloride | f | alkaline ion in depilatory |
| ______ | 7. disulfide linkage | g | used in deodorants |
| ______ | 8. lanolin | h | surface skin |
| ______ | 9. polyvinylpyrrolidone | i | skin disinfectant |
| ______ | 10. thioglycolic acid | j | emollient |

# Part IV—Automotive Products

A significant amount of chemistry is involved in the production and operation of the automobile. Before an automobile can exist, metals must be separated from their ores; plastics must be synthesized and fabricated; paints have to be formulated; sulfuric acid for the battery must be made; rubber must be synthesized, formed into shape, and vulcanized; and glass has to be mixed, fired, molded, and cut. Many chemical reactions are required to power the automobile. The two most familiar are the combustion of the fuel and the electrochemistry of the battery. In its wake the automobile leaves its chemical exhaust to undergo a variety of chemical reactions, some of which lead to smog formation. The chemistry of fuels and exhaust gases has been discussed in previous chapters. In this section, we shall discuss some of the chemicals that consumers buy for their automobiles.

## Miscellaneous Gasoline Additives

In addition to tetraethyllead and/or aromatic compounds that reduce preignition or "knock" in an automobile engine, numerous other chemicals are added to gasoline to improve its properties.

***Deposit modifiers*** are added to gasoline to prevent the ignition of new fuel by glowing particles from a previous ignition. Phosphorus compounds such as tricresyl phosphate (one trade name is TCP) and, more recently, boron compounds, have been used for this purpose. These reduce the melting point of the lead oxide residue (by forming a lead borate or phosphate glass), and the resulting hot liquid is blown out with the exhaust. The phosphorus compounds also prevent spark plug deposits from becoming so electrically conductive that the charge leaks away instead of firing the plug.

$(H_3C-C_6H_4-O)_3PO$

*Tricresyl phosphate*

***Antioxidants,*** such as phenylenediamine, aminophenols, dibutyl-*p*-cresol, and *ortho*-alkylated phenols, are added to prevent the formation of peroxides that lead to knock and gum formation. About two or three pounds of these additives are added to every 1,000 barrels of gasoline. The mechanism of antioxidation of automotive oils is very similar to the mechanism described for BHA and BHT in food additives. Because copper ions catalyze gum formation, metal ion scavengers such as ethylenediamine are added to chelate the trace amounts of copper ions and render them ineffective. The copper gets into the gasoline from the copper tubing used for fuel lines and from brass parts of the engine.

**Brass is an alloy of copper and zinc.**

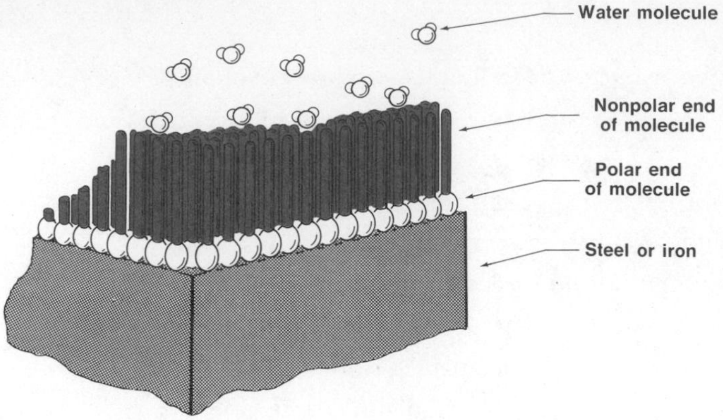

**Figure 18–22** The action of surface active agents, such as rust inhibitors, mild antiwear agents, and some deicing agents.

To inhibit water from corroding and rusting storage tanks, pipelines, tankers, and fuel systems of engines, ***antirust agents*** are added. Four compounds used to prevent corrosion are trimethyl phosphate, sodium and calcium sulfonates, and N,N′-di-sec-butyl-*p*-phenylenediamine. All these compounds have a polar or ionic end and a nonpolar end in the molecule. These agents coat metal surfaces with a very thin protective film that keeps water from contacting the surfaces, as shown in Fig. 18–22. This also helps prevent gummy deposits in the carburetor and combats carburetor icing during cold weather.

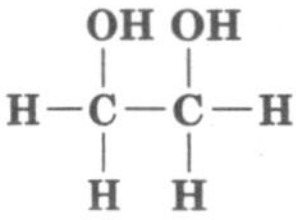

***Antiicing agents*** coat metal surfaces, as do antirust agents, and thus prevent ice particles from accumulating on surfaces, and/or depress the freezing point. The small ice particles pass harmlessly through the carburetor and into the engine where the heat converts them to water vapor that eventually exits with the exhaust gases. The freezing point depressants, which include alcohols and glycols, act in the same manner as the antifreeze in the engine's cooling system.

OH OH
H—C—C—H
H H

*Ethylene glycol*

***Detergents,*** which include alkylammonium dialkyl phosphates, are added to prevent the accumulation of high-boiling components on the walls of the carburetor. These deposits interfere with the air flow into the carburetor and cause rough idling, frequent stalls and increased fuel consumption. The effectiveness of these detergents stems from their surface active properties, as shown in Figure 18–22. The film of detergent provides a thin, nonpolar coating on the metal surfaces that prevents high molecular weight, nonpolar gums from forming thick deposits on the surfaces.

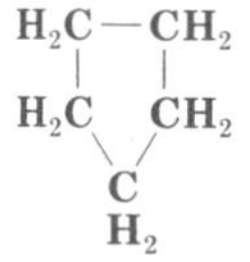

***Upper cylinder lubricants*** are sometimes blended into gasoline. These are usually light mineral oils or low viscosity naphthenic distillates (such as cyclopentane) that lubricate the moving metal parts and dissolve away deposits in the intake system from the carburetor, cylinders, top piston rings, and valves. (Nonpolar deposits tend to dissolve in nonpolar solvents.)

$H_2C—CH_2$
$H_2C\quad CH_2$
C
$H_2$

*Cyclopentane*

At one time ***dyes*** were added to distinguish antiknock "ethyl" gasolines from nonantiknock gasolines. Always present in very small amounts, they are now used to distinguish grades and brands.

## Lubricants and Greases

***Lubricating oils*** from petroleum consist essentially of complex mixtures of hydrocarbon molecules. These generally range from low viscosity oils, having molecular weights as low as 250, to very viscous lubricants with molecular weights as high as about 1,000. The viscosity of an oil can often determine its use. For example, if the oil is too viscous, it offers too much resistance to the moving metal parts. On the other hand, if the oil is not viscous enough, it will be squeezed out from between the metal

**"Viscous" means resistant to flow, like molasses.**

**TABLE 18–8 Viscosity Data at −18° C and 99° C for the SAE Method of Rating Motor Oils**

| Motor Oil | Viscosity (SUS)* −18° C | | 99° C | |
|---|---|---|---|---|
| | *Min.* | *Max.* | *Min.* | *Max.* |
| 5W | | 4,000 | 39 | |
| 10W | 6,000 | 12,000 | 39 | |
| 20W | 12,000 | 48,000 | 39 | |
| 20 | | | 45 | 58 |
| 30 | | | 58 | 70 |
| 40 | | | 70 | 85 |
| 50 | | | 85 | 110 |

*SUS is the Saybolt Universal Second, which is the time in seconds required for 60 ml of oil to empty out of the cup in a Saybolt viscometer through a carefully specified capillary opening. Note the extremely shortened time for the outflowing of the hotter oil.

surfaces, and consequently offer insufficient lubricating action. For these and other reasons, motor oil is often a mixture of oils with varying viscosities. The common 10W-30 oil, for instance, combines the low-temperature viscosity of the Society of Automotive Engineers (SAE) 10W classification (Table 18–8) for easy low-temperature starting with SAE 30 high-temperature viscosity for better load capacity in bearings at the normal engine running temperature.

The usefulness of motor oil is improved by the addition of substances such as antiwear agents, oxidation inhibitors, rust inhibitors, detergents, viscosity improvers, and foam inhibitors.

A number of synthetic oil blends have shown superior performance in that the oil molecules are more stable to heat. Breakdown of oil is the result of the thermal decomposition of the oil molecules to smaller ones, and also the partial oxidation of hydrocarbons at high temperatures.

***Greases*** are essentially lubricating oils thickened with a gelling agent such as fatty acid soaps. The soaps form a network of fibers that entrap the oil molecules within the interlacing fiber structures. Carbon black, silica gel, and clay are also used to thicken petroleum greases. Chemical additives similar to those used in lubricating oils and gasolines are added to greases to improve oxidation resistance, rust protection, and extreme pressure properties. Synthetic greases are being developed that deteriorate so slowly that longer intervals between grease jobs are now possible. Silicone greases have a useful life of up to 1,000 hours at 450°F (232°C). Unfortunately, silicone greases provide relatively poor lubrication for gears and other sliding devices. Diester greases such as di(2-ethylhexyl) sebacate have found extensive use among synthetic greases. Lithium soaps dissolve well in the diester oil and form a grease with equal or better lubrication characteristics and a considerably longer useful life than petroleum greases. Blends of silicone oil and diester oil provide greases with good low-resistance lubricating power even at low temperatures (−73°C).

$CH_3$
$(CH_2)_3$
$CH-C_2H_5$
$CH_2$
$O$
$C=O$
$(CH_2)_8$ *Diester*
$C=O$
$O$
$CH_2$
$CH-C_2H_5$
$(CH_2)_3$
$CH_3$

*Di(2-ethylhexyl) sebacate*

## Solid Lubricants

Perhaps the most tenacious and wear-resistant solid lubricant is molybdenum disulfide, $MoS_2$. Like graphite, a common solid lubricant, $MoS_2$ has a layered structure that enables one layer to float over another and provide lubrication (Fig. 18–23). Strong bonds exist within each S—Mo—S layer, while weak S—S bonds between the layers allow easy sliding of one layer over another. $MoS_2$ is extensively used in

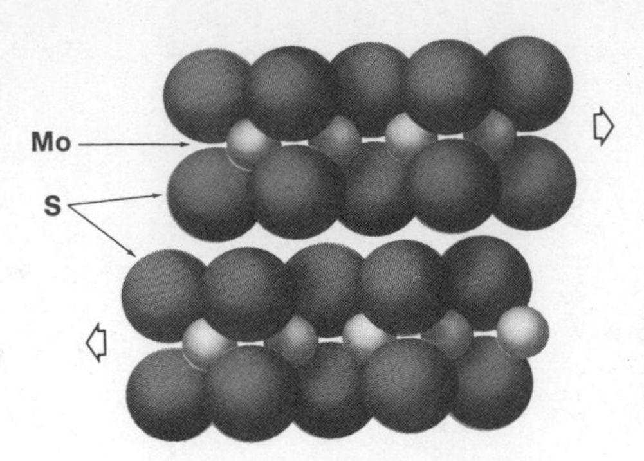


**Figure 18–23** The structure of molybdenum disulfide, $MoS_2$. The small spheres represent molybdenum atoms. Note how the sulfides are adjacent to sulfides in every other layer. This allows one layer to slide over another, accounting for the slippery nature of $MoS_2$.

greases for automotive chassis lubrication. When used alone, $MoS_2$ suffers like any other solid lubricant from having poor wear resistance and being unable to heal any breaks in its surface coating.

## Antifreeze

**Ethylene comes from cracking petroleum. Ethylene glycol is prepared from ethylene, $CH_2{=}CH_2$, by an oxidation reaction**

$$2CH_2{=}CH_2 + O_2 \longrightarrow 2\underset{\text{O}}{CH_2{-}CH_2}$$

*Ethylene oxide*

**followed by hydrolysis:**

$$\underset{\text{O}}{CH_2{-}CH_2} + H_2O \longrightarrow \underset{\text{OH}\quad\text{OH}}{CH_2{-}CH_2}$$

An antifreeze is a substance that is added to a liquid, usually water, to lower its freezing point. Although various substances have been used as antifreezes in the past, nearly all of the current market is supplied by ethylene glycol and methyl alcohol.

```
  OH OH            OH
  |  |             |
H—C—C—H          H—C—H
  |  |             |
  H  H             H
```

ETHYLENE GLYCOL  METHYL ALCOHOL

More than 95% of the antifreeze on the market is "permanent" antifreeze, having ethylene glycol as the major constituent. The largest use of antifreeze is in protection of water-cooled engines. Water has two serious disadvantages as an engine coolant. First, it has a relatively high freezing point and second, under normal operating conditions, it is corrosive. Modern antifreeze mixtures effectively counteract these problems.

Antifreeze protection charts always show the temperatures at which the first ice crystals form. Below this temperature the antifreeze solution turns to slush. If the slush is unable to circulate through the radiator, overheating, boiling, and engine damage can result. For this reason it is best to add sufficient antifreeze to prevent the formation of the first ice crystals even at the lowest anticipated temperature (Fig. 18–24).

**The density of an antifreeze solution can be used to determine its freezing temperature.**

A service station attendant measures the effectiveness of the antifreeze in your car by reading the position at which the hydrometer floats in a sample of a radiator solution. He is really measuring the solution's density, which varies with the amount of antifreeze present, since ethylene glycol has a density greater than that of water.

**Good auto antifreeze formulations also contain rust inhibitors.**

The prevention of corrosion is the second most important job of antifreeze. Metals in the cooling system that are subject to corrosion are copper, steel, cast iron, aluminum, solder (lead and tin), and brass (copper and zinc). The presence of oxygen, along with high temperatures, pressures, and flow rates, increases the possibility of general corrosion. Although corrosion can "eat" through the walls of the cooling system, the most general trouble is overheating caused by flakes of metal oxide clogging the radiator. Some additives inhibit corrosion by acting as reducing agents (e.g., nitrites), some as ion scavengers (e.g., phosphates), and some as surface active agents (detergents; e.g., triethanolamine). Most antifreezes contain two or more inhibitors for the different metals. All inhibitors are depleted with use.

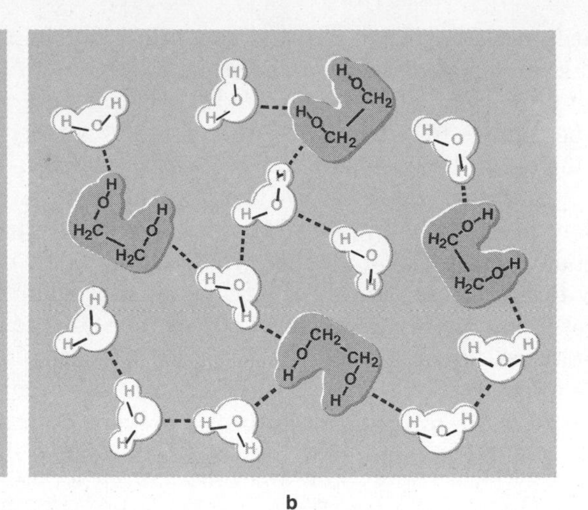


**Figure 18–24** How foreign molecules prevent water from forming its normal crystal structure. For example, by forming hydrogen bonds to water molecules, ethylene glycol (*b*) prevents water molecules from assuming their places in the ice structure (*a*).

Radiator sealants have been on the market for years. Modern sealants in antifreeze include asbestos fiber and polystyrene spheres. When a radiator springs a leak, the coolant penetrates the crack because the pressure inside the cooling system is greater than atmospheric pressure. Asbestos fibers are often too big to squeeze through and thus get caught in the crack, plugging the hole in the radiator. On the other hand, the larger polystyrene particles initially plug up most of the leak while the smaller ones build up behind them. The spheres fuse together under the pressure and temperature conditions that exist and form a solid plug in the crack (Fig. 18–25). This is effective in stopping up holes or cracks up to 0.5 millimeter width, which includes about 90% of all radiator leaks.

Foaming is caused by one or more of several factors: air leaks in hoses, water pump, or radiator; exhaust gas leaking into the cooling system; failure to drain out cooling system cleansers; or extended use of antifreeze. Foaming can be corrected by tightening the system or by use of antifoam additives, such as silicones, polyglycols, mineral oils, high molecular weight alcohols, organic phosphates, alkyl lactates, castor oil soaps, and calcium acetate. These substances reduce the surface tension so bubbles cannot hold together.

Corrosion begins in the cooling system when the pH of the coolant drops below 7. Most commercial antifreeze contains an extra amount of alkali to maintain the pH above 7. As time passes this alkali is exhausted and corrosion begins. A check of your

**Pink litmus—acidic.
Blue litmus—basic.**

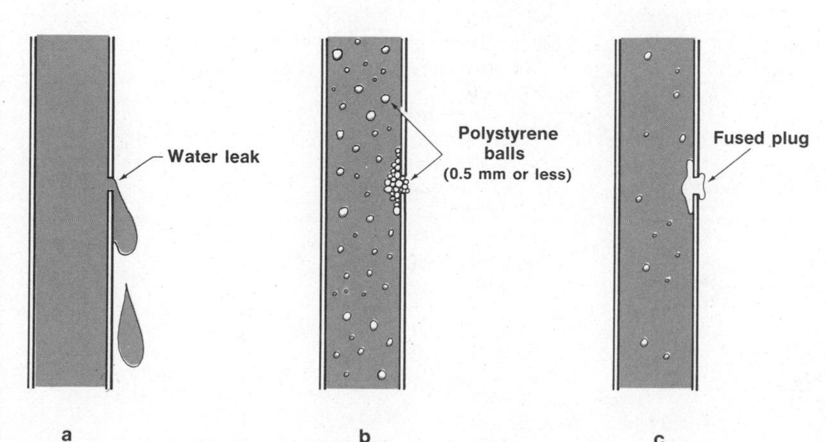


**Figure 18–25** How polystyrene works to plug radiator leaks. Details are given in the text.

antifreeze's corrosion-fighting capability can be made with a piece of litmus paper. Pink means acid and trouble through corrosion. Adding a can of rust inhibitor (which contains an alkali) will adjust the pH back to a value above 7. A retest with litmus paper shows blue (basic). Your car is now protected as long as the mixture composition doesn't change due to boil-away or leaks.

***Deicing fluids*** are a type of antifreeze, but they also will melt ice and frost. They are chiefly employed in removing ice and frost from parked aircraft and car windows. The glycol-alcohol type used on cars came on the market in 1959. The better formulations contain the following ingredients: ethylene or propylene glycol (for protection against refogging); one or more of the lower alcohols, such as 2-propanol, denatured alcohol, and so on (for low viscosity and good spray pattern); water (to minimize inside fogging by diminishing the evaporative-cooling effect of the formulation); nitrite or other inhibitor (to prevent corrosion of the container); and wetting agent, such as ethanolamine (to break surface tension so fluid will attack the ice readily).

**Deicers lower the freezing point of water.**

A simple alcohol-water mixture works satisfactorily for quick frost removal and to defog the inside of glass, but it evaporates quickly and refogging occurs unless

**TABLE 18–9 Typical Composition of Some Automotive Products**

**Engine and Motor Cleaners**

| | |
|---|---|
| Ethylene dichloride | 63% |
| *p*-Cresol | 25% |
| Oleic acid | 7.2% |
| Potassium (or sodium) hydroxide | 1.4% |
| Water | 3.0% |

**Radiator Cleaners**

| | |
|---|---|
| (1) Oxalic acid | 40% |
| Boric acid | 60% |
| (2) Sodium carbonate | 85% |
| Potassium dichromate | 18% |
| May also contain: | |
| Sulfamic acid | |
| 1-Butanol (n-butyl alcohol) | |
| 2-Propanol (isopropyl alcohol) | |

**Radiator Stop Leak**

| | |
|---|---|
| Dextrin | 5–10% |
| Cellulose gum | 0.5% |
| Asbestos | 5% |
| Sodium carbonate | 0.8% |
| Isopropyl alcohol | 10–15% |
| Water | to 100% |

**Shock Absorber Fluids**

| | |
|---|---|
| (1) Petroleum ether | 97% |
| Kerosene | 3% |
| (2) Mineral oils | 90–100% |
| Fatty oils | 0–5% |
| May also contain: | |
| Viscosity improvers (polymethacrylate esters) | 0–5% |
| Dyes | 0–150 ppm |

**Automatic Transmission Fluids**

| | |
|---|---|
| Mineral oils | 75–100% |
| Oxidation inhibitors and detergents | 0–20% |
| Viscosity improvers | 0–5% |
| Polymethacrylate esters | |
| Polyisobutylenes | |
| Antiwear agents | 0–2% |
| Organic borates | |
| Antifoam agents | 0–15% |
| Polysiloxanes | |
| Sealant | to 100% |
| Triaryl phosphate | |
| Dyes | 0–200 ppm |

**Brake Fluids**

| | |
|---|---|
| Lubricant | 20–25% |
| Castor oil | |
| Butyl or glyceryl ether of polyoxyethylene propylene glycol | |
| Polypropylene glycol | |
| Solvent | 80–85% |
| Methyl, ethyl, and butyl ethers of ethylene glycol and related glycols | |
| May also contain: | |
| Inhibitors | |
| Amine soaps | |
| Potassium soaps | |
| Borax | |
| Antioxidants | |
| Hydroquinone | |
| Dyes | |

glycol is present. Before a deicer formulation is applied, the snow should be removed and the ice should be scored to permit faster penetration.



### Miscellaneous Automotive Products

Consumers spend millions of dollars each year on a large variety of specialized automotive products. The formulations of some typical examples of these products are given in Table 18–9.

## Self-Test 18-E

1. Antirust agents such as trimethyl phosphate in gasoline prevent the formation of rust by: (a) reacting with oxygen; (b) coating metal surfaces; (c) converting rust ($Fe_2O_3$) into iron metal; (d) absorbing water. Select one answer.
2. Gum in the automobile fuel system is formed by __________ of hydrocarbons which must have one or more __________ bonds per molecule.
3. Since gums are nonpolar, a __________ solvent such as cyclopentane is used to dissolve the gum in an engine.
4. Antioxidants such as aminophenols in gasoline prevent oxygen from reacting with gasoline prior to ignition by destroying __________, which have unpaired electrons.
5. Lubricating oils with a rating of 10W-30: (a) have one type of molecule; (b) have the same viscosity at high and low temperatures; (c) are a mixture of oils with varying viscosities. Select one answer.
6. Molybdenum disulfide ($MoS_2$) is an excellent solid lubricant because it consists of __________, which slide over each other.
7. The viscosity of oils and greases is increased by adding linear polymers because the chains of the polymers become __________.
8. Automotive greases are essentially thickened (or gelled) motor oils. (True/False)
9. The development of what two kinds of greases have made it possible to go longer between automotive lubrications? __________ and __________
10. Ethylene glycol makes it more difficult for water to boil because the —OH groups on an ethylene glycol molecule form hydrogen __________ with water molecules and detain them from escaping into the atmosphere.

## Matching Set

| | | |
|---|---|---|
| ______ | 1. tricresyl phosphate | **a** phenylenediamine |
| ______ | 2. antioxidant | **b** oil gelled in soap |
| ______ | 3. antifreeze | **c** in some antifreeze mixtures to stop leaks |
| ______ | 4. grease | **d** permanent antifreeze mixture |
| ______ | 5. $MoS_2$ | **e** lightweight motor oil |
| ______ | 6. polystyrene | **f** ethylene glycol |
| ______ | 7. greater density than water | **g** deposit modifier |

| | | | |
|---|---|---|---|
| ______ | 8. 5W | h | added to gasoline to keep carburetor clean |
| ______ | 9. dyes | i | used to distinguish grades of gasoline |
| ______ | 10. detergent | j | solid lubricant |

## Part V—The Chemistry of Photography

### Black and White Photography

Photochemistry deals with the chemical changes produced by absorbed radiant energy.

Many chemical substances are known to be photosensitive. The chloride, bromide, and iodide salts of silver can illustrate the useful ability of storing an image of the absorbed light. The various processes known as ***photography*** are based upon this property.

J. H. Schulze observed in 1727 that a mixture of silver nitrate and chalk darkened on exposure to light. The first permanent images were obtained in 1824 by Nicéphore Niepce, a French physicist, by means of glass plates coated with a coal derivative (called bitumen) containing silver salts. In the early 1830's, Niepce's partner, Louis Daguerre, discovered by accident that mercury vapor was capable of developing an image from a silver-plated copper sheet that had been sensitized by iodine vapor. The ***daguerreotype*** image was rendered permanent by washing the plate with hot concentrated salt solution. In 1839 Daguerre demonstrated his photographic

Pronunciation of daguerreotype: (dä-ger′-ō-tīp).

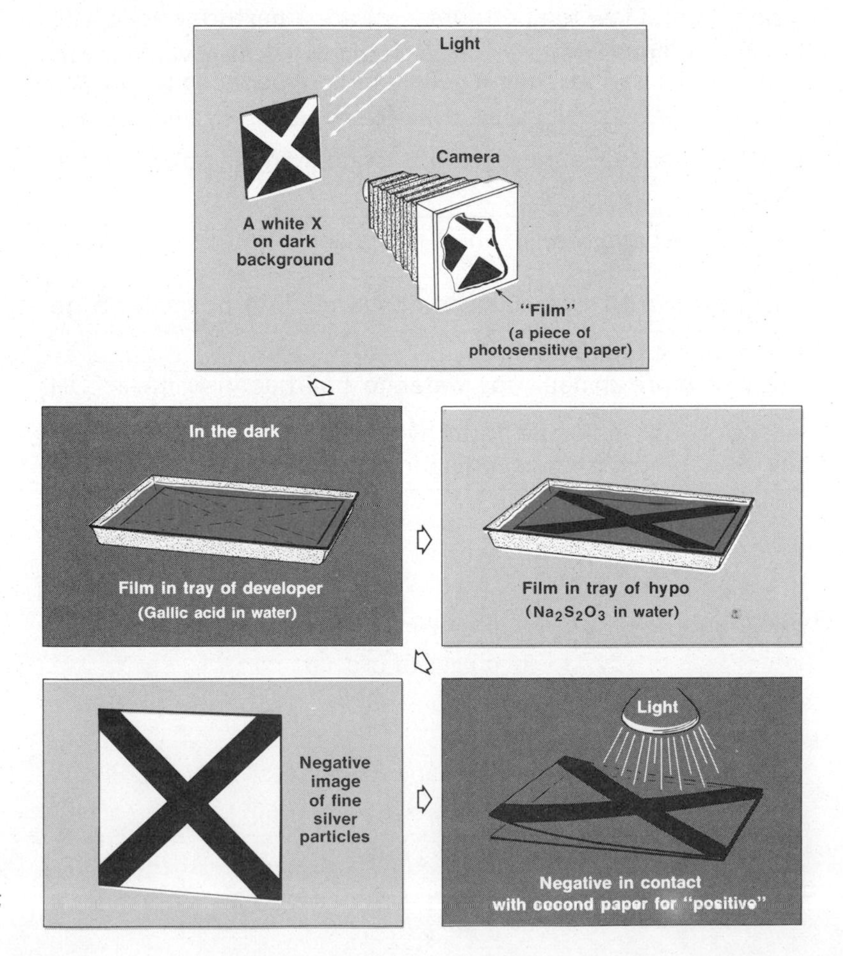


**Figure 18–26** Talbot process.

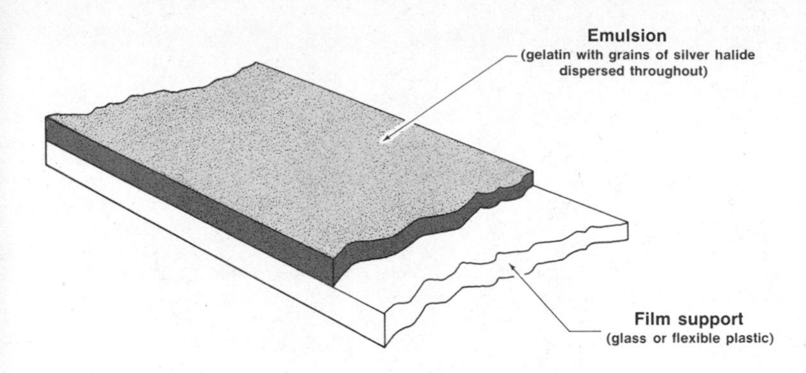


**Figure 18–27** A modern photosensitive gelatin emulsion.

process to the Academy of Sciences in Paris. The process was improved by using sodium thiosulfate to wash off the unexposed silver salts.

In 1841, an Englishman, William Henry Fox Talbot, announced the calotype process. The Talbot process (Fig. 18–26) involved a paper made sensitive to light by silver iodide. The light-sensitive paper could be developed into a negative image with gallic acid in a development process essentially the same as is used today. When made with semitransparent paper, Talbot's negatives could be laid over another piece of photographic paper which, when exposed and developed, yielded a "positive," or direct copy of the original. Although the Talbot process required less time than the Daguerre process, the Talbot images were not sharp. It was obvious that some way of holding the silver halides on a transparent material would have to be devised.

At first, the silver salts were held on glass with egg white as binder. This provided sharp, though easily damaged, pictures. By 1871, the problem had been solved by an amateur photographer and physician, Dr. R. L. Maddox. He discovered a way to make a gelatin emulsion of silver salts and apply it to glass. In 1887, George Eastman introduced the Kodak, a camera using film made by attaching a gelatin emulsion to a plastic (cellulose nitrate) base (Fig. 18–27). The camera could take 100 pictures and then camera and film had to be sent to Rochester, New York for processing. The age of modern photography had arrived.

**Cellulose acetate replaced easily combustible cellulose nitrate as the film support in 1951.**

## Photochemistry of Silver Salts

To understand the chemistry of photography, we must first look at the photochemistry of silver salts. A typical photographic film contains tiny crystallites called ***grains*** (Fig. 18–28), which are composed of a slightly soluble silver salt, such as silver bromide, AgBr. The grains are suspended in gelatin, and the resulting gelatin emulsion is melted and applied as a coating on glass plates or plastic film.

When light of an appropriate wavelength strikes one of the grains, a series of reactions begins that leaves a small amount of free silver in the grain. Initially, a free bromine atom is produced when the bromide ion absorbs the photon of light:

$$Ag^+Br^- \xrightarrow{\text{light absorption}} Ag^+ + Br^0 + e^-$$

The silver ion can combine with the electron to produce a silver atom.

$$Ag^+ + e^- \longrightarrow Ag^0$$

Association within the grains produces species such as $Ag_2^+$, $Ag_2^0$, $Ag_3^+$, $Ag_3^0$, $Ag_4^+$, and $Ag_4^0$. The presence of this free silver in the exposed silver bromide grains provides the ***latent image,*** which is later brought out by the development process. The grains containing the free silver in the form of $Ag_4^0$ are readily reduced by the developer to form relatively massive amounts of free silver; hence a dark area appears at that point on the film. The unexposed grains are not reduced by the developer under the same critical conditions.

**In order for an exposed AgBr grain to be developable, it will need a minimum of four silver atoms as $Ag_4^0$.**

**The latent image is the "invisible developable image" stored in the silver halide grains.**

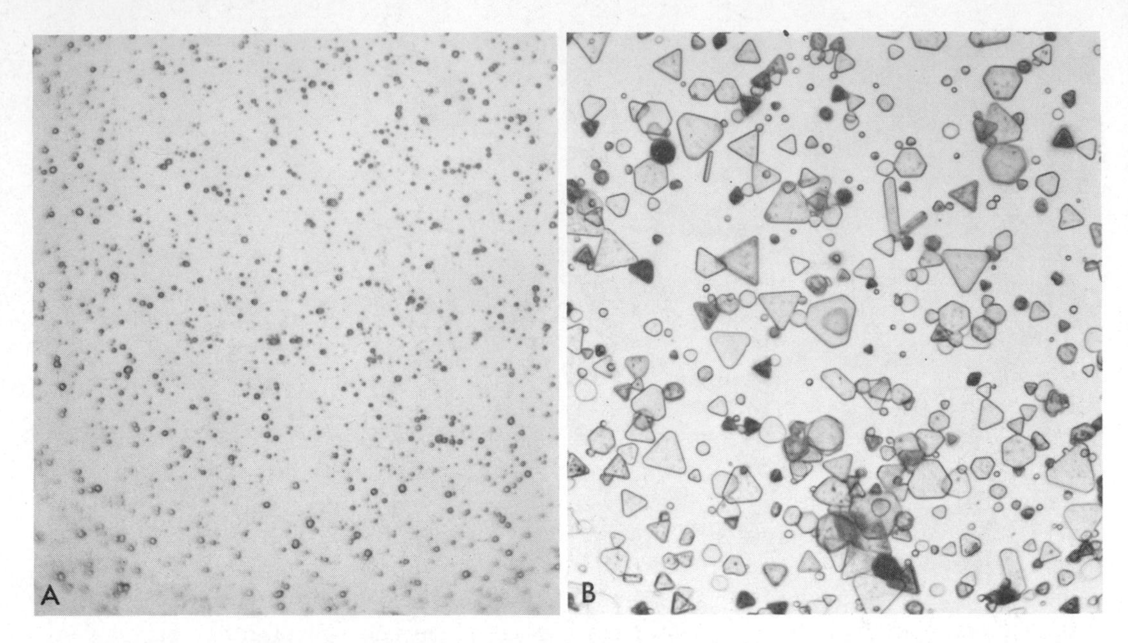


**Figure 18–28** (*a*) Photomicrograph of the grains of a slow positive emulsion. (*b*) The grains of a high speed negative emulsion at the same magnification.

**Film sensitivity is rated on the American Standards Association (ASA) scale. The larger the number, the more sensitive the film is to light.**

Film sensitivity is related to grain size and to the halide composition. As the grain size in the emulsion increases, the effective light sensitivity of the film increases (up to a point; Fig. 18–28). The reason for this is that the same number of silver atoms are needed to initiate reduction of the entire grain by the developer despite the grain size.

## Amplification of the Latent Image—Development

Silver halides are not the most photosensitive materials known. Why, then, are they effective image producers? The answer lies in the fact that the impact of a single photon on a silver halide grain produces a nucleus of at least four silver atoms, and this effect is amplified as much as a billion times by the action of a proper reducing agent ***(developer).***

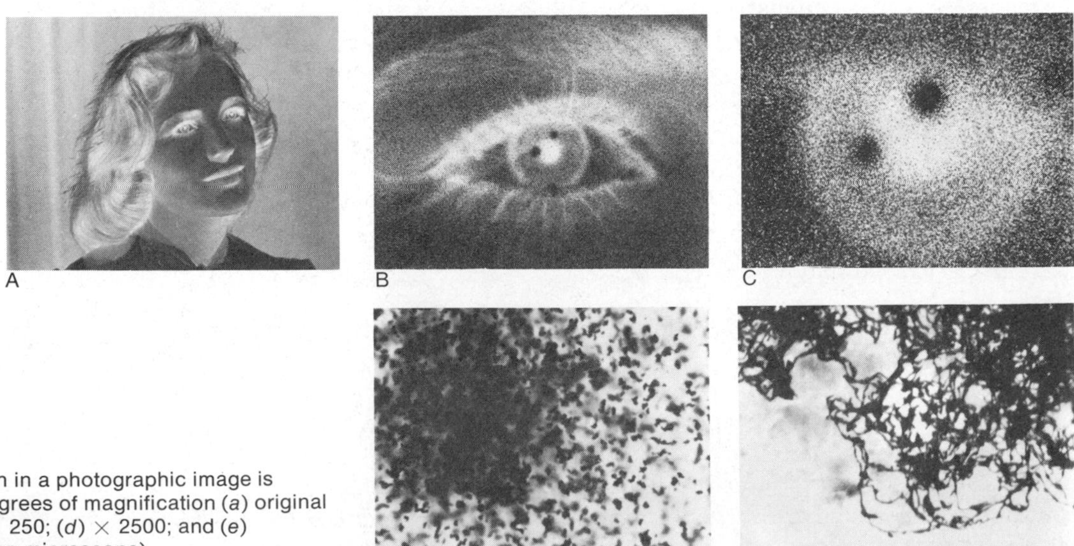


**Figure 18–29** Grain in a photographic image is illustrated by five degrees of magnification (*a*) original size; (*b*) × 25; (*c*) × 250; (*d*) × 2500; and (*e*) × 25,000 (by electron microscope).

**TABLE 18–10 Some Compounds Used as Photographic Developers**

| Name | Formula |
|---|---|
| Gallic acid | 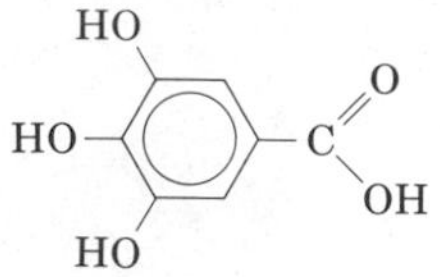 HO, HO, HO on benzene ring; —C(=O)OH |
| *o*-Aminophenol | 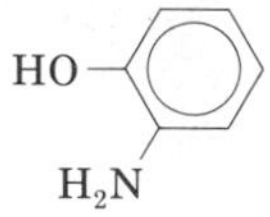 HO—, $H_2N$— on benzene ring |
| Hydroquinone | HO—(benzene ring)—OH |
| *p*-Methylaminophenol (Metol) | 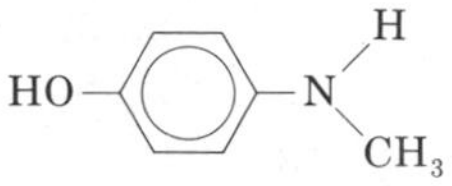 HO—(benzene ring)—N(H)($CH_3$) |
| 1-Phenyl-3-pyrazolidone (Phenidone) | 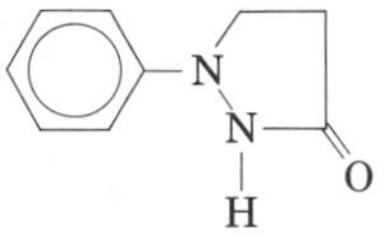 (benzene ring)—N, N—H, =O (pyrazolidone ring) |

When an exposed film is placed in developer, the grains that contain silver atom nuclei are reduced faster than those grains that do not. The more nuclei present in a given grain, the faster the reaction. The reduction reaction is

$$\underset{\text{IN GRAINS CONTAINING } Ag^0_4}{Ag^+} + e^- \longrightarrow Ag^0$$

Factors such as temperature, concentration of the developer, pH, and the total number of nuclei in each grain determine the extent of development and the intensity of free silver (blackness) deposited in the film emulsion.

**The blackness on the negative is due to free silver atoms, $Ag^0$.**

Not only must the developer be capable of reducing silver ions to free silver, but it must be selective enough not to reduce the unexposed grains, a process known as "fogging." Table 18–10 lists some substances that are used as developing agents.

Most developers used for black and white photography are composed of hydroquinone and Metol or hydroquinone and Phenidone. A typical developer consists of a developing agent (or two), a preservative to prevent air oxidation, and an alkaline buffer to prevent the actual reduction reaction from being retarded (Table 18–11). Other chemicals might be added but they are not absolutely necessary.

When hydroquinone acts as a developer, quinone is formed. Two hydrogen ions are also produced for every two silver atoms, as shown in the margin.

OH (benzene ring) OH

HYDROQUINONE

$$+ 2Ag^+ \rightleftharpoons \text{(quinone: O=ring=O)} + 2Ag^0 + 2H^+$$

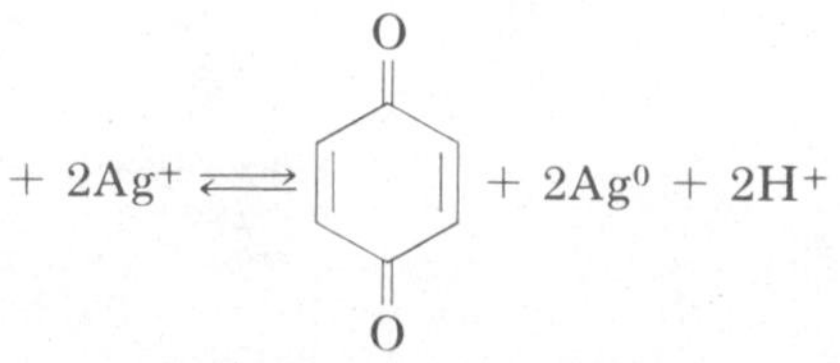

QUINONE

**TABLE 18–11 Formula for a Typical Developer for Black and White Films**

| | |
|---|---|
| 750 ml water at 50°C; dissolve in this water: | |
| Metol | 2.0 g |
| Hydroquinone | 5.0 |
| Sodium sulfite | 100.0 |
| Borax, $Na_2B_4O_7 \cdot 10H_2O$ | 2.0 |
| Add cold water to make 1000 ml of solution. | |

Since this reaction is reversible, a buildup of either hydrogen ions or quinone would impede the development process. The sodium sulfite reacts with quinone and destroys its ability to revert back to hydroquinone.

$$C_6H_4O_2 \text{ (quinone)} + SO_3^{2-} + H_2O \longrightarrow C_6H_3(OH)_2SO_3^- + OH^-$$

The hydrogen ions are neutralized effectively by the hydroxide ions.

$$H^+ + OH^- \longrightarrow H_2O$$

If development proceeds either too long or at a higher temperature than recommended, sufficient fogging occurs to render a negative useless. Since the rates of the development reactions increase with increasing temperature, the photographer usually controls the temperature of the development bath very carefully.

The development process is terminated by a ***stop bath.*** The stop bath usually contains a weak acid such as acetic acid, which decreases the pH. The action of a stop bath is to build up the amount of hydrogen ions, which effectively stops the hydroquinone $\longrightarrow$ quinone reaction.

## Fixing

One of the principal problems in the early days of photography was the lack of permanence of the image. If development only produces free silver where the light intensity was greatest and nothing further is done to the negative, the undeveloped silver halide will be exposed the instant it is taken into the light. After that, almost any reducing agent will completely fog the negative. In order to overcome this problem, a suitable substance had to be found to remove the unreduced silver halides. The most commonly used ***fixing agent*** in black and white photography is the thio-

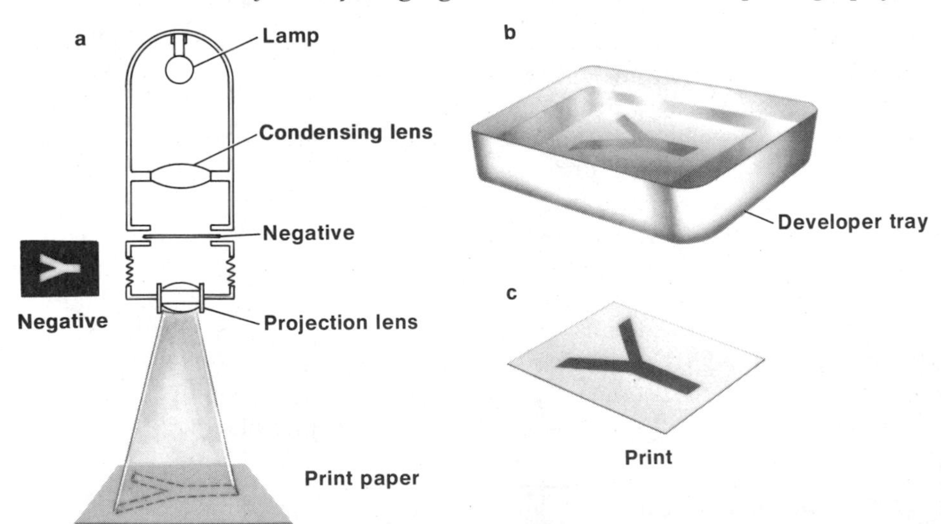


**Figure 18–30** Black and white enlargement printing involves choosing a negative and placing it in an enlarger (*a*). The print paper is exposed for a length of time necessary to produce the desired print (usually by trial and error) and then placed in developer (*b*). After development, the print is fixed with hypo and washed thoroughly to produce the finished product (*c*), which is a direct opposite of the negative.

sulfate ion ($S_2O_3^{2-}$). Thiosulfate ions form stable complexes with silver ions in aqueous solution:

$$\underset{\substack{\text{INSOLUBLE}\\\text{SALT}\\\text{(UNDEVELOPED)}}}{AgBr(s)} + \underset{\substack{\text{FROM}\\\text{"HYPO"}\\\text{SOLUTION}}}{2S_2O_3^{2-}} \longrightarrow \underset{\substack{\text{WATER-SOLUBLE}\\\text{COMPLEX}}}{Ag(S_2O_3)_2^{3-}} + Br^-$$

Solutions of sodium thiosulfate, $Na_2S_2O_3$, known as "hypo," were first used by Sir J. W. F. Herschel to "fix" negatives.

## Instant Black and White Pictures

In 1947, Dr. Edwin H. Land introduced his invention of a process that produced a finished picture in one minute. Since their introduction, the Polaroid and similar processes have been popular for just this unique feature.

After exposure, a Polaroid film is brought into contact with a piece of receiver paper; at the same time a pod containing both a developer and a silver solvent is broken and the pasty mixture is spread over the film. As the developer reduces the exposed silver halide grains in the film emulsion, the silver solvent picks up the unexposed silver ions, which then diffuse across the boundary onto the receiver paper. There, in contact with minute grains of silver already in the paper, the developer reduces the silver in the hypo complex to free silver and a ***positive*** image. This positive image forms since black areas in the original object photographed exposed no silver grains in the emulsion, and it was this silver, as $Ag^+$ ions, which was then carried onto the receiver paper and reduced (Fig. 18–31).

## Spectral Sensitivity

Probably the most important ingredients in a black and white photographic emulsion, other than the silver halide salts themselves, are the spectral sensitizing dyes. Silver halides are most sensitive to blue light or higher energy electromagnetic radiation such as ultraviolet light (Fig. 18–32). A film manufactured with only silver halides as the photosensitive agents will be only blue-sensitive and will not "see" reds, yellows, greens, and so on, as ordinary colors.

By adding spectral sensitizing dyes, photographic emulsions can be made that are sensitive to selected regions of the spectrum with wavelengths from 100 nanometers (ultraviolet) to 1,300 nanometers (infrared).

In 1873, while trying to eliminate light scattering problems in photographing the solar spectrum, W. H. Vogel, a German chemist, added a yellow dye to his emulsion. To his surprise he discovered that he could now record images in the green region of the visible spectrum. Later, in 1904, another German, B. Homolka, discovered a dye, pinacyanol, which when added to a silver halide emulsion rendered it sensitive to the entire visible spectrum (Fig. 18–33). Films of this type are called ***panchromatic*** or "pan" films.

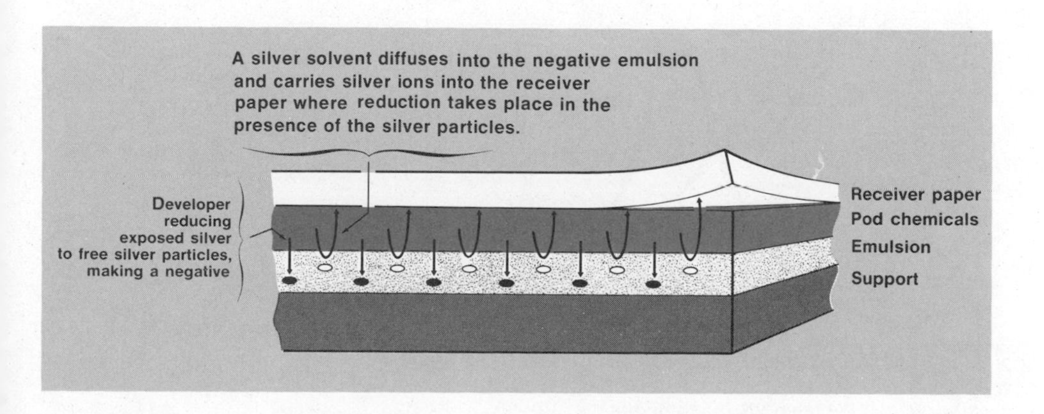
A silver solvent diffuses into the negative emulsion and carries silver ions into the receiver paper where reduction takes place in the presence of the silver particles.
Developer reducing exposed silver to free silver particles, making a negative
Receiver paper
Pod chemicals
Emulsion
Support

**Figure 18–31** A schematic diagram of the chemistry of the Polaroid process.

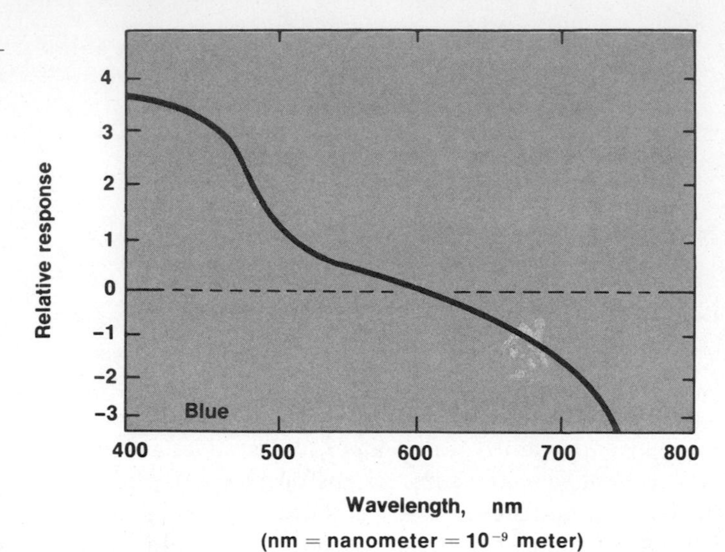


**Figure 18–32** Spectral sensitivity of a typical AgBr emulsion.

The mechanism by which a dye molecule can impart spectral sensitivity to silver halide grains seems to involve initially the absorption of a photon of light by the dye molecule. Next, the excited molecule ejects an electron into the silver halide grain, where a free silver atom is formed. The electron-deficient dye molecule then oxidizes the bromide ion, producing a bromine atom:

**A panchromatic film is sensitive to the entire range of visible wavelengths.**

$$\text{Dye} \xrightarrow{\text{light}} \text{Dye}^+ + e^-$$

$$Ag^+ + e^- \longrightarrow Ag^0 \longrightarrow \text{Silver nuclei}$$

$$\text{Dye}^+ + Br^- \longrightarrow \text{Dye} + Br^0$$

Thus, the process is effectively the same as a photon striking the bromide in the grain itself, the dye serving as a catalyst.

## Color Photography

The chemistry of color photography, which dates back to 1861, is more complicated than that of black and white photography. James Clerk Maxwell, the famous English scientist, was the first to photograph an object in color. He used three ex-

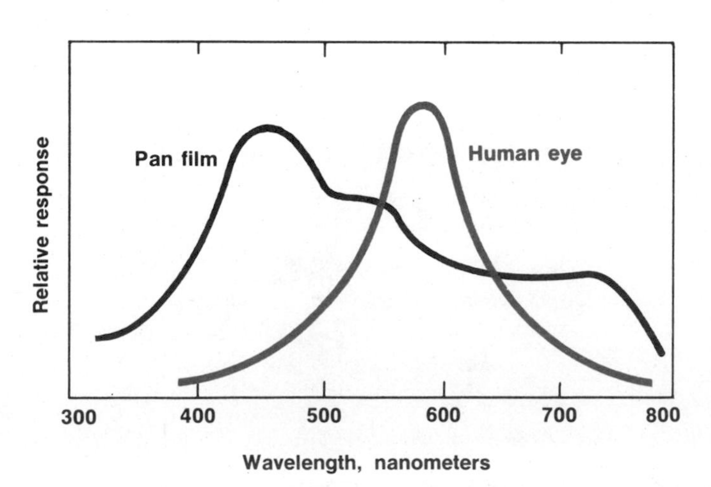


**Figure 18–33** Spectral sensitivity of a panchromatic film compared to that of the human eye.

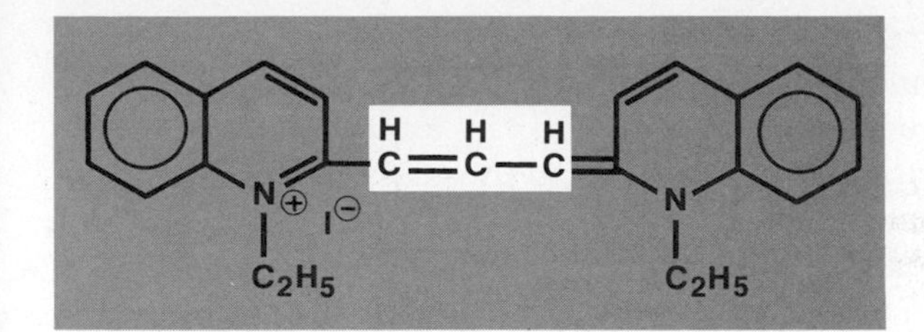


**Figure 18–34** Pinacyanol, a cyanine dye. The conjugated group (double bond, single bond, etc.—one sequence in the white block) produces the color of the dye. Other cyanine dyes have more CH groups, absorb longer wavelengths of light, and shift the film sensitivity toward the red.

posures through three primary color filters and then resynthesized the color of the object by projecting the image through the same filters. This experiment actually predated panchromatic film, but because of a peculiarity of his particular emulsion, it was essentially panchromatic nonetheless. The important point is that the results were consistent with the then emerging theory of color vision and thus led to other more significant results.

## Additive and Subtractive Primary Colors

As early as 1611, De Dominis showed that the visible spectrum is composed of three fundamental colors: red, green, and blue (known as ***additive primaries***). This concept has since proved useful in the development of color vision theory and in color photography. After 1861 the idea slowly evolved that in order to reproduce color images, a film would have to be made with three different layers, each layer sensitive to one of the three primary colors. After the discovery of color-sensitizing dyes and panchromatic black and white film, several different techniques for color photography were developed; but not until 1935, when Kodachrome was placed on the market, did the products reach the consumer. The Kodachrome process produces transparencies (or slides) that are viewed with transmitted light.

If additive primary colors are used to form an image by superposition, such as we might expect in a color transparency, problems with light transmittance arise. Combinations of additive primary-color filters produce black. In order to overcome these problems and obtain a color slide (or color negative, for that matter), another system of primary-color filters was developed, known as ***subtractive primary*** colors. These colors are produced by dyes that absorb the additive primary colors. Thus a dye that absorbs red light transmits or reflects the remainder of the spectrum and appears greenish-blue (*cyan*). Absorption of blue light renders a dye yellow, and absorption of green light makes the dye appear bluish-red (*magenta*).

**Blue light + Green light** } **Cyan**

**Green light + Red light** } **Yellow**

**Blue light + Red light** } **Magenta**

When the proper mixture of subtractive primary dyes is formed in a photographic emulsion during the development process, an image is produced with the desired color. For example, a mixture of magenta and cyan dyes would appear blue, since the magenta dye absorbs green light and the cyan dye absorbs red light, leaving only blue to be transmitted out of the three components of white light.

The use of subtractive primaries in color photography was suggested as early as 1869, but it was much later before the chemistry was worked out in enough detail to yield good results. The problem is to get the right amount of the correct subtractive primary dye in the right place to reproduce the correct true-to-life color. White is produced by the absence of all three subtractive primaries, and black is produced by an equal balance of all three.

## Color Film

Generally, a color film consists of a support and three color-sensitive emulsion layers. The blue-sensitive layer is usually on the top since silver halides are inherently blue-

The thickness of the entire emulsion of color film is only about 0.0254 mm (0.001 inch).

sensitive. Next, a yellow colored filter layer is added. This layer absorbs blue light and serves to protect the lower emulsion layers from blue light. A green-sensitive layer is added and followed by a red-sensitive layer and the support (Fig. 18–35). These layers are rendered color-sensitive by dyes similar to those in the cyanine class, which render black and white film panchromatic. It should be realized, however, that the color-sensitizing dyes are *not* generally involved in producing the final primary colors responsible for the color of the image. It is the final processing of the color film that yields the color image.

## Color Development

Most color films are developed with the aid of a dye-forming color process first introduced by a German chemist, R. Fischer, in 1912. The basis for this process is the oxidation of the developer to a dye-forming substance, which is then allowed to react with a molecule called a ***coupler*** to form the dye.

Color developers are generally substituted amines and as such are reducing agents. An example is N,N-diethyl-*p*-phenylenediamine:

$$H_2N-C_6H_4-N(C_2H_5)_2$$

N,N-DIETHYL-*p*-PHENYLENEDIAMINE
(A COLOR DEVELOPER)

To form a cyan dye during the development process, a phenol compound such as α-naphthol acts as a coupler:

$$C_{10}H_7OH + H_2N-C_6H_4-N(C_2H_5)_2 + 4Ag^+ \longrightarrow$$

COUPLER (α-NAPHTHOL) DEVELOPER

$$O=C_{10}H_6=N-C_6H_4-N(C_2H_5)_2 + 4Ag^0 + 4H^+$$

A CYAN DYE (ABSORBS AT 630 nm)

Light with a wavelength of 630 nm is red.

Thus, in the development of an exposed silver halide grain in the red-sensitive emulsion layer, a small amount of cyan dye is produced. The free silver must be bleached out prior to finishing.

## The Kodachrome Process

An interesting example of a widely used color photography system is the Kodachrome process of the Eastman Kodak Company. The Kodachrome process is a ***reversal*** process; this means colors are reproduced in terms of their correct values and not their negative or complementary colors. The first developer in the Kodachrome pro-

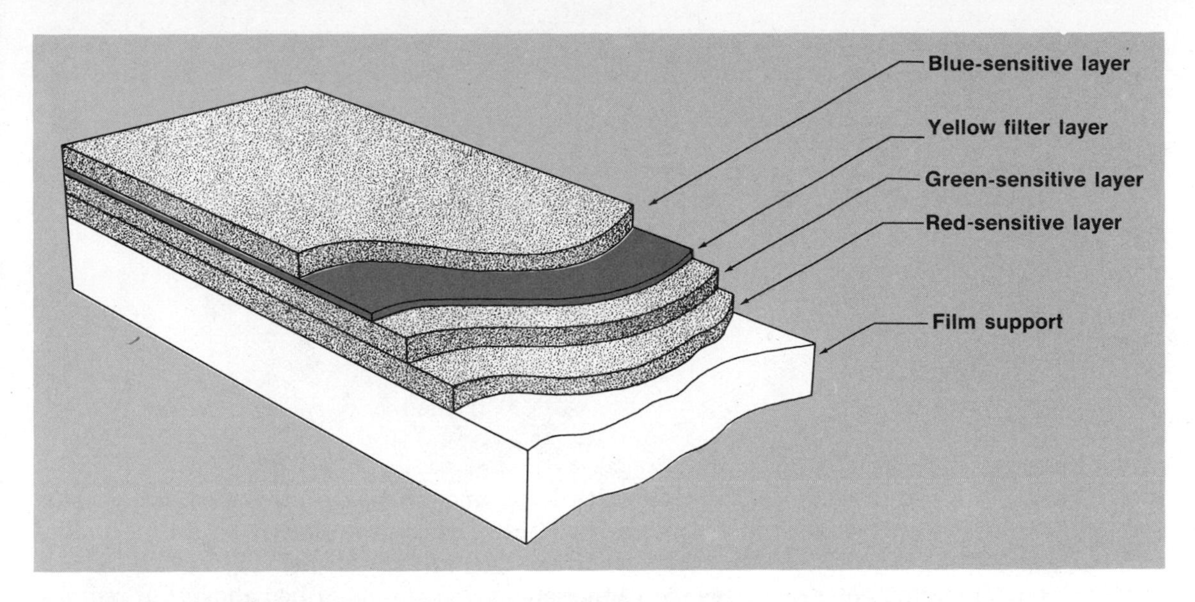


**Figure 18–35** A typical arrangement of color-sensitive emulsion layers in a color film.

cess is a black and white developer. By careful temperature control, development of the exposed silver halide is made essentially complete.

The remaining unexposed silver halide in the three color-sensitive emulsions is a positive record of the original exposure. For example, red light striking the film would, upon black and white development, leave free silver in the red-sensitive layer (Fig. 18–36). Since no other color-sensitive layers were exposed by the original image, they contain no information. Now, selective reexposure and color development will produce free silver throughout the emulsion layers, along with the colored dyes, *except* where the red light originally struck the film. No dye forms there since the silver was previously reduced with a black and white developer.

**All the silver is bleached out of color negative and reversal films during processing.**

Next, all the silver in the three emulsion layers, as well as the yellow-colored protective layer, is bleached with an oxidant such as cyanoferrate ion, $Fe(CN)_6^{3-}$:

$$Ag^0 + Fe(CN)_6^{3-} \longrightarrow Ag^+ + Fe(CN)_6^{4-}$$

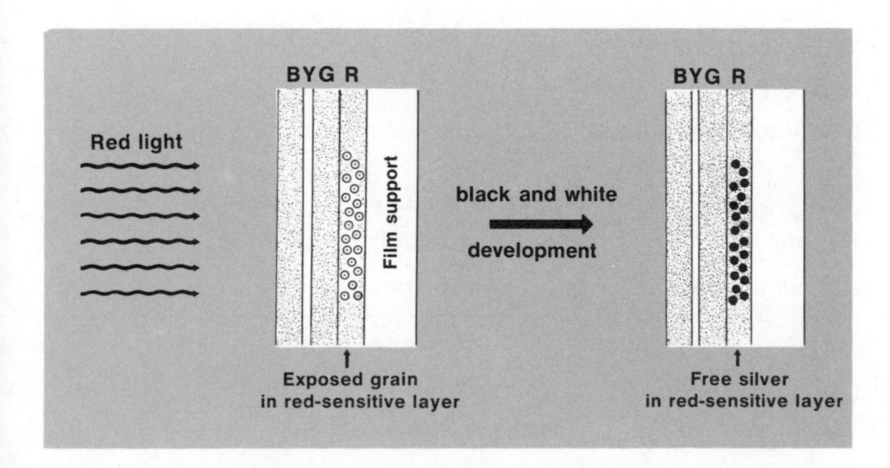


**Figure 18–36** Simplified color image-forming process. B—blue-sensitive layer; Y—yellow filters; G—green-sensitive layer; R—red-sensitive layer.

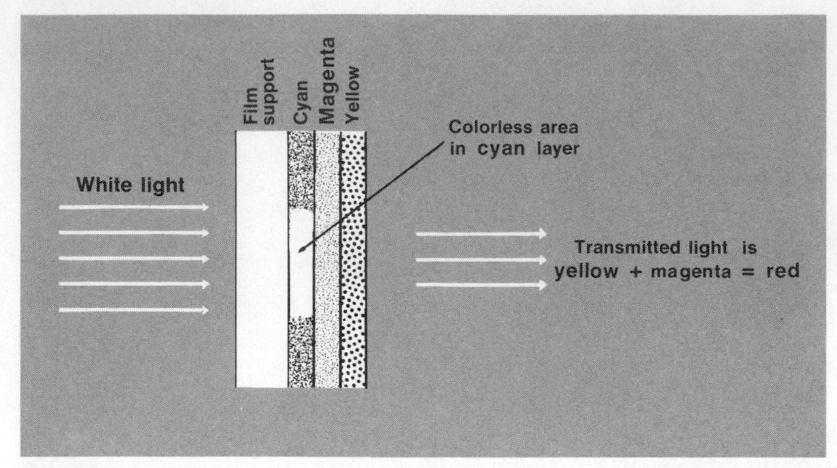


**Figure 18–37** White light passes through a three-layer transparency. Since blue light and green light are absorbed, the transmitted light is red.

Once oxidized, the silver is treated with hypo and washed from the emulsion. The resulting emulsion is colored, but transparent. Considering that red light originally exposed the film, we see that the transmitted light will appear red (Fig. 18–37).

"Instant" color pictures involve the same principles described above, but require a delicate balance of light exposure, photochemical reagents, dyes, developers, and couplers.

## Self-Test 18-F

1. The branch of chemistry that deals with the changes in matter produced by absorbed radiant energy is __________.
2. What chemical composes the dark regions on a negative? __________
3. Name a chemical that can be used as a black and white developer. __________
4. In the development process, silver is (oxidized/reduced).
5. Complete the following equation for the fixing process.

   $Ag^+ + 2S_2O_3^{2-} \longrightarrow$ __________
6. Which are more sensitive to light, very small or somewhat larger grains of AgBr?

   __________
7. Photographic developers are (oxidizing/reducing) agents.
8. Hypo is another name for __________ __________.
9. a. cyan + magenta = __________

   b. cyan + yellow = __________

   c. magenta + yellow = __________
10. The primary additive colors are __________, __________, and __________.
11. A Kodachrome color transparency has no silver in it. (True/False)
12. In the formation of a dye in color film, silver is (oxidized/reduced).
13. Film sensitive to all light in the visible range is known as __________ film.

## Matching Set

| | | |
|---|---|---|
| ______ | 1. negative | **a** introduced Kodak |
| ______ | 2. larger grain size | **b** developer |

| | | | |
|---|---|---|---|
| _______ | 3. daguerreotype | c | stop bath |
| _______ | 4. George Eastman | d | more light sensitive |
| _______ | 5. hydroquinone | e | additive primary colors |
| _______ | 6. acid solution | f | early photographic form |
| _______ | 7. sodium thiosulfate | g | a cyanine dye |
| _______ | 8. panchromatic film | h | introduced instant photography |
| _______ | 9. red, green, and blue | i | exposed areas are dark |
| _______ | 10. pinacyanol | j | sensitive to broad band of wavelengths |
| _______ | 11. Edwin H. Land | k | fixer |

## Questions on Part I—Chemicals in Foods

**1.** What are the primary reasons that food spoils?

**2.** How does salt preserve food?

**3.** Which of the following food additives should be avoided? Give your reasons.

a. butter yellow c. glycerin
b. propionic acid d. sodium cyclamate

**4.** Why does it take less time to cook food in a pressure cooker than in an open pot of boiling water?

**5.** Why does cooking aid digestion?

**6.** What would happen to your ability to digest protein if you kept the acid in your stomach neutralized all the time? Is acid bad for your stomach?

**7.** Why is saccharin preferable to chloroform as a sweetener?

**8.** A label on a brand of breakfast pastries lists the following additives: dextrose, glycerin, citric acid, potassium sorbate, vitamin C, sodium iron pyrophosphate, and BHA. What is the purpose of each substance?

**9.** What is a common flavor enhancer? How do flavor enhancers work?

**10.** How do BHA and BHT prevent potato chips from becoming rancid?

**11.** Choose a label from a food item and try to identify the purpose of each additive.

**12.** Describe some of the chemical changes that occur during the cooking of

a. a carbohydrate
b. a protein
c. a fat

**13.** What causes fat in foods to become rancid? How can this be avoided?

**14.** What causes bread to rise?

**15.** What do the letters FDA represent, and what does this governmental agency do?

**16.** What are the pros and cons of eating "natural" foods as opposed to foods containing chemical additives?

**17.** Why were cyclamates taken off the market?

**18.** Do you think it is wise to use animals in safety tests for drugs and food additives? Should mental patients and prisoners be used for this purpose?

**19.** Many consumer products are almost identical in chemical composition but are sold at widely different prices under different trade names. Do you think the products should be identified by their chemical names or their trade names? Why?

**20.** What is the GRAS list?

**21.** What foods have you eaten during the past week that did not have chemicals added or applied to them?

**22.** See what you can find out about correlation between taste and smell. Are they the same sensation? Are they independent of each other?

**23.** Do you think it would be possible to live on a diet of entirely synthetic foods?

## Questions on Part II—Medicines

**1.** Explain the need for the chemical action of an antacid.

**2.** What is an analgesic? Give three examples, two of which are habit-forming.

**3.** What are enkephalins? How are they related to drug use?

**4.** What is the common name for procaine? In what area is it often used?

**5.** Define the following terms: antacid, analgesic, antipyretic, generic name of a drug.

**6.** What is the chemical formula for aspirin?

**7.** What is aspirin's greatest known danger?

**8.** How do sulfa drugs kill bacteria?

**9.** Name three common antiseptics.

**10.** Acetaminophen (Tylenol) and acetylsalicylic acid (aspirin) are analgesic and antipyretic. Based on their structures shown in the text, do you expect their mechanisms of action to be similar or different?

**11.** What chemical has taken the "sting" out of iodine?

**12.** Select the least used analgesic in prescriptions: morphine, codeine, heroin, demerol.

**13.** How does penicillin kill bacteria?

**14.** Which was the first to be widely used: the sulfa drugs or the penicillins?

**15.** What is the common molecular feature of the steroids?

**16.** What are the active chemicals in birth control pills?

**17.** What are the essential structural differences between the female hormone, progesterone, and the male hormone, testosterone?

**18.** What is the purpose of pyribenzamine in medical use?

**19.** What medicines are in a class known as the tetracyclines?

**20.** Write a paragraph to give your evaluation of diet pills.

## Questions on Part III–Beauty Aids

**1.** You read in a newspaper about a new compound that will break disulfide bonds in proteins. What potential use might it have?

**2.** What do skin, hair, and nails have in common?

**3.** What is the major difference between keratin in the hair and other proteins?

**4.** a. What is the purpose of an emulsifier?
b. In which of the following cosmetics is an emulsifier important: suntan lotion, hair spray, cold cream?

**5.** Why do the lips dry so easily?

**6.** What is the structure of the monomer unit in polyvinylpyrrolidone?

**7.** What is the purpose of each of the following:
a. polyvinylpyrrolidone in hair sprays
b. aluminum chloride in deodorants
c. *p*-aminobenzoic acid in suntan lotion

**8.** Describe in chemical terms what happens when a person gets a permanent.

**9.** What specific substance is broken down during the bleaching of hair?

**10.** What three types of chemical bonds hold hair proteins together?

**11.** Is hair curl the result of primary, secondary, or tertiary protein structure?

**12.** Use the structures of the constituents of lanolin to justify its ability to emulsify face creams.

**13.** Hydrogen bonding is a very handy theoretical tool. Name three applications of hydrogen bonding in consumer products.

**14.** What happens to the solar energy absorbed by *p*-aminobenzoic acid in a suntan lotion or oil?

**15.** If you were going to formulate a suntan lotion, what particular spectral property would you look for in choosing the active compound?

**16.** If a substance is astringent, what is its action?

**17.** What is the purpose of talc in face powder?

**18.** Name an astringent widely used in deodorants.

**19.** What is the major ingredient in lipstick?

**20.** What is the chemical action of hydrogen peroxide on hair?

**21.** Commercial lanolin comes from what animal?

## Questions on Part IV – Automotive Products

**1.** Write a balanced equation for the complete combustion of 2,2,4-trimethylpentane.

**2.** What is meant by high detergency (HD) gasoline? Specifically, what do the detergents do for a gasoline engine?

**3.** What is the purpose of each of these substances in gasoline?
a. ethylenediamine
b. TCP
c. ethylene glycol
d. dyes

**4.** What is the meaning of 10W-30 oil?

**5.** How do the antiwear agents in motor oils work?

**6.** What is the difference in the composition of motor oils and vegetable oils, such as corn oil or palm oil?

**7.** What is the basic difference in the composition of motor oils and automotive greases?

**8.** Referring to the structure of ethylene glycol, explain why this compound is so soluble in water.

**9.** How does antifreeze lower the freezing point of water?

**10.** Why would it be undesirable to have a methanol-type antifreeze in your radiator during the summer?

**11.** Explain how a deicer melts ice.

**12.** Why is the methanol-type antifreeze called temporary and the ethylene glycol-type called permanent antifreeze?

**13.** Write a chemical equation for the rusting of iron in a radiator.

**14.** Would you purchase an antifreeze mixture containing radiator sealants for a new car?

**15.** List three typical additives in antifreeze formulations and describe the action of each.

## Questions on Part V—The Chemistry of Photography

**1.** What would be the effect of lower pH on a typical developer?

**2.** What is the chemical explanation for fogging on film?

**3.** Explain the term fixing. What is a fixer and why is it important in photography?

**4.** Describe what happens chemically from the time the shutter is opened on a camera until the latent image is developed on the film.

**5.** Write a chemical equation for the development of silver ions by Metol.

**6.** Explain what the stop bath solution does in the development process.

**7.** What are the subtractive primary colors?

**8.** Explain how a red dot would be photographed with a color reversal film such as Kodachrome.

**9.** What color is produced by the superposition of the following subtractive primary colors?

a. magenta and yellow
b. cyan and yellow
c. magenta and cyan
d. magenta, cyan, and yellow

**10.** Is all of the silver reduced in the development of a black and white film? Explain.

**11.** What different support materials have been used for the light-sensitive emulsion in photography?

**12.** What is the purpose of sodium sulfite in black and white film development?

**13.** Where are the developer and fixer in the black and white Polaroid film?

**14.** What was done to increase the spectral range of black and white film?

**15.** Explain the action of a coupler in color photography.

APPENDIX

# A The International System of Units (SI)

Since 1960, a coherent system of units known as the Système International (SI system), bearing the authority of the International Bureau of Weights and Measures, has been in effect and is gaining acceptance among scientists. It is an extension of the metric system that began in 1790, with each physical quantity assigned a unique SI unit. An essential feature of both the older metric system and now the newer SI is a series of prefixes that indicate a power of ten multiple or submultiple of the unit.

## Units of Length

The standard unit of length is the ***meter.*** It was originally meant to be one ten-millionth of the distance along a meridian from the North Pole to the equator. However, the lack of precise geographical information necessitated a better definition. For a number of years the meter was defined as the distance between two etched lines on a platinum-iridium bar kept at 0°C (32°F) in the International Bureau of Weights and Measures at Sèvres, France. The inability to measure this distance as accurately as desired prompted a recent redefinition of the meter as being a length equal to 1,650,763.73 times the wavelength of the orange-red spectrographic line of $^{86}_{36}Kr$.

The meter (39.37 inches) is a convenient unit with which to measure the height of a basketball goal (3.05 meters), but it is unwieldy for measuring the parts of a watch or the distance between continents. For this reason, prefixes are defined in such a way that, when placed before the meter, they define distances convenient for our particular purposes. Some of the prefixes with their meanings are:

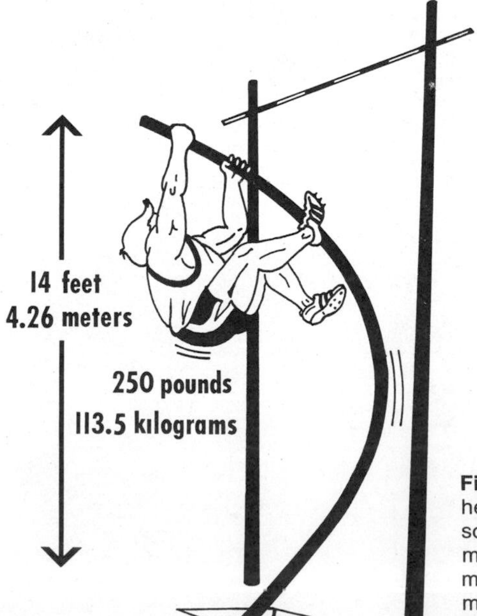


**Figure A-1** The pole-vaulter is easily recognized as hefty when described by 250 pounds and his jump something less than a record 14 feet. As Americans move closer to the use of the system of international measurements, the 113.5 kilograms and the 4.26 meters will produce similar conceptualizations related to previous experience.

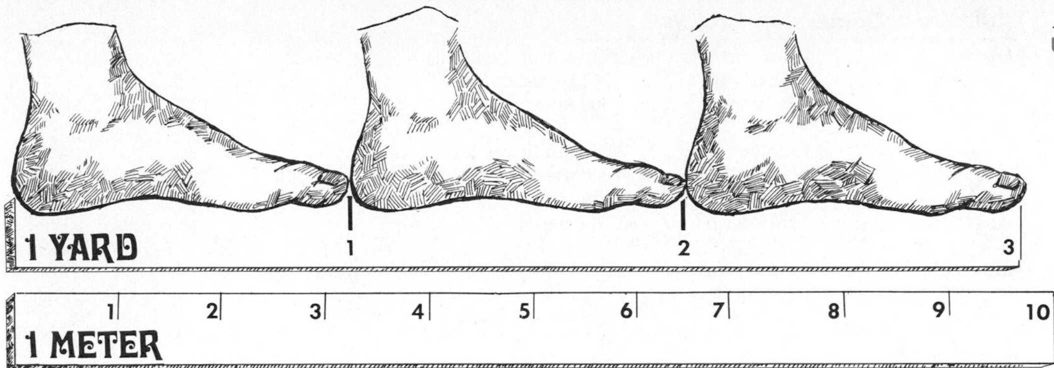


**Figure A-2** A meter equals 1.094 yards.

nano—1/1,000,000,000 or 0.000000001
micro—1/1,000,000 or 0.000001
milli—1/1,000 or 0.001
centi—1/100 or 0.01
deci—1/10 or 0.1
deka—10
hecto—100
kilo—1,000
mega—1,000,000

The corresponding units of length with their abbreviations are the following:

nanometer (nm)—0.000000001 meter
micrometer ($\mu$m)—0.000001 meter
millimeter (mm)—0.001 meter
centimeter (cm)—0.01 meter
decimeter (dm)—0.1 meter
meter (m)—1 meter
dekameter (dam)—10 meters
hectometer (hm)—100 meters
kilometer (km)—1,000 meters
megameter (Mm)—1,000,000 meters

Since the prefixes are defined in terms of the decimal system, the conversion from one metric length to another involves only shifting the decimal point. Mental calculations are quickly accomplished.

How many centimeters are in a meter? Think: Since a centimeter is the one-hundredth part of a meter, there would be 100 centimeters in a meter.

Conversion of measurements from one system to the other is a common problem. Some commonly used English-SI equivalents (conversion factors) are given in Table A–1.

## Units of Mass

The primary unit of mass is the ***kilogram*** (1000 grams). This unit is the mass of a platinum-iridium alloy sample deposited at the International Bureau of Weights and Measures. One pound contains a mass of 453.6 grams (a five-cent nickel coin contains about 5 grams).

Conveniently enough, the same prefixes defined in the discussion of length are used in units of mass, as well as in other units of measure.

**TABLE A–1 Conversion Factors***

| | | |
|---|---|---|
| Length: | 1 inch (in.) | = 2.54 centimeters (cm) |
| | 1 yard (yd.) | = 0.914 meter (m) |
| | 1 mile (mi.) | = 1.609 kilometers (km) |
| Volume: | 1 ounce (oz.) | = 29.57 milliliters (ml) |
| | 1 quart (qt.) | = 0.946 liter (l) |
| | 1.06 quart (qt.) | = 1 liter (l) |
| | 1 gallon (gal.) | = 3.78 liters (l) |
| Mass (weight)†: | 1 ounce (oz.) | = 28.35 grams (g) |
| | 1 pound (lb.) | = 453.6 grams (g) |
| | 1 ton (tn.) | = 907.2 kilograms (kg) |

*Common English units are used.

†Mass is a measure of the amount of matter, whereas weight is a measure of the attraction of the earth for an object at the earth's surface. The mass of a sample of matter is constant, but its weight varies with position and velocity. For example, the space traveler, having lost no mass, becomes weightless in earth orbit. Although mass and weight are basically different in meaning, they are often used interchangeably in the environment of the earth's surface.

## Units of Volume

The SI unit of volume is the ***cubic meter*** ($m^3$). However, the volume capacity used most frequently in chemistry is the liter, which is defined as 1 cubic decimeter (1 $dm^3$). Since a decimeter is equal to 10 centimeters (cm), the cubic decimeter is equal to $(10\ cm)^3$ or 1,000 cubic centimeters (cc). One cc, then, is equal to one milliliter (the thousandth part of a liter). The ml (or cc) is a common unit that is often used in the measurement of medicinal and laboratory quantities. There are then 1,000 liters in a kiloliter or cubic meter.

## Units of Energy

The SI unit for energy is the ***joule*** (J), which is defined as the work performed by a force of one newton acting through a distance of one meter. A newton is defined as that force which produces an acceleration of one meter per second per second when applied to a mass of one kilogram. Conversion units for energy are:

1 calorie = 4.184 joules
1 kilowatt-hour = $3.5 \times 10^6$ joules

## Other SI Units

Other SI units are listed below.

| | |
|---|---|
| Time | second (s) |
| Temperature | Kelvin (K) |
| Electric current | ampere (A) = 1 coulomb per second |
| Amount of molecular substance | mole (mol) = $6.023 \times 10^{23}$ molecules |
| Pressure | pascal (Pa) = 1 newton per square meter |
| Power | watt (W) = 1 joule per second |
| Electric charge | coulomb (C) = $6.24196 \times 10^{18}$ electronic charges |
| | = $1.036086 \times 10^5$ faradays |

Further information on SI units can be obtained from "SI Metric Units—An Introduction," by H. F. R. Adams, McGraw-Hill Ryerson Ltd., Toronto, 1974.

# Temperature Scales

APPENDIX

# B

The system of measuring temperature that is used in scientific work is based on the ***centigrade*** or ***Celsius*** temperature scale. This temperature scale was defined by Anders Celsius, a Swedish astronomer, in 1742. The Celsius scale is based upon the expansion of a column of mercury that occurs when it is transferred from a cold standard temperature to a hot standard temperature. The cold standard temperature is the temperature of melting ice and is defined as 0°C. The hot standard temperature is the temperature at which water boils under standard conditions of pressure; it is defined as 100°C. The expansion of the mercury is assumed to be linear over this range, and the distance through which the mercury column expands between 0°C and 100°C is divided into 100 equal parts, each corresponding to one degree.

On the scale commonly used in the United States, the Fahrenheit scale (developed by and named after Gabriel Daniel Fahrenheit, a German-Dutch physicist of the early eighteenth century), the freezing mark is 32° and the boiling mark is 212°. Obviously, the ice is no hotter when a Fahrenheit thermometer is stuck into the system than when a Celsius thermometer is used. The marks on the tubes are simply different names for the same thing.

The ideal gas law predicts another temperature scale based on the temperature-volume behavior of a gas (Charles' law). Lord Kelvin reasoned that if an ideal gas lost $\frac{1}{273}$ of its volume during a temperature decrease from 0°C to −1°C, then at −273°C, theoretically at least, there should be no volume at all. All gases liquefy before this temperature is reached. However, this temperature, −273°C, appears to be a zero temperature defined by nature. Thus the Kelvin temperature scale was developed. Minus 273°C becomes 0 Kelvin and 0°C becomes 273 Kelvin. (More accurately, 273.15 K.) Note that current SI usage does not write °K, but K.

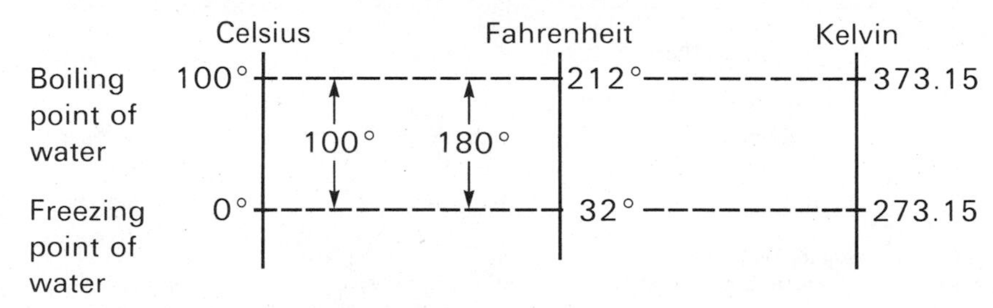


Because all these systems are now in common use, it is necessary at times to convert from one system to another.

Note that 100 Celsius degrees = 180 Fahrenheit degrees or
1 C degree = 1.8 ( or $\frac{9}{5}$) F degrees.

But if we wish to convert temperature from one scale to another, we must remember that 0°C is the same as 32°F. Therefore, 32 must be added to the calculated number of degrees Fahrenheit in order to revert to the start of the counting in the Fahrenheit system.

Therefore,

$$°F = \frac{9}{5}(°C) + 32$$

which can be arranged to

$$°C = \frac{5}{9}(°F - 32)$$

**Example 1**
Convert 50°F to the corresponding Celsius temperature:

$$\begin{aligned} °C &= \frac{5}{9}(°F - 32) \\ &= \frac{5}{9}(50 - 32) \\ &= \frac{5}{9}(18) \\ &= 10°C \end{aligned}$$

**Example 2**
Convert 25°C to the corresponding Fahrenheit temperature:

$$\begin{aligned} °F &= \frac{9}{5}(°C) + 32 \\ &= \frac{9}{5}(25) + 32 \\ &= 45 + 32 \\ &= 77°F \end{aligned}$$

To convert Celsius to Kelvin:

$$K = °C + 273$$

**Example 3**
Convert 23°C to Kelvin:

$$\begin{aligned} K &= 23° + 273 \\ &= 296\ K \end{aligned}$$

---

Rather than memorizing formulas, many students prefer to figure out temperature conversions using common sense. A common-sense approach is shown in Table B–1.

**TABLE B–1 Common-Sense Method for Converting Temperatures**

Starting point: { freezing point (fp) of water is 0°C = 32°F; boiling point (bp) of water is 100°C = 212°F }

Other information needed: 100 C degrees = 180 F degrees

| General Procedure | Example 1 (50°F = ?°C) | Example 2 (25°C = ?°F) | Example 3 (98.6°F = ?°C) |
|---|---|---|---|
| 1. Select a reference point. | 1. fp of water | 1. fp of water | 1. bp of water |
| 2. Determine relationship of temperature of interest to reference point. | 2. 50°F is 18 F degrees above fp of water | 2. 25°C is 25 C degrees above fp of water | 2. 98.6°F is 113.4 F degrees below bp of water |
| 3. Convert the number of degrees. | 3. 18 F degrees is $18\ F° \times \frac{100\ C°}{180\ F°} =$ 10 C degrees | 3. 25 C degrees is $25\ C° \times \frac{180\ F°}{100\ C°} =$ 45 F degrees | 3. 113.4 F degrees is $113.4\ F° \times \frac{100\ C°}{180\ F°} =$ 63.0 C degrees |
| 4. Express temperature, taking into account the selected reference point. | 4. The temperature is 10 C degrees above the fp of water (which is 0°C), so the temperature is *10°C* | 4. The temperature is 45 F degrees above the fp of water (which is 32°F), so the temperature is 45 + 32 = *77°F* | 4. The temperature is 63.0 C degrees below the bp of water (which is 100.0°C), s the temperature is 100.0 − 63.0 = *37.0°C* |

# Factor-Label Approach to Conversion Problems

APPENDIX

# C

For converting a measurement from one system of units to another, the following factor-label method of solution is straightforward and does not require the decision of whether to divide or multiply to obtain the proper answer. This method makes the decision for you, a decision that is sometimes difficult when dealing with new units with which you have had little experience.

**Example**

How many liters are in 6 quarts?

1. Write down the unit to which you are converting. The question mark indicates the number of liters to be determined.

   1. ? liters =

2. On the right-hand side of the equal sign, write down the quantity given. Write both the number and the name or label.

   2. ? liters = 6 quarts

3. Now look at the two units. Recall a conversion between these two units. These conversions must be learned or looked up.

   3. 1 liter = 1.06 quarts, 1 liter per 1.06 quarts

   $$\text{or } \frac{1 \text{ liter}}{1.06 \text{ quarts}}$$

   $$\text{or } \frac{1.06 \text{ quarts}}{\text{liter}}$$

4. Write the conversion factor on the right-hand side so that the unwanted units will cancel; that is, a unit in the numerator will cancel the same unit in the denominator, and only the unit you want will remain. Do not, of course, cancel the numbers, just the units.

   4. ? liters =

   $$6 \cancel{\text{quarts}} \times \frac{1 \text{ liter}}{1.06 \cancel{\text{quarts}}}$$

5. Do the indicated multiplication and division. The line, of course, means divided by. Check the units on both sides of the equation. The units should be the same and in the same position (numerator or denominator).

   $$5.\ ?\text{ liters} = \frac{6}{1.06} \text{ liters}$$

   $$= 5.66 \text{ liters}$$

---

Now, suppose you needed to convert 3 pints to ml, and you do not know a conversion factor that will make the conversion in one step.

Proceed as before:

? ml = 3 pints

Note this is a volume conversion from the English system to the metric system as before. Recall the volume conversion factor that you know between these systems.

1 liter = 1.06 quarts

Convert in the system of the given quantity until you reach the unit in the conversion factor for this system. Then write down the conversion factor between systems so that like units in separate factors will cancel.

$$? \text{ ml} = 3 \cancel{\text{pints}} \times \frac{1 \cancel{\text{quart}}}{2 \cancel{\text{pints}}} \times \frac{1 \text{ liter}}{1.06 \cancel{\text{quarts}}}$$

Cancel units and recall a conversion factor that converts the units you have left to the unit you want. Do the indicated multiplication and division.

$$? \text{ ml} = \frac{3 \times 1 \times 1 \cancel{\text{liter}}}{2 \times 1.06} \times \frac{1000 \text{ ml}}{1 \cancel{\text{liter}}}$$

$$= \frac{3 \times 1 \times 1 \times 1000}{2 \times 1.06 \times 1} \text{ ml}$$

$$= 1420 \text{ ml}$$

In brief, the method is very simple:

1. Write the units you want, an equal sign, and the quantity you have given.
2. Write conversion factors so the unwanted factors will cancel.
3. Keep in one system until you come to units in a familiar intersystem conversion factor.
4. Do the indicated arithmetic and check the units.
5. Examine the size of the answer for reasonableness.

*Note:* This approach to problem solving has wide applicability to many other types of problems. It is especially useful in problems pertaining to weight relationships in chemical reactions. See Appendix D.

# Calculations with Chemical Equations

APPENDIX

# D

The bases for calculations with chemical equations were presented in Chapter 5. Problems of a more complex nature and a systematic approach to their solution are presented in the following examples. Finally, a list of exercise problems is given for further study.

**Example 1**

## Balanced Equations Express Number Ratios for Particles

In the reaction of hydrogen with oxygen to form water, how many molecules of hydrogen are required to combine with 19 oxygen molecules?

*Solution:* A chemical equation can be written for a reaction only if the reactants and products are identified and the respective formulas determined. In this problem, the formulas are known and the unbalanced equation is:

$$H_2 + O_2 \longrightarrow H_2O$$

It is evident that one molecule of oxygen contains enough oxygen for two water molecules and the equation, as written, does not account for what happens to the second oxygen atom. As it is, the equation is in conflict with the conservation of atoms in chemical changes. This conflict is easily corrected by balancing the equation:

$$2H_2 + O_2 \longrightarrow 2H_2O$$

Now all atoms are accounted for in the equation and it is obvious that two hydrogen molecules are required for each oxygen molecule. In other words, two hydrogen molecules are equivalent to one oxygen molecule in their usage. This can be expressed as follows:

2 hydrogen molecules {are equivalent to} 1 oxygen molecule;

or,

$$2H_2 \text{ molecules} \sim O_2 \text{ molecule};$$

or,

$$\frac{2H_2 \text{ molecules}}{O_2 \text{ molecule}},$$

which can be read as two hydrogen molecules per one oxygen molecule.

Using now the factor-label approach developed in Appendix C, the solution is readily achieved.

$$? \; H_2 \text{ molecules} = 19O_2 \text{ molecules}$$

$$\begin{aligned} ? \; H_2 \text{ molecules} &= 19\cancel{O_2 \text{ molecules}} \times \frac{2H_2 \text{ molecules}}{\cancel{O_2 \text{ molecule}}} \\ &= 38H_2 \text{ molecules} \end{aligned}$$

*Note:* The reader is likely to say at this point that the method is cumbersome and that he can quickly see the answer to be $38H_2$ molecules without "the method." However, problems to follow are made much easier if a systematic method of approach is used.

### Example 2

## Laboratory Mole Ratios Identical with Particle Number Ratios

How many moles of hydrogen molecules must be burned in oxygen (the reaction of Example 1) to produce 15 moles of water molecules (about a glassful)?

*Solution:* The balanced equation

$$2H_2 + O_2 \longrightarrow 2H_2O$$

tells us that two molecules of hydrogen produce two molecules of water; or,

2 molecules hydrogen $\sim$ 2 molecules water,

and therefore,

1 molecule hydrogen $\sim$ 1 molecule water.

It is obvious then that the number of water molecules produced will be equal to the number of hydrogen molecules consumed regardless of the actual number involved. Therefore,

$6.02 \times 10^{23}$ molecules of hydrogen $\sim$ $6.02 \times 10^{23}$ molecules of water

Since $6.02 \times 10^{23}$ is a number called the mole, it follows that one mole of hydrogen molecules will produce one mole of water molecules. The general conclusion, then, is the following: the ratio of particles in the balanced equation is the same as the ratio of moles in the laboratory. The solution to the problem logically follows:

$$\left.\begin{array}{r}\text{? moles of hydrogen}\\ \text{molecules}\end{array}\right\} = \left\{\begin{array}{c}15\ \cancel{\text{moles}}\\ \cancel{\text{water molecules}}\end{array}\right\} \times \frac{1\ \text{mole hydrogen molecules}}{1\ \cancel{\text{mole water molecules}}}$$

$$= 15\ \text{moles hydrogen molecules}$$

*Note:* Again the solution to the problem looks simple enough without resorting to the factor-label method. However, in Examples 3 and 4, the numbers become such that a quick mental solution is not readily achieved by most students.

### Example 3

## Mole Weights Yield Weight Relationships

How many grams of oxygen are necessary to react with an excess of hydrogen to produce 270 grams of water?

*Solution:* From the balanced equation

$$2H_2 + O_2 \longrightarrow 2H_2O$$

the mole ratio between oxygen and water is immediately evident and is one mole of oxygen molecules per two moles of water molecules, or

$$\frac{1\ \text{mole oxygen molecules}}{2\ \text{moles water molecules}}$$

This mole ratio can be changed into a weight ratio since the mole weight can be easily calculated from the atomic weights involved. One molecule of oxygen ($O_2$) weighs 32 amu (16 amu for each oxygen atom). Therefore, a mole of oxygen molecules weighs 32 grams. Similarly, two moles of water weigh 36 grams [2(16 + 1 + 1)]. Therefore, the weight ratio is:

$$\frac{1\ \cancel{\text{mole oxygen molecules}} \times \dfrac{32\ \text{grams oxygen}}{\cancel{\text{mole oxygen molecules}}}}{2\ \cancel{\text{moles water molecules}} \times \dfrac{18\ \text{grams water}}{\cancel{\text{mole water molecules}}}}$$

or,

$$\frac{32\ \text{grams oxygen}}{36\ \text{grams water}}$$

This weight relationship is exactly the conversion factor needed to answer the original question:

$$\begin{aligned} ?\ \text{grams oxygen} &= 270\ \cancel{\text{grams water}} \times \frac{32\ \text{grams oxygen}}{36\ \cancel{\text{grams water}}} \\ &= 240\ \text{grams oxygen} \end{aligned}$$

*Note:* It should be observed that a weight relationship could be established between any two of the three pure substances involved in the reaction, regardless of whether they are reactants or products.

**Example 4**

How many molecules of water are produced in the decomposition of 8 molecules of table sugar? The unbalanced equation is as follows:

$$C_{12}H_{22}O_{11} \longrightarrow C + H_2O$$

*Solution:* Balance the equation

$$C_{12}H_{22}O_{11} \longrightarrow 12C + 11H_2O$$

$$\begin{aligned} ?\ \text{molecules of water} &= 8\ \cancel{\text{molecules sugar}} \times \frac{11\ \text{molecules water}}{1\ \cancel{\text{molecule sugar}}} \\ &= 88\ \text{molecules of water} \end{aligned}$$

**Example 5**

How many grams of mercuric oxide are necessary to produce 50 grams of oxygen? Mercuric oxide decomposes as follows:

$$2HgO \longrightarrow 2Hg + O_2$$

*Solution:*

Weight of two moles of HgO = 2(201 + 16) = 2(217) = 434 g

Weight of 1 mole of $O_2$ = 2(16) = 32 g

$$\begin{aligned} ?\ \text{g HgO} &= 50\ \cancel{\text{g oxygen}} \times \frac{434\ \text{g mercuric oxide}}{32\ \cancel{\text{g oxygen}}} \\ &= 678\ \text{g mercuric oxide} \end{aligned}$$

### Example 6

How many pounds of mercuric oxide are necessary to produce 50 pounds of oxygen by the reaction:

$$2HgO \longrightarrow 2Hg + O_2$$

*Solution:* Note that the problem is the same as Example 5 except for the units of chemicals. Also note that the conversion factor of Example 5

$$\frac{434 \text{ g mercuric oxide}}{32 \text{ g oxygen}}$$

can be converted to any other units desired:

$$\frac{434 \cancel{\text{g}} \text{ mercuric oxide} \times \dfrac{1 \text{ pound}}{454 \cancel{\text{g}}}}{32 \cancel{\text{g}} \text{ oxygen} \times \dfrac{1 \text{ pound}}{454 \cancel{\text{g}}}} = \frac{434 \text{ pounds mercuric oxide}}{32 \text{ pounds oxygen}}$$

It is evident that the ratio, $\frac{434}{32}$, expresses the ratio between weights of mercuric oxide and oxygen in this reaction regardless of the units employed.

$$? \text{ pounds mercuric oxide} = 50 \cancel{\text{ pounds oxygen}} \times \frac{434 \text{ pounds mercuric oxide}}{32 \cancel{\text{ pounds oxygen}}}$$
$$= 678 \text{ pounds of mercuric oxide}$$

## Problems

**1.** What weight of oxygen is necessary to burn 28 g of methane, $CH_4$? The equation is:

$$CH_4 + 2O_2 \longrightarrow CO_2 + 2H_2O$$

***Ans.*** 112 g oxygen

**2.** Potassium chlorate, $KClO_3$, releases oxygen when heated according to the equation:

$$2KClO_3 \longrightarrow 2KCl + 3O_2$$

What weight of potassium chlorate is necessary to produce 1.43 g of oxygen? ***Ans.*** 3.65 g $KClO_3$

**3.** $Fe_3O_4$ is a magnetic oxide of iron. What weight of this oxide can be produced from 150 g of iron? ***Ans.*** 207 g oxide

**4.** Steam reacts with hot carbon to produce a fuel called water gas; it is a mixture of carbon monoxide and hydrogen. The equation is:

$$H_2O + C \longrightarrow CO + H_2$$

What weight of carbon is necessary to produce 10 g of hydrogen by this reaction? ***Ans.*** 60 g carbon

**5.** Iron oxide, $Fe_2O_3$, can be reduced to metallic iron by heating it with carbon.

$$2Fe_2O_3 + 3C \longrightarrow 4Fe + 3CO_2$$

How many tons of carbon would be necessary to reduce 5 tons of the iron oxide in this reaction? ***Ans.*** 0.56 ton carbon

**6.** How many grams of hydrogen are necessary to reduce 1 pound (454 g) of lead oxide (PbO) by the reaction:

$PbO + H_2 \longrightarrow Pb + H_2O$

***Ans.*** 3.91 g hydrogen

**7.** Hydrogen can be produced by the reaction of iron with steam.

$4H_2O + 3Fe \longrightarrow 4H_2 + Fe_3O_4$

What weight of iron would be needed to produce one-half pound of hydrogen?

***Ans.*** 10.5 lb iron

**8.** Tin ore, containing $SnO_2$, can be reduced to tin by heating with carbon.

$SnO_2 + C \longrightarrow Sn + CO_2$

How many tons of tin can be produced from 100 tons of $SnO_2$?

***Ans.*** 79 tons tin

# Answers to Self-Tests and Matching Sets

## Chapter 1

### Self-Test 1-A

1. a. air
   b. concrete
   c. vegetables
   d. paint
2. operational definition
3. a. table sugar
   b. distilled water
   c. diamond
   d. copper wire
4. false

### Self-Test 1-B

1. (a) metals and (b) nonmetals

| metals: | nonmetals |
|---|---|
| a. iron | a. oxygen |
| b. copper | b. silicon |
| c. gold | c. carbon |
| d. silver | d. nitrogen |
| e. chromium | e. chlorine |
| f. magnesium | f. fluorine |

2. 106
3. false
4. a. burning of coal
   b. dissolving of iron ore in an acid
   c. making steel from iron ore
   d. making of aspirin
5. a. cutting diamond
   b. blowing glass
   c. molding plastic
   d. slicing bread
6. chemical substance
7. submicroscopic
8. a. macroscopic
   b. microscopic
   c. molecular
9. a. theories
   b. laws
   c. facts
10. a. a sodium atom, or a mole of sodium
    b. two sodium atoms, or two moles of sodium
    c. a hydrogen chloride molecule, or a mole of hydrogen chloride
    d. reacts to form
    e. a hydrogen molecule, or a mole of hydrogen
    f. two "molecules" of sodium chloride, or two moles of sodium chloride

### Matching Set

| | | |
|---|---|---|
| 1. b | 5. c | 9. g |
| 2. e | 6. f | 10. h |
| 3. d | 7. i | 11. j |
| 4. a | 8. k | |

## Chapter 2

### Self-Test 2-A

1. Leucippus and Democritus
2. b. philosophy
3. gained in a chemical reaction
4. CO and $CO_2$
5. a. a new compound
   b. 2:4:1
   c. law of multiple proportions
6. d. atoms are recombined in different arrangements
7. a. the same, b. atoms
8. repel; attract
9. alpha ($\alpha$), beta ($\beta$), gamma ($\gamma$), gamma ($\gamma$)

### Self-Test 2-B

1. protons, neutrons
2. small
3. nucleus
4. electrons, protons, neutrons
5. electrons and protons
6. atomic
7. about 1,836
8. 33, 33, and 42
9. different, identical
10. electrons
11. electrons and protons
12. false

### Self-Test 2-C

1. particles, waves
2. spectrum
3. farther from, closer to
4. d. wave nature of the electron
5. wavelength
6. 18, 2
7. a. no, not the paths
   b. in a way, yes; they are representations of the space in which we can expect to find the electron with 90% certainty.
   c. of a given type, yes, with 90% certainty

### Matching Set

1. e
2. a
3. j
4. i
5. d
6. n
7. b
8. f
9. m
10. g
11. h
12. c
13. k
14. l

## Chapter 3

### Self-Test 3-A

1. IIA, VIII, VIIA, IIIA, IB
2. 2, 5, 7, 4
3. R, R, T, N, I
4. Mendeleev
5. atomic numbers
6. periodic
7. metals

### Self-Test 3-B

1. 3, $GaCl_3$; 2, $BaCl_2$; 2, $SeCl_2$; 1, ICl
2. metalloid, metal, nonmetal, metal, metal, nonmetal, nonmetal, metalloid
3. 1, 2, 7, 7, 6, 3, 4
4. ionization
5. He, F, Br, S
6. K, Br, S, In
7. IA, IVA, IIA

### Matching Set

1. i
2. c
3. e
4. a
5. d
6. g
7. b
8. h
9. f
10. k

## Chapter 4

### Self-Test 4-A

1. ions
2. ionic
3. one
4. He, Cs
5. $CaI_2$
6. $Cl^-$
7. valence
8. losing
9. gaining
10. Rb, one electron lost
    S, two electrons gained
    Ca, two electrons lost
    Mg, two electrons lost
    K, one electron lost
    Br, one electron gained

### Self-Test 4-B

1. $H_2$, HF
2. six
3. three
4. fluorine
5. a. eight
   b. octet
   c. most of the time
6. fluorine
7. N, O, F
8. $H_2O$

### Self-Test 4-C

1. $\sigma$, $\pi$
2. $\sigma$, $\pi$
3. s, p
4. nonbonding, bonding
5. single
6. three
7. linear, planar, octahedral

### Matching Set I

1. l
2. d
3. j
4. f
5. i
6. g
7. b
8. e
9. k
10. c
11. h
12. a
13. m

### Matching Set II

1. d
2. g
3. h
4. f
5. a
6. e
7. b
8. c

### Matching Set III

1. a
2. c
3. d
4. e
5. b

## Chapter 5

### Self-Test 5-A

1. temperature, because rate of bacterial growth slows down, as does the rate of any chemical reaction, with decreasing temperature
2. hydrogen and oxygen
3. hemoglobin and oxygen, and calcium oxide (CaO) and water
4. a. double replacement
   b. combination or synthesis
   c. single replacement
   d. decomposition
5. a. $SeO_2$ and $SeO_3$
   b. $SeF_4$ and $SeF_6$
   c. $H_2SeO_3$ and $H_2SeO_4$
6. a. $2Na + F_2 \longrightarrow 2NaF$
   b. $Mg + Cl_2 \longrightarrow MgCl_2$
   c. $2Ra + O_2 \longrightarrow 2RaO$
   d. $CaCl_2 + 2AgNO_3 \longrightarrow Ca(NO_3)_2 + 2AgCl$
   e. $2K + H_2CO_3 \longrightarrow K_2CO_3 + H_2$
   f. $2Li + 2HClO_4 \longrightarrow 2LiClO_4 + H_2$
   g. $Ba + S \longrightarrow BaS$
   h. $Mg(OH)_2 + 2HCl \longrightarrow MgCl_2 + 2H_2O$
7. (2) flour in dust form
8. hemoglobin oxygen uptake and release
9. formation of salt from sodium and chlorine, formation of $SO_2$ from S and $O_2$, and derusting of iron

### Self-Test 5-B

1. absorbs
2. 136 kcal
3. a. $2Mg + O_2 \longrightarrow 2MgO$
   b. $Si + 2Cl_2 \longrightarrow SiCl_4$
   c. $4Al + 3O_2 \longrightarrow 2Al_2O_3$
4. a. $2H_2O \longrightarrow 2H_2 + O_2$
   b. 18 grams of water
   c. 16 grams of oxygen
   d. 18 tons of water

### Matching Set I

1. h
2. f
3. e
4. b
5. c
6. d
7. a
8. g

### Matching Set II

1. c
2. e
3. a
4. f
4. b
6. d

## Chapter 6

### Self-Test 6-A

1. an electrolyte
2. a base, an acid
3. a base, an acid
4. amphiprotic
5. neutral
6. water
7. an acid reacts with a base

### Self-Test 6-B

1. 3 (a) acidic, 10 (b) basic, 7 (c) neutral
2. 1,400 g or 1.4 kg
3. $NH_4^+ + OH^-$
4. $H_3O^+ + Cl^-$
5. pH = 2, more acidic than pH = 6
6. yes
7. insoluble
8. $CO_2$
9. fertilizer manufacture; explosives manufacture

### Self-Test 6-C

1. gains, gets reduced; loses, gets oxidized
2. reducing, oxidizing
3. reduced; cathode
4. $2Ag^+ + Cu^{2+}$ = no
   $F_2 + Zn$ = yes
   $Fe^{2+} + Ag^+$ = yes
   $Mg^{2+} + 2Fe^{2+}$ = no
5. a reduction
6. an oxidation

### Matching Set

1. h
2. d
3. k
4. a
5. f
6. l
7. j
8. b
9. c
10. e
11. g
12. i

## Chapter 7

### Self-Test 7-A

1. oxygen
2. aluminum
3. Minnesota
4. limestone
5. copper
6. copper, aluminum, magnesium
7. slag
8. reduced
9. cathode
10. magnesium
11. positive ions

### Self-Test 7-B

1. oxygen
2. lead
3. ammonia
4. potassium, nitrogen, and phosphorus
5. petroleum hydrocarbons
6. sulfuric acid
7. KCl
8. iron compounds
9. $MgCl_2$

### Matching Set

| | |
|---|---|
| 1. i | 6. e |
| 2. j | 7. f |
| 3. b | 8. a |
| 4. g | 9. c |
| 5. h | 10. d |

## Chapter 8

### Self-Test 8-A

1. organic
2. tetrahedral
3. 13, no
4. false
5. two (a or b or d, and e or f)
6. double bond
7. 2,4-dimethylhexane
8. $-C_2H_5$

### Self-Test 8-B

1. four
2. rotation of plane polarized light
3. 32
4. delocalized
5. 12 (6 C, 6 H)
6. 1,2,4-trimethylbenzene
7. false

### Matching Set

| | | |
|---|---|---|
| 1. e | 4. c | 7. b |
| 2. g | 5. f | 8. h |
| 3. a | 6. i | 9. d |

## Chapter 9

### Self-Test 9-A

1. c
2. a. methanol
   b. ethanol
   c. methanoic acid or formic acid
   d. 2-butanol
   e. ethanoic acid or acetic acid
   f. ethylene glycol
   g. acetaldehyde
   h. 2-bromopropane
3. a. fermentation of starch
   b. hydration of ethylene
4. stearic acid, palmitic acid, or oleic acid, among others
5. a. alcohol
   b. carboxylic acid
   c. aldehyde
   d. ketone

### Self-Test 9-B

1. $CH_3CH_2CH_2O\overset{\overset{O}{\|}}{C}CH_2CH_3$ (an ester) + $H_2O$
2. a salt of a long-chain fatty acid
3. a. A fat is solid at room temperature, while an oil is liquid. The fat molecule has fewer double bonds.
   b. By hydrogenation (adding hydrogen to C=C double bonds).
4. false
5. c
6. carbolic

### Matching Set

| | | |
|---|---|---|
| 1. g | 4. i | 7. b |
| 2. e | 5. a | 8. c |
| 3. h | 6. d | 9. f |

## Self-Test 10-A

1. monomers
2. a. $H_2C{=}CH(CH_3)$ (CH₃ on the CH carbon)

   b. 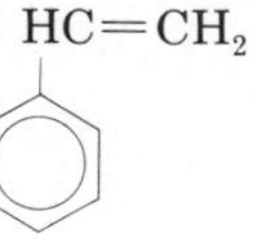 $HC{=}CH_2$ with a benzene ring on the HC carbon

   c. $F_2C{=}CF_2$
3. polyester
4. isoprene
5. copolymer
6. polyamide or condensation
7. water
8. 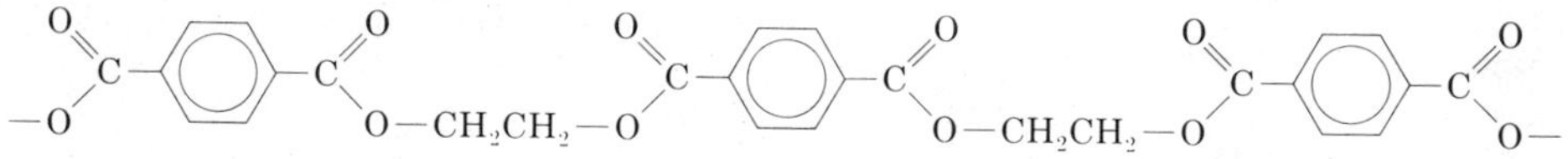 $-O-C(=O)-C_6H_4-C(=O)-O-CH_2CH_2-O-C(=O)-C_6H_4-C(=O)-O-CH_2CH_2-O-C(=O)-C_6H_4-C(=O)-O-$
9. condensation
10. thinner, pigment
11. linseed
12. emulsions
13. alcohols, acids

## Self-Test 10-B

1. 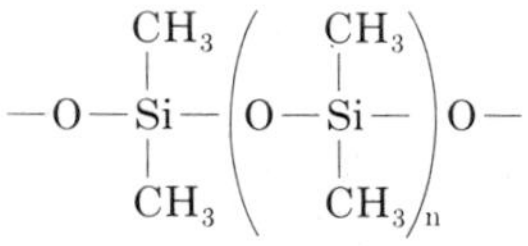 $-O-Si(CH_3)_2-\left(O-Si(CH_3)_2-\right)_n O-$
2. ultraviolet light
3. plasticizer
4. O (oxygen); it is a polymer held together by a network of Si—O bonds
5. HCl
6. ultraviolet
7. polyethylene

## Matching Set

| | | |
|---|---|---|
| 1. l | 6. g | 11. h |
| 2. d | 7. b | 12. m |
| 3. j | 8. c | 13. e |
| 4. k | 9. f | 14. n |
| 5. a | 10. i | |

## Chapter 11

### Self-Test 11-A

1. carbon, hydrogen, and oxygen
2. monosaccharide
3. glucose, fructose
4. D-glucose
5. D-glucose
6. prostaglandins

### Self-Test 11-B

1. amino acids
2. essential amino acids
3.
```
      H
      |
   —N—C—
        ||
        O
```
4.
```
R—CHCOOH
   |
   NH2
```
5.
```
   H  O  H
   |  ||  |          O
H—C—C—N—CH2—C//
   |              \
   NH2             OH
```
6. a. sequence of amino acids
   b. helical structure in which the amino acid chains are coiled
   c. the way in which the helical sections are themselves folded
   d. the aggregation of subunits
7. a. 27
   b. 6

### Self-Test 11-C

1. catalyst
2. key, lock
3. niacin
4. enzyme
5. enzyme
6. coenzyme
7. niacin
8. active site
9. a. ribose
   b. deoxyribose
10. phosphoric acid, a sugar (ribose or deoxyribose), and a nitrogenous base
11. double helix
12. true, when it invades a host cell

### Matching Set

1. d
2. g
3. e
4. i
5. k
6. b
7. c
8. a
9. h
10. l
11. m
12. j
13. f

## Chapter 12

### Self-Test 12-A

1. the sun
2. ATP
3. ADP, phosphate or phosphoric acid, energy
4. free
5. $CO_2$, $H_2O$, energy
6. NADPH
7. 3-phosphoglyceric acid
8. glucose

### Self-Test 12-B

1. hydrolysis
2. lipids
3. emulsifying agents
4. ATP, lactic acid
5. $CO_2$, $H_2O$
6. coupled

### Self-Test 12-C

1. a. DNA
   b. transfer RNA
2. hydrogen
3. ATP
4. adenine: thymine or uracil
   cytosine: guanine
   guanine: cytosine
   thymine: adenine
   uracil: adenine
5. false
6. false

### Matching Set

1. i
2. h
3. a
4. f
5. g
6. c
7. e
8. b
9. j
10. d
11. k

## Chapter 13

### Self-Test 13-A

1. observed experimental facts
2. same
3. rate
4. change

5. radioactive
6. food

### Matching Set

| | | |
|---|---|---|
| 1. f | 5. a | 9. d |
| 2. g | 6. b | 10. e |
| 3. h | 7. c | 11. i |
| 4. j | 8. d | |

## Chapter 14

### Self-Test 14-A

1. about 22%
2. natural gas
3. 42
4. over 50%
5. CO, $H_2$, $N_2$
6. 33%
7. coal, petroleum, natural gas
8. oxygen, water
9. true
10. work
11. quantitatively
12. entropy
13. quality
14. calorie, joule, BTU
15. watt, kilowatt
16. aluminum, aluminum

### Self-Test 14-B

1. fission
2. critical mass
3. uranium-235
4. fusion
5. containment of reactants at high temperature
6. tritium
7. plutonium-239, uranium-233
8. beta
9. intensity
10. ${}^{85}_{35}Br$, ${}^{242}_{94}Pu$

### Matching Set

| | | | |
|---|---|---|---|
| 1. f | 5. n | 9. g | 13. k |
| 2. l | 6. c | 10. a | 14. m |
| 3. b | 7. d | 11. p | 15. o |
| 4. j | 8. e | 12. h | 16. i |

## Chapter 15

### Self-Test 15-A

1. dehydration, hydrolysis
2. oxidizing
3. hemoglobin, oxygen
4. false
5. cyanide ($CN^-$), cytochrome oxidase, oxygen
6. fluoroacetic acid
7. heavy metal poisons, complex
8. true

### Self-Test 15-B

1. neurotoxins
2. synapses
3. acetylcholine
4. genes, chromosomes
5. false
6. chimney sweeping
7. tetrahydrocannabinol (THC)
8. acetaldehyde
9. cancer, heart disease
10. teratogen

### Matching Set

| | | |
|---|---|---|
| 1. b | 5. j | 9. c |
| 2. f | 6. d | 10. k |
| 3. h | 7. g | 11. a |
| 4. i | 8. e | |

## Chapter 16

### Self-Test 16-A

1. water
2. chlorination
3. biological oxygen demand
4. oxygen, liter
5. limnology
6. sodium tripolyphosphate

### Self-Test 16-B

1. chlordan
2. toxic
3. electrical
4. calcium, magnesium
5. chlorine
6. electrical
7. two
8. more pure than
9. settling, filtration
10. metal, organics

### Matching Set

1. e
2. h
3. i
4. a
5. c
6. b
7. g
8. j
9. d
10. f

## Chapter 17

### Self-Test 17-A

1. parts per million, ten thousand, 920
2. carbon monoxide, hydrocarbons, nitrogen oxides, ozone, sulfur dioxide, etc.
3. peroxy acyl nitrate, aldehydes, ketones, ozone; sulfuric acid, sulfurous acid, $SO_3$
4. aerosols
5. surface area
6. Cottrell precipitator
7. nitrogen dioxide
8. warm, cool
9. sulfur dioxide, nitrogen dioxide
10. sulfur dioxide
11. fuels
12. first
13. other elements
14. endothermic
15. oxides
16. false
17. false
18. $SO_2$, $SO_3$, $NO_2$

### Self-Test 17-B

1. limited
2. hemoglobin
3. false
4. benzo($\alpha$)pyrene (BaP)
5. ozone, $O_3$
6. polar
7. eight
8. air
9. lead
10. yes
11. false
12. hydrocarbons, carbon monoxide, nitrogen oxides
13. platinum (or palladium)
14. increases, increases

### Matching Set

1. b
2. o, g
3. l
4. j
5. h
6. c
7. e
8. m
9. i
10. d
11. o
12. a
13. k
14. h

## Chapter 18

### Self-Test 18-A

1. synergism
2. salting, drying
3. false
4. volatile
5. more
6. sequestrant
7. false
8. flavor enhancer
9. dehydrating
10. free radicals

### Self-Test 18-B

1. saccharin
2. double bonds and aromatic rings
3. citric and lactic acids
4. carbon dioxide
5. flavor enhancer
6. breaks
7. generally recognized as safe
8. active
9. acid

### Matching Set (Food)

1. b
2. c
3. d
4. e
5. f
6. a

### Self-Test 18-C

1. hydrochloric acid
2. pain killer
3. morphine
4. acid, ester
5. fever
6. Alexander Fleming
7. true
8. sulfa drugs, antihistamines

### Matching Set (Medicine)

1. g
2. f
3. h
4. a
5. i
6. b
7. d
8. c
9. e

### Self-Test 18-D

1. 4
2. cystine
3. disulfide, ionic
4. ionic
5. colorless
6. flexible

7. adsorbent
8. fatty acids
9. sheep
10. false

## Matching Set (Beauty Aids)

| | |
|---|---|
| 1. h | 6. g |
| 2. d | 7. a |
| 3. f | 8. j |
| 4. i | 9. e |
| 5. c | 10. b |

## Self-Test 18-E

1. b
2. polymerization, double
3. nonpolar
4. free radicals
5. c
6. layers
7. entangled
8. true
9. silicone and diester greases with lithium soaps
10. bonds

## Matching Set (Automotive Products)

| | |
|---|---|
| 1. g | 6. c |
| 2. a | 7. d |
| 3. f | 8. e |
| 4. b | 9. i |
| 5. j | 10. h |

## Self-Test 18-F

1. photochemistry
2. silver
3. hydroquinone
4. reduced
5. $Ag(S_2O_3)_2^{3-}$
6. somewhat larger
7. reducing
8. sodium thiosulfate
9. a. blue
   b. green
   c. red
10. red, blue, and green
11. true
12. reduced
13. panchromatic

## Matching Set (Photography)

| | | |
|---|---|---|
| 1. i | 5. b | 9. e |
| 2. d | 6. c | 10. g |
| 3. f | 7. k | 11. h |
| 4. a | 8. j | |

# Credits

## Chapter 1

**page 4** Courtesy of Dr. Spurny, Institut für Aerobiologie.

## Chapter 2

**page 26** From Ihde, A. J.: *The Development of Modern Chemistry.* New York, Harper and Row, 1964.

**Figure 2–13** From Masterton, W. L., and Slowinski, E. J.: *Chemical Principles.* Philadelphia, Saunders College Publishing, 1973.

## Chapter 3

**Figure 3–2** Modified from Hein, Morris, et al.: *Foundations of Chemistry in the Laboratory,* 4th ed. Dickenson Publishing Co., 1977.

**Figure 3–4** Modified from Peters, Edward I.: *Introduction to Chemical Principles,* 3rd ed. Philadelphia, Saunders College Publishing, 1982.

**Figure 3–8** From Peters, Edward I.: *Introduction to Chemical Principles,* 3rd ed. Philadelphia, Saunders College Publishing, 1982.

## Chapter 4

**Figure 4–9** From Berlow, Peter P., Burton, Donald J., and Routh, Joseph I.: *Introduction to the Chemistry of Life.* Philadelphia, Saunders College Publishing, 1982.

## Chapter 5

**Figure 5–6** Modified from Clark, M. E.: *Contemporary Biology.* Philadelphia, Saunders College Publishing, 1973.

## Chapter 6

**Figure 6–11** Courtesy of GRT Record Pressing, Nashville, TN.

## Chapter 7

**Figure 7–3** From Lee, G., Van Orden, H. O., and Ragsdale, R. O.: *General and Organic Chemistry.* Philadelphia, Saunders College Publishing, 1971.

**Figure 7–10** Courtesy of Union Carbide Corporation, Linde Division, Danbury, CT.

**Figure 7–12** Courtesy of the Corning Glass Company.

**Figure 7–14** Courtesy of Intel Corporation.

**Figure 7–15** From Turk, A., et al.: *Environmental Science.* 2nd ed. Philadelphia, Saunders College Publishing, 1978.

**Table 7–2** U. S. Standard Atmosphere Tables.

## Chapter 9

**Figure 9–1** From Routh, J. I., Eyman, D. P., and Burton, D. J.: *A Brief Introduction to General, Organic and Biochemistry.* Philadelphia, Saunders College Publishing, 1971.

**Figure 9–6** Courtesy of the DuPont Company, Textile Fiber Department

## Chapter 10

**Figure 10–2** Courtesy of Sears, Roebuck and Co., Chicago, IL.

**Table 10–6** *Chem. Tech.,* September 1980, p. 550.

## Chapter 11

**Figure 11–12** Courtesy of M. F. Perutz and *Science, 140*:863, 1963.

## Chapter 12

**Figure 12–5** Adapted from Routh, J. I.: *Introduction to Biochemistry.* Philadelphia, Saunders College Publishing, 1971, p. 100.

**Figure 12–6** Adapted from Routh, J. I.: *Introduction to Biochemistry.* Philadelphia, Saunders College Publishing, 1971, p. 102.

**Table 12–2** Adapted from Routh, J. I., Eyman, D. P., and Burton, D. J.: *Essentials of General, Organic, and Biochemistry,* 3rd ed. Philadelphia, Saunders College Publishing, 1977.

**Figure 12–14** From *Chemical and Engineering News,* January 11, 1971.

## Chapter 13

**Figure 13–1** Reproduced with permission of Yale University Press.

**Figure 13–2** From *Chemical and Engineering News,* August 2, 1971.

**Figure 13–4** Courtesy A. E. Gunther and the University Press, Oxford.

**Figure 13–5** From "The Realm of Chemistry," Econ-Verlag Gmb H, Düsseldorf, 1965.

**Figure 13–6** Adapted from Turk, A., Turk, J., and Wittes, J. T.: *Ecology, Pollution, and Environment.* Philadelphia, Saunders College Publishing, 1972.

**Figure 13–7** From *Radioactive Wastes.* Courtesy of the U. S. Atomic Energy Commission, Washington, D.C.

**Figure 13–8** From Turk, A., Turk, J., Wittes, J.: *Ecology, Pollution, and Environment.* Philadelphia, Saunders College Publishing Co., 1972.

**Figure 13–10** Courtesy of Chief Douglas K. Burrows, Peel Regional Police Force, Brampton, Ontario.

**Figure 13–11** Courtesy of Dr. D. L. Brockway, Federal Water Quality Administration.

**Figure 13–13** Courtesy of Varian, Palo Alto, CA.

**Table 13–1** By permission from *Industrial Research and Development,* October, 1978.

**Table 13–4** From *Scientific Aspects of Pest Control,* National Academy of Science Publication No. 1402.

**Table 13–3** United States Food and Drug Administration, June 1979.

## Chapter 14

**Figure 14–1** Source: Exxon Corporation.

**Figure 14–2** From Turk, A., et al.: *Environmental Science.* 2nd ed. Philadelphia, Saunders College Publishing, 1978.

**Figure 14–3** From *Chemical and Engineering News,* January 10, 1972.

**Figure 14–5** From *Science,* April 19, 1974.

**Figure 14–9** Source: Exxon Corporation.

**Figure 14–15** Courtesy of the Department of Energy, 1982.

**Figure 14–26** After Masterton, W. L., and Slowinski, E. J.: *Chemical Principles.* Philadelphia, Saunders College Publishing, 1977.

**Table 14–2** *Chem Tech,* September 1980, p. 550

## Chapter 15

**Figure 15–2** From Meyer, E.: *Chemistry of Hazardous Materials.* Englewood Cliffs, N.J., Prentice-Hall, Inc., 1977.

**Table 15–8** From Sax, N.I.: *Cancer-Causing Chemicals.* New York, Van Nostrand Reinhold, 1981.

## Chapter 16

**Figure 16–2** From Singer, S. F.: "Federal Interest in Estuarine Zones Builds." *Environmental Science and Technology, 3*:2, 1969.

**Figure 16–3** From Turk, A., et al.: *Environmental Science.* 2nd ed. Philadelphia, Saunders College Publishing, 1978.

**Figure 16–4** From Tidwell, P.: "Anti-pollution: A Fish Management Priority." *The Tennessee Conservationist, 37*:4, 1971.

**Figure 16–8** From *The Living Waters.* U. S. Public Health Service Publication No. 382.

**Figure 16–15** Courtesy of Robert L. Lawrence Jr. Filtration Plant, Nashville, TN.

**Table 16–1** U. S. Department of Commerce.

**Table 16–4** Environmental Protection Agency.

## Chapter 17

**Figure 17–2** From Turk, A., et al.: *Environmental Science.* 2nd ed. Philadelphia, Saunders College Publishing, 1978.

**Figure 17–3** Courtesy of *Fortune* Magazine, May 4, 1981.

**Figure 17–5** Courtesy of American Airlines.

**Figure 17–7** Courtesy of Wide World Photos, Inc.

**Figure 17–16** From Turk, A., et al.: *Environmental Science.* 2nd ed. Philadelphia, Saunders College Publishing, 1978.

**Table 17–1** From U. S. Environmental Protection Agency, Research Triangle Park, North Carolina, March, 1981.

**Table 17–3** From U. S. Environmental Protection Agency.

**Table 17–6** From U. S. Environmental Protection Agency.

## Chapter 18

**Figure 18–8** From an article by Fleming, *British Journal of Experimental Pathology,* June, 1929.

**Figure 18–13** Courtesy of E. Bernstein and C. B. Jones, *Science, 166*:252–253, 10 October, 1969. Copyright 1969 by the American Association for the Advancement of Science.

# Index

Note: Numbers in *italics* indicate a figure; d following a page number indicates a definition; s indicates a molecular structure; and t indicates a table.


Abbeville, Louisiana, 403
Abbot Laboratories, 439
Abietic acid, 454s
Absorption, 393d
Academy of Sciences, Paris, 463
Acetal, 227s
Acetaldehyde, 195
  product of ethyl alcohol oxidation, 364
Acetamide, uses of, 179t
Acetaminophen, 432
Acetate ion, 120
Acetic acid, 179t, 195, 342
  from ethanol, 197
  hydrolysis product of acetylcholine, 351
  ionization, 120
  relative acid strength, 121t
  use of, 179t
Acetone, 179t
  from 2-propanol, 195
  use of, 179t
Acetyl coenzyme A, *272*
  in alcohol oxidation, *364*
Acetylcholine, 349s, 351
  cycle, *350*
  hydrolysis, *352*
Acetylene, 172s
  structure and bonding, *94*
Acetylsalicylic acid (aspirin), *205,* 432s
Acid rain, 369, 397, *398*
  carbon dioxide and, 408
  nitrogen oxides and, 403
  tall stacks and, 400
Acid(s)
  amino. See *amino acids.*
  and bases, 117
    strengths, 120, 121t
  Bronsted, 118
  carboxylic group, 179t
  examples of, 113
  fatty, 238
  in foods, 427t
  organic, 165d
  rebound, 429
  strength, 121t
Acidic solution, 119
Acidity of oceans, 408
Acidulant, 424
Aconitic acid, *272*
Acrilan, 227t
Acropolis, decay of, 398
Acrylamide, 227s
Acrylic, 227t
Acrylonitrile
  carcinogen, 359t
  monomer, 212t
Activated sludge, 381
Active sites in enzymes, 249d
Activity series, 100, 133
Addition polymerization, mechanism, 211
Addition polymers, 210
  and paints, 216
  ethylene derivatives, 212t
Addition reactions, 170
  mechanism, 204
  organic, 203
Additive primary colors, 469
Adenine, 254s, 278, 356
Adenosine, 263
Adenosine diphosphate, 261, 262
Adenosine monophosphate, 261, 262
Adenosine triphosphate, 261, 262. See also *ATP.*
Adipic acid in nylon, 20
Adrenalin. See *Epinephrine.*
Adsorption, *393d*
Aeration of water, 368
Aerobic glucose metabolism, 271
Aerosols
  air pollution and, 393
  smog and, 397
Aflatoxins, 355t
  carcinogens, 359t
Agriculture, 154d
Air pollution
  aerosols and, 393
  air quality standards and, 393t
  automobile and, 409
  carbon dioxide, 408
  carbon monoxide, 404
  costs of, 413
  discharges from, *392,* 392t
  electrostatic precipitation of, *395*
  fatalities from, 391
  future of, 413
  history of, *292,* 390
  hydrocarbons, 405
  nitrogen oxides, 400–403
  ozone, 406
  particulates and, 394
  primary pollutants, 396d
  secondary pollutants, 396d
  smog, 396, 401
  sulfur dioxide as, 395–400
  thermal inversion, *397*
Airplanes, air pollution and, *394*
Alanine, 175s, 242t
  optical isomers of, 175
Alanylglycine, 243s
Alcohol(s), 189, 190t
  absorption by body, 365
  effect of drugs on use of, 441
  functional group, 177, 178t
  lauryl, 202
  nomenclature of, 177
  polyhydric, 425
  solubility in water, 195t
  toxicity of, 364
  used in synthesis, 194
Alcohol group in carbohydrates, 233
Alcoholic beverages, 192
Aldehyde, functional group, 179t
Aldehyde group in carbohydrates, 233
Aldrin, 376s
Aliphatic, 172d
Alkali metals, 55
Alkalies in foods, 427t
Alkaline cell, 137t
Alkaline earths, 56
Alkaloid neurotoxins, 354t
Alkaloids, 430
Alkanes, 170d
Alkenes, 170d
  in petroleum cracking, 188
Alkyd paint, 219, 220s
Alkyl-, 405d
Alkynes, 170d
  unsaturation and, 170
Allergen, 437d
Allergy, 437d
Allied Chemical Corporation, 428
Alloy, 142d
Allura Red AC, 428
Alpha particles, 27–28, 34
  penetration of, *325*
Aluminum, 109
  electronic structure of, 43t
  energy for production of, 311t
  energy required to manufacture, 145, 230t
  from ore, 130t
  production, 144, *145*
  properties, 50t
  uses, 145
Aluminum chlorhydrate, 449
Aluminum chloride, 449
Aluminum hydroxide, 430t
Aluminum oxide, 109
Aluminum sulfate, 449
Amalgam, 346d
Amber, 27
American Standards Association (ASA), 464
Amide(s), functional group, 179t
Amide linkage, 220
Amine(s), functional group, 179t
Amino acids, 241, 242t
  essential, 241
  in protein synthesis, 278
  messenger RNA codes for, 279t
p-Aminobenzoic acid, 434, 451
  ultraviolet light absorption, *452*
Aminocaproic acid, 221
p-Aminodiphenylaminesulfonic acid, 446s
Aminopeptidases, 268t
Aminophenols, 455
Ammonia, 108, 403
  and First World War, 293
  covalent bonds in, 75
  production by Haber Process, 157
  solutions of, 121
Ammonium carbamate, 159
Ammonium chloride, 185
Ammonium cyanate, 185
Ammonium nitrate, 127, 403
  as fertilizer, 158
Amount of substance, SI unit, 478

Amphetamines
diet control with, 440
hallucinogen, 362t
Amphiprotic species, 118
Ampicillin, 429t
Amu, 321d
Amygdalin, 342
beta-Amylase, 249
action of, 250
Amylopectin, 236
dextrins and, *237*
hydrolysis of, *237*
Amylose, 236, *237*
*Anabaena cylindrica,* 331
Anabuse, mechanism of action, *364,* 365s
Analgesics, 429, 430
local, 431
Anemia, sickle cell, 246
Anerobic glucose metabolism, 271
Anesthetics, 429
Anhydride, 227d
Aniline, carcinogen, 358
Anion exchangers, 383
Anode, 29d
in electrolysis, 131
Anorexigenic drugs, 440d
Antacids, 429
chemistry of, 430t
Antibiotics, 429, 433
Antibody, 433d
Anticaking agents, 424
Anticholinesterase poisons, 350
Anticodon, 279
Antifoam agents in antifreeze, 459
Antifreeze, 458
mechanism of action, *459*
Antigen E, 438
Antihistamines, 437–440
Antiicing agents, 456
Antimicrobial medicines, 433
Antimicrobial preservatives, 418
Antioxidants
in foods, 416, 419, 427t
in gasoline, 455
in rubber formulation, 214
Antirust agents in gasoline, 455
Antiseptics, 432t
Antitoxins, 429
Apoenzymes, 250d
Apple seeds, 342
Apple smell ester, 198t
Aqueous solutions, 114
Arachidonic acid, 239s
Arginine, 242t
Argon, 56
electronic structure of, 43t
percentage in air, 148
properties, 50t
Argonne National Laboratory, 56
Aristotle, 21
Arnel, 227t
Aromatic compounds, substitution reactions, 204
Aromatic hydrocarbons, 179–181
Arsenic
air pollutant, 394
carcinogen, 359t
FDA tolerances in foods, 344
in solar battery, 332
teratogen, 358t
toxicity, 343–345
Artificial sweeteners, 422
Asbestos
cancer and, 394
carcinogen, 359t
Ascorbic acid, 426t
Aspartame, 423s
Aspartic acid, 242t
Asphalt, 187t
Aspirin, 429, 431
lethal dose, 337
stomach bleeding and, 432
synthesis, 205
Astatine, 56
Astringents, 448, 449
Asymmetric atom, 174d, *174*
D- and L- and, 175
mirror images and, *175*
Atherosclerosis, 238
Atmosphere
carbon dioxide effect on, 408
carbon dioxide emissions, *300*
composition, 148
content of, 390
stratosphere, 406d
troposphere, 406d
Atmospheric oxidation, of foods, 418
Atom(s), 12d
ancient definition, 20d
Dalton's theory of, 23–24
nucleus of, 34–35
wave mechanical model of, 42–46
weights of, 24
Atomic bomb
effects of, 320
hydrogen bomb and, 327
mechanism of, *320*
Atomic mass, 35d
Atomic number, 35d
electrons, protons, and, 35
Atomic pile, contents of, 322
Atomic radii, *59*
Atomic sizes, relative to ionic sizes, *73*
Atomic theory
and the Periodic Table, 60
Greek influence, 20
Atomic volumes, periodic trends, 58, *59*
Atomic Weight Order, 52
Atomic weight scale(s), 26
Atomic weights
molecular weights and, 108
weighted average of isotopes, 37
ATP, 261, 262, 265, 266
from light reaction, 267
in Krebs cycle, 272
Atropine, 351, 354s
Aureomycin, 436s
Authority, refusal to accept, 287
Automobile
air pollution by, 409–411
air-fuel ratio of, 411
Baker Electric Coupe, 412
Borland Electric Delux, 412
catalytic muffler for, *410*
combustion chamber of, *409*
emission standards for, 410t
hydrocarbon emissions of, 406
hydrogen powered, 331
new kinds of, 411
PCV valve of, 410
pollution controls on, *411*
sources of air pollution of, *409*
Automotive products, 455–461

B vitamins, 426t
Bacon, Francis, 21
Bakelite, 227t
BAL. See *British Anti-Lewisite.*
Balanced equations, calculations and, 483
Ballistic separation of solid wastes, 229
Banana smell ester, 198t
Barium-141, radioactivity of, 320
Barium sulfate, use in X-ray diagnosis, 128
Barrel
of crude oil, 187
unit of volume, 312d
Bartlett, Neil, 56
Base(s)
and acids, 117
Bronsted, 118
examples, 113
nitrogenous, 254t
Basic solution, 119
Battery(ies), 135, 137t
solar, 331–332
Bauxite, 144
Beauty aids, 442
Becquerel, Henri, 27
Beer, 192t
Benzene, 179, *180,* 189
carcinogen, 359t
in aspirin synthesis, 205
Benzidine, carcinogen, 359t
Benzo(alpha) pyrene, 405s
carcinogen, 355t, 358, 359t
Benzocaine, 451
in diet pills, 440
Benzoic acid in foods, *417*
Bertollet, Comte Claude Louis, 23
Beryllium
carcinogen, 359t
electronic structure of, 43t
properties, 50t
Beta particles, 27–28, 320
penetration of, *325*
Beverages, alcoholic, 192
BHA, 416, 419s
in gasoline additives, 455
BHT, 419s
in food production, *417*
in gasoline additives, 455
Bicarbonate of soda. See *Sodium hydrogen carbonate.*
Bile salts, 269
Binding energy, 321d, *322*
Biochemical energy, 261
Biochemical Oxygen Demand (BOD), 370, 371d
Biochemistry, 233d
Biodegradable, 369d
Biogenetic engineering, 283
Biological half-life, 376d
Biotin, 426t
Biphenyl, 184s
PCB's and, 184
Birds of prey, effects of insecticides on, 297
Birth control pills, 437
Biuret from urea manufacture, 159
Bladder cancer, 359, 422
Blast furnace, *141*
Bleaching hair, 445
Blindness from alcohol, 191
Block copolymer, 215
Blood
cells, renewal rate, 251
clotting, prostaglandins, and, 240
plasma, 429
pressure, prostglandins, and, 240
sugar in, 234
sugar. See glucose.
BOD, 371
characteristic levels, 372
Body odor, 449
Bohr, Niels, *39,* 41
atomic theory of, 38–42
Boiling points
of alcohols, 193t
of the halogens, *82*
of the hydrogen halides, *83*
periodic trends, *58*
straight chain vs. branched chain hydrocarbons, 168

Bolts and nuts, *99*
Bomb
  atomic, 327
  hydrogen, 327
  neutron, *328*
Bond(s)
  carbon, polarity of, 168
  coordinate covalent, 76d
  covalent, 74
  dash and, 165d
  disulfide in proteins, 246
  double, in hydrocarbons, 169
  hydrogen. See *Hydrogen bonding.*
  ionic, 123
  metallic, 85
  peptide, 242d
  pi, in benzene, *180*
  rotation about single, 166
  sigma, in benzene, 179, *180*
  triple, in hydrocarbons, 169
  types of, 65
Bond energy(ies), 77d
  and electronegativity, 79
Bond lengths, 77t
  and energies, 77t
Bones, collagen and, 244
Boron, 43st
  in solar battery, 332
  plastic reinforcing agent, 228t
  properties, 50t
Boron trichloride, structure, 90
Boron trifluoride, covalent bonds in, 75
Bottle gas, 186
Bottled water, 380
Boyle, Robert, 21
Brake fluids, 460t
Branched polymer chain, 211
Brandy, 192t
Brass, 455
Bread, leavened, 425
Breeder reactors, 323, *324*
Brewer's wort, 191
Brick, energy for production of, 311t
British Anti-Lewisite(BAL), 345s
British Thermal Unit, 16d
Brittleness, ionic compounds, 71
Bromine, 56
1-Bromo-2-methylpropane, 173
Bromoethyne, 173
1-Bromopropane, 173
2-Bromopropane, 173, 195
Bronchial asthma, 438
Brönsted acid and base, 118
Brönsted, J.N., 117
Brownies, recipe, 15t
BTU, 16d
Buffers, 424
  in foods, 427t
Bullet-proof plastics, 221
Buna rubber, 213
  uses of, 212t
1, 3-Butadiene, 212t
n-Butane, 167s, *168*, 171s
  properties of, 167
1, 4-Butanediol,
  in polyurethane synthesis, 226
1-Butene, 169s, 172s
  properties of, 169
2-Butene, 172s. See also *trans-2-Butene.*
t-Butyl alcohol, 189
Butylated hydroxyanisole, 416, 419t
Butylated hydroxytoluene, 419t
  in food production, *417*
1-Butyne, 169s, 172s
  properties of, 169
2-Butyne, 169s, 172s
  properties of, 169
Cadmium, 322
  carcinogen, 359t
  teratogen, 358t
  toxicity, 343
Caffeine, 354s, 355t
  in diet pills, 440
  teratogen, 358t
Calcium
  electronic structure of, 43t
  from ore, 130t
  properties, 50t
Calcium carbonate, 430t
  Iceland spar as, 175
Calcium dihydrogen phosphate in baking powder, 425
Calcium fluoride, 56
Calcium hydroxide, 56
Calcium ion in hard water, 201
Calcium silicate. See *Slag.*
Calcium sulfide, 448
Calcium thioglycolate, 448
California crude, 189
Calorie, 16d
  food, 238d, 308d
  intake, 440
Calvin, Melvin, 266
Canal rays, 32, *33*
Cancer, 358
  and smoking, 361
  asbestos and, 394
  benzo(alpha) pyrene and, 405
  bladder, 358, 422
  cyclamates and, 358
  frequency in parts of the world, 360
  initiation, 360, *361*
  liver, 359
  radioactivity and, 326
  treatment of, 327
Cannizzaro, Stanislao, *26*
Caprolactam, 221s
Caprolan, 227t
Captan, 355t
Carbohydrases, action of, 250
Carbohydrate(s), 233ff
  digestion and absorption, 269
Carbon
  electronic structure of, 43t
  properties, 50t
  uses of, 164
Carbon-12
  basis for atomic weight scale, 26
Carbon black, UV stabilizer, 228t
Carbon bonds, polarity of, 168
Carbon dioxide, 25
  air pollution and, 408
  greenhouse effect and, 408
  in baking, 425
  percentage in air, 148
  role in photosynthesis, 264
  solution in water, 127
  sublimation of, 66
Carbon monoxide, 25
  air quality standards for, 393t
  as air pollutant, 404
  metabolic poison, 340
  quantity discharged into air, *392*, 392t
  reaction with hemoglobin, 341
Carbon tetrachloride, 407
  carcinogen, 359t
Carbonate structure in polycarbonates, 222
Carbonic acid, 127, 408
Carboxyl functional group, 165d
  acidity of in acids, 195
Carboxylic acids, functional group, 179t
Carboxypeptidases, 268t
Carbyl, *353*
Carcinogens, 358
beta-Carotene, 423
Carothers, Wallace, and nylons, 220
Carrageenin, 424
Cartilage, collagen and, 244
Catalase, action of, 251
Catalysis by metal ions in paint drying, 218
Catalyst(s)
  enzymes as, 248–253
  for making alcohol, 190
Catalytic
  mufflers on automobiles, 410
  reforming, 188
Cathode rays, 28
Cathode, 29d
  in electrolysis, 131
Cathodic protection, 134
Cation exchangers, 383
Cell
  damage, by radiation, 325
  electrochemical, 135
  human, *255*
Cellophane, cellulose and, 238
Cellulose, 234, *238*
  digestibility of, 238
Cellulose acetate in photography, 463
Celsius degree, 15
Celsius temperature scale, 479d
Celsius, Anders, 479
Centi-, 477d
Centigrade temperature scale, 479d
Centrifugal separation, *395*
Cetyl palmitate, 453
Chadwick, James, 33
Chain reaction, nuclear, 320
Change
  chemical, 1
  physical, 1
Charcoal, 186
Charge
  electrical, *27*
  electrostatic, *27*
  law of, 27
Charles' Law, 479
Chelate, *420*
Chelating agent, 345
Chemical bonds, types of, 65
Chemical change, 1, 6
Chemical coefficients in balancing equations, 107
Chemical dumps, 377
Chemical equation(s), 12d
  balancing of, 107
  calculations with, 483
Chemical formula(s), 12d
  coefficients, 13
  condensed, 173d
  subscripts in, 13
Chemical hazards, 298
Chemical landfills, 377
Chemical properties, 7, 49
Chemical reactions
  addition with alkenes and alkynes, 170
  combination, 99
  decomposition, 99
  combustion of hydrocarbons, 170
  double arrows in, 105d
  double replacement, 101
  driving forces of, 101
  energy and, 106
  metathetical, 101
  nonmetal-nonmetal, 98
  rates of, 101, 104
  reversibility of, 101
  single replacement, 100
  synthesis, 99
  weight relationships and, 107

Chemical spill, clean-up costs, 377
Chemical symbol, 12d
Chemicals
  control of, 301
  hazardous, 298
Chemistry, 2d
  in the kitchen, 425
  organic, 164d
  science of, 1
Chicago, University of, 322
Chimney, tallest, *400*
Chinese Restaurant Syndrome, 421
Chloracne, 296
Chlordan, 376s
Chlorides, metal, *57*
Chlorination
  of water, 386
  trihalomethanes and, 387
Chlorine, 56
  electronic structure of, 43t
  in magnesium production, 147
  properties, 50t
  reaction with potassium, 98
  reaction with sodium, 107
  transportation accident, *298*
Chlorobenzene in aspirin synthesis, 205
Chloroorganic compounds in water, 387
Chlorophyll A and B, 265
Chloroprene, 355t
Cholesterol, 436s
  and atherosclerosis, 238
  in lanolin, 449
Choline, 351s
Cholinergic nerves, *349*
Cholinesterase, 350
Chromium
  carcinogen, 359t
  toxicity, 343
Chromosomes in humans, 353
Chronic poisoning, 346
Chymotrypsin, 250, 268t
Cigarette smoking
  cancer and, 361
  pollution and, 405
cis-Aconitic acid, *272*
Citric acid, *272,* 416
  in foods, 419
Civetone, 452s
Clean Air Act, 391
Cloud chamber, 32
Coal gas, 315
Coal gasification, 314, *315*
Coal, 186
  as a source of plastics, 230
  consumption of, *306, 311*
  energy content of, 312t
  minable, 314d
  production of, *314*
  sulfur content of, 396
  tar, 186
Coalescing agents in latex paints, 217t
Cobalamin, 426t
Cobalt, teratogen, 358t
Cocaine, 354t, 431s
Codeine, 354s, 430s
Codon, 279
Coenzyme, 249d
  action of, *250, 251*
Coke, 186
  in iron reduction, 140
  in silicon production, 153
Cold cream, composition of, 450
Collagen, 244, *245*
Colloids, 450d
  in latex paints, 217t
Color
  development, 471
  film, 469–471
  image-forming process, *471*
  photography, 468
Colors in foods, 423, 427t
Combustion
  hydrocarbons and, 170
  of octane, 411
Common elements, 5t
Common metals and their minerals, 140
Compound(s), 5d
  aliphatic, 172d
  and mixtures, 6
  number of carbon, 164
  of carbon, hydrogen, and oxygen, 189
Concentration, reaction rates and, 103
Condensation polymers
  polyamides, 220
  polyesters, 218
  polycarbonates, 221
  silicones, 223, 224s
Conductance of solutions, 115
Configurations, molecular structure and, 168
Conformations, 167, 168
  eclipsed, *166*
  in proteins, 243
  staggered, *166*
Conjugate acid and base, 119d
Conjugated molecules, 423
Constant composition, law of, 23
Consumer chemistry, 416
Conversion problems, 481
Cooking, chemistry of, 426
Cooling water, needs in U.S., 373t
Coordinate covalent bond, 76d
Copolymer
  block polymer, 215
  homopolymer, 214
Copper
  effects in gasoline, 455
  energy for production of, 230t, 311t
  from ore, 130t
  in automobile, 455
Copper carbonate, 22–23
Copper reduction, 132
Copper-zinc electrochemical cell, 136
Coppertown, Tennessee, *390*
Corey, R.B., 243
Corneal layer, 443
Corrosion
  by sulfur dioxide, 399t
  of iron, 143
  prevention by antifreeze, 458
  prevention of, 143
Corrosive poisons, 338
beta-Corticotropin, synthetic, 280
Cortisone, 436s
Cosmetics, 442
Cotton, *239*
Cottrell precipitator, *395*
Coupled reactions, 275
Coupler in color development, 470
Covalent bonds, 74
Cracking, thermal, of oil, 187
Creams, 450
Creslan, 227t
Crick, Francis H.C., *257*
Critical mass, 320d
Crookes, Sir William, 291
Crop loss due to insects, 296
Crosslinking in paint drying, 218
Crude oil, 187
  California, 189
  Pennsylvania, 189
Cryogenic fluids, 149
Cryolite, 145
Cryosurgery, 149
Crystallization, 126
  as means of purifying water, 368
Cubic decimeter, 478
Cubic meter, 478d
Curare, 349, 351, 354s
Curie
  Marie, 27, 325
  unit of radiation, 325
Cuticle softeners, 454
Cyan dye, 470
Cyanide ion
  as a base, 118
  toxicity, 337, 342
Cyanoferrate ion, 471
Cyclamates, 422
  and bladder cancer, 422
Cyclobutane, 170s
  properties of, 170
Cyclohexane, 181s
1,4-Cyclohexane dimethanol, 227s
Cyclopentane, 456
Cysteine, 242t
Cystine, 242t
  in skin and hair, 443
Cytochrome C, 342
Cytosine, 254s, 278, 356s

Dacron, 219, 227t
Daguerre, Louis, 462
Daguerreotype, 462
Dalton, John, 21, *23,* 24
  atomic theory of, 23–24
Dark reaction in photosynthesis, 264, 266
Darvon, 431s
Davisson, Clinton, 43
DDT
  in water, 369
  Lake Michigan and, 376
  use of, 297
  water pollution and, 376
de Broglie, Louis, 43
de Chancourtois, A. Beguyer, 50
De Dominis, 469
de Voeux, Harold, 396
Death caused by disease, 433t
n-Decane, 171s
Deci-, 477d
Definitions
  operational, 3
  theoretical, 3
Defoamers in latex paints, 217t
Dehydrogenases, action of, 251
Deicing fluids, 460
Deka-, 477d
Delrin, 227t
Demerol, 431s
Democritus, *20*
Denaturant in alcohol, 191d
Denkenwalter, Robert G., 280
Densities, alkali metals, 55
Density, 49d
Deodorants, 449
alpha-2-deoxy-D-ribose, 254s
  in DNA, 254–258
Deoxyribonucleic acids (DNA), 254d, *257*
  secondary structure of, 256
Department of Transportation (DOT), chemical hazard classes, 298t
Depilatories, sulfide ion in, 448
Deposit modifiers, 455
Destructive distillation of wood, 191
Detergent builders, 374d

Detergents
in gasoline additives, 456
synthetics, 201
Deuterium, 327
Development, in photography, 464–466
Dextrins, 237d
uses of, 237
Dextrose. See also *Glucose.*
in diet pills, 440
DFP, 352s
Di (2-ethylhexyl) sebacate, 457s
1,2-Diaminoethane in polyimide synthesis, 227
Diammonium phosphate as fertilizer, 160
Diastase in alcohol production, 191
1,2-Dibromoethane, 410
1,2-Dibromopropane, 173
Dibutyl-p-cresol, 455
Dichlorodiethyl sulfide, 355t
1,1-Dichloroethane, 178
1,2-Dichloroethane, 410
cis and trans isomers of, 177s
properties of, 177
1,2-Dichloroethylene, 204
Dieldrin, 376s
Diet pills, 440
Diethyl ether. See *Ethyl ether.*
N,N-Diethyl-p-phenylenediamine, 470s
Diethylene glycol monomethyl ether, 453s
Diethylstilbestrol
carcinogen, 359t
teratogen, 358t
Diets, fats and, 238
Diffraction
of electrons, *44*
of light, *44*
of x rays, *44*
Digestion, 268
carbohydrate, 269
Digestive enzymes, 268t
Digitalis, 429
Digitoxin, 429t
Diglycerides, 425
Dihydroxy acetone phosphate, *267*, *271*
Dihydroxyaluminum sodium carbonate, 430t
5,6-Dihydroxyindole, *447*
Dilution as means of purifying water, 368
Dimethyl sulfate, 355t
1,2-Dimethylbenzene, 181. See also *o-xylene.*
1,3-dimethylbenzene, 181. See also *m-xylene.*
1,4-dimethylbenzene, 181. See also *p-xylene.*
Dimethyldichlorosilane in silicone synthesis, 224
2,2-Dimethylpropane, 168s
Dinitrogen oxide, 127
Dioctyl adipate, plasticizer, 228s
Dioctyl phthalate (DOP)
as polymer additive, 226
plasticizer, 228s
Dipeptidases, 268t
1,3-Diphosphate glyceric acid, *271*
Dirac, Paul, 42
Disaccharides, 235d, 269
Discovery, methods of, 289–290
Disease
death by, 433t
water-borne, 386
Disinfectants, 432, 449
Dispersing agents in latex paints, 217
Dissolution of ions, 114
Distillation, 4
as means of purifying water, 368
of sea water, 382
Disulfide linkages, 443
Disulfiram, 365s
Divinyl, monomer, 212t
DNA
effect of mutagens on, 353
radioactivity and, 326
replication of, 276, 277
DOA. See *Dioctyl adipate.*
Dobereiner, Johann Wolfgang, 50, 52
Dolomite, water hardness, 386
Domagk, Gerhard, 434
Donora, Pennsylvania, *391*
DOP. See *Dioctyl phthalate.*
Doping of silicon crystals, 153
Dose, 337
Double bond, 77d
Drugs
in combination, 440–441
prescription, 429t
Dry cell battery, 137t
Dry skin, 449
Drying
of foods, 417
of latex paint, 216
of oil-based paints, 217, 218
Du Pont Company, and nylons, 221
du Vigneaud, Vincent, 280
Dubos, René, 436
Ducktown, Tennessee, *390*
Duggan, B.M., 436
Dust collector, *394*
Dyes
for hair, 445–446
in gasoline additives, 456
in lipsticks, 450–451
in photographic film, 467–472
Dynel, 227t

Earth, cooling by particulates, 395
Eastman, George, 463
Edison storage cell, 137t, 289
EDTA, 348s, 419, 420
iron chelate, 162
Efficiency, 310d
electrical production and, 317
Ehrlich, Paul, 434
Eight electrons in valence shell, 60
Einstein, Albert, 308
equation of, 321
Ekaaluminum, gallium, 52
Ekaboron, scandium, 52
Ekasilicon, germanium, 52t
Elastomer, 213
Electric cars, *412*
Electric charge, SI unit, 478
Electric current, SI unit, 478
Electrical conductivity, ionic compounds, 71
Electrical energy consumption in U.S., thermal pollution and, 373t
Electricity
conductance of in solutions, 115
from nuclear reaction, *329*
from solar energy, *331*
nuclear energy and, 323
production of, 316
Electrochemical cell, 135
Electrodes in electrolysis, 131
Electrodialysis, purification of sea-water, *384*
Electrolysis, 111d, 131
of water, *104*
Electrolytes, 114
strong and weak, 120
Electrolytic solutions, 113
conductance of, 115
Electron pair repulsion, molecular structure and, 90
Electron transfer reactions, 129
Electron volt, 67d
Electron(s)
arrangements of in atoms, 43t
orbitals and, 43t, 45–46
wave nature of, *44*
Electronegativity, 79d
of carbon, bonding and, 165
Electronic configurations. See *Electronic structures.*
Electronic structures
of atoms, 60, *61*
of ions, 69t
Electrons
cathode-ray tubes and, 29
charge of, 30, 32d, 34
delocalized, 179d
localized, 179d
energy of, 38
mass of, 30–31, 32d, 34
number per atom, 35
placement of in atoms, 38–43, *43*
quanta and, 38
spectra and, 39
Electroplating, 131
Electroscope, *27*
Electrostatic precipitators, 327, *395*
Elemental properties, 50t
Elementary Treatise on Chemistry, 21
Elements, 5d
abundance in earth's crust, *140*
natural, number of, 164
new, predicted by Periodic Table, 61
Embden-Meyerhof Pathway, 271, 273
Emollient, 449
Emphysema, air pollution and, 393
Emulsifier, 425
Emulsion, 450d
oil-in-water, 217
on color film, 470, *471*
paints, 217
soap in oil-in-water, 200
water-in-oil, 217
Endothermic, 111d
Energy units, 478
Energy, 308d
activation, 248
binding, 321d
biochemical, 261
carbohydrate storage of, 234
chemical reactions and, 106
consumption of, *306*, *311*
efficiency, 310, *317*
equivalences, 308t
examples, 1
from fats, 238
from glucose metabolism, 274, 275
fuel equivalents, 323
fuel pie, *317*
in natural gas, 186
kinetic, 38d
law of conservation of, 308d
loss of, 317, *318*
nuclear, 319
potential, 38d
primary sources of, 310d
required to manufacture plastics, 230t

requirements for product production, 311
secondary sources of, 310d
societal needs, *307*
solar, 330
steel making, use in, 142
units of, 308t
use per capita, *307*
useful, 309
zero growth of, 323
Engine cleaners, 460t
English-SI conversions, 478t
Enhanced radiation weapon, *328*
Enkephalins, 431
Enterokinase, 268t
Entropy, 309d
Environmental Protection Agency (EPA), 374, 376, 378
gasoline and, 189
Environmental toxic substances, 337
Enzyme(s), 248–253
action of, *248*
active sites on, 249
and heredity, 270
apoenzyme, 250
as meat tenderizers, 251
coenzymes, 249d
digestive, 278t
for cellulose, 238
globular protein and, 249
lock and key analogy, *249*
naming of, 250
oil spills and, 251
substrate and, 249d
Eosin, 451s
EPA. See *Environmental Protection Agency*
Epinephrine, 175s
optical isomers of, 175
Epoxy, 227t
Equilibrium, ionic, 120
Erythromycin, 429t
Ester(s), 189, 198t
functional group, 179t
hydrolysis of by trypsin, 249
nucleic acids as, 254
Estradiol, 437s
Estrogens, 437
Estrone, 437s
Ethane, *166,* 171s
1,2-Ethanediol, 189t
Ethanoic acid. See *Acetic acid.*
Ethanol, 190t, 191
of glucose fermentation, 191
solubility, in water, 195t
synthetic, 192
toxicity of, 364
Ethene, geometrical isomers and, 177. See also *Ethylene.*
Ether functional group, 179t
Ethyl acetate, 179t, 199s
use of, 179t
Ethyl alcohol. See *Ethanol.*
Ethyl ethanoate. See *Ethyl acetate.*
Ethyl ether, 179t
Ethyl-, 178d
Ethylene glycol, 193
in antifreeze, 458
in deicing fluids, 460
in gasoline additives, 456
in polyesters, 219
Ethylene oxide, carcinogen, 359t
Ethylene, 172s
addition polymer of, 210
monomer, 212t
reaction with chlorine, 204
structure and bonding, 94
used to make ethanol, 192
Ethylenediaminetetraacetic acid, 348s
in foods, 419, 420
Ethyne. See *Acetylene.*
Eutrophication, 370d
Eye makeup, 453
Eye shadow, 453

Face powder, composition of, 452
Factor-label approach to problems, 481
Fahrenheit temperature scale, 479
Fahrenheit, Gabriel Daniel, 479
Families, in Periodic Table, 53
Fats, 238
digestion and absorption, 269
edible, 313
into Krebs cycle, 275
metabolism, *272*
oils and soaps, 199
oxidation of in foods, 418
saturated, 199
unsaturated, 199
Fats, oils, and soaps, 199
Fatty acid(s), 197, 199
essential, 238
saturated, 238, 239t
unsaturated, 239t
Fermi, Enrico, 321
Fertilizers, 154, 155t
and water pollution, 374
history of, 291
labels on, 160
prices related to petroleum costs, 156
usage in United States, 156
use of, 291
Fetus, damage to by chemicals, 357
Fibril, 239d
Field-effect transistor, 154
Filler in rubber formulation, 214
Film, color, 469
Filtration, 4
as means of purifying water, 368
Firestone chemists, role in making rubber, 214
Fischer, R., 470
Fish kill, *372*
and BOD, *371*
Fish, trash, as animal feed, 251
Fission, 319
Fixing in photography, 466
Flavin adenine dinucleotide (FAD), 253s
Flavor enhancers in foods, 427t
Flavorings in foods, 420, 427t
Fleming, Alexander, 434, *435*
Flour, 7
Fluorides, physiological effects of, 394
Fluorine, 56
covalent bond in, 75
electronic structure of, 43t
properties, 50t
Fluoroacetic acid, 343
Fluorocarbon-11, 407t
Fluorocarbon-12, 407t
Fluoropolymers in paints, 217
Foam rubber, 226
Foaming agents, polymer additives, 226
Folic acid, 426t
Food and Drug Administration (FDA), 423
ban on sodium nitrite, 356
GRAS list, 428
limits on arsenic, 344
PCB tolerances in foods, 296t
saccharin ban, 422
Food colors, 423, 427t
Food preservation, radioactivity and, 327
Food supply, cellulose and, 238
Foods
drying of, 417
PCB's in, 296t
preservation of, 102
Formaldehyde, product of methyl alcohol oxidation, 365
use of, 179t
Formation of ions, 71t
Formic acid
product of methyl alcohol oxidation, 365
synthesis, 196
used in making perfume, 453
Formica, 227t
Formulas, 12d
chemical, condensed, 173d
Fortrel, 227t
Fossil fuels
consumption of, *306*
energy content of, 312t
rate of using, *311*
thermodynamics and, 309
Fractional distillation of petroleum, 187
Francium, 55
Franklin, Benjamin, 27
Free energy
from glucose metabolism, 275
from hydrolysis of ATP, 263
Free radicals
in addition polymerization, 211
in food oxidation, 419
produced by radiation, 325
Freeze-thaw additives, 217t
Freezing of sea water, 382
Fresh water from the sea, 381
Frisch, Otto, 319
Fructose, *235,* 269
Fructose 1,6-diphosphate, *267, 271*
Fructose 6-phosphate, *267, 271*
Fruits
decomposition of, 102
sugar in, 234
Fuel cell(s), *331*
Fuels
equivalent energy of, 323
low sulfur, 399
use of, related to technology, 291
Fumaric acid, *272*
Functional groups, 165d, 179t
-R, 178d
carboxyl, 165d
isocyanate group, 225
Fungal protease, 426
Fusion, 327
controlled, 329
Future elements, Periodic Table and, 61

Gabon Republic, 321
Galactose, 235, *236*
Galileo, 21
Gallium
Periodic chart, 52
teratogen, 358t
Galvanizing, 143d
Gamma rays, 27–28
Gas
natural, 315–316
coal, 315
natural, 315–316
consumption of, *306*
power, 314d
synthesis, 315

Gases
  noble, 56
Gasification of coal, 314, *315*
Gasoline, 186t
  additives for, 455–456
  composition of, 108
  ethyl, 409
  lead-free, 410
  octanes and, 108
  shortage of, 307
Gasoline engine, knocking in, 188
Gastric lipase, 268t
Geiger counter, 32
General Electric Research Laboratory, 224
Genetic code, 283
Genetic effects, radioactivity and, 326
Genetic engineering, 283
Geometrical isomers, 177–178
  trans-2-Butene, 169
Geraniol, 453s
Geranyl formate, 453s
German chemists making alcohol, 190
Germanium, 52
Germicide, 432
Germs, killing of, 449
Giant molecule, 209
Gibbs, J. Willard, 263
Glass, 150d
  as plastic reinforcing agent, 228t
  colored, 151
  specialty, 152
Glucose, 177s, *234*, 269, *271*, 273. See also *Dextrose.*
  as optical isomer, 177
  cotton and, *239*
  from dark reaction, 267
  in alcohol production, 191
  in amylopectin, 237
  in amylose, 236
  in cellulose, 238
  in dextrins, 237
  in leavening bread, 425
  in photosynthesis, 264
  in starch, 236
  metabolism, 271
  oxidation of, 248
  to ethanol, 192
Glucose 1-phosphate, *271*
Glucose 6-phosphate, *271*
Glue solvents, 355t
Glutamic acid, 242t, 444
Glutathione, *345*
Glyceraldehyde 3-phosphate, *271*
Glycerin. See *Glycerol.*
Glycerol, 190t, 193, 199, 269
  in alkyd paints, 220
Glycine, 241s, 242t
  in silk, 241
  sweetness of, 422
Glycocholic acid, sodium salt of, 270
Glycogen, 236, 269, *271*
  physiological function of, 237
Glycylglycine, 243s
Gold, 20
  14-carat, 23
Goldstein, E., 32
Goodyear scientists, role in making rubber, 214
Goodyear, Charles, 212
Goulian, Mehran, 283
Grain alcohol. See *Ethyl alcohol.*
Grains, for making alcohol, 191
  in photographic, emulsion, 463, *464*
Grape sugar. See *Glucose.*
Graphite
  as plastic reinforcing agent, 228t
  atomic pile moderator, 322
GRAS list, 427t, 428
Greases, 456–458
  soaps and, 457
Greenhouse effect, *408*
Ground state, 42d
Groundwater, 368
Group properties in Periodic Table, 55
Groups in Periodic Table, 53
Guanine, 254s, 278, 356s
Gutta-percha, 214
Gutte, Bernd, 280

Haber, Fritz, 293
Haber process, 157
Hahn, Otto, 319
Hair sprays, 447, *448*
Hair, *443*
  bleaching, 445
  coloring, 444
  dyeing of, *446*
  permanent wave, *445*
  waving, 444
Half-life, 324d, 325
  of pesticide in environment, 376
Hallucinogens, 361
Halogens, 56
  boiling points, *82*
Hard water, 201, 385
Hardness, 49
  ionic compounds, 71
Hardwoods, source of methanol, 190
Hay fever, 438
Hazardous chemicals, 298
Hazardous wastes, 377t
Haze, cause of, 403
Heart
  disease, atherosclerosis and, 238
  human, *105*
Heat capacity of water, 373
Heat of vaporization of water, 373
Heavy distillates, 187t
Heavy metals, toxicity of, 343
Hecto-, 477d
Heisenberg, Werner, 42
Helium, 56
  electronic structure of, 43t
  percentage in air, 148
  properties, 50t
Helix, *244*
Hemoglobin, 246, *247*, 340
  oxygen and, *105*
Heptachlor, 376s
n-Heptane, 171s, 188, 189
Heptane, isomers of, 169t
Herculon, 227t
Heroin, 352, 430s
Herschel, Sir J.W.F., 467
*Hevea brasilieusis* tree, 212
Hexamethylene diisocyanate, 226
Hexamethylenediamine in nylon, 220
n-Hexane, 171s, 189
Hexane, isomers of, 169t
HGH. See *Human growth hormone.*
Hiroshima, Japan, 328
Hirschmann, Ralph F., 280
Histamine, 438s
Histidine, 242t
Homopolymer, 214
Homulka, B., 467
Hormones, 429
Hughes, John, 431
Human growth hormone, synthesis, 282
Humidity, relative, 397d
Hybrid atomic orbitals, 87
  designation, 87
  shapes, *88*
  spatial orientations, *88*
Hybridization, benzene and, 179–180
Hydrides, Group VA, 111
Hydro energy, *306*
Hydrocarbons, 165d
  air quality standards for, 393t
  aliphatic, 172d
  alkanes, 170
  alkenes, 170
  alkynes, 170
  aromatic, 179–181
  as air pollutants, 405
  halogenated, emissions of, 407t
  halogenated, ozone and, 406
  halogenated, production of, 407t
  in eye makeup, 453
  number of isomers of, 169t
  quantity discharged into air, *392*, 392t
  saturated, 183d
  uses of, 185
Hydrochloric acid, 118
  from phosgene, 339t
  of stomach, 429
  relative acid strength, 121t
Hydrofluoric acid, relative acid strength, 121t
Hydrogen, 25
  atomic theory and, 41t
  electronic structure of, 43t
  for the Haber process, 158
  isotopes of, 37
  position in periodic table, 56
  properties, 50t
  rocket fuels and, 109
  spectrum of, *39*
Hydrogen atom
  structure of, *35*
  transfer in biological oxidation-reduction, 252
Hydrogen bomb, 327
Hydrogen bonding, 82, *84*
  in alcohols, 193
  in cotton, *239*
  in DNA, *257*
  in nylon, 221, *222*
  in proteins, 243, *244*, *245*
  sugars and, 234
Hydrogen chloride gas, 116
  in solution, 116
Hydrogen cyanide, 118
Hydrogen economy, 331
Hydrogen fluoride molecule, 75
Hydrogen halides, boiling points, *83*
Hydrogen iodide, 104
Hydrogen ion, 113
  concentrations, 121
  hydrated, 116
Hydrogen molecule, bond in, 74
Hydrogen peroxide, 444
  catalysis of, 251
  in hair bleaching, dyeing, 446, *447*
Hydrogen sulfide ion, 118
Hydrolases, action of, 250
Hydrolysis
  carbon unaffected by, 165
  of ATP, 261, 262
  of nucleic acids, 253
  of proteins, 241
  of sugars, 233
Hydrometer, test of coolant, 458
Hydronium ion, 116
  relative acid strength, 121t
Hydroquinone, 465s

Hydroxide ion
as base, 118
relative acid strength, 121t
Hydroxyl-, 178d
Hydroxyproline, 242t
Hygroscopic substances, 424
Hypertonic, 417
Hypo, 467
Hypochlorite ion, corrosive poison, 339t
Hypothesis, 12d
Hypoxanthine, 356s

I.G. Farbenindustrie, 434
Ice, structure and hydrogen bonding, *84*
Iceland spar, 175
IgE antibody, 438
Industrial categories, related to toxic pollutants, 378t
Industrial chemistry, 139
Industrial revolution, related to technology, 290
Industrial smog, 396
Industrial wastes and water pollution, 377t, 378t
Inner transition elements, 53
Inosinic acid, 421s
Insecticides
cholinesterase poisons, *353*
containing arsenic, 344t
use of, 296
Instant color pictures, 472
Institute of Medicine of the National Academy of Science, 422
Insulation, R-values for, 310d
Insulin, 269
synthetic, 280
Integrated circuits, 154
International Bureau of Weights and Measures, 13, 476
International System of Units, 476
Intestinal lipase, 268t
Inventions, incubation intervals, 290t
Inversion, air pollution and, *397*
Iodine, 56, 433
corrosive poison, 339t
periodic chart, 52
test for starch, 236
Iodophor, 433
Ion, 32, 66d
hydrogen, 113
Ion exchange
purification of sea water, *382*
to purify hard water, 386
Ion migration in solution, 115
Ion sizes, 72, *73*
Ionic bond(s), 66
in proteins, 443, 444
Ionic bonding, 123
Ionic compound(s), 66d
compared with covalent compounds, 82t
properties, 71
Ionic dissociation, 114
Ionic equilibrium, 120
Ionic solutions, 114
Ionization
in solution, 116
of HCl, 116
Ionization energy, 41, *60*
of elements, 58
of selected gaseous atoms, 67t
related to ion formation, 67
Ions
names of, 123
properties, 70
solvation of, 124
Iron, 130t
and steel, 140
from ores, 141
as a trace nutrient, 161
chelated, in soil, 161
EDTA chelate, *162*
from ore, 130t
role in cyanide poisoning, 343
rusting of, *99,* 103, 107
Iron (III) hydroxide, 107
Iron oxide, 129
Iron pyrite, 396, 399
Iron-phospate complex, 374
Isobutane. See *Methylpropane.*
Isocitric acid, *272*
Isocyanate group, 225s
Isoleucine, 242t
Isomers, 165d
benzene and, 180
from digits, 165
geometrical, 177–178
optical, 174–177
structural, 167
sugars, 235
Isooctane, 188
Isopentane. See *2-Methylbutane.*
Isoprene in natural rubber, 212
Isopropyl alcohol, 190t
Isotope(s), 36d
half-life of, 324d, 325
number of, 37
of Francium, 55
IUPAC (International Union of Pure and Applied Chemistry), nomenclature of hydrocarbons and, 170

James River, 377
Jet fuel, 187t
Johnson, Samuel, 398
Joule, 16d, 478d

Karlsruhe, Germany, 26
Kelvin temperature scale, 479
Kepone, 377
Keratin, 443
Kerosene, 187t, 312
alpha-Ketoglutaric acid, *272*
Ketone, functional group, 179t
Ketone group in carbohydrates, 233
Khorana, Gobind, 283
Kilo-, 477d
Kilogram, 477
Kinetic energy, 38d
Kinetic Molecular Theory, reaction rates and, *103,* 111
King on the Mountain, 100
Kitchen chemistry, 425
Knocking in gasoline engines, 188
Kodachrome film, 469
Kodachrome process, 470
Kodak, 463
Kodel, 227t
Kornberg, Arthur, 283
Kosterlitz, Hans, 431
Krebs cycle, *272,* 273
blocked by fluoroacetate ion, 343, *344*
Krypton, 56
Superman and, 48
Krypton-86 isotope, 476
Krypton-92, radioactivity of, 320

Lactase, 268t, 270
action of, 250
Lactic acid, *271, 272,* 273
fatigue and, 175
optical isomers of, 175
Lactose, *236,* 269
hydrolysis of, *236*
intolerance, 270
source of, 235
Lake (dye), 450d
Lake Michigan, water pollution and, 374
Land, Edwin H., 467
Lanolin, 449
Lasers, fusion and, 329, *330*
Latent image, 463
Latex paints, 216
Laughing gas, 127
Lauryl alcohol, 202
Lauryl hydrogen sulfate, 202
Lavoisier, Antoine, 13, 21, 23, 100
Law
Conservation of Matter, 21, *22, 25,* 100, 107
Constant Composition, 23, *25*
Multiple Proportions, 23, *25*
of Octaves, 51
Periodic, 50
scientific, 9d
LD50, 337d
Lead
air pollutant, 394
air quality standards for, 393t
equilibrium in body, *347*
in paints, 216
storage cell, 137t
teratogen, 358t
toxicity, 343, *347*
Lead acetate, sweetness of, 422
Lead arsenate, 344t
Leavened bread, 425
Length units, 476
Lethal dose, 337
Leucine, 242t
Leucippus, 20
Leukemia, radioactivity and, 326
Lexan, 221
Li, Choah Hao, 282
Lidocaine, 431s
Light
atomic cause of, *40*
polarized, 175d, *176*
quantum theory of, *40*
wave theory of, *40*
Light reaction in photosynthesis, 264, 265
Lime, 56, 104, 399
related to iron in soil, 161
slaked, 107
Limestone, 399
decay of, 398
water hardness, 386
Limnology, 374d
Linear polyethylene molecule, *211*
Linoleic acid, 239s
Linolenic acid, 239s
Linseed oil in paints, 217
Lipase(s), action of, 251
Lipid, 199, 269
Lipstick, 450–451
composition, 451
dyes for, 451
Liquid air, 148
Liquid nitrogen, 149
fertilizer, 159
Liquid oxygen, 149, *150*
Lister, Sir Joseph, 449
Liter, 14
Liter flask, *117*
Lithium
electronic structure of, 43t
properties, 50t
radioactivity shield, 329
teratogen, 358t
Lithium chloride, 55

Lithium deuteride, 327
Liver cancer, 358
Liver extract, 429
Living systems, synthesis of, 276
Local analgesics, 431t
London, England, air pollution and, 391
Los Angeles, California, *402*
ozone and, 406
pollutants in, 403t
Love Canal, 377
Lowry, T.M., 117
LOX, 149, *150*
LSD, 355t
hallucinogen, 362t
mechanism of action, *363*
Lubricants, 456–458
solid, 457
upper cylinder, 456
Lucite, 227t
Lucretius, 20
Lungs, human, *105*
Lymphocytes, *326*
Lysergic acid diethylamide, hallucinogen, 362t
Lysine, 242t, 444s

Macromolecule, 209d
Macronutrient, 154, 155t
Macroscopic structure, 9
Mad Hatter, 346
Maddox, R.L., 463
Magnesium
electronic structure of, 43t
from ore, 130t
from sea, 139, 146
ore, 146
production, *146*
properties, 50t
reaction with oxygen, 99
Magnesium ion, hard water, 201
Magnesium oxide, 430t
formation of, 99
Magnesium silicate, 424
Magnetic bottle, 329
Malathion, *353*
Maleic acid, *272*
Maleic hydrazide, 355t
Maltase, 249, 268t
action of, 250
in alcohol production, 191
Maltol, 421s
Maltose, *236*
action on by maltase, 249
hydrolysis of, *236*
in alcohol fermentation, 191
source of, 235
D-mannitol, 425s
Marble, decay of, 398
Marihuana, hallucinogen, 362t
Marsh gas, 405
Mascara, 453
Mass
atomic, 26
critical, 320d
units of, 477
weight and, 26
Mass defect, 321d, *322*
Mass spectrometer, *36*
Mast cells, *439*
Matter
charged, *27*
conservation of, 21
duality of, 43d
law of conservation of, 100
Mayer, Julius Robert, 308
Measurement, 13
importance of, 21
Meat tenderizers, 270, 426
enzymes and, 251
Mechanism of cyanide poisoning, 342
Medflies, 353
Medicines, 428, 429
Mega-, 477d
Meiosis, 277
Meitner, Lise, 319
Melamine, 227s
Melanin, 445, *447*, 451
Melmac, 227t
Melting points
of ionic compounds, 71
periodic trends, *58*
Mendeleev, Dmitri Ivanovich, 51
Meperidine, 352s, 431s
Meprobamate, 429t
Mercaptans, 396
Mercury, 21
from ore, 130t
ozone and, 406
teratogen, 358t
toxicity, 343, *346*
Mercury cell, 137t
Mercury (II) oxide, 21
Merlon, 221
Merrifield, Robert B., 280
research notes, *289*
Mescaline, 362t
Messenger RNA, 278
Meta-, 181d
Metabolic poisons, 340
Metabolism of glucose, 271
Metal chlorides, *57*
Metallic bonding, 85
Metalloids, *57*
Metallurgy, 139d
Metals, 50, *57*
and their preparation, 139
from ores, 130t
Metathetical, 101d
Meter, 13, 476d
Methadone, 352, 353s
Methanal. See *Formaldehyde.*
Methane, *165*, 171s
energy content of, 312t
energy from, 319
in magnesium production, 147
percentage in air, 148
Methanoic acid. See *Formic acid.*
Methanol, 179t, 190t. See also *Methyl alcohol.*
use of, 179t
Methaqualone, 362t
Methionine, 242t
Methyl alcohol, 190t
from wood, 190
in antifreeze, 458
lethal dose, 337
synthetic, 190
toxicity of, 364
Methyl cellulose in diet pills, 440
Methyl chloride, 224
Methyl chloroform, 407
Methyl group, 171d
Methyl methacrylate, monomer, 212t
Methyl salicylate, synthesis, 205
2-Methyl-2-pentene, 172s
2-Methyl-2-propanol, 189
Methylamine, 179t
2-Methylbutane, 168s, 189
2-Methylpentane, 172s
3-Methylpentane, 172s
Methylpropane, 167s, *168*
Metol, 465
Meyer, Julius Lothar, 53
Micro-, 477d
Microballoons, fusion and, 329, *330*
Microcomputer, *154*
Microfibril, 239d
Micron, 393d
Micronutrient, 154, 155t
Microorganisms, in foods, 417
Microprocessor, 154
Microscopic structure, 9
Milk of magnesia, 430t
Milli-, 477d
Millikan, Robert Andrews, 30–32, *31*
Mineral oil, carcinogen, 359t
Mineral spirits in paints, 217
Minerals, food supplements, 427
Minimum structural change, principle of, 202
Mitochondrion, *255*
Mitosis, 277
Mixtures, and compounds, 6
separation of, 3
Model compounds, to test theory of macromolecules, 209
Moisturizers, on skin, 443
Molar solutions, 116
Molarity of solutions, 117
Mole, 12d, 108d
chemical equations and, 107
examples of, 12
Mole ratios, 484
Molecular formula, 12d
Molecular structure(s), 86
cyclic, 170
predicted by atomic orbitals, 86
predicted by orbital hybridization, 87
Valence-Shell Electron-Pair Repulsion Theory, 90
Molecular weight(s)
atomic weights and, 108
of lubricating oils, 456
Molecule(s), 12d
useful energy and, 309
Molina, M.J., 407
Molybdenum disulfide, 457, *458*
Mono-, in saccharides' naming, 233d
Monomer, 210d
Monosaccharides, 233d, 234
Monosodium glutamate, 421s, 426, *417*
Monosodium methanearsenate, 344t
Morphine, 351, 352, 354s, 430s
lethal dose, 337
Motion, perpetual, 308
Motor cleaners, 460t
Motor oil, viscosity of, 457t
MSG. See *Monosodium glutamate.*
Mt. St. Helens, 394, 395
Mucilage, 237
Mulder, G.T., 241
Multiple bonds, 76
dot structures, *77*
line structures, *77*
Multiple Proportions, law of, 23
Mustard gas, 355t
Mutagens, 353
in plants and animals, 355t
Mutations
DNA structure and, 284
radioactivity and, 326
Mylar, 219, 227t

$NAD^+$ in Krebs cycle, 272
$NADP^+$, 265
in Krebs cycle, 272
NADPH, 265, 266
Nail polish, 453–454
composition of, 455

Nano-, 477
Nanometer, 40d, 393d
alpha-Naphthol, 470s
2-Naphthylamine
  bladder cancer, 359
  carcinogen, 385s
Naphthylamines, carcinogenic properties, 360
National Academy of Science, Institute of Medicine, 422
Natural gas, 186, 315–316
  as fuel for automobile, 411
  energy content of, 186, 312t
  supply of, *316*
Natural rubber, 212
Negative ions, sizes relative to atoms, *73*
Neon, 56
  electronic structure of, 43t
  isotopes of, *37*
  percentage in air, 148
  properties, 50t
Neopentane. See *2,2-Dimethylpropane.*
Neoprene, 227t
Nerve gas
  DFP, 352s
  Sarin, *353*
  Tabun, *353*
Neurotoxins, 349
  use in medicine, 351
Neutral solution, 119
Neutralization of acid with base, 117, 119
Neutron(s)
  charge of, 33, 34
  damage to metals, 325
  fission and, 320
  mass of, 33, 34, 321
  number per atom, 36
Neutron bomb, *328*
New elements, 61
New York, N.Y.
  air pollution and, 391
  haze over, *396*
Newlands, John, 51
Newton, Isaac, 21
Niacin, 252, 252s, 429t
Niagara Falls, N.Y., 377
NiCad battery, 137t
Nickel-cadmium (NiCad) cell, 137t
Nicotinamide, 426t
Nicotinamide adenine dinucleotide phosphate ($NADP^+$), 252s, 265
Nicotinamide adenine dinucleotide ($NAD^+$), 252s
Nicotine, 354s
Nicotinic acid, 429t
Niepce, Nicéphore, 462
Nitrate ion, 119
Nitrate salts, 127
Nitric acid, 119, 403
  relative acid strength, 121t
  used to make ammonium nitrate, 159
Nitrocellulose, 453
Nitrogen
  chemistry, 156
  diatomic, 164
  electronic structure of, 43t
  fixed, 157, 401d
  percentage in air, 148
  properties, 50t
Nitrogen cycle, *401*
Nitrogen dioxide
  corrosive poison, 339t
  formation of, 108
  physiological effects of, 403
  plant growth and, 403
Nitrogen oxides
  air quality standards for, 393t
  as air pollutants, 400–403
  quantity discharged into air, *392,* 392t
Nitrogen pool, 277
Nitrogenous bases, 254t
Nitroglycerin, 429t
Nitrous acid, 355t, 356, 403
Noble gases, 53, 56
  reactions of, 56
Noble, Richard, 283
Nomenclature of hydrocarbons, 170, 171t
n-Nonane, 171s
Nonbonding electrons, 89
Nonelectrolytes, 114
Nonmetals, 50, *57*
Nonpolar covalent bond, 79, 168
Nonpolar molecules, 80
Norepinephrine, *349,* 363s
Novocaine, 431s
Nuclear energy, 319
  binding energy, 321
  breeder reactors for, 323
  consumption of, *306*
  electricity from, *329*
  fission, 319–320
  fusion, 327
  power plants for, *323*
  radiation damage from, 324
Nuclear power plant, *323*
Nuclear power reactors, location of, *324*
Nuclear wastes, burial of, *294*
Nucleases, 268t
Nucleic acids, 253–258
  as esters, 254
  synthetic, 283
Nucleotidase, 268t
Nucleotide, 255s
5′-Nucleotides, 416, 421
Nucleus, discovery of, *34, 35*
Nutrient supplements, 427
Nutrient, plant. See *Fertilizer.*
Nuts and bolts, *99*
Nylon 501, 227t
Nylon 6, 221
Nylon 66, 220
  peptide bonds and, 242
Nylons
  as polyamides, 220
  commercialization of, 221
Nytril, 227t

Objective knowledge, 287
Oceans, carbon dioxide effect on, 408
Octane, 108
  combustion of, 411
  energy content of, 312t
  isomers of, 169t
  rating, 188
n-Octane, 171s
Octaves, Law of, 51
Octet of electrons, 74
Octet Rule
  exceptions, 75
  hydrocarbons and, 169
Odor
  of esters, 198t
  of perfumes, 452
Oil(s), 199
  consumption of, *306*
  crude, 187
  digestion and absorption, 269
  lubricating, 456
  recoverable, 312d
  undiscovered, 312
Oil of wintergreen, 205
Oil shale, 314
Oil spills, 251
Oily skin, 452
Oleic acid, 197, 269
Oligo-, in saccharides' naming, 233d
Oligonucleotides, 254d
Oligosaccharides, 233d, 235
OPEC, 307
Open hearth furnace, *142*
Operational definition, 3d
Opium, 430
Optical isomers, 174–177
  in sugars, 235
  number of, 177
Orbitals, 44d
  electronic arrangements and, 43t, 45–46
  probability and, 45
  s, p, d, f designations of, 44
  shapes of, 45
  spatial orientation of, 45
Orbits, atomic, 41d
Order, elemental, 49
Ore, 139d
Organic acids, 189, 195. See also *Acids.*
  functional group, 165
Organic chemistry, 164d
Organic matter in water, 371
Organic synthesis, 185
  useful products, 202
Organophosphorus insecticides, 377
Organosilanes, 224
Orlon, 227t
Ortho-, 181d
Osmosis, 385d, 417
Ovulation, *438*
  and the pill, 437, *438*
Oxalic acid, corrosive poison, 339t
Oxaloacetic acid, *272*
Oxalosuccinic acid, *272*
Oxidases, action of, 251
Oxidation
  as means of purifying water, 368
  in foods, 418
  of fats in foods, 418
  single replacement reactions as, 100
Oxidation-reduction, 113, 129
  hydrogen atom transfer and, 252
  niacin and, 252
Oxide catalyst for making alcohol, 190
Oxidizing agents, 129
  strengths, 132
Oxygen
  atmospheric height of, 390
  atomic, 402
  diatomic, 164
  discovery of, 100
  electronic structure of, 43t
  in water, 371
  percentage in air, 148
  properties, 50t
  reaction with hemoglobin, 341
  use in steel making, 142
Oxyhemoglobin, 341
  oxygen and, *105*
Ozone, 355t, 406
  air quality standards for, 393t
  corrosive poison, 339t
  halogenated hydrocarbons and, 406

PABA, 451
Paint(s)
  additives, 217
  baked-on, 219

binders, 216
composition, 216
curing by radiation, 327
drying of latex, 216
drying of oil-based, 217
pigments, 216
toxicity of, 347
Palmitic acid, 197, 269
Palmitooleostearin, 269
Pan (panchromatic) films, 467
Pancreatic amylase, 268t
Pancreatic lipase, 268t
Pantothenic acid, 426t
Papain, 426
Paper chromatography, 4
Paper, cellulose and, 238
Para-, 181d
Paraffin, 33
Paraoxon, *353*
Parathion, *353*
Paregoric, 429t
Paris green, 344t
Part per million (ppm), 397d
Particle number ratios, 484
Particulates
air quality standards for, 393t
concentration in atmosphere, 393
damage done by, 394
quantity discharged into air, *392*, 392t
removal from air, 395
Pauling, Linus, 79, *243*
Pavesi, Angelo, 26
PCB's (polychlorinated biphenyls), 184
teratogen, 358
PCP, hallucinogen, 362t
Peach, decomposition of, *102*
Pellagra, 251
Penicillin G, potassium, 429t, 434s
Penicillin VK, 429t
Penicillins, 434–435
*Penicillium notatum,* 435
Pennsylvania crude, 189
Pentane, 11s
as foaming agent, 228t
n-Pentane, 168s, 171s, 189
properties of, 168
1-pentanol, 178s
2-pentanol, 178s
3-pentanol, 178s
Pentazocine, 431s
Pepsin, 250, 268t, 270
Pepsinogen, 270
Peptide bond(s), 242d
Perborates, 444
Perchloric acid, relative acid strength, 121t
Perchloroethylene, 407
Perfume, 452
Periodic Law, 50, 53
Periodic Table, 49, 53, *54*
Atomic Theory, 60
in the future, 61
Mendeleev and, 51
predicting chemical formulas and, 101
uses of, 55
Periods in Periodic Table, 53
Permanent wave, 444
Peroxide ion, corrosive poison, 339t
Peroxyacyl nitrate (PAN), 401
Perpetual motion, 308
Persistent chemicals in water cycle, 369
Persistent pesticides, 376
Perspiration, 449
Pesticides and water pollution, 376
Petrochemicals, 313
Petroleum, 186
and plastics, 230
consumption of, *306, 311*
discovery of, 312
energy content of, 312t
sulfur content of, 396
supply of, *312, 313*
Pfizer Laboratories, 436
PGA. See *3-Phosphoglyceric acid.*
pH, 121
and corrosion in radiator, 459
control in latex paints, 217t
in digestion, 269
in foods, 423
limits for life, 398
of stomach, 429
Phencyclidine (PCP), hallucinogen, 362t
Phenidone, 465
Phenobarbitol, 429t
Phenol
in aspirin synthesis, 205
in disinfectants, 449s
Phenyl salicylate, 228s
Phenylalanine, 242t
p-Phenylenediamine, 446s
Phenylenediamine in gasoline, 455
p-Phenylenediaminesulfonic acid, 446s
Phosgene, 339t
in polycarbonate synthesis, 222
Phosphate ion, 76
Phosphate rock as fertilizer, 159
3-Phosphate-glyceric acid, *271*
Phosphates and water pollution, 374
Phosphoenol pyruvic acid, *271,* 271s
3-Phosphoglyceric acid, 266, *267*
Phosphorus
chain structure and, 164
electronic structure of, 43t
properties, 50t
Phosphorus pentachloride, 99s
Phosphorus trichloride, 99s
Photochemical smog, 396, 401
mechanism of, *401*
Photographic developers, 465t
formulation, 465t
Photographic image, *464*
Photography
chemistry of, 462–472
color, 468
Photon(s), 38d
photosynthesis and, 233
Photosynthesis, 264
and water pollution, 374
electricity from, *331*
glucose and, 233
rate of, 309
sugar production and, 233
Phthalic acid in alkyd paints, 220
Physical change, 1, 6
Physical properties, 7d, 49
Phytoplankton, 408
Pi bond, 93d
Pica, 348
Pigments
in hair, 445
in skin, 451
Pill, the, 437
Pinacyanol, *469*
Pineapple, smell ester, 198t
Pitch, 186
Plasma, atomic, 327
Plasticizers, polymer additives, 226
Plastics, 209d
and petroleum, 230
energy for production of, 311t
energy requirements to manufacture, 230t
for recycling, *229*
from coal, 230
future sources, 230
reasons for being, 209
waste disposal, 229
Platinum in catalytic mufflers, 410
Plato, 21
Plexiglas, 227t
Plutonium-239, 319, 323
Poisons
corrosive, 338
metabolic, 340
Polar bond(s), 79d, *80*
carbon and, 168
Polar molecules, 80
Polar solvent, water as, 124
Polarity, enzymatic action and, 249
Polarization in organic molecules, 204
Polarized light, *176*
Polaroid photography, 467
Polaroid sheet, polarized light and, 175
Pollen, 438
Pollution. See also specific type, e.g., *Air pollution.*
thermal, 373
water, 367
Poly-, in saccharides, naming of, 233d
Poly-cis-isoprene, *213*
synthetic, 214
Poly-trans-isoprene, *213*
Polyacrylonitrile, uses of, 212t
Polyatomic ion, 76d
names and formulas, 76t
Polycarbonates, 221, 222s
Polychlorinated biphenyls (PCB's)
fate of, 295
history of use, 295s
in foods, 296t
Polyesters, 218
of terephthalic acid, 219
Polyethylene
energy required to manufacture, 230t
high density, 211
how made, 211
properties, 211
uses of, 212t
Polyhydric alcohols, 425
in foods, 427t
Polyimide, 227s
Polymer(s), 210d
carbohydrates, 233
cellulose, 237
examples, 209
future of, 226
nucleic acids, 253–258
polynucleotides as, 254
proteins, 242
starches, 236
trade names, 227t
Polymer additives, 228t
Polymerization, 210d
Polymethyl methacrylate, 212t
Polynucleotides, 254
Polypeptide. See *Protein.*
Polysaccharides, 233d, 235
cellulose, 237
glycogen, 236
starches, 236
Polysorbate 80, 425
Polystyrene
energy required to manufacture, 230t
in antifreeze, 459
use in making synthetic protein, *282*
uses of, 212t

Polytetrafluoroethylene, 212t
Polythene, 227t
Polyurethane, 226s
Polyvinylacetate, 212t
Polyvinylchloride (PVC), 212t
  energy required to manufacture, 230t
Polyvinylpyrrolidone (PVP), 447s, *433*
  in hair sprays, 447
Portland cement, energy for production of, 311t
Positive ions, sizes, *73*
Potable water, 381
Potash, fertilizer, 159
Potassium p-aminobenzoate in diet pills, 440
Potassium acid tartrate, 423, 424
Potassium bromate, 444
Potassium hydrogen tartrate in baking powder, 425
Potassium nitrate, 127
Potassium
  electronic structure of, 43t
  properties, 50t
  reaction with chlorine, 98
Potential energy, 38d
Potentiation, 416
Power gas, 314d
Power, 308d, 316d
  of automobile, 411
  SI unit, 478
  units of, 361t
Power plants
  electrical efficiency of, *318*
  nuclear, *323*
  sulfur dioxide from, 396
ppm (parts per million), 397d
Precipitate, 101d
Prednisone, 429t
Pregnancy, termination with prostaglandins, 240
Prescription drugs, 429t
Preservatives, food, 418, 427t
Pressor amines, 441
Pressure, SI unit, 478
Priestly, Joseph, 100
Primary colors, 469
Primary water treatment, 379, *380*
Principle of Minimum Structural Change, 202
Printing techniques, development of, 287
Printing in photography, *466*
Probability, electronic structures and, 45
Procaine, 431t
Products, chemical, 98d
Progesterone, 437s
Proline, 242t
Prontosil, 434
Propane, *166,* 171s, 186
  transportation accident, *298*
1,2,3-Propanetriol, 189t
1-Propanol, 190t
2-Propanol, 190t
  solubility, 195t
Propanols, 192
Propanone. See *Acetone.*
Propene, 172s
Properties
  chemical, 49
  of Elements, periodic, 53
  physical, 49
Propoxyphene, 431s
n-Propyl alcohol, 190t
Propyl gallate, food preservative, 419s
Prostaglandins, 240s
Protective coatings, 143
Protein synthesis, research notes, *289*
Protein(s), 241–253
  amino acids, 241
  as apoenzymes, 250
  digestion and absorption, 270
  enzymes as, 248–253
  fibrous, 244
  globular, *246*
  helical structure, *245*
  hydrolysis of, 241
  into Krebs cycle, 275
  number of, 243
  peptide bonds and, 242
  primary structure of, 243d
  quaternary structure of, 246d
  secondary structure of, 243d
  sheet structure, *244*
  synthesis of, 277
  synthetic, 280
  tertiary structure of, 246d
  viruses and, 258
Proton(s)
  charge of, 32, 34
  discovery of, 32
  mass of, 32, 34, 321
  number per atom, 35
Proton transfer, 117
Proust, Joseph Louis, 21–23
Ptyalin, 268t
  action of, 250
Public Health Service, 370t
Pure substances, 2d
  examples, 2
Pure water, 368d
Purification of mixtures, 4
Purification of water, natural, 368
Purine, 181s
PVP, 447
Pyribenzamine, 440s
Pyridine, 181s
Pyridoxine, 426t
Pyrimidine, 181s
Pyromellitic anhydride, 227s
Pyruvic acid, *271, 272*

Quaalude, 362t
Quanta, photosynthesis and, 233
Quantum, 38d
Quaternary ammonium chloride, 432
Quicklime, 104
Quinidine sulfate, 429t
Quinone, 465s

R-values of insulation, 310d
Rad, unit of radiation, 326
Radiation
  alpha, 27–28
  beta, 27–28
  gamma, 27–28
  separation of, *28*
Radiation damage, 324
  human beings and, 326
Radiation effects, 326
Radiator, 459
  cleaners, 460t
  sealants, 459
  stop leak, 460t
Radical, free, 419
Radii, atomic, *59*
Radioactive isotopes, 55
Radioactive waste, disposal of, 324
Radioactivity, 27d
  elemental, 49
Radiotherapy, 327
Radium, 27
Radon, 56
Ragweed pollen, 438–439
  map, *439*
Rat poison, 343
Raynor, Gilbert S., 397
Rayon, 238
Reactants, chemical, 98d
Reaction rate, 102d
Reactive site(s), 249d
  in addition polymerization, 211
Reactors, breeder, 323
Rearrangement polymers, 225
Recombinant DNA technology, 284
Recoverable oil, 312d
Recrystallization, 4
Red Dye No. 2, 428
Redox, 113
Redox reactions, 129
Reducing agents, 129
  strengths, 132
Reduction
  of metals, 130
  single replacement reactions as, 100
Reduction-oxidation, 113, 129
Reforming by catalysis, 188
Relative abundance of elements, 7
Relative humidity, 397d
Replication of DNA, 276, 277
Representative elements, 53
Reserpine, 429t
Reversal process, 470
Reverse osmosis, *385*
Reversibility, chemical reactions and, 104
Rhizobium bacteria, 157
Rhodanase, 343
Riboflavin, 253s, 426t
Ribonuclease A, 280, *281*
Ribonucleic acids (RNA), 254d
  messenger, 256
  molecular weights of, 256
  ribosomal, 256
  transfer, 256, 278
  types of, 256
Ribose 1,5-diphosphate, 266
Ribose 5-phosphate, *267*
alpha-D-Ribose, 254s
  in RNA, 254–258
Ribosome, in protein synthesis, *278*
Ribulose 1,5-diphosphate, *267*
Ribulose 5-monophosphate, *267*
Ribonucleic acids (RNA), 254d
  function, 255
RNA
  effect of mutagens on, 353
  messenger, 278
  transfer, 278
Rochow, E.G., 224
Rocket fuels, 109
Roentgen, unit of radiation, 325
Roentgen, William, 325
Rolaids, 430t
Roland, F.S., 407
Rubber
  buna, 213
  foamed polyurethane, 226
  formulation, 214t
  in World War II, 213
  supplies in World War II, 215
Rubbing alcohol. See *Isopropyl alcohol.*
Rum, 192t
Rum smell ester, 198t
Rust, *99,* 129
  formation of, 107
Rust inhibitors, 458
Rusting, 142
Rutherford, Ernest, *34*
  discovery of atomic nucleus, *35*

Saccharin, 422s
SAE (Society of Automotive Engineers), 457
Safe Drinking Water Act, 378
Safrole, 421s
Salicylic acid in aspirin synthesis, 205
Salt(s), 71d
  content in sea water, 124
  in solution, 124
  names of, 123
  of organic acids, 196
  preparation of, 122
  solubility of, 126
  uses of, 126
Salt solutions, 122
Saran, 227t
Sarin, *353*
Saturated fats, 199
Saturated hydrocarbons, 183d
Saturated organic compounds, 170d
Saturated solution, 125
SBR rubber, 215
Schrödinger, Erwin, 42
Schulze, J.H., 462
Science, 288d
Scientific fact, 9d
Scientific journals, 288
Scientific method, 288d
Scrubbers, *395*
Sea water, 124
  salt content, 124
  source of fresh water, 381
  source of magnesium, 146
Sealants, in antifreeze, 459
Searle, G.D., Co., 423
Secondary water treatment, *380*
Sedimentation in purifying water, 368
Sedoheptulose 1,7-diphosphate, *267*
Selenium, 106
Semiconductors, 57, *333*
Semimetals, *57*
Sequesterants in foods, 419d, 427t
Serine, 242t
Serotonin, 363s
Sevin, *353*
Sewage, treatment by radiation, 327
Shale oil, 314
Shock absorber fluids, 460t
SI (system of measurement), 13, 476
Sickle cell anemia, 246
Sigma bond, 93d
Silicon tetrachloride, 91s
Silicon
  alloys, 153
  chain structure and, 164
  electronic structure of, 43t
  in solar battery, 332
  production, 153
  properties, 50t
  zone-refining, 153
Silicone oils, 224
Silicone rubber, 224s
  room-temperature-vulcanizing, 224
  uses, 224, *225*
Silk, 244
  glycine and, 241
  nylon and, 221
Silly Putty, 225
Silt as water pollutant, 379
Silver
  from ore, 130t
  ozone and, 406
Silver bromide, 463
Silver cell, 137t
Silver chloride, 56, 185
Silver cyanate, 185
Silver salts, photochemistry of, 463
Silver tarnish, 112
Single covalent bond, 74
Skin, *442*
  preparations for, 449–454
Slag in blast furnace, 141
Slaked lime, 56, 107
Smell, *420*
Smog
  industrial, 396
  photochemical, 396, *401*
Smoking
  cigarette, 405
  damage to fetus, 358
Snyder, Solomon, 431
Soaps, 189, 199
Society of Automotive Engineers (SAE), 457
Sodium
  electronic structure of, 43t
  from ore, 130t
  properties, 50t
  reaction with chlorine, 99, 107
Sodium acetate, 342
Sodium acid pyrophosphate in baking powder, 425
Sodium benzoate in food preservation, 418
Sodium bicarbonate (baking soda), 425
  as antacid, 430t
Sodium bromate, 444
Sodium carbonate, 399
Sodium chloride, 56, 99, 126
  formation of, 107
  structure, *72*
Sodium citrate, 430t
  detergent builder, 375
Sodium cyanide, 342
Sodium cyclamate, 422s
Sodium fluoroacetate, 343
Sodium formate, 196
Sodium glycocholate, 270
Sodium hydrogen carbonate, 127
Sodium hydroxide, 55
  corrosive poison, 339t
Sodium lauryl sulfate, 202
Sodium nitrate, 127
Sodium nitrite, 356
Sodium oxide, 55
  in solution, 118
Sodium perborate, corrosive poison, 339t
Sodium phenoxide in aspirin synthesis, 205
Sodium propionate in food preservation, 418
Sodium salicylate in aspirin synthesis, 205
Sodium sulfide, 448
Sodium sulfite, 399
Sodium tripolyphosphate, 374
Solar battery, 331–332, *333*
Solar energy
  collectors of, *332*
  electricity from, *331*
  U.S. reception of, *330*
  use of, *306*
Solar radiation, *452*
Solubility of salts, 126
Solute, 113d
Solution(s), 113d
  acidic and basic, 119
  ammonia in water, 121
  aqueous, 114
  conductance of, 115
  electrolytic, 114
  ionic, 114
  molar, 116
  molecular, 114
  salts, 122
  saturated, 125
Solvation of ions, 124
Solvent(s), 113d
  in glue, 355t
Somatic effects of radioactivity, 326
Soot, carcinogen, 359t
Sorbitan monostearate, 425
D-Sorbitol, 425s
Spandex, 227t
Spectral sensitivity, 467, *468*
  of human eye, *468*
  of pan film, *468*
Spectrum, *39*, 39d
  bright-line emission, 39d
  continuous, 39d
  emission, 39d
  hydrogen, *39*
Spermaceti wax, 453
Spiegelman, Sol, 283
Stabilizers in foods, 424, 427t
Stainless steel, 144
*Staphylococcus aureus*, 434
Starch, 236
  digestibility of, 238
  hydrolysis of, *426*
  used to make alcohol, 191
Staudinger, Hermann, 209
Steam stripping of water, *382*
Stearic acid, 197
Steel, energy for production of, 311t
Stereochemical control in making rubber, 212
Stereochemical model of smell, *420*
Stereoisomers, 174d
  geometrical, 177–178
  optical, 174–177
Steroids, 436s
Stomach digestion, 270
Stomach
  bleeding caused by aspirin, 432
  pH in, 429
Stop bath, 466
Strassman, Fritz, 319
Stratosphere, 406d
Stratum corneum, 443
Strengths
  of acids and bases, 121t
  of oxidizing agents, 134t
  of reducing agents, 134t
Streptomycin, 436
Strong electrolytes, 120
Strontium sulfide, 448
Structure of matter, discussion, 9
Structure(s), electronic, 43t
  molecular, 167–168
    branched chain, 167d
    configurations versus conformations, 168
    straight chain, 167d
  of proteins, 243–256
    primary, 243d
    quaternary, 246d
    secondary, 243d, 256
    tertiary, 246d
Strychnine, 349
Styrene, 210
  monomer, 212t
Styrene-butadiene rubber, 215
Submicroscopic structure, 9
Substitution reactions, 204
Substrate in biochemical reactions, 249d
Subtractive primary colors, 469
Succinic acid, *272*
Succinyl coenzyme A, *272*
Sucrase, 249, 268t
  action of, 250

Sucrose, *236*, 269
annual production of, 235
composition of, 23
hydrolysis of, *236*
source of, 235
Sugar(s), 233
alpha-D-ribose, 254s
alpha-2-deoxy-D-ribose, 254s
from oil, 313
Sugar solution, *113*
Sulfa drugs, 434
Sulfanilamide, 434
Sulfate ion, 76
Sulfhydryl group and heavy metal poisoning, 344
Sulfide ion, 118
as base, 118
Sulfite ion, 118
corrosive poison, 339t
Sulfonamides, 429
Sulfur
burning of, 102
chain structure and, 164
electronic structure of, 43t
formula of, molecular, 164
in vulcanization of rubber, 212
properties, 50t
Sulfur dioxide, 102, 395
effects of, *390*
physiological effects of, 399t
source of, 396
Sulfur hexafluoride, 92s
Sulfur oxides, 392
air quality standards for, 393t
quantity discharged into air, *392*, 392t
Sulfur trioxide, 396
reactions with water, 99
Sulfuric acid
air pollution and, 397
corrosive poison, 338, 339t
formation of, 99
Sulfurous acid, 398
Sunlight, damage to plastics, 228
Sunscreen, 451
ozone as, 406
Suntan lotions, 451
Superphosphate, 159
Supplements, nutrient, 427
Surface active agents, 425
in foods, 427t
in gasoline additives, 456
Sweat glands, 449
Sweeteners, 235
artificial, 422
Sweetness, 422
for sugars, 235
Symmetry
asymmetric and, 174d
mirror images and, *175*
optical isomers and, 174
Synapse, 350
Syndets, 201
Synergism, 416
between bacteria and plants, 157
Synthesis
of living systems, 276
of nucleic acids, 283
of proteins, 277
organic, 185
with alcohols, 194
Synthesis gas, 315
Synthetic detergents, 201
Synthetic protein, 280

Table, Periodic, 49
Tabun, *353*
Talbot, William Henry Fox, 463
Talbot process, *162*, 163
Tall stacks, *400*
Talwin, 431s
Tar, carcinogen, 359t
Tartaric acid, 423
TCP, 455s
Technological development, 299
Technology, 287d
and the human environment, 294
control of, 300
triumphs and problems, 291
Teeth, collagen and, 244
Teflon, 227t
in paints, 217
Tellurium, 52
Temperature
effect on reaction rate, 102
effects on aquatic life, 373
effects on oxygen in water, 371
SI unit, 478
Temperature conversion, 480t
Temperature inversion, *397*
Temperature scales, 15, 479
Tenderizers, 270
meat, 426
Tendons, collagen and, 244
Teratogens, 357
Teratology, 357
Terephthalic acid, 218
Terramycin, 436s
Tertiary water treatment, 380
Terylene, 219, 227t
Testosterone, 437s
Tetrabromofluorescein, 451s
Tetracycline HCl, 429t
Tetracyclines, 436
Tetraethyllead, 189
and automobile, 409
toxicity of, 347
Tetrafluoroethylene, monomer, 212t
Tetrahydrocannabinol, 362t
TFE. See *Teflon.*
Thalidomide, 357s
Thallium, 343
teratogen, 358t
Thames River, thermal pollution of, 373
Theoretical definition, 3d
Theoretical model, 10d
Theory, 12d
vital force, 185
Thermal cracking of oil, 187
Thermal inversion, *397*
Thermal pollution, 373
Thermodynamics, 308d
first law, 308
second law, 309
Thermoplastic, 210d
Thermosetting plastic, 210d
Thiamine, 426t
Thickeners
in foods, 424, 427t
in latex paints, 217t
Thiocyanate ion, 343
Thioglycolic acid, 444
Thiosulfate ion, 343, 476
Thomson, George, 43
Thomson, Joseph John, 30–31, *31*
Threonine, 242t
Thymine, 254s, 278
Thyroid, drug, 429t
Time, SI unit, 478
Tin, 164
Tincture, 433d
Titanium, energy for production of, 311t
Titanium dioxide
in eye makeup, 453
in paints, 216
Tobacco, 351
Tobacco mosaic virus, *258*
alpha-Tocopherol, 426t
Toluene, 189, 405s
Toone, Tennessee, 377
Toxic pollutants, 378t
Toxic substances, 337
Toxic Substances Control Act, 296
trans-2-Butene, 169s
properties, 169
Transfer RNA, 278
Transistor, 153
Transition elements, 53
Transmission fluids, 460t
Transportation accidents involving hazardous chemicals, 298
Trends in periods of Periodic Table, 56
Triads, elemental, 50
Trichloroacetic acid, 195
Trichlorobenzenes, 181s
Tricresyl phosphate, 455s
Triethylenemelamine, 355t
Triglyceride, 269
Trihalomethanes, 378
2,2,4-Trimethylpentane, 188
Triolein, 200
Triple bond, 77d
Trisodium phosphate, 339t
Tristearin, 200
Tritium, 327
Troposphere, 406d
Trypsin, 249
digestive enzyme, 268t
protein hydrolysis by, 240
Tryptophan, 242t
Turpentine, 217
Tylenol, 432s
Tyrosinase, *447*
Tyrosine, 242t, *447*

Ultraviolet light, sunburn and, 451–452
Unsaturated fats, 199
Unsaturated organic compounds, 170d
Upper cylinder lubricants, 456
Uracil, 254s, 356s
Uranium, 27, 319
discovery of fission, and, 319
enriched, 323
fission of natural sample, 320
fission of, 320
isotopic mixture of, 320
Uranium-233, 323
Urea, 185
as fertilizer, 159
synthesis, 185
Urethane linkage, 225s
Urine, lead in, 348
Useful products from organic synthesis, 202
UV stabilizers as polymer additives, 226

Vaccines, 429
Valence electrons, 65d
Valence shell, 56, 60d
Valence-shell electron-pair repulsion theory, 90
double bonds and, 92
triple bonds and, 93
Valine, 242t
Van der Waals' attractions, 85
carbon and, 168
in alcohols, 193
Vanishing cream, 450
Vectra, 227t
Vegetables, damage by sulfur dioxide, 399

Velon, 227t
Vinegar, 179t
Vinyl acetate, 227s
  monomer, 212t
Vinyl chloride, 212t
  carcinogen, 359t
Vinyl pyridine, 227s
Vinylidine dinitrilè, 227s
Virus, 258d
Viscosity, 151
  of glass, 151
  of oils, 456, 457t
Vital force theory, 185
Vitamin(s), 251–253
  as coenzymes, 249, 251–253
  as medicine, 429
  B2. See *Riboflavin.*
  B3. See *Niacin.*
  in diet pills, 440
  in food production, *417,* 427
  table of, 426t
Vitamin A, 423, 426t
Vitamin C, 426t
Vitamin D, 426t
Vitamin E, 426t
Vitamin K, 426t
Vogel, W.H., 467
Volcanoes, 391
  list of eruptions, 394
Volume units, 478
Vulcanization, 212
Vulcanized rubber, *213*

Waksman, Selman, 436
Water, 23, 25
  and hydrogen bonding, 83
  as polar solvent, 124
  decomposition of, *104*
  hard, 201
  hardness, 385
  hydrogen bonds with sugars, 234
  relative acid strength, 121t
  reuse, 367
  usage, 367, 368t
Water cycle, *369*
Water molecule, bonds in, 75
Water pollution, 367
  by detergent builders, 374
  by fertilizers, 374
  by silt, 379
  classes of pollutants, 370t
  scope, 369
Water purification, 379
  home treatment, 384
Water solubilities,
  of alcohols, 195t
Water treatment, 379
  primary, 379
  secondary, *380*
  tertiary, 380
Watson, James D., *257*
Watt, power unit, 316d
Wave mechanical model of the
    atom, 42–46
Waving hair, 444
Weak electrolytes, 120
Weight, 26
  atomic, 26
  chemical reactions and, 107
  mass and, 26
Weight relationships, 484
Wein, W., 32
Whiskey, 192t
White lead in paints, 216
Wilkins, Maurice H.F., 257
Wine, 192t
Wöhler, Fredrich, 185
Wood alcohol. See *Methyl alcohol.*
Wool grease, 449
Wooley, Dilworth, 363
World population, 293
Wort in alcohol production, 191

X rays, 28
  dental, 325
  radiation from dental, 325
  using barium sulfate, 128
Xanthine, 356s
Xenon, 56
m-Xylene, 181. See also *1,3-Dimethylbenzene.*
o-Xylene, 181. See also *1,2-Dimethylbenzene.*
p-Xylene, 181. See also *1,4-Dimethylbenzene.*
Xylenes, 181
Xylocaine, 431s

Yeast in breadmaking, 425
Yonashiro, Donald, 283

Zantrel, 227t
Zeolite, detergent builder, 375
Zero Energy Growth, 323
Zinc, 130t
  in automobile, 455
  teratogen, 358t
Zinc oxide in eye makeup, 453
Zone-refining, *153*
Zymase, action of glucose, 425

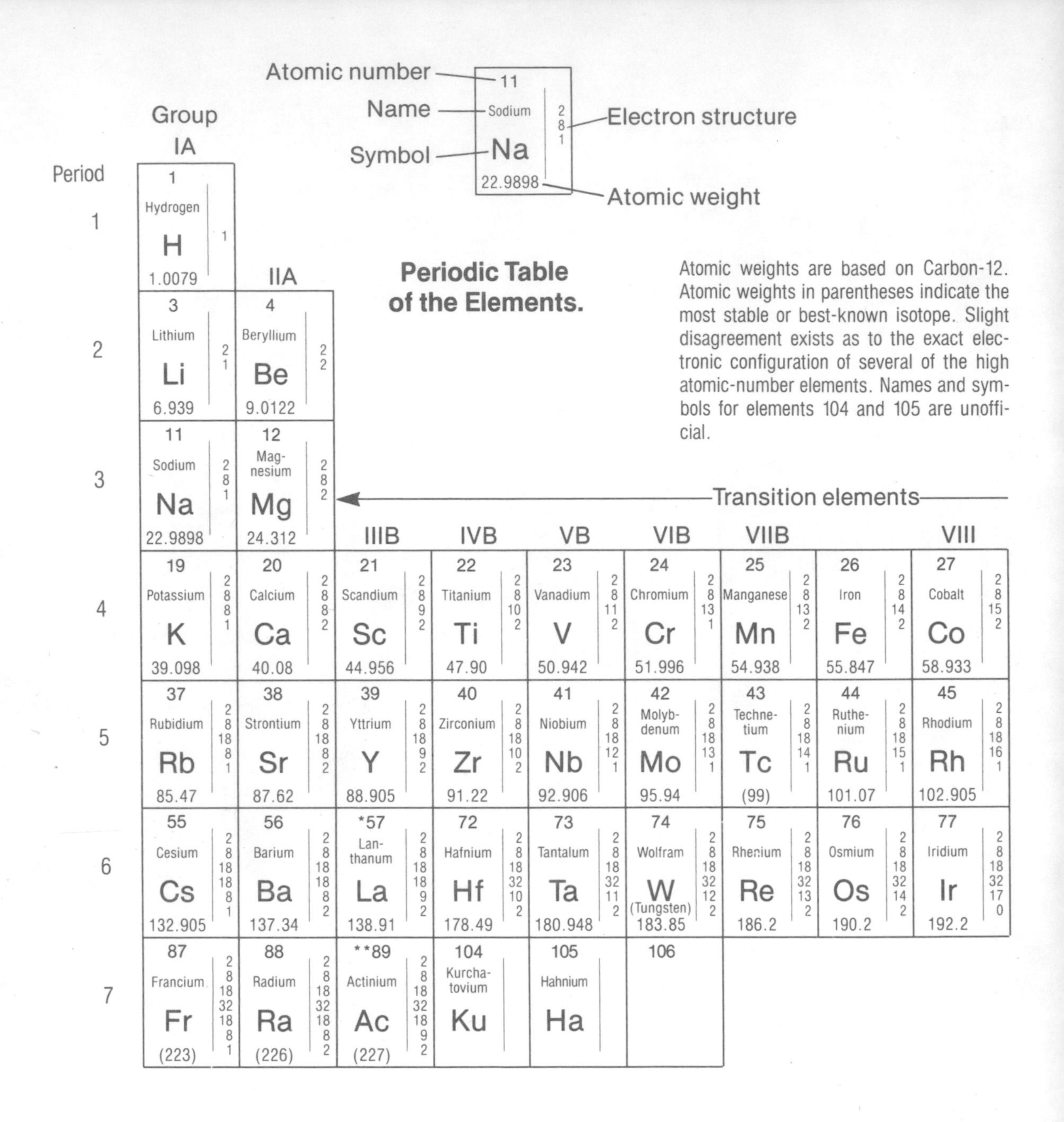


**Periodic Table of the Elements.**

Atomic weights are based on Carbon-12. Atomic weights in parentheses indicate the most stable or best-known isotope. Slight disagreement exists as to the exact electronic configuration of several of the high atomic-number elements. Names and symbols for elements 104 and 105 are unofficial.

| Period | Group IA | IIA | IIIB | IVB | VB | VIB | VIIB | | VIII |
|---|---|---|---|---|---|---|---|---|---|
| 1 | 1<br>Hydrogen<br>H<br>1.0079<br>1 | | | | | | | | |
| 2 | 3<br>Lithium<br>Li<br>6.939<br>2 1 | 4<br>Beryllium<br>Be<br>9.0122<br>2 2 | | | | | | | |
| 3 | 11<br>Sodium<br>Na<br>22.9898<br>2 8 1 | 12<br>Mag-nesium<br>Mg<br>24.312<br>2 8 2 | | | | | | | |
| 4 | 19<br>Potassium<br>K<br>39.098<br>2 8 8 1 | 20<br>Calcium<br>Ca<br>40.08<br>2 8 8 2 | 21<br>Scandium<br>Sc<br>44.956<br>2 8 9 2 | 22<br>Titanium<br>Ti<br>47.90<br>2 8 10 2 | 23<br>Vanadium<br>V<br>50.942<br>2 8 11 2 | 24<br>Chromium<br>Cr<br>51.996<br>2 8 13 1 | 25<br>Manganese<br>Mn<br>54.938<br>2 8 13 2 | 26<br>Iron<br>Fe<br>55.847<br>2 8 14 2 | 27<br>Cobalt<br>Co<br>58.933<br>2 8 15 2 |
| 5 | 37<br>Rubidium<br>Rb<br>85.47<br>2 8 18 8 1 | 38<br>Strontium<br>Sr<br>87.62<br>2 8 18 8 2 | 39<br>Yttrium<br>Y<br>88.905<br>2 8 18 9 2 | 40<br>Zirconium<br>Zr<br>91.22<br>2 8 18 10 2 | 41<br>Niobium<br>Nb<br>92.906<br>2 8 18 12 1 | 42<br>Molyb-denum<br>Mo<br>95.94<br>2 8 18 13 1 | 43<br>Techne-tium<br>Tc<br>(99)<br>2 8 18 14 1 | 44<br>Ruthe-nium<br>Ru<br>101.07<br>2 8 18 15 1 | 45<br>Rhodium<br>Rh<br>102.905<br>2 8 18 16 1 |
| 6 | 55<br>Cesium<br>Cs<br>132.905<br>2 8 18 18 8 1 | 56<br>Barium<br>Ba<br>137.34<br>2 8 18 18 8 2 | *57<br>Lan-thanum<br>La<br>138.91<br>2 8 18 18 9 2 | 72<br>Hafnium<br>Hf<br>178.49<br>2 8 18 32 10 2 | 73<br>Tantalum<br>Ta<br>180.948<br>2 8 18 32 11 2 | 74<br>Wolfram<br>W<br>(Tungsten)<br>183.85<br>2 8 18 32 12 2 | 75<br>Rhenium<br>Re<br>186.2<br>2 8 18 32 13 2 | 76<br>Osmium<br>Os<br>190.2<br>2 8 18 32 14 2 | 77<br>Iridium<br>Ir<br>192.2<br>2 8 18 32 17 0 |
| 7 | 87<br>Francium<br>Fr<br>(223)<br>2 8 18 32 18 8 1 | 88<br>Radium<br>Ra<br>(226)<br>2 8 18 32 18 8 2 | **89<br>Actinium<br>Ac<br>(227)<br>2 8 18 32 18 9 2 | 104<br>Kurcha-tovium<br>Ku | 105<br>Hahnium<br>Ha | 106 | | | |

Inner transition elements

| | | | | | | | |
|---|---|---|---|---|---|---|---|
| *<br>Lanthanide series<br>6 | 58<br>Cerium<br>Ce<br>140.12<br>2 8 18 20 8 2 | 59<br>Praseo-dymium<br>Pr<br>140.907<br>2 8 18 21 8 2 | 60<br>Neo-dymium<br>Nd<br>144.24<br>2 8 18 22 8 2 | 61<br>Prome-thium<br>Pm<br>(147)<br>2 8 18 23 8 2 | 62<br>Samarium<br>Sm<br>150.35<br>2 8 18 24 8 2 | 63<br>Europium<br>Eu<br>151.96<br>2 8 18 25 8 2 | 64<br>Gado-linium<br>Gd<br>157.25<br>2 8 18 25 9 2 |
| **<br>Actinide series<br>7 | 90<br>Thorium<br>Th<br>232.038<br>2 8 18 32 18 10 2 | 91<br>Protac-tinium<br>Pa<br>(231)<br>2 8 18 32 20 9 2 | 92<br>Uranium<br>U<br>238.03<br>2 8 18 32 21 9 2 | 93<br>Nep-tunium<br>Np<br>(237)<br>2 8 18 32 22 9 2 | 94<br>Plu-tonium<br>Pu<br>(242)<br>2 8 18 32 23 9 2 | 95<br>Ameri-cium<br>Am<br>(243)<br>2 8 18 32 25 8 2 | 96<br>Curium<br>Cm<br>(247)<br>2 8 18 32 25 9 2 |